P9-AFQ-271

FIFTH EDITION
Natural
Hazards &
Disasters

DONALD HYNDMAN
University of Montana

DAVID HYNDMAN
Michigan State University

CENGAGE

Australia • Brazil • Canada • Mexico • Singapore • United Kingdom • United States

Natural Hazards and Disasters, Fifth Edition
Donald Hyndman, David Hyndman

Product Director: Dawn Giovaniello

Product Manager: Morgan Carney

Content Developer: Rebecca Heider

Associate Content Developer: Kellie N. Petruzzelli

Product Assistant: Victor Luu

Market Development Manager: Julie Schuster

Content Project Manager: Hal Humphrey

Senior Designer: Michael C. Cook

Manufacturing Planner: Becky Cross

Production Service: Jill Traut, MPS Limited

Photo Researcher: Nazveena Begum Syed, Lumina Datamatics

Text Researcher: Ganesh Krishnan, Lumina Datamatics

Copy Editor: Heather McElwain

Text Designer: Liz Harasymczuk

Cover Designer: Michael Cook

Cover and Title Page Image: Photo by Donald and David Hyndman, Hurricane Ike aftermath, Crystal Beach, northeast of Galveston, Texas, September 2008

Compositor: MPS Limited

For product information and technology assistance, contact us at
Cengage Customer & Sales Support, 1-800-354-9706

For permission to use material from this text or product,
submit all requests online at **www.cengage.com/permissions**
Further permissions questions can be e-mailed to
permissionrequest@cengage.com

Library of Congress Control Number: 2015944018

Student Edition:

ISBN: 978-1-305-58169-2

Cengage
200 Pier 4 Boulevard
Boston, MA 02210
USA

Cengage is a leading provider of customized learning solutions with employees residing in nearly 40 different countries and sales in more than 125 countries around the world. Find your local representative at **www.cengage.com**.

To learn more about Cengage platforms and services, register or access your online learning solution, or purchase materials for your course, visit **www.cengage.com**.

Printed in the United States of America
Print Number: 06   Print Year: 2021

To Shirley and Teresa
for their endless encouragement and patience

About the Authors

DONALD HYNDMAN is an emeritus professor of Geosciences at the University of Montana, where he has taught courses in natural hazards and disasters, regional geology, igneous and metamorphic petrology, volcanology, and advanced igneous petrology. He continues to lecture on natural hazards. Donald is co-originator and co-author of six books in the Roadside Geology series and one on the geology of the Pacific Northwest; he is also the author of a textbook on igneous and metamorphic petrology. His B.S. in geological engineering is from the University of British Columbia, and his Ph.D. in geology is from the University of California, Berkeley. He has received the Distinguished Teaching Award and the Distinguished Scholar Award, both given by the University of Montana. He is a Fellow of the Geological Society of America.

DAVID HYNDMAN is a professor in and the chair of the department of Geological Sciences at Michigan State University, where he has taught courses in natural hazards and disasters, the dynamic Earth, and advanced hydrogeology. His B.S. in hydrology and water resources is from the University of Arizona, and his M.S. in applied earth sciences and Ph.D. in geological and environmental sciences are from Stanford University. David was selected as a Lilly Teaching Fellow and has received the Ronald Wilson Teaching Award. He was the 2002 Darcy Distinguished Lecturer for the National Groundwater Association, is the 2016 Chair of the Board of Directors for the Consortium of Universities for the Advancement of Hydrologic Science, and is a Fellow of the Geological Society of America.

Brief Contents

Contents

Preface

The further you are from the last disaster, the closer you are to the next.

Why We Wrote This Book

In teaching large introductory environmental and physical geology courses for many years—and, more recently, natural hazards courses—it has become clear to us that topics involving natural hazards are among the most interesting for students. Thus, we realize that employing this thematic focus can stimulate students to learn basic scientific concepts, to understand how science relates to their everyday lives, and to see how such knowledge can be used to help mitigate both physical and financial harm. For all of these reasons, natural hazards and disasters courses appear to achieve higher enrollments, have more interested students, and be more interesting and engaging than those taught in a traditional environmental or physical geology framework.

A common trend is to emphasize the hazards portions of physical and environmental geology texts while spending less time on subjects that do not engage the students. Students who previously had little interest in science can be awakened with a new curiosity about Earth and the processes that dramatically alter it. Science majors experience a heightened interest, with expanded and clarified understanding of natural processes. In response to years of student feedback and discussions with colleagues, we reshaped our courses to focus on natural hazards.

Students who take a natural hazards course greatly improve their knowledge of the dynamic Earth processes that will affect them throughout their lives. They should be able to make educated choices about where to live and work, how to better recognize natural hazards, and to deal with those around them. Perhaps some who take this course will become government officials or policy makers who can change some of the current culture that contributes to major losses from natural disasters.

Undergraduate college students, including nonscience majors, should find the writing clear and stimulating. Our emphasis is to provide them a basis for understanding important hazard-related processes and concepts. This book encourages students to grasp the fundamentals while still appreciating that most issues have complexities that are beyond the current state of scientific knowledge and involve societal aspects beyond the realm of science. Students not majoring in the geosciences may find motivation to continue studies in related areas and to share these experiences with others.

Natural hazards and disasters can be fascinating and even exciting for those who study them. Just don't be on the receiving end!

Living with Nature

Natural hazards, and the disasters that accompany many of them, are an ongoing societal problem. We continue to put ourselves in harm's way, through ignorance or a naïve belief that a looming hazard may affect others but not us. We choose to live in locations that are inherently unsafe.

The expectation that we can control nature through technological change stands in contrast to the fact that natural processes will ultimately prevail. We can choose to live *with* nature or we can try to fight it. Unfortunately, people who choose to live in hazardous locations tend to blame either "nature on the rampage" or others for permitting them to live there. People do not often make such poor choices willfully, but rather through their lack of awareness and understanding of natural processes. Even when they are aware of an extraordinary event that has affected someone else, they somehow believe "it won't happen to me." These themes are revisited throughout the book, as we relate principles to societal behavior and attitudes.

People often decide on their residence or business location based on a desire to live and work in scenic environments without understanding the hazards around them. Once they realize the risks, they often compound the hazards by attempting to modify the environment. Students who read this book should be able to avoid such errors. Toward the end of the course, our students sometimes ask, "So where is a safe place to live?" We often reply that you can choose hazards that you are willing to deal with and live in a specific site or building that you know will minimize impact of that hazard.

It is our hope that by the time students have finished reading this textbook, they should have the basic knowledge to critically evaluate the risks they take and the decisions they make as voters, homeowners, and world citizens.

Our Approach

This text begins with an overview of the dynamic environment in which we live and the variability of natural processes, emphasizing the fact that most daily events are small and generally inconsequential. Larger events are less frequent, though most people understand that they can happen. Fortunately, giant events are infrequent; regrettably, most people are not even aware that such events can happen. Our focus here is on Earth and atmospheric hazards that appear rapidly, often without significant warning.

The main natural hazards covered in the book are earthquakes and volcanic eruptions; extremes of weather, including hurricanes; and floods, landslides, tsunami, wildfires, and asteroid impacts. For each, we examine the nature and processes that drive the hazard, the dangers associated with it, the methods of forecasting or predicting such events, and approaches to their mitigation. Throughout the book, we emphasize interrelationships between hazards, such as the fact that building dams on rivers often leads to greater coastal erosion. Similarly, wildfires generally make slopes more susceptible to floods, landslides, and mudflows.

The book includes chapters on dangers generated within the Earth, including earthquakes, tsunami, and volcanic eruptions. Society has little control over the occurrence of such events but can mitigate their impacts through a deeper understanding that can afford more enlightened choices. The landslides section addresses hazards influenced by a combination of in-ground factors, human actions, and weather, a topic that forms the basis for many of the following chapters. A chapter on sinkholes, subsidence, and swelling soils addresses other destructive in-ground hazards that we can, to some extent, mitigate and that are often subtle yet highly destructive.

The following hazard topics depend on an understanding of the dynamic variations in weather, thunderstorms and tornadoes, so we begin with a chapter to provide that background. The next two chapters on climate change address the overarching atmospheric changes imposed by increasing carbon dioxide and other greenhouse gases that affect weather and many hazards described in the following chapters. Chapters on streams and floods begin with the characteristics and behavior of streams and how human interaction affects both a stream and the people around it. Chapters follow on wave and beach processes, hurricanes and nor'easters, and wildfires. The final chapter addresses asteroid impacts on Earth.

The book is up-to-date and clearly organized, with most of its content derived from current scientific literature and from our own personal experience. It is packed with relevant content on natural hazards, the processes that control them, and the means of avoiding catastrophes. Numerous excellent and informative color photographs, many of them our own, illustrate scientific concepts associated with natural hazards. Diagrams and graphs are clear, straightforward, and instructive.

Extensive illustrations and Case in Point examples bring reality to the discussion of principles and processes. These cases tie the process-based discussions to individual cases and integrate relationships between them. They emphasize the natural processes and human factors that affect disaster outcomes. Illustrative cases are placed at the chapter end to not interrupt continuity of the discussion. Coverage of natural hazards is balanced with excellent examples across North America and the rest of the world. As our global examples illustrate, although the same fundamental processes lead to natural hazards everywhere, the impact of natural disasters can be profoundly different depending on factors such as economic conditions, security, and disaster preparedness.

End-of-chapter material also includes Critical View photos with paired questions, a list of Key Points, Key Terms, Questions for Review, and Critical Thinking Questions.

New to the Fifth Edition

With such a fast-changing and evolving subject as natural hazards, we have extensively revised and added to the content, with emphasis not only on recent events but also on those that best illustrate important issues. We have endeavored to keep material as up-to-date as possible, both with new Cases in Point and in changes in governmental policy that affect people and their hazardous environments. New to this edition is a Survival Guide feature that highlights risk, preparedness, and safety information related to relevant hazards. To make space for new Cases in Point, some older cases have been moved online, where they can be accessed in the CourseMate available at cengagebrain.com.

In recognition of the rapid advances in understanding of climate change and its increasing importance, we now present this important topic in two separate chapters. That material is thoroughly reorganized, rewritten, and revised, with numerous new graphs and photos. Graphs have been updated with the most recent available information.

In addition to these overall changes, some significant additions to individual chapters include the following:

- **Chapters 3 and 4, Earthquakes**, include new coverage of the giant 2015 Nepal earthquake that killed more than 8600 people, destroyed most of the capital, Kathmandu and surrounding cities, and flattened most of its priceless ancient temples. We have added new insights on earthquakes associated with fracking, the latest way to drill for oil and gas. The moderate-size but destructive 2014 earthquake near Napa, California's iconic wine-growing area provides another wake-up call for this region.

- **Chapters 6 and 7, Volcanoes**, include an update on Hawaii's lava flows, which continued into 2015.

- **Chapter 8, Landslides**, features a new Case in Point on the tragic Oso landslide in western Washington, which occurred in a known hazard area that permitted building of a new subdivision.

- **Chapter 10, Weather, Thunderstorms, and Tornadoes**, has been significantly updated and revised. We have added coverage of the polar vortex, a process that is now better understood and more relevant to the public after millions of people in the northeastern United States lived through the bitterly cold winter of 2014. A new Case in Point focuses on the 2013 EF5 tornado that struck Moore, Oklahoma (the fourth in 14 years), killing many people who had no tornado shelters, in spite of federal support to partially pay for them. Another new Case is devoted to the severe California drought.

- **Chapters 11 and 12, Climate Change**, breaks the existing climate change coverage into two updated and expanded chapters. Chapter 11 focuses on processes related to climate change, whereas Chapter 12 focuses on the impacts of climate change and mitigation strategies. Coverage has been significantly expanded to encompass new data and illustrations from the 5th Intergovernmental Panel on Climate Change (IPCC).

- **Chapters 13 and 14, Streams and Floods**, features a new Case in Point about the disastrous 2013 flash floods in the Rocky Mountain foothills near Denver that provided a reminder of the Big Thompson canyon event almost 40 years before.

- **Chapter 16, Hurricanes and Nor'easters**, includes coverage of Hurricane Sandy in late 2012, which was a major wake-up call for those who view a "weak" hurricane as a minor inconvenience.

- **Chapter 17, Wildfires**, includes new Cases in Point about two large fires near Colorado Springs and the tragic Yarnell Hill fire in Arizona that killed 14 professional firefighters.

- **Chapter 18, Asteroid and Comet Impacts**, includes a new Case on the 2013 Chelyabinsk meteor in Russia, which was a frightening near miss that nearly became a catastrophe.

Ancillaries

Instructor Resources

Cengage Learning Testing Powered by Cognero

The Test Bank is offered through Cengage Learning Testing Powered by Cognero and contains multiple-choice, true/false, matching, and discussion exercises. Cengage Learning Testing is a flexible, online system that allows you to author, edit, and manage test bank content, create multiple test versions, and deliver tests from your LMS, your classroom, or wherever you want.

Instructor Companion Site

On the Instructor Companion Site you can access Microsoft PowerPoint™ lecture presentations, the Instructor's Manual, images, videos, animations, and more, all updated for the 5th edition.

Student Resources

Earth Science CourseMate with eBook
ISBN: 9781305866560
Make the most of your study time by accessing everything you need to succeed in one place. Read your text, take notes, review flash cards, watch videos, take practice quizzes, and more, all online with CourseMate.

Virtual Field Trips in Geology, Hazards Edition
ISBN: 9781111668891
The *Virtual Field Trips in Geology*, by Dr. Parvinder S. Sethi from Radford University, are concept-based modules that teach students geology by using famous locations throughout the United States. The Hazards Edition includes geologic hazard concepts, including Mass Wasting, Earthquakes and Seismicity, Volcano Types, Desert Environments, and Running Water. Designed to be used as homework assignments or lab work, the modules use a rich array of multimedia to demonstrate concepts. High-definition videos, images, panoramas, quizzes, and Google Earth layers work together in *Virtual Field Trips in Geology* to bring concepts to life.

Virtual Field Trips in Geology: Complete Set of 15
ISBN: 9780495560692
The *Virtual Field Trips in Geology*, by Dr. Parvinder S. Sethi from Radford University, are concept-based activities that teach you geology by using famous locations throughout the United States. Designed to be used as homework assignments or lab work, the field trips use a rich array of multimedia to demonstrate concepts. High-definition videos, images, panoramas, quizzes, and Google Earth layers work together in *Virtual Field Trips in Geology* to bring concepts to life. Topics include: Desert Environments, Geologic Time, Hydrothermal Activity, Running Water, Sedimentary Rocks, Earthquakes & Seismicity, Glaciers & Glaciation, Igneous Rocks, Mass Wasting, Volcano Types, Plate Tectonics, Mineral Resources, Metamorphism & Metamorphic Rocks, Groundwater, and Shorelines & Shoreline Processes.

Global Geoscience Watch
ISBN: 9781111429058
Updated several times a day, the Global Geoscience Watch is an ideal one-stop site for current events and research projects for all things geoscience! Broken into the four key course areas (Geography, Geology, Meteorology, and Oceanography), you can easily find the most relevant information for the course you are taking. You will have access to the latest information from trusted academic journals, news outlets, and magazines. You also will receive access to statistics, primary sources, case studies, podcasts, and much more!

Acknowledgments

We are grateful to a wide range of people for their assistance in preparing and gathering material for this book, far too many to list individually here. However, we particularly appreciate the help we received from the following:

■ We especially wish to thank Rebecca Heider, Development Editor who not only expertly managed and organized all aspects of recent editions but suggested innumerable and important changes in the manuscript. In large measure, the enhancements are in response to her insight, perception, and skillful editing.

■ For editing and suggested additions: Dr. Dave Alt, University of Montana, emeritus; Ted Anderson; Tony Dunn (San Francisco State University); Shirley Hyndman; Teresa Hyndman; Dr. Duncan Sibley, Dr. Kaz Fujita, and Dr. Tom Vogel from Michigan State University; Dr. Kevin Vranes, University of Colorado, Center for Science and Technology Policy Research.

■ For information and photos on specific sites: Dr. Brian Atwater, USGS; Alexis Bonogofsky, National Wildlife Federation, Billings, MT; Dr. Rebecca Bendick, University of Montana; Michael Burnside, Missoula, MT; Karl Christians, Montana Department of Natural Resources and Conservation; Susan Cannon, USGS; Michael Cline, USGS; Jack Cohen, Fire Sciences Laboratory, U.S. Forest Service; Dr. Paul Cole, Earth Sciences, Plymouth University, UK; Dr. Joel Harper, University of Montana; Bretwood Hickman, University of Washington; Peter Bryn, Hydro.com; Dr. Dan Fornari, Woods Hole Oceanographic Institution, MA; Dr. Kaz Fujita, Michigan State University; Ney Grant, West Coast Flying Adventures; Colin Hardy, Program Manager, Fire, Fuel, Smoke Science, U.S. Forest Service; Dr. Benjamin P. Horton, University of Pennsylvania, Philadelphia; Dr. Roy Hyndman, Pacific Geoscience Center, Saanichton, British Columbia; Sarah Johnson, Digital Globe; Walter Justus, Bureau of Reclamation, Boise, ID; Ulrich Kamp, Geography, University of Montana; Bob Keane, Supervisory Research Ecologist, U.S. Forest Service; Dr. M. Asif Khan, Director, National Center of Excellence in Geology, University of Peshawar, Pakistan; Karen Knudsen, executive director, Clark Fork Coalition; Kristy Kroeker, University of California, Santa Cruz, CA; Dr. Ian Lange, University of Montana; Martin McDermott, McKinney Drilling Co.; Dr. Ian Macdonald, Texas A&M University, Corpus Christi; Andrew MacInnes, Plaquemines Parish, LA; Dr. Jamie MacMahan, Naval Postgraduate School, Monterey, CA; Curtis McDonald, LiveHailMap.com; Tanya Milligan, Zion National Park, UT; Dr. Alan Benimoff, College of Staten Island, NY; Alan Mockridge, Intralink America; Andrew Moore, Kent State University; Jenny Newton, Fire Sciences Laboratory, U.S. Forest Service; Dr. Mark Orzech, Naval Postgraduate School, Monterey, CA; Jennifer Parker, Geography, University of Montana; Dr. Barclay Poling, North Carolina State University; Peter Reid, SCI-Fun, The University of Edinburgh; Dr. Stanley Riggs, Institute for Coastal Science and Policy, East Carolina University; Dr. Vladmir Romanovsky, University of Alaska; Dr. Dave Rudolf, University of Waterloo; Dr. Steve Running, Numerical Terradynamic Simulation Group, University of Montana; Todd Shipman, Arizona Geological Survey; Dr. Duncan Sibley, Michigan State University; Stephen Slaughter, Washington Dept. Natural Resources; Robert B. Smith, University of Utah; Dr. Seth Stein, Northwestern University; Dr. Bob Swenson, Alaska Division of Geological and Geophysical Surveys; Donald Ward, Travis Co., TX; Karen Ward, Terracon, Austin, TX; John M. Thompson; Dr. Ron Wakimoto, Forest Fire Science, University of Montana; Dr. Robert Webb, USGS, Tucson, AZ; Dr. Jeremy Weiss, University of Arizona; Vallerie Webb, Geoeye.com, Thornton, CO; Ann Youberg, Arizona Geological Survey.

■ For providing access to the excavations of the Minoan culture at Akrotiri, Santorini: Dr. Vlachopoulos, head archaeologist, Greece.

■ For assisting our exploration of the restricted excavations at Pompeii, Italy: Pietro Giovanni Guzzo, the site's chief archaeologist.

■ For logistical help: Roberto Caudullo, Catania, Italy; Brian Collins, University of Montana; and Keith Dodson, Brooks/Cole's earth sciences textbook editor at the time of the first edition.

In addition, we appreciate chapter reviewers who suggested improvements that we made in the Fifth Edition: Jennifer Rivers Cole, Northeastern University, Joan E. Fryxell, California State University, James Hibbard, North Carolina State University, David M. Hondula, University of Virginia, Steve Kadel, Glendale Community College, David Mrofka, Mt. San Antonio College.

Donald Hyndman and David Hyndman

■ During Hurricane Sandy the east end of the Mantoloking Bridge, at left, was submerged, and houses on the barrier island battered. Some like the one in the lower left floated off their foundations.

Natural Hazards and Disasters

1

Living in Harm's Way

Why would people choose to put their lives and property at risk? Large numbers of people around the world live and work in notoriously dangerous places—near volcanoes, in floodplains, or on active fault lines. Some are ignorant of potential disasters, but others even rebuild homes destroyed in previous disasters. Sometimes the reasons are cultural or economic. Because volcanic ash degrades into richly productive soil, the areas around volcanoes make good farmland. Large floodplains attract people because they provide good agricultural soil, inexpensive land, and natural transportation corridors. Some people live in a hazardous area because of their job. For understandable reasons, such people live in the wrong places. Hopefully they recognize the hazards and understand the processes involved so they can minimize their risk.

But people also crowd into dangerous areas for frivolous reasons. They build homes at the bases or tops of large cliffs for scenic views, not realizing that big sections can give way

> Those who cannot remember the past are condemned to repeat it.
>
> —George Santayana (Spanish philosopher), 1905

in landslides or rockfalls. They build beside picturesque streams without realizing they have put themselves in a flood zone. Far too many people build houses in the woods because they enjoy the seclusion and scenery of this natural setting without understanding their risk from wildfires. Others choose to live along edges of sea bluffs where they can enjoy ocean views, or on the beach to experience the ocean more intimately. But in these locations they also expose themselves to coastal storms. In October 2012, the devastating effects of Hurricane Sandy, only a Category 1 storm, reminded many people of the hazards of living on the Atlantic coast.

Some natural catastrophe experts say these people have chosen to live in "idiot zones." But people don't usually reside in hazardous areas knowingly—they generally don't understand or recognize the hazards. However, they might as well choose to park their cars on a rarely used railroad track. Trains don't come frequently, but the next one might come any minute.

Catastrophic natural hazards are much harder to avoid than passing freight trains; we may not recognize the signs of imminent catastrophes because these events are infrequent. So many decades or centuries may pass between eruptions of a large volcano that most people forget it is active. Many people live so long on a valley floor without seeing a big flood that they forget it is a floodplain. The great disaster of a century ago is long forgotten, so folks move into the path of a calamity that may not arrive today or tomorrow, but it is just a matter of time.

Catastrophes in Nature

Geologic processes, like erosion, have produced large effects over the course of Earth's vast history, carving out valleys or changing the shape of coastlines. While some processes operate slowly and gradually, infrequent catastrophic events have sudden and major impacts.

Although streams may experience a few days or weeks of flooding each year, major floods occurring once every few decades do far more damage than all of the intervening floods put together. Soil moves slowly downslope by creep, but occasionally a huge part of a slope may slide. Pebbles roll down a rocky slope daily, but every once and a while a giant boulder comes crashing down (**FIGURE 1-1**). Mountains grow higher, sometimes slowly, but more commonly by sudden movements. During an earthquake, a mountain can abruptly rise several meters above an adjacent valley.

Some natural events involve disruption of a temporary *equilibrium*, or balance, between opposing influences. Unstable slopes, for example, may hang precariously for thousands of years, held there by friction along a slip surface until some small perturbation, such as water soaking in from a large rainstorm, sets them loose. Similarly, the opposite sides of a fault may stick until slowly building stress finally tears them loose, triggering an earthquake. A bulge may form on a volcano as molten magma slowly rises into it, then it collapses as the volcano erupts. The behavior of these natural systems is somewhat analogous to a piece of plastic wrap that can stretch up to a point, until it suddenly tears.

FIGURE 1-1 The Unexpected

On December 12, 2013, a huge mass of sandstone separated from a prominent cliff above homes along Highway 9 in the community of Rockville, Utah, instantly killing the two home owners. This hazardous area of homes was highlighted in a Utah Geological Survey report in 2013. The same home was pictured as in a dangerous location in the previous edition of this textbook printed in late 2012, one year before the disaster.

People watching Earth processes move at their normal and unexciting pace rarely pause to imagine what might happen if that slow pace were suddenly punctuated by a major event. The fisherman enjoying a quiet afternoon trout fishing in a small stream can hardly imagine how a 100-year flood might transform the scene. Someone gazing at a serene, snow-covered mountain can hardly imagine it erupting in an explosive blast of hot ash followed by destructive mudflows racing down its flanks. Large or even gigantic events are a part of nature. Such abrupt events produce large results that can be disastrous if they affect people.

Human Impact of Natural Disasters

When a natural process poses a threat to human life or property, we call it a **natural hazard**. Many geologic processes are potentially hazardous. For example, streams flood as part of their natural process and become a hazard to those living nearby. A hazard is a **natural disaster** when the event causes significant damage to life or property. A moderate flood that spills over a floodplain every few years does not often wreak havoc, but when a major flood strikes, it may lead to a disaster that kills or displaces many people. When a natural event kills or injures large numbers of people or causes extensive property damage, it is called a **catastrophe**.

The potential impact of a natural disaster is related not only to the size of the event but also to its effect on the public. A natural event in a thinly populated area can hardly pose a major hazard. For example, the magnitude 7.6 earthquake that struck the southwest corner of New Zealand on July 15, 2009, was severe but posed little threat because it happened in a region with few people or buildings. In contrast, the much smaller January 12, 2010, magnitude 7.0 earthquake in Haiti killed more than 46,000 (**FIGURE 1-2**). In another example, the eruption of Mt. St. Helens in 1980

caused few fatalities and remarkably little property damage simply because the area surrounding the mountain is sparsely populated. On the other hand, a similar eruption of Vesuvius, in the heavily populated outskirts of Naples, Italy, could kill hundreds of thousands of people and cause property damage beyond reckoning.

You might assume that more fatalities occur as a result of dramatic events, such as large earthquakes, volcanic eruptions, hurricanes, or tornadoes. However, some of the most dramatic natural hazards occur infrequently or in restricted areas, so they cause fewer deaths than more common and less dramatic hazards such as floods or droughts. **FIGURE 1-3** shows the approximate proportions of fatalities caused by typical natural hazards in the United States.

In the United States, heat and drought together account for the largest numbers of deaths. In fact, there were more U.S. deaths from heat waves between 1997 and 2008 than from any other type of natural hazard. In addition to heat stress, summer heat wave fatalities can result from dehydration and other factors; the very young, the very old, and the poor are affected the most. The same populations are vulnerable during winter weather, the third most deadly hazard in the United States. Winter deaths often involve hypothermia, but some surveys include, for example, auto accidents caused by icy roads.

Flooding is the second most deadly hazard in the United States, accounting for 16 percent of fatalities between 1986 and 2008. Fatalities from flooding can result from

FIGURE 1-2 A Disaster Takes a High Toll

Searchers dig for survivors of the Haiti earthquake of January 12, 2010, which killed more than 316,000, mostly in concrete and cinder block buildings with little or no reinforcing steel.

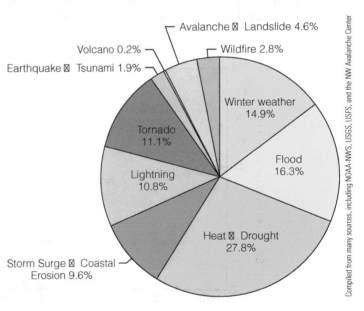

FIGURE 1-3 Hazard-Related Deaths

Approximate percentages of U.S. fatalities due to different groups of natural hazards from 1986 to 2008, when such data are readily available. For hazardous events that are rare or highly variable from year to year (earthquakes and tsunami, volcanic eruptions, and hurricanes), a 69-year record from 1940 to 2008 was used.

hurricane-driven floods; some surveys place them in the hurricane category rather than floods.

The number of deaths from a given hazard can vary significantly from year to year due to rare, major events. For example, there were about 1800 hurricane-related deaths in 2005 when Hurricane Katrina struck, compared with zero in other years. The rate of fatalities can also change over time as a result of safety measures or trends in leisure activities. Lightning deaths were once among the most common hazard-related causes of death, but associated casualties have declined significantly over the past 50 years, due in part to satellite radar and better weather forecasting. In contrast, avalanche deaths have increased significantly over a similar period, a change that seems to be associated with increased snowmobile use and skiing in mountain terrains.

Some natural hazards can cause serious physical damage to land or man-made structures, some are deadly for people, and others are destructive to both. The type of damage sustained as a result of a natural disaster also depends on the economic development of the area where it occurs. In developing countries, there are increasing numbers of deaths from natural disasters, whereas in developed countries, there are typically greater economic losses. This is because developing countries show dramatic increases in populations relegated to marginal and hazardous land on steep slopes and near rivers. Such populations also live in poorly constructed buildings and have less ability to evacuate as hazards loom; many lack transportation and financial ability to survive away from their homes.

For an example of this phenomenon, in 2010, earthquakes of similar sizes (magnitude 7.0) struck Haiti, a poor, developing country, and New Zealand, a prosperous, developed country. In Haiti, between 46,000 and 316,000 people were killed (U.S. government versus Haitian government estimates), mostly in the collapse of poorly built masonry buildings. Total damages were estimated to be about U.S. $7.8 billion. In contrast, only 185 people died in the New Zealand earthquake, which also occurred near a populous area. New Zealand's buildings were generally well constructed. Despite this, damages were still estimated to be about U.S. $6.5 billion.

The average annual cost of natural hazards has increased dramatically over the last several decades (**FIGURE 1-4**). This is due in part to the increase in world population, which doubled in the 40 years between 1959 and 1999. By July 2015, it reached 7.3 billion. It is also a function of the increased value of properties at risk and to human migration to more hazardous areas. Overall losses have increased even faster than population growth. Population increases in urban and coastal settings result in more people crowding into land that is subject to major natural events. In effect, people place themselves in the path of unusual, sometimes catastrophic events. Economic centers of society are increasingly concentrated in larger urban areas that tend to expand into regions previously considered undesirable, including those with greater exposure to natural hazards.

The reality of climate change adds an additional dimension to these problems; it is one of the greatest challenges facing the human race. Scientists agree that global temperatures are rising. As world population grows and large numbers of people become more affluent and use larger amounts of resources, greenhouse gas emissions increase dramatically.

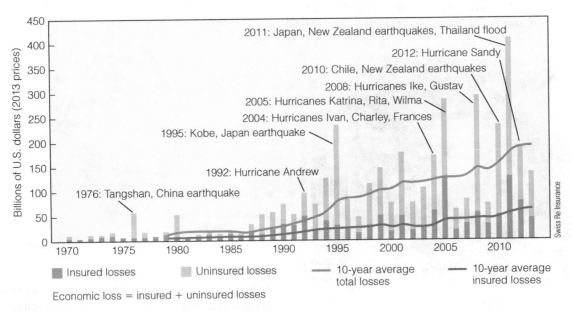

Economic loss = insured + uninsured losses

FIGURE 1-4 Increasing Costs of Natural Hazards

The cost of natural hazards is increasing worldwide. The 2011 earthquake and tsunami in Japan alone caused losses of about $235 billion.

FIGURE 1-5 Homes at Risk

Homes along channels in the Mekong River delta in southern Vietnam are almost in the water under normal circumstances. They would be washed away in the next major cyclone.

Our generation of greenhouse gases seems likely to cause population collapse in some parts of the world, especially in poor areas most affected by natural hazards. People's living conditions will be severely disrupted. Millions will die from increased incidence of storms and coastal flooding, heat stroke, dehydration, famine, disease, and wars over water, food, heating fuel, and other resources.

Climate change is expected to lead to more rapid erosion of coastlines, along with more extreme weather events that cause landslides, floods, hurricanes, and wildfires. Some small islands in the Indian Ocean, far from the 2004 Sumatra earthquake's epicenter, were completely overwashed by tsunami waves. As sea level continues to rise, such low-lying islands will gradually submerge, even without a catastrophic event. Extensive low-lying coastal regions of major river deltas in Southeast Asia feed and are homes to millions of poor people. Deltas of the Ganges and Brahmaputra Rivers in Bangladesh, the Irrawaddy River in Myanmar, and the Mekong River in Vietnam and Cambodia are subject to 2-m ocean tides more than 200 km upstream (**FIGURE 1-5**). Major storms can submerge most of the deltas, including all of their rice fields and homes, under more than 2 m of water, with storm waves on top of that. Sea-level rise with climate change is expected to worsen those effects, killing thousands in major typhoons. The number of hurricanes has not increased significantly, but since 1990 the annual number of the most intense storms—Categories 4 and 5—nearly doubled to 18 worldwide in 2005 although the future trend remains unclear. Hurricane development and intensity depend on energy provided by higher sea-surface temperatures.

Predicting Catastrophe

A catastrophic natural event is unstoppable, so the best way to avoid it would be to predict its occurrence and get out of the way. Unfortunately, there have been few well-documented cases of accurate prediction, and even the ones on record may have involved luck more than science. Use of the same techniques in similar circumstances has resulted in false alarms and failure to correctly predict disasters.

Many people have sought to find predictable cycles in natural events. Those that occur at predictable intervals are called *cyclic events*. However, most recurrent events are not really cyclic; too many variables control their behavior. Even with cyclic events, overlapping cycles make resultant extremes noncyclic, which affects the predictability of a specific event. So far as anyone can tell, most episodes, large and small, occur at seemingly random and essentially unpredictable intervals.

Although scientists cannot predict exactly when an event will occur, based on past experience they can often **forecast** the chance that a hazardous event will occur in a region within a few decades. For example, they can forecast that there will be a large earthquake in the San Francisco Bay region over the next several decades, or that Mt. Shasta will likely erupt sometime in the next few centuries. In many cases, their advice can greatly reduce the danger to lives and property.

Ask a stockbroker where the market is going, and you will probably hear that it will continue to do what it has done during recent weeks. Ask a scientist to forecast an event, and he or she will probably look to the geologically recent past and forecast more of the same; in other words, *the past is the key to the future*. Most forecasts are based on linear projections of past experience. However, we must be careful to look at a long enough sample of the past to see prospects for the future. Many people lose money in the stock market because *short-term* past experience is not always a good indicator of what will happen in the future.

Similarly, statistical forecasts are simply a refinement of past recorded experiences. They are typically expressed as **recurrence intervals** that relate to the probability that a natural event of a particular size, or **magnitude**, will happen within a certain period of time, or with a certain **frequency**. For example, the history of movement along a fault may indicate that it is likely to produce an earthquake of a certain size once every hundred years on average.

A recurrence interval is not, however, a fixed schedule for events. Recurrence intervals can tell us that a 50-year flood is likely to happen sometime in the next several decades but *not* that such floods occur at intervals of 50 years. Many people do not realize the inherent danger of an unusual occurrence, or they believe that they will not be affected in their lifetimes because such events occur infrequently. That inference often incorrectly assumes that the probability of another severe event is lower for a considerable length of time after a major event. In fact, even if a 50-year flood occurred last year, that does not indicate that there will not be another one this year or for the next ten years.

To understand why this is the case, take a minute to review probabilities. Flip a coin, and the chance that it will come up heads is 50%. Flip it again, and the chance is again 50%. If it comes up heads five times in a row, the next flip still has a 50% chance of coming up heads. So it goes with floods and many other kinds of apparently random natural events. The chance that someone's favorite fishing stream will stage a 50-year flood this year and every year is 1 in 50, regardless of what it may have done during the last few years.

As an example of the limitations of recurrence intervals, consider the case of Tokyo. This enormous city is subject to devastating earthquakes that for more than 500 years came at intervals of close to 70 years. The last major earthquake ravaged Tokyo in 1923, so everyone involved awaited 1993 with considerable apprehension. The risk steadily increased during those years as the strain across the fault zone grew, as did the size of the population at risk. More than 20 years later, no large earthquake has occurred. Obviously, the recurrence interval does not predict events at equal intervals, in spite of the 500-year Japanese historical record. Nonetheless, the knowledge that scientists have of the pattern of occurrences here helps them assess risk and prepare for the eventual earthquake. Experts forecast that there is a 70% chance that a major quake will strike that region in the next 30 years.

To estimate the recurrence interval of a particular kind of natural event, we typically plot a graph of each event size versus the time interval between sequential individual events. Such plots often make curved lines that cannot be reliably extrapolated to larger events that might lurk in the future (**FIGURE 1-6**). Plotting the same data on a logarithmic scale often leads to a straight-line graph that can be

extrapolated to values larger than those in the historical record. Whether the extrapolation produces a reliable result is another question.

The probability of the occurrence of an event is related to the magnitude of the event. We see huge numbers of small events, many fewer large events, and only a rare giant event (**By the Numbers 1-1:** Relationship between Frequency and Magnitude). The infrequent occurrence of giant events means it is hard to study them, but it is often rewarding to study small events because they may well be smaller-scale models of their uncommon larger counterparts that may occur in the future.

Many geologic features look the same regardless of their size, a quality that makes them **fractal**. A broadly generalized map of the United States might show the Mississippi River with no tributaries smaller than the Ohio and Missouri Rivers. A more detailed map shows many smaller tributaries. An even more detailed map shows still more. The number of tributaries depends on the scale of the map, but the general branching pattern looks *similar across a wide range of scales* (**FIGURE 1-7**). Patterns apparent on a small scale quite commonly resemble patterns that exist on much larger scales that cannot be easily perceived. This means that small events may provide insight into huge ones that occurred in the distant past but are larger than any seen in historical time; we may find evidence of these big events if we search. The geologic record provides evidence for massive natural catastrophes in the Earth's distant past, such as the impact of a large asteroid that caused the extinction of the dinosaurs. We need to be aware of the potential for such extreme events in the future.

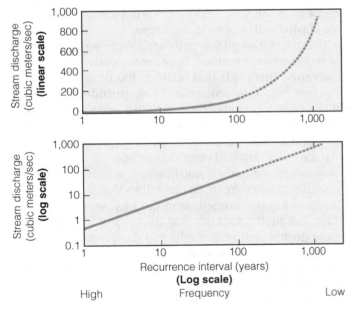

FIGURE 1-6 Recurrence Interval

If major events are plotted on a linear scale (top graph, vertical axis), the results often fall along a curve that cannot be extrapolated to larger possible future events. If the same events are plotted on a logarithmic scale (bottom graph), the results often fall along a straight line that can use historical data to forecast what to expect in future events.

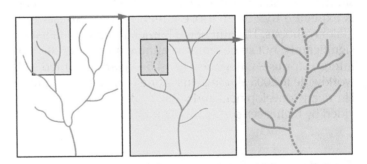

FIGURE 1-7 Fractal Systems

The general *pattern* of a branching stream looks similar regardless of scale—from a less-detailed map on the left to the most detailed map on the right.

It is impossible in our current state of knowledge to predict most natural events, even if we understand in a general way what controls them. The problem of avoiding natural disasters is like the problem drivers face in avoiding collisions with trains. They can do nothing to prevent trains, so they must look and listen. We have no way of knowing how firm the natural restraints on a landslide, fault, or volcano may be. We also do not generally know what changes are occurring at depth. But we can be confident that the landslide or fault will eventually move or that the volcano will erupt. And we can reasonably understand what those events will involve when they finally happen.

Relationships among Events

Although randomness is a factor in forecasting disasters, most natural events do not occur as randomly as tosses of a coin. Some events are directly related to others—formed as a direct consequence of another event (**FIGURE 1-8**). For example, the slow movement of Earth's huge outer layers colliding or sliding past one another clearly explains the driving forces behind volcanic eruptions and earthquakes. Heavy or prolonged rainfall can cause a flood or a landslide. But are some events unrelated? Could any of the arrows in Figure 1-8 be reversed?

Past events can also create a contingency that influences future events. It is certainly true, for example, that sudden movement on a fault causes an earthquake. But the same movement also changes the stress on other parts of the fault and probably on other faults in the region, so the next earthquake will likely differ considerably from the last. What if, after an earthquake movement on a fault, one side of the fault is now across from a very slippery area of rock on the other side of the fault. Might then the fault break more easily and the next movement on the fault come sooner? Similar complex relationships arise with many other types of destructive natural events.

Some processes result in still more rapid changes—a **feedback effect**. For example, global warming causes more rapid melting of Arctic sea ice. The resulting darker sea water absorbs more of the Sun's energy than the white ice, which in turn causes even more sea ice melting. Similarly, global warming causes faster melting of the Greenland and Antarctic ice sheets. More meltwater pours through fractures to the base of the ice, where it lubricates movement, accelerating the flow of ice toward the ocean. This leads to more rapid crumbling of the toes of glaciers to form icebergs that melt in the ocean.

In other cases, an increase in one factor may actually lead to a decrease in a related result. Often as costs of a product or service go up, usage goes down. With increased costs of hydrocarbon fuels, people conserve more and thus burn less. A rapid increase in the price of gasoline in 2008 led people to drive less and to trade in large SUVs and trucks for smaller cars. In some places, commuter train, bus, and bicycle use increased dramatically. With the rising cost of electricity, people are switching to compact fluorescent bulbs and using less air conditioning. These changes had a noticeable effect on greenhouse gas emissions and their effect on climate change (discussed in Chapter 12).

Sometimes major natural events are preceded by a series of smaller **precursor events**, which may warn of the impending disaster. Geologists studying the stirrings of Mt. St. Helens, Washington, before its catastrophic eruption in 1980 monitored swarms of earthquakes and decided that most of these recorded the movements of rising magma as it squeezed upward, expanding the volcano. Precursor events alert scientists to the potential for larger events, but events that appear to be precursors are not always followed by a major event.

The relationships among events are not always clear. For example, an earthquake occurred at the instant Mt. St. Helens exploded, and the expanding bulge over the rising magma collapsed in a huge landslide. Neither the landslide nor the earthquake caused the formation of molten magma, but did they trigger the final eruption? If so, which one triggered the other—the earthquake, the landslide, or the eruption? One or more of these possibilities could be true in different cases.

Events can also overlap to amplify an effect. Most natural disasters happen when a number of unrelated variables overlap in such a way that they reinforce each other to amplify an effect. If the high water of a hurricane storm surge happens to arrive at the coast during the daily high tide, the two reinforce each other to produce a much higher storm surge (**FIGURE 1-9**). If this occurs on a section of coast that happens to have a large population, then the situation can become a major disaster. Such a coincidence caused the catastrophic hurricane that killed 8000 people in Galveston, Texas, in 1900.

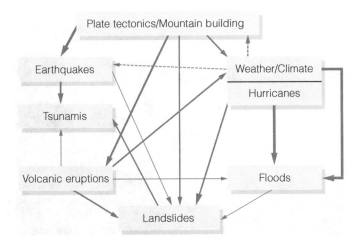

FIGURE 1-8 Interactions among Natural Hazards

Some natural disasters are directly related to others. The bolder arrows in this flowchart indicate stronger influences. Can you come up with words to describe these influences?

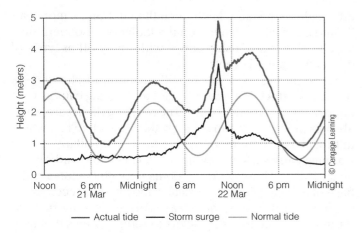

FIGURE 1-9 **Amplification of Overlapping Effects**

If events overlap, their effects can amplify one another. In this example, a storm surge (black line) can be especially high if it coincides with high tide (red line). The blue line shows the much higher tide that resulted when the tide overlapped with the storm surge.

Mitigating Hazards

Because natural disasters are not easily predicted, it falls to governments and individuals to assess their risk and prepare for and mitigate the effects of disasters. **Mitigation** refers to efforts to prepare for a disaster and reduce its damage. Mitigation can include engineering projects such as levees, as well as government policies and public education efforts. "Soft" solutions for hazardous areas include zoning to prevent building in certain regions and strict building codes, which minimize damage and are much less expensive in the long run. "Hard" alternatives, including levees on rivers and riprap along coasts, are expensive, often short-term, and create other problems. Throughout this book, we examine mitigation strategies related to specific disasters.

Land-Use Planning

One way to reduce losses from natural disasters is to find out where disasters are likely to occur and restrict development there, using **land-use planning**. Ideally, we would prevent development along major active faults by reserving that land for parks and natural areas. We should also limit housing and industrial development on floodplains to minimize flood damage and along the coast to reduce hurricane and coastal erosion losses. Limiting building near active volcanoes and the river valleys that drain them can curtail the hazards associated with eruptions.

It is hard, however, to impose land-use restrictions in many areas because such imposition tends to come too late. Many hazardous areas are already heavily populated, perhaps even saturated with inhabitants. Many people want to live as close as they can to a coast or a river and resent being told that they cannot; they oppose attempts at land-use restrictions because they feel it infringes on their property rights. Almost any attempt to regulate land use in the public interest is likely to ignite intense political and legal opposition.

Developers, companies, and even governments often aggravate hazards by allowing—or even encouraging—people to move into hazardous areas. Many developers and private individuals view restrictive zoning as an infringement on their rights to do as they wish with their land. Developers, real estate agents, and some companies are reluctant to admit the existence of hazards that may affect a property for fear of lessening its value and scaring off potential clients (**FIGURE 1-10**). Most local governments consider news of hazards bad for growth and business. They shun restrictive zoning or minimize possible dangers for fear of inhibiting improvements in their tax base. As in other venues, different groups have different objectives. Some are most concerned with economics, others with safety, still others with the environment.

Should landowners be permitted to do whatever they wish with their property? Property rights advocates often say yes. If a governmental entity permits building on land within its jurisdiction, should the taxpayers in the district shoulder the responsibility if there is a disaster? Should the government inform a buyer that a property is in a hazardous location and what the hazards are? Should a landowner be prevented from developing a piece of property that might be subjected to a disaster? If so, has the government effectively taken the landowner's anticipated value without compensation, a taking characterized in the courts as **reverse condemnation?**

FIGURE 1-10 **Risky Development**

Some developers seem unconcerned with the hazards that may affect the property they sell. High spring runoff floods this proposed development site in Missoula, Montana.

What about personal responsibility? As adults we like to think that we are responsible for our actions. That assumes, of course, that we know what we are doing and understand the consequences of our actions. If we build in the forest, surrounded by brush and trees, who should be responsible for fighting a forest fire (**FIGURE 1-11**)? Who should pay if we suffer loss? If we decide to build our home close to a stream, do we really understand enough about the natural behavior of streams to safely and responsibly do that? If the government were to restrict us from building on our own property, would they be infringing on our rights? If a future major flood wipes out our investment or causes severe damage to a downstream neighbor who sues us because of our construction, then what? Are we then likely to blame the government for permitting us to build there in the first place? If we demand personal rights, we need to be responsible for the consequences of our actions.

If you buy a property you later decide is at risk of a hazard, what responsibility do you have to a potential buyer? In many aspects of society, property sellers are held responsible if they are aware of some aspect of a property that is dangerous or damaged. Home owners commonly blame—and often sue—others for damages to property they have purchased. However, perhaps they should have remembered the old adage, "Buyer beware."

Insurance

Some mitigation strategies are designed to help with recovery once a disaster occurs. **Insurance** is one way to lessen the financial impact of disasters after the fact. People buy property insurance to shield themselves from major losses

Texas Parks and Wildlife

FIGURE 1-11 Who Should Pay?
Remains of a home surrounded by brush and trees that was destroyed by the Bastrop fire in Texas, 2011.

they cannot afford. Insurance companies use a formula for risk to establish premium rates for policies. **Risk** is essentially a hazard considered in the light of its recurrence interval and expected costs (**By the Numbers 1-2**: Assessing Risk). The greater the hazard and the shorter its recurrence interval, the greater the risk.

In most cases, a company can estimate the cost of a hazard event to a useful degree of accuracy, but they can only guess at its recurrence interval, and therefore the level of risk. The history of experience with a given natural hazard in any area of North America is typically less than 200 years. Large events recur, on average, only every few decades or few hundred years or even more rarely. In some cases, most notably floods, the hazard and its recurrence interval are both firmly enough established to support a rational estimate of risk. But the amount of risk and the potential cost to a company can be so large that a catastrophic event would put the company out of business.

The uncertainties of estimating risk make it impossible for private insurance companies to offer affordable policies that protect against many kinds of natural disasters. As a result, insurance is generally available for events that present relatively little risk, mainly those with more or less dependably long recurrence intervals. In high-risk areas for a particular hazard, for example Florida or Louisiana for hurricanes and sinkholes, insurance companies may either charge very high insurance premiums to cover their risks or refuse to cover damages from such hazards. In those states, nonprofit state programs have been formed to provide insurance that is not otherwise available. In California, where the risks and expected costs of earthquake damages are very high, insurance companies are required by law to provide earthquake coverage. As a result, companies now make insurance available through the California Earthquake Authority, a consortium of companies, in order to spread out their risks.

Insurance for some natural hazards is simply not available. Landslides, most mudflows, and ground settling or swelling are too risky for companies, and each potential hazard area would have to be individually studied by a scientist or engineer who specialized in such a hazard. The large number of variables makes the risk too difficult to quantify; it is too expensive to estimate the different risks for the relatively small areas involved.

People who lose their houses in landslides may not only lose what they have already paid into the mortgage or home loan, but can be obligated to continue paying off a loan on a house that no longer exists. In some states, such as California, there are laws preventing what are called "deficiency judgments" against such mortgage holders. This permits home owners to walk away from their destroyed homes, and the bank cannot go after them for the remainder of the loan. Banks and others that make loans or provide insurance on property need to therefore be aware of natural processes and risks to their investments.

The Role of Government

The U.S. and Canadian governments are involved in many aspects of natural hazard mitigation. They conduct and sponsor research into the nature and behavior of many kinds of natural disasters. They attempt to forecast hazardous events and mitigate the damage and loss of life they cause. Governmental programs are split among several agencies.

The U.S. Geological Survey (USGS) and Geological Survey of Canada (GSC) are heavily involved in earthquake and volcano research, as well as in studying and monitoring stream behavior and flow. The National Weather Service monitors rainfall and severe weather and uses this and the USGS data to try to forecast storms and floods.

The Federal Emergency Management Agency (FEMA) was created in 1979, primarily to bring order to the chaos of relief efforts that seemed invariably to emerge after natural disasters. After the hugely destructive Midwestern floods of 1993, it has increasingly emphasized hazard reduction. Rather than pay victims to rebuild in their original unsafe locations, such as floodplains, the agency now focuses on relocating them. Passage of the Disaster Mitigation Act in 2000 signals greater emphasis on identifying and assessing risks before natural disasters strike and taking steps to minimize potential losses. The act funds programs for hazard mitigation and disaster relief through FEMA, the U.S. Forest Service, and the Bureau of Land Management.

To determine risk levels and estimate loss potential from earthquakes, federal agencies such as FEMA consider potential hazards, inventories of the hazards, direct damages, induced damages, direct economic and social losses, and indirect losses. To address the hazards and potential damages, they need to understand the hazards and processes that drive them.

Unfortunately, some government policies can be counterproductive, especially when politics enter the equation. In some cases, disaster assistance continues to be provided without a large cost-sharing component from states and local organizations. Thus, local governments continue to lobby Congress for funds to pay for losses but lack incentive to do much about the causes. FEMA is charged with rendering assistance following disasters; it continues to provide funds for victims of earthquakes, floods, hurricanes, and other hazards. It remains reactive to disasters, as it should be, but it is only beginning to be proactive in eliminating the causes of future disasters. Congress continues to fund multimillion-dollar Army Corps of Engineers projects to build levees along rivers and replenish sand on beaches. The Small Business Administration disaster loan program continues to subsidize credit to finance rebuilding in hazardous locations. The federal tax code also subsidizes building in both safe and hazardous sites. Real estate developers benefit from tax deductions, and ownership costs, such as mortgage interest and property taxes, can be deducted from income. A part of uninsured "casualty losses" can still be deducted from a disaster victim's income taxes. Such policies do not discourage future damages from natural hazards.

The Role of Public Education

Much is now known about natural hazards and the negative impacts they have on people and their property. It would seem obvious that any logical person would avoid such potential damages or at least modify their behavior or their property to minimize such effects. However, most people are not knowledgeable about potential hazards, and human nature is not always rational.

Unfortunately, a person who has not been adversely affected in a serious way is much less likely to take specific steps to reduce the consequences of a potential hazard. Migration of the population toward the Gulf and Atlantic coasts accelerated in the last half of the twentieth century and still continues. Most of those new residents, including developers and builders, are not very familiar with the power of coastal storms. Even where a hazard is apparent, people are often slow to respond. Is it likely to happen? Will I have a major loss? Can I do anything to reduce the loss? How much time will it take, and how much will it cost? Who else has experienced such a hazard?

Several federal agencies have programs to foster public awareness and education. The Emergency Management Institute—in cooperation with FEMA, the National Oceanic and Atmospheric Administration (NOAA), USGS, and other agencies—provides courses and workshops to educate the public and governmental officials. Some state emergency management agencies, in partnership with FEMA and other federal entities, provide workshops, reports, and informational materials on specific natural hazards. Where the risk of earthquakes is high, the government places emphasis on preparing the public through drills and education programs (**FIGURE 1-12**).

Some people are receptive to making changes in the face of potential hazards. Some are not. The distinction depends partly on knowledge, experience, and whether they feel vulnerable. A person whose house was badly damaged in an earthquake is likely to either move to a less earthquake-prone area or live in a house that is well braced for earthquake

FIGURE 1-12 Earthquake Drill

The Great ShakeOut Earthquake Drill, October 16, 2014. Middle school students in Reston, Virginia, participate.

resistance. People losing their homes to a landslide are more likely to avoid living near a steep slope. The best window of opportunity for effective hazard reduction is immediately following a disaster of the same type. Studies show that this opportunity is short—generally, not more than two or three months.

Different Ground Rules for the Poor

In countries where poverty is widespread, the forces that drive many people's behavior are different than those in prosperous countries. Food and shelter dictate where they live. In Guatemala and Nicaragua, giant corporate farms now control most of the fertile valley bottoms, leaving peasants little choice but to work for them in the fields at minimal wages and to provide their own shelter on the steep landslide-prone hillsides. Their choice of steep hillsides as a place to live is a hazardous one, but they have little choice to survive. Compounding the problem is that fertility rates and population growth in desperately poor countries are among the highest in the world, so more people are forced to live in less suitable areas.

The disaster differences between poor and more affluent countries were accentuated in early 2010 by the earthquakes in Haiti and Chile. The magnitude 7.0 Haiti earthquake, which struck near the capital city of Port-au-Prince, killed more than 159,000 people and left 1.3 million homeless, whereas the magnitude 8.8 earthquake in Chile killed about 800. Although the Chile earthquake was about 500 times stronger at its epicenter, it caused far fewer fatalities. The greatest difference was that Haiti has no building codes, and even if it did, most people have no money

*World Bank, 2014.

to follow them. The annual per capita income for Haiti is only $834*, so most people live on only about $2 per day. Most are uneducated and cannot even read or write. Chile, however, is the wealthiest country in Latin America with an annual per capita income of $14,520*. Being along one of the world's most active subduction zones, Chile is very familiar with earthquakes; its building codes, upgraded after the 1960 earthquake (the world's largest), are among the strongest in the world.

The 2015 magnitude 7.8 earthquake in Nepal provided a severe reminder of the hazards of a small, poor country living in the collision zone of two major tectonic plates. The ancient Indian plate's ongoing collision with the Asian plate crumples the country to raise the Himalayas, the highest mountain range in the world.

Walls of most buildings in the country were constructed from loose stone or bricks with little or no mortar. Heavy poured concrete floors of two- to four-story buildings were supported on thin concrete posts with little or no reinforcing. Shaking from that strong earthquake on such heavy, weak, and rigid structures collapsed many like a tall stack of kid's blocks. Large aftershocks brought down many buildings weakened by the main shocks. More than 9000 people died, most crushed under tons of masonry (**FIGURE 1-13**).

The example of Hurricane Katrina, in New Orleans in 2005, provides additional understanding of the effects of a catastrophic event on poor people, even in an affluent country (**FIGURE 1-14**). Despite clear prediction of a dangerous storm approaching, many residents had no means to leave the city and nowhere to go even if they could have left. Ten days after the storm, an estimated 10,000 people still refused to leave their homes in flooded and seriously contaminated

FIGURE 1-13 Earthquake Aftermath

Buildings in Kathmandu and most other cities in Nepal collapsed into piles of bricks and stone, providing occupants little chance of survival.

FIGURE 1-14 Failed to Evacuate

Survivors of Hurricane Katrina wait on a roof for rescue. Why did they not evacuate when warned? Lack of transportation? Previous false alarms?

areas. The predominantly poor people from the eastern part of the city, who were unable to evacuate before the storm's arrival, made up most of the flood survivors who crowded into the Superdome and the Convention Center during the storm. Before Katrina, 23% of New Orleans residents lived below the poverty line. Many survived from day to day, had no savings or working cars, and lived in rented homes or apartments. Less than half of New Orleans residents had flood insurance.

A major problem for many families—especially the poor—was separation of family members during evacuation and the storm. For example, a mother or father evacuated with children but left behind an elderly parent who couldn't be moved. In some cases, parents were separated from their children during evacuation, or one parent left to help a relative, friend, or neighbor and was unable to return. Because their home was no longer a point of reference, communication became difficult or impossible.

In the case of Hurricane Katrina, evacuees' medical problems were compounded by the storm. Patients could not find their doctors and doctors could not find their patients. Patients on prescription medications or undergoing specialized treatments often could not remember the names of their medicines or the details of their treatments. Years of medical records were lost.

For such poor societies, the reduction of vulnerability to natural hazards does not depend much on strengthening the zoning restrictions against living in dangerous areas or improving warning systems, though education can help. More so, it depends on cultural, economic, and political factors. Because more affluent individuals and corporations control many of those factors, the poor are left to fend for themselves.

Living with Nature

Catastrophic events are natural and normal processes, but the most common human reaction to a current or potential catastrophe is to try to stop ongoing damage by controlling nature. In our modern world, it is sometimes hard to believe that scientists and engineers cannot protect us from natural disasters by predicting them or building barriers to withstand them.

Unfortunately, we cannot change natural system behaviors, because we cannot change natural laws. Most commonly, our attempts tend only to temporarily hinder a natural process while diverting its damaging energy to other locations. In other cases, our attempts cause energy to build up and produce more severe damage later.

If, through lack of forethought, you find yourself in a hazardous location, what can you do about it? You might build a levee to protect your land from flooding. Or you might build a rock wall in the surf to stop sand from leaving your beach and undercutting the hill under your house.

If you do any of these things, however, you merely transfer the problem elsewhere, to someone else, or to a later point in time. For example, if you build a levee to prevent a river from spreading over a floodplain and damaging your property, the flood level past the levee will be higher than it would have been without it. Constricting river flow with a levee also backs up floodwater, potentially causing flooding of an upstream neighbor's property. Deeper water also flows faster past your levee, so it may cause more erosion of a downstream neighbor's riverbanks.

Overall, as a society, we seem to be following trends recognized by Jared Diamond in his book *Guns, Germs, and Steel: The Fates of Human Societies* (1999). Diamond recounts the collapse of ancient civilizations that failed to anticipate a problem, failed to recognize an existing problem, or neglected to fix the problem until it was too late. For gradual problems, many even denied that there was a problem. In some cases, they denied that they themselves were the cause of the problem and therefore could not see a solution. Does this sound like today? Throughout this book, we address natural hazards that affect us, some sudden and unexpected (for example, earthquakes, some landslides); some repeated but which we address with temporary short-term fixes (floods, hurricanes), and some slow and insidious that we deny exist or that we contribute to (climate change).

Individually and as a society, we must learn to live with nature, not try to control it. Mitigation efforts typically seek to avoid or eliminate a hazard through engineering. Such efforts require financing from governments, individuals, or groups likely to be affected. Less commonly, but more appropriately, mitigation requires changes in human behavior. Behavioral change is usually much less expensive and more permanent than the necessary engineering work. In recent years, governmental agencies have begun to learn this lesson, generally through their own mistakes, but then they often regress because of local pressure from groups of affected residents or from groups wanting to develop an affected site. In a few places along the Missouri and Sacramento Rivers, for example, some levees are being reconstructed back from the riverbanks to permit water to spread out on floodplains during future floods.

Where are specific natural hazards most widespread? Where can we live to avoid them? As **FIGURE 1-15** shows, there is no easy answer: There are earthquakes along the West Coast, especially in California; wildfires throughout the western states, especially in southern California; tornadoes in the Midwest and farther east; hurricanes and floods along the Gulf Coast and central Atlantic coast; flash floods in southern California; sinkholes in Florida; and landslides and floods happen throughout the country, including in the Great Plains. These hazards can be found almost anywhere, although they may vary in severity. Volcanoes are active in the Pacific Northwest, and tsunamis can affect almost any coast. How do we stop the rise of global warming and learn to live with its worsening effects? You may believe you will not be impacted by a major hazard, but you may well be wrong. The following chapters help you decide what natural hazards are most likely to affect you wherever you live—and how to avoid them.

Given the constraints of health, education, and livelihood, we can minimize living in the most hazardous areas. We can avoid one type of hazard while tolerating a less ominous one. Above all, we can educate ourselves about natural hazards and their controls, how to recognize them, and how to anticipate increased chances of a disaster. Although prediction

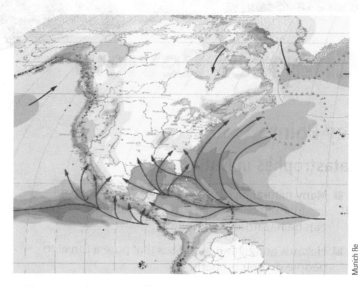

FIGURE 1-15 Living with Tropical Storms

Tropical storms (green) sweep north along the East Coast and up through the Gulf of Mexico, bringing sometimes severe weather to the central and eastern United States (no map color). Earthquakes (yellow to red) concentrate near the West Coast and Intermountain West.

may not be realistic, we can forecast the likelihood of certain types of occurrence that may endanger our property or physical safety. This book provides the background you need to be knowledgeable about natural hazards.

Chapter Review

Key Points

Catastrophes in Nature

- Many natural processes that we see are slow and gradual, but occasional sudden or dramatic events can be hazardous to humans.

- Hazards are natural processes that pose a threat to people or their property.

- A large event becomes a disaster or catastrophe only when it affects people or their property. Large natural events have always occurred but do not become disasters until people place themselves in harm's way.

- More common and less dramatic hazards, such as heat, cold, and flooding, often have higher associated fatalities than rare but dramatic hazards, such as earthquakes and volcanoes. **FIGURE 1-3**.

- The cost of natural hazards is increasing worldwide as a result of growth in population and development. Developed countries suffer greater financial losses in a major disaster; poor countries suffer more fatalities. **FIGURE 1-4**.

- Climate change and global warming potentially increase the severity of weather-related hazards.

Predicting Catastrophe

- Events are often neither cyclic nor completely random.

- Although the precise date and time for a disaster cannot be predicted, understanding the natural processes that control them allows scientists to forecast the probability of a disaster striking a particular area.

- Statistical predictions or recurrence intervals are average expectations based on past experience. **FIGURE 1-6**.

- There are numerous small events, fewer larger events, and only rarely a giant event. We are familiar with the common small events but, because they come along so infrequently, we tend not to expect the giant events that can create major catastrophes. **By the Numbers 1-1**.

- Many natural features and processes are fractal—that is, they have similarities across a broad range of sizes. Large events tend to have characteristics that are similar to smaller events. **FIGURE 1-7**.

Relationships among Events

- Different types of natural hazards often interact with or influence one another. **FIGURE 1-8**.

- Natural processes can sometimes trigger other, more rapid changes in a feedback effect.

- Overlapping influences of multiple factors can lead to the extraordinarily large events that often become disasters. **FIGURE 1-9**.

Mitigating Hazards

- Mitigation involves efforts to avoid disasters rather than merely dealing with the resulting damages.

- Land-use planning can prevent development of hazardous areas, but it often faces opposition.

- Insurance can help people recover from a disaster by providing financial compensation for losses. Risk is proportional to the probability of occurrence and the cost from such an occurrence. **By the Numbers 1-2**.

- People need to be educated about natural processes and how to learn to live with and avoid the hazards around them.

- In poor societies, the reduction of vulnerability to natural hazards does not depend on zoning restrictions or improving warning systems but on cultural, economic, and political factors.

Living with Nature

- Erecting a barrier to some hazard will typically transfer the hazard to another location or to a later point in time.

- Humans need to learn to live with some natural events rather than trying to control them.

Key Terms

catastrophe, p. 3

feedback effect, p. 7

forecast, p. 5

fractal, p. 6

frequency, p. 5

insurance, p. 9

land-use
   planning, p. 8

magnitude, p. 5

mitigation, p. 8

natural disaster, p. 3

natural hazard, p. 3

precursor event, p. 7

recurrence
   interval, p. 5

risk, p. 9

reverse condemnation p. 8

Questions for Review

1. What are some of the reasons people live in areas prone to natural disasters?

2. Is the geologic landscape controlled by gradual and unrelenting processes or intermittent large events with little action in between? Provide an example to illustrate.

3. Some natural disasters happen when the equilibrium of a system is disrupted. What are some examples?

4. Contrast the general nature of catastrophic losses in developed countries versus poor countries. Explain why this is the case.

5. What are the three most deadly natural hazards in the United States?

6. What are the main reasons for the ever-increasing costs of catastrophic natural events?

7. Why are most natural events not perfectly cyclic, even though some processes that influence them are cyclic?

8. What is the difference between a prediction and a forecast?

9. Give an example of a feedback effect in natural processes.

10. Give an example of a fractal system.

11. Describe the general relationship between the frequency and magnitude of an event.

12. If the recurrence interval for a stream flood has been established at 100 years and the stream flooded last year, what is the probability of the stream flooding again this year?

13. What are the two main factors insurance companies use to determines the cost of an insurance policy for a natural hazard?

14. When people or governmental agencies try to restrict or control the activities of nature, what is the general result?

Critical Thinking Questions

1. If people should not live in especially dangerous areas, what beneficial uses could there be for those areas? What are some examples?

2. What responsibility does the government have to ensure that its citizens are safe from natural hazards? Conversely, what freedom should individuals have to choose where they want to live?

3. A small town suffering economic losses from the closure of a factory considers a plan to build a new housing development in an area where there is a record of infrequent flooding. Make a case for and against this development. In your case for the development, describe what measures need to be taken to minimize hazards.

4. Should people be permitted to build in hazardous sites? Should they expect government help in case of a disaster? Should they be required to pay for all costs incurred in a disaster?

5. What are the main natural hazards in the region where you currently live or where you grew up?

6. If I sell a house that is later damaged by landsliding, who is responsible, that is, what are the main considerations?

7. When the federal government provides funds to protect people's homes from floods or wildfires, individual home owners benefit at the expense of taxpayers. What are the two main types of alternatives to eliminate taxpayer expense for natural hazard losses?

8. What is meant by the expression "Those who ignore the past are condemned to repeat it"? Provide an example related to natural hazards.

9. When we note that people need to take responsibility for their own actions when living in a hazardous environment, what is different about the behavior of poor people living in underdeveloped countries?

■ The Himalayas of Nepal mark the collision zone between the Eurasian tectonic plate and the northward-moving plate containing India.

2

Plate Tectonics and Physical Hazards

The Big Picture

Why is the Pacific coast of North America so rugged and mountainous, with steep slopes prone to landslides, whereas the Atlantic coast is relatively flat with gently sloping beaches? Why are most active volcanoes and large earthquakes concentrated near the edge of the Pacific? Why are most tsunamis in the Pacific Ocean rather than the Atlantic? Why do the continents on the edges of the Atlantic Ocean look like they would fit if you were to push them together? The answers to all of these questions and many others lie in the theory known as plate tectonics.

We now know that giant blocks of the upper layers of Earth move around, grind sideways, collide, or sink into the hot interior of the planet where they cause melting of rocks and formation of volcanoes. Those collisions between tectonic plates squeeze up high mountain ranges, even as landslides and rivers erode them away. To understand these processes and associated hazards, we need to understand the forces that drive them. Natural hazards of all kinds are ultimately driven by mountains, oceans, coasts, and the weather and climate that they control.

Earth Structure

At the center of Earth is its **core**, surrounded by the thick **mantle** and covered by the much thinner **crust** (**FIGURE 2-1**). The distinction between the mantle and the crust is based on rock composition. We also distinguish between two zones of Earth based on rock rigidity or strength. The stiff, rigid outer rind of Earth is called the **lithosphere**, and the inner, hotter, more easily deformed part is called the **asthenosphere** (**FIGURE 2-1B**). Continental lithosphere includes silica-rich crust 30 to 50 km thick, underlain by the upper part of the mantle (see Appendix 2 online for detailed rock compositions). Oceanic lithosphere is generally only about 60 km thick; its top 7 km are a low-silica basalt-composition crust. Continental crust is largely composed of high-silica-content minerals, which give it the lowest density (2.7 g/cm³) of the major regions on Earth. Oceanic crust is denser (3 g/cm³) because it contains more iron- and magnesium-rich minerals.

Because we do not have direct observations of crustal thickness, scientists measure the gravitational attraction of Earth (which is greater over denser rocks) and analyze the velocity and timing of seismic waves as they radiate away from earthquake epicenters to provide indirect evidence of the density, velocity, and thickness of subsurface materials. The boundary between Earth's crust and mantle has been identified as a major difference in density called the *Mohorovičić discontinuity* or *Moho*. It marks the base of the continental crust.

Deeper in the mantle, the next major change in material properties occurs at the boundary between the strong, rigid lithosphere above and the weak, deformable asthenosphere below. This boundary was first identified as a near-horizontal zone of lower velocity earthquake waves that move at several kilometers per second. The so-called low-velocity zone is concentrated at the top of the asthenosphere and may contain a small amount of molten basalt over a zone a few hundred kilometers thick. The cold, rigid lithosphere rides on that asthenosphere made weak by its higher temperatures and perhaps also by small melt contents.

Earth's topography clearly shows the continents standing high relative to the ocean basins (**FIGURE 2-2**). The thin lithosphere of the ocean basins stands low; the continents with their thick lower-density lithosphere float high and sink deep into the asthenosphere.

The elevation difference between the continental and oceanic crusts is explained by the concept of **isostacy**, or buoyancy. A floating solid object displaces an amount of liquid with the same mass. Although Earth's mantle is not liquid, its high temperature (above 450°C or 842°F) permits it to flow slowly as if it were a viscous liquid. As a result, the proportion of a mass immersed in the liquid can be calculated from the density of the floating solid divided by the density of the liquid (**By the Numbers 2-1: Height of a Floating Mass**). Similarly, where the weight of an extremely large glacier is added to a continent, the crust and upper mantle slowly sink deeper into the mantle. That happened during the last ice age when thick ice covered most of Canada and the northern United States.

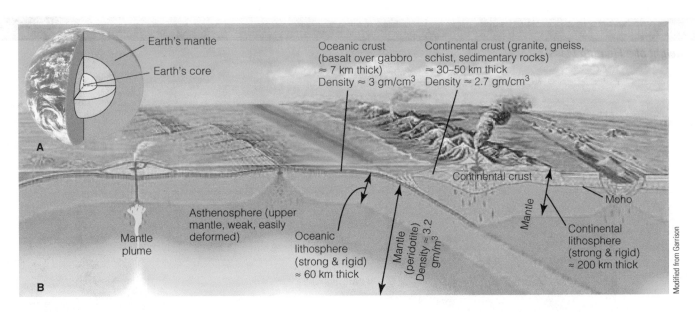

FIGURE 2-1 Earth Structure

A (whole-Earth inset). A slice into Earth shows a solid inner core and a liquid outer core, both composed of nickel-iron. Peridotite in Earth's mantle makes up most of the volume of Earth. Earth's crust, on which we live, is as thin as a line at this scale.
B (main diagram). This expanded view shows the relative thickness and density of different parts of Earth's mantle and crust. The boundary between the mantle and crust is called the *Moho*.

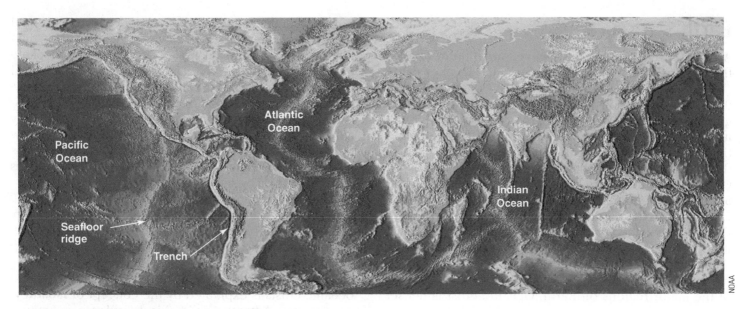

FIGURE 2-2 Mountain Ranges

This shaded relief map shows the continents standing high. Mountain ranges in red tones concentrate at some continental margins. Dark blue areas are the deep ocean floor. Light blue seafloor ridges in oceans are mountain ranges on the ocean floor. Dark blue trenches are deep subsea valleys.

As the ice melted, these areas gradually rose back toward their original heights.

About 150 years ago, measurements showed that gravitational attraction of the huge mountain mass of the Himalayas pulled plumb bobs of very precise surveying instruments toward the mountain range more than would be expected based on the height of the mountains above sea level. A scientist, George Airy, inferred that the mountains must be thicker than they appeared, not merely standing on Earth's crust but extending deeper into it. Based on measurements of the density and velocity of earthquake waves through the crust, it now seems that many major mountain ranges do in

By the Numbers 2-1

Height of a Floating Mass

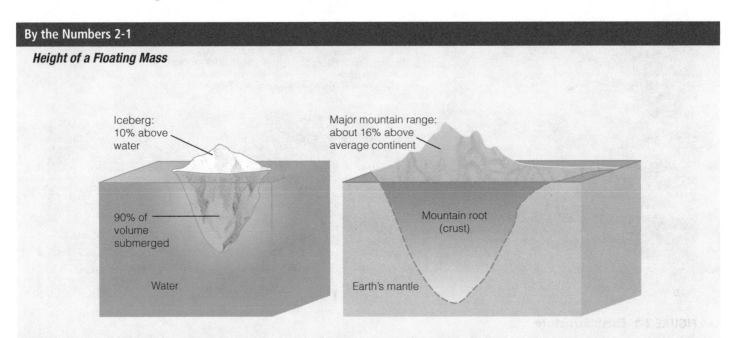

The height to which a floating block of ice rises above water depends on the density of water compared with the density of ice. For example, when water freezes, it expands to become less dense (ice density is 0.9 g/cm^3, liquid water is 1.0 g/cm^3). Thus, 90% of an ice cube or iceberg will be underwater. Similarly, for many large mountain ranges, approximately 84% of a mountain range of continental rocks (2.7 g/cm^3) will submerge into the mantle (3.2 g/cm^3) as a deep mountain root. Note that 2.7/3.2 = 0.84 or 84%. That is, Earth's crust rocks are about 84% as dense as Earth's mantle rocks.

fact have roots, and their crust is much thicker than adjacent older crust. As the mountains grew higher, their roots sank deeper into a fluid Earth, like a block of wood floating in water.

The temperature of the crust also affects its elevation. The crust that makes up the mountains of western Canada and the northwestern United States is no thicker than the 40-km-thick continental crust to the east, and in some areas it is even thinner. Why then does it rise higher above sea level? Measurement of the temperature of the deep crust shows it to be hotter than old, cold continental crust to the east. In certain mountain ranges such as this one, the hot, more expanded, crust of the mountain range is less dense, so it floats higher than old, dense continental crust on the underlying asthenosphere. Heat in the thin crust may have been provided from the hot underlying mantle asthenosphere that stands relatively close to Earth's surface.

Plate Movement

The lithosphere is not continuous like the rind on a melon. It is broken into a dozen or so large **lithospheric plates** and about another dozen much smaller plates (**FIGURE 2-3**).

Even though they are uneven in size and irregular in shape, the plates fit neatly together almost like a mosaic that covers entire surface of Earth. Most plates consist of a combination of continent and ocean areas. About half of the South American Plate lies under the Atlantic Ocean and half is continent. Even the Pacific Plate, which is mostly ocean, includes a narrow slice of western California and part of New Zealand.

The lithospheric plates move over the weak, deforming asthenosphere at rates up to 8 cm (3.2 in.) per year, as confirmed by satellite Global Positioning System (GPS) measurements. Many move in roughly an east–west direction, but some don't. **Plate tectonics** is the big picture theory that describes the movements of Earth's plates. We will present the evidence for plate tectonics at the end of this chapter.

Some plates separate, others collide, and still others slide under, over, or past one another (**FIGURES 2-3** and **2-4**). In some cases, their encounters are head on; in others, the collisions are more oblique. Plates move away from each other at **divergent boundaries**. Plates move toward each other at collision or **convergent boundaries**. In cases where one or both of the converging plates are oceanic lithosphere, the denser plate will slide

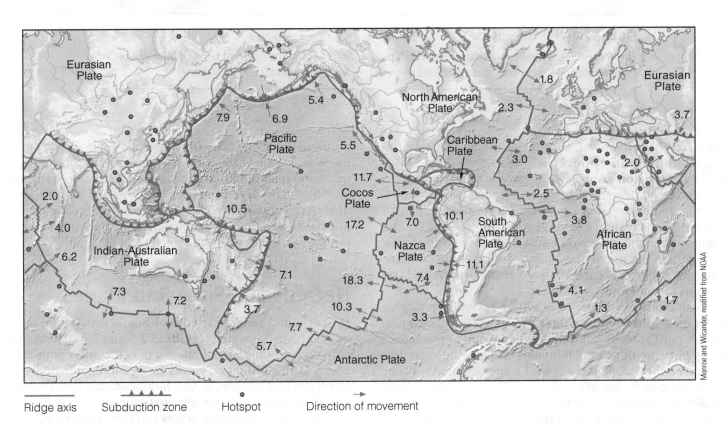

Ridge axis   Subduction zone   Hotspot   Direction of movement

FIGURE 2-3 Lithospheric Plates
Most large lithospheric plates consist of both continental and oceanic areas. Although the Pacific Plate is largely oceanic, it does include parts of California and New Zealand. General direction and velocities of plate movement (compared with hotspots that are inferred to be anchored in the deep mantle), in centimeters per year, are shown with red arrows.

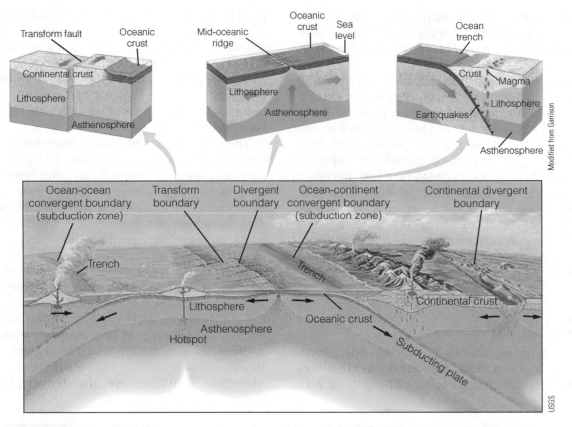

FIGURE 2-4 Plate Boundaries

This three-dimensional cutaway view shows a typical arrangement of the different types of lithospheric plate boundaries: transform, divergent (spreading ridge), and convergent (subduction zone).

down, or be *subducted*, into the asthenosphere, forming a **subduction zone**. When two continental plates collide (**collision zone**), neither side is dense enough to be subducted deep into the mantle, so the two sides typically crumple into a thick mass of low-density continental material. This type of convergent boundary is where the largest mountain ranges on Earth, such as the Himalayas, are built. In the remaining category of plate interactions, two plates slide past each other at a **transform boundary**, such as the San Andreas Fault.

Plate motion is driven by **seafloor spreading**. Magma wells up at **mid-oceanic ridges** to form new oceanic crust. As the crust spreads out from the ridge, older crust moves away from the ridge until it finally sinks into the deep oceanic **trenches** along the edges of some continents. Plates continue to drift apart at the Mid-Atlantic Ridge, for example, making the ocean floor wider and moving North America and Europe farther apart. In the Pacific Ocean, the plates diverge at the East Pacific Rise; their oldest edges sink in the deep ocean trenches near the western Pacific continental margins.

In some places, different types of plate edges intersect. For example, at the Mendocino *triple junction* just off the northern California coast, the Cascadia subduction zone at the Washington-Oregon coast joins both the San Andreas transform fault of California and the Mendocino transform fault that extends offshore. The north end of the same subduction zone joins both the Juan de Fuca spreading ridge and the Queen Charlotte transform fault at a triple junction just off the north end of Vancouver Island.

Hazards and Plate Boundaries

Most of Earth's earthquake and volcanic activity occurs along or near plate boundaries (**FIGURES 2-5** and **2-6**). Most of the convergent boundaries between oceanic and continental plates form subduction zones along the Pacific coasts of North and South America, Asia, Indonesia, and New Zealand. Collisions between continents are best expressed in the high mountain belts extending across southern Europe and Asia. Most rapidly spreading divergent boundaries follow oceanic ridges. In some cases, slowly spreading continental boundaries, such as the East African Rift zone, pull continents apart. Each type of plate boundary has a distinct pattern of natural events associated with it.

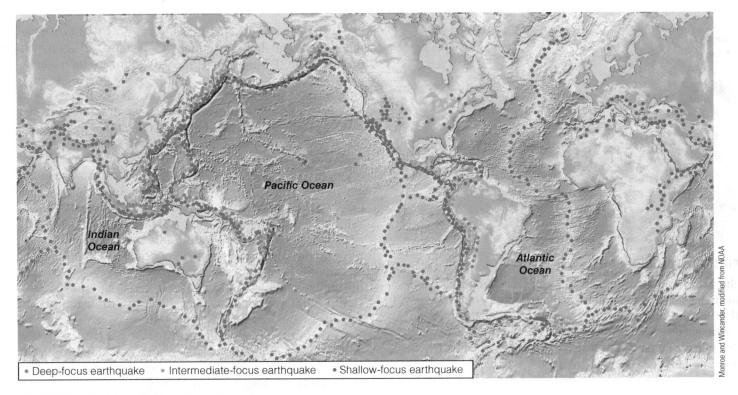

• Deep-focus earthquake • Intermediate-focus earthquake • Shallow-focus earthquake

FIGURE 2-5 Earthquakes at Plate Boundaries

Most earthquakes are concentrated along boundaries between major tectonic plates, especially subduction zones and transform faults, with fewer along spreading ridges. The depth at which the earthquake occurred is called the focus, and is shown in different colors.

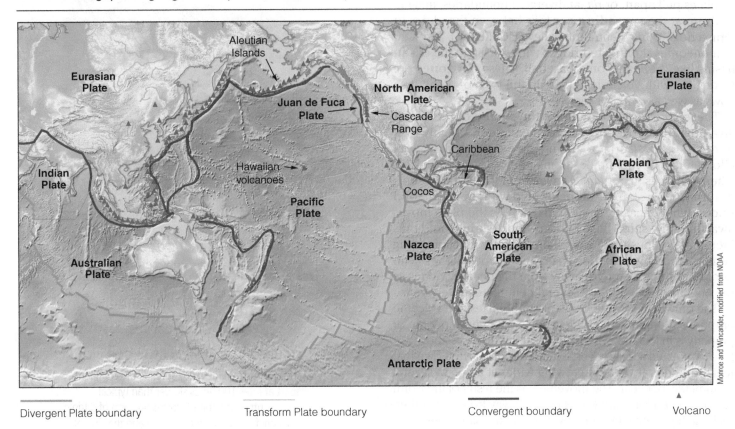

Divergent Plate boundary   Transform Plate boundary   Convergent boundary   ▲ Volcano

FIGURE 2-6 Volcanoes Near Plate Boundaries

Most volcanic activity also occurs along plate tectonic boundaries. Eruptions tend to be concentrated along the continental side of subduction zones and along divergent boundaries, such as rifts and mid-oceanic ridges.

FIGURE 2-7 Mid-Oceanic Ridge

The spreading Mid-Atlantic Ridge, fracture zones, and transform faults are dramatically exhibited in this topography of the ocean floor. Fracture zones across the ridge are parallel to the spreading-movement direction but not necessarily perpendicular to the ridge. Transform faults are offsets (e.g., of ridge crests) along fracture zones.

Divergent Boundaries

As plates pull apart, or *rift*, at divergent boundaries, magma wells up at the **spreading centers** between the plates to form a ridge with a central rift valley (**FIGURE 2-7**). A system of more-or-less connected ridges winds through the ocean basins like the seams on a baseball. These **rift zones** are associated with volcanic activity in the form of basalt lava flows as well as earthquakes.

These spreading centers are the source of the basalt lava flows that cover the entire ocean floor, roughly two-thirds of Earth's surface, to an average depth of several kilometers. The molten basalt magma rises to the surface, where it comes in contact with water. It then rapidly cools to form pillow-shaped blobs of lava with an outer solid rind initially encasing molten magma. As the plate moves away from the spreading center, it cools, shrinks, and thus increases in density. This explains why the hot spreading centers stand high on the subsea topography. New ocean floor continuously moves away from the oceanic ridges as the oceans grow wider by several centimeters every year.

The only place where frequent earthquakes and volcanic eruptions along oceanic ridges pose a danger to people or property is in Iceland, where the oceanic ridge rises above sea level. Repeated surveys over several decades have shown that Iceland's central valley is growing wider at a rate of several centimeters per year. The movement is the result of the North American and Eurasian Plates pulling away from each other, making the Atlantic Ocean grow wider at this same rate.

Iceland's long recorded history shows that a broad fissure opens in the floor of its central valley every 200 to 300 years. It erupts a large basalt lava flow that covers as much as several thousand square kilometers. The last fissure opened in 1821. Finally in April 2010, rifting under a glacier again erupted basalt magma. The hot magma melted the ice causing flooding and an immense ash cloud that spread over most of northern Europe, curtailing air traffic for days. Another such event could happen at any time. Fortunately, the sparse population of the region limits the potential for a great natural disaster.

Spreading centers in the continents pull apart at much slower rates and do not generally form along plate boundaries. The East African Rift zone that extends north–south through much of that continent (**FIGURE 2-8**) may be the early stage of a future ocean. Continental rifts, such as the Rio Grande Rift of New Mexico and the Basin and Range of Nevada and Utah, spread so slowly that they cannot split the continental plate to form new ocean floor (**FIGURE 2-9**). Continental spreading was responsible for creating the Atlantic Ocean long ago (**FIGURE 2-10**).

Continental spreading centers experience a few earthquakes—sometimes large—and volcanic eruptions. Volcanic activity is varied, ranging from large rhyolite calderas in the Long Valley Caldera of the Basin and Range region of southeastern California and the Valles Caldera of the Rio

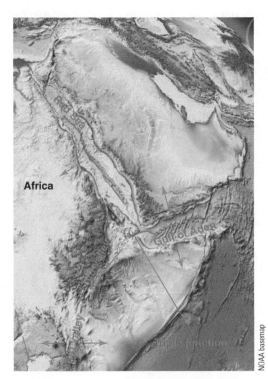

FIGURE 2-8 Beginning of an Ocean

The East African Rift Valley spreads the continent apart at rates 100 times slower than typical oceanic rift zones. This rift forms one arm of a triple junction, from which the Red Sea and the Gulf of Aden form along more rapidly spreading rifts.

FIGURE 2-9 Continental Spreading

The Basin and Range terrain is found southwest of Salt Lake City, Utah. This broad area of spreading in the western United States is marked by prominent basins between mountain ranges. Centered in Nevada and western Utah, it gradually decreases in spreading rate to the north across the Snake River Plain, near its north end. Its western boundary includes the eastern edge of the Sierra Nevada Range, California, and its main eastern boundary is at the Wasatch Front near the east side of Great Salt Lake, Utah. An eastern branch includes the Rio Grande Rift of central New Mexico.

Grande Rift of New Mexico, to small basaltic eruptions at the edges of the spreading center.

Most of the magmas that erupt in continental rift zones are either ordinary rhyolite or basalt with little or no intermediate andesite (see Table 6-1). But some of the magmas, as in East Africa, are peculiar, with high sodium or potassium contents. Some of the rhyolite ash deposits in the Rio Grande Rift and in the Basin and Range provide evidence of extremely large and violent eruptions of giant rhyolite volcanoes. But those events appear to be infrequent, and much of the region is sparsely populated, so they do not pose much of a volcanic hazard.

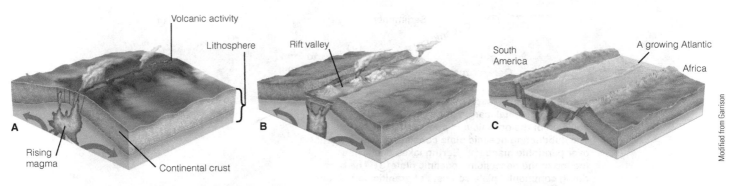

FIGURE 2-10 Evolution of a Spreading Ridge

A spreading center forms as a continent is pulled apart to form new oceanic lithosphere. This process separated the supercontinent of Pangaea into South America and Africa, thereby forming the Atlantic Ocean. A, B, and C show progressive stages in opening of the ocean, beginning with swelling of the hot lithosphere. This spreading eventually results in an ocean basin, as shown in C.

Convergent Boundaries

Convergent boundaries, where plates come together, consist of both subduction zones and continental collision zones. Both zones are associated with earthquakes; volcanoes are more common at subduction zones.

SUBDUCTION ZONES As Earth generates new oceanic crust at boundaries where plates pull away from each other, it must destroy old oceanic crust somewhere else. It swallows this old crust in subduction zones, where one plate slides beneath the other and dives into the hot interior. The plate that sinks is the denser of the two, the one with oceanic crust on its outer surface. It absorbs heat as it sinks into the much hotter rock beneath. The subduction of one plate under another results in volcanic activity and earthquakes (**FIGURE 2-11**).

Where an oceanic plate sinks in a subduction zone, a line or *arc* of picturesque volcanoes rises inland from the trench. The process begins at the oceanic spreading ridge, where fractures open in the ocean floor. Seawater penetrates the dense peridotite of the upper mantle, where the two react to make a greenish rock called serpentinite. That altered ocean floor eventually sinks through an oceanic trench and descends into the upper mantle, where the serpentinite heats up, breaks down, releases its water, and reverts back to peridotite. The water rises into the overlying mantle, which it partially melts to make basalt magma that rises toward the surface. If the basalt passes through continental crust, it can heat and melt some of those rocks to make rhyolite magma. The basalt and rhyolite may erupt separately or mix in any proportion to form andesite and related rocks, the common volcanic rocks in stratovolcanoes. The High Cascades volcanoes in the Pacific Northwest are a good example; they lie inland from an oceanic trench, the surface expression of the active subduction zone (**FIGURE 2-12**).

Recall that most mountain ranges stand high. They stand high because they are either hot volcanoes of the volcanic arc or part of the hot *backarc*, the area behind the arc, above the descending subduction slab. The backarc environment stands high because it weakens, perhaps due to circulating hot water-bearing rocks of the asthenosphere that spread, expand, and rise. In some cases, an oceanic plate descends beneath another section of oceanic plate attached to a continent. The same melting process described previously generates a line of basalt volcanoes because there is no overlying continental crust to melt and form rhyolite.

Volcanoes above a subducting slab present hazards to nearby inhabitants and their property. Deterring people from settling near these hazards can be difficult, because volcanoes are very scenic, and the volcanic rocks break down into rich soils that support and attract large populations. Volcanoes surrounded by people are prominent all around the Pacific basin and in Italy and Greece, where the African Plate collides with Europe.

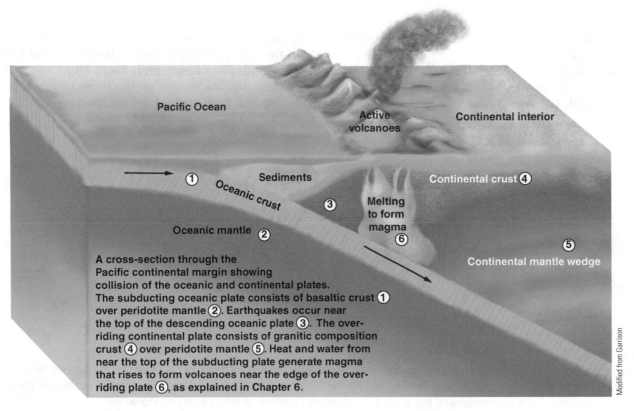

FIGURE 2-11 Subduction Zone Processes

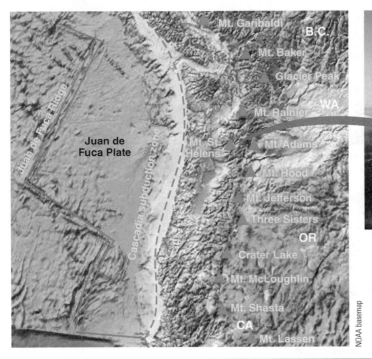

FIGURE 2-12 Volcanoes Near Subduction Zones
The Cascade volcanic chain forms a prominent line of peaks parallel to the oceanic trench and 100 to 200 km inland. Mt. St. Helens (in foreground) and Mt. Rainier (behind, to the north) are two of the picturesque active volcanoes that lie inland from the Cascadia subduction zone.

The sinking slab of lithosphere also generates many earthquakes, both shallow and deep. Grinding rock against rock, the slippage zone sticks and occasionally slips, with an accompanying earthquake. Earth's largest earthquakes are generated along subduction zones; some of these cause major natural catastrophes such as the 2011 earthquake in Japan. Somewhat smaller—but still dangerous—earthquakes occur in the overlying continental plate between the oceanic trench and the line of volcanoes.

Sudden slippage of the submerged edge of the continental plate over the oceanic plate during a major earthquake can cause rapid vertical movement of a lot of water, which creates a huge tsunami wave. The wave both washes onto the nearby shore and races out across the ocean to endanger other shorelines.

COLLISION ZONES Where two continental plates collide, called a continental collision zone, neither plate sinks, so high mountains, such as the Himalayas, are pushed up in fits and starts, accompanied by large earthquakes (**FIGURE 2-13**). During the continuing collision of India against Asia to form the Himalayas, and between the Arabian Plate and Asia to form the Caucasus range farther west, earthquakes regularly kill thousands of people such as in the 2015 earthquake in Nepal. These earthquakes are distributed across a wide area because of the thick, stiff crust in these mountain ranges.

Transform Boundaries

At transform boundaries, or transform faults, plates simply slide past each other without pulling apart or colliding. Some transform boundaries offset the mid-oceanic ridges. Because the ridges are spreading zones, the plates move away from them.

The section of the fault between the offset ends of the spreading ridge has significant relative movement (**FIGURE 2-14A**). Lateral movement between the ridge ends occurs in the opposite direction compared to beyond the ridges, where there is no relative movement across the same fault. Note also that the offset between the two ridge segments does not indicate the direction of relative movement on the transform fault.

Oceanic transform faults generate significant earthquakes without causing casualties because no one lives on the ocean floor. On continents it is a different story. The San Andreas Fault system in California (**FIGURE 2-14B**) is a well-known continental example. The San Andreas Fault is the dominant member of a swarm of more-or-less parallel faults that move horizontally. Together, they have moved a large slice of western California, part of the Pacific Plate, north more than 350 km so far.

Transform plate boundaries typically generate large numbers of earthquakes, a few of which are catastrophic. A sudden movement along the San Andreas Fault caused the devastating San Francisco earthquake of 1906, with its large toll of casualties and property damage. The San Andreas system of faults passes through the metropolitan areas south of San Francisco and just east of Los Angeles. Both areas are home to millions of people, who live at risk of major earthquakes that have the potential to cause enormous casualties and substantial property damage with little or no warning. Even moderate earthquakes in 1971 and 1994 near Los Angeles, in 1989 near San Francisco, and in 2003 near Paso Robles, between them, killed almost 200 people. The threat of such sudden havoc in a still larger event inspires much public concern and major scientific efforts to find ways to predict large earthquakes.

For reasons that remain mostly unclear, some transform plate boundaries are also associated with volcanic activity.

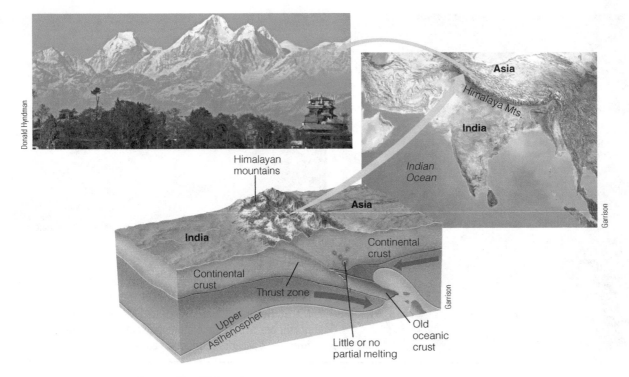

FIGURE 2-13 Continental Collision Zones

The Himalayas, which are the highest mountains on any continent, were created by collision between the Indian and Eurasian Plates. Collision of two continental plates generally occurs after subduction of oceanic crust. The older, colder, denser plate may continue to sink, or the two may merely crumple and thicken. Collision promotes thickening of the combined lithospheres and growth of high mountain ranges.

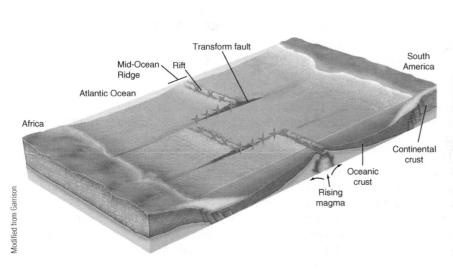

FIGURE 2-14 Transform Fault

A. In this perspective view of an oceanic spreading center, earthquakes (stars) occur along spreading ridges and on transform faults offsetting the ridge.

B. The San Andreas Fault, indicated with a yellow, dashed line, is an example of a continental transform fault. Shown here is the heavily populated area that straddles the fault just south of San Francisco.

Several large volcanic fields have erupted along the San Andreas system of faults during the last 16 million or so years. One of those, in the Clear Lake area north of San Francisco, erupted recently enough to suggest that it may still be capable of further eruptions.

Hotspot Volcanoes

Despite being remote from any plate boundary, **hotspot volcanoes** provide a record of plate tectonic movements. Hotspots are the surface expressions of hot columns of partially molten rock anchored (at least relative to plate movements) in the deep mantle. Their origin is unclear, but many scientists infer that they arise from deep in the mantle, perhaps near the boundary between the core and the mantle. At a hotspot, plumes of abnormally hot but solid rock rising within Earth's mantle begin to melt as the rock pressure on them drops. Wherever peridotite of the asthenosphere partially melts, it releases basalt magma that fuels a volcano on the surface. If the hotspot is under the ocean floor, the basalt magma erupts as basalt lava. If the hot basalt magma rises under continental rocks, it partially melts those rocks to form rhyolite magma; that magma often produces violent eruptions of ash.

The melting temperature of basalt is more than 300°C hotter than rhyolite, so a small amount of molten basalt can melt a large volume of rhyolite. The molten rhyolite rises in large volumes, which may erupt explosively through giant rhyolite calderas, such as those in Yellowstone National Park in Wyoming and Idaho, Long Valley Caldera in eastern California, and Taupo Caldera in New Zealand.

The rising column or plume of hot rock appears to remain nearly fixed in its place as one of Earth's plates moves over it, creating a track of volcanic activity. The movement of the plate over an oceanic hotspot is evident in a chain of volcanoes, where the oldest volcanoes are extinct, and possibly submerged, while newer, active volcanoes are created at the end of the chain. Mauna Loa and Kilauea, for example, erupt at the eastern end of the Hawaiian Islands, a chain of extinct volcanoes that become older westward toward Midway Island (**FIGURE 2-15**). Beyond Midway, the Hawaiian-Emperor chain doglegs to a more northerly course. It continues as a long series of defunct volcanoes that are now submerged. They form seamounts to the western end of the Aleutian Islands west of Alaska. So far as anyone knows, the hotspot track of dead volcanoes will continue to lengthen until eventually the volcanoes and the plate carrying them slide into a subduction zone and disappear.

Hotspot volcanoes leave a clear record of the direction and rate of movement of the lithospheric plates. Remnants of ancient hotspot volcanoes show the direction of movement in the same way that a saw blade cuts in the opposite direction of movement of a board being cut. The ages of those old volcanoes provide the rate of movement of the lithospheric plate. The assumption, of course, is that the mantle

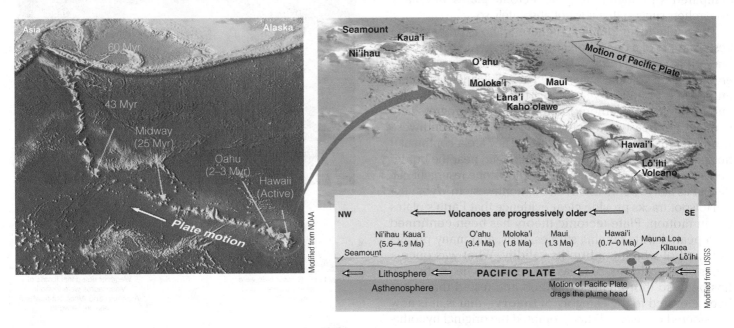

FIGURE 2-15 Oceanic Hotspots

The relief map of the Hawaiian-Emperor chain of volcanoes clearly shows the movement of the crust over the hotspot that is currently below the Big Island of Hawaii, where there are active volcanoes. Two to three million years ago, the part of the Pacific Plate below Oahu was over the same hotspot. The approximate rate and direction of plate motion can be calculated using the common belief that the hotspot is nearly fixed in the Earth for millions of years. The distance between two locations of known ages divided by the time (age difference) indicates a rate of movement of about 9 cm per year. The lithospheric plate, moving across a stationary hotspot in Earth's mantle (moving to the left in this diagram), leaves a track of old volcanoes. The active volcanoes are over the hotspot.

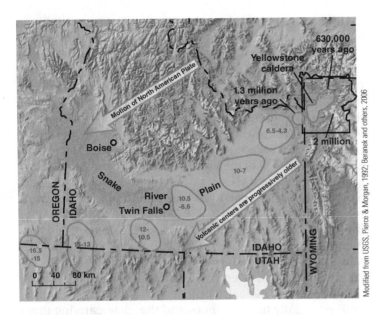

FIGURE 2-16 Continental Hotspots

This shaded relief map of the Snake River Plain shows the outlines of ancient resurgent calderas leading northeast to the present-day Yellowstone caldera. Caldera ages are shown in millions of years before present.

containing the hotspot is not itself moving. Comparison of different hotspots suggests that this is generally valid compared with migration of the tectonic plates, but many researchers suggest that it is not absolutely so.

The Snake River Plain of southern Idaho is probably the best example of a continental hotspot track. Along this track is a series of extinct *resurgent calderas*, depressions where the erupting giant volcano collapsed. Those volcanoes began to erupt some 14 million years ago. They track generally east and northeast in southern Idaho, becoming progressively younger northeastward as the continent moves southwestward over the hotspot (**FIGURE 2-16**). They are a continental hotspot track that leads from its western end near the border between Idaho and Oregon to the Yellowstone resurgent caldera at its active northeastern end in northwestern Wyoming.

Hotspot tracks provide clear evidence that Earth's plates are in motion. Plate tectonic theory has been confirmed by repeated wide-ranging studies, tests, and many predictions, all of which confirm its validity. What was once a series of **hypotheses**, or ideas that remain to be confirmed, has been so thoroughly examined and tested that it has been elevated to the category of **theory**—that is, it is now considered to be *fact*. What prompted the original hypotheses and how did it finally lead to the present understanding?

Development of a Theory

When you look at a map of the world, you may notice that the continents of South America and Africa would fit nicely together like puzzle pieces. In fact, as early as 1596,

Abraham Ortelius, a Dutch mapmaker, noted the similarity of the shapes of those coasts and suggested that Africa and South America were once connected and had since moved apart. In 1912, Alfred Wegener detailed the available evidence and proposed that the continents were originally part of one giant supercontinent that he called **Pangaea** (**FIGURE 2-17**). Wegener noted that the match between the shapes of the continents is especially good if we use the real edge of the continents, including the shallowly submerged continental shelves.

To test this initial hypothesis, Wegener searched for connections between other aspects of geology across the Atlantic Ocean: mountain ranges, rock formations and their ages, and fossil life forms. Continued work showed that ancient rocks, their fossils, and their mountain ranges also matched on the other side of the Atlantic. This analysis is similar to what you would use to put a jigsaw puzzle together; the pieces fit and the patterns match across the reconnected pieces. With confirmation of former connections, he hypothesized that the continents had moved apart; North and South

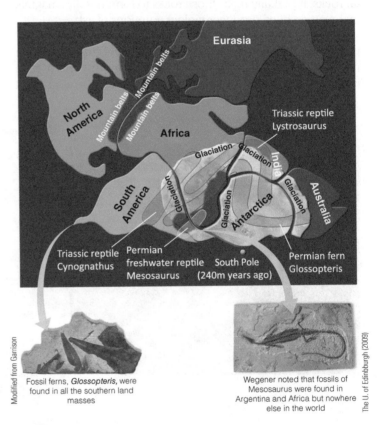

Fossil ferns, *Glossopteris*, were found in all the southern land masses

Wegener noted that fossils of Mesosaurus were found in Argentina and Africa but nowhere else in the world

FIGURE 2-17 Continents Once Fit Together

Before continental drift a few hundred million years ago, the continents were clustered together as giant supercontinent Pangaea. The Atlantic Ocean had not yet opened. The continents match almost perfectly at the continental shelves, which are part of the continents. Some distinctive fossils and mountain ranges lie in belts across the Atlantic and Indian oceans.

America separated from Europe and Africa, widening the Atlantic Ocean in the process. He suggested that the continents drifted through the oceanic crust, forming mountains along their leading edges. This hypothesis, called **continental drift**, remained at the center of the debate about large-scale Earth movements into the 1960s.

As research has continued, other lines of evidence supported the continental drift hypothesis. Exposed surfaces of ancient rocks in the southern parts of Australia, South America, India, and Africa show grooves carved by immense areas of continental glaciers (**FIGURE 2-18**). The grooves show that glaciers with embedded rocks at their bases may have moved from Antarctica into India, eastern South America, and Australia. The rocks were once buried under glacial ice,

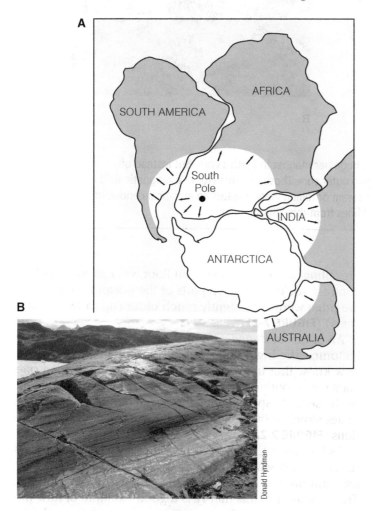

FIGURE 2-18 Glaciation in Warm Areas
A. Continental masses of the southern hemisphere appear to have been parts of a supercontinent 300 million years ago, from which a continental ice sheet centered on Antarctica spread outward to cover adjacent parts of South America, Africa, India, and Australia. After separation, the continents migrated to their current positions. **B.** The inset photo shows glacial grooves in Torres del Paine, Chile, part of the largest remaining ice field in South America.

yet many of these areas now have warm to tropical climates. In addition, the remains of fossils that formed in warm climates are found in areas such as Antarctica and the present-day Arctic: coal with fossil impressions of tropical leaves, the distinctive fossil fern *Glossopteris*, and coral reefs.

Despite this evidence, many scientists rejected Wegener's whole hypothesis because they could show that his proposed mechanism was not physically possible. English geophysicist Harold Jeffreys argued that the ocean floor rocks were far too strong to permit the continents to plow through them. Others who were willing to consider different possibilities eventually came up with a mechanism that fit all of the available data.

The first step in understanding how the continents were separating was to learn more about the topography of the ocean floor, what it looked like, and how old it was. Oceanographers from Woods Hole Oceanographic Institution in Massachusetts, who were measuring depths from all over the Atlantic Ocean in the late 1940s and 1950s, found an immense mountain range down the center of the ocean, extending for its full length—a mid-oceanic ridge. Later, scientists recognized that most earthquakes in the Atlantic Ocean were concentrated in that central ridge.

Although the anti-continental drift group dominated the scientific literature for years, in 1960 Harry Hess of Princeton University conjectured that the ocean floors acted as giant conveyor belts carrying the continents. Hess calculated the spreading rate to be approximately 2.5 cm (1 in.) per year across the Mid-Atlantic Ridge. If that calculation was correct, the whole Atlantic Ocean floor would have been created in about 180 million years.

Confirmation of seafloor spreading finally came in the mid-1960s through work on the magnetic properties of ocean floor rocks. We are all aware that Earth has a **magnetic field** because a magnetized compass needle points toward the north magnetic pole. Slow convection currents in Earth's molten nickel-iron outer core are believed to generate that magnetic field (**FIGURE 2-19**). Because of changes in those currents, this field reverses its north–south orientation every 10,000 to several million years (every 600,000 years on average).

The ocean floor consists of basalt, a dark lava that erupted at the mid-oceanic ridge and solidified from molten magma. Iron atoms crystallizing in the magma orient themselves like tiny compass needles, pointing toward the north magnetic pole. As a result, the rock is slightly magnetized with an orientation like the compass needle. When the magnetic field reverses, that reversed magnetism is frozen into rocks when they solidify. A compass needle at the equator remains nearly horizontal but one at the north magnetic pole points directly down into Earth. At other latitudes in between, the needle points more steeply downward as it approaches the poles. Thus we can tell the latitude at which the rock formed when it solidified by the inclination of its magnetism.

British oceanographers Frederick Vine and Drummond Matthews, studying the magnetic properties of ocean-floor rocks in the early 1960s, discovered a striped pattern

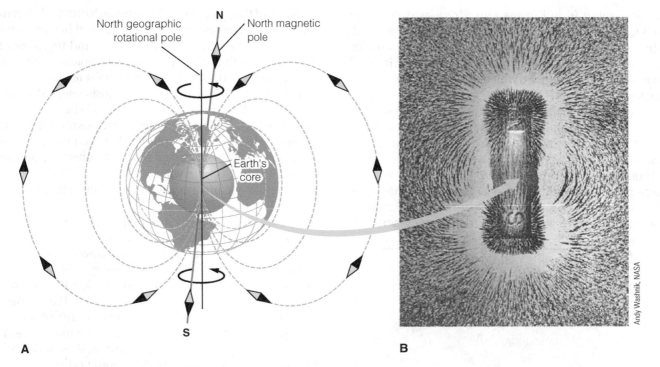

FIGURE 2-19 Earth's Magnetic Field

A. The shape of Earth's magnetic field suggests the presence of a huge bar magnet in Earth's core. But instead of a magnet, Earth's rotation is thought to cause currents in the liquid outer core. Those currents create a magnetic field in a similar way in which power plants generate electricity when steam or falling water rotates an electrical conductor in a magnetic field. **B.** Metal filings align with the magnetic field lines from this "bar magnet".

parallel to the mid-oceanic ridge (**FIGURE 2-20**). Some of the stripes were strongly magnetic; adjacent stripes were weakly magnetic. They realized that the magnetism was stronger where the rocks solidified while Earth's magnetism was oriented parallel to the present-day north magnetic pole. Where the rock magnetism was pointing toward the south magnetic pole, the recorded magnetism was weak—it was partly canceled by the present-day magnetic field. Because it reversed from time to time, Earth's magnetic field imposed a pattern of magnetic stripes as the basalt solidified at the ridge. As the ridge spread apart, ocean floor formed under alternating periods of north- versus south-oriented magnetism to create the matching striped pattern on opposite sides of the ridge.

These magnetic anomalies provide the relative ages of the ocean floor; their mapped widths match across the ridge, and the rocks are assumed to get progressively older as they move away from mid-oceanic ridges. Determination of the true ages of ocean-floor rocks eventually came from drilling in the deep-sea floor by research ships of the Joint Oceanographic Institutions for Deep Earth Sampling (JOIDES), funded by the National Science Foundation. The ages of basalts and sediments dredged and drilled from the ocean floor showed that those near the Mid-Atlantic Ridge were young (up to 1 million years old) and had only a thin coating of sediment. Both results contradicted the prevailing notion that the ocean floor was extremely old. In contrast, rocks from deep parts of the ocean floor far from the ridge were consistently much older (up to 180 million years) (**FIGURE 2-21**).

All of this evidence supports the modern theory of plate tectonics, the big picture of Earth's plate movements. We now know that the world's landmasses once formed one giant supercontinent, called Pangaea, 225 million years ago. As the seafloor spread, Pangaea began to break up, and the plates slowly moved the continents into their current positions (**FIGURE 2-22**).

As it turns out, Wegener's hypothesis that the continents moved apart was confirmed by the data, although his assumption that they plowed through the ocean was not. The evolution of this theory is a good example of how the scientific method works.

The **scientific method** is based on logical analysis of data to solve problems. Scientists make observations and develop tentative explanations—that is, hypotheses—for their observations. A hypothesis should always be testable, because science evolves through continual testing with new observations and experimental analysis. Alternate hypotheses should be developed to test other potential explanations for observed behavior. If observations are inconsistent with a hypothesis, it can either be rejected or revised. If a hypothesis continues to be supported by all available data over a

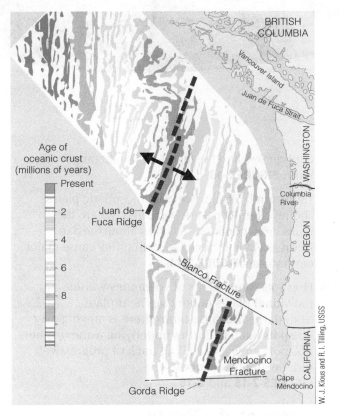

FIGURE 2-20 **Magnetic Record of Ocean-Floor Spreading**

The magnetic polarity, or orientation, across the Juan de Fuca Ridge in the Pacific Ocean shows a symmetrical pattern, as shown in this regional survey (a similar nature of stripes exists along all spreading centers). Basalt lava erupting today records the current northward-oriented magnetism right at the ridge; basalt lavas that erupted less than 1 million years ago recorded the reversed, southward-oriented magnetic field at that time. The south-pointing magnetism in those rocks is largely canceled out by the present-day north-pointing magnetic field, so the ocean floor shows alternating strong (north-pointing) and weak (south-pointing) magnetism in the rocks.

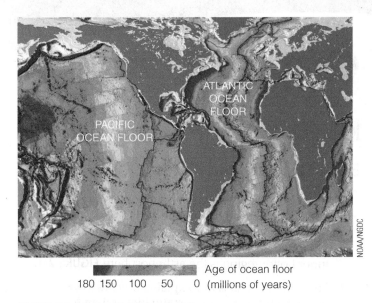

Age of ocean floor
180 150 100 50 0 (millions of years)

FIGURE 2-21 **Ages of Ocean Floor**

Ocean-floor ages are determined by their magnetic patterns. Red colors at the oceanic spreading ridges grade to yellow at 48 million years ago, to green 68 million years ago, and to dark blue some 155 million years ago.

long period of time, and if it can be used to predict other aspects of behavior, it becomes a theory.

After a century of testing, Wegener's initial hypothesis of continental drift was modified to be the foundation for the modern theory of plate tectonics. Plate tectonics is supported by a large mass of data collected over the last century. Modern data continue to support the concept that plates move, substantiate the mechanism of new oceanic plate generation at the mid-oceanic ridges, and support the concept of plate destruction at oceanic trenches. This theory is a fundamental foundation for the geosciences and important for understanding why and where we have a variety of major geologic hazards, such as earthquakes and volcanic eruptions.

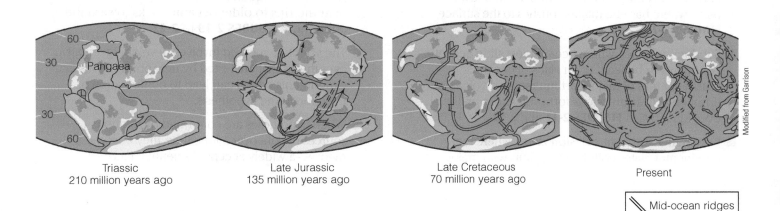

| Triassic 210 million years ago | Late Jurassic 135 million years ago | Late Cretaceous 70 million years ago | Present |

Mid-ocean ridges
Oceanic trench

FIGURE 2-22 **Continents Spread Apart**

The supercontinent Pangaea broke up into individual continents starting approximately 225 million years ago.

Chapter Review

Key Points

Earth Structure and Plates

- Earth is made up of an inner and outer core, surrounded by a thick mantle and covered by a much thinner crust. The crust and stiff outer part of the underlying mantle is called the lithosphere. The inner, hotter region is the asthenosphere. **FIGURE 2-1.**

- The concept of isostacy explains why the lower-density continental rocks stand higher than the higher-density ocean-floor rocks and sink deeper into the underlying mantle. This behavior is analogous to ice (lower density) floating higher in water (higher density). **FIGURE 2-2** and **By the Numbers 2-1.**

- A dozen or so nearly rigid lithospheric plates make up the outer 60 to 200 km of Earth. They slowly slide past, collide with, or spread apart from each other. **FIGURES 2-3** and **2-4.**

Hazards and Plate Boundaries

- Much of the tectonic action, in the form of earthquakes and volcanic eruptions, occurs near the boundaries between the lithospheric plates. **FIGURES 2-5** and **2-6.**

- Where plates diverge from each other, new lithosphere forms. If the plates are continental material, a continental rift zone forms. As this process continues, a new ocean basin can develop, and the spreading continues from a mid-oceanic ridge, where basaltic magma pushes to the surface. **FIGURES 2-7** to **2-10.**

- Subduction zones, where ocean floors slide beneath continents or beneath other slabs of oceanic crust, are areas of major earthquakes and volcanic eruptions. These eruptions form volcanoes on the overriding plates. **FIGURES 2-11** and **2-12.**

- Continent–continent collision zones, where two continental plates collide, are regions with major earthquakes and the tallest mountain ranges on Earth. **FIGURE 2-13.**

- Transform faults involve two lithospheric plates sliding laterally past one another. Where these faults cross continents, such as along the San Andreas Fault through California, they cause major earthquakes. **FIGURE 2-14.**

- Hotspots form chains of volcanoes within individual plates rather than near plate boundaries. Because lithosphere is moving over hotspots fixed in Earth's underlying asthenosphere, hotspots grow as a trailing track of progressively older extinct volcanoes. **FIGURES 2-15** and **2-16.**

Development of a Theory

- The hypothesis of continental drift was supported by matching shapes of the continental margins on both sides of the Atlantic Ocean, as well as the rock types, deformation styles, fossil life forms, and glacial patterns. **FIGURES 2-17** and **2-18.**

- Continental drift evolved into the modern theory of plate tectonics based on new scientific data, including the existence of a large ridge running the length of many deep oceans, matching alternating magnetic stripes in rock on opposite sides of the oceanic spreading ridges, and age dates from oceanic rocks that confirmed a progressive sequence from very young rocks near the rifts to older oceanic rocks toward the continents. **FIGURES 2-19** to **2-22.**

- The scientific method involves developing tentative hypotheses that are tested by new observations and experiments, which can lead to confirmation or rejection of the hypothesis.

- When hypotheses are confirmed by multiple sources of data over a long time, they become a theory—a widely accepted scientific fact.

Key Terms

asthenosphere, p. 17

collision zone, p. 20

continental drift, p. 29

convergent boundary, p. 19

core, p. 17

crust, p. 17

divergent boundary, p. 19

hotspot volcano, p. 27

hypothesis, p. 28

isostacy, p. 17

lithosphere, p. 17

lithospheric plate, p. 19

magnetic field, p. 29

mantle, p. 17

mid-oceanic ridge, p. 20

Pangaea, p. 28

plate tectonics, p. 19

rift zone, p. 22

scientific method, p. 30

seafloor spreading, p. 20

spreading center, p. 22

subduction zone, p. 20

theory, p. 28

transform boundary, p. 20

trench, p. 20

Questions for Review

1. Describe differences between Earth's crust, lithosphere, asthenosphere, and mantle.

2. What does oceanic lithosphere consist of and how thick is it?

3. What are the main types of lithospheric plate boundaries described in terms of relative motions? Provide a real example of each (by name or location).

4. Why does oceanic lithosphere almost always sink beneath continental lithosphere at convergent zones?

5. Along which type(s) of lithospheric plate boundary are large earthquakes common? Why?

6. Along which type(s) of lithospheric plate boundary are large volcanoes most common? Provide an example.

7. What direction is the Pacific Plate currently moving, based on FIGURE 2-15? How fast is this plate moving?

8. Before people understood plate tectonics, what evidence led some scientists to believe in continental drift?

9. If the coastlines across the Atlantic Ocean are spreading apart, why isn't the Atlantic Ocean deepest in its center?

10. What evidence confirmed seafloor spreading?

11. Why are high volcanoes such as the Cascades found on the continents and in a row parallel to the continental margin?

12. Explain how the modern theory of plate tectonics developed in the context of the scientific method.

13. How does the height of a mountain range compare with the thickness of the crust or lithosphere below the mountain? Relate this to the percentage of an iceberg above the water line.

Critical Thinking Questions

1. Explain the role of Earth material densities with respect to Earth's features such as mountains and mid-oceanic ridges. For example, why is the top of basaltic crust below sea level while the surface of granitic crust is generally above sea level?

2. The Basin and Range region of Nevada and Utah is a continental spreading zone. Because it is pulling apart, why isn't its elevation low, rather than as high as it is? Why isn't it an ocean?

3. The scientific community initially rejected Wegener's hypothesis of continental drift and remained skeptical for decades. Does skepticism help or hinder scientific progress?

4. In common usage, *hypothesis* often indicates a guess or hunch, whereas a scientific *theory* is based in evidence tested over a long time and is considered to be scientific fact. What other scientific issues have caused broad debate in social and political culture? Are they well-supported scientific theories or just hypotheses?

A cellar worker at Kieu Hoang Winery in Napa tallies damage after the earthquake.

3

Earthquakes and Their Causes

Public Ignorant of Bay Area Earthquake Risk

In 2005, residents in California's wine country, some 110 km north of San Francisco, remarked to us that "we only get a few small quakes here, nothing to worry about." The individuals we talked to lacked critical information about the region's earthquake risk. Although the San Andreas Fault is approximately 40 km (25 mi) to the west, Napa and Sonoma Valleys are dissected by two active faults, which together have higher risk of large earthquakes than the main San Andreas Fault in the region. At 3:20 a.m. on August 24, 2014, Napa was rocked by a magnitude 6 earthquake. Although the loss of life was low, the economic damages were significant, including loss of wine from many of the producers in the area who did not adequately secure their barrels and bottles (**Case in Point:** Recent San Francisco Bay Earthquakes, 1989 and 2014, Chapter 4, p. 85).

Faults and Earthquakes

To understand why earthquakes happen, remember that the plates of Earth's crust move, new crust forms, and old crust sinks into subduction zones. These movements give rise to earthquakes, which form along **faults**, or ruptures, in Earth's crust. Faults are simply fractures in the crust along which rocks on one side of the break move past those on the other. Faults are measured according to the amount of displacement along the fractures. Over several million years, for example, the rocks west of the San Andreas Fault of California have moved at least 450 km north of where they started. Thousands of other faults have moved much less than 1 km in the same period.

Some faults produce earthquakes when they move; others produce almost none. Some faults have not moved for such a long time that we consider them inactive; others are clearly still active and potentially capable of causing earthquakes. Earthquakes are common in the mountainous western parts of North America, where the rocks are deformed into complex patterns of faults and folds. Active faults are rare in regions such as the American Midwest and central Canada, where the continental crust has been stable for hundreds of millions of years. Such stable regions contain many faults that geologists have yet to recognize. Some of these first announce their presence when they cause an earthquake; others are marked by the line of a recent break near the base of a mountainside called a *fault scarp*.

Types of Faults

Faults can be classified according to the way the rocks on either side of the fault move in relation to each other (**FIGURE 3-1**). **Normal faults** move on a steeply inclined surface. Rocks above the fault surface slip down and over the rocks beneath the fault. Normal faults move when Earth's crust pulls apart, during crustal extension. **Reverse faults** move rocks on the upper side of a fault up and over those below. **Thrust faults** are similar to reverse faults, but the fault surface is more gently inclined. When thrust faults don't break the surface, they are called **blind thrusts**. Blind thrusts are dangerous because they often remain unknown until they cause an earthquake (**Case in Point:** A Major Earthquake on a Blind Thrust Fault—Northridge Earthquake, California, 1994). Reverse and thrust faults move when Earth's crust is pushed together, during crustal compression. **Strike-slip faults** move horizontally as rocks on one side of a fault slip laterally past those on the other side. If rocks on the far side of a fault move to the right, it is a right-lateral fault. If they moved to the left, it would be a left-lateral fault.

The orientations of rock layers and faults are described in terms of *strike* and *dip*. *Strike* is the compass orientation of a horizontal line on a rock surface. *Dip* is the inclination angle (perpendicular to the strike direction) down from horizontal to the rock surface (**FIGURE 3-2**).

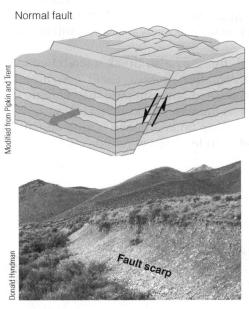
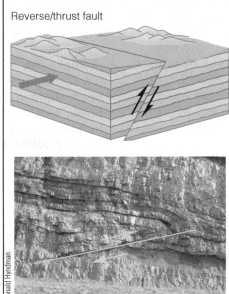

FIGURE 3-1 Types of Fault Movement

A. A normal fault near Challis, Idaho, moved in the 1983 earthquake. The rock mass above the fault slipped downward relative to the mass below.

B. A small thrust fault (reverse fault movement, where one slab of rock moves up and over another, see arrows) east of Vail, Colorado.

C. A strike-slip (lateral slip) fault offset the road after the Landers earthquake in California.

FIGURE 3-2 Strike and Dip

In this example, the strike orientation is about 30° west of north (if north is parallel to the edge of the photo). The dip is 45° down from a horizontal plane to the surface of the rock layer.

Causes of Earthquakes

At the time of the great San Francisco earthquake of 1906, the cause of earthquakes was a complete mystery. The governor of California at the time appointed a commission to ascertain the cause of earthquakes. The director of this commission, Andrew C. Lawson, was a distinguished geologist and one of the most colorful personalities in the history of California. Lawson and his students at the University of California (UC)–Berkeley had already recognized the San Andreas Fault and mapped large parts of it, but until the 1906 event they had no idea that it could cause earthquakes. During their investigation, members of the commission found numerous places where roads, fences, and other structures had broken during the 1906 earthquake just where they crossed the San Andreas Fault. In every case, the side west of the fault had moved north as much as 7 m (23 ft). That led to the theory of how fault movement causes earthquakes.

The earthquake commission hypothesized that as Earth's crust moved, the rocks on opposite sides of the fault had bent, or *deformed*, instead of slipping, over many years. As the rocks on opposite sides of the fault bent, they accumulated energy. When the stuck segment of the fault finally slipped, the bent rocks straightened with a sudden snap, releasing energy in the form of an earthquake (**FIGURE 3-3**). Imagine pulling a bow taut, bending it out of its normal shape, and then releasing it. It would snap back to its original shape with a sudden release of energy capable of sending an arrow flying. This explanation for earthquakes, called the **elastic rebound theory**, has since been confirmed by rigorous testing.

We now know enough about the behavior of rocks in response to stress to explain why faults either stick or slip. We think of rocks as brittle solids, but rocks are elastic, like a spring, and can bend when a force is applied. We use the term **stress** to refer to the forces imposed on a rock and **strain** to refer to the change in shape of the rock in response to the imposed stress. The larger the stress applied, the greater the strain.

Rocks deform in broadly consistent ways in response to stress. Typical rocks will deform *elastically* under low stress, which means that they revert to their former shape when the applied force is relieved. At higher stress, these rocks will deform *plastically*, which means they permanently change shape or flow when forces are applied. Deformation experiments show that most rocks near Earth's surface, where they are cold and not under much pressure from overlying rocks, deform elastically when affected by small forces. Under other conditions, such as deep in Earth where they are hot and under high pressure imposed by the overlying load of rocks, it is much more likely that rocks will deform plastically.

Rocks can bend, but they also break if stretched too far. In response to smaller stresses, rocks may merely bend, while in response to large stresses, they fracture or break. As stress levels increase, rocks ultimately succumb to *brittle failure*, causing fault slippage during an earthquake (**FIGURE 3-4**). Under these conditions, a fault may

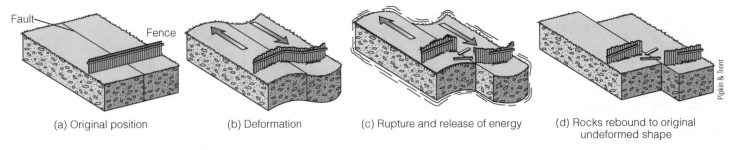

(a) Original position   (b) Deformation   (c) Rupture and release of energy   (d) Rocks rebound to original undeformed shape

FIGURE 3-3 Elastic Rebound Theory

Rocks near a fault (a) are slowly deformed elastically (b) until the fault breaks during an earthquake (c), when the rocks on each side slip past each other, relieving the stress. After the earthquake the rocks regain their original, undeformed shape, but in a new position (d).

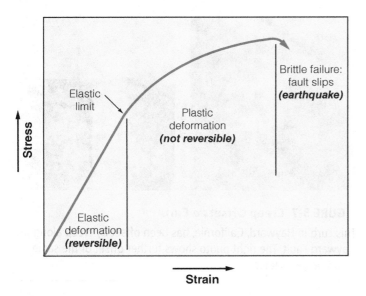

FIGURE 3-4 Stress and Strain

With increasing stress, a rock deforms elastically, then plastically, before ultimately failing or breaking in an earthquake. A completely brittle rock fails at its elastic limit.

FIGURE 3-5 Slickensides

Scratches on a fault surface in the French Alps (horizontal in this photo) indicate movement orientation.

begin to fail, with smaller slips, called **foreshocks**, preceding the main earthquake. It then continues to adjust with smaller slips called **aftershocks** after the event. In a few cases aftershocks can be large and devastating. About 7 weeks after the magnitude 7.8, 2015 earthquake in Nepal that killed almost 9000 people, a magnitude 7.3 aftershock killed an additional 218.

When brittle failure occurs, rocks break in a predictable direction. Deformation of a rock by compression generally results in slippage diagonal to the direction of compression (**By the Numbers 3-1**: Compression and Rock Shear). In Figure 3-1A, for example, Earth's gravity is pulling straight down, but the rock breaks along a dipping fault. Along a fault, differential plate motions apply stresses continuously. Because those plate motions do not stop, elastic deformation progresses to plastic deformation within meters to kilometers of the fault, and the fault finally ruptures in an earthquake. When stress on a section of a fault releases as slippage during a large earthquake, some of that stress is often transferred to increasing stress on a part of the fault beyond the slip zone or to adjacent faults. That makes those adjacent areas more prone to slip than they were before.

The orientation of movement along a fault can often be determined from surface scratches—*slickensides*—imposed by sliding of one surface of the fault against the other (**FIGURE 3-5**).

The size of an earthquake is related to the amount of movement on a fault. The displacement, or **offset**, is the distance of movement across the fault, and the **surface rupture length** is the total length of the break (**FIGURE 3-6**). The largest earthquake expected for a

particular fault generally depends on the *total fault length*, or the longest segment of the fault that typically ruptures.

This relationship between fault-segment length and earthquake size puts a theoretical limit on the size of an earthquake at a given fault. A short fault only a few kilometers long can have many small earthquakes but

By the Numbers 3-1

Compression and Rock Shear

Experimental study of compression of a cylinder of rock from the top and bottom breaks the rock on diagonal shear planes. Shear is generally on one plane only, as shown by the red line. The maximum principal stress is σ_1.

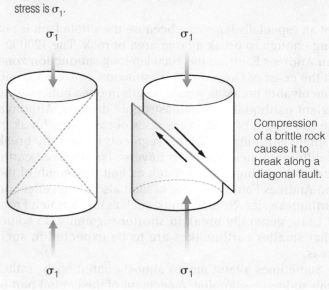

Compression of a brittle rock causes it to break along a diagonal fault.

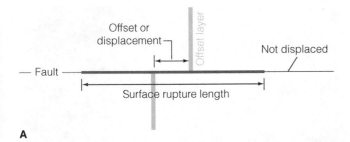

A

FIGURE 3-6 Offset

A. This diagram shows offset and surface rupture length on a fault. Beyond the ends of the rupture, the fault does not break or offset. **B.** A fence near Point Reyes, north of San Francisco, was offset approximately 2.6 m in the 1906 San Francisco earthquake on the San Andreas Fault (indicated with the yellow dashed line).

FIGURE 3-7 Creep Offsets a Curb

This curb in Hayward, California, has been offset by creep along the Hayward Fault. The right photo shows further offset of the same curb two years later.

the San Andreas Fault south of Hollister slips slowly and nearly continuously without causing significant earthquakes. In this zone, strain in the fault is released by **creep** and thus does not accumulate to cause large earthquakes. Why that segment of the fault slips without causing major earthquakes, whereas other segments stick until the rocks break during a tremor, is not entirely clear. Presumably, the rocks at depth are especially weak, such as might be the case with shale or serpentine; or perhaps water penetrates the fault zone to great depth, making it weak. The Hayward Fault also creeps but can produce large earthquakes (**FIGURE 3-7**). The continuously creeping section of the fault almost halfway between San Francisco and Los Angeles seems to be a zone of soft rocks that would be unable to build up a significant stress. Perhaps that means that no more than half of the length of the fault is likely to break in one sudden movement. Slip of only half the length of the fault could still generate a catastrophic earthquake.

not an especially large one because the whole fault is not long enough to break a large area of rock. The 1200 km San Andreas Fault, or the 1000-km-long subduction zone off the coast of Oregon and Washington, however, could conceivably break its whole length in one shot, causing a giant earthquake and catastrophic damage. Although the full length of such faults does occasionally break in a single earthquake, shorter segments commonly break at different times. Scientists have so far observed earthquakes breaking only as much as half the length of the San Andreas Fault. The type of fault also has an effect on earthquake size. Normal faults, such as the Wasatch Front of Utah, generally break in shorter segments, so somewhat smaller earthquakes are to be expected in such areas.

Sometimes a fault moves almost continuously, rather than suddenly snapping. A segment of the central part of

Tectonic Environments of Faults

Because earthquakes are triggered by the motion of Earth's crust, it follows that earthquakes are associated with plate boundaries. The sense of motion during a future earthquake is dictated by the relative motion across a plate boundary— large strike-slip faults commonly move along transform boundaries, thrust faults are typically associated with convergent boundaries, including subduction zones and continent–continent collision boundaries; normal faults are most prominent in divergent boundaries, or spreading zones.

This section discusses examples of faults in these major tectonic environments, as well as fault systems isolated from plate boundaries. Chapter 4 will explore the human impact of earthquake activity in some of these earthquake zones.

Transform Faults

The most important example of a transform fault in the United States is the San Andreas Fault, which slices through a 1200 km length of western California, from near the Mexican border to Cape Mendocino in northern California (**FIGURE 3-8**). The trace of the fault appears from the air and on topographic maps as lines of narrow valleys, some of which hold long lakes and marshes, that have eroded because rocks along the fault are crushed by its movements.

The San Andreas Fault is a continental transform fault in which the main sliding boundary marks the relative motion between the Pacific Plate, which moves northwest, and the North American Plate, which moves slightly south of west. As shown in **FIGURE 3-9**, the total motion of the westernmost slice of California moves more than the slices closer to the continental interior. Areas where there is the greatest *difference* between those arrow lengths, which represent movement rates across a fault, have the greatest likelihood of new fault slippage or an earthquake.

The west side of the northwest-trending fault moves northwestward at an average rate of 3.5 cm per year, or 3.5 m every 100 years, relative to the east side. The rupture length for a magnitude 7 earthquake for a 3.5 m offset would be 50 km; release of all the strain accumulated in 100 years would require a series of such earthquakes along the length of the fault. However, earthquakes frequently occur in clusters separated by periods of relative seismic inactivity.

The San Andreas Fault stretches from the San Francisco Bay area southward past Los Angeles to the north end of the Gulf of California. The northward drag of the Pacific Plate against the continent is slowly crushing the Los Angeles basin northward at roughly 7 mm per year, a small part of the overall plate movement. The sedimentary formations buckle into folds and break along thrust faults, both of which shorten the basin as they move.

The San Andreas Fault appears to have accumulated a total displacement of 235 km in the approximately 16 million years since it began to move. The fault has been stuck along

FIGURE 3-8 Transform Fault

The San Andreas Fault and other major faults nearby appear as a series of straight valleys slicing through the Coast Ranges in this shaded relief map of California. In the inset view from the air, streams jog abruptly (yellow arrows) where they cross the San Andreas Fault in the Carrizo Plain north of Los Angeles. The 1857 Fort Tejon earthquake caused 9.5 m of this movement.

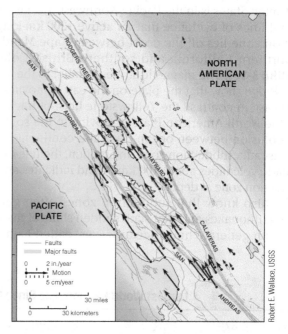

FIGURE 3-9 Rate of Fault Movement Along the San Andreas Fault

The San Andreas Fault system is a wide zone that includes nearly the entire San Francisco Bay area. Based on Global Positioning System measurements, the black arrows are proportional to the rates of ground movement relative to the stable continental interior. Energy builds up when there is a differential movement, as can be seen across each of the major faults shown in orange.

its big bend, south of Parkfield, since the Fort Tejon earthquake of 1857. More recent earthquakes near the southern San Andreas Fault are associated with blind thrusts, which means that some of the crustal movement is being taken up in folding and thrust faults near the main fault instead of in slippage along the main fault itself.

Subduction Zones

Subduction zones present another tectonic environment in which earthquakes occur, including the largest recorded earthquakes since AD 1700 (**Table 3.1**).

The subduction zone along the western coast of South America has given rise to some of a largest earthquakes in history. For example, a magnitude 9.5 earthquake struck the coast of Chile in 1960 and another of magnitude 8.8 in February 2010. In 1868, a magnitude 9 event in Peru (now in northern Chile) killed several thousand. In 2001, a magnitude 8.4 earthquake on the same subduction zone may have increased stress on nearby parts of the boundary. It was followed on August 15, 2007, by a magnitude 8.0 event that struck the coast of Peru and killed more than 510 people, many from collapse of their adobe-brick homes.

The most important example of subduction-zone faults in the United States is in the Pacific Northwest. We know from several lines of evidence that an active 1200-km-long subduction zone lies off the coast between Cape Mendocino in northern California and southern British Columbia (**FIGURE 3-10**). The magnetic stripes parallel to the Juan de Fuca Ridge show that the plate on the east is moving to the southeast; in contrast, the Yellowstone hotspot track shows that the North American Plate is moving to the southwest. The collision between ocean floor and continent is along the Cascadia subduction zone. In addition, the line of active Cascade volcanoes about 100 km inland indicates an active subduction zone at depth.

We also know that subduction zones often generate giant earthquakes and that such sudden shifts of the ocean floor can generate huge ocean waves, called tsunami. A

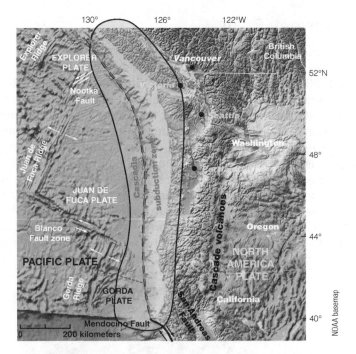

FIGURE 3-10 Giant Quake in the Northwest

The Cascadia oceanic trench dominates the Pacific continental margin off Washington, Oregon, and southern British Columbia. The December 2004 magnitude 9.15 earthquake rupture near Sumatra (superimposed here with yellow shading) is comparable to the size of the potential Cascadia subduction zone slip.

comparable zone in Sumatra generated a giant earthquake and tsunami in December 2004 that killed about 230,000 people (see Chapter 5 for details). Several earthquakes of magnitudes 6.3 to 8.4 occurred along the Sumatra subduction zone in the following years. In 2011, a subduction-zone earthquake and tsunami devastated Japan (**Case in Point:** Giant Subduction-Zone Earthquake—Sendai (Tōhoku) Earthquake, Japan, 2011).

Slabs of oceanic lithosphere sinking through an oceanic trench at subduction-zone boundaries typically generate earthquakes from as deep as several hundred kilometers, so the apparent absence of those deep earthquakes in the Pacific Northwest has worried geologists for years. Several lines of evidence now show that major earthquakes do indeed happen but at such long intervals that none have struck within the 300 or so years of recorded Northwest history. Radiocarbon dating of leaves, twigs, and other organic matter in the buried soils at Willapa Bay, Washington, indicates seven deep earthquakes in the past 3500 years, an average of one per 500 years. Elsewhere along the coast, the records show that 14 have occurred in the last 6500 years, since the eruption of Mt. Mazama in Oregon. The last one was a little more than 300 years ago, so the next could come at any time.

Radiocarbon dating of the peat and buried trees in the Pacific Northwest helped place the last major earthquake in the region within a decade or two of the year 1700. In a separate analysis, careful counting of tree rings from killed and

Table 3-1	Largest World Earthquakes Since 1700	
EARTHQUAKE	DATE	MAGNITUDE
Chile	May 22, 1960	9.5
Anchorage, Alaska	Mar. 28, 1964	9.2
Northern Sumatra	Dec. 26, 2004	9.15
Tōhoku, Japan	March 11, 2011	9.0
Kamchatka	Nov. 4, 1952	9.0
Cascadia	Jan. 26, 1700	9.0
Arica, Chile	Aug.13,1868	9.0
Ecuador	Jan. 13,1906	8.8
Maule, Chile	Feb. 27, 2010	8.8

damaged trees indicates that the event happened shortly after the growing season of 1699.

In a clever piece of sleuthing, geologists of the Geological Survey of Japan found old records with an account of a great wave 2 m high that washed onto the coast of Japan at midnight on January 27, 1700. No historical record tells of an earthquake at about that time on other Pacific margin subduction zones, in Japan; Kamchatka, Alaska; or South America. That leaves the Northwest coast as the only plausible source. Correcting for the day change at the international date line and the time for a wave to cross the Pacific Ocean, the earthquake would have occurred on January 26, 1700, at approximately 9 p.m.

Coastal Indians in the Pacific Northwest have oral traditions that tell of giant waves that swept away villages on a cold winter night. Archaeologists have now found flooded and buried Indian villages strewn with debris. These many lines of data help confirm the timing of the last giant earthquake in this area.

The historical record of major earthquakes in that region could be an indicator of future events. The oceanic plate sinking through the trench off the Northwest coast is now stuck against the overriding continental plate. The continental plate is bulging up, as shown by precise surveys (**FIGURE 3-11**). The locked zone is 50 to 120 km off the coasts of Oregon, Washington, and southern British Columbia. Just inland, the margin is now rising at a rate between 1 and 4 mm per year and shortening horizontally by as much as 3 cm per year. Collapse of the coastal bulge during a giant earthquake drops the low coastal area below sea level, as happened, for example, in the 1964 Alaska earthquake and the 2011 Japan earthquake. In Japan the coastline shifted 5.2 m horizontally and sank as much as 1.2 m.

If the fault broke again as it seems to have broken in 1700, along this entire 1200 km length of coast, it would likely correspond to an earthquake of about magnitude 9, similar to that of the December 2004 Sumatra earthquake. Such an enormous earthquake offshore would generate a wave large enough to cause considerable damage all around the Pacific Ocean. Tsunami waves could arrive at the North American coast within 15 minutes of the earthquake, leaving little time to evacuate.

We now know that giant earthquakes in the southern part of the subduction zone occur at intervals as short as 480 years in the north, 230 years to the south—less than the time since the AD 1700 earthquake! It has also become apparent that earthquakes on major faults can trigger earthquakes on adjacent faults.

Recent research indicates an *average* recurrence interval of 500 to 530 years for a giant magnitude 8.6 to 9.3 event that breaks along the whole 1200-km-long margin at one time. However, even at a single location, the time between such events varies widely. At Willapa Bay in southwestern Washington, for example, the last five large earthquakes appear to have occurred in about 1000 BC, 600 BC, AD 300, AD 700, and

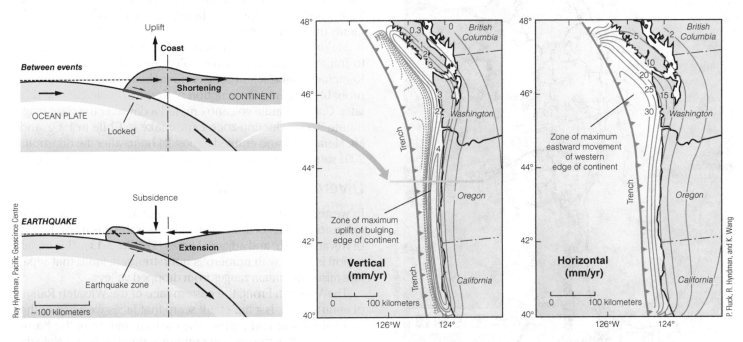

FIGURE 3-11 Flexing the Continental Edge

A. Denser oceanic plate sinks in a subduction zone. As strain accumulates, a bulge rises above the sinking plate while an area landward sinks. Those vertical displacements reverse when the fault slips, causing an earthquake.

B. The subduction zone is locked east of the oceanic trench in the area near the coast. Precise surveying shows the bulge growing near the continental margin, both rising (map on left) and spreading inland (map on right).

AD 1700. Although specific dates and events may change with further research, this illustrates that intervals between events are highly variable. Breaks of shorter fault segments tend to have shorter time gaps and somewhat smaller earthquakes. This research indicates there is an 80% chance of a giant subduction-zone earthquake along the southern part of the fault off southern Oregon and northern California in the next 50 years! A major earthquake on the northern San Andreas could possibly trigger one on the southern Cascadia subduction zone—and vice versa, a disturbing thought. Studies reported in 2010 show that earthquake ground motions of 45 cm/s are expected in Seattle, 15 cm in Tacoma, and 7 cm in Portland. These velocities along with as much as five minutes of low-frequency shaking suggest that high-rise buildings, often considered to be the strongest and best built, are at greatest risk of collapse.

At depths on the subducted slab greater than the locked zone that generates the giant earthquakes, higher temperatures permit the zone to slip continuously without earthquakes. At still greater depths, near the base of the continental crust (near the Moho), is a zone of episodic tremors about 14 months apart, separated by gradual slip (**FIGURE 3-12**).

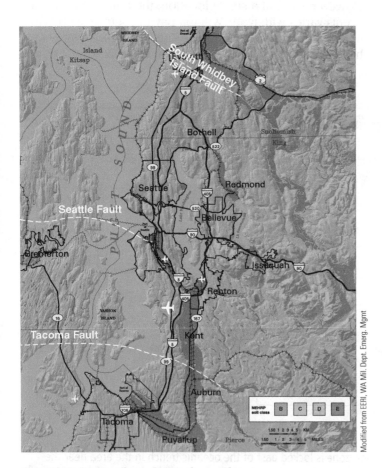

FIGURE 3-12 Earthquake Under a City

This map of the Seattle, Washington, area shows the Seattle Fault and related major recent active fault zones. The Seattle Fault runs east–west almost through the interchange of I-90 and I-5 at the southern edge of Seattle. Red colors = greatest shaking.

EARTHQUAKES ABOVE THE SUBDUCTION ZONE In contrast, most of the earthquakes in the Puget Sound area of northwestern Washington State do not involve slip on the collision boundary at the oceanic trench offshore. Instead they accompany movement at shallow depth on faults that trend west or northwest and straddle Puget Sound (see FIGURE 3-12). Every three or four years, the Puget Sound area feels the jolt of a moderate to large earthquake with a Richter magnitude of 5 to 7.

The Seattle Fault is the best known, and perhaps most dangerous, inland fault in the region. It trends east through the southern end of downtown Seattle, almost through the interchange between Interstate 5 and Interstate 90. Seventy kilometers of the fault are mapped; the part that reaches the surface dips steeply down to the south. Studies show that the rocks south of the fault rose 15.6 m in a large earthquake about AD 900 to 930. That movement generated large tsunami waves in the water in Puget Sound and caused landslides into Lake Washington at the eastern edge of downtown Seattle (also discussed in Chapter 5). In 2001, movement on a related fault not far to the south during the Nisqually earthquake caused more than $2.45 billion* in property damage.

Shaking during the 2001 Nisqually earthquake produced cracks in concrete supports of the double-decker Alaska Viaduct that skirted the west edge of Seattle. Concern for its safety prompted building of a replacement north–south highway tunnel under Seattle. The viaduct has now been demolished and replaced by a ground-level road; the remaining 3.2 km will be a tunnel now being bored through glacial sediments under the waterfront of downtown Seattle.

Major subduction-zone earthquakes appear occasionally to initiate eruption of some volcanoes. The magnitude 9.0 Kamchatka subduction-zone earthquake in 1952, for example, probably triggered the eruption of Karpinski volcano one day later. Cordón-Caulle volcano erupted a day after the giant magnitude 9.5 subduction-zone earthquake in Chile in 1960, and Mt. Merapi volcano erupted almost 24 hours after the disastrous 2004 subduction-zone earthquake in Sumatra, killing 28 people.

Divergent Boundaries

In western North America, the best-known area of continental extension and associated normal faults is the Basin and Range of Nevada, Utah, and adjacent areas (**FIGURE 3-13**). This broad region is laced with numerous north-trending faults that separate raised mountain ranges from dropped valleys.

The Wasatch Front, the eastern face of the Wasatch Range of central Utah, is a high fault scarp that faces west across the Salt Lake basin and defines the eastern margin of the Basin and Range. It is the eastern counterpart to the Sierra Nevada front of California. The Wasatch Front overlooks the deserts of Utah in the same way that the Sierra Nevada overlooks those of Nevada. Many small earthquakes shake the Wasatch Front, but none of any consequence have been felt since Brigham Young's party founded Salt Lake City in 1847.

---

* For the sake of comparison, 2010 dollars are used in discussion of earthquake damages.

Susan Rhea, USGS

Donald Hyndman

Magnitude   ≤5.9   ○ 6   ○ 7   ○ 8   ○ 9

FIGURE 3-13 **Spreading Center Earthquakes**
The north–south faults of the Basin and Range of Nevada, western Utah, and adjacent areas occupy a spreading zone accompanying the northwestward drag of the Pacific Ocean floor. That drag also causes shear to form the San Andreas Fault. The western margin of the Basin and Range is marked by the precipitous eastern edge of the Sierra Nevada; the eastern margin is the equally precipitous Wasatch Front at Salt Lake City. Most of the earthquake activity of the Basin and Range is concentrated along the east face of the Sierra Nevada and the Wasatch Front.

One way to interpret the modest size of many deposits of stream sand and gravel at the mouths of canyons at the base of the Wasatch Front is to suggest that the fault movement has dropped the valley relative to the Wasatch Range during the geologically recent past, probably within tens of thousands of years. That would roughly correspond to the time in which the Sierra Nevada last rose. In fact, both faults remain active as their ranges rise. The active fault zone extends from central Utah, north to southeastern Idaho. The central section near Weber, Salt Lake City, Provo, and Nephi is the most active, but even the end segments are capable of causing magnitude 6.9 earthquakes.

Intraplate and Eastern North American Earthquakes

Earthquakes occasionally strike without warning in places that are remote from any plate boundary and lack any recent record of seismic activity. These intraplate, or within-continent, earthquakes can be devastating, especially because most local people are unaware of their threat. Some of these isolated earthquakes are enormous, easily capable of causing a major natural catastrophe. Although many geologists have offered tentative explanations for these earthquakes, their causes remain generally obscure.

The intraplate earthquakes that struck southeastern Missouri in 1811 and 1812 were among the most severe to hit North America during its period of recorded history. The three great earthquakes that struck near New Madrid, Missouri, in December 1811, January 1812, and February 1812 were felt throughout the eastern United States, toppling chimneys in Ohio, Alabama, and Louisiana and causing church bells to ring in Boston. Once thought to have magnitudes more than 8, those values have been revised to 7.7, 7.5, and 7.7, respectively. USGS researchers now infer that their magnitudes were more likely around 7.

Although there has not been another large earthquake in the region since then, the area is seismically active enough that people as far away as St. Louis and Memphis, the nearest big cities, occasionally hear the ground rumble as their dishes and windows rattle (**FIGURE 3-14**). A repetition of an earthquake in this magnitude range could kill many people and cause major property damage in Memphis, St. Louis, Louisville, Little Rock, and many smaller cities that have older masonry buildings. Few of the buildings in such cities are designed or built to resist significant earthquakes.

EASTERN NORTH AMERICAN EARTHQUAKES Most people do not think of eastern North America as earthquake country, but earthquakes there are not uncommon and some have caused considerable damage. A few earthquakes in this region have had magnitudes of 6.0 to 7.3, and any future such events could have tragic consequences for old brick

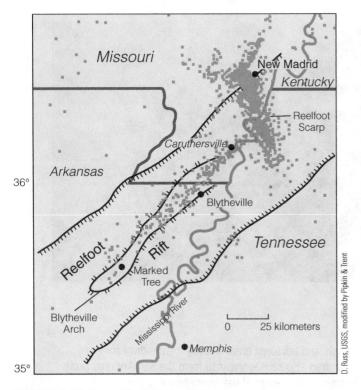

FIGURE 3-14 New Madrid Fault Zone

Recent microearthquake epicenters in the New Madrid region appear to outline three fault zones responsible for the earthquakes of 1811 and 1812. Two northeast-trending lateral-slip faults are offset by a short fault that pushed the southwestern side up over the northeastern side.

and stone buildings. The largest historic earthquakes in the region include events near the big cities of Philadelphia, New York, Boston, Toronto, and Montreal, among the oldest and most vulnerable locations in the northeast (**FIGURE 3-15**).

A magnitude 5.8 earthquake occurred 61 km northwest of Richmond, Virginia, on August 23, 2011, as reverse-fault movement on a north or northeast-trending plane in the Central Virginia seismic zone; it was felt throughout the eastern United States from Georgia to Maine. Nuclear power plants were immediately shut down and travel was disrupted all along the East Coast. Although strong shaking startled people, sending many running into the street, it caused only light damage to buildings, some as far away as New York City. Total damages were estimated between $200 and 300 million, less than half of which was insured.

The strongest earthquakes in the region are concentrated in southern Quebec and adjacent Ontario, and the New Madrid zone between St. Louis and Memphis. Strong shocks, however, have occurred in New Hampshire, near Boston, and in Charleston, South Carolina. In 1886, an earthquake of about magnitude 7.3 caused severe damage to Charleston, South Carolina. Structural brick walls failed and some chimneys fell all the way through the houses. It killed 60 people and caused $5 to 6 million in damages. Other large earthquakes in New England and southern Quebec occurred in 1638, 1663, 1732, and 1755. A magnitude 5 earthquake occurred north of Ottawa, Ontario, on June 23, 2010.

Although large earthquakes are more frequent in the West, the few large ones that have occurred in eastern North America have been much more damaging because Earth's crust in the East transmits earthquake waves more efficiently, with less loss of energy, than that in the West, which is hotter and more broken along faults. That explains why the area of significant damage for an earthquake of a given size is greater in the East than in the West (**FIGURE 3-16**). Good land-use planning and building codes for new structures cost little and may someday avert enormous loss of life and property damage in a city that does not now suspect it is living dangerously.

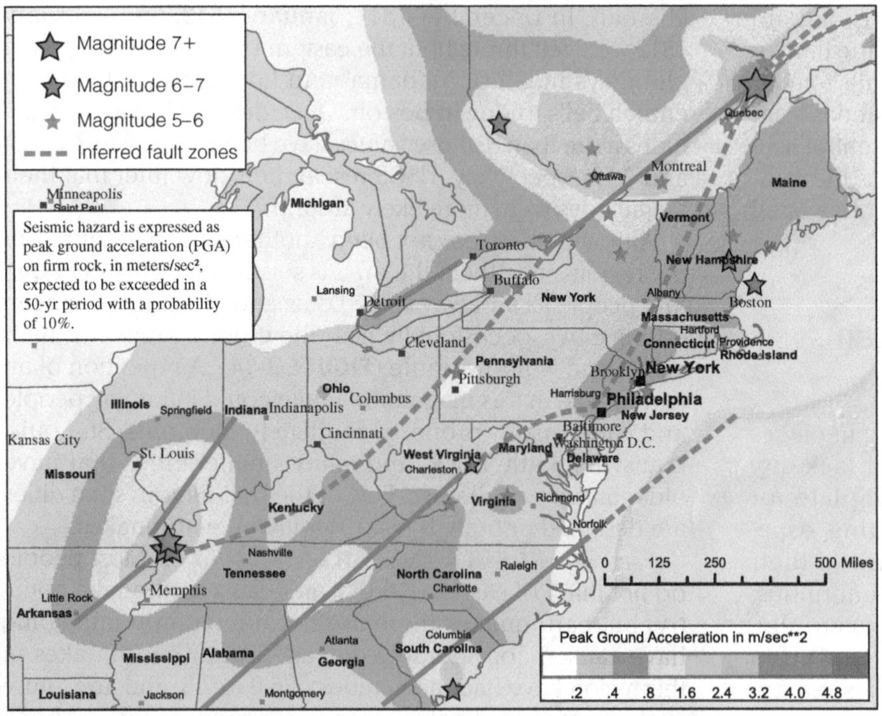

FIGURE 3-15 East Coast Earthquake Hazard

Fault zones in this region (blue lines) are not precisely mapped but appear to lie near ancient continental-margin boundaries, approximately parallel to the present continental margin. The youngest of these is at the coast; those to the northwest are tens of millions of years older. Areas of greatest seismic accelerations (violence) are shaded orange to red. The largest historical earthquakes are indicated by red stars.

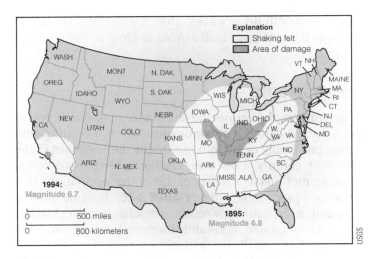

FIGURE 3-16 Shaking and Damage by Region

A comparison of similar magnitude earthquakes shows that the damage would be much greater for an earthquake in the Midwest than in the mountainous West.

Earthquake Waves

When a fault slips, the released energy travels outward in **seismic waves** from the place where the fault first slipped, called the **focus**, or hypocenter, of the earthquake. The **epicenter** is the point on the map directly above the focus (**FIGURE 3-17**). The behavior of earthquake waves explains both how we experience earthquakes and the types of damage they cause.

Types of Earthquake Waves

Observant people have noticed for centuries that many earthquakes arrive as a distinct series of shakings that feel different. The different types of shaking are a result of the

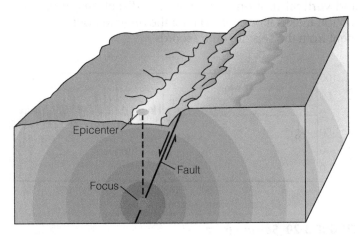

FIGURE 3-17 Epicenter and Focus

The epicenter of an earthquake is the point on Earth's surface directly above the focus where the earthquake originated.

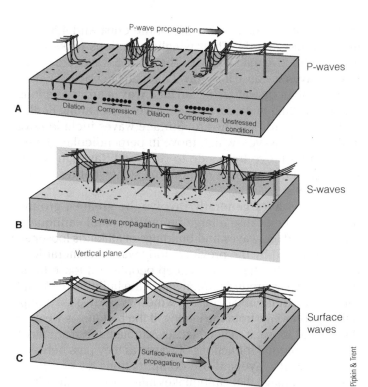

FIGURE 3-18 Types of Earthquake Waves

A. P waves travel with alternating compression and extension as they move through the rock. **B.** S waves travel in a wiggling motion perpendicular to the direction of wave travel (only the horizontal direction is shown here). **C.** In the rolling motion of surface waves, individual particles at Earth's surface move in a circular motion (opposite the direction of travel) both in vertical and horizontal planes.

different types of earthquake waves (**FIGURE 3-18**). The first event is the arrival of **P waves**, the primary or compressional waves, which come as a sudden jolt. People indoors might wonder for a moment whether a truck just hit the house. P waves consist of a train of compressions and expansions. P waves travel roughly 5 to 6 km/s in the less-dense continental crust and 8 km/s in the dense, less-compressible rocks of the upper mantle. People sometimes hear the low rumbling of the P waves of an earthquake. Sound waves are also compressional and closely comparable to P waves, but they travel through the air at only 0.34 km/s.

After the P waves comes a brief interval of quiet while the cat heads under the bed and plaster dust sifts down from cracks in the ceiling. Then come the **S waves** (secondary, or shear, waves), moving with a wiggling motion—like that of a rhythmically shaking rope—and making it hard to stand. Chimneys may snap off and fall through floors to the basement. Streets and sidewalks twist and turn. Buildings jarred by the earlier P waves distort and may collapse. S waves are slower than P waves, traveling at speeds of 3.5 km/s in the crust and 4.5 km/s in the upper mantle. Their wiggling motions make them more destructive than P waves. The P and S waves are called **body waves** because they travel through the body of Earth.

After the body waves, the **surface waves** arrive as a long series of rolling motions. Surface waves travel along Earth's surface and fade downward. Surface waves include Love and Rayleigh waves, which move in perpendicular planes. Love waves move from side to side, and Rayleigh waves move up and down in a motion that somewhat resembles ocean swells.

Surface waves generally involve the greatest ground motion, so they cause a large proportion of all earthquake damage. Surface waves find buildings of all kinds loosened and weakened by the previous body waves, vulnerable to a final blow. Inertia tends to keep people and loose furniture in place as ground motion yanks the building back and forth beneath them. Shattering windows spray glass shrapnel as plaster falls from the ceiling. If the building is weak or the ground loose, it may collapse. Although there are more complex, internal refractions of waves as they pass between different Earth layers, those complications do not much affect the damage that earthquakes inflict because the direct waves are significantly stronger.

The differences people feel during this series of earthquake waves can be explained by the different characteristics of those waves. To describe the vibrations of earthquake waves, we use a variety of terms (**FIGURE 3-19**). The time for one complete cycle between successive wave peaks to pass is the **period**; the distance between wave crests is the **wavelength**; and the amount of positive or negative wave motion is the **amplitude**. The number of peaks per second is the **frequency** in cycles per second, or Hertz (Hz).

When you bend a stick until it breaks, you hear the snap and feel the vibration in your hands. When Earth breaks along a fault, it vibrates back and forth with the frequency of a low rumble, although the frequencies of earthquake waves are generally too low to be heard with the human ear. P and S waves generally cause vibrations in the frequency range between 1 and 30 cycles per second (1 to 30 Hz). Surface waves generally cause vibrations at much lower frequencies, which dissipate less rapidly than those

associated with body waves. That is why they commonly damage tall buildings at distances as great as 100 km from an epicenter.

Seismographs

A **seismograph** records the shaking of earthquake waves on a record called a **seismogram**. When recording seismographs finally came into use during the early part of the twentieth century, it became possible to see those different shaking motions as a series of distinctive oscillations that arrive in a predictable order (**FIGURE 3-20**). Imagine the seismograph as an extremely sensitive mechanical ear clamped firmly to the ground, constantly listening for noises from the depths. It is essentially the geologist's stethoscope.

We normally stand firmly planted on solid ground to watch things move, but how do we stand still and watch Earth move? Seismographs consist of a heavy weight suspended from a rigid column that is firmly anchored to the ground. The whole system moves with the earthquake motion, except the suspended weight, which stays relatively still due to its inertia. In seismographs designed to measure horizontal motion, the weight is suspended from a wire, whereas in those designed for vertical motion, it is suspended from a weak spring (**FIGURE 3-21**).

The first seismograph used in the United States was at UC Berkeley in 1887. It used a pen attached to the suspended weight to make a record on a sheet of paper that was attached to the moving ground. Most modern seismographs work on the same basic principle but detect and record ground motion electronically.

Seismograms can help scientists understand more about how a fault slipped, as well as where it slipped and how much. Faults with different orientations and directions of movement generate various patterns of motion. Specialized seismographs are designed to measure those various directions of earthquake vibrations—north to south, east to west, and vertical motions. Knowing the directions of ground motion makes it possible to infer the direction of fault movement from the seismograph records.

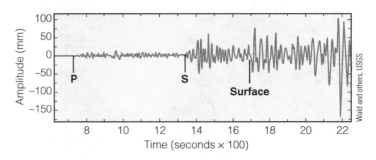

FIGURE 3-20 Seismogram
A seismogram for the 1906 San Francisco earthquake shows arrival of the main seismic waves in sequence—first P, then S, and finally surface waves.

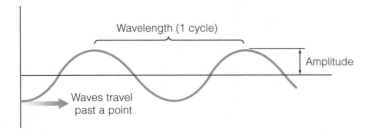

FIGURE 3-19 Wave Forms
The definition of wavelength and amplitude for earthquake waves.

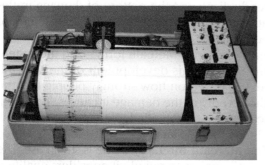
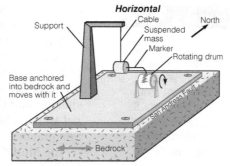
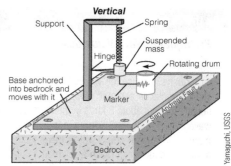

FIGURE 3-21 Seismographs

A. Although many seismograph stations now record earthquake waves digitally, a recording drum seismograph, like this one, is especially useful for visualizing the nature of earthquake waves: their amplitude, wavelength, and frequency of vibration.
B. Some seismographs record the side-to-side motion of a mass suspended from a cable.
C. Other seismographs record the up-and-down motion of a mass suspended from a spring.

Locating Earthquakes

The time interval between the arrivals of P and S waves recorded by a seismograph can also help scientists locate the epicenter of an earthquake. Imagine the P and S waves as two cars that start at the same place at the same time, one going 100 km/h, the other 90 km/h. The faster car gets farther and farther ahead with time. An observer who knows the speeds of the two cars could determine how far they are from their starting point simply by timing the interval between their passage. In exactly the same way, because we know the velocity of the waves, the time interval between the arrivals of the P and S waves reveals the approximate distance between the seismograph and the place where the earthquake struck (**FIGURE 3-22**). This calculation is explained in greater detail in **By the Numbers 3-2**: Earthquake-Wave Velocities.

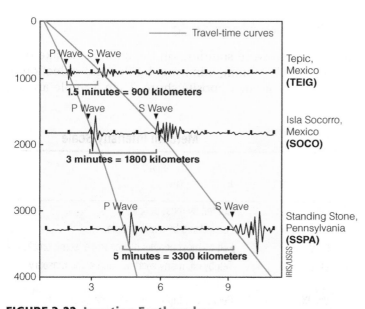

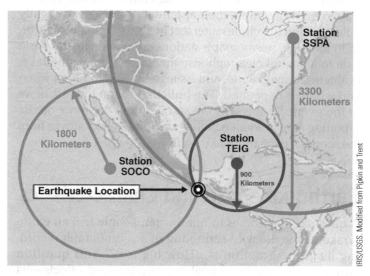

FIGURE 3-22 Locating Earthquakes

A. S waves arrived only 1.5 minutes after P waves at Tepic, Mexico (Station TEIG), indicating that the station is 900 km from the earthquake epicenter. S waves arrived 5 minutes after P waves in Pennsylvania (Station SSPA), more than 3000 km away. This method indicates the distance but not the location of the earthquake from each seismograph.

B. Using the calculated distance of the earthquake as the radius, circles are plotted around each of three seismograph stations. The point where all three circles intersect is the earthquake's epicenter in the Mexico trench.

Earthquake-Wave Velocities

Surface-wave arrival times increase linearly with distance from an earthquake because the waves travel with nearly constant velocity in shallow rocks. In contrast, P waves travel at 5 to 6 km/s in the continental crust but about 8 km/s in the more dense rocks of the mantle. S waves travel at about 3.5 km/s in the crust but about 4.5 km/s in the mantle. Because P and S waves travel faster deeper in Earth, waves at greater depths can reach a seismograph faster along those curving paths.

Calculate distance from an earthquake using this simple formula:

$$D = T/(1/V_S - 1/V_P)$$

where V_S = S-wave velocity

V_P = P-wave velocity

The arrival times of P and S waves at a single seismograph indicate how far from the seismograph an earthquake originated, but it does not indicate in which direction the earthquake occurred. This means the earthquake could have happened anywhere on the perimeter of a circle drawn with the seismograph at its center and the distance to the earthquake as its radius. In order to better pinpoint the location of the earthquake, this same type of data is needed from at least three different seismograph stations. The three circles will intersect at only one location, and that is where the earthquake struck. In fact, because earthquake waves travel at slightly different velocities through different rocks on their way to a seismograph, their apparent distances are slightly different, and the circles intersect in a small triangle of error.

In practice, seismograph stations communicate the basic data to a central clearinghouse that locates the earthquake, evaluates its magnitude, and issues a bulletin to report when and where it happened. The bulletin is often the first news of the event. That is why we so often find the news media reporting an earthquake before any information arrives from the scene of the earthquake itself.

Earthquake Size and Characteristics

A question that comes to mind when people feel an earthquake or see the wild scribbling of a seismograph recording its ground motion is, "How big is it?" This question can be answered by describing its perceived effects—its intensity—or by measuring the amount of energy released—its magnitude.

Earthquake Intensity

After the great Lisbon earthquake of 1755, the archbishop of Portugal sent a letter to every parish priest in the country asking each to report the type and severity of damage in his parish. Then the archbishop had the replies assembled into a map that clearly displayed the pattern of damage in the country. Jesuit priests have been prominent in the study of earthquakes ever since.

Italian scientist Giuseppe Mercalli formalized the system of reporting in 1902 with his development of the Mercalli Intensity Scale. It is based on how strongly people feel the shaking and the severity of the damage it causes. The original Mercalli Scale was later modified to adapt it to construction practices in the United States.

The **Modified Mercalli Intensity Scale** is still in use. The USGS sends questionnaires to people it considers qualified observers who live in an earthquake area and then assembles the returns into an earthquake intensity map, on which the higher Roman numerals record greater intensities (**Table 3-2**).

Mercalli Intensity Scale maps reflect both the subjective observations of people who felt the earthquake and an objective description of the level of damage. They typically show the strongest intensities in areas near the epicenter and areas where ground conditions cause the strongest shaking (**FIGURE 3-23**).

The map shown in FIGURE 3-23 is an example of recently developed computer-generated maps of ground motion called **ShakeMaps**, which show the distribution of maximum acceleration for many potential earthquakes. Such maps are useful in land-use planning because they help forecast the pattern of shaking in future earthquakes along the same fault, so they can be used to infer the likely level of damage. A real-time ShakeMap can help send emergency-response teams quickly to areas that have likely suffered the greatest damage.

Earthquake Magnitude

Suppose you were standing on the shore of a lake on a perfectly still evening admiring the flawless reflection of a mountain on the opposite shore. Then a ripple arrives,

Table 3-2	Mercalli Intensity Scale
INTENSITY AT EPICENTER	**EFFECT ON PEOPLE AND BUILDINGS**
I–II	Not felt by most people.
III	Felt indoors by some people.
IV–V	Felt by most people; dishes rattle, some break.
VI–VII	Felt by all; many windows and some masonry cracks or falls.
VIII–IX	People are frightened; most chimneys fall; major damage to poorly built structures.
X– XI	People panic; most masonry structures and bridges are destroyed.
XII	Nearly total damage to masonry structures; major damage to bridges, dams; rails are bent.
>XII	Near total destruction; people see ground surface move in waves; objects are thrown into air.

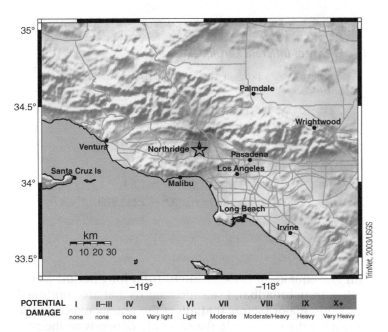

POTENTIAL DAMAGE

I	II–III	IV	V	VI	VII	VIII	IX	X+
none	none	none	Very light	Light	Moderate	Moderate/Heavy	Heavy	Very Heavy

FIGURE 3-23 Intensity of Shaking

This ShakeMap shows the distribution of shaking during the 1994 Northridge earthquake. The intensity of shaking in the Los Angeles area is nearly concentric around the area of fault that ruptured. In contrast, the San Francisco area has earthquakes with intensities that are elongated parallel to valleys, because these valleys are filled with soft, wet sediments (See Figure 3-29).

momentarily marring the reflection. Did a minnow jump nearby, or did a deranged elephant take a flying leap into the lake from the distant opposite shore? Nothing in the ripple as you would see it could answer that question. You need to know how far it traveled and spread before you saw it, because the size of the wave decreased with distance. Useful as it is, the Mercalli Intensity Scale measures only how an earthquake is experienced (like the ripple), not the actual size of the event (minnow or elephant). That is the problem that Charles Richter of the California Institute of Technology addressed when he first devised a new earthquake magnitude scale in 1935.

RICHTER MAGNITUDE Richter developed an empirical scale, called the Richter Magnitude Scale, based on the maximum amplitude of earthquake waves measured on a seismograph of a specific type, the Wood-Anderson seismograph. Although wave amplitude decreases with distance, Richter designed the magnitude scale as though the seismograph were 100 km from the epicenter.

Seismograms vary greatly in amplitude for earthquakes of different sizes. To make that variation more manageable, Richter chose to use a logarithmic scale to compare earthquakes of different sizes. At a given distance from an earthquake, an amplitude 10 times as great on a seismograph indicates a magnitude difference of 1.0—an earthquake of magnitude 6 sends the seismograph needle swinging 10 times as far as one of magnitude 5 (**FIGURE 3-24**).

Seismographs, like buildings and people, sense shaking at different frequencies. Tall buildings, for example, sway back and forth more slowly than short ones—they have longer periods of oscillation. P waves, S waves, and surface waves have different amplitudes and different periods. With this variability in earthquake waves, Richter chose to use as the standard waves with periods, or back-and-forth sway times, of 0.1 to 3.0 seconds. The **Richter magnitude** is now known as M_L, for *local magnitude*.

Distant earthquakes travel through Earth's interior at higher velocities and frequencies. To work with distant earthquakes, Beno Gutenberg and Charles Richter developed two more-specific magnitude scales in 1954. M_S, the surface-wave magnitude, is calculated in a similar manner to that described for M_L. The number quoted in the news media is generally the surface-wave magnitude, as it is in this book, unless specified otherwise. Surface waves with a period of 20 seconds or so generally provide the largest amplitudes on seismograms. Special seismographs record earthquake waves with such long periods. M_B, the body-wave magnitude, is measured from the amplitudes of P waves.

To estimate the magnitude of an earthquake, we need the amplitude (from the S wave or surface wave). Because the amplitude of shaking decreases with distance, we also need the distance to the epicenter (from the P minus S time). These calculations can be made using a graphical method, the earthquake nomogram, on which a straight line is plotted between the P – S time and the S-wave amplitude (**FIGURE 3-25**). This line intersects the central line at the approximate magnitude of the earthquake.

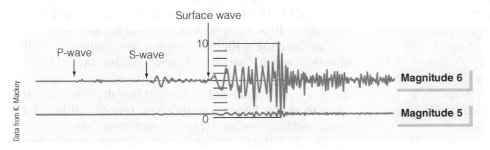

FIGURE 3-24 The Richter Scale

An earthquake of magnitude 6 has 10 times the amplitude as an earthquake of magnitude 5 from the same location and on the same seismograph. That difference is an increase in 1 on the Richter scale. The horizontal axis on these seismograms is time, and the vertical axis is the ground motion recorded.

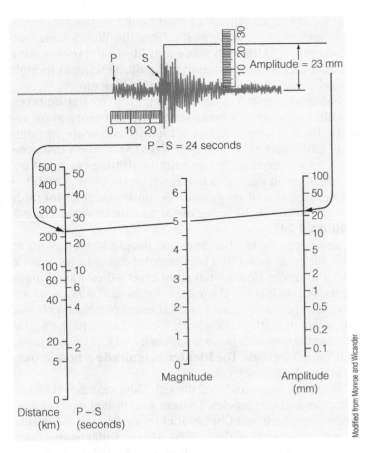

FIGURE 3-25 Estimating Earthquake Magnitude
A nomograph chart uses the distance from the earthquake (P – S time in seconds) and the S-wave amplitude (in mm) to estimate the earthquake magnitude.

By the Numbers 3-3

Energy of Different Earthquakes

To compare energy between different earthquakes, a Richter magnitude difference of:

0.2 is ~ 2 times the energy

0.4 is ~ 4 times the energy

0.6 is ~ 8 times the energy

1.0 is ~ 32 times the energy

2.0 is > 1000 times the energy ($32 \times 32 = 1024$)

4.0 is > 1,000,000 times the energy ($32^4 \approx 1,148,576$)

For earthquakes with M_L above 6.5, the strongest earthquake oscillations, which have a lower frequency, may lie below the frequency range of the seismograph. This may cause saturation of earthquake records, which occurs when the ground below the seismograph is still going in one direction while the seismograph pendulum, which swings at a higher frequency, has begun to swing back the other way. Then the seismograph does not record the maximum amplitude. So the Richter magnitude becomes progressively less accurate above M_L 6.5, and a different scale becomes more appropriate.

An earthquake of magnitude 6 indicates ground motion or seismograph swing 10 times as large as that for an earthquake of magnitude 5, but the amount of energy released in the earthquakes differs by a factor of about 32 (**By the Numbers 3-3**: Energy of Different Earthquakes). Below M_L 6 or 6.5, the various measures of magnitude differ little; but above that, the differences increase with magnitude. For larger earthquakes, the energy released is a better indicator of earthquake magnitude than ground motion.

MOMENT MAGNITUDE **Moment magnitude, M_W**, is essentially a measure of the total energy expended during an earthquake. It is determined from long-period waves taken from broadband seismic records that are controlled by three major factors that affect the energy expended in breaking the rocks. Calculation of M_W depends on the *seismic moment*, which is determined from the shear strength of the displaced rocks multiplied by both the surface area of earthquake rupture and the average slip distance on the fault. The largest of these variables, and the one most easily measured, is the offset or slip distance.

Small offsets of a fault release small amounts of energy and generate small earthquakes. If the length of fault and the area of crustal rocks broken is large, then it will cause a large earthquake. Because the relationships are consistent, it is possible to estimate the magnitude of an ancient earthquake from the total surface rupture length. For typical rupture thicknesses, a fault offset of 1 m would generate an earthquake of approximately M_W 6.5, whereas a fault offset of 13 m would generate an earthquake of approximately M_W 9. If you find a fault with a measurable offset that occurred in a single earthquake, then you can infer the approximate magnitude of the earthquake it caused (**FIGURE 3-26**).

MAGNITUDE AND FREQUENCY In 1954, Gutenberg and Richter worked out the relationship between frequency of occurrence of a certain size of earthquake and its magnitude. Recall from Chapter 1 that there are many small events, fewer large ones, and only rarely a giant event. Quantitatively, that translates as a *power law*. Plotted on a graph of earthquake frequency versus magnitude, the power law can be plotted as a log scale: 10^1 or 10 to the power of 1 is 10; 10^2 or 10 to the second power is $10 \times 10 = 100$; $10^3 = 10 \times 10 \times 10 = 1,000$; and so on.

The Gutenberg-Richter frequency–magnitude relationship tells us that if we plot all known earthquakes of a certain size against their frequency of occurrence (on a logarithmic axis), we get a more or less straight line that we can extrapolate to events larger than those on record (**FIGURE 3-27**). Small earthquakes are far more numerous than large earthquakes, and giant earthquakes are extremely rare, which is why we have not had many in the historical record.

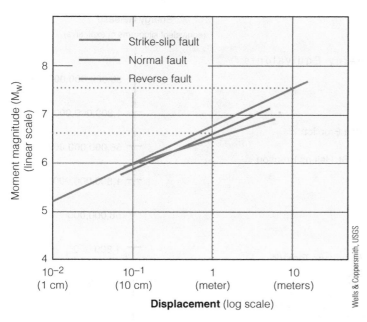

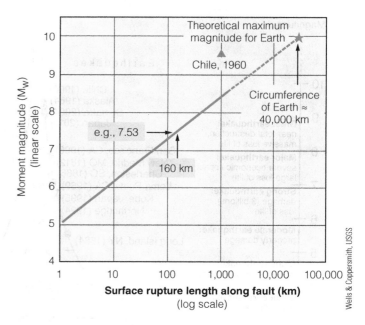

FIGURE 3-26 Moment Magnitude

These graphs show the relationships between the magnitude of the earthquake and the maximum fault offset (during earthquakes on all types of faults), the surface rupture length.

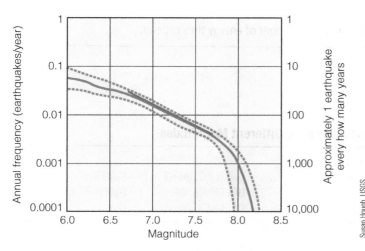

FIGURE 3-27 Earthquake Magnitude and Frequency

This graph plots the Gutenberg-Richter frequency–magnitude relationship for the San Francisco Bay region. The logarithm of the annual frequency of earthquakes plotted against their Richter magnitude generally plots as a straight line, or nearly so (red line). The curved line (green) provides the best fit to the data, which mostly plot between the dashed blue lines. Different faults would plot as somewhat different lines.

Most of the total energy release for a fault occurs in the few largest earthquakes (**FIGURE 3-28**). Each whole-number increase in magnitude corresponds to an increase in energy release of approximately 32 times. Thus,

32 magnitude 6 earthquakes would be necessary to equal the total energy release of 1 magnitude 7 earthquake. And more than 1000 earthquakes of magnitude 6 would release energy equal to a single earthquake of magnitude 8 ($32 \times 32 = 1024$).

Ground Motion and Failure during Earthquakes

How much and how long the ground shakes during an earthquake is related to how much and where the fault moves. **Table 3-3** summarizes the relationship between earthquake magnitude and ground motion. Local conditions can also amplify shaking and increase damage.

Ground Acceleration and Shaking Time

Sometimes it helps to think of ground motion during an earthquake as a matter of acceleration, that is, the strength of the shaking. **Acceleration** is normally designated as some proportion of the acceleration of gravity (g); 1 g is the acceleration felt by a freely falling body, such as what you feel when you step off a diving board. Most earthquake accelerations are less than 1 g; a few are more. A famous photograph taken after the San Francisco earthquake of 1906 shows a statue of the eminent nineteenth-century scientist Louis Agassiz stuck headfirst in a courtyard on the campus of Stanford University, its feet in the air. Perhaps the statue was

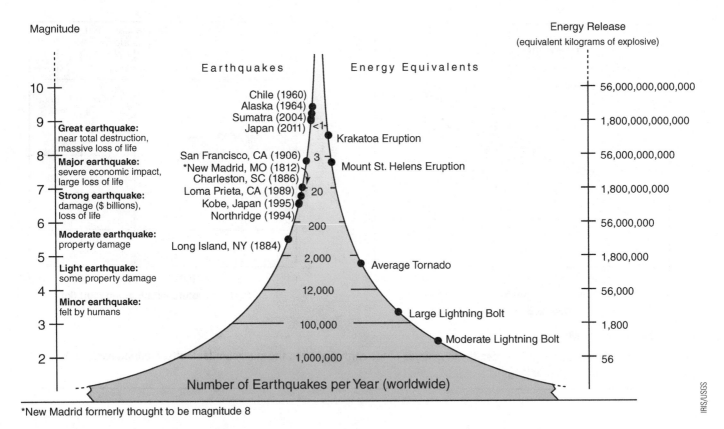

*New Madrid formerly thought to be magnitude 8

FIGURE 3-28 Magnitude and Energy Release
The average number of earthquakes per year, worldwide, and a comparison of the amount of energy they expend.

Table 3-3	Characteristics of Earthquakes of Different Magnitudes				
RICHTER MAGNITUDE (M_L)	MAXIMUM ACCELERATION (APPROX. % OF g)[a]	MAXIMUM VELOCITY OF BACK-AND-FORTH SHAKING[a]	APPROXIMATE TIME OF SHAKING NEAR SOURCE (sec)	DISPLACEMENT DISTANCE, OR OFFSET	SURFACE RUPTURE LENGTH (km)
<2	0.1–0.2				
~3	<1.4	<1	0–2		
~4	1.4–9	1–8	0–2		
~5	9–34	8–31	2–5	~1 cm	1
~6	34–124	31–116	10–15	60–140 cm	~8
~7	>124	>116	20–30	~2 m	50–80
~8	>124	>150	~50	~4 m	200–400
~9+	>124		>80	>13 m	>1200
M_W = 9.7–9.8					Earth's circumference

[a]g = 9.8 m/sec[2]

tossed off its pedestal at a moment when the vertical ground acceleration was greater than 1 g. It is commonplace after a strong earthquake to find boulders tossed a meter or more.

The duration of strong shaking in an earthquake depends on the size of the earthquake. The time that the ground moves in one direction during an earthquake, before the oscillation moves back in the other direction, is similar to the time of initial fault slip in one direction. The total duration of motion is longer because the ground oscillates back and forth.

An increase in magnitude above 6 does not cause much stronger shaking; rather, it increases the area and total time of shaking. Earthquakes of magnitude 5 generally last only 2 to 5 seconds; those of magnitude 7 from 20 to 30 seconds; and those of magnitude 8 almost 50 seconds (see Table 3-3). A magnitude 6 earthquake, shaking only 10–15 seconds, provides only a short time to evacuate. A magnitude 7 earthquake provides more time, but evacuation is harder to do, with accelerations approaching 1 g. The longer shaking lasts, the more damage occurs; structures weakened or cracked in the first few seconds of an earthquake are commonly destroyed with continued shaking. Because there is almost no time to evacuate, and because running outside can result in death by falling debris, it is generally best to duck under a sturdy desk or lie next to a very sturdy piece of furniture for protection.

The amount of shaking also relates to distance from an earthquake's focus. Waves radiating outward from an earthquake source show a significant decrease in violence of shaking with distance, especially in bedrock and firmly packed soil. For this reason, earthquakes that occur deep underground may cause less property damage than smaller earthquakes that occur near Earth's surface. The focus for most earthquakes is generally at depths shallower than 100 km, because rocks at greater depths behave plastically and slip continuously.

Shaking severity is also affected by the type of material waves are traveling through. For example, upon reaching an area of loose, water-saturated soils, such as old lakebed clay or artificial fill at the edge of bays, earthquake waves are strongly amplified to accelerations many times greater than nearby waves in bedrock (**FIGURE 3-29**). The violence of shaking depends on the frequency of the earthquake waves compared with the frequency of the ground oscillation. The lower-frequency oscillations of surface waves often correspond to the natural oscillation frequency of loose, water-saturated ground.

Secondary Ground Effects

Earthquakes often trigger landslides (see Chapter 8). If you place a pile of loose, dry sand on a table, then sharply whack the side of the table, some of the sand will immediately slide down the pile. In nature, if sand or soil in the ground is saturated with water, the quick back-and-forth acceleration from a quake has a pumping effect on

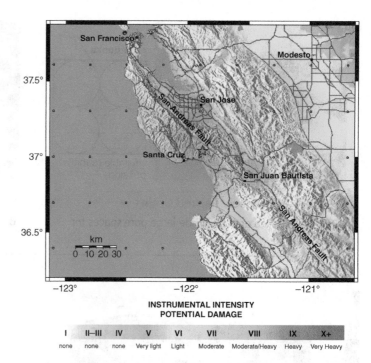

**INSTRUMENTAL INTENSITY
POTENTIAL DAMAGE**

I	II–III	IV	V	VI	VII	VIII	IX	X+
none	none	none	Very light	Light	Moderate	Moderate/Heavy	Heavy	Very Heavy

FIGURE 3-29

This ShakeMap of earthquake intensity for the 1989 Loma Prieta earthquake near Santa Cruz, California, shows a Mercalli Intensity VIII at the epicenter (indicated with a star) northeast of Santa Cruz, around the south end of the San Francisco Bay, and in the two big valleys to the south, where wet muds amplified the shaking. See Figure 3-23 for a comparison with a Los Angeles Area earthquake in a region without soft sediments in valleys.

the water between the grains. Water is forced into spaces between the grains with each pulse of an earthquake. This sudden increase in water pressure in the pore spaces can effectively push the grains apart and permit the mass to slide downslope.

Earthquakes can also cause **liquefaction**—in which soils that ordinarily seem perfectly stable become almost liquid when shaken and then solidify again when the shaking stops. Many soft sediment deposits consist of extremely loose sand or silt grains with water-filled pore spaces between them. An earthquake can shake these deposits down to a much tighter grain packing, expelling water from the pore spaces (**FIGURE 3-30**). The escaping water carries sediment along as it rapidly flows to the surface, creating *sand boils* and *mud volcanoes* that are typically a few meters across and several centimeters high.

Liquefaction can cause significant damage to buildings and roads on soft sediment (**FIGURE 3-31**). During the 1964 Alaska earthquake, the ground in Anchorage, 100 km from the epicenter, began to shake and continued for 72 s. Clays liquefied in the Turnagain Heights district, causing bluffs up to 22 m high to collapse along 2.8 km of coastline. The

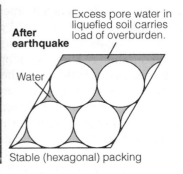

FIGURE 3-30 Loosely Packed Grains

A. Loosely packed grains provide large pore spaces for water.

B. If the grains collapse to a tighter arrangement, much of the water must be squeezed out to cause liquefaction.

FIGURE 3-31 Liquefaction

Liquefaction of the foundation under the left side of this building during the August 17, 1999, earthquake in Izmit, Turkey, caused settling of the left section and destruction of the middle section.

swiftly flowing liquid clay carried away many modern frame houses that were as much as 300 m inland. The $3.8 billion in property damage (2010 dollars) included roads, bridges, railroad tracks, and harbor facilities.

Shaking of the 1971 San Fernando Valley earthquake near Los Angeles induced liquefaction of the upper face of the Van Norman Dam, nearly causing its failure just upstream from the homes of tens of thousands of people.

The deep fill of soft sediments and high groundwater levels in the Salt Lake Valley, Utah, have a significant likelihood of liquefaction during earthquakes. Together, these factors may also amplify ground motion more than 10 times. The ground is saturated with water only a few meters below the surface in Salt Lake City. Liquefaction of wet clays would cause loss of bearing capacity and downslope flow.

In the next chapter, we use the principles and related discussions from Chapter 3 to consider the possibilities of earthquake forecasts and prediction, and we discuss how to avoid or minimize damages caused by earthquakes. A few prominent examples illustrate results of some of those effects.

Giant Subduction-Zone Earthquake
Sendai (Tōhoku) Earthquake, Japan, 2011 ▶

On Friday, March 11, 2011, at 2:46 p.m. local time, in Sendai, Japan, a professor's cell phone sounded a distinctive warning that indicated an imminent large earthquake. He and his students immediately dove under their sturdy desks just before the building began shaking violently. Cracks formed in the walls and plaster began falling from the ceiling; the lights went out as the power failed. Others took cover in steel-frame doorways but were unable to remain standing in the powerful shaking as a giant magnitude 9.0 earthquake hit northeastern Japan.

In Japan, the world's most advanced earthquake warning system is triggered by the shaking of P waves that arrive first but do minimal damage. The system automatically alerts factories, schools, hospitals, radio, and television of the impending earthquake. In this earthquake, the system gave residents a critical 32 seconds of warning time before damaging S waves reached Sendai, enough time to take cover or shut down critical instruments or machines. About 2 minutes later the

earthquake location and magnitude were determined; less than one minute after that the tsunami warning sirens blared. Residents had only 9 to 24 minutes after the earthquake to try to evacuate to higher ground before arrival of the first waves of a devastating tsunami (discussed further in Chapter 5).

The Pacific Plate moves relatively westward and under the Eurasian Plate at 8.3 cm per year. The long-stuck subduction zone finally slipped about 70 km offshore, from the nearest coast, 130 km east of Sendai, Japan, under

the continental shelf, at a depth of 24.4 km. Foreshocks, including four with magnitudes greater than 6, began on March 9. Aftershocks included a magnitude 7.4 earthquake about 30 minutes after the main event. The giant earthquake, the fifth largest ever recorded, filled a seismic gap along the subduction zone. Previous historic giant earthquakes in the region were in the years 869, 1611, 1896, 1933. The 2011 quake caused shaking at Mercalli Intensity VII and buildings to sway for several minutes in Tokyo, 373 km to

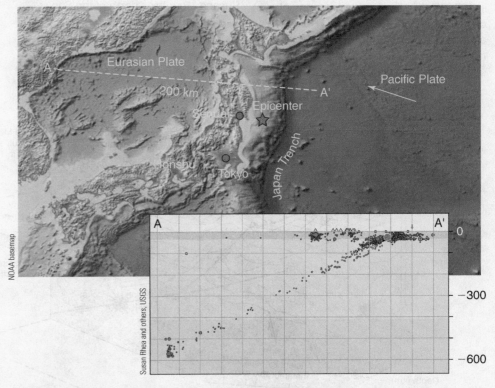

■ The Pacific plate is subducted under the island of Honshu, forming the Japan Trench. The epicenter of the magnitude 9.0 2011 earthquake is indicated with a star. The inset diagram shows a record (1900–2007) of earthquake depth along the subducting plate, in the area extending from the coast of North Korea (A) to the Pacific Ocean off the east coast of Japan (A'). The size of the circles indicates earthquake magnitude and the color indicates their depth. Yellow triangles indicate active volcanoes.

■ *This building was damaged during the Sendai earthquake.*

the southwest. Although the shaking did topple filing cabinets and anything loose in offices, stores, and homes, the huge earthquake did remarkably little structural damage because of stringent building codes in a country that sees many earthquakes. Although 15,826 people died in the event and 3810 remained missing several months later, 90% of the casualties resulted not from the earthquake, but from the tsunami, described in Chapter 5.

A Major Earthquake on a Blind Thrust Fault
Northridge Earthquake, California, 1994 ▶

On January 17, 1994, at 4:31 a.m., a magnitude 6.7 earthquake struck Northridge, California; its epicenter was 20 km southwest of that of the San Fernando Valley earthquake of similar size in 1971. Both accompanied movement on faults north of Los Angeles. The earthquake was caused when a thrust fault slipped at a depth of 10 km. The offset reached to within 5 km of the surface but did not break it, so the fault is a blind thrust. The fault, unknown before the earthquake, is now known as the Pico thrust fault. The fault movement raised Northridge 20 cm (8 in.); the Santa Susana Mountains north of the San Fernando Valley rose 40 cm.

Ground acceleration reached almost 1 g in many areas and approached 2 g at one site in Tarzana, and ground velocity locally exceeded 1 m/s. A few people felt the ground motion as much as 300 to 400 km away from the epicenter. Local topography and bedrock structure may have amplified ground shaking. Anomalously strong shaking occurred on some ridgetops and at the bedrock boundaries of local basins.

Damage inflicted during the Northridge earthquake reached Mercalli Intensity IX near the epicenter. Intensity V effects were noted as far as 120 km to the north, 180 km to the west, and 200 km to the southeast.

The Northridge earthquake caused $63.5 billion in property damage (2015 dollars). Some 3000 buildings were closed to all reentry; another 7000 buildings were closed to occupation. Sixty-one people were killed. In fact, the toll would probably have been in the thousands had the earthquake happened later in the day, when more people would have been up and about. Most buildings and parking garages that collapsed

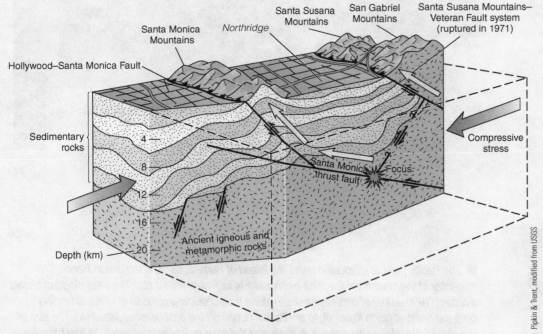

Pipkin & Trent, modified from USGS

■ This cutaway block diagram shows the blind thrust fault movement that caused the 1994 Northridge earthquake. Yellow arrows indicate the relative direction of movement.

Collapsed first-story garages

D.L. Carver, USGS

■ *Many first-floor garages under apartment buildings were not well braced, so the buildings collapsed on the cars below.*

were essentially unoccupied. Freeway collapse occurred at seven locations, 170 bridges were damaged, and a few unlucky people were on the road.

Many California faults are capable of causing large earthquakes, tens of times larger than those in the San Fernando Valley in 1971 and Northridge in 1994. Future earthquakes could cause stronger shaking that would last longer and extend over broader areas than those previous California earthquakes. The consequences in densely populated areas would be tragic.

CRITICAL THINKING QUESTIONS

1. In what ways are blind thrust faults more dangerous in the Los Angeles area than the San Andreas Fault?
2. What features of the fault that caused the Northridge earthquake provide clues to the nature of future earthquakes in the Los Angeles area?

Earthquake in a Continent-Continent Collision Zone
Nepal Earthquake, 2015 ▶

On Saturday, April 25, 2015 at 11:56 a.m. local time, a major magnitude 7.8 earthquake struck the southern slopes of the Himalayas, 77 km northwest of Kathmandu, Nepal. It was the second strongest event ever recorded in Nepal. Slip estimated at 2 m occurred on the main frontal thrust fault 15 km beneath the surface, in the collision zone between the India Plate and the Eurasian Plate, which are converging at a rate of 4.5 cm/year. At least 8856 died in Nepal along with 78 in nearby parts of India and at least 26 in Tibet. Damages could exceed $10 billion.

Shaking in Kathmandu, with a population of about 1.4 million, was severe, reaching Mercalli Intensity VIII. Smaller cities nearby, Bharatpur with 107,000 people and Patan with 183,000 people, were subjected to similar shaking. Weak structures suffered heavy damage, and even earthquake-resistant structures had moderate to heavy damage. A magnitude 8.0 earthquake in 1934, 240 km southeast of the present event, and 180 km southeast of Kathmandu, killed about 10,600 people.

Numerous old buildings in cities and towns in the region collapsed, many of them built from structural brick and lacking reinforcement. People were crushed by falling brick walls and heavy floor beams. A popular six-story tourist hotel in Kathmandu collapsed, killing at least 50. The brick, cylindrical nine-story Dharahara tower with a spiral staircase to a lookout platform collapsed, killing at least 180. Fortunately, many people were outside or at work when the earthquake occurred, or the toll would have been higher. The earthquake triggered a massive snow avalanche at Mt. Everest, killing at least 17.

In the immediate aftermath of the quake, most survivors were afraid to reenter their damaged houses even to retrieve warm clothing, plastic sheets for shelter, or food, even with most stores closed. Large numbers of people, especially in outlying areas, went without adequate food and clean water for days. Tens of thousands slept outside without shelter in overnight temperatures of 12° to 14°C (mid-50°s F) with showers and thunderstorms, fearing that the many aftershocks as strong as magnitude 6.7 would cause further collapse.

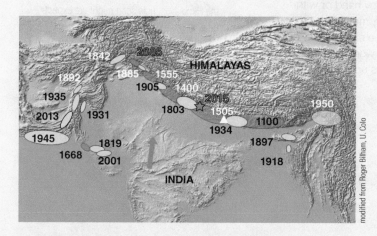

modified from Roger Bilham, U. Colo

■ *Dates of historical large earthquakes in the Himalayan collision zone between the Indian and Eurasian Plates. The main subduction-zone thrust fault follows the southern edge of earthquake ovals. The 2015 Nepal earthquake is at the red star; Kathmandu at the white triangle.*

Prakash Mathema/Stringer/Getty Images

Nepal rescue usaid

■ *Buildings constructed with lightly reinforced concrete floor posts with brick infill collapsed in piles of rubble that would have left little chance of survival. The rigid construction was unable to flex in the strong earthquake shaking.*

Although the first earthquake relieved some stress on the fault that ruptured, seismologists noted that it increased stress on nearby faults and could cause them to fail. On May 12, two weeks after the main quake, a major aftershock of magnitude 7.3, 83 km east of Kathmandu, collapsed weakened buildings and killed at least 117 more people in Nepal and 17 in India. Even this large aftershock did not relieve all of the pent-up stress in the overall zone, and other earthquakes are likely.

Poor infrastructure and remote locations made an international aid response in the region difficult. Tents, water, and food began arriving, but much more was needed. The small international airport at Kathmandu could accommodate only a limited number of flights so some relief flights had to turn back. Aid distribution to outlying towns remained problematic because of closed roads and weather too dangerous for helicopter flights. An unfortunate problem that slowed relief was that the Nepal government wanted to tax the shipments as ordinary imports.

Outside the main cities, small towns suffered the same problems and were isolated by landslides, mudflows, and collapsed roads. In this very poor country, hospital facilities and emergency services are severely limited. Both locals and emergency workers struggled to find and dig out victims mostly by hand or with hand tools; heavy equipment was in short supply. A week after the earthquake, one or two people per day were still being dug out of fortunate cavities under buildings.

People attempted to identify missing relatives among the recovered casualties. Most of the dead were cremated in outdoor funeral pyres, mostly in open fields.

CRITICAL THINKING QUESTIONS

1. Why were so many people killed in the Nepal earthquake? What could be done to minimize future casualties given the lack of resources to rebuild as we might in North America?

2. Given past earthquakes along this whole continental collision zone between India and the Himalayas, how soon might we expect another major earthquake in this region? Where would you most expect such an earthquake and why?

Critical View

A This central Idaho irrigation ditch was offset during an earthquake in the 1980s (see arrows).

G. Reagor, USGS

1. What type of fault movement is illustrated?
2. What tectonic region is this in?
3. What regional movement causes this type of fault?

B This industrial area is at the edge of San Francisco Bay, near the airport, built on mud and artificial fill.

NOAA/NGDC and University of Colorado

1. What fault near here might cause an earthquake?
2. In case of a large earthquake, what processes would affect the buildings?
3. Compared with a location a little closer to the fault, would damages here be greater or less? Why?

C This view in southwestern Montana shows three ridges with their ends lopped off to form triangles. (One of the ridges is dotted and its end is outlined in red.)

Donald Hyndman

1. What type of fault movement would have lopped off the ridge ends (red triangle)?
2. Does that same movement explain why the mountain range stands high? Why?

D This southern California view from above shows recent movement on a fault (between arrows).

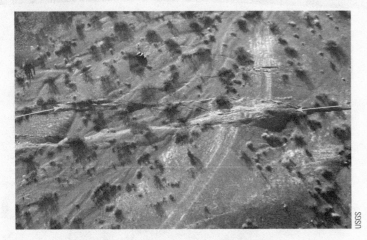

USGS

1. What type of fault movement is illustrated?
2. What direction of slip is illustrated? Explain why.
3. Assuming the tire tracks are from a car, approximately how much offset occurred during the earthquake?
4. How would you determine the size of earthquake that caused the offset?

F This pile of sand with a crater in its top is about 1 m across. It shows signs of water having spilled out of the crater and flowed down the pile's sides.

J. R. Tinsley, USGS

1. What would cause the sand pile to form?
2. Where did the sand come from?
3. Why did the sand move? Explain how the process works.

E This fault (between arrows) just south of San Francisco, California, shows movement that bent the layers.

Donald Hyndman

1. Which way did the fault move (right side went which way)?
2. What type of fault movement is this called?
3. What direction of force (direction of push) would be required to cause this movement?

G This curb is in the San Francisco Bay area.

Kious & Tilling, USGS

1. What type of fault movement is illustrated?
2. What direction of slip is illustrated? Explain why.
3. What specific fault or fault zone is this likely to be?

Chapter Review

Key Points

Faults and Earthquakes

- Faults move in both vertical and lateral motions controlled by earth stresses in different orientations. **FIGURE 3-1** and **By the Numbers 3-1**.

- Earthquakes are caused when stresses in Earth deform or strain rocks until they finally snap. Small strains may be elastic, where the stress is relieved and the rocks return to their original shape; plastic, where the shape change is permanent; or brittle, where the rocks break in an earthquake. **FIGURES 3-3** and **3-4**.

- The size of an earthquake is related to the surface rupture length, and the offset indicates how much the fault moved. Faults can move continuously, through creep, or a sudden snap. **FIGURES 3-6** and **3-7**.

Tectonic Environments of Faults

- Earthquakes typically occur along plate boundaries. Strike-slip faults move along transform boundaries such as along the San Andreas Fault in western California; thrust faults are typically associated with subduction zones such as in the Pacific Northwest and continent–continent collision boundaries; and normal faults move in spreading zones such as in the Basin and Range area. **FIGURES 3-8** to **3-13**.

- Intraplate earthquakes, though less frequent, can also be quite large as in the case of the New Madrid, Missouri, events of 1811–1812. **FIGURES 3-14** and **3-15**.

- Large earthquakes along the eastern fringe of North America are less frequent but can be significant and highly damaging. **FIGURES 3-15** and **3-16**.

Earthquake Waves

- When a fault first slips in an earthquake, energy travels outward from the focus of the earthquake in the form of seismic waves. The epicenter is the point on the surface of Earth directly above the focus. **FIGURE 3-17**.

- Earthquakes are felt as a series of waves: first the compressional P waves, then the higher-amplitude shear-motion S waves, and finally the slower surface waves. **FIGURES 3-18** and **3-24** and **By the Numbers 3-2**.

- Earthquakes are measured using a seismograph, which is essentially a suspended weight that remains relatively motionless as Earth moves under it. **FIGURES 3-20** and **3-21**.

- We can estimate the distance to an earthquake using the time between arrival of the S and P waves to a seismograph and then locate the earthquake by plotting the distances from at least three seismographs in different locations. **FIGURE 3-22**.

Earthquake Size and Characteristics

- The size of an earthquake, as estimated from the degree of damage at various distances from the epicenter, is recorded as the Mercalli Intensity. **FIGURE 3-23** and **Table 3-2**.

- The strength of an earthquake, recorded as the Richter magnitude, can be determined from its amplitude on a seismograph. An increase of one Richter magnitude corresponds to a 10-fold increase in ground motion and about a 32-fold increase in energy. **FIGURES 3-24** and **3-25**, and **By the Numbers 3-3**.

- Small, frequent earthquakes are caused by short fault offsets and rupture lengths; larger earthquakes are less frequent, with longer offsets and rupture lengths; and giant earthquakes are infrequent and caused by extremely long fault offsets and rupture lengths. **FIGURES 3-26** to **3-27**.

Ground Motion and Failure during Earthquakes

- The rigidity of the ground has a large effect on damage to buildings. Soft sediments with water-filled pores shake more violently than solid rocks. The ground may fail by liquefaction or landsliding. **FIGURES 3-30** and **3-31**.

Key Terms

acceleration, p. 51

aftershock, p. 37

amplitude, p. 46

blind thrust, p. 35

body wave, p. 46

creep, p. 38

elastic rebound theory, p. 36

epicenter, p. 45

fault, p. 35

focus, p. 45

foreshock, p. 37

frequency, p. 46

liquefaction, p. 53

Modified Mercalli Intensity scale, p. 48

moment magnitude (M_w), p. 50

normal fault, p. 35

offset, p. 37

period, p. 46

P wave, p. 45

reverse fault, p. 35

Richter magnitude (M_L), p. 49

seismic wave, p. 45

seismogram, p. 46

seismograph, p. 46

ShakeMap, p. 48

strain, p. 36

stress, p. 36

strike-slip fault, p. 35

surface rupture length, p. 37

surface wave, p. 46

S wave, p. 45

thrust fault, p. 35

wavelength, p. 46

Questions for Review

1. Explain how earthquakes occur according to the elastic rebound theory. Draw sketches to support your explanation.

2. Give examples of four significant and active earthquake zones in North America. Tell what type of fault characterizes each zone and what kind of earthquake activity is typical for each zone.

3. Why does the ground near the coast drop dramatically during a major subduction-zone earthquake? Draw a diagram to illustrate.

4. In what sequence do different types of earthquake waves occur?

5. Which type or types of earthquake waves move through the mantle of Earth?

6. Which type of earthquake waves shake with the largest amplitudes (largest range of motion)?

7. What is the approximate highest frequency of vibration (of back-and-forth shaking) in earthquakes?

8. What is the difference between the focus of an earthquake and its epicenter?

9. How is the distance to an earthquake epicenter determined?

10. What does the Richter Magnitude Scale measure?

11. On a seismogram, how much higher is ground motion from a magnitude 7 earthquake than a magnitude 6 earthquake?

12. Name three factors that help determine the strength of shaking during an earthquake. What combination of factors would produce the most intense shaking?

13. Explain the role water can play in ground failure during an earthquake.

Earthquake Critical Thinking Questions are at the end of Chapter 4.

The Haiti earthquake on January 12, 2010, destroyed the capital city of Port-au-Prince and surrounding areas, killing 159,000 people.

US Air Force Sgt. Jeremy Lock

Earthquake Predictions, Forecasts, and Mitigation

4

A City Unprepared for the Worst

At 4:53 in the afternoon on January 12, 2010, the sprawling capital of Haiti, Port-au-Prince, shook violently in a magnitude 7 earthquake that killed about 159,000 people out of a population of 1.2 million.* Survivors described the earthquake arriving with a sound like a freight train or a jet engine, immediately followed by a sudden jolt and continuous violent shaking for about 35 seconds. Clouds of dust billowed up from collapsing buildings, turning the sky gray. The power failed so the city was mostly dark after the sun went down and communications were almost nonexistent (see **Case in Point: Deadly Collapse of Poorly Constructed Heavy Masonry Buildings—Haiti, January 2010, p. 89**).

---

*The government of Haiti claimed 316,000 deaths, but foreign government agencies suggest these figures may have been inflated to appeal for more foreign aid.

Predictions and Short-Term Forecasts

Charles Richter, the developer of the magnitude scale, once remarked, "Only fools, charlatans, and liars predict earthquakes." In fact, psychics, astrologers, crackpots, assorted cranks, and miscellaneous prophets regularly predict earthquakes, but never with any success. A prediction of a new big earthquake between St. Louis and Memphis on the New Madrid Fault, which suffered three major quakes in 1811–1812, caused considerable concern and mobilization of rescue resources in 1990 but there was no earthquake. The self-proclaimed earthquake expert who made the prediction actually had no background in earthquake studies.

Not even scientists can effectively predict the date and time when an earthquake will strike, although they do understand which regions are likely to experience them. It is perfectly reasonable to say that earthquakes are far more likely to strike California than North Dakota. However, it is still just as impossible to know exactly when earthquakes will strike California as it is to know in July the dates of next winter's blizzards in North Dakota.

One main objective of earthquake research has been to provide people with some reasonably reliable warning of impending events. Scientists have been able to identify some short-term earthquake indications, and there is a possibility of developing a short-term forecasting system for them. However, until such forecasts are more reliable, they raise a host of complex ethical and political issues for governments. For now, most efforts are going into development of early warning systems for areas away from epicenters, to provide warning after an earthquake starts but before shaking reaches that area.

Earthquake Precursors

One main avenue of research in earthquake forecasting has been to explore phenomena that warn of an imminent earthquake. Although few of those efforts have proved useful to date, that may be changing.

Some researchers hope that tracking the movement of a fault will allow them to anticipate when the fault will break. Global Positioning System (GPS) stations set up across a fault can measure deformation at an accuracy of 1 to 2 mm horizontal distance and about 1.5 cm vertical. Straight lines surveyed across an active fault gradually bend before the next earthquake and fault slip (**FIGURE 4-1**). More than 250 permanent GPS stations, for example, have been set up in the Los Angeles Basin and surrounding areas to monitor ongoing ground movements.

Tectonic plates appear to move at constant speeds, and it seems reasonable to suppose that rock strength in a major fault zone could be nearly constant through time. If so, stresses might accumulate enough to break a stuck fault loose at a predictable time. Unfortunately, it is fairly common to find that a stress applied at a constant rate does not produce results at a constant rate. Broadly speaking, this is because the natural situation is far more complicated than our mental image. More specifically, every time a fault slips, for example, it juxtaposes different rocks of different strengths, thus changing the terms of the problem of predicting when it will slip again. A large fault movement will change the stress patterns along other faults, which dramatically changes their chances of future movement.

Some hope persists that swarms of minor earthquakes, or **foreshocks**, may warn of major earthquakes if they announce the onset of fault slippage. Foreshocks precede

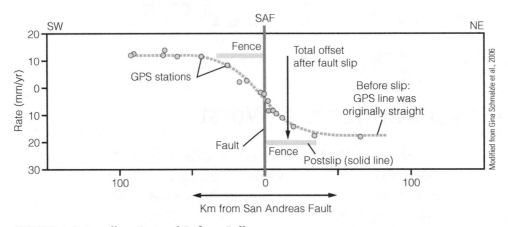

FIGURE 4-1 Bending Ground Before Failure

GPS stations placed in a straight line across the San Andreas Fault become curved as Earth's crust on one side bends elastically with respect to the crust on the other side. Eventually, an earthquake causes a break along the fault, causing the tan fence in this diagram to be offset by the fault. Compare this to Figure 3-6, the fence offset during the 1906 San Francisco earthquake.

one-third to one-half of all earthquakes. However, large earthquakes often occur with no foreshocks. Even when foreshocks do precede large earthquakes, they are rarely distinguishable from the ongoing background of common small earthquakes.

Recently, it became possible to determine precisely the focus of large numbers of minute earthquakes, or *microearthquakes*. This can reveal the presence of a previously unknown fault system, as it did in the New Madrid, Missouri, area. The changes with time in the detailed distribution of tiny crackling movements along a fault may ultimately lead to some kinds of predictions. Unfortunately, no one knows until after the fact whether the preceding small earthquake was a foreshock or the main event.

A change in levels of the radioactive gas radon is another possible earthquake precursor. Radon is one step that uranium atoms pass through in their long decay chain, which finally ends with the formation of lead. Radon is one of the rare gases. It forms no chemical compounds and remains in rocks only because it is mechanically trapped within them. Uranium is a widely distributed minor element, so most rocks contain small amounts of radon, which escapes along fractures. Radon is responsible for nearly all of the background radiation in well water and long-term exposure is a hazard in many parts of the world.

Geologists in the Soviet Union reported during the 1970s that they had observed a strange set of symptoms in the days immediately before an earthquake. These included a rise of a few centimeters in the surface elevation above the fault, a drop in the elevation of the water table, and a rise in background radioactivity in wells. Evidently, the rocks along the fault were beginning to break. The water table dropped because water was draining into the new fractures, and radon was escaping through the fractures. It all seemed to make sense as a set of precursor events.

A change in ground elevation, such as that observed in the Soviet Union, is another possible indication of an impending earthquake. In 1975, geologists of the U.S. Geological Survey (USGS) noticed that the surface elevation had risen in the area around Palmdale, California, near Los Angeles. In addition, the groundwater level was dropping while the level of background radiation was rising. Furthermore, Palmdale is on a part of the fault that had not shown a recent large earthquake. Many members of the geologic community began to hold their breath waiting for a major earthquake, but there has been no major earthquake, at least not yet.

High zones of fluid pressure in a fault zone may also localize fault movement. Detailed study of the Parkfield area of the San Andreas Fault approximately halfway between San Francisco and Los Angeles suggests that fluid pressure may be an important factor there. In fact, there have been cases where the injection of water underground inadvertently triggered earthquakes. In the early 1960s, the U.S. Army pumped waste fluids into an ancient inactive fault zone in Colorado, which apparently triggered a series of earthquakes in an area not noted for them. When the Army stopped pumping in fluids, the earthquake frequency dropped from 20 to fewer than 5 per month. In Basel, Switzerland, more than 100 small earthquakes, including 2 larger than magnitude 3, occurred from December 2006 through January 2007 following injection of water into deep hot rocks for geothermal power production (**FIGURE 4-2**). Following the events, a local prosecutor began an investigation to determine whether the company injecting the water should pay damages.

Injection of fluids in wells to increase recovery of oil and gas can generate small earthquakes up to magnitude 5+. In a populated area these can cause some damage. Further documentation of the connection between fracking fluid injection and earthquakes has come from a Transportable Array of seismographs installed in Oklahoma that

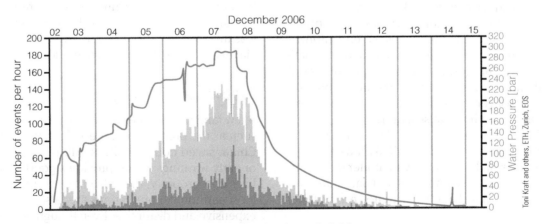

FIGURE 4-2 Earthquakes Correlate with Injection of Fluids
Injection of fluids for geothermal power production triggered a series of earthquakes at Basel, Switzerland, in December 2006. Red is large earthquakes; gray is small earthquakes (left scale). Blue line is water pressure injected (right scale; bar = atmospheric pressure). As water was pumped in, earthquake frequency increased.

shows earthquakes migrating away from injection wells. In November 2011, a magnitude 5.6 earthquake 70 km east of Oklahoma City damaged several houses; research suggests that it was caused by fracking. We will return to the discussion of fracking in Chapter 12.

From these correlations between fluid pressure and earthquakes, geologists infer that the addition of fluid increases pressure between mineral grains until the rocks break in an earthquake. The situation is analogous to the way water addition can trigger landslide movement (see Chapter 8). Some people have suggested deliberately injecting water into a fault to cause movement and initiate modest-size earthquakes. This might relieve stress on the fault and preempt the occurrence of a giant earthquake. However, as we saw in Chapter 3 (By the Numbers 3-3), each magnitude differs from the next lower magnitude by about 32 times the energy. Thus it would take about 32 magnitude 6 earthquakes to relieve the stress from a magnitude 7 earthquake. Also, the potential damage from that many magnitude 6 events suggests that may not be a good idea!

Early Warning Systems

Earthquake early warning (EEW) systems are designed to provide warning before the strong shaking of an earthquake arrives. P waves travel fast but are weak so they generally cause little damage. S waves travel at about half that speed but have much stronger shaking and do most of the damage in an earthquake. There are two main types of EEW systems: a single station and networks of stations. In a single-station system, a seismograph sensor is placed near critical facilities such as power plants, pumping stations, trains, and schools. This sensor first detects the faster P wave, providing seconds to tens of seconds of warning before the damaging S waves arrive (**FIGURE 4-3**). This advance warning could allow time to quickly shut down machinery or trains. Students in a school would have time to take cover.

In a network of stations, numerous sensors are spread around earthquake-prone areas. Many sensors send information of the earthquake to a central location, which automatically locates the earthquake, estimates its size, and sends out warnings to critical facilities. Networks are more complex and may take a little longer to provide warning, but they give more accurate warning with few false alarms.

More than 2142 seismometers, with plans for 7000, the Advanced National Seismic System (ANSS), has recently been installed in areas of the country with moderate to high seismic risk, including near Los Angeles and San Francisco Bay, the Puget Sound, Salt Lake City, Reno, Anchorage, Memphis, and St. Louis.

Although an EEW system provides a little advance warning of earthquake shaking, it will not provide enough time to evacuate a building, so building codes and earthquake preparedness, discussed later in this chapter, are also important.

EEW systems are now being developed and tested in the western United States, primarily in western California and

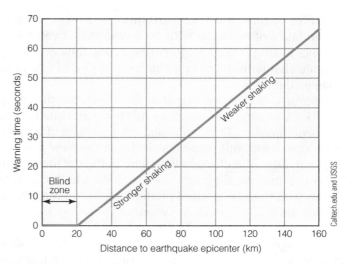

FIGURE 4-3 Warning Time for EEW

About 10 seconds are required to detect the event and issue a warning. This leaves a blind zone close to the epicenter that shakes before an alert is received. Locations within about 20 km of the earthquake epicenter receive no warning, whereas more distant locations have more time to prepare.

Washington State. The USGS already issues rapid automatic alerts for moderate to large earthquakes and information using text messages, email, social media, and the Internet. In Canada, British Columbia has a similar network that is being expanded. The magnitude 6 earthquake near Napa, California, in August 2014, occurred while an EEW system was being tested by the USGS, Berkeley Seismological Lab, and others; it functioned as designed. Mexico and Japan have the only fully operational and widespread EEW networks. Mexico's network, installed in 1991 after the disastrous 1985 Mexico City earthquake, has sensors along the Pacific coast; most warning systems are located in Mexico City and a few coastal cities. Japan's countrywide network was installed in October 2007. They alert the general public via text message, radio, and television.

Prediction Consequences

A few famous instances of successful earthquake predictions, such as the prediction of the 1976 earthquake in Haicheng, China, saved countless lives. Although short-term forecasts or early warnings would provide a good opportunity to save lives and property, they also raise complex political issues. Most earthquake predictions are false alarms, which can be expensive and disruptive. Even though the successful prediction of the Haicheng earthquake saved many lives, the Chinese government clamped down on unofficial predictions in the late 1990s after 30 false alarms severely curtailed industry and business operations.

Imagine yourself the governor of a state, perhaps California. Now imagine that your state geological survey has just given you a perfectly believable prediction that an earthquake of magnitude 7.3 will strike a large city next Tuesday afternoon. What would you do?

If you were to appear on television to announce the prediction, you would run the risk of causing a hysterical public reaction that might result in major physical and economic damage before the date of the predicted earthquake. Imagine the consequences if you broadcast the warning and nothing happened. If you do not announce the prediction, you could stand accused of withholding vital information from the public. The problem is a political and ethical minefield.

However, the potential benefits in saved lives and minimized economic damage make short-term predictions a desirable goal. If, for example, warning of an impending earthquake could be given several minutes before a large event, people could evacuate buildings, bridges and tunnels could be closed, trains stopped, critical facilities prepared, and emergency personnel mobilized.

With the current state of knowledge, earthquakes seem as inherently unpredictable in the short term as the oscillations of the stock market. Until prediction methods become sounder, governments will have to weigh the consequences of a false prediction carefully before taking action. On the other hand, officials may face accountability in the wake of deadly earthquakes if they don't act on predictions.

An Italian technician predicted that an earthquake would occur in Sulmona, Italy, on March 30, 2009, based on a series of small tremors, along with increased levels of radon emitted from the ground. The government brought together a commission to evaluate the risks of the increased earthquake activity and found that "A major earthquake in the area is unlikely but cannot be ruled out." In a news conference following the meeting, the chair of the committee told a local reporter that the earthquake activity posed "no danger" and "in fact it is a favorable situation, that is to say a continuous discharge of energy." Then, at 3:32 a.m. on April 6, an earthquake of moment magnitude 6.3 struck near L'Aquila, Italy, about 60 km northwest of Sulmona, killing 309 people and injuring 1500 more.

The technician's predicted timing was off by a week and his location was off by 60 km. Although the correlation between radon levels and earthquakes is not well documented, changes in radon levels, along with the frequent earthquakes in January through March could have provided cause for concern. If officials had taken the prediction more seriously, would lives have been saved? In fact, if Sulmona had been evacuated as recommended, many evacuees would probably have been housed in the larger town of L'Aquila—where the earthquake actually caused most of the deaths and damage. In September 2011, six Italian scientists and one government official appointed to *evaluate* earthquake risks to the population were charged with manslaughter. Prosecutors insisted that the charge was not that the scientists did not *predict* the earthquake, but that they did not properly convey the dangers to the public. The scientists were finally acquitted on appeal in 2014.

Scientists should minimize the chance that the media and public will infer that an event will (or will not) happen in a certain period of time or a specific place, by making clear that they cannot predict earthquakes. They should emphasize that earthquake forecasts indicate only an approximate percent chance of an earthquake in a general area over several decades.

A case can probably be made for explaining the symptoms, and possible precautions, to the endangered public in a straightforward, honest way and letting people decide individually what action they wish to take. Perhaps the potential threat could be expressed as a percent chance of an event of a certain size within a certain time period—much as is done for storms by the National Weather Service.

Earthquake Probability

Although scientists cannot make specific predictions about when and where an earthquake will occur, they can make reasonably reliable forecasts about where and how frequently events of different sizes *are likely* to occur. Being aware of the tectonic environments that control earthquakes helps scientists quantify risks for different regions and predict generally what earthquakes in those regions would be like. Establishing a record of past events helps determine the likely frequency of future events.

Forecasting Where Faults Will Move

Understanding the plate tectonic movements that control earthquakes allows scientists to identify the probable locations for future earthquakes. Remember from Chapter 3 that the faulting that leads to earthquakes most commonly occurs near plate boundaries, and different types of boundaries lead to different types of faults. By establishing a pattern of movement on a particular fault, scientists get a better idea of where the next earthquake is likely to strike.

Through **paleoseismology**, the study of prehistoric fault movements, seismologists can establish fault movement trends that predate written records. Some basins and ranges in western North America show distinctive white stripes along the lower flanks of the ranges. Many of these stripes mark the fault scars where the basin dropped and the range rose during a past earthquake. They are most obvious in the arid Southwest, where vegetation does not obscure fault scarps.

Trenches dug across segments of an active fault generally expose the fault, along with sediments deposited across it (**FIGURE 4-4**). Earthquakes displace sediment layers, so older sedimentary layers are more offset because they have experienced multiple fault movements. Because the amount of offset during fault movement is generally proportional to earthquake magnitude, trenches also provide a record of the magnitudes of prehistoric earthquakes. Evidence of past earthquakes can also be found in features that indicate soil liquefaction, sand boils (see Chapter 8), and sediment compaction.

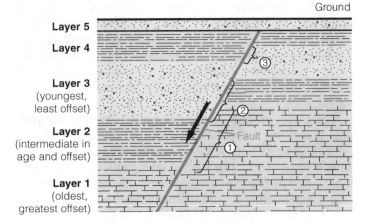

FIGURE 4-4 Earthquake History Exposed in a Trench

This diagram of a trench dug across a fault shows the relative displacement of layers offset by movements at different times. Offset (top of brick pattern) during first earthquake (1); offset (top of gray layer) during second earthquake (2); offset (top of orange layer) during last earthquake (3). Note that more sediment is deposited, in each stage, on the down-dropped side of the fault, and the first offset (1) gets progressively greater during each following offset.

Major faults are segmented by smaller faults, called *cross faults*, that run at an angle to the main fault (**FIGURE 4-5**). By studying the activity along different segments of a fault, geologists can understand which segments are likely to move next and how big the resulting earthquake might be. Cross faults limit the size of an earthquake that can occur on the main fault because typically only one segment of the fault breaks at a time. Recall from Chapter 3 that earthquake magnitude depends on the surface rupture length.

The pattern of earthquakes along fault segments also provides clues into future earthquake activity. Segments of a major fault that have less earthquake activity than neighboring segments represent a **seismic gap**. Experience shows that these seismic gaps are far more likely to be the locations of large earthquakes than the more active segments of the same fault. The magnitude 6.9 Loma Prieta earthquake of 1989 filled a seismic gap in the San Andreas Fault that geologists had identified as an area statistically likely to produce an earthquake within a few decades (**FIGURE 4-6**). The length of a seismic gap provides an indication of the possible length of a future break. Because the length of fault rupture during an earthquake is proportional to the magnitude of the accompanying earthquake, the maximum likely length of the rupture of a fault provides an indication of the magnitude of an earthquake in that segment.

Earthquakes are known to migrate along certain major faults, which can also help scientists anticipate future movements. One of the best examples of **migrating earthquakes** is along the North Anatolian Fault in Turkey, which trends east for 900 km. It closely resembles the San Andreas Fault in California and is comparable in length and slip rate. Earthquakes along the North Anatolian Fault are caused as the Arabian and African Plates jam northward against Eurasia at a rate of 1.8 to 2.5 cm per year. Between 1939 and

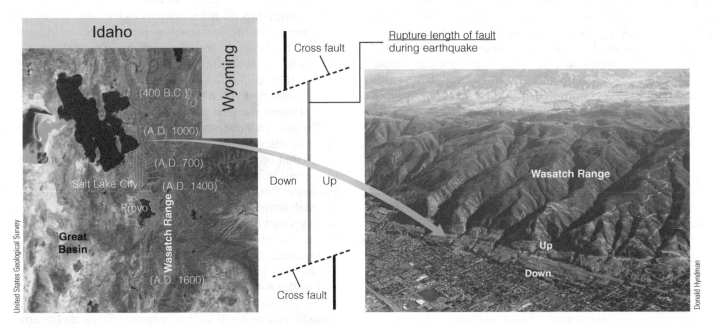

FIGURE 4-5 Earthquake Activity on a Segmented Fault

The Wasatch Fault near Salt Lake City is broken into segments by east–west cross faults (not shown on map). Individual fault segments typically break one at a time; longer segments can generate larger earthquakes. The historical record of earthquake activity along the north–south fault suggests an earthquake of magnitude greater than 6.5 every 350 years on average.

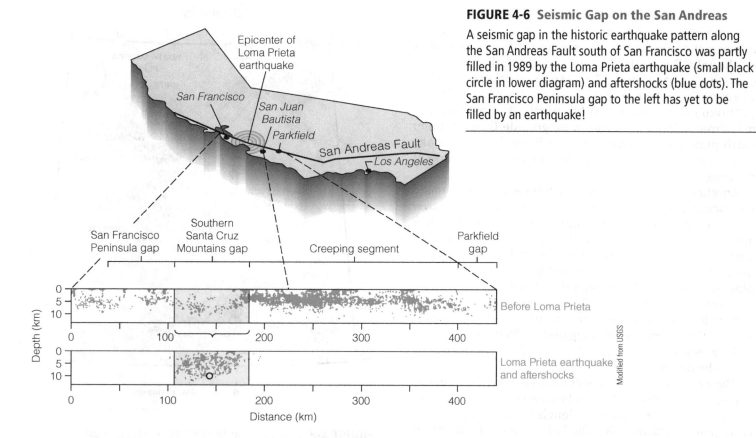

FIGURE 4-6 Seismic Gap on the San Andreas

A seismic gap in the historic earthquake pattern along the San Andreas Fault south of San Francisco was partly filled in 1989 by the Loma Prieta earthquake (small black circle in lower diagram) and aftershocks (blue dots). The San Francisco Peninsula gap to the left has yet to be filled by an earthquake!

1999, earthquakes moved sequentially westward along the North Anatolian Fault (p. 87), including the 1999 earthquake that devastated Izmit. If that pattern of earthquakes continues, the next one should be a little farther west (**Case in Point:** One in a Series of Migrating Earthquakes— Izmit Earthquake, Turkey, 1999, p. 86). Is Istanbul the next major disaster?

Forecasting When Faults Will Move

Although scientists cannot predict the date on which an earthquake will occur, they can make forecasts about the likelihood of an earthquake striking during a given time frame, for example, over a period of decades. Making forecasts about earthquake frequency generally requires knowledge of past earthquakes along a fault. That makes it possible to estimate the **recurrence interval** for earthquakes of various sizes. Once scientists have established a recurrence interval for a certain area, they can statistically estimate the probability that an earthquake greater than a given magnitude will strike within a specified period in the future.

A few faults move at somewhat regular intervals, although this kind of behavior is so uncommon, and the time series so unreliable, that it has never been a useful method of earthquake prediction. One of the most reliable cases of equal-interval earthquakes was on the San Andreas Fault, near Parkfield, north of Los Angeles. Until recently, a series

of six moderate-size earthquakes occurred about every 22 years on average. After that, there was a 38-year gap to the most recent moderate earthquake on September 28, 2004.

For more than 500 years, especially strong earthquakes devastated Tokyo, Japan, at intervals of roughly 70 years, most recently in 1923, when 105,000 people died (**FIGURE 4-7**). Many people, especially those in the

FIGURE 4-7 Tokyo Leveled in 1923

A magnitude 8.2 earthquake in 1923 destroyed Tokyo from an epicenter 90 km away.

insurance business, watched the approach of 1993 with extreme concern. No earthquake has yet occurred, but it may just come later than expected. Tokyo University's Earthquake Research Institute, in 2012, estimated a 70% chance of a magnitude 7 or larger earthquake within 75 km of Tokyo before 2016! Since the March 2011 Sendai (Tōhoku) earthquake 300 km north, the Institute has measured a large increase in magnitude 6 and smaller earthquakes near Tokyo. A magnitude 7.3 earthquake there could kill 10,000 people and cause US $1 trillion in damages.

Another recently recognized pattern is that earthquakes may occur in groups. A major fault can be quite active over many decades and then lie quiet for many more before it again becomes active. In North America, our written history may be a bit short to demonstrate such patterns clearly, but in Turkey the historical record is much longer. Beginning in AD 967, the North Anatolian Fault moved to cause earthquakes at intervals of 7 to 40 years, followed by no activity for 204 years. The fault saw more earthquakes from 1254 to 1784, then another long hiatus. Then again in 1939, activity picked up at intervals from 1 to 21 years and continues to the present day.

The coast of northern California, Oregon, Washington, and southern British Columbia has a record of strong earthquakes but is currently experiencing a long gap in such earthquakes (**FIGURE 4-8**). The last major earthquake here was just over 315 years ago. As with some other subduction zones, the expected magnitude 8 to 9 earthquake would make up for that long interval.

Scientists look for clues to establish dates for earthquakes that occurred before recorded history. Radiocarbon dating of organic matter, most commonly charcoal, in the sedimentary layers broken along a fault can reveal the maximum dates of fault movements. Radiocarbon dates on offset sediments exposed in a trench across the Reelfoot Fault scarp in Tennessee reveal that the New Madrid Fault system has moved three times within the last 2400 years, which, in combination with other data, suggests that the recurrence interval for earthquakes in the New Madrid area may be 500 to 1000 years, although the uncertainty is large. One limitation of radiocarbon dating is that it can be used to date events that happened only in the last 40,000 years.

A combination of such clues indicates that Utah should be prepared for a major earthquake. The intermountain seismic zone through central Utah, Yellowstone Park, and western Montana feels more earthquakes than one might expect given the significant distance to any plate boundary. No major movement has occurred on the main fault near Brigham City, north of Salt Lake City, for at least 1300 years. But that does not offer much reassurance to the people of the Salt Lake area. Nor does the extremely fresh appearance of the Wasatch Front. The lack of large deposits of eroded material at the base of the Wasatch Front suggests that the range is rising rapidly, presumably with

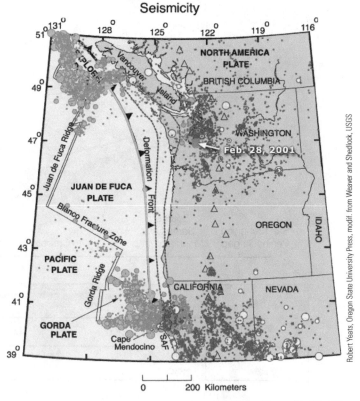

FIGURE 4-8 Seismic Gap in the Pacific Northwest

The seismic gap in Cascadia subduction-zone earthquakes between southern British Columbia and northern California is marked by a double arrow along the "deformation front." Red dots are earthquakes. Open circles are shallow earthquakes with larger circles indicating larger earthquakes; triangles are volcanoes.

accompanying earthquakes. The people of the Salt Lake area should consider themselves at high risk for major earthquakes.

An obvious fault scarp shows evidence of a large earthquake in Little Cottonwood Canyon in the southeastern part of Salt Lake City (**FIGURE 4-9**). Radiocarbon dates on scraps of charcoal show that the scarp rose approximately 9000 years ago. More radiocarbon dates on offset sedimentary layers exposed in trenches cut across the fault reveal evidence of movements 5300, 4000, 2500, and 1300 years ago. The average recurrence interval appears to be about 1350 years, with an average offset of 2 m per event. It is not reassuring to see that developers continue to encourage building homes next to the fault.

Movement on the Wasatch scarp appears to occur in segments. Within a 6000-year record, 17 fault offsets exposed in trenches suggest an earthquake of magnitude greater than 6.5 every 350 years on average. Maximum segment lengths and offsets found in some trenches suggest earthquake magnitudes up to 7.5. Despite these warning signs, development

FIGURE 4-9 Fault Scarps and Development

Wasatch Front fault scarp cuts a 19,000-year-old glacial moraine from Cottonwood Canyon near Salt Lake City. The scarp (between arrows) formed during a large earthquake since that time.

continues near, and even within, the fault zone. Recent studies indicate that a magnitude 7 earthquake on the Wasatch Fault could generate local horizontal accelerations of 1 g and higher.

Monitoring ground movements using GPS networks is now widely used in earthquake forecasting. Hundreds of

GPS stations have now been installed in California, southern Canada, Mexico, Europe, and elsewhere around the world. Continuous monitoring movements across active faults can lead to better estimates of the most likely locations, magnitudes, and probabilities of future earthquakes.

Populations at Risk

Seismologists use what they have learned about probabilities of where and when an earthquake will strike to estimate the risk for a given area on a **risk map** (**FIGURES 4-10** and **4-11**). These maps are based on past activity measured in terms of both frequency and magnitude. They are invaluable when choosing sites for major structures—such as dams, power plants, public buildings, and bridges—and important for insurance purposes.

Notice on the map in FIGURE 4-11 that one of the highest-risk areas in the United States is the heavily populated zone near the coast of California. California gets far more than its share of North American earthquakes because it is along the boundary between the Pacific and North American tectonic plates. That part of the plate boundary is marked by the San Andreas Fault, one of Earth's largest and most active transform faults. This well-researched earthquake zone provides a good example of how forecasting allows scientists to estimate the probability of earthquakes and their likely levels of damage.

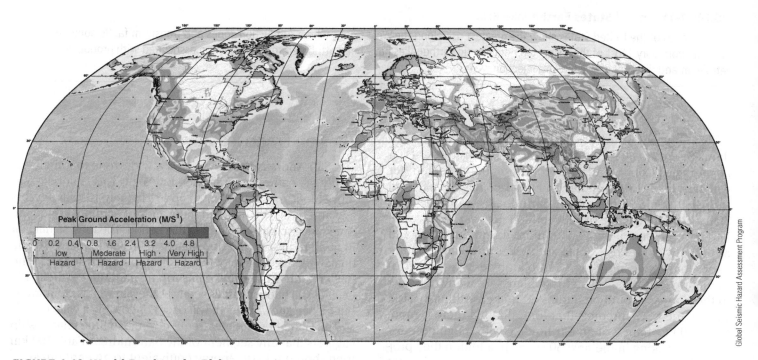

FIGURE 4-10 World Earthquake Risk

Peak accelerations (darker red colors) correspond to regions with strong earthquakes, typically along subduction zones such as those along the west coasts of North and South America and continental collision zones such as in southern Europe and Asia.

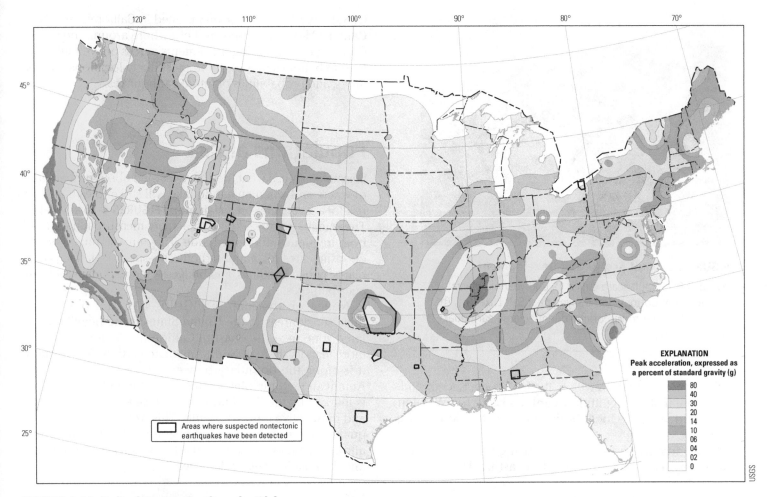

FIGURE 4-11 United States Earthquake Risk

Seismic zones of the United States show the greatest risk along the West Coast, where subduction zones and transform faults dominate the plate boundary. Most of the suspected nontectonic earthquakes have been attributed to fluid injection associated with production of petroleum and disposal of wastewater injection wells.

Roughly one-third of California's future earthquake damage will probably result from a few large events in Los Angeles County. A large proportion of the remainder will most likely occur in the San Francisco Bay area. Before the 1989 Loma Prieta event, earthquakes in the United States cost an average of $230 million per year. Costs escalate as more people move into more dangerous areas and as property values rise. Some authorities now expect future losses to average more than $4.4 billion per year, with 75% of that in California. Given these high stakes, any taxpayer in these areas should be interested in the earthquake risk. In fact, the USGS predicts that for California there is more than a 99% chance of an earthquake with magnitude 6.7 or larger striking in the next 30 years. In addition they project a 46% chance that a much larger magnitude 7.5 event will occur somewhere in the state in the next 30 years (**FIGURE 4-12**).

The San Francisco Bay Area

The San Andreas Fault system is a wide zone that includes nearly the entire San Francisco Bay area. This fault began to move along most of its length approximately 16 million years ago and has likely inflicted thousands of earthquakes on the San Francisco Bay area during this period. The most significant in modern history, the devastating 1906 San Francisco earthquake, reduced that city to ruins. The people who live in the Bay area dread the next catastrophic earthquake, their "Big One." It will happen just as surely as the sun will rise tomorrow, but no one knows when it will occur or how big it will be.

The Gutenberg-Richter frequency–magnitude relationship suggests that any segment of the San Andreas Fault 100 km long should release energy equivalent to an earthquake of magnitude 6 on average every 8 years, one of magnitude 7 on average every 60 years, and one of magnitude 8 on

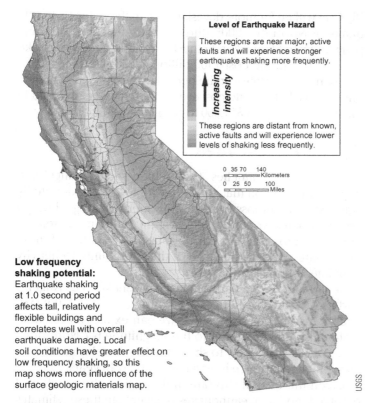

Level of Earthquake Hazard

These regions are near major, active faults and will experience stronger earthquake shaking more frequently.

Increasing intensity

These regions are distant from known, active faults and will experience lower levels of shaking less frequently.

0 35 70   140
                Kilometers
0 25 50   100
                Miles

Low frequency shaking potential: Earthquake shaking at 1.0 second period affects tall, relatively flexible buildings and correlates well with overall earthquake damage. Local soil conditions have greater effect on low frequency shaking, so this map shows more influence of the surface geologic materials map.

USGS

FIGURE 4-12 Chance of the Next Big California Earthquake

California is almost certain (>99% probability) to have a major earthquake in the next 30 years. The Los Angeles and San Francisco regions, respectively, have a 63 and 67% chance of suffering a major earthquake in that time.

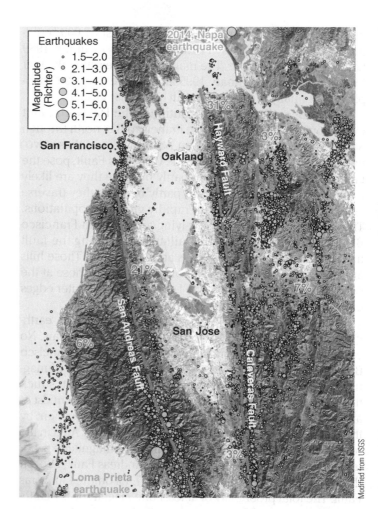

FIGURE 4-13 San Francisco Area Earthquakes

The San Francisco Bay area faults show probabilities of magnitude 6.7 or larger earthquakes on the main faults before 2030. The Loma Prieta epicenter is shown as a larger orange dot. The 1906 earthquake broke the fault from near the Loma Prieta event to far north of this map.

average about every 700 years (recall FIGURE 3-37). If this is correct, then several fault segments are overdue. Energy builds up across the fault until an earthquake occurs, so the chance of a major event grows as the time since the last large earthquake increases.

Researchers have calculated different overall slip rates on the fault using different types of data. Using a relatively low slip rate of 2 cm per year, strain should accumulate on stuck faults at a rate that would cause an earthquake of moment magnitude (M_w) 7.5 every 120 to 170 years. Alternatively, there could be six earthquakes of magnitude 6.6 in the same period. Because the actual number has been far fewer, this again suggests that the area is overdue.

Although the San Andreas is the main fault, others in that fault system also pose risk to populations in the area. The Hayward Fault splays north from the San Andreas and runs along the base of the hills near the east side of San Francisco Bay through a continuous series of cities, including Hayward, Oakland, Berkeley, and Richmond. The Hayward Fault has not caused a major earthquake since 1868, but many geologists believe it is one of the most dangerous faults in California and may be ready to break (**FIGURE 4-13**). A moderate

earthquake (magnitude 5.6) struck the Hayward Fault northeast of San Jose on October 30, 2007, without doing much damage, although it did serve as a reminder of the hazard. A magnitude 6 earthquake on the West Napa Fault shook the town of Napa and vicinity at 3:20 a.m. on August 24, 2014, 8 km SSW of Napa. It lasted 10 to 20 seconds and injured at least 120 people, 3 of them critically. Most injuries were from falling debris, including chimneys. It ruptured many water lines and caused gas leaks, causing 8 to 10 large fires. In the older part of town, brick façades and cornices collapsed onto the sidewalks. Damages are likely to be as much as $300 million (see **Case in Point:** Recent San Francisco Bay Earthquakes, 1989 and 2014 p. 85).

An offset of as much as 3 m on the Hayward Fault would probably cause a magnitude 7 earthquake that would cause shaking for 20 to 25 seconds. The USGS estimates that several thousand people might be killed and at least 57,000 buildings would be damaged, 11 times as many as in the 1989 Loma Prieta earthquake. The magnitude 6.8 earthquake that hit Kobe, Japan, in January 1995 killed more than 6000 people in an equally modern city built on similar ground.

The Rodgers Creek Fault continues the Hayward Fault trend from the north end of San Francisco Bay north through Santa Rosa. Along with the San Andreas Fault, these two, and their southward extension, the Calaveras Fault, pose the greatest hazards in the region, partly because they are likely to cause large earthquakes and partly because they traverse large metropolitan areas with rapidly growing populations. The risk is not only in the low-lying areas of San Francisco Bay muds, or in the heavily built-up cities along the fault trace, but in the precipitous hills above the fault. Those hills are blanketed with expensive homes, including those at the crest of the range along Skyline Drive, where the outer edges of many are propped on spindly-looking stilts.

In the 70 years before the 1906 San Francisco event, earthquakes of magnitude 6 to 7 occurred every 10 to 15 years. No earthquakes greater than magnitude 6 struck the San Francisco Bay area between 1911 and 1979. Four earthquakes of magnitudes 6 to 7, including the 1989 Loma Prieta event, struck the region between 1979 and 1989. We may be in the midst of another cluster of strong earthquakes. A recent assessment by the USGS of the earthquake probabilities in the Bay Area in the next 30 years suggests a 21% probability of a magnitude 6.4 or larger earthquake on the northern San Andreas Fault, 29% on the Hayward-Rodgers Creek Fault, and 7.4% on the Calaveras Fault. The total probability of at least one magnitude 6.7 earthquake in the next 30 years somewhere in the San Francisco Bay area is 72%. The same size earthquake caused severe damage in the Los Angeles area in 1994.

A 2014 analysis of historic earthquakes in the San Francisco Bay Area shows that large earthquakes of magnitude 6.6 to 7.8 sometimes occur in clusters. From about 1690 to 1776, six large earthquakes occurred on 5 different faults. This would have released much of the regional stress built up by tectonic movement between the Pacific Plate and the North American Plate. In the following 86 years without major earthquakes, stress between the two plates was gradually building. Instead of planning only for an eventual "Big One" like the magnitude 7.9 event in 1906, perhaps we should be more concerned with the possibility of a cluster of somewhat smaller events on different faults—a combination that would release a similar total amount of energy but could present greater problems than a single large event.

Just as the risk of a large earthquake increases as more time passes without one, so does the potential damage such an earthquake would cause. People build in areas close to or even on top of faults, right on potential future epicenters. In some areas immediately south of San Francisco, developers have filled fault-induced depressions and built subdivisions right across the trace of the fault. Within San Francisco itself, zoning prevents building in some areas near the fault. All those people and all that development guarantee that the next

Big One will be far more devastating than the last major earthquake. A November 2005 study by Allstate Insurance Company indicated that if San Francisco had the same size quake as in 1906, it could cost $400 billion to rebuild the city. For comparison, the state's entire budget in 2004 was $164 billion. This means that it could cost every man, woman, and child in California more than $11,000 to rebuild San Francisco.

The Los Angeles Area

The San Andreas Fault lies 50 km northeast of Los Angeles. A large earthquake on that part of the fault would undoubtedly cause significant loss of life and property damage, but the prospect of lesser earthquakes on any of the many related faults that cross the Los Angeles basin is a matter of more immediate concern because of their proximity to heavily populated areas (**FIGURE 4-14**).

Several moderate earthquakes have shaken Los Angeles and the Transverse Ranges area during the last 150 years. Such a history guarantees that we can expect more. Moderate earthquakes similar to the magnitude 6.7 Northridge event of 1994 are likely to shake the Los Angeles basin with an average recurrence interval of less than 10 years.

Given the overall slip rate in the region, the number of observed moderate earthquakes is fewer than these estimates would lead us to expect. A larger earthquake of magnitude 7.5, requiring an approximately 160 km long rupture should occur on average once every 300 or so years. Far too few moderate earthquakes have occurred in the Los Angeles area to relieve the observed amount of strain built up across the San Andreas Fault between 1857 and 2007. Releasing the accumulated strain would have required 17 moderate earthquakes, but only two (1971 and 1994) have been greater than magnitude 6.7. The reason for the deficiency is unclear. Elsewhere in the world, clusters of destructive earthquakes have occurred at intervals of a few decades. Perhaps several faults that were mechanically linked ruptured at roughly the same time to generate a large earthquake. Such combination ruptures appear to have occurred in the Los Angeles area in the past.

Seismic activity has been notably absent along the segment of the San Andreas Fault south of the Fort Tejon slippage of 1857, almost to the Mexican border. Fault segments with such seismic gaps are far more likely to experience an earthquake than fault segments that have recently moved.

Trenches dug across the San Andreas Fault northeast of Los Angeles exposed sand boils that record sediment liquefaction during nine large earthquakes during the 1300 years before the Fort Tejon earthquake of 1857. They struck at intervals of between 55 and 275 years, the average interval between these events being 160 years. Based on this history, the area appears due for a major earthquake.

A magnitude 7.4 earthquake would cause the ground to shake over a much larger area and for a longer duration than the Northridge earthquake did in 1994, and the greater area and longer shaking would result in far more casualties and property damage. Studies of the fault in 2006 showed

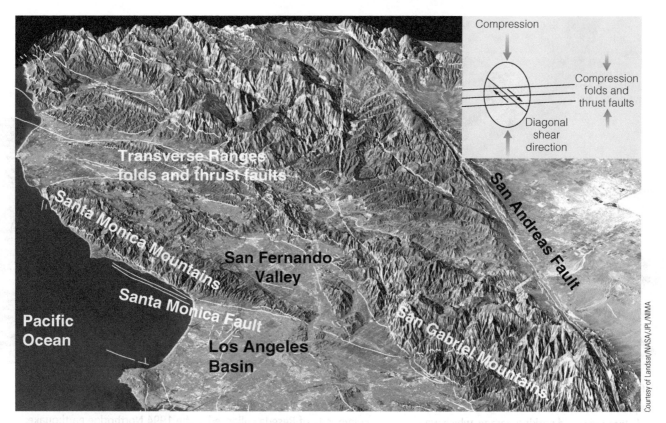

Courtesy of Landsat/NASA/JPL/NIMA

FIGURE 4-14 Major Faults in the Los Angeles Area

The Los Angeles area faults (thin white lines) are largely oriented northwest—southeast parallel to the San Andreas Fault, with east—west or east-southeast—west-northwest blind thrusts in between, especially in the area north of Los Angeles near the "big bend" of the San Andreas. (*Modified from NASA basemap*)

that it was sufficiently stressed to break with a magnitude 7 earthquake but could, in the future, see a magnitude 8 (the Northridge and San Fernando Valley earthquakes were both magnitude 6.7). The nearby, lesser-known San Jacinto Fault, which runs northwest through San Bernardino and Riverside, is being stressed about twice as fast as formerly thought. It is capable of generating magnitude 7 earthquakes.

Relationships between nearby parallel faults suggest that some may be codependent. Recent studies suggest that the San Jacinto Fault, about 50 km west of the San Andreas Fault in southernmost California, began taking up some of the regional displacement about 1 million years ago, finally becoming dominant as the San Andreas declined. About 90,000 years ago, displacement on the San Andreas again began taking over regional movement, which continues today. The overall regional offset rate seems to remain nearly unchanged.

Minimizing Earthquake Damage

Throughout history, earthquakes have had devastating effects, destroying cities and decimating their populations. Given the high probability of a destructive earthquake in a major urban area, what is the potential damage from such an earthquake, and how can it be mitigated?

The primary cause of deaths and damage in an earthquake is the collapse of buildings and other structures, which we explore in greater detail in the following material. But the shaking during an earthquake can trigger other damage as well. In the aftermath of the earthquake, fire becomes a serious hazard. The fires sparked in the 1906 San Francisco earthquake caused more damage than the initial shaking. During an earthquake, electric wires fall to an accompaniment of great sparks that are likely to start fires. Release of gas or other petroleum products in port areas, such as Los Angeles, could spark a firestorm. Buckled streets, heaps of fallen rubble, and broken water mains would hamper firefighters' efforts.

Without immediate outside help, the days after a major earthquake commonly bring epidemics caused by impure water, because broken sewer mains leak contaminants into broken water mains. Decomposing bodies buried in rubble also contribute to the spread of disease. It is not uncommon for many people to die from diseases following a major earthquake, especially in the poorest countries and warm climates. Meanwhile, fires continue to burn and expenses mount.

Steel plates welded on floor beam and ledge

Precast floor beam

Support beam

Mehmet Celebi USGS

James W. Dewey USGS

FIGURE 4-15 Failure of Joins and Beams

A. The precast floor of the Northridge Fashion Center's three-year-old parking garage, which was supported on the ledge of a supporting beam, failed during the 1994 Northridge, California, earthquake.

B. The exterior façade loosely hung on a building framework in the community of Reseda collapsed in the 1994 Northridge earthquake. Directional shaking detached one wall while leaving the perpendicular wall intact.

Scientists know a number of ways to mitigate earthquake damage, but these mitigation efforts are expensive and hard to enforce. The costs should be weighed against the large number of lives that could be saved if these dollars were spent on improved medical care or mitigation of other hazards.

Types of Structural Damage

Earthquakes don't kill people—falling buildings and highway structures do. Load-bearing masonry walls of any kind are likely to shake apart and collapse during an earthquake, dropping heavy roofs on people indoors. Some bridge decks and floors of parking garages are not strongly anchored at their ends because they must expand and contract with changes in temperature. A strong earthquake can shake them off their supports. Many external walls are loosely attached to building frameworks. They may break free during an earthquake and collapse onto the street below (**FIGURE 4-15**).

Reinforced concrete often breaks in large earthquakes, leaving the formerly enclosed reinforcing steel free to buckle and fail (**FIGURE 4-16**). Segments of the Interstate 5/California Highway 14 interchange collapsed while under construction in the 1971 San Fernando Valley earthquake

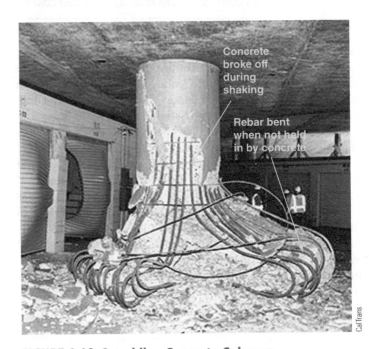

Concrete broke off during shaking

Rebar bent when not held in by concrete

CalTrans

FIGURE 4-16 Crumbling Concrete Columns

The steel reinforcing bars collapsed when the rigid concrete of this freeway support column shattered during the 1994 Northridge earthquake.

when concrete support columns failed. Highway officials failed to learn from this mistake and rebuilt them to the same specifications so that they collapsed again during the 1994 Northridge earthquake. Reinforced concrete fares much better if it is wrapped in steel to prevent crumbling. Much of the strengthening of freeway overpasses in California during the 1990s involved fitting steel sheaths around reinforced concrete columns. New construction often involves wrapping steel rods around the vertical reinforcing bars in concrete supports.

Weak floors at any level of a building are a major problem during an earthquake. Upper floors move back and forth during shaking, leading to the collapse of either individual floors or the whole building (**FIGURE 4-17**). Garages and storefronts commonly lack the strength to resist major lateral movements, as do commercial buildings with too many unbraced windows on any floor. The addition of diagonal beams can provide the support needed to resist lateral movement. Heavy reinforced concrete floors and corner supports with unreinforced fill-in walls of brick are very common

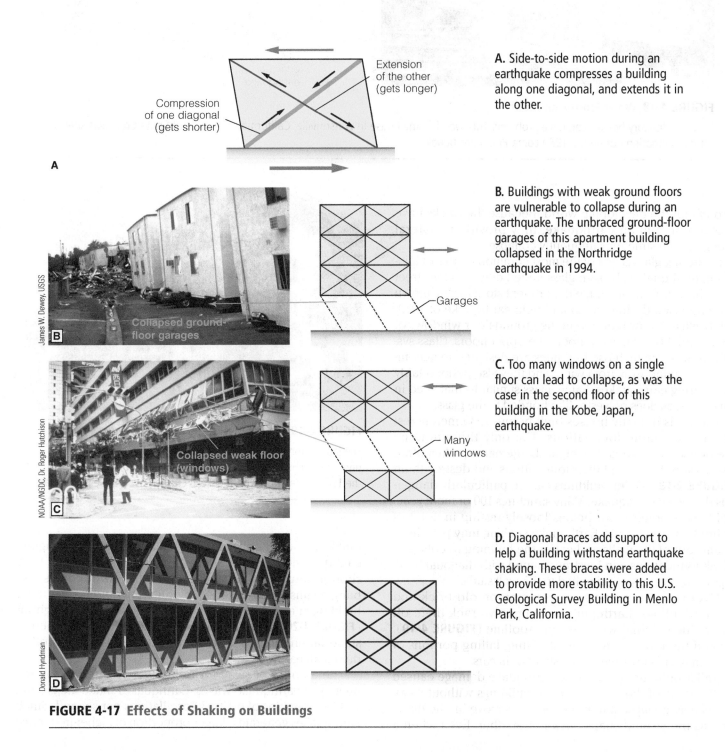

A. Side-to-side motion during an earthquake compresses a building along one diagonal, and extends it in the other.

B. Buildings with weak ground floors are vulnerable to collapse during an earthquake. The unbraced ground-floor garages of this apartment building collapsed in the Northridge earthquake in 1994.

C. Too many windows on a single floor can lead to collapse, as was the case in the second floor of this building in the Kobe, Japan, earthquake.

D. Diagonal braces add support to help a building withstand earthquake shaking. These braces were added to provide more stability to this U.S. Geological Survey Building in Menlo Park, California.

FIGURE 4-17 Effects of Shaking on Buildings

Carl W. Stover, USGS

FIGURE 4-18 Weak Floor Connections

Even single-story houses can have problems. This wood-frame house in Watsonville, California, southeast of Santa Cruz, was shaken off its foundation during the 1989 Loma Prieta earthquake.

throughout the developing world. Because the bricks have little strength in an earthquake, they provide no lateral strength; such structures frequently collapse.

Shattering glass is one of the most common causes of injuries in earthquakes. Broken glass rained down on the street from the windows of a large department store in downtown San Francisco during the Loma Prieta earthquake of 1989. Safety glass is now required in the ground-floor windows of commercial buildings but not in the upper floors. Glass systems in modern high-rise buildings are designed to accommodate routine sway from wind, and they also perform fairly well during earthquakes. Safety glass similar to that used in cars helps, as does polyester film bonded to the glass.

The walls of many houses built before 1935 merely rest on their concrete foundations. The only thing holding the house in place is its weight. Large earthquakes often shake older houses off their foundations and destroy them (**FIGURE 4-18**). Older buildings can be particularly dangerous during an earthquake. Many structures 100 or more years old have wooden floor beams loosely resting in notches in brick or stone walls. Earthquake shaking may pull these beams out of the walls and cause the building to collapse. Weak foundations can be strengthened with diagonal bracing or sheets of plywood nailed to the wall studs.

Overhanging parapets are common on old brick and stone buildings. Earthquakes commonly crack them off where the external wall joins the roofline (**FIGURE 4-19**). Even if the building remains standing, falling portions of walls may crush people on the street or in cars.

Taller buildings experience earthquake damage caused by the sway of the building. Even buildings without weak floors may collapse if their upper floors move in one direction as the ground snaps back in the other. Even when a

Edgar V. Leyendecker USGS

FIGURE 4-19 Structural Brick and Overhangs

The fourth-story wall and overhanging brick parapet of an unreinforced building in San Francisco collapsed onto the street during the 1989 Loma Prieta earthquake. It crushed five people in their cars.

building doesn't collapse, its oscillation may cause significant damage. Taller buildings sway more slowly than their shorter neighbors. Adjacent tall and short buildings can bang against each other. This commonly breaks the taller building at about the level of the top of its shorter neighbor (**FIGURE 4-20**). Buildings should be either firmly attached or stand far enough apart that they do not bash one another during an earthquake.

The amount of sway for buildings is related to the frequency of earthquake waves. Earthquakes shake the ground with frequencies of 0.1 to 30 oscillations per second. Small earthquakes generate a larger proportion of higher-frequency

FIGURE 4-20 Unsynchronized Sway

Adjacent buildings of different heights will sway at different frequencies, so they can collide during earthquakes. During the 1985 Mexico City earthquake, the shorter building on the left hammered its taller neighbor, collapsing one story at the point of impact.

Frequency of Building Vibration

Buildings of different heights sway at different frequencies, like inverted pendulums, as shown in the following figure.

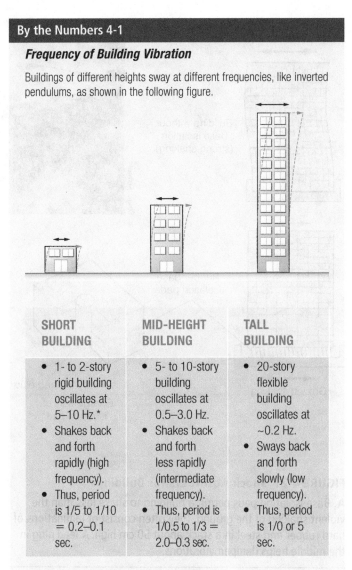

SHORT BUILDING	MID-HEIGHT BUILDING	TALL BUILDING
• 1- to 2-story rigid building oscillates at 5–10 Hz.* • Shakes back and forth rapidly (high frequency). • Thus, period is 1/5 to 1/10 = 0.2–0.1 sec.	• 5- to 10-story building oscillates at 0.5–3.0 Hz. • Shakes back and forth less rapidly (intermediate frequency). • Thus, period is 1/0.5 to 1/3 = 2.0–0.3 sec.	• 20-story flexible building oscillates at ~0.2 Hz. • Sways back and forth slowly (low frequency). • Thus, period is 1/0 or 5 sec.

*Hz = Hertz = cycles of back-and-forth motion per second.

vibrations. If the natural oscillation frequency of a building is similar to that of the ground, it may resonate in the same way as a child on a swing pulls in resonance with the oscillations of the swing. In both cases, the resonance greatly amplifies the motion. In general, soft mud shakes at a low frequency whereas solid rock shakes at a high frequency. Tall, heavy structures, like some raised freeways built across mud-filled bays, frequently collapse unless their supports are deeply anchored and designed to minimize low-frequency shaking. Buildings and other structures need to be designed to avoid matching their natural vibration frequency with that of the shaking ground beneath. (See **Case in Point:** Shaking Amplified in Soft Mud and Clays—Recent San Francisco Bay Earthquakes, 1989 and 2014, p. 85.)

Tall buildings are most vulnerable to lower-frequency vibrations. To understand why, you can simulate the swaying of a tall building by dangling a weight at the end of a string and moving the hand holding the string. For example, if you move your hand back and forth 30 cm (1 ft) each second, the weight at the end of a string 30 cm long will cause a large swing back and forth. This swing is in resonance with the oscillation period of the pendulum. On the other hand, if you move your hand back and forth three times per second, the weight will hardly move. Similarly, at three times per second, the weight will hardly move if the string is much longer than 30 cm. A tall building sways back and forth more slowly than a short one, so low-frequency earthquake vibrations are more likely to damage tall buildings (**By the Numbers 4-1:** Frequency of Building Vibration). Short buildings up to several stories high vibrate at high frequencies and do not sway much.

The ground also vibrates at different frequencies depending on the sediment type. Soft sediment vibrates at low frequencies; it makes a dull thud when hit with a hammer. In contrast, bedrock vibrates at high frequency, so it rings when hit with a hammer. Buildings sustain the most damage when they oscillate at a frequency similar to that of the ground. Because short, rigid buildings have high frequencies, they survive earthquakes better when built on soft sediment (without taking into account secondary ground effects such as liquefaction and landslides). A tall, flexible building, which sways at a low frequency, often survives an earthquake well on high-frequency bedrock.

Fortunately, buildings can be built to withstand a severe earthquake well enough to minimize the risk to people inside. Houses framed with wood generally have enough flexibility to bend without shattering. A house may bounce off its foundation and be wrecked during an earthquake, but it is unlikely to collapse and kill anyone. Commercial buildings with frames made of steel beams welded or bolted together are also generally flexible enough to resist collapse. Most building codes now

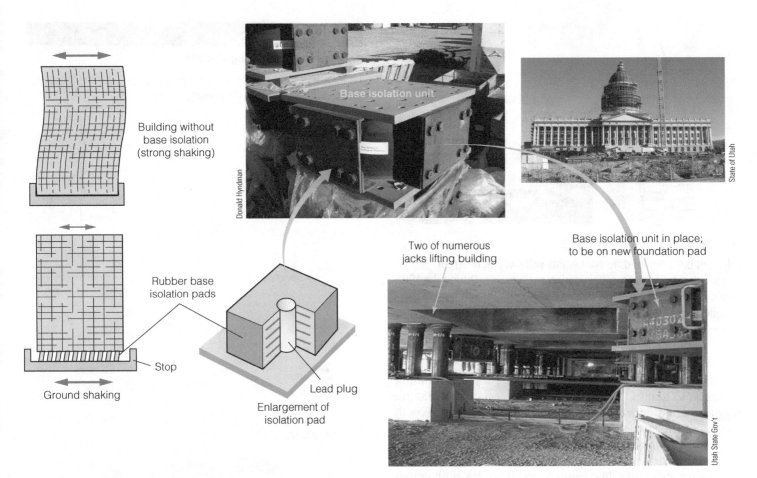

FIGURE 4-21 **Shock Absorbers for Buildings**

A. Base isolation pads permit a building to shake less than the violent shaking of the ground. They often consist of laminations of hard rubber and steel in a stack about 50 cm high. A lead plug in the middle helps dampen vibrations.

B. In a large project retrofit, base isolation pads are emplaced under jacked-up Utah State Capitol building (inset) in 2006.

require builders to bolt the framing to the foundation—an excellent example of an inexpensive change that can make an enormous improvement in the ability of a house to withstand an earthquake.

Modifying existing buildings to minimize damage during strong earthquake motion, or **retrofitting**, can provide additional protection. Large earthquakes have a low probability of occurrence, but they may happen during the average life of a building, which is considered to be 50 to 150 years. Retrofitting existing buildings to survive these large but rare events is extremely expensive; it is much less expensive to construct new buildings to a higher standard. This leaves scientists and policymakers unsure of the best strategy when developing codes for new or retrofitted buildings.

Guidelines for new construction as well as retrofitting of older buildings can reduce damage to buildings during an earthquake. The Federal Emergency Management Agency

(FEMA) responded in 2001 with new guidelines for steel construction in areas prone to earthquakes. Most high buildings built in the last 30 years have welded steel frames designed to resist earthquake motions. In recent years, seismic engineers have tried to minimize the shaking, and therefore the damage, by isolating buildings from the shaking ground in a procedure called **base isolation** (**FIGURE 4-21**). They place the building on thick flexible pads, which act like a car's springs and shock absorbers that isolate us from many bumps in the road.

Base isolation pads are generally installed during initial construction; however, historic buildings can be retrofitted with them. In 1989, the historic City and County Building in Salt Lake City, five floors of stone masonry construction, was detached from its foundation, reinforced, and fitted with base isolation pads. That building is now designed for an earthquake of Richter magnitude 5 or 6. Seismic risk maps indicate a 10% probability of such an earthquake

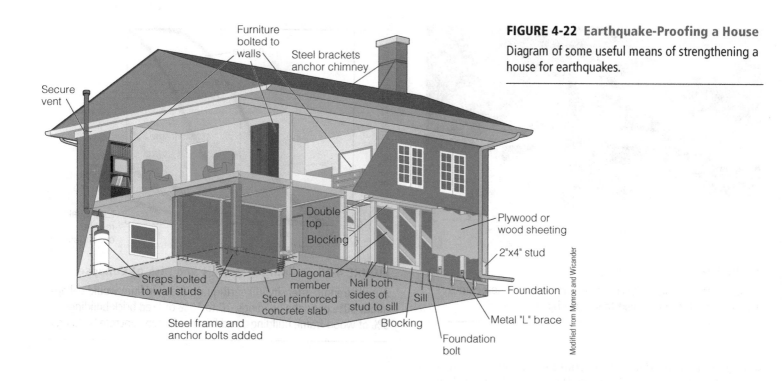

Furniture bolted to walls

Steel brackets anchor chimney

Secure vent

Double top

Blocking

Plywood or wood sheeting

2"x4" stud

Straps bolted to wall studs

Diagonal member

Nail both sides of stud to sill

Sill

Foundation

Steel reinforced concrete slab

Steel frame and anchor bolts added

Blocking

Metal "L" brace

Foundation bolt

Modified from Monroe and Wicander

within 50 years. In a still larger project in the mid-2000s, the Utah State Capitol building in Salt Lake City was also jacked up and placed on rubberized pads about 50 cm thick (**FIGURE 4-21B**).

Earthquake Preparedness

Preparing homes for an earthquake and knowing what to do when an earthquake occurs can reduce damage and save lives (see the Earthquake Survival Guide at the end of this chapter). Evaluating structural weaknesses in a home and retrofitting are the first steps (**FIGURE 4-22**). In general, walls of all kinds should be well anchored to floors and the foundation. Other recommendations include bolting bookcases and water heaters to walls and securing chimneys and vents with brackets. Earthquakes commonly break gas mains and electric wires, which start fires that firefighters cannot readily combat if water mains are also broken (**FIGURE 4-23**). Sparks from broken wires are likely to ignite gas in the air. It helps if water and gas mains are flexible and if stoves, refrigerators, and television sets are well anchored to floors or walls.

If you live in an area with significant earthquake risk, you should consider purchasing earthquake insurance. Although most well-built wood-frame houses will not collapse during an earthquake, damage may make the house uninhabitable and worthless. Because a house is typically the largest investment for most people, total loss of its value can be financially devastating. Basic homeowners insurance does not cover such damage. Earthquake insurance can cover only the cost of replacing the house, not

FIGURE 4-23 **Weak Utility Lines**
The Northridge earthquake severed gas lines and caused fires.

the land, which can be a big difference, as in the expensive real estate of western California.

Plan ahead of time what you will do when an earthquake occurs. Most earthquakes last less than a minute. People who find themselves inside a building during an earthquake are well advised to stay there because the earthquake will probably end before they can get out. Remember from TABLE 3-3 that for earthquakes with magnitudes less than 6, the shaking time is so short that by the time you realize what is happening, there is little time to move to safety. For greater magnitudes, there may be enough time, but the accelerations are

FIGURE 4-24 Survival Spaces?
Collapse of buildings with concrete-slab floors can sometimes leave triangular spaces next to sturdy objects.

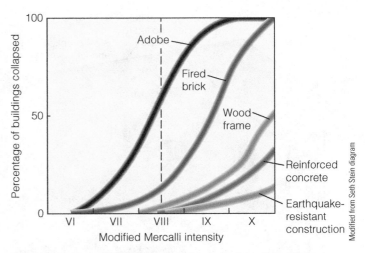

FIGURE 4-25 Building Materials and Risk of Collapse
A mid-strength intensity VIII earthquake (dashed line) would collapse about 60% of adobe buildings, 10 to 15% of fired brick buildings, 5% of wood-frame buildings, and few reinforced concrete buildings.

so high that it is hard to stay on your feet. If you do attempt to leave a building, avoid elevators. People outdoors and away from buildings are much safer than those indoors because no roof is overhead to collapse on them. A car parked next to a building provides little safety from falling debris. Glass and other debris falling from nearby buildings might make it advisable to run for open ground.

To increase your chances of surviving the collapse of a building during an earthquake, the latest guidelines suggest exiting and moving away from the building if there is time. If not, lie down next to a sturdy object that will not completely flatten if the building collapses. Even falling concrete can leave triangular spaces next to a heavy and sturdy object. In a modest-size earthquake, taking refuge under a sturdy table or strong piece of furniture can protect you from smaller falling objects (**FIGURE 4-24**).

Land-Use Planning and Building Codes

Governments also have a role in preparing for earthquakes and mitigating their damage. Land-use planning and building codes are the best defenses against deaths, injuries, and property damage in earthquakes.

Building codes in areas of likely earthquake damage should require structures that are framed in wood, steel, or appropriately reinforced concrete. They should forbid masonry walls made of brick, concrete blocks, stone, or mud that support roofs, which are much more susceptible to collapse in an earthquake (**FIGURE 4-25**). The Uniform Building Code provides a seismic zonation map of the United States that indicates the level of construction standards required to provide safety for people inside a building (compare FIGURE 4-11).

Enforcement of building codes is one reason the largest earthquakes do not necessarily kill the most people.

Typically, large earthquakes in developed countries with modern construction codes—and strong enforcement of those codes—cause significant damage but fewer deaths. In poor countries with substandard construction or little enforcement of existing construction codes, such earthquakes result in high death tolls (see **Cases in Point:** Collapse of Poorly Constructed Buildings that Did Not Follow Building Codes, Wenchuan [Sichuan], China, Earthquake, 2008 and Deadly Collapse of Poorly Constructed Heavy Masonry Buildings—Haiti Earthquake, 2010, pp. 89–90). Contrast these death tolls (tens of thousands) with 486 killed in the February 2010 magnitude 8.8 earthquake, the 7th or 8th largest on record in Chile, a country with very strong earthquake building codes.

Most high death tolls from earthquakes come from countries notable for poor-quality building construction or unsuitable building sites (**Table 4-1**). The high death toll from the January 2001 earthquake in San Salvador was the result of both a huge landslide in a prosperous part of the capital and poorly built adobe houses in the poor areas. The tens of thousands of deaths in the January 2001 earthquake in Bhuj, India, derived mainly from the collapse of houses that were poorly built with heavy materials.

A cursory examination of TABLE 4-1 also shows that death tolls from earthquakes are not significantly declining with time. More than 200 years ago, and even in the past 50 years, many tens of thousands of people, and occasionally hundreds of thousands, died in major earthquakes. Unfortunately, there are now far more people living in crowded conditions and often in poorly constructed buildings (**FIGURE 4-26**). Developed countries tend to be better off in this respect but can still experience devastating loss of life in an earthquake.

Zoning should strictly limit development in areas along active faults, on ground prone to landslides, or on soft mud or fill. Parks and golf courses are better uses for such areas. If people did not live near faults, their sudden shifts would not create problems, but cities and towns grew in those areas for reasons that had nothing to do with Earth's movements. Now with millions of people residing in hazardous environments, societies are beginning to realize that we have a worsening problem, both for individuals and civilization as a whole. In order to deal with the hazards, we need to know more about what creates each danger.

Table 4-1	Some of the Most Catastrophic Modern Earthquakes in Terms of Casualties*		
EARTHQUAKE	**DATE**	**MAGNITUDE (M_S)**	**CASUALTIES**
Iquique, Chile	April 1, 2014, 8:46 p.m.	8.2	6
Japan, Sendai	March 11, 2011, 2:46 p.m.	9.0	18,000—mostly in tsunami caused by the quake
Haiti, Port-au-Prince	Jan. 12, 2010, 4:53 p.m.	7.0	159,000—mostly in collapse of poor-quality buildings
China, Wenchuan	May 12, 2008, 2:28 p.m.	7.9	87,587—collapse of recent, poorly built schools and apts.
Kashmir	Oct. 8, 2005, 8:50 a.m.	7.6	87,351—mostly in collapse of schools and apartment buildings
Sumatra	Dec. 26, 2004, 7:59 a.m.	9.15	297,000—including from the tsunami caused by the quake
Iran, Bam	Dec. 26, 2003, 5:26 a.m.	6.7	~26,000—mostly in buildings of adobe and brick
China, Tangshan	July 27, 1976	7.6	250,000—mostly in collapsed adobe houses
Japan, Kwanto	Sept. 1, 1923	8.2	143,000—including deaths in fire started by the quake
China, Kansu	Dec. 16, 1920	8.5	180,000
Italy, Sicily, Messina	Dec. 28, 1908	7.5	120,000

*Information gathered from various sources. M_s = surface-wave magnitude.

FIGURE 4-26 **Hidden Flaws Behind Plaster**

A. What appears to be a nice, strong stone wall may be merely textured plaster over weak masonry. South of Madrid, Spain.

B. Houses that were damaged in old parts of Athens, Greece, in a magnitude 5.9 earthquake in September 1999, were shoddily constructed from local rocks weakly cemented together. The collapsing houses killed 143 people. Would you stay in a hotel with this type of construction? Could you tell the type of construction if its walls were covered with plaster or stucco? Red "X" placed after earthquake to designate "no re-entry."

EARTHQUAKE SURVIVAL GUIDE

U.S. High Hazard Area	■ U.S. Pacific Coast: British Columbia, Washington, Oregon, California. Mid-continent Mississippi River area near the intersection of Missouri, Illinois, Kentucky, Tennessee, and Arkansas. ■ East Coast of the Carolinas, Philadelphia to Maine.
Preparation	■ Check with state and local emergency preparedness offices and schools about appropriate preparations. Review USGS guidelines for surviving an earthquake and their "Seven Steps." ■ Identify potential hazards in your home: whether the building is structurally stable, what heavy items may topple or fall, and begin to fix them. Know the locations of electrical and gas shutoffs. Anchor chimneys, water gas appliances. Where in the house will you take shelter if there is no time to run outside? Review what to expect and what you should do if in a high-rise building. ■ Prepare a survival kit appropriate to your location: fire extinguisher, three days of drinking water (1 gal./person/day), emergency food for a few days or weeks, dust mask and goggles, medications, spare contact lenses/glasses, first aid kit, flash light and batteries, critical papers, portable radio, cell phone and charging cord, some extra cash. ■ Prepare with family what to do in evacuation and, in case of separation, where to meet and who and how to contact them (in another state). Educate family and friends.
Warning Signs	■ Be aware of significant earthquake faults in your area and USGS projections of chance of a significant earthquake in your area.
During the Event	■ Protect yourself from falling objects or building collapse during earthquake shaking.
Aftermath	■ After the quake, check for injuries and damage. Be aware of ongoing hazards such as gas leaks, electrical wires, and fire. Leave the building, since it may not be structurally stable in aftershocks. ■ When safe, continue to follow your disaster-preparedness plan.

Cases in Point*

Shaking Amplified in Soft Mud and Clays
Recent San Francisco Bay Earthquakes, 1989 and 2014 ▶

On October 17, 1989, Game 3 of the World Series between the San Francisco Giants and the Oakland Athletics was about to begin. At 5:04 p.m., the sudden shock of an earthquake jarred everyone to a stop. That was the P wave arriving. Ten seconds later, the shaking suddenly intensified enough to knock a few people off their feet. That was the S wave arriving. The 10-second interval between the P and S waves indicated that the earthquake was some 80 km away. Then the light towers began swaying as the surface waves arrived. The San Andreas Fault moved near Loma Prieta in the Santa Cruz Mountains 80 km southeast of the stadium. It was magnitude 6.9.

The fault slippage that caused the Loma Prieta earthquake began at a depth of 18 km, where rocks west of the fault moved 1.9 m (6.2 ft) north and 1.3 m up. The offset did not break the surface or even the upper 6 km of the

crust. Most of the fatalities and damages resulted from the liquefaction of ground materials that undermined structures.

When the quake occurred, cars swerved and traffic stopped on a two-and-a-quarter-mile double-decked stretch (the Cypress structure) of Interstate 880 in Oakland. At first, some drivers thought they had flat tires. Initial excitement turned to terror for those on the lower deck, when chunks of concrete popped out of the support columns and the upper deck collapsed onto their vehicles.

Soft mud along the edge of San Francisco Bay had amplified the ground motion under the freeway by a factor of 5 to 8, despite the 90-km distance from the epicenter. It was especially unfortunate that the sediments reverberated with the vibration frequency of the elevated freeway (2 to 4 lateral cycles per second) because that greatly amplified

the damage. When the structure collapsed, motorists on the lower level were crushed or trapped in their cars. Forty-two motorists died, accounting for two-thirds of the fatalities from the quake. If not for the fact that many people were at home to watch the start of the baseball game, many more people would have died on I-880.

Damages in San Francisco were also associated with liquefaction of loose sediment. Buildings in the Marina district were especially vulnerable because much of the construction on the edge of San Francisco Bay stands on fill that amplified ground motion. The water-saturated sediments of the landfill turned to a mushy fluid, causing settling that broke gas lines and set fires. It also broke water mains, forcing firefighters to string hoses to pump seawater from the bay.

■ *Severe shaking of the Interstate 880 double-deck freeway in Oakland sheared off heavily reinforced concrete supports, and much of the upper deck collapsed onto the lower deck. Note that the heavy column in mid-photo, for example, failed at the level of the lower deck, where the two parts were joined during construction. Seismographs show that the shaking was much stronger on mud than on bedrock (see upper-right inset diagram).*

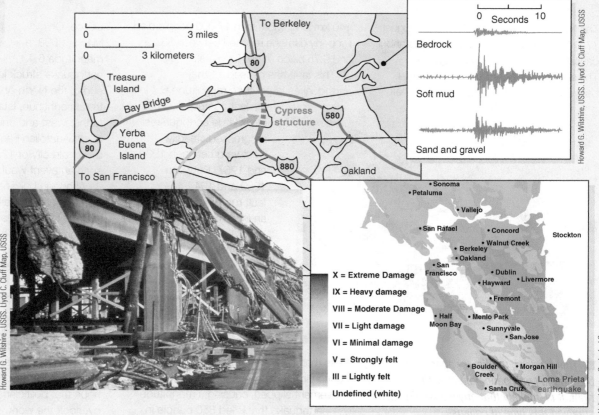

* Visit the book's Website for these additional Cases in Point: Devastating Fire Caused by an Earthquake—San Francisco, CA, 1906; *Predicted Earthquake Arrives on Schedule*— Haicheng, China, Earthquake, 1975

Earthquake Predictions, Forecasts, and Mitigation **85**

In the final tally, collapsing structures killed 62 people and injured 3757. Some 12,000 people were displaced from their homes. The earthquake destroyed 963 homes and inflicted $6 billion in property damage ($11.4 billion in 2015 dollars).

In the early morning hours of August 24, 2014, very strong shaking from a magnitude 6 earthquake, on the north–south trending West Napa Fault, shook the area around Napa, just north of San Francisco Bay. It was the largest California earthquake since Loma Prieta. Two hundred people were injured by falling debris, including chimneys, three of them critically. One woman died from a subdural hematoma when an old TV set on a stand fell on her. The earthquake ruptured many water and gas pipes, causing eight to ten large fires. In the older part of town, brick façades and cornices collapsed onto the sidewalks. Several old buildings in Napa were severely damaged, and nearby

J.K. Nakata , USGS

■ *This apartment building, constructed on the artificial fill of the Marina district of San Francisco, collapsed during the Loma Prieta earthquake in 1989. Note that the second-floor balcony behind the car is now at street level.*

wineries suffered major losses. Damages reached at least $362 million.

An earthquake early-warning system being developed at the Berkeley Seismological Lab, 45 km south of Napa, detected the initial P waves, but the 5-second delay before arrival of the stronger S waves provided too little time to warn area residents.

CRITICAL THINKING QUESTIONS

1. What mistakes did planners make when they built the Cypress structure of the freeway?
2. Now that the Cypress structure is removed, what are three things that Bay Area officials could do to prepare for the next earthquake?

One in a Series of Migrating Earthquakes
Izmit Earthquake, Turkey, 1999 ▶

The North Anatolian Fault slipped on August 17, 1999, in the area 100 km east of Istanbul, causing a magnitude 7.4 earthquake. The surface displacement was horizontal, between 1.5 and 2.7 m along a fault break

Mehmet Celebi, USGS

■ *This modern, multistory apartment building in Izmit pancaked over to the right. It is unlikely that anyone would have survived this type of collapse.*

140 km long. More than 17,000 people died. Property damage reached between $3.25 and 9.75 billion (in 2015 dollars).

This fault has caused 11 major earthquakes larger than magnitude 6.7 in the last century. A series of six large earthquakes progressed steadily westward between 1939 and 1957. Between 1939 and 1944, the fault ruptured along an incredible 600 km of continuous length. Meanwhile, three other earthquakes occurred near both ends of the fault, beyond the sequence of six. The earthquake of August 17, 1999, filled a seismic gap. Less than three months later, a 43-km section immediately east of the previous movement slipped, causing a magnitude 7.1 earthquake that killed 850 people in

Düzce. On May 1, 2003, a magnitude 6.4 earthquake struck just south of the east end of the North Anatolian Fault. If current trends continue, Istanbul, a short distance farther west and only 20 km north of the North Anatolian Fault, may be next. For a very old city of 12.6 million inhabitants, such an event would be a catastrophe.

Although building codes in Turkey match those of the United States with respect to earthquake safety, enforcement is poor and much of the construction is shoddy. Because buildings are taxed on the area of the street-level floor, developers favor the erection of buildings with second stories that extend over the street. Many of these buildings collapsed during the earthquakes, and liquefaction caused buildings to tilt or fall over. In addition, uppermost floors are sometimes added illegally during election years when builders hope that politicians and inspectors will overlook the work.

■ A group of older apartment buildings in Izmit lies in ruins after the poorly braced lower floors of most of them collapsed. Poor-quality construction contributed. Diagonal braces and shear walls would have prevented lateral shift.

Collapsed lower floors

Mehmet Celebi, USGS

CRITICAL THINKING QUESTIONS

1. What are two major reasons for many buildings in Izmit falling over to one side? For each of these reasons, what could be done to retrofit such buildings?
2. What is a third major difference between buildings in Izmit compared with most comparable-size buildings in the United States or Canada?
3. If earthquake sequence along the North Anatolian Fault continues as it has in recent decades (see FIGURE 4-6), when and where do you think the next major earthquake will occur and how should people prepare for it?

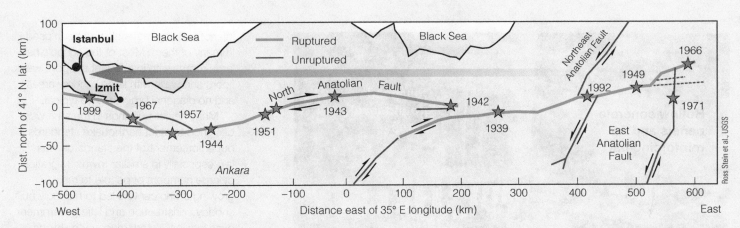

Ross Stein et al., USGS

■ The North Anatolian Fault in Turkey shows earthquake migration with time. The red arrow in the top map shows earthquake migration (not the sense of fault movement) from 1939 to 1999. Is Istanbul next?

Collapse of Poorly Constructed Buildings that Did Not Follow Building Codes
Wenchuan (Sichuan), China, Earthquake, 2008 ▶

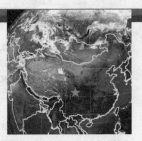

At 2:28 p.m. on May 12, 2008, sudden movement along almost 300 km of the northeast-trending Longmenshan thrust belt moved the eastern high Tibetan Plateau, northeast and upward against the Sichuan Basin on its east. The upper plate, on the west, shifted as much as 4.9 m east and 6.5 m up, a total offset of more than 8 m. The resulting magnitude

7.9 earthquake began 19 km below the surface and 75 km west-northwest of Chengdu, a city in southwestern China with a population of about 10 million.

A sudden jolt and violent back-and-forth shaking threw people off their feet. It felt like free fall except that they fell sideways rather than down. People were both stunned and unable to run because

the floor under them moved rapidly, first one way and then the other. Plaster fell from ceilings and walls buckled. As buildings rocked back and forth for as long as two minutes, much of the rigid masonry construction cracked and ultimately collapsed. The rigidity of the terrain, shallow depth of the earthquake

■ *Roofs and structural brick walls of old buildings in Mianyang collapsed into narrow streets, crushing people who ran outside.*

Ying Ying Huang, USGS

Rebecca Bendick

■ *Structural brick and masonry walls and heavy concrete floors in Bechuan collapsed.*

Hollow concrete panels and no reinforcing bar

Collapsed ground-level stores

Kelin Wang photos; Pacific Geoscience Center/Courtesy of Natural Resources of Canada, Geological Survey of Canada

■ *Masonry construction, with little or no reinforcing bar, crumbled easily. Stores and apartments occupying ground floors and garages had no diagonal bracing. Being "weak floors," they collapsed during the earthquake. Wenchuan, China.*

focus, and proximity of such a large population amplified damages and deaths.

Cities and rural villages lie in narrow, steep-sided valleys where people live in multistory concrete and brick apartment complexes built since the 1950s or in small one- and two-story homes built of brick or concrete blocks. In older homes a sparse framework of timber beams and wood slats supported heavy tile roofs.

Most children were in school when the earthquake leveled most of the buildings near the epicenter, killing thousands of students and teachers. Thirteen hundred died in one middle school built in 1998; at least 1000 were buried when another school, seven stories high and 160 km from the epicenter, collapsed into a 2 m high pile of rubble. Examination of collapsed buildings quickly showed that

inferior building materials had been used in many of them. Most of the schools had large rooms with insufficient support; weak floors and walls with large window areas and no diagonal bracing were typical.

Many newer buildings in the area were designed to strict earthquake standards but enforcement of the standards was lax, especially in smaller towns. Migration of large numbers of people to rapidly growing urban centers led to hurriedly built shoddy construction and little government oversight. Multistory masonry buildings used insufficient mortar and too few steel reinforcing bars.

Engineering reports indicated that school construction problems in rural areas were aggravated by poor funding and exploitation by government and school officials.

Highways and rail lines into the area were buried by landslides or had collapsed, so rescue teams were delayed. Bridges, power lines, and cell phone towers fell, blocking communication, especially to rural towns in rugged valleys. Immediately following the earthquake, rescue teams rushed to the larger towns where the largest numbers of people would need help, often bypassing small communities where people urgently needed help. They lacked shelter, warm clothes, blankets, food, and clean water; cold spring rains added to their misery. Rains brought down more landslides. Even where damaged buildings could provide some shelter, hundreds of aftershocks threatened to collapse them so most survivors remained outside. Without external help, the only rescue came from local people who had no heavy equipment to move the rubble that covered roads and to lift heavy concrete slabs.

Although many people were rescued from the rubble initially, by eight days after the earthquake the numbers dwindled to only one or two. Without food, they can survive for up to 60 days but without water few can last for more than five to seven days. A few survived by drinking their own urine; doctors say this is safe because it is naturally antiseptic. The only help for some people buried in collapsed buildings came from family members or coworkers. In many cases they could not reach victims because concrete slabs were too heavy to move by hand or blocks were linked by steel reinforcing bars. Former residents, who had left to work in industrial cities to the east, returned to search for relatives. They asked friends and searched nearby shelters, in many cases without success; a lot of survivors were taken to larger shelters many miles away.

The total confirmed dead reached 87,500, including thousands long missing. More than 5 million remained homeless. Total costs of reconstruction from the earthquake were estimated as $160 billion (2015 dollars).

CRITICAL THINKING QUESTIONS

1. Why did so many school buildings collapse in this earthquake, in contrast to otherwise similar buildings?
2. Why did so many buildings collapse despite strong earthquake standards for construction of buildings?
3. Many older homes in rural areas collapsed. What was so dangerous about these homes? What inexpensive changes could be made in rebuilding or retrofitting those homes to make them more earthquake resistant?

Deadly Collapse of Poorly Constructed Heavy Masonry Buildings
Haiti Earthquake, January 12, 2010 ▶

Haiti's rampant corruption and poverty, the worst in the Western Hemisphere, led to shoddy construction that was unsafe even before the magnitude 7.0 earthquake that struck on January 12, 2010 when thousands of people were crushed in collapsing buildings. Haiti has virtually no construction standards. Walls were commonly poor quality, composed of handmade and lightweight cinder blocks with minimal cement and little or no reinforcing steel. Thin, poor-quality concrete floors often had only small amounts of lightweight reinforcing bar. The central cathedral, the best hotel in the city, and the local United Nations Development Program headquarters collapsed, and the Presidential Palace was severely damaged.

At least 1.5 million residents were left homeless, approximately 159,000 were killed, and damages reached about $11 billion. The earthquake contrasts strikingly with the February 27, 2010, Chile earthquake, magnitude 8.8 (about 500 times stronger), that killed fewer than a thousand people. In Chile, construction standards are enforced and the country is much more affluent.

When the dust settled, the city was in chaos. Buildings everywhere were destroyed; bodies lay in the street or

Anthony Crone, USGS

■ *No one could have survived pancaking of lower floors of this house with heavy concrete floors and weak walls.*

USN Senior Chief Mass Communication Specialist Spike Call. Via #55

■ *Large crowded areas of poorly built houses collapsed.*

protruded from under debris. People immediately began searching for relatives, coworkers, and neighbors. Many children were separated from their parents or left orphans. With relief workers focused on search and rescue, and most streets blocked by debris, dozens of bodies remained stacked in piles. With no reasonable burial sites, the marginally functional government removed tens of thousands of bodies with bulldozers and trucks, dumping them in mass burial sites in swamps at the edge of the city. More than 150,000 ended up there, mostly unidentified. When people gave up any hope the authorities would come to take the decaying corpses of relatives, some burned the bodies in impromptu funeral pyres using wood crates, scrap wood, and old tires.

Aid came slowly. Aid groups already on the ground, from UN countries, charities, and religious organizations, provided initial searches as best they could, but they had also lost many of their members in the earthquake. Water lines failed, leading to contaminated water, and food distribution channels were disrupted. Port-au-Prince's seaport, its main conduit for food imports, was destroyed in the earthquake. Its single-runway international airport was damaged, and the main highway—of poor quality to begin with—was further hindered by damage and debris. Many cars and trucks were damaged or destroyed. The airport was quickly repaired but its single runway, limited aircraft parking, and limited supply of aircraft fuel made it an aid bottleneck. Except for medical help to the injured, clean water is most important. People can survive much longer without food than without water. Aid flights brought in bottled water, and the desalination plant of a U.S. aircraft carrier standing off the coast soon provided more than 750,000 liters of fresh water per day.

By the end of January, 700,000 homeless Haitians were living in scattered, chaotic, unsanitary tent camps. Some tent camps had only one portable toilet for 2000 people! Most were forced to use a nearby gutter. Earthquake relief teams were also leveling ground outside the city to move half of those in squalid tent settlements to safer areas—where tents could be spread out on drier ground, latrines could be constructed, and clean water and food more easily provided.

Although the main rainy season begins in May, early heavy rains in mid-February flooded low areas and turned streets and the ground under tent camps to mud. Many remained without even plastic sheets for cover. Poor sanitation and widespread contaminated water led to concern about diarrhea, typhoid, cholera, and malaria, which would increase in the rainy season and with higher summer temperatures. Malaria and dengue fever are endemic in Haiti. Some hospitals reported that half of the children being treated had malaria. In the two years since the earthquake, 8300 Haitians died from cholera. In 2011, the UN concluded that much of the problem was caused when Nepalese peacekeepers improperly discharged untreated sewage into the Artibonite River, which many Haitians use as their source of drinking water.

Following the 2010 earthquake, and in spite of aid, its economy has stagnated. Major tropical storms in 2012 didn't help.

By the end of 2012, nongovernment organizations had departed and the economy had expanded by only 2.5%. Much of the billions of dollars of pledged international aid either didn't materialize or was provided to a few international bodies that spent most of it on temporary tents, water distribution, and salaries of temporary foreign workers. Donors are reluctant to support projects that lack appropriate paperwork.

Cleanup and reconstruction, hopefully with better-built buildings, will take years, even with massive external aid. Although Haiti imports a large percentage of its food, its agricultural sector largely escaped damage. Unfortunately much of the imported food is subsidized rice from the United States; that undercuts and discourages local rice production.

CRITICAL THINKING QUESTIONS

1. In the wake of the earthquake, countries around the world pledged money to aid in the relief and rebuilding efforts. What responsibility do rich countries have in providing financial support when poor countries experience a natural disaster? Do they have any responsibility to mitigate the damages before the next disaster occurs?

2. What positive and negative scenarios might Haiti face as a result of the international attention the earthquake brought to the situation there?

3. Given the extreme poverty of countries like Haiti, what could the residents and/or aid groups do to build safer homes at minimal expense?

A This building in western China was destroyed during the 2009 Wenchuan earthquake.

B These buildings in Machu Picchu Village, Peru, lie east of the South American subduction zone. Most are constructed from cinder blocks cemented together, and upper floors sometimes overhang the ground floor. Plaster covers some outer walls to cover the rough construction.

Brick walls

Collapsed wall

Concrete floor slab and concrete posts

USGS

Cinder-block construction

Plaster covering cinder blocks

Overhanging second floor

Concrete floor slab

Reinforcing bars cemented into cinder blocks

Donald Hyndman

1. What tectonic environment controlled the location of that earthquake?
2. What aspects of this building construction would lead to damage during an earthquake? Why?
3. During a major earthquake, what aspects of the building are likely to pose a hazard to people inside the building?
4. What hazards would affect people outside the building and why?

1. What aspects of this building construction would lead to building damage during an earthquake?
2. During a major earthquake, what aspects of the building are likely to pose a hazard to people inside the building?
3. What hazards would affect people outside the building and why?

C This modern overpass in downtown Philadelphia, Pennsylvania, has floor-supporting beams resting on heavy cross-supports.

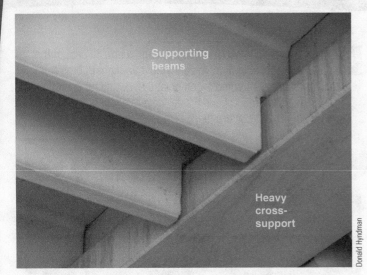

Supporting beams

Heavy cross-support

Donald Hyndman

1. Is there any earthquake danger in this city and, if so, why?
2. Would this type of construction be more hazardous in St. Louis, Missouri, than here? Why?
3. In case of a moderate-size earthquake, describe what might happen to the structure and why.

D This building is in downtown Portland, Oregon, about 200 km from the Cascadia subduction zone. Note the heavy-duty diagonal steel I-beams. Note also the display windows on the ground floor.

Donald Hyndman

1. Explain what role these I beams play in stabilizing the building during an earthquake.
2. Does it appear that the I beams were added later, after completion of the building? Why?
3. During a major earthquake, what aspects of the building are likely to pose an indoor hazard?
4. What hazards would affect people outside the building and why?

E This building in Talca, Chile, was destroyed in the February 27, 2010, magnitude 8.8 earthquake.

Donald Hyndman

1. What tectonic environment would have caused this Chile earthquake?
2. Describe the apparent construction materials, whether there was any structural strengthening, and if so, what it was.
3. Is this building material inherently strong or weak? What makes it so?
4. What structural features make the structure especially weak or especially strong?
5. Before the earthquake, what could have been done to strengthen the structure to minimize this type of damage?

Chapter Review

Key Points

Predicting Earthquakes

■ Precursors that suggest an imminent earthquake include foreshocks, changes in ground elevation, radon gas emissions, and changes in groundwater levels.

■ Early warning systems are triggered when a fault moves, so they provide less than a minute to prepare for the arrival of an earthquake.

■ The lifesaving potential of short-term earthquake forecasts and early warnings must be weighed against the political and economic costs of false alarms.

Earthquake Probability

■ Earthquake forecasts, which are now fairly reliable, specify the probability of an earthquake in a magnitude range within a region over a long time period, such as a few decades.

■ Paleoseismology, the study of prehistoric fault movements, helps us understand future fault behavior. Past fault movements are visible in offset sedimentary layers exposed in trenches and fault scars. **FIGURE 4-4.**

■ Examining trends in past fault movements helps scientists determine where future earthquakes are likely to occur. Seismic gaps, where movement has not occurred on a fault segment, are likely areas for future fault movements. Earthquakes can also migrate along a fault over time. **FIGURES 4-5, 4-6**, and **4-8; CIP**, p. 87.

■ The magnitude of future earthquakes can be estimated based on the lengths of fault segments that are likely to break. **FIGURES 4-5** and **4-8**.

■ Establishing a recurrence interval for earthquakes helps scientists determine when earthquakes are likely to occur. Some faults slip at regular intervals. Others may be very active and then experience a gap in earthquake activity. **FIGURE 4-8**.

Populations at Risk

■ Seismologists use risk maps to visualize areas of high earthquake risk. The Pacific Northwest is the highest risk area for very large earthquakes in the United States. **FIGURES 4-10** and **4-11**.

■ The probability of a major earthquake on the San Andreas Fault can be assessed from the relationship between frequency and magnitude or strain accumulated by the overall slip rate on the fault for the San Francisco Bay area or the Los Angeles area.

■ The San Andreas transform fault, running much of the length of coastal California, is the dominant earthquake fault in North America. Its most dangerous areas are the large population centers around San Francisco Bay and Los Angeles. Blind thrusts occur off the main fault. **FIGURES 4-12, 4-13**, and **4-14**.

Minimizing Earthquake Damage

■ Earthquake damages include collapsed buildings and other structures, as well as fire and, over the longer term, disease. **FIGURES 4-15, 4-17, 4-19, 4-20, 4-24**, and **4-26**.

■ People generally die in earthquakes because things fall on them, not because of the shaking itself. Therefore, weak, rigid buildings and highway overpasses are most dangerous. Flexible, well-built wood-frame houses may be damaged but are not likely to collapse on people. **FIGURES 4-15, 4-16**, and **4-18**.

■ Weak floors are likely to collapse, as are floors without lateral bracing. Walls not anchored to floors and roofs can separate and fall. Brick parapets can collapse onto the street below. **FIGURES 4-17** to **4-19**.

■ Where the frequency of vibration or back-and-forth oscillation is the same in the building as in the ground under it, the shaking is strongly amplified and the building is more likely to fall. Back-and-forth shaking can cause adjacent buildings of different heights to collide. **By the Numbers 4-1** and **FIGURE 4-20.**

■ Earthquake damage can be mitigated through building codes, retrofitting, and land-use planning, as well as educating the public about earthquake preparedness. **FIGURES 4-21** to **4-26.**

Key Terms

base isolation, p. 80

earthquake early warning (EEW) systems p. 66

foreshock, p. 64

migrating earthquake, p. 68

paleoseismology, p. 67

recurrence interval, p. 69

retrofitting, p. 80

risk map, p. 71

seismic gap, p. 68

Questions for Review

1. To what extent can earthquakes be predicted?
2. List several of the precursors that may indicate an imminent earthquake.
3. What can a trench dug across an active fault show about past fault movement? Use a sketch to illustrate your answer.
4. What is a seismic gap, and what is its significance in determining future fault activity?
5. What information indicates the probable magnitude of future earthquakes along a specific fault segment?
6. An earthquake on the North Anatolian Fault in Turkey caused more than 30,000 deaths in 1999. What North American fault is it similar to and in what way?
7. Why does the North Anatolian Fault kill many more people than its North American counterpart?
8. What kinds of structural materials make walls dangerously weak during an earthquake?
9. What type of wall strengthening is commonly used to prevent a building from lateral collapse during an earthquake?
10. Why do the floor or deck beams of parking garages and bridges often fail and fall during an earthquake?
11. When a tall building stands next to a short building, why is the tall building often damaged during an earthquake? Where on the tall building does the damage occur?
12. What feature is sometimes used to prevent a building from shaking too much during an earthquake?
13. Name three things that would help prepare a home for an earthquake.
14. What should you do if you are in a building, in the event of an earthquake?

Critical Thinking Questions

1. When water was injected into deep hot rocks to generate geothermal power in Switzerland it caused many small earthquakes (see discussion related to FIGURE 4-2). Who should be held responsible: the geothermal power company, the drilling company, the governmental agency that provided the drilling permit, or someone else and if so, who?
2. If injection of fluids into a fault zone can trigger earthquakes, why not trigger some small earthquakes instead of waiting for the "Big One"? Consider numbers of earthquakes of a given size, consequences for public safety, what might go wrong, and who should be liable.
3. The section on Prediction Consequences discusses the pros and cons of making predictions. If you were the mayor of Los Angeles and U.S. Geological Survey earthquake experts advised you an earthquake of magnitude 7.5 will strike the city the day after tomorrow, what would you do? What are the consequences of notifying the public? What would you tell them to do? What are the consequences of withholding that information?
4. Many older buildings were not designed for safety in earthquakes. If you were a policymaker for an earthquake-prone area such as California, would you design laws that require expensive retrofitting or require that buildings be demolished and rebuilt to a safer standard? Why? Why would some groups argue against such a law?

■ Fishing boat lies among the tsunami rubble in Kesunnuma City, Japan, on March 31, 2011.

Tsunami

Swept Away

5

On March 11, 2011, when a giant magnitude 9 earthquake struck off the coast of Japan, the vertical movement of the ocean floor during the earthquake pushed a massive amount of water into broad waves in the open ocean. In the coastal area near Sendai, tsunami sirens sounded, warning residents to seek higher ground. Strong shaking lasted for 6 minutes. Only 20 minutes later, 10-m-high tsunami waves swept ashore, pouring over tsunami barriers. The flood of water moved faster than a person could run, as far as 6 km inland, collecting everything in its path. The tsunami swept away everything including buildings, cars, and boats. Eighteen thousand people died, mostly by drowning in the tsunami. The roiling seawater carried mud, sewage, oil, gasoline, and chemicals—a noxious mess.

Bullet trains and subways shut down, a coastal train was swept off its track, construction cranes swayed wildly, and roads cracked and separated; the airport in nearby Sendai, a city of about 1 million people, was inundated with water and rendered inoperable. People who had retreated to the roof of the four-story city hall found themselves standing waist-deep in water;

80% of the people who worked in the building died. Broken fuel lines caused widespread fires in buildings, in a natural gas storage plant, and even in Tokyo.

The tsunami also disabled the emergency cooling systems of the Fukushima nuclear reactors at the coast, 150 km from the epicenter. Those systems ultimately failed, resulting in a partial meltdown and major radioactive contamination of the surrounding area. This necessitated evacuations and severe complications to the already catastrophic situation (described in Chapter 12). See also **Case in Point:** Massive Tsunami from a Subduction Zone Earthquake—Sendai, Japan, March 2011, p. 113.

Tsunami Generation

Tsunami means *harbor wave* in Japanese, referring to the fact that the waves rise highest where they are focused into bays or harbors (**FIGURE 5-1**). Although tsunami are sometimes called tidal waves, this term is misleading because tsunami are not related to tides.

Tsunami are most commonly generated by earthquakes, but they can also be caused by other mechanisms that suddenly displace large volumes of water. These can include volcanic eruptions, landslides or rockfalls, volcano flank collapses, and asteroid impacts.

Earthquake-Generated Tsunami

Most tsunami are generated during shallow-focus underwater earthquakes associated with the sudden rise or fall of the seafloor, which displaces a large volume of water. Earthquake-generated tsunami occur most commonly by displacement of the ocean bottom on a reverse- or thrust-fault movement on a subduction-zone fault (and occasionally on a normal fault). Strike-slip earthquakes seldom generate tsunami because they do not displace much water.

In a subduction zone, recall that oceanic lithosphere typically slides under continental lithosphere. Although the plates move at a nearly constant rate, the boundary between two plates sticks for many years. Where these plates stick, the continental edge is pulled downward and toward the continent as the subducting plate moves under it. This causes the overlying plate to flex upward in a **coastal bulge**, in the same way that a piece of paper bulges upward if you pull its far edge toward you (**FIGURE 5-2**). When the stuck zone, which commonly stretches a considerable length parallel to the coast, finally ruptures in an earthquake, the edge of the continent snaps up and oceanward. The bulge itself drops, often submerging low-lying coastal areas (**FIGURE 5-3**). Such areas remain submerged until crustal forces gradually bulge them upward again; compare bulging in Figure 5-2. This suddenly displaces a huge volume of water, creating tsunami waves that move in both directions (toward and away from the continent) from the location where they are generated (**FIGURE 5-4**).

The height of a tsunami wave depends on the magnitude of the shallow-focus earthquake, area of the rupture zone, rate and volume displaced, sense of ocean floor motion, and depth of water above the rupture. Vertical movement on a fault, as during a subduction-zone earthquake, causes a large displacement of water.

The size of an earthquake-generated tsunami in the open ocean is limited to the maximum displacement or offset on a fault. Based on the relationship between displacement and fault magnitude, an earthquake of moment magnitude (M_w) 8 from vertical displacement on a normal fault—the most likely type to displace significant water—could have a vertical offset of 15 m. A thrust-fault movement might have greater offset, but its gentler dip would likely cause less vertical displacement of water. Because the tsunami wave height approximates the vertical displacement on a fault, the maximum wave height from an earthquake is about 15 m, which would increase as waves are pushed into shallow water and bays. The most vulnerable parts of the United States and Canada are Hawaii and the Pacific coast (California, Oregon, Washington, British Columbia, and Alaska). On average, a major tsunami forms somewhere around the Pacific Ocean roughly once a decade; and once every 20 years, a 30-m-high wave hits.

John Russell/AFP/Getty Images

FIGURE 5-1 A Tsunami Floods Ashore
Startled people near the beach react to tsunami sweeping ashore in Koh Raya, Thailand.

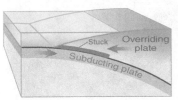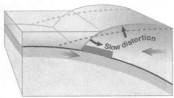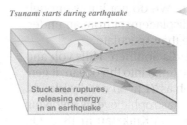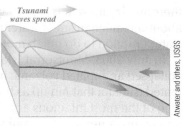

Tsunami starts during earthquake

Tsunami waves spread

Stuck
Overriding plate

Subducting plate

Slow distortion

Stuck area ruptures, releasing energy in an earthquake

———————— Between earthquakes ———————— During an earthquake   Minutes later

Atwater and others, USGS

FIGURE 5-2 Tsunami Formation

A subduction-zone earthquake snaps the leading edge of a continent up and forward, displacing a huge volume of water to produce a tsunami. The coastal bulge (second diagram) abruptly drops (third diagram).

FIGURE 5-3 Collapse of Coastal Bulge

A. During the 1964 earthquake in Anchorage, Alaska, the coastal bulge collapsed, flooding inland areas, remaining under water for years. **B.** After the quake, the bulge began to slowly rise again. This photo shows the same area 27 years later.

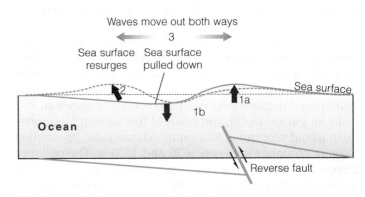

Waves move out both ways
3
Sea surface resurges   Sea surface pulled down
Sea surface
1a
1b
Ocean
Reverse fault

FIGURE 5-4 Initiation of a Tsunami

The sequence of events that creates a tsunami generated by a subsea reverse- or thrust-fault in the ocean floor are: (1a) seafloor snaps up, pushing water with it; (1b) sea surface drops to form a trough; (2) displaced water resurges to form wave crest; and (3) gravity restores water level to its equilibrium position, sending waves out in both directions.

Subduction-zone earthquakes off Japan, Kamchatka, the Aleutian Islands/Gulf of Alaska, Mexico, Peru, and Chile are the most frequent culprits. The subduction zone off the coast of Washington and Oregon is like a tightly drawn bow waiting to be released.

The magnitude 7 earthquake in Haiti in 2010 moved enough water to cause a 3-m-high tsunami. It destroyed several houses and killed 3 people who went to watch it from the shore. In this case, motion on the strike-slip fault had partial vertical motion, which pushed water up to the west.

Even in southern California, south of the Cascadia subduction zone, a nearby earthquake poses a potential problem. A tsunami from an earthquake on the Santa Catalina Fault offshore from Los Angeles would reach the community of Marina Del Rey, just north of the Los Angeles Airport, in only *eight minutes*. Given the large population and near sea-level terrain, the results could be tragic.

Tsunami Generated by Volcanic Eruptions

Tsunami are also caused by volcanic processes that displace large volumes of water. Water is driven upward or outward by fast-moving flows of hot volcanic ash or submarine volcanic explosions into a large body of water. Volcanoes can also collapse in a giant landslide, as addressed below. More than one of these mechanisms can occur at an individual volcano. Tsunami generated by volcanic eruptions are occasionally catastrophic, but they are poorly understood;

their maximum size is unknown. We do not know enough about the mechanism of water displacement from an underwater eruption to do much more than speculate.

In July 2003, Montserrat Island's Soufrière Hills volcano collapsed and spilled volcanic material into the ocean; it generated tsunami that ran up as high as 21 m on nearby islands. One of the most infamous and catastrophic events involving a volcano-generated tsunami was at Krakatau in 1883. On August 27, the mountain exploded in an enormous eruption, the climax of activity that had been going on for several months. Thirty-five minutes later, a series of waves as high as 30 m (almost 100 ft) flattened the coastline of the Sunda Strait between Java and Sumatra, including its palm trees and houses. Only a few who happened to be looking out to sea saw the incoming wave in time to race upslope to safety. More than 35,000 people died. Studies of the distribution of pyroclastic flow deposits and seafloor materials in the Sunda Straits between Krakatau and the islands of Java and Sumatra suggest that seawater seeping into the volcano interacted with the molten magma to generate huge underwater explosions and upward displacement of a large volume of seawater.

In an earlier event, the cone of the eastern Mediterranean island volcano of Santorini collapsed into its erupting magma chamber between 1630 and 1550 BC The collapse displaced a series of huge tsunami waves that washed ashore in Crete to the south, southwestern Turkey to the east, and Israel, at the east end of the Mediterranean. Some researchers correlate this event to the legend of Atlantis, the city that disappeared under the sea.

Tsunami from Fast-Moving Landslides or Rockfalls

When major fast-moving rockfalls or landslides enter the ocean, they can displace immense amounts of water and generate tsunami. You might expect that the height of the tsunami depends primarily on the volume of the mass that displaces water. However, a more important parameter is the height of fall; higher falls displace more water because the material is moving faster when it hits the water. A striking example is the 150-m-high tsunami in Lituya Bay, Alaska, which was generated when a nearby earthquake caused a large section of cliff to slide into a coastal fjord (**Case in Point:** Immense Local Tsunami from a Landslide—Lituya Bay, Alaska, 1958, p. 116). Although it killed only two people, this is the highest tsunami in the historical record. A similar event in southern Chile on April 21, 2007 was also triggered by a magnitude 6.2 earthquake. A large landslide plunged into a narrow fjord and caused 7.6-m-high waves that swept away 10 people at a beach and destroyed some boats.

Submarine landslides, those that occur underwater, can also generate tsunami. Eight thousand years ago, the giant submarine Storegga landslide offshore from Norway, caused a tsunami 11 m high that ran up on the coasts of eastern Scotland and Norway (**FIGURE 5-5**). The slide moved 800 km out into the deep ocean floor and affected

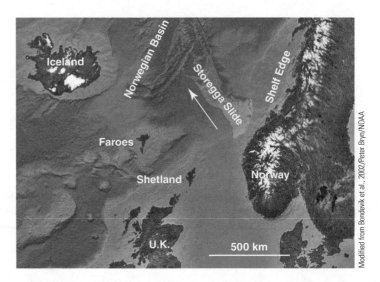

FIGURE 5-5 Continental Shelf Collapse
The Storegga Slide collapsed the continental shelf off Norway, causing tsunami that inundated nearby coasts.

95,000 km^2, an area larger than Scotland and nearly the size of Virginia. The slide, at the end of the last ice age, may have been triggered by an earthquake that destabilized frozen methane-ice layers (discussed in Chapter 12) in the continental shelf sediments.

Submarine landslides from the outer continental shelf off the eastern United States and Canada can also generate tsunami. On November 18, 1929, tsunami from the magnitude 7.2 Grand Banks earthquake killed 27 people on the southern coast of Newfoundland. The epicenter 250 km south of Newfoundland and 610 km east of Halifax, Nova Scotia, triggered a submarine landslide from the edge of the continental shelf that broke 12 submarine telegraph cables on the ocean floor. A series of three waves, which were 3 to 8 m high, arrived at the coast at 105 km/h (65 mph), amplifying to run up as high as 13 m at the head of narrow bays. The waves were recorded in Newfoundland, the east coast of Canada, and the United States as far south as Charleston, South Carolina, almost six hours later. They were even recorded across the Atlantic Ocean in Portugal. Similar submarine slides all along the edge of the Atlantic continental shelf are documented in the ocean-floor record. The recurrence interval for a magnitude 7 earthquake off the New England coast is 600 to 3000 years. Any of those could cause catastrophic slope failure and a major tsunami. Recently discovered fractures along a 40-km stretch of the continental shelf 160 km off Virginia and North Carolina suggest the possibility of a future undersea landslide. Such a slide could generate a tsunami like the one that occurred 18,000 years ago just south of those fractures.

A major subduction-zone earthquake in the Caribbean—for example, north of Puerto Rico—may well trigger a large tsunami that could inundate low-lying areas of the Gulf Coast and East Coast states, even at a considerable distance from the epicenter (**FIGURE 5-6**). Tsunami-deposited sand layers

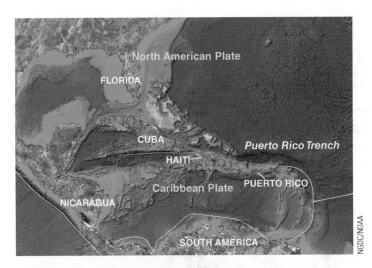

FIGURE 5-6 Future Tsunami in Puerto Rico

The east–west Puerto Rico Trench could generate a major subduction-zone earthquake and tsunami that would inundate much of the east coast of the United States and Canada.

have also been discovered at several sites on islands west of Norway. A smaller, more recent subsea slide in 1998 generated a tsunami that killed 2200 people in Papua New Guinea. The giant Sendai, Japan, earthquake of 2011 triggered large landslides, enhancing the size of the resulting tsunami.

Tsunami can also occur in large, deep lakes. On the west side of Lake Tahoe, on the California–Nevada border, a broad shelf of lake sediments collapsed some time after the glaciers from the last ice age melted (**FIGURE 5-7**). It caused a catastrophic landslide and tsunami. Giant blocks as large as 0.5 to 1 km across dropped 0.5 km and moved 10 to 15 km across the bottom of the lake. Although the north half of the sediment shelf, including much of Tahoe City, did not collapse, it may yet do so in the future. Study of lake-floor sediments suggests magnitude 7 earthquakes every 2000 to 3000 years; such earthquakes could trigger slumps below the lake surface.

Coral limestone boulders up to 9 m across in Tonga, in the southwest Pacific Ocean, appear to have been swept ashore less than 7000 years ago by tsunami generated by huge submarine landslides from adjacent submarine volcanoes. The boulders are 10 to 20 m above sea level and 100 to 400 m inland.

Tsunami from Volcano Flank Collapse

The flanks of many major oceanic volcanoes, including those of the Hawaiian Islands in the Pacific Ocean and the Canary Islands in the Atlantic Ocean, apparently collapse on occasion and slide into the ocean, suddenly displacing thousands of cubic kilometers of water. Resultant tsunami can be hundreds of meters high.

Volcanoes, such as those that make up the Hawaiian Islands, grow from the seafloor for 200,000 to 300,000 years before breaking sea level, then build a reasonably solid sloping dome above sea level, called a lava shield, for a similar time. Mega-landsliding occurs near the end of the

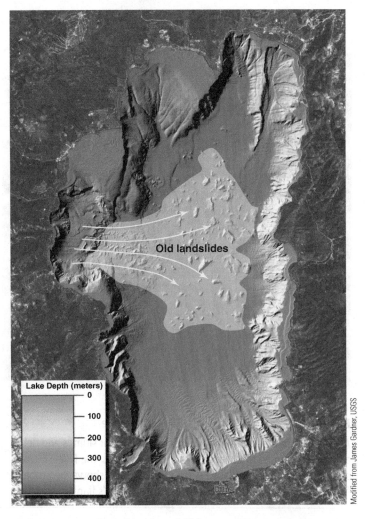

Old landslides

Lake Depth (meters)
0
100
200
300
400

FIGURE 5-7 Lake Tahoe Landslides and Tsunami

A detailed survey of the bottom of Lake Tahoe shows clear topography of old large landslides, which could cause tsunami in the lake.

shield-building stage, when the growth rate is fastest, heavy rock load on top is greatest, and slopes are steepest and thus least stable. The lower part of each volcano, below sea level, consists largely of loose volcanic rubble formed when the erupting lava chilled in seawater and broke into fragments. It has little mechanical strength.

The three broad ridges that radiate outward from the top of a volcano spread slightly under their own immense weight, producing rift zones along their crests. The volcano eventually breaks into three enormous segments that look on a map like a pie cut into three slices of approximately equal size. One or more of the three volcano segments may begin to move slowly seaward.

The rifts between segments provide easy passage for molten magma rising to the surface. Those rifts that become the sites of most eruptions also form weak vertical zones in the volcano. There is a long history of one or more volcano segments breaking loose to slide into the ocean, sometimes catastrophically.

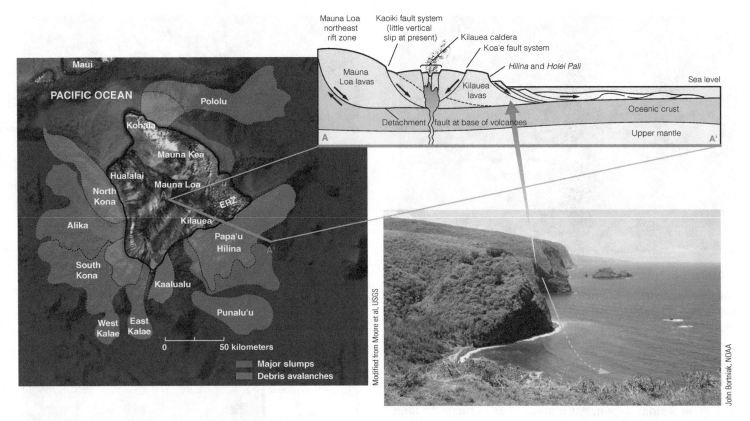

FIGURE 5-8 Hawaii Volcano Flank Collapse

A. This map of the island of Hawaii shows the major slumps and debris avalanches formed by collapse of the island's flanks. The NW–SE cross-section of Mauna Loa-Kilauea volcano from A to A' shows the probable failure surfaces that lead to collapse of the volcano's flanks.

B. Giant cliffs, or pali, amputate the lower slopes of the big island of Hawaii.

Studies of the ocean floor using side-scanning radar around the Hawaiian Islands reveal 68 giant **debris avalanche** deposits, each more than 20 km long (**FIGURE 5-8**). Some extend as far as 230 km from their source and contain several thousand cubic km of volcanic debris. At least some of those deposits are the remains of debris avalanches that raised giant tsunami waves, which washed high onto the shores of the Hawaiian Islands as one of the enormous pie segments of an active volcano plunged into the ocean.

Similar situations are now known to exist on volcanoes of the Lesser Antilles in the Caribbean, Mount Etna in the Mediterranean Sea, and the Marquesas Islands in the Pacific Ocean. Specific sites on the north end of a volcano on Dominica in the Caribbean could fail and produce tsunami that would inundate tourist beaches and a population of 30,000 people on the island of Guadeloupe. A flank collapse of Reunion Island in the Indian Ocean could cause tsunami inundation and destruction of many coastal areas, including the dense sea-level populations of Bangladesh. Collapse of a flank of the Canary Islands, off the northwestern coast of Africa, could generate a giant tsunami that would cross the Atlantic Ocean to decimate cities on the eastern coast of North America and, perhaps, those in western Europe (see details, p. 111).

Although earthquake-generated tsunami are more common, there is a limit to the size of those waves. It seems likely that a catastrophic tsunami, many times larger than any in the historical record, is likely to come from the flank collapse of an oceanic volcano. Our geologic record of such events is clear enough to show that they have happened, and they will again.

Tsunami from Asteroid Impact

The impact of a large asteroid into the ocean would generate huge tsunami radiating outward from the impact site—much as happens with any other tsunami (discussed further in Chapter 18, Asteroid Impacts). The average frequency of such events is low, but a 1-km-diameter asteroid falling into a 5-km-deep ocean might generate a 3-km-deep cavity. Cavity walls would collapse at speeds up to supersonic, sending a plume high into the atmosphere. Initial kilometer-high waves would crest, break, and interfere with one another. Waves with widely varying frequencies would radiate outward. The behavior of such complex waves is not well understood, but they are thought to decrease fairly rapidly in size away from the impact site. The different wave frequencies would, however, interfere and locally pile up on one another to cause immense run-ups at the shore.

The chance of a 1-km asteroid colliding with Earth is not great—only about one per million years, so such a hazard is not likely in our lifetime. However, the chance of a catastrophic tsunami from the flank collapse of an oceanic island such as one of the Hawaiian or Canary Islands is perhaps 10 times as great, or one per 100,000 years.

Tsunami Movement

A wave can be described by its wavelength, height, and period (**FIGURE 5-9**). **Wavelength** is the distance between two waves, measured from the highest point of each wave, called the *crest*. **Period** is the time between the passage of two successive wave crests. Wave height is measured from the shallowest point of the wave, its *trough*, to the top of the crest.

Tsunami wave heights in the open ocean are small. Tsunami generated by ocean earthquakes are often no more than a meter or two high far out in the ocean, with a maximum of about 15 m near an earthquake epicenter. The average wavelength of a tsunami wave is 360 km, so the slopes on the wave flanks are extremely gentle. The time between waves, or the period, can be half an hour. For example, for a wave with a 30-minute period, a ship would go from wave trough to crest and back to trough in 30 minutes. A ship at sea would not even notice such a gentle wave.

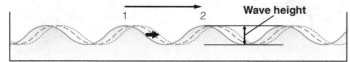

Wavelength (distance from peak to peak)
Period (time for passage of one wave)

Wave height

FIGURE 5-9 Characteristics of Waves

Waves can be described in terms of their wavelength, height, and period.

Water particles in waves travel in a circular motion that fades downward. At depths of less than approximately half their wavelengths, this circular motion reaches the seafloor and waves are said to *feel bottom*. As a wave drags on the bottom, it slows and the waves become much shorter, perhaps one-sixth of their former wavelength. Because wave volume remains the same, its height must rise dramatically, perhaps to six times its open-ocean height. For example, a 3-m-high wave in the open ocean could rise in shallow water to 18 m!

The mechanisms that drive tsunami waves are different from those that drive typical waves, which are driven by wind. Wind-driven waves have relatively short wavelengths so they feel bottom only near shore, then lean forward and break on the shore (discussed in Chapter 14). Tsunami waves, however, have extremely long wavelengths, so they drag on the bottom everywhere in the ocean. Because a tsunami feels bottom far from shore, it gains amplitude to become a steep front of water; as it rides over the continental shelf and approaches land, it continues to flow forward like a flash flood (**FIGURE 5-10**).

This same relationship between wavelength and depth affects the speed of a wave. In the open ocean, tsunami can travel as fast as 870 km/h, the speed of modern jet aircraft, but as they reach shallower water, they slow because their circular motions at depth drag even more strongly on the ocean bottom. Thus, water depth is related to wave velocity (**FIGURE 5-11** and **By the Numbers 5-1:** Velocity of Tsunami Waves). On the continental shelf, a 1-m open-ocean tsunami wave may slow to 150 to 300 km/h (90 to 180 mph). Clearly this is too fast for escape after you see it coming, unless there is high ground nearby.

Tsunami on Shore

In some cases, tsunami may appear much like ordinary breaking waves at the coast, except that their velocities are much greater and they are much larger. Some come in as a high breaking wave that destroys everything in its path.

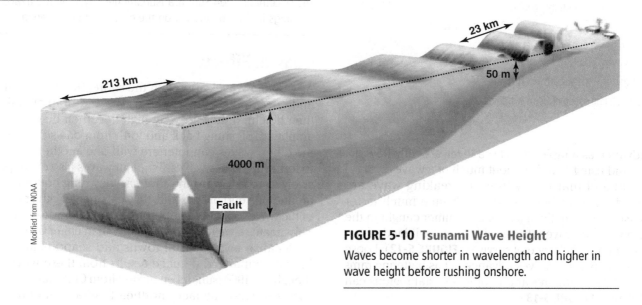

FIGURE 5-10 Tsunami Wave Height

Waves become shorter in wavelength and higher in wave height before rushing onshore.

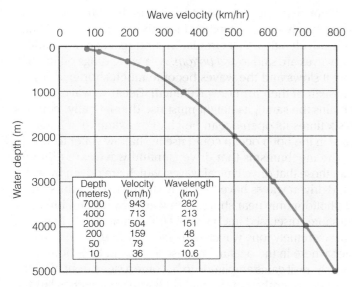

FIGURE 5-11 Wave Velocity and Water Depth

Waves are faster in deeper water and slower in shallower water. As the wave slows down in shallower water, the wavelength is reduced. Because the water volume in a wave remains the same, shortening its wavelength forces its height to rise.

Depth (meters)	Velocity (km/h)	Wavelength (km)
7000	943	282
4000	713	213
2000	504	151
200	159	48
50	79	23
10	36	10.6

By the Numbers 5-1

Velocity of Tsunami Waves

The velocity of tsunami waves depends on the water depth and gravity

$$C = \sqrt{gD}$$

where

C = velocity in meters per second (m/s)

D = depth in meters (m)

g = gravitational acceleration (9.8 m/sec²)

Thus, $C = 3.13\sqrt{D}$

For example, if D = 4,600 m (deep ocean):

$C = 3.13 \times \sqrt{4,600}$ m/s = 3.13 × 67.8 m/s = 212 m/s,

or 763 km/h (the speed of some jet aircraft!).

If D = 100 m (near shore):

$C = 3.13\sqrt{100} = 3.13 \times 10 = 31.3$ m/s,

or 112.7 km/h (the speed of freeway traffic).

Others advance as a rapid rise of sea level, a swiftly flowing, churning, and rising "river" without much of a wave.

Even tsunami that rise without a breaking wave are extremely dangerous because they advance much faster than a person can run. Even a strong swimmer caught in the swift current as the wave retreats will be swept inland or out to sea, with minimal chance of survival (**FIGURE 5-12**). Loose debris picked up as the waves advance act as battering rams that impact both structures and people. Coastal regions can be annihilated (**FIGURE 5-13**).

FIGURE 5-12 Incoming Wave

A massive tsunami surges into Khao Lak, Thailand, carrying sand, debris, and struggling people.

FIGURE 5-13 Tsunami Destruction

The December 2004 Sumatra tsunami destroyed all the near-shore buildings in this community on the east coast of Sri Lanka.

Coastal Effects

As tsunami waves approach the shore, the mouths of rivers and coastal bays funnel the waves and dramatically raise their height. If they arrive at high tide, their height is further amplified. Sloshing back and forth from one side of a bay to the other, waves can interfere with one another, combining to form still higher waves. Tsunami waves also curve progressively to face toward the shore as they drag bottom; moving around one end of an island, they can merge with another part of the wave as it curves around the opposite side of the island. This creates a wave up to 70% larger than either of the parts.

Because most coastal towns and seaports are located in bays, enhanced damage results from these waves. Even though the 1960 tsunami emanated from Chile, far to the southeast, and Hilo Bay faces northeast, refraction of the waves around the island left the head of the bay vulnerable to waves 4 m above sea level. On December 12, 1992, a magnitude 7.5 earthquake in Indonesia generated a tsunami in the Flores Sea. The southern coast of the small island of Babi, situated opposite the direction from which the waves came, was hit by 26-m-high waves, twice as high as the northern coast. In this case, the waves reaching the northern coast split and refracted around the circular island, interfering and combining with one another on the opposite coast. More than 1000 people died.

The coastal geography of other regions can protect them from tsunami. Many low-lying Pacific and Caribbean islands are surrounded by offshore coral reefs that drop steeply into deep water. Thus, tsunami waves are forced to break on the reef, providing some protection for the islands themselves.

Run-Up

When a wave reaches shore, we talk about the characteristics of its **run-up**, or the height that a wave reaches as it rushes onshore. Run-up is greater than incoming-wave height; it varies depending on distance from a fault rupture and whether the wave strikes the open coast or a bay. For the largest earthquakes, such as the 1964 subduction event in Alaska and the 1960 earthquake in Chile, run-up heights were generally 5 to 10 m above normal tide level. Local run-up reached as high as 30 m in Chile.

Water levels can change rapidly, as much as several meters in a few minutes. A wave will typically run up onto shore in a direction perpendicular to the orientation of the wave crest; the wave then drains back offshore straight downslope. Driftwood, trees, and the remains of boats, houses, and cars commonly mark the upper limit of tsunami run-up. Even long after the fact, evidence of past tsunami can be seen in a **trimline**, or the line across a mountainside where tall trees upslope are bordered by distinctly shorter trees downslope.

Period

The mound of water suddenly appearing at the sea surface in response to a major event generates a series of waves that may cross the entire Pacific Ocean. Because the initial mound of water oscillates up and down a few times before fading away, it generates a series of waves, just like a stone thrown into a pond. Thus the arrival of a giant wave is followed by others that are often larger than the first. As the initial wave slows, those following catch up and thus arrive more frequently. At a velocity of 760 km/h and a wavelength of 200 km offshore, a wave would pass any point in the ocean every 15 minutes.

What seems like a calm sea or a sea in retreat can be the trough before the next wave (**FIGURE 5-14**). Survivors of tsunami often report an initial withdrawal of the sea with a hissing or roaring noise. In many cases, curious people drown when they explore the shoreline as the sea recedes before the first big wave or subsequent waves (**FIGURE 5-15**). In 1946, people in Hilo, Hawaii, assuming the danger had past, went out to see the wide, exposed beach littered with stranded boats and sea creatures. There they were caught in the second, larger wave.

As a wave recedes into the trough before the next wave, its current and load of debris flowing back offshore are almost as fast and dangerous to people as the initial tsunami. Because the time between tsunami waves at the coast is often more than a half hour, the wave trough is well offshore; people and wreckage are carried out to sea.

Tsunami waves may continue for several hours, and the first wave is commonly not the highest (**FIGURE 5-16**). In harbors, tsunami have wave periods of 10 to 35 minutes, and the series of large waves may continue arriving for up to six hours. In Hawaii in 1960, in spite of several hours of tsunami warning, people went down to the beach to watch the spectacular wave, only to be overwhelmed by it.

FIGURE 5-14 Series of Tsunami Waves

A. At Kalutara, Sri Lanka, on December 26, 2004, the first wave of the tsunami surrounds all of the houses in this satellite view. Here it begins to drain back to the ocean.

B. A view of a larger area shows the broad offshore beach exposed after the first wave drains back offshore. The red dotted line is the normal beach edge.

FIGURE 5-15 From Shock to Panic

People on the beach at Krabi, Thailand, were stunned to see a giant wave breaking on the horizon and headed their way. Within a few seconds the tsunami was on them.

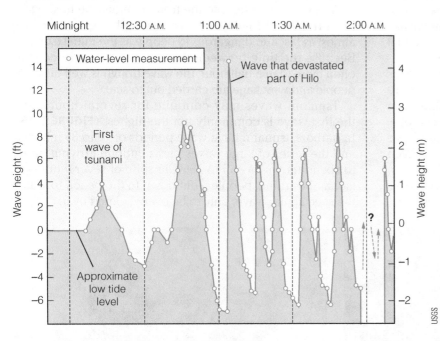

FIGURE 5-16 Tsunami Tide Gauge

This tide gauge record shows the tsunami waves in Hilo, Hawaii, May 23, 1960, following the Chilean earthquake. In this case, the first wave was relatively low, followed by successively higher waves that rose to more than 4 m above the low tide that preceded the tsunami. After the first couple of waves, the frequency increased.

Tsunami Hazard Mitigation

Tsunami hazards can be mitigated by land-use zoning that limits building to elevations above those that would potentially be flooded by a tsunami. In Hilo, Hawaii, the waterfront area at the head of the bay where the worst damage occurred from disastrous tsunami in 1946 and 1960 was converted into a park to minimize future damage.

If lower elevations are to be developed, potential tsunami impact should be taken into account during planning. Coastal developments that orient streets and buildings perpendicular to wavecrests tend to survive better than those aligned parallel to the shore, because waves can dissipate as they flow through open streets, limiting debris impact. Structures should also be designed to resist erosion and scour. Landscaping with vegetation that resists wave erosion and scour can help. Trees can slow waves while permitting water to flow between them, but they need to be well rooted or they can become projectiles. A large ditch or reinforced concrete wall placed in front of houses can reduce the impact of the first wave and may provide a little extra evacuation time. Both options are locally used in Japan. Although tsunami barriers provide some protection, large tsunami waves can overtop or even destroy them (**FIGURE 5-17**).

FIGURE 5-17 Battered Tsunami Barrier

A battered tsunami wall in Otsuchi, Japan, slowed the 2011 tsunami and collected large amounts of tsunami debris behind it.

Tsunami Warnings

Depending on their distance from an earthquake's epicenter, tsunami are most likely to appear within a few minutes to several hours after an earthquake that involves major vertical motion of the seafloor. Tsunami warning systems have been perfected for *far-field tsunami,* those far from the source that generated them, although many regions failed to invest in such systems until after the 2004 tsunami (**Case in Point:** Lack of Warning and Education Costs Lives—Sumatra Tsunami, 2004, p. 114). Tsunami-warning buoys on the Pacific off Japan are seaward of the oceanic subduction zone. After transmitting a signal to warning offices, they provide too little time to warn the public. Japanese earthquake

warning systems are much faster but depend on people being aware of the potential for a following tsunami.

A tsunami warning network around the Pacific Ocean monitors large earthquakes and ocean waves and signals the possibility of tsunami generation and arrival time to 26 participating countries. A world network of seismographs locates the epicenters of major earthquakes, and the topography of the Pacific Ocean floor is so well known that the travel time for a tsunami to reach a coastal location can be accurately calculated (**FIGURE 5-18**). In addition, environmental satellites take readings from tidal sensors along the coasts, and ocean bottom sensors detect ocean surface heights as waves radiate outward across the Pacific Ocean (**FIGURE 5-19**). This information now permits prediction of tsunami arrival times within five minutes, at any coastal location around the Pacific Ocean. Pacific tsunami warning centers are located in Honolulu, Hawaii, and Palmer, Alaska. Some low-lying areas, such as parts of Hawaii, are equipped with sirens mounted on high poles to warn people in dangerous coastal areas.

The Pacific Tsunami Warning System has two levels: a **tsunami watch** and a **tsunami warning**. A watch is issued when an earthquake of magnitude 7 or greater is detected somewhere around the Pacific Ocean. If a significant tsunami is identified from the buoy system, the watch is upgraded to a warning, and civil defense officials order evacuation of low-lying areas that are in jeopardy.

A new warning system for detection of tsunami waves is likely to come from satellites. Because wavelengths of tsunami are very long and the speed of tsunami waves is so much faster, they differ from other disturbances on the ocean surface. Those differences can be detected by satellites. The main limitation is providing real-time data to scientists so that warnings can be issued in time to warn vulnerable population centers.

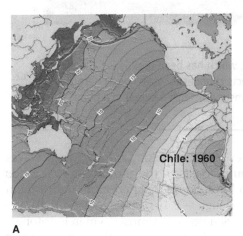
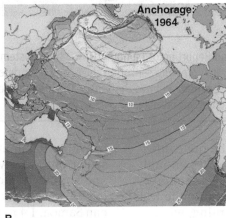
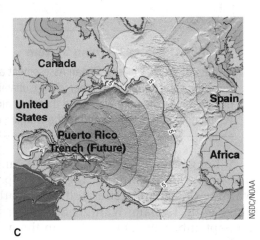

FIGURE 5-18 Travel Times for Different Earthquake Zones

Estimated tsunami travel times across the Pacific and Atlantic oceans from **A.** Chile, 1960 earthquake; **B.** Alaska, 1964 earthquake; and **C.** possible future Puerto Rico Trench subduction-zone earthquakes. Concentric arcs are travel-time estimates in hours after each earthquake.

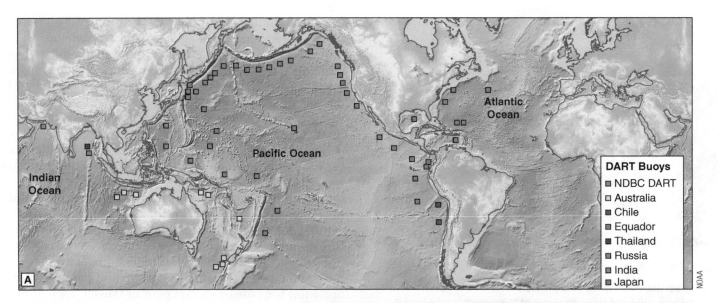

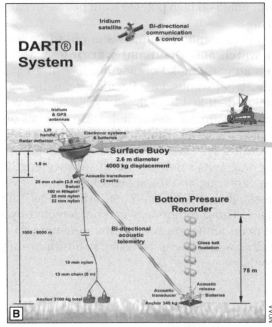

FIGURE 5-19 Buoy Warning System

A. Map of the 2014 Deep-Ocean Assessment and Reporting of Tsunami (DART) buoy system, which has been greatly expanded since the catastrophic December 2004 tsunami. **B.** A pressure sensor on the ocean floor detects changes in wave height because a higher wave puts more water and therefore more pressure above the sensor. The pressure sensor transmits a signal to a buoy **(C)** floating at the surface and to the warning center via satellite.

Although tsunami warning systems can be quite effective for far-field tsunami, they rarely provide enough warning of tsunami generated by a nearby earthquake. The Pacific Tsunami Warning System was put to the test on September 29, 2009, at 6:48 in the morning, when a major earthquake struck the westward-dipping Tonga subduction zone near Samoa, halfway between Hawaii and Australia. Shaking lasted two to three minutes. At a depth of 18 km and in deep water, it generated a tsunami that inundated Samoa, American Samoa, and other nearby islands.

Within 10 minutes after the shaking stopped, huge waves swept onto low-lying coasts of the mountainous islands and amplified into bays. Some reached more than 1.5 km inland. Waves rose as high as 3.14 m at Pago Pago, in American Samoa, 1.4 m in Samoa, and almost 0.5 m in Rarotonga. Coastal houses and tourist resorts were flattened or swept off their foundations and demolished; cars were lifted and smashed into buildings, and some, along with people, were swept out to sea (**FIGURE 5-20**). Many people drowned or were crushed by debris floating in the surging water. Some could not swim or were trapped under water. The Samoa

FIGURE 5-20 Samoa Tsunami

A. The second wave into the harbor at Pago Pago carried cars and the shattered remnants of houses.

B. Survivors search for belongings in the remnants of their homes.

tsunami killed 149 people in Samoa, and 34 in American Samoa, about 200 km from the epicenter, and 9 on Tonga.

The Pacific Tsunami Warning Center in Hawaii sent out an alert when the earthquake struck, but the local population had only about 10 minutes to respond before the arrival of the first wave. Many did not realize how urgent the situation was. Reportedly, some radio stations did not interrupt their music to provide any warning. Although Samoa has a text-message system for warning residents, mobile phone service is scattered in more remote areas, and some people's phones were not turned on. With such a brief window of time for evacuation, warning dissemination measures taken by local authorities, as well as the education of the local population about how to respond to those warnings, is key to saving lives.

Surviving a Tsunami

As was the case in Samoa in 2009, a nearby earthquake allows almost no time for official tsunami warning. Whenever coastal residents feel an earthquake, they should move immediately to higher ground. Tsunami warning signs are often provided (**FIGURE 5-21**). After a nearby subduction-zone tremor, there is little time before the first wave arrives, possibly only 10 to 20 minutes. You need to get to high ground or well inland immediately. Roads heading directly inland are escape routes, but blocked roads and traffic jams are likely. Climb a nearby slope as far as possible.

Even if there are no nearby hills, quickly moving inland can still help, because the energy, height, and speed of a tsunami dissipates on land. Moving to an upper floor of a well-reinforced building away from the beach may also help. When determining how far to evacuate, keep in mind that a wave reaching shore may either break on the beach or rush far onshore with a steep front. Large tsunami can reach more than a kilometer inland in low-lying areas.

FIGURE 5-21 Tsunami Zone Signs

This sign warns of potential tsunami along the Oregon coast.

In addition to drowning in an incoming wave, tsunami dangers include death or injury from being thrown against solid objects or hit by debris, severe abrasions from dragging along the ground at high speeds, and being carried out to sea when a wave recedes. Hazardous fragments can include boards and other remains of houses, trees, and cars. Even when a wave slows to, say, 55 km/h as it drags on shallow bottom, it is much too fast to outrun.

Even if you don't live in a coastal area at risk for tsunami, you might be vacationing in such an area when a tsunami hits, as many tourists in Thailand were during the 2004 Sumatra tsunami. See the Tsunami Survival Guide at the end of this chapter for safety information for tsunami.

Future Giant Tsunami

The largest tsunami that are likely to impact people are caused by giant earthquakes in subduction zones and, less frequently, flank collapse of an island shield volcano. We describe three cases of known major hazards that are likely to affect North America in the future:

1. Giant tsunami following a huge subduction-zone earthquake in the Pacific Northwest,
2. Catastrophic flank collapse of a shield volcano on the big island of Hawaii, and
3. Catastrophic flank collapse of a shield volcano on the Canary Islands in the Atlantic Ocean.

Pacific Northwest: Historical Record of Giant Tsunami

The lack of recent earthquakes along the Pacific Northwest coast from northern California through Oregon and Washington to southern British Columbia is more of a concern than a comfort. As discussed in Chapter 4, the last major earthquake in the area was in January 1700, but there have been many other giant earthquakes at intervals of a few hundred years. Although the subduction zone was known to exist, and almost all such zones are marked by major earthquakes, evidence of major earthquakes here remained elusive.

Finally, in the 1980s, Brian Atwater of the U.S. Geological Survey (USGS) found the geologic record of giant earthquakes in marshes at the heads of coastal inlets. Evidence of giant tsunami generated by the earthquakes included a consistent and distinctive sequence of sedimentary layers. A bed of peat, consisting of partially decayed marsh plants that grew just above sea level, lies at the base of the sequence (**FIGURE 5-22A**). Above the peat lies a layer of sand notably lacking the sort of internal layering contained in most sand deposits. Above the sand is a layer of mud that contains the remains of saltwater plants. That sequence tells a simple story that begins with peat accumulating in a salt marsh barely above sea level. It appears that a large earthquake caused huge tsunami that rushed up on shore and into tidal inlets, carrying sand swept in from the continental shelf. The sand covered the old peat soils in low-lying ground inland from the bays as the salt marsh suddenly dropped as much as 2 m below sea level. Then the mud, with seaweed (later fossilized), accumulated on the sand. The sequence of peat, sand, and mud is repeated over and over. In some cases, forests were drowned by the invading salt water or trees were snapped off by a huge wave (**FIGURE 5-22B, C**). Huge tsunami-flattened forests in low-lying coastal inlets are found all down the Pacific coast from British Columbia to southern Oregon. These stumps are now at and below sea level because the coastal bulge dropped during the 1700 earthquake.

Sand sheets were deposited at elevations up to 18 m above sea level. Tsunami of this size expose coastal communities to extreme danger. For an earthquake that breaks the entire 1100-km length of the subduction zone, mathematical models suggest that Victoria's harbor, at the southern end of Vancouver Island, could see 4-m-high waves 80 minutes after the earthquake. The harbor area of Seattle could see 1-m-high waves three hours after the earthquake and 2-m-high waves after six hours. Calculations suggest that Portland, Oregon, would likely not be at significant tsunami risk from such a subduction-zone earthquake because it is well up the shallow water of the Columbia River; waves are expected to largely dissipate before reaching it. Tsunami from an earthquake on the Seattle Fault pose a greater danger for Seattle. A magnitude 7.6 event could generate a wave of up to 6 m in the Seattle harbor area.

Communities on the open coast or smaller coastal bays, however, are in real peril. An earthquake near the coast could generate a tsunami wave that could sweep ashore in less than 20 minutes, leaving too little time for warning and evacuation of danger areas. The small coastal community of Seaside, Oregon, west of Portland, for example, swells with 40,000 summer residents and visitors, many of whom are unaware of the hazard or what to do when it comes. Feeling an earthquake along the coast, people should immediately move inland to higher ground. Following the March 2011 tsunami in Japan, a tsunami warning was triggered in Seaside and other coastal communities. Many people evacuated, but the waves from the distant earthquake were small. The first indication along the coast of Oregon, Washington, or British Columbia may be an unexpected rise or fall of sea level. If the compressional bulge where the continental plate overrides the subduction zone is under water offshore (for example, off the coast of Oregon; see Figure 3-11), the bulge will drop, pulling water down with it. Thus the first sign at the coast will be a drop in water level as the water pulls back offshore. If the compressional bulge is onshore (for example, in Washington), the bulge will drop but the toe of the plate, under water, thrusts up. The first sign of a tsunami at the Washington coast may be a rapid rise in water level. Later waves are likely to be larger than this first wave. Commonly the initial wave withdraws because the bulge is offshore underwater.

Based on past events, estimates indicate about 1 chance in 10 for such a giant earthquake and tsunami in the next 50 years. That chance may seem small but, as with the weather, the event may come at any time. Recent research shows three to four major tsunami per 1000 years between 4600 and 2800 years ago, a gap of about 1000 to 1200 years—then four during the next 1000 years, followed by another gap of 670 to 750 years. In AD 1700, a giant earthquake and tsunami occurred. Perhaps most disturbing is that although the average recurrence time in the northern part of the fault is 525 years, the average in the southern part is only 278 years. That area may be overdue for the next major earthquake and tsunami.

The sudden drop of the coastal area will raise relative sea level there. Thus, the tsunami will rush ashore to higher levels than would otherwise be expected. Given the length of the subduction zone, comparable in size to the one that caused the catastrophic 2004 earthquake and tsunami in Sumatra, the next event is likely to be disastrous. Calculations suggest

FIGURE 5-22 The Ghost Forest

A. Tsunami sand from a mega-thrust earthquake deposited in January 1700 over dark brown peat in a British Columbia coastal marsh. The scale is in tenths of a meter (10 cm). **B.** This ancient Sitka spruce forest in the bay at Neskowin, Oregon, was felled by a giant tsunami following the same giant earthquake. Stumps of the giant trees punctuate low tide at this beach 25 km north of Lincoln City. The forest—with trees as old as 2000 years—was suddenly dropped into the surf during such a mega-thrust earthquake and then felled by the huge tsunami that followed. **C.** Simplified sketch showing tsunami sand deposited immediately after a subduction earthquake when a tidal marsh suddenly drops below sea level.

that tsunami up to 16 m high may invade some coastal bays. The record of the last event indicates that waves rose as high as 20 m where they funneled into some inlets. Computer models estimate that the heights of those waves will be approximately 10 m high offshore. These heights would be amplified by a factor of two to three in some bays and inlets. Port Alberni, at the head of a long inlet on the west coast of Vancouver Island, British Columbia, in the 1964 Alaska earthquake, for example, saw the waves amplified by a factor of three compared with the open ocean. Inland, where the wave sloshed up against a slope, the run-up reached 25 to 30 m (80 to 100 ft). Although approximate, similar numbers are obtained from studies of onshore damage.

Coastal communities in the earthquake zone are endangered. Cannon Beach and Seaside, in northern Oregon, are well aware of the problem and are beginning to address solutions (**FIGURE 5-23**).

A subduction-zone earthquake 160 km offshore could shake for 5 minutes, and generate tsunami that would reach Seaside in 20 minutes with 12-m-high (40-ft) waves.

Westport, Washington, also built largely on a low-lying barrier bar, is very susceptible to major tsunami from a nearshore subduction-zone earthquake. The town of 2300 people is addressing tsunami preparedness by designing a new elementary school that allows water to pass under the structure. In addition to sheltering students and staff, a refuge built atop the school would accommodate 700 other people.

Kilauea, Hawaii: Potentially Catastrophic Volcano-Flank Collapse

Hawaiian geologists wondered for years about blocks of coral and other shoreline materials strewn across the lower slopes of the islands more than 6 km inland and to elevations up to 400 m above sea level. The flanks of some of the islands have lost much or all of their soil up to a similar elevation. It now seems clear that both the displaced coral and the scrubbed slopes are evidence of monstrous waves that washed up on the flanks of the islands as one of the enormous pie segments

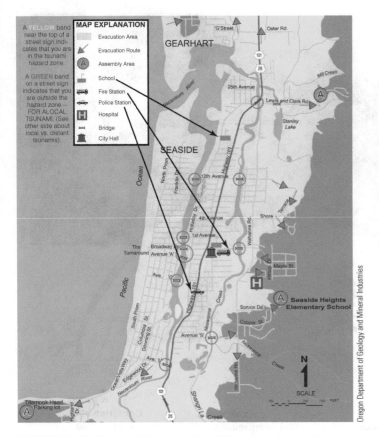

FIGURE 5-23 Tsunami Evacuation Map

Tsunami hazard map of Seaside, Oregon, showing the area of possible inundation in yellow. Bridges (circled) are likely to fail in an earthquake of magnitude greater than 8.5. Evacuation time is too short for schools, and emergency response of firefighters and police is endangered because their headquarters are in the hazard zone. The area west of the north–south channel is especially endangered because bridge failures are likely, making it difficult to evacuate.

of an active volcano plunged into the ocean. The head scarps of such collapsed segments become gigantic coastal cliffs, some more than 2000 m high (see FIGURE 5-8B).

Tsunami formed by island-flank collapse are documented from deposits as high as tens of meters above sea level. On Molokai, tsunami left cemented fragments of limestone reef and basalt 70 m above the sea; on Lanai, they left blocks of coral up to 326 m above sea level. An eventual repetition of those events seems inevitable, and much of the population of the Hawaiian Islands lives below these levels.

The landslides appear to occur during major eruptive cycles and have a recurrence interval of roughly 100,000 years. Head scarps of the slides are the giant pali, or cliffs, that mark one or more sides of each of the Hawaiian Islands. Despite the existence of such evidence, the frequency of these horrifying events remains unclear. If we can judge from the age of coral fragments washed onto the flanks of several islands, the most recent slide detached a large part of the island of Hawaii 105,000 years ago. That slide, on the island

of Lanai, raised tsunami waves to the above-mentioned 326 m above sea level. Mauna Loa, the gigantic volcano on the big island of Hawaii, the youngest and largest of the Hawaiian Islands, has collapsed repeatedly to the west. Two of these collapses were slumps and two were debris avalanches. Most collapses were submarine, though the head scarp of the North Kona slump grazes the west coast of Hawaii. Another deposit of coral fragments about 110 thousand years ago, near the north end of the Big Island of Hawaii, ran up to more than 400 m above sea level at the time and 6 km inland.

Kilauea, the youngest and most active volcano in Hawaii, is now slowly slumping. The potential for future collapse of its flank is emphasized in the Hilina scarps. A mass 100 km wide and 80 km long is moving seaward at 10 to 15 cm per year, sometimes suddenly (**FIGURE 5-24**). On November 19, 1975, the southern flank of the volcano moved more than 7 m seaward and dropped more than 3 m during a magnitude 7.2 earthquake. On June 4, 2013, a magnitude 5.6 earthquake centered about 55 km off the southeast coast of the Big Island of Hawaii (compare Figure 5-8) originated at a depth of about 40 km. Its modest shaking caused little damage. The resulting relatively small tsunami drowned two people nearby, destroyed coastal houses, and sank boats in Hilo Bay on the northeastern side of the island. What this portends for further movement is not clear. Will the flank of the volcano continue to drop incrementally at unpredictable intervals or could it fail catastrophically?

If this huge slump suddenly collapses into the sea, perhaps accompanying a large earthquake or major injection of magma, it could generate tsunami more than 100 m high. Many low-lying coastal communities in Hawaii would be obliterated with little or no warning. However, to put these numbers in perspective, if 100,000 people were killed in such an event every 100,000 years, the average would be one person per year. Although unimaginably catastrophic when

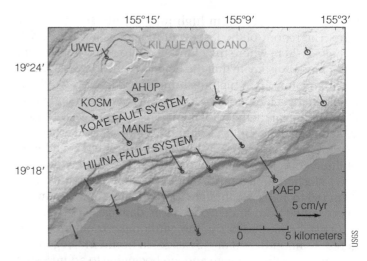

FIGURE 5-24 Kilauea Sliding Seaward

The south flank of Kilauea volcano is slowly slumping seaward. The arrows indicate directions and rates of movement as measured by Global Positioning System.

it does happen, there are certainly greater dangers, on average, in one person's lifetime.

The danger is not limited to Hawaii. If the flank of Kilauea, now moving seaward, should fail catastrophically, it could generate a tsunami large enough to devastate coastal populations all around the Pacific Ocean. Those in Hawaii would have little warning. The Pacific coast of the Americas would get several hours. It remains to be seen how many people could be warned and how many of those would heed the warning. The transportation networks of major urban centers, such as San Francisco and Los Angeles, could not accommodate enough traffic to permit evacuation. We hope that the next event will not be anytime soon, but we have no way of knowing.

Canary Islands: Potential Catastrophe in Coastal Cities across the Atlantic

Like other large basaltic island volcanoes, Tenerife—in the Canary Islands off the northwestern coast of Africa—shows evidence of repeated collapse of its volcano flanks. Tenerife reaches an elevation of 3718 m, almost as high as Mauna Loa. It is flanked by large-volume submarine debris deposits that have left broad valleys on the volcano flanks. Lava filling these valleys is as thick as 590 m and overlies volcanic rubble along an inferred detachment surface that dips seaward at about 9°.

Collapse may have been initiated by subsidence of the 11- to 14-km-wide summit caldera into its active magma chamber. The most recent caldera and island flank collapse was 170,000 years ago. That event carried a large debris avalanche from the northwestern coast of Tenerife onto the ocean floor. It carried 1000 km³ of debris, some of which moved 100 km offshore. The much larger El Golfo debris avalanche detached 15,000 years ago from the northwestern flank of El Hierro Island. It carried 400 km³ of debris as much as 600 km offshore.

An average interval of 100,000 years between collapse events on the Canary Islands may be long, and some scientists argue that it is unlikely to happen, but the consequences of such an event would be catastrophic. And the interval is merely an average. The next collapse could come at any time, and the giant tsunami caused by such a collapse would not only catastrophically inundate heavily populated coastal areas around the northern Atlantic Ocean but also reach coastal Portugal in only two hours, England in little more than three hours, and the east coasts of Canada and the United States in six to nine hours (**FIGURE 5-25**). Because large populations live in low-lying coastal cities and on unprotected barrier islands along the coast, millions would be at risk. Predicted wave heights in Florida, for example, would reach 20 to 25 m, more than the height of a five- to seven-story building! Even if warning were to reach endangered areas as much as six hours before arrival of the first wave, we know from experience with hurricanes that evacuation would likely take much longer that that. Imagine hundreds of thousands of people trying to evacuate without a well-thought-out plan and in traffic that is heavy under normal circumstances. What about congestion on the single two-lane bridges that link most barrier islands to the mainland? How many would ignore the warning, not realizing the level of danger? The death toll could be staggering.

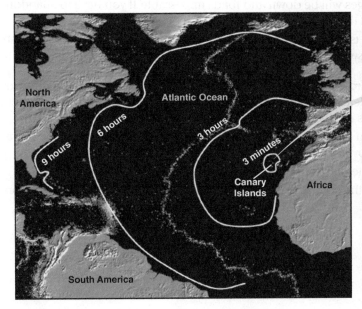
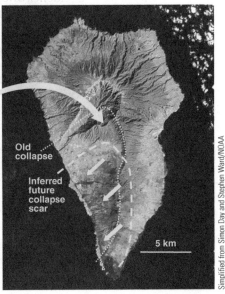

FIGURE 5-25 Potential for a Tsunami on the Atlantic Coast

A. A large landslide from La Palma, Canary Islands, could generate immense tsunami waves that would fan out into the Atlantic Ocean. Computer simulations suggest that huge waves would reach the east coast of North America in six to seven hours.

B. La Palma, Canary Islands, showing a major old collapse scar and a possible future collapse site inferred by W. J. McGuire and others.

TSUNAMI SURVIVAL GUIDE

Be aware that this list does not include everything. Use your own common sense.

U.S. High-Hazard Area	■ Pacific Coast, near shore, especially in low-lying areas. After a large earthquake off the coast of Oregon, Washington, British Columbia, or northernmost California, waves may arrive onshore within 10 to 15 minutes. ■ Inlets or bays are more hazardous because waves tend to be funneled to higher levels. ■ East Coast and Gulf Coast are at risk to a lesser extent but tsunami are still possible.
Preparation	■ Have a plan for where you would go to be above the largest waves, including their run-up height against the shore. ■ Plan for lack of availability of clean drinking water (especially) and food, as well as important medications.
Warning Signs	■ You feel a very large earthquake (generally magnitude > 7 and lasting more than 10 to 15 seconds; larger earthquakes shake for longer times). ■ The sea unexpectedly rises or drops (recedes well offshore) or an unusually large wave approaches. ■ You hear loud, roaring sound from offshore. ■ You hear a report of a very large earthquake elsewhere near a related coastline; large waves can arrive within minutes or hours. Warnings should be announced by reliable sources. ■ You see a very large landslide into a nearby body of water.
During the Event	■ Protect yourself from the effects from falling debris and other effects of the earthquake. ■ Immediately after the shaking ceases, run uphill as far as possible, at least 100 m above sea level; if that is not possible, run inland at least 1 km; or take cover on an upper floor or roof of a well-built building; or climb a large tree. Save yourself, not your belongings. ■ Do not count on your car; roads will be quickly jammed with other traffic and accidents; bridges will be down and roads impassable. If you have to abandon your car, move it well off the road. ■ If caught in the tsunami, take hold of a large floating object and avoid being struck by other debris. If swept offshore, continue holding the floating object until help arrives.
Aftermath	■ After the first large wave, do not come down or back to the shore area until authorities sound the all clear. Typically there are several waves separated by up to an hour or more; the first is not the largest. ■ Do not go down to the shore to watch the waves; they move onshore too fast and traffic jams hinder movement. ■ Stay tuned to the radio, television, text alerts. ■ After the event, the coastal area may end up submerged below sea level by more than 2 m

Massive Tsunami from a Subduction Zone Earthquake
Sendai, Japan, March 2011 ▶

The subduction zone offshore from Sendai, Japan, broke in a massive magnitude 9 earthquake at 2:46 p.m. on March 11, 2011. Damage from the earthquake was minimal due to the distance from the epicenter and excellent earthquake engineering in Japan. The tremendous damage and loss of life was a result of the tsunami generated by the earthquake. As discussed in the chapter opener, the vertical motion of the seafloor displaced a huge volume of water into a series of large tsunami waves, which reached shore in as little as 20 minutes.

Although the tsunami warning system in Japan is the best in the world, residents of coastal cities may have believed they were safe behind the tsunami walls and barriers. Many were overwhelmed when 10-m waves overtopped and toppled the walls. More than 15,826 people died in Japan and 2700 remain missing. Hundreds of thousands were left homeless. Power and drinking water were lost to 4 million homes.

In the aftermath, supplies couldn't be replenished because factories had shut down, fuel ran out, and roads were blocked or washed out. For days the damaged region remained in chaos. Survivors ran short of food and drinking water, stranded because of blocked roads and lack of fuel for their vehicles. They suffered from freezing overnight temperatures and snow, many without fuel for heating or warm enough clothes. The military sent 100,000 troops to open roads, hunt for survivors in the rubble, and distribute water and food.

More than 452,000 found shelter in school gymnasiums, sports arenas, and other improvised locations; even those ran short of water, food, and medical supplies. It took months to find temporary housing for the tens of thousands of evacuees. The natural orderliness of the Japanese was evident in shelters as evacuees separated into small groups, elected a captain for each, and discussed the best ways to cope with managing distribution of limited supplies, cooking, medical care, and sanitation.

Officials estimate that cleaning up, disposing of the incredible mess, and rebuilding could take five years and cost U.S. $300 billion. Economic losses came to about $210 billion, with insured losses of $30 billion.

Although a geologist who had studied tsunami deposits in the vicinity of Sendai had clearly shown that similar earthquakes and tsunami have ravaged the same coast of Japan every 450 to 800 years, his results published seven months before were sadly largely ignored. Also largely ignored were hundreds of very old stone markers reminding people of past tsunami and warning to not build below a certain point on the coast.

In Wakuya, Japan, the tsunami obliterated everything along the coast.

■ *A fishing boat lies among the remains along a tree-lined street of Ofunato, Japan. Rooftop debris on a sturdy building at left indicates the surge height.*

CRITICAL THINKING QUESTIONS

1. People sometimes ignore history at their peril. For this tsunami, what did they ignore that might have prepared them better?

2. For people who were too far away from steep slopes to which they could retreat, what could they have done to find a safe place from the tsunami?

3. Japan, being very familiar with earthquakes and tsunami, has systems to help keep the public safe. What are they and how effective were they in this case? Be specific.

* Visit the book's website for these additional Cases in Point: Subduction-Zone Earthquake Generates a Major Tsunami—Anchorage, Alaska, 1964; *An Ocean-Wide Tsunami from a Giant Earthquake*—Chile Tsunami, 1960.

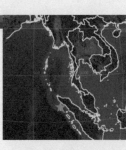

Lack of Warning and Education Costs Lives
Sumatra Tsunami, 2004 ▶

The Indian Plate is moving northeast at 6 cm per year relative to the Burma Plate. In the 10 years preceding this event, there were 40 events larger than magnitude 5.5 in the area, but none generated tsunami. In the last 200 or so years, several other earthquakes larger than magnitude 8 have generated moderate-sized tsunami that have killed as many as a few hundred people. Paleoseismic studies show that giant events occur in the region on an average of once every 230 years.

The subduction boundary had been locked for hundreds of years, causing the overriding Burma Plate to slowly bulge like a bent stick; it finally slipped to cause a magnitude 9.15 earthquake on December 26. Given the size of the earthquake, offset on the thrust plane was some 15 m, with the seafloor rising several meters. The subduction zone broke suddenly, extending north over approximately 1200 km of its length, shaking violently for as long as eight minutes.

The sudden rise of the ocean floor generated a huge wave that moved away from the earthquake source at speeds of more than 700 km/h; it reached nearby shores within 15 minutes. A short time later, tsunami waves 5 m high struck northernmost Sumatra, wiping out 25 km^2 of the provincial capital of Banda Aceh. Locally, the wave swept inland as far as 8 km; it had a 24-m-high run-up on one hill almost a kilometer inland.

Most people in this steeply mountainous country live in low-lying coastal areas, around river mouths, for example; they had little or no warning of the incoming wave. Most people were preoccupied with the earthquake, and few were aware of even the possibility of tsunami. For some who did not happen to be looking out to sea, the first indication was apparently a roaring sound similar to fast-approaching locomotives. Elsewhere there was no sound as the sea rose.

In less than two hours, the first of several tsunami waves crashed into western Thailand, the east coast of Sri Lanka, and shortly thereafter the east coast of India; seven hours later, it reached Somalia on the east coast of Africa. In Sri Lanka, a coastal train carrying 1000 passengers was washed off the tracks into a local swamp; more than 800 bodies were recovered. By January 13, more than 297,000 people were dead and tens of thousands more remained missing. Even with many hours between the earthquake and the first waves,

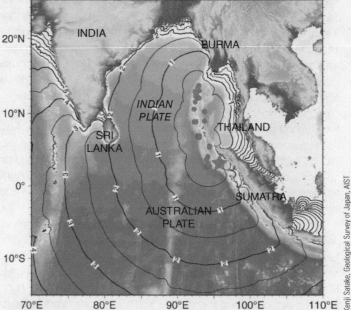

■ *Tsunami wave-front travel times (in hours) are shown emanating from the rupture zone, which spans from the earthquake epicenter (red star) through the area of aftershocks (red dots).*

Kenji Satake, Geological Survey of Japan, AIST

hundreds of people died in Somalia on the northeastern African coast.

At least 31,000 died in Sri Lanka; 10,750 in India; and 5400 in Thailand. In Indonesia, at least 168,000 were dead or missing. In Banda Aceh alone, 30,000 bodies may remain in the area in which no buildings were left standing. Relief organizations were overwhelmed by the unprecedented scale of the disaster encompassing 11 countries. Five million people in the region lost their homes; hundreds of thousands of survivors huddled in makeshift shelters. This tsunami was the most devastating natural disaster of its kind on record.

Although the massive earthquake was recorded worldwide, people along the affected coasts were not notified of the possibility of a major tsunami. Unfortunately, the Indian Ocean did not have a tsunami warning network. The

■ *Because they float and are sturdily built, some fishing boats, though battered, survived the tsunami.*

E. Schneider, UN photo

■ **A.** *The northern part of Banda Aceh, Sumatra, on June 23, 2004, before the tsunami.* **B.** *The same area on December 28, 2004, after the tsunami. Note that virtually all of the buildings were swept off the heavily populated island. The heavy rock riprap along the north coast of the island before the tsunami remains only in scattered patches afterward. A large part of the island south of the riprap has disappeared, as has part of the southern edge, between the two bridges, where closely packed buildings were built on piers in the bay.*

Pacific Tsunami Warning Center in Hawaii alerted member countries around the Pacific and tried to contact some countries around the Indian Ocean about potential tsunami that might have been generated. Because tsunami in the Indian Ocean are infrequent, no notification framework was in place to rapidly disseminate the information between or within countries. By the summer of 2006, a warning system was finally up and working.

An official in West Sumatra recorded the earthquake and spent more than an hour unsuccessfully trying to contact his national center in Jakarta. An official in Jakarta later sent email notices to other agencies but did not call them. A seismologist in Australia sent a warning to the national emergency system and to Australia's embassies overseas but not to foreign governments because of concern for breaking diplomatic rules. Officials in Thailand had up to an hour's notice but apparently failed to disseminate the warning. Among the public, few people had any knowledge of tsunami or that earthquakes could produce them. Ironically, the country's chief meteorologist, now retired, had warned in the summer of 1998 that the country was due for a tsunami. Fearing a disaster for the tourist economy, government officials had labeled him crazy and dangerous. He is now considered a local hero.

Although scientists had expressed concern about the lack of a warning system in the Indian Ocean, most officials in Thailand and Malaysia viewed tsunami as a Pacific Ocean problem, and the tens of millions of dollars it would cost to set up a network left it a low priority in a region with limited finances. However, by 2009 they had placed ocean-floor sensors in the Indian Ocean for detection of any future tsunami.

Even if there had been an Indian Ocean warning system in place, it would not have been able to save most of the lives in the most devastated region of Sumatra because the time between the earthquake and the wave arrival was short. Compounding the problem was the time delay in determining the size of the earthquake. The location of the quake was determined quickly and automatically from the arrival times of seismic waves from several locations. However, the magnitude was initially estimated by Indonesian authorities at only 6.6, a size that would not generate a significant tsunami. It often takes more than an hour to determine the magnitude

■ *A lone mosque remains standing in Banda Aceh.*

■ *Few coastal-area homes, or their foundations, near Banda Aceh survived the tsunami.*

of a giant earthquake because the surface waves they generate travel slowly. By that time, it would have been too late for a local warning in Sumatra; however, it could have saved many lives in more distant locations, such as Sri Lanka and India.

As with all hazards, public education could have done much to save lives. There was a lack of knowledge, even among officials, that a large earthquake could generate large tsunami. On the other hand, a 10-year-old girl who had recently learned about tsunami in school saw the sea recede before the first wave and yelled to those around her to run uphill. A dock worker on a remote Indian island had seen a television special on tsunami, felt the earthquake, and ran to warn those in a nearby community that giant waves were coming. Together, these two saved more than 1500 people. Knowledge of hazard processes can save lives.

Although improvements have been made to the tsunami warning systems in this region, more recent tsunami in 2005, 2006, and 2007 have also claimed lives.

CRITICAL THINKING QUESTIONS:

1. For people living along low-lying coastal plains where a subduction zone lies only 100 km or so offshore, what can people do to protect themselves from an earthquake-generated tsunami?
2. Approximately how long do people have between the earthquake and arrival of the first wave?
3. What are the best refuges from being killed by an incoming tsunami wave?

Immense Local Tsunami from a Landslide
Lituya Bay, Alaska, 1958 ▶

One of the most spectacular tsunami resulting from a land-based rockfall occurred in Lituya Bay, Alaska, in 1958. Lituya Bay, a deep fjord west of Juneau, Alaska, and at the western edge of Glacier Bay National Park, was the site of one of the highest tsunami run-ups ever recorded. On July 9, 1958, 60 million m³

■ **A.** *A huge rockfall into the head of Lituya Bay, Alaska, generated a giant tsunami wave that stripped the forest and soil from a ridge. This view to the northeast shows the broad areas of forest that the tsunami swept from the fringes of the bay. The scarp left by the rockfall is visible at the head of the bay (arrow).* **B.** *The rockfall crashed into the head of the bay, displacing water up to 524 m, over a ridge and back into the bay.* **C.** *Detailed view of the felled forest and stripped soil at the crest of that ridge.*

of rock and glacial ice, loosened by a nearby magnitude 7.5 earthquake on the Fairweather Fault, fell into the head of Lituya Bay. The displaced water created a wave 150 m high—the height of a 50-story building. It surged to an incredible 524 m over a nearby ridge and removed forest cover up to an average elevation of 33 m and up to 152 m over large areas. This was a huge wave compared with common tsunami, which may be 10 to 15 m high; it swept through Lituya Bay at between 150 and 210 km/h.

Although three fishing boats, with crews of two each, were in the bay at the time, only those on one boat died, when their boat was swept into a rocky cliff. On another boat, on the south side of the island in the center of the bay, Howard Ulrich and his seven-year-old son hung on as their boat was carried high over a submerged peninsula into another part of the bay. They were actually able to motor out of the bay the next day. On the third boat, Mr. and Mrs. William Swanson, anchored on the north side of the bay, were awakened as the breaking wave lifted their boat bow first and snapped the anchor chain. The boat was carried at a height of 25 m above the tops of the highest trees, over the bay-mouth bar, and out to the open sea. Their boat sank, but they climbed onto a deserted skiff and were rescued by another fishing boat two hours later.

1874 wave

1936 wave

1958 wave

Lituya Bay

Donald Miller, USGS

■ *This view of Lituya Bay shows trimlines from two previous tsunami that were even larger than the 1958 event. Do narrow, steep-sided inlets elsewhere show similar healed trimlines? Could they provide hazard information for future rockfall tsunami?*

Examination of the forested shorelines of Lituya Bay, above the trimline of the 1958 event, shows two, much higher trimlines produced by earlier tsunami. Scientists from the USGS examined trees at the level of these higher trimlines and found severe damage caused by the earlier events. Counting tree rings that had grown since then, they determined that the earlier tsunami had occurred in 1936, 1874, and 1853 or 1854.

Glacier Bay, 50 km east of Lituya Bay, is in a similar spectacular environment. Could it be the site of the next big

landslide-generated tsunami? It is a deep fjord bounded by precipitous rock cliffs and glaciers. It lies between two major active strike-slip faults, the Fairweather Fault and the Denali Fault, each 50 or 60 km away, and each capable of earthquakes of magnitudes greater than 7. Glacier Bay is a prominent destination for cruise ships touring from Seattle or Vancouver to Alaska, so a tsunami generated by a large landslide into the bay is a concern. Study by USGS geologists suggests that an unstable rockslide mass on the flank of a tributary

Glacier Bay

Wave heights from cliff collapse:

~ 4 m

~ 20 m

Side channel

main Glacier Bay

Gerald F. Wieczorek, USGS

■ *View is from the apex of the slide with the tidal inlet in the foreground; two cruise ships in Glacier Bay are visible in mid-photo. The tidal inlet landslide mass next to Glacier Bay, Alaska, includes the rock face from the new higher scarp to below water.*

inlet to Glacier Bay would generate waves with more than 100-m run-ups near the source and tens of meters within the inlet.

In the deepwater channel of the western arm of Glacier Bay, the wave amplitude would decrease with distance into the bay. Ships near the mouth of the tributary inlet could encounter a 10-m-high wave only 4 minutes after the slide hit the water, then 20-m-high waves after 20 minutes. The waves would likely strike the cruise ships broadside (left photo). If the ships kept to this central channel, the largest waves would likely be approximately 4 m high. The response of a ship to waves near the mouth of the tributary inlet would depend on the wave heights, the wave frequencies relative to the ship's rocking frequency, and the height of the lowest open areas on the ship. Because cruise operators are now aware of this risk, they avoid the dangerous near-inlet waters.

Critical View

A This damage is in the coastal area of Samoa after the 2009 earthquake and tsunami.

Casey Deshong, FEMA

1. Is this damage likely from the earthquake or the tsunami? Why?
2. If you were caught in the incoming tsunami, what would cause injuries or death?
3. Would you be better off in a vehicle or outside it? Why?
4. Where would be a safer location if you had five minutes of warning?

C This damage is in the coastal area of Sri Lanka after the December, 2004 Sumatra earthquake.

USGS

1. Where did the loose sand and the scattered bricks come from and how did they get here?
2. What would severely injure or kill you in this event?
3. If you had only two or three minutes as the wave approached, what would be your safest plan of action?

B This damage is in the Sukuiso area of Japan, after the 2011 tsunami.

Mass Communication Specialist 3rd Class Dylan McCord/Released/U.S. Navy

1. Why were the buildings in the foreground obliterated, while the ones in the upper left of the photo remained largely undamaged?
2. Given the level of devastation, how is it possible that the white building on the coastline survived?

D This damage is in the coastal area of Minato, Japan, one week after the 2011 tsunami.

Lance Cpl. Ethan Johnson/U.S. Marine Corps/U.S. Navy

1. One week after the tsunami, this coastal area remained submerged. Why?
2. Like many coastal towns, Minato is at the mouth of a river valley. Where would be safe areas that might be reached in 15 to 20 minutes after feeling a major earthquake and why?
3. Given that most roads either follow the coast or along river valleys inland, how would you reach such a safe area?

Chapter Review

Key Points

Tsunami Generation

- Tsunami are caused by any large, rapid displacement of water, including earthquake offsets or volcanic eruptions underwater; landslides; and asteroid impacts into water.

- In coastal subduction zones, the leading edge of the continental plate slowly bulges upward before suddenly dropping during an earthquake. The displacement of water caused by this rupture is a common source of tsunami. **FIGURES 5-2 to 5-4.**

- The height of an earthquake-generated tsunami is related to the magnitude of the earthquake, which in turn depends on the length of the rupture and the vertical extent of offset underwater.

- Volcanic processes—such as an underwater eruption, lava flow into the sea, or volcano flank collapse—can all trigger tsunami.

- Landslides into the ocean as well as landslides that occur underwater can trigger tsunami. The height of the waves produced depends in large part on the height of the fall. **FIGURES 5-5 to 5-8.**

- The impact of a large asteroid into the ocean, although rare, would displace a huge amount of water and generate a massive tsunami.

Tsunami Movement

- Tsunami have such long wavelengths that they always drag on the ocean bottom. As they drag on bottom in shallow water, they slow, and their wave height increases. **FIGURE 5-9.**

- Wave velocity depends on water depth. Tsunami waves in the deep ocean are low and far apart but move at velocities of several hundreds of kilometers per hour. They slow and build much higher in shallow water near the coast. **FIGURES 5-10 to 5-12 and By the Numbers 5-1.**

Tsunami on Shore

- The effects of tsunami are amplified by coastal bays, river mouths, and tides.

- Tsunami run-up is the height the water reaches when it sweeps up on shore.

- Tsunami reach shore as a series of waves, often tens of minutes apart, and may continue for hours. The third and subsequent waves are often the largest. **FIGURE 5-16.**

Tsunami Hazard Mitigation

- Tsunami damage can be mitigated by land-use planning and appropriate considerations for development. Structures and vegetation can also reduce tsunami impact. **FIGURE 5-17.**

- Warning systems are able to predict tsunami travel time fairly accurately, even to coastlines far from an earthquake epicenter. **FIGURES 5-18 and 5-19.**

- Coastal residents should be aware of danger signals for tsunami, including earthquakes and a rapid rise or fall in sea level. Once these signs are noticed, inhabitants should run upslope or drive directly inland. The safest areas are a kilometer or more inland and several tens of meters above sea level.

Future Giant Tsunami

- The record of subduction-zone tsunami is based on sand sheets over felled forests and marsh vegetation pushed into coastal bays. **FIGURE 5-22.**

- Tsunami from major Pacific Coast subduction earthquakes occur every few hundred years and reach shore within 20 minutes of the quake, destroying coastal communities, particularly those in bays and inlets. **FIGURE 5-23.**

- A volcano flank collapse that suddenly moves an enormous amount of water, such as in Hawaii or the Canary Islands, could generate giant tsunami that would be catastrophic for the coasts of North America, especially low-lying coastal communities. **FIGURES 5-24 and 5-25.**

Key Terms

coastal bulge, p. 96

debris avalanche, p. 100

period, p. 101

run-up, p. 103

submarine landslides, p. 98

trimline, p. 103

tsunami, p. 96

tsunami warning, p. 105

tsunami watch, p. 105

wavelength, p. 101

Questions for Review

1. What are three of the main causes of tsunami?

2. Of the three main types of fault movements—strike-slip faults, normal faults, and thrust faults—which can and which cannot cause tsunami? Why?

3. How high are the largest earthquake-caused tsunami waves in the open ocean compared to when they reach shore?

4. How fast do tsunami waves tend to move in the deep ocean compared to when they reach shore?

5. Do tsunami speed up or slow down as they approach the coast? Why?

6. How does the height of a tsunami wave change as it reaches shore? Provide a sketch to illustrate your explanation.

7. Why should you stay away from the beach even after a tsunami wave has retreated? How long should you stay away?

8. Why is even the side of an island facing away from the source earthquake not safe from a tsunami?

9. Name three ways to mitigate damages from tsunami.

10. In what situations would a tsunami warning system not be effective for evacuating a population at risk?

11. Name three signs of an imminent tsunami.

12. If you are in a coastal area, what should you do to increase your chances of surviving a tsunami?

13. For a subduction-zone earthquake off the coast of Oregon or Washington, how long would it take for a tsunami wave to first reach the coast?

14. What specific evidence is there for multiple tsunami events having struck coastal bays of Washington and Oregon?

15. Because the Atlantic coast experiences fewer large earthquakes, what other specific event could generate a large tsunami wave that would strike the Atlantic coast of North America?

Critical Thinking Questions

1. Given the past record of major earthquakes and tsunami along the Pacific coasts of northern California, Oregon, Washington, and British Columbia, what should be done about people currently living in especially hazardous areas such as:

   a. Those living in bays along the coast–for example, Seaside and Newport, Oregon?

   b. Those living on barrier bars along the coast–for example, bay-mouth bars in southwestern Washington State (Westport, Long Beach)?

   Consider possibilities such as moving endangered houses or even whole towns, building major barriers to slow the advance of a tsunami, building safe evacuation structures, building safe evacuation routes away from the coast, or doing nothing.

2. Given the dangers noted in question 1, what should be done to prevent more people from moving into such hazardous areas?

3. Because both permanent residents and visitors occupy such hazardous areas and thousands of people from inland cities visit the same coastal areas, who should pay for the costs of mitigation or removal?

■ Mount Hood looms over the city of Portland, Oregon. People who live here admire the scenic view and rarely remember that these Cascades volcanoes are active.

6

Volcanoes: Tectonic Environments and Eruptions

Cascade Range Volcanoes Are Active

Until the 1970s, most of the Cascades volcanoes were thought by almost everyone to be extinct or at least dormant. The rule of thumb was that if a volcano had significant glacial erosion that had not been erased by later eruptions, it probably had not erupted since the last ice age some 10,000 years ago. The inference was that because it had not erupted in such a long time, it was unlikely to erupt again. It seems that our human time frame colored our view of what a volcanic time frame should be. Mt. Lassen, in northern California, staged a series of eruptions that started in 1914 and intermittently continued with declining vigor until 1921, but most geologists regarded it as an unusual event. This eruption did not persuade them that the High Cascades were an active volcanic chain. With the new understanding of plate tectonics and the fact that active subduction zones are associated with earthquakes and overlying active volcanoes, geologists began to wonder whether in fact the Cascades volcanoes might really be active. If a subduction zone is active, based on either measured convergence of two plates or major earthquakes, the chain of volcanoes above it is almost certainly active. As the term is now used by volcanologists, an *active volcano* is one likely to erupt again. Evidence for such activity is generally provided by documentation of its past eruptions and their average frequency. Nonspecialists or the news media may refer to a volcano as dormant because it has not erupted for hundreds or thousands of years, but some volcanoes have even longer periods between eruptions. We now know that several Cascade volcanoes are probably active. Glacier Peak, Mt. Rainier, Mt. St. Helens, Mt. Adams, Mt. Hood, Three Sisters, Mt. Shasta, and Mt. Lassen have all erupted in the last 200 years. Mt. Baker and Crater Lake erupted somewhat earlier.

Introduction to Volcanoes: Generation of Magmas

A **volcano** is typically a cone-shaped hill or mountain formed at a vent from which molten rock, called **magma,** or magmatic gases reach Earth's surface and erupt. Once the magma reaches the surface, it is called **lava.** Because most of Earth beneath the surface is not molten, volcanoes erupt in a limited number of geologic settings where magma is generated at depth and can rise to erupt at the surface.

To understand the generation of magmas, it is important to appreciate the basic difference between states of matter: solid, liquid, and gas, which are related to the movement of their particles. Molecules in a solid are tightly bound together in a rigid shape, so they hardly move. Those in a liquid are loosely held together by flexible bonds that permit them to move fluidly. Molecules in a gas are far apart, free to move and fill available space. With very few exceptions, solids are most dense, liquids are less dense, and gases are least dense (**FIGURE 6-1**).

As temperature and pressure change, substances undergo a change from one state to another. For example, as you compress a gas, it first becomes a liquid and then a solid. The reverse is also true: as you decrease the pressure on a hot solid rock, it may expand and melt into a liquid and then expand even more to vaporize into a gas. These principles govern the transition from rock to magma and the release of gases from magma. A very hot rock may melt if pressure decreases. A gas dissolved in a liquid may expand and separate with decreasing pressure. The **melting temperature**, which controls when a rock becomes magma, depends on pressure and the availability of water. A hot rock deep within Earth may melt if temperature rises, pressure falls, or water is added; addition of water shifts the melting curve to lower temperature (**By the Numbers 6-1:** Melting Temperature of a Rock). The same principle governs when magma becomes solid, or crystallizes. Newly formed magmas may crystallize by cooling or loss of water before reaching the surface. A water-rich magma will lose water and begin to crystallize as it rises.

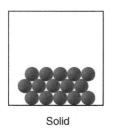
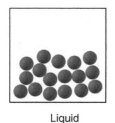
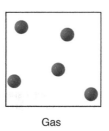

Solid   Liquid   Gas

FIGURE 6-1 Solid, Liquid, and Gas
Molecules in a solid are tightly packed, whereas molecules in a liquid are held together loosely, and molecules in a gas can spread out to fill a container.

By the Numbers 6-1

Melting Temperature of a Rock

The melting temperature of a rock depends on its depth within Earth and the amount of available water. The red boundary on the graph below shows the melting point for a particular rock above a subduction zone, at different temperatures and pressures. For example, as shown by the green arrows from point A, the rock may melt to become magma as a result of an increase in temperature (due to regional heating by injection of another magma nearby) or a decrease in pressure (as the rock moves toward the surface). The addition of water would also shift the melting boundary so that the rock would melt at lower temperatures, as shown by the lower green arrow.

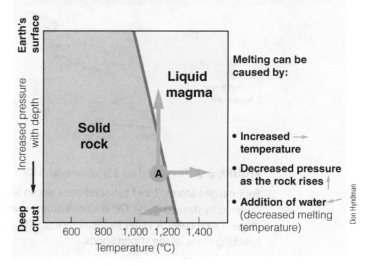

Masses of magma rise through the crust because they are less dense than the surrounding rocks, much as an iceberg rises into the air because ice has lower density than the water surrounding it. Magmas may even rise into rocks of lower density; as long as the column of magma is less dense than the surrounding rock, the magma will float in it. Magma may rise through cracks, sometimes breaking off and incorporating pieces of the adjacent rocks. **Magma chambers** are large masses of molten magma that rise through Earth's crust, often erupting at the surface to build a volcano. The behavior of earthquake waves, minute irregularities in Earth's gravitational attraction, or a slight rise or tilt of the ground surface may indicate an expanding magma chamber. Erosion eventually exposes and dissects old volcanoes and their magma chambers, which are now crystallized into masses of rock solidified from magma.

Magma Properties and Volcanic Behavior

No two volcanoes are quite alike, nor are any two eruptions from the same volcano. What happens during an eruption depends mainly on how fluid the magma is (its viscosity), the quantity of water vapor and other volcanic gases it contains (its volatiles), and the type and amount of magma that erupts (its volume).

VISCOSITY The nature of a volcanic eruption depends in part on the composition of the magma, including its content of water and other gases, and its temperature, both of which affect its viscosity. **Viscosity** refers to how fluid magma is; high-viscosity magmas are thick and pasty, whereas low-viscosity magmas are thinner and more fluid. The viscosity of a magma depends on its chemical composition—the internal arrangement of its atoms and molecules. By far the most abundant atoms in a magma are oxygen and, to a lesser extent, silicon and aluminum. Almost every silicon atom in its natural state is surrounded by four oxygen atoms arranged in the shape of a tetrahedron (**FIGURE 6-2**). All silicate rocks and minerals, including the common volcanic rocks, consist of an array of submicroscopic silica tetrahedra that generally are linked to atoms of aluminum, iron, magnesium, calcium, potassium, sodium, and other elements. The chemical bonds between silicon and oxygen atoms are too strong to bend or break easily, so silicate structures are rigid, like an assemblage of Tinkertoy parts.

Differences in viscosity among the major magma types (basalt, andesite, and rhyolite) are due mainly to different percentages of silica. In general, the higher the percentage of silica, the more viscous the magma. As shown in Table 6-1, the composition spectrum ranges from low-viscosity (fluid) basalt magmas, which have around 50% silica, to high-viscosity rhyolite magmas, which have around 70% silica. Andesite magmas fall in between these two extremes, with around 60% silica. (See Appendix 2 online.)

At the more fluid end of the spectrum, **basalt** is black or brownish black, and its magma is about as fluid as cold molasses. Dark magmas have fewer silica tetrahedra, and these either are not linked directly to others or share two oxygen atoms to form long chains of tetrahedra. The tetrahedra chains link to others with weaker bonds provided by charged atoms of calcium, magnesium, or iron; they can

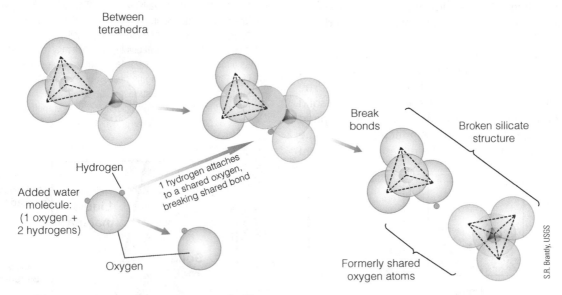

FIGURE 6-2 Tetrahedral Structure of Silica
Four oxygen atoms (blue) surround each silicon atom (gray) to make a silica tetrahedron (shape outlined by dashed lines). Other silicon atoms share oxygen atoms (blue-green atom shared) to form mineral molecules. A water molecule reacts with a silicate structure and its shared oxygen atoms, breaking some of the strong bonds.

Table 6-1

Characteristics of the Common Magmas or Lavas*

MAGMA	BASALT GRADATIONAL TO →	ANDESITE TO DACITE GRADATIONAL TO →	RHYOLITE
Viscosity	Fluid (low viscosity)	Medium viscosity	Thick and pasty (high viscosity)
Rock color	Black	Dark to medium shades of gray, green, reds, and other intermediate colors	Pale colors, including white, pink, yellow, green
Composition	Magnesium, iron, and calcium-rich	Gradational to →	Potassium and silica-rich
Silica content	≈ 50%	≈ 60%	≈ 70%
Temperature	1100° to 1200°C (yellow to red-hot)	Gradational to →	800° and 900°C (dull red)
Volatiles	Little water content unless magma encounters groundwater; may contain significant carbon dioxide	Gradational to →	Generally more water
Erupted material**	Mostly lava flows	Lava flows, ash, broken fragments (rubble)	Mostly ash

*All of these rock types are gradational. ** Discussed later in the chapter.*

wiggle about in magma like worms in a can, which explains why dark magmas are so fluid (**FIGURE 6-3**).

Rhyolite comes in white or pale shades of gray, yellow, pink, green, and lavender. Rhyolite magma is extremely viscous, causing it to flow stiffly, if at all. Flows of rhyolite glass, obsidian, are rare; the few that exist are so viscous that thick flows move and solidify to form flow fronts 50 to 100 m high with 45° slopes. Pale magmas, such as rhyolite, contain more silicon atoms to attach to oxygen atoms. Most of the silica tetrahedra share oxygen atoms to build exceptionally rigid frameworks of tetrahedra linked in all directions. The other atoms, including aluminum, sodium, and potassium, do not provide much freedom of movement, so pale magmas are extremely viscous.

FIGURE 6-3 Red-Hot Lava
Lava erupting as a bright-red curtain from a Hawaiian rift zone.

The viscosities of **andesite** magmas are intermediate, falling between low-viscosity basalt magmas and high-viscosity rhyolite magmas. Andesite magma is the most abundant type, erupting from subduction-zone volcanoes like those in the Cascades. These magmas contain enough silica to be quite viscous, and as a result they flow very slowly, if at all; they solidify to form thick flows. Because of its high viscosity, andesite lava solidifies on steep slopes, resulting in steep-sided volcanoes. The ash and broken-andesite rubble tumble down the same slopes and come to rest at angles near 30°. Gas content of the dark andesite magmas is generally quite low, and these magmas generally erupt in sluggish flows that are considerably thicker than basalt flows, or they blow out chunks of solidified lava or cinders. Dacite, with compositions between andesite and rhyolite, tend to erupt more vigorously to make clouds of ash and fields of broken rubble.

Temperature also plays a role in viscosity, because as magma cools, more bonds link between atoms and molecules and the magma becomes more and more viscous. At temperatures reaching 1100°C to 1200°C, red-hot basalt lava generally pours down stream valleys or spreads across flat ground like pancake batter across a griddle, making relatively thin flows that commonly cover dozens of square kilometers. In contrast, the more viscous rhyolite magmas erupt at temperatures between 800°C and 900°C, a dull red heat.

VOLATILES The **volatiles** of a volcano refer to the dissolved gases it contains. Almost all volcanoes emit gases, often quietly but sometimes violently. The gases rise either from seemingly random locations high on the volcano or at an eruptive vent, sometimes during an eruption, sometimes long after. Water vapor is the most abundant volcanic gas and the most important in governing what happens when magma erupts. Under volcanic conditions, water in magmas exists mostly as vapor hot enough to glow and instantly ignite anything flammable in its path. Carbon dioxide is normally second to water in abundance among the volcanic gases, but it has much less influence on the explosive nature of an eruption. It is relatively more abundant in basalt magmas than in those closer to rhyolite in composition.

Water content is critical because this is mainly what drives volcanoes to violence. Water at the high temperature of magma expands to form steam at low pressures near Earth's surface. Pressure holds in any dissolved water, so the amount of water a magma can contain decreases dramatically as it rises into levels where the rock pressure is lower. Rhyolite magma with 2% water at a depth of 20 km could hold only half that much at a depth of 2 or 3 km. By the time the magma reaches the low pressure of Earth's surface, it can dissolve virtually no water or other gases; any gases that were dissolved at depth must separate from the magma and bubble out (**FIGURE 6-4**). It is those separating gases that drive explosive volcanic eruptions. Even if magma stops rising, steam can separate and build up pressure to drive an eruption. Most of the new crystals growing in the magma contain little or no water. As they grow, they displace the water or

steam into an ever-smaller volume of the magma. This may eventually drive an eruption.

Magmas that contain little water erupt quietly as lavas. Basalt magmas, for example, are fluid enough to let their small amounts of water and carbon dioxide escape without causing much commotion. Those that contain large amounts of dissolved water, however, are likely to explode unless they are fluid enough to let the steam fizz quietly away. Rhyolite magmas contain 0 to 10% water by weight and much more than that by volume. When rhyolite magma reaches the surface nearly devoid of water, it erupts quietly as a dome that may grow to the size of a small mountain. When it arrives at the surface with a heavy charge of water, rhyolite magma explodes into clouds of steam full of foamy pumice and white rhyolite ash.

Because basalt magmas feeding a volcano above a subduction zone stream up from deep in Earth's mantle, they may rise into a water-rich rhyolite magma chamber. The extremely hot basalt heats the rhyolite, causing its water to boil into steam. Rapid expulsion of this steam near the surface causes explosive eruption of the rhyolite magma. Dark inclusions of basalt sometimes found in pale rhyolite help document this process. Such magma interactions may be a frequent trigger for explosive eruptions; the process adds to the complications of predicting the style and timing of eruptions.

VOLUME Viscosity and volatiles of a magma are the properties that determine the nature of an eruption, and volume is the property that determines its size or magnitude. The volume of magma expelled in a single eruption has a significant bearing on the degree of hazard. It affects the size of a lava flow, as well as the volume, areal extent, and time span of an ash eruption. Recognition of various types of volcanoes from their size and the slopes of their flanks permits us to interpret their behavior, even from a distance.

For andesite to rhyolite gas-driven explosive eruptions, the volume of fragmental material erupted depends on the volume of magma reaching Earth's surface, the viscosity of the magma, and the proportion of dissolved gas it contains. Because the dissolved gases tend to rise toward the top of the magma chamber and the expanding gases drive the eruption, the upper gas-rich part of the magma (the upper third or so of the magma chamber) erupts explosively. The remaining magma stays underground, or a little may erupt as a lava flow late in the eruption process. Because these magmas are highly viscous, it may take decades to centuries for more gas to collect in the upper part of the magma chamber to again drive an eruption.

The composition and total volume of magma rising under a subduction-zone volcano depends on the subduction rate; the temperature and water content of the descending slab; the composition, temperature, and water content of the overlying crustal rocks; the ease with which the magma can rise through the crust; and other intangible factors. For an individual volcano, most of these factors are unlikely to change much over time. Thus, the behavior of an individual volcano is likely to be more or less predictable.

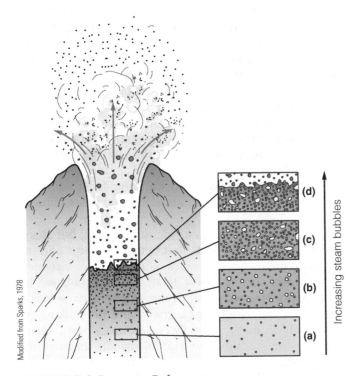

Modified from Sparks, 1978

Increasing steam bubbles

FIGURE 6-4 Pressure Release

(a) and (b) represent steam separating from magma in open bubbles. (c) and (d) show that the bubbles grow and the magma begins to froth, expand, and rise. That pushes magma upward, and its pressure decreases, launching a chain reaction of increasingly rapid bubble formation that leads to an explosive eruption.

Tectonic Environments of Volcanoes

Plate tectonic environments, where Earth's tectonic plates spread apart or converge (see FIGURE 2-3), are common locations of changes in temperature, pressure, or water content. Most of the world's volcanoes are along these plate boundaries. Spreading zones, where plates pull apart, and subduction zones, where one plate pushes under another, are the most common locations for eruptions. Hotspots produce large volcanoes even though they are not on plate boundaries. Few volcanoes occur in continent–continent collision zones, such as in the Himalayas. Transform boundaries (faults that slide laterally past one another) rarely show volcanic eruptions.

The nature of an eruption is influenced by the tectonic environment in which it forms. Relatively peaceful eruptions are characteristic of spreading zones, while violent eruptions are characteristic of subduction zones.

Spreading Zones

Ocean-floor rifts or spreading zones, such as in the Pacific and Atlantic Oceans, erupt basalt lavas that spread out on the adjacent ocean floor. These don't provide much hazard to humans except in Iceland, where the Mid-Atlantic Ridge extends above sea level. There the ridge crest makes a broad valley that follows a dogleg course across the island, its opposite walls moving an average of a few centimeters farther away from each other per year. Every few hundred years, a long fissure or crack opens in the floor of the valley and erupts a large basalt flow. Until 2010, the most recent such occasion was in 1783, when the Laki fissure erupted 12 km³ of basalt lava that spread 88 km down a gentle slope. Given the events of the last 1000 years of written Icelandic record it is not surprising that the spreading zone erupted again, producing basalt lava, along with ash that drifted over much of Europe. In that case water to drive the ash eruption came from the overlying glacier melted by the basalt magma.

In Iceland, as everywhere along the oceanic ridge system, peridotite, the hot dark rock of Earth's mantle, rises at depth to fill the gap between lithospheric plates that separate at an oceanic ridge. As the dry mantle rock slowly rises, the lessening load of overlying rock subjects it to progressively lower pressures that permit it to partly melt to make basalt magma. The lower density of the magma lets it rise to erupt at the ridge. Its low viscosity (high fluidity) permits it to spread out as a thin basalt lava flow. Similar flows erupt in the crest of the oceanic ridge system along its entire length, to build the basalts that make up the entire ocean floor.

Watch wax dribbling down a candle and think about molten basalt lava erupting on the ocean floor. The cooling dribble of wax soon acquires a thin skin of nearly solid wax. Then the molten wax within bursts through that skin and dribbles down a new path. As that happens again and again, the original dribble of wax forms a network of dribbles. Basalt lava

FIGURE 6-5 Pillow Basalts

A thick pile of pillow basalts from uplifted ocean floor in Oman.

erupting into water behaves the same way. A chilled skin of solid basalt forms on the outside of the flow; then the molten basalt within bursts out and pours off in a new direction. The result is sometimes a pile of basalt cylinders, about the size of small barrels, called **pillow basalts**. Exposed in a cliff or road cut, they look like a pile of oversized pillows in dull shades of greenish to brownish black (**FIGURE 6-5**).

On continents, the largest **flood-basalt** flows are simply giant lava flows with volumes of several hundred cubic kilometers, more than 100 times those of ordinary basalt flows. Individual flows may cover tens of thousands of square kilometers to depths of 30 to 40 m. Some in the Pacific Northwest of Washington, Oregon, and western Idaho cover areas almost the size of Maine.

Continental rifts or spreading zones, such as the Basin and Range of Nevada and vicinity, the Rio Grande Rift of New Mexico, and the East African Rift Zone, move apart at much slower rates than those on the ocean floor. The mantle rocks under the Basin and Range and Rio Grande Rift rise more slowly, so their decrease in pressure is slower, and they melt at lower rates and produce much less magma. Much of the magma erupted through the continents in the last million years or so rose along rift-zone faults. Fewer but much larger and more violent eruptions spread incinerating flows of rhyolite ash over large areas; those would have been catastrophic for any nearby populations.

Subduction Zones

Subduction zones, the locations where oceanic plates slide under either oceanic or continental plates, are widespread and spawn most active volcanoes. Volcanoes formed in this environment are the most spectacular and most hazardous on Earth. Here, cold ocean-floor lithosphere (basaltic crust and hydrated upper mantle) collides with and descends beneath warmer, lower-density ocean-floor or continental rocks. Cold ocean-floor rocks formed tens of millions of years ago at oceanic rift zones; ocean water percolating deep in cracks warms up at depth, circulates widely, and over millions of years combines with the minerals in the crustal basalt and mantle peridotite. By the time these altered rocks descend into the oceanic trench, they contain enough water to make a difference in the way magmas are produced.

The descending oceanic plate or slab slowly heats up and, at depths of 100 km or so, begins to boil off some of the water. The water rises into hot mantle rocks under the continental crust. In that environment, the melting boundary for the hot mantle rocks moves to lower temperatures because of the added water.

Whenever mantle peridotite melts, it forms basalt magma; this new lower-density magma rises into the overlying crustal rocks, most commonly continental crust. Basalt magma is hot enough to melt granitic composition rocks of the continental crust to form rhyolite magma. Thus the basalt and rhyolite magmas intermingle. Either or both can rise to Earth's surface to erupt in a volcano. If the two magmas mingle and mix, they form magma of intermediate composition—andesite—that rise to erupt in the same volcano. Thus, subduction-zone volcanoes, including those in the Cascade Range of western North America, lie on the continental side of oceanic trenches and are characterized by basalt, andesite, and rhyolite compositions. Proportions of those rocks depend on a variety of factors, including the composition and thickness of continental crust. Magmas are commonly generated near a depth of 100 km, above the inclined subducted slab. Volcanoes above that location will be farther inland if the inclination of the slab is gentler, and closer to the trench if the inclination is steeper.

Because magma in the subduction-zone environment is generated due to the release of water from the subducted slab, the magmas formed generally contain some water. That water contributes to the eruption behavior of the magmas. Instead of flowing out quietly as lavas, water in the magma near the surface separates into gas bubbles that expand rapidly under the lower pressure. The expanding bubbles blast the magma into fragments and ash in a violent eruption.

Hotspots

Hotspot volcanoes are far fewer in number but generally produce large volumes of magma. Instead of appearing at plate boundaries, they grow within tectonic plates at what appear to be random locations. Because an active hotspot volcano lies at the end of a series of older inactive volcanoes, the source of the magma must be in the underlying relatively stationary asthenosphere, rather than in the moving lithospheric plate, as discussed in Chapter 2. Melting of mantle peridotite under an ocean basin produces dry basalt

magma that rises through peridotite and basalt-composition rocks, so it continues to the surface to erupt as basalt lava. Basalt magma derived from deep, dry mantle peridotite is not inherently explosive because it contains no water. Thus a hotspot volcano in an ocean basin, such as in Hawaii, typically erupts basalt lavas that flow out quietly without blasting into fragments. If the magma picks up some groundwater en route, the water can vaporize and blast the magma into fragments during eruption.

Although few hotspot volcanoes appear on the continents, Yellowstone is a prominent example. As with oceanic hotspot volcanoes, the basalt magma source must be in the mantle, but this magma rises through thick continental crust. Because it is dense, little of the basalt makes it all the way to the surface, but its intense heat melts the silica-rich continental crust that has a lower melting temperature. That new magma rises to erupt at the surface as light-colored rhyolite ash; its dissolved water drives explosive eruptions.

Volcanic Eruptions and Products

Characteristics of volcanoes depend on the same factors that control the nature of eruptions. The magma volume, viscosity, and volatile content control the size of the volcano, the steepness of its slopes, and its eruption products. Depending on its style of eruption, a volcano can produce lava, pyroclastic materials (air-fall ash, pumice, and pyroclastic flow deposits), and lahars (volcanic mudflows) (**FIGURE 6-6**). **Table 6-2** gives an overview of volcanic materials.

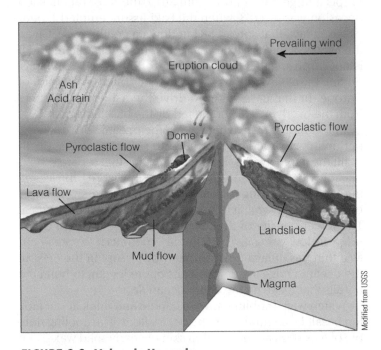

FIGURE 6-6 Volcanic Hazards
Most volcanoes produce a broad range of hazards.

	LAVA	PYROCLASTIC MATERIAL (ASH, PUMICE, OTHER FRAGMENTS)	LAHARS
DEFINITION	Molten magma that flows out and onto Earth's surface	Fragments of solidified magma blown out of a volcano; may be deposited by a pyroclastic flow or by air-fall ash	Volcanic ash and other fragments transported downslope with water (mudflows)
GENERAL CHARACTERISTICS	Solidifies as coherent sheets or broken jumbles of volcanic rock	Fragments range from less than 2 mm ash to tens of cm across; larger pieces may be broken from older volcanic rocks on the sides of the vent	Angular to rounded; unsorted particles from mud to boulders

Nonexplosive Eruptions: Lava Flows

Most basalt magmas do not explode, because they are fluid and contain modest amounts of gas, mainly water and carbon dioxide, which bubbles harmlessly out of the melt. Instead, basalt magma tends to spill out of a volcano and flow down its sides in the form of a **lava flow**. Lava flows fed by a hidden magma chamber may spill out from a central

crater (a depression created as ash blasts out) or pour from a spreading crack or rift on the flank of the cone.

Where fluid basalt lava flows dominate eruptions, the flows are sometimes called *Hawaiian-type lava*, after one of the places that best characterizes them. Some of those basalt flows have a smooth, ropy, or billowy surface generally called **pahoehoe,** a Hawaiian term pronounced *pah-hoy-hoy* (**FIGURE 6-7A**). Pahoehoe surfaces develop on lavas rich

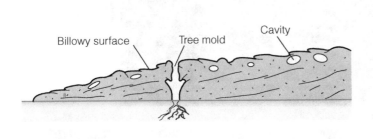

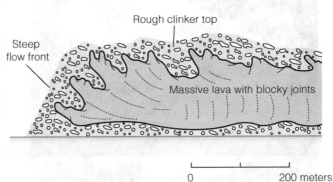

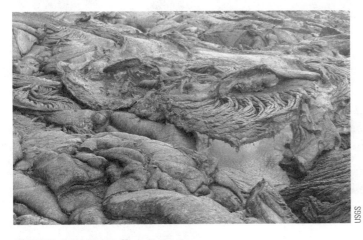

FIGURE 6-7 Hawaiian-Type Lavas

A. Smooth-topped pahoehoe lava in cross-section (top) and flowing on Kilauea (bottom).

B. A ragged, clinkery surface of aa lava in cross-section (top) and from Mauna Loa Volcano, Waikoloa, Hawaii (bottom).

in steam and other volcanic gases. The ropy form develops when fluid lava drags a thin cooling skin into small wrinkles or folds. These flows commonly develop open passages as molten lava escapes from under a solidified crust several centimeters to a meter or two thick. This crust is important because it makes naturally insulated pipes through which lava can flow for several kilometers without much cooling. The molten lava can then break out downslope and again flow on the surface. People walking across recently erupted but solidified lava flows risk breaking through a thin crust and falling into the molten lava. Because of their low viscosity, basalt flows spread out easily and solidify on gentle slopes.

Basalt lavas that are charged with less steam and other gases, and that have partially crystallized, develop an extremely rubbly and clinkery surface called **aa**, another Hawaiian term, pronounced *ah-ah* (**FIGURE 6-7B**). The rubble tumbles down the slowly advancing front of the flow, which runs over it the way a bulldozer lays down and runs over its tread. Aa lavas flow more slowly and form thicker flows than pahoehoe.

Explosive Eruptions: Pyroclastic Materials

Explosive eruptions produce fragments of solidified magma known generally as **pyroclastic material**. The more viscous the magma and the greater its gas content, the more likely it is to explode. Steam bubbles separate with difficulty from rhyolite or dacite magma as it approaches the surface; the froth expands to form a foam of glassy bubbles called pumice. The bubbles expand as the magma continues to rise until they burst into **volcanic ash**, which consists mostly of curving shards of glass, formerly the walls of bubbles resembling soap bubbles. When the pumice bursts into ash, the steam escapes, and the whole mass explodes into a cloud of volcanic gases and suspended ash (**FIGURE 6-8A, B**).

Volcanic ash may rain down from an eruption cloud or be blown downwind as air-fall ash. Heavier parts of the eruption cloud may collapse onto the volcano flank as a **pyroclastic flow** (**FIGURE 6-8C**). Pyroclastic flows may also spill downslope directly from the crater rim. Many pyroclastic flows develop in the collapse of an

FIGURE 6-8 Volcanic Ash

A. This explosive eruption of Mt. St. Helens on July 22, 1980, pushed an ash cloud to a height of 14 km, dwarfing the mountain visible in the lower right. **B.** A fragment of volcanic ash from the main Mt. St. Helens eruption in 1980, magnified about 3000 times. **C.** A pyroclastic flow races down the flank of Mt. Merapi, Java.

expanding volcanic dome. When a dome collapses to form a pyroclastic flow, it is extremely dangerous for anything in its path. Pyroclastic flows sometimes travel distances of 20 km—in a few cases, much farther, destroying everything in their path. Continuing expansion and expulsion of steam, and heating and expansion of air trapped below the ash cloud, maintain the internal turbulence and high speeds of pyroclastic flows. If ash and fragment deposits on a volcano's flank soak up sufficient water from rain or melting snow, the mixture may pour downslope as a volcanic mudflow called a **lahar** (Case in Point: Deadly Lahar—Mt. Pinatubo, Philippines, 1991, p. 139).

On March 18, 2007, partial collapse of Mt. Ruapehu's crater rim, on the north island of New Zealand, released a large lahar. Seepage through the crater rim following several days of heavy rain caused its collapse, releasing about 1.3 million m^3 of water flowing at 2500 m^3/s in a valley 7 km downstream. Because the area is almost uninhabited, there was little damage.

The size of an ash eruption depends on many factors, including the amount of magma, the magma viscosity, and its water-vapor content. Steam explosions eject particles of magma at velocities that depend upon the amount and pressure of the steam and the narrowness of the vent. Those factors dictate the rate at which the steam can expand and escape. Escaping gas slows as it expands and cools. Erupting steam commonly moves between 400 and 700 km/h but may exceed the speed of sound in air—1200 km/h. Steam explosions may throw large blocks and blobs of magma, called *bombs*, as far as 5 to 10 km. Smaller particles drift farther downwind.

Styles of Explosive Eruptions

Styles of eruptions range from frequent and mild to infrequent and violent, as classified by the **volcanic explosivity index (VEI)**, which crudely quantifies eruption size, volume, and violence (Table 6-3). Hawaiian eruptions are the most frequent and least violent. Many of the more explosive, less frequent types of eruptions are typified by Italian volcanoes that provide their names.

PHREATIC AND PHREATOMAGMATIC ERUPTIONS **Phreatic eruptions** are violent, steam-driven explosions generated by vaporization of shallow water in the ground, a nonvolcanic lake, a crater lake in a volcano, or shallow sea. The result is often a maar, a broad, bowl-shaped crater encircled by a low rim that commonly rises only slightly above the surrounding terrain. Steam dominates; magma is not erupted. If magma incorporates groundwater, it causes a *phreatomagmatic eruption*. In either case, such water-rich eruptions can be especially dangerous; the erupting column of steam and ash can collapse to form a base surge that sweeps rapidly outward. The surge carries hot sand and rock fragments that can sandblast and uproot trees and overwhelm and kill people.

STROMBOLIAN ERUPTIONS Stromboli, off the west coast of Italy, is fed by magma that interacts with groundwater or

Table 6-3		**Volcanic Explosivity Index (VEI) for Different Styles of Individual Eruptions**				
VEI	**VOLUME OF EJECTA (KM³)**	**ERUPTION COLUMN HEIGHT (KM)**	**ERUPTION STYLE**	**DURATION OF CONTINUOUS BLAST (HRS)**	**ERUPTION FREQUENCY (APPROXIMATE)**	**EXAMPLE ERUPTION**
0	<0.0001	<0.1 (100 m)	Hawaiian	<1		Kilauea, Hawaii
1	0.0001–0.001	0.1–1	Hawaiian Strombolian	<1	100 per year	Kilauea, Hawaii Stromboli, 1996
2	0.001–0.01	1–5	Strombolian Vulcanian	<1	15 per year	Unzen, Japan, 1994
3	0.01–0.1	3–15	Vulcanian	1–6	2–3 per year	Nevado del Ruiz, Columbia, 1985
4	0.1–1	10–25	Vulcanian to Plinian	6–12	1 / 2 years	Iceland, 2010 El Chichon, 1982
5	1–10	>25	Plinian	>12	1 / 10 years	Mt. St. Helens, 1980
6	10–100	>25	Plinian	>12	1 / 40 years	Krakatau, 1883 Pinatubo, 1991 Thera (Santorini), 1600 BC
7	100–1000	>25	Plinian	>12	1 / 200 years	Tambora, Indonesia, 1815
8	>1000	>25	Yellowstone	Off scale	1 / 2000 to 1 / 1,000,000 years	Yellowstone, 600,000 BC Long Valley, California, 730,000 BC

Increasing volume → · Increasing eruption height → · Increasing duration → · Decreasing frequency →

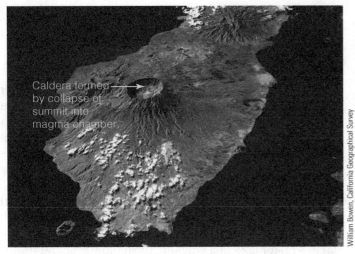

Caldera formed by collapse of summit into magma chamber.

William Bowen, California Geographical Survey

Dr. Haraldur Sigurdsson, University of Rhode Island

FIGURE 6-9 **Plinian Eruption at Mt. Tambora**

A. Pyroclastic flows from Mt. Tambora in Indonesia killed more than 10,000 people in a giant eruption in 1815.

B. Artifacts excavated from ash from the 1815 eruption. People and their communities were buried in the eruption.

seawater. Rapidly expanding steam bubbles in the magma blow it into cinders and bomb-size blocks that fall around the vent and tumble down steep slopes to form a cinder cone. This is a **Strombolian eruption**. The fluid nature of the magma is associated with generally mild eruptions.

VULCANIAN ERUPTIONS Vulcano, an island off the north coast of Sicily, is fed by highly viscous, andesite magmas that are rich in gas. Dark eruption clouds blow out blocks of volcanic rock, along with ash, pea-size *lapilli*, and bombs. With a **Vulcanian eruption**, ash falls may dominate, but pyroclastic flows and lateral-blast eruptions can develop with—or follow—the ash fall.

PELÉAN ERUPTIONS Mt. Pelée in Martinique, the West Indies, erupted in 1902 to obliterate the town of St. Pierre and its 28,000 inhabitants (see Online **Case in Point:** Deadly Pyroclastic Flows—Mt. Pelée, Martinique, West Indies). Its rhyolite, dacite, or andesite eruptions can be violent, especially early in the eruption, when high columns of ash may be ejected. **Peléan eruptions** are characterized by giant ash columns that collapse to form incandescent pyroclastic flows. Most distinctive in the 1902 eruption, however, was the growth of Pelée's huge, steep-sided spine of solidified magma that the magma pushed up from the vent. At intervals, the sides of the expanding dome collapsed to form searing hot block and pyroclastic flows. Occasionally, the magma below slowly pushed a viscous lava spine outward from the surface of the dome, sometimes as much as 300 m high.

PLINIAN ERUPTIONS The great eruptions of Vesuvius in AD 79 and Mt. St. Helens in 1980 produced powerful continuous blasts of gas that carried huge volumes of pumice high into the atmosphere. Even larger than Peléan eruptions, **Plinian eruptions** can be truly catastrophic for any nearby

population. Silica-rich ash falls accompany incinerating pyroclastic flows, including those consisting mostly of pumice flows. Ejection of a large volume of magma often causes collapse of the magma chamber to form a **caldera**, as in the case of Mt. Mazama (Crater Lake), Oregon, 7700 years ago; Santorini, Greece, in approximately 1620 BC (**Case in Point:** A Long History of Caldera Eruptions—Santorini, Greece, p. 140); Krakatau in Indonesia in 1883; and Tambora in Indonesia in 1815 **(FIGURE 6-9)**.

Small eruptions typically form craters at the eruptive vent; a depression or crater forms where the vent is excavated by the violent eruption. The distinction between crater and caldera is not so much size but the mechanism of depression formation. A cinder cone blows out vent material to form a crater. A giant Plinian eruption, such as the one that formed Crater Lake, Oregon, 7700 years ago, collapses into the emptying magma chamber to form a caldera. To be precise, Crater Lake should be called Caldera Lake. Similarly, the eruption of the Bishop tuff and collapse of the Long Valley caldera about 760,000 years ago was a giant Plinian eruption.

Types of Volcanoes

Volcanoes range from giant and gently sloping to small and steep-sided **(FIGURE 6-10)**. Based on the appearance of a volcano, primarily its size and the slopes of its flanks, you can infer the magma composition that produced it and its volatile content; both reflect the eruptive style and the types of associated hazards and risks (**Table 6-4**). In this section, we describe the main differences between volcano types, using examples to illustrate the range of behaviors and hazards.

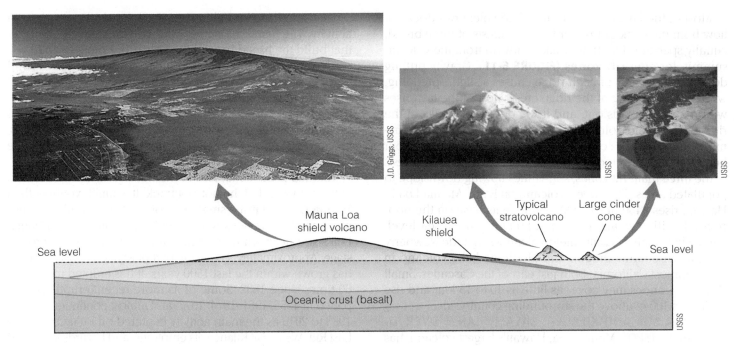

FIGURE 6-10 Volcanoes in All Sizes

The different types of volcanoes have dramatically different sizes (or volumes). Mauna Loa, the giant shield volcano in Hawaii (left photo), is roughly 220 km in diameter and rises some 9450 m above the seafloor; the 4169 m above sea level is less than half its total height. The gentle slopes of Mauna Loa are typical of giant shield volcanoes. In comparison, a typical subduction-zone volcano such as Mt. Rainier in Washington State (center photo), on a base near sea level is only 2000 to 3000 m high. Cinder cones such as Sunset Crater near Flagstaff, Arizona (right photo), are still smaller.

Shield Volcanoes

Persistent basalt eruptions of very fluid basalt lavas within a small area eventually build a gently sloping pile of thin flows. More than a century ago, geologists decided to call those piles **shield volcanoes** because of a fancied resemblance to the shape of an ancient Roman shield. The flows are characterized by their low viscosity, low volatile content, broad and gently sloping sides, and large to giant volumes.

Table 6-4			General Characteristics of Common Volcanoes			
VOLCANO TYPE	**TECTONIC ENVIRONMENT**	**VISCOSITY**	**SLOPE OF FLANKS**	**VOLATILES**	**VOLUME (SIZE)**	**ERUPTION STYLE**
Shield volcano (basalt), Figs. 6-10, 11	Oceanic hotspots, some volcanic chains	Low	Gentle	Low	Large to giant	Quiet lavas
Cinder cone (basalt), Figs. 6-10, 14 to 16	Flanks of shield volcanoes, continental rifts	Low	Steep	Moderate	Small	Explosive but not very dangerous
Stratovolcano (andesite), Figs. 6-9, 10, 17	Above active subduction zones	Moderate	Moderate	Moderate to high	Moderate	Violent, dangerous pyroclastic flows and air-fall ash
Lava dome (rhyolite)	Above active subduction zones	High	Steep	Low to moderate	Small to moderate	Dangerous dome collapse and pyroclastic flows
Resurgent caldera (rhyolite), Fig. 6-19	Continental hotspots, some continental rift zones	High	Very gentle	Moderate	Giant	Violent, dangerous pyroclastic flows and ash

Most of the lava erupted from shield volcanoes does not flow from the peak but rather from the crests of three broad, equally spaced ridges that radiate outward from the volcano summit, forming **rift zones (FIGURE 6-11)**. Gravity pulling down on the flanks of each ridge causes a slight spreading, which causes rifting of the ridge crest. Most of the lava rises within a rift, spreads out, and flows down the flanks of the ridge. The shield volcano may have a caldera in the summit where the surface rocks sank into the magma chamber below. Many have cinder cones or lava cones along the crests of the three main rift zones, sometimes flowing downslope to populated areas. The largest volcano on Earth, Mauna Loa in Hawaii, rises 9450 m (31,000 ft) above its base on the floor of the Pacific Ocean, 4270 m (14,000 ft) above sea level. Somewhat smaller basalt shield volcanoes include Newberry volcano in central Oregon and Medicine Lake volcano in northernmost California, both just east of the Cascades. Small shield volcanoes may cover as little as 10 to 15 km² and rise as little as 100 m above the surrounding countryside.

MAUNA LOA AND KILAUEA: BASALT GIANTS OVER AN OCEANIC HOTSPOT Mauna Loa, Hawaii's largest volcano, has erupted 33 times since 1843. Recent eruptions include those in 1926, 1940–1942, 1949–1950, and 1975. Seven of its flows, including one in 1984, came within 6.5 km of Hilo. Mauna Loa and Kilauea exhibit the mildest of eruptions, with VEIs of 0 to 1.

Oceanic hotspot volcanoes go through three stages of activity. The first is a long series of eruptions below sea level that build the broad base of the volcano, a great heap of volcanic rubble with little mechanical strength. In the second or main stage, eruptions produce basalt lava flows that build the main, visible mass of the volcano. The late stage of activity comes as the volcano moves off the hotspot. Eruptions become smaller and less frequent. Before major eruptions, volcanoes slowly swell at a few centimeters per year with the pressure of inflating magma chambers.

Kilauea is the successor to Mauna Loa, the new volcano at the active end of the hotspot track. It is much younger than Mauna Loa, still just a small fraction of its size, and produces much smaller eruptions. Kilauea looks like a big ledge on Mauna Loa's southeastern flank. Kilauea is now in its main growth stage. Approximately 95% of the part of Kilauea above sea level has grown within the last 1500 years. It has erupted more than 60 times since 1832. Some of its more recent eruptions happened in 1955, 1961–1974, 1977, and almost continuously from 1983 to 2015 as the main activity centered on Pu'u O'o on the East Rift. We revisit Kilauea, its eruptions, and hazards at the end of Chapter 7, **Case in Point: Kilauea's East Rift Eruptions Continue**. Most erupted material is lava, but cinder cones blast out bubbly chunks of basalt that tumble downslope.

Even though Kilauea is still young, its edifice has already split into pie-slice segments that have begun to spread—only

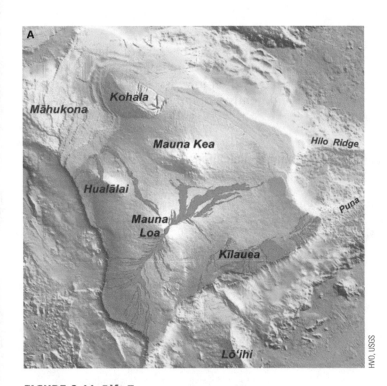

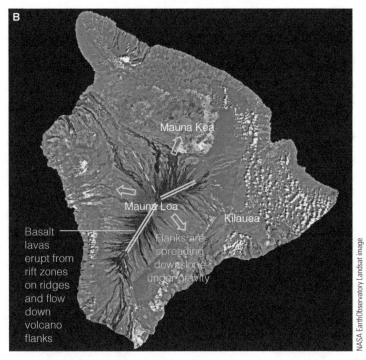

FIGURE 6-11 Rift Zones

Mauna Loa, the dominant volcano on the Big Island of Hawaii, erupts along three spreading rift zones. **A.** Hawaii above sea level is shown in gray; recent lava flows erupt along rift zones of each volcano, as shown in red. **B.** The sagging flanks (three arrows) of the crudely three-sided volcano maintain spreading at the three rifts, permitting basalt lavas to erupt at the crests of the rifts and flow, like dark fingers, downslope from the rifts.

two, because one side is buttressed against Mauna Loa. Most of its eruptions are along the rift zones that radiate from the summit and separate the segments. The southern flank of Kilauea broke along a scarp more than 500 m high, Hilina Pali, and has begun slowly sliding toward the ocean.

Like many volcanoes, Kilauea announces an impending eruption with swarms of small earthquakes that originate at shallow depths, along with harmonic tremors that record magma movement. Sensitive tiltmeters may detect an inflation of the volcano summit as magma pressure increases. In some cases, these symptoms raise false alarms.

Kilauea's eruptions typically produce only basalt. Although lava flows cross and burn roads (**FIGURE 6-12**) and can torch buildings, they rarely injure or kill anyone. But hot gases, presumably water vapor, sulfur dioxide, and carbon dioxide, did drive a surge in 1790 that killed 80 warriors from King Keoua's army as they crossed a high flank of Kilauea. In February 2007, 30 scientists and naturalists exploring underground tunnels in a Canary Islands' shield volcano off northwestern Africa got lost in the maze of caverns; six suffocated, probably from breathing carbon dioxide, and six were hospitalized. In March, 2008, a gas vent on the wall of Halema'uma'u Crater, the lava lake near the summit of Kilauea volcano, became active and exploded, ejecting steam and ash. Hazardous sulfur dioxide gas in "vog" (volcanic gas and fog) required evacuation of nearby communities. Lava was seen sloshing in the vent in the floor of Halema'uma'u Crater in December 2009.

MT. ETNA, SICILY The huge mass of Mt. Etna broods over Catania and other cities in eastern Sicily. It is 3315 m high, the largest continental volcano on Earth. Except for Stromboli, it is also the most active volcano in Europe, typically erupting basalt lava flows and cinders **(FIGURE 6-13)**. Etna stands on an unstable base of soft muds deposited from seawater onto oceanic crust.

FIGURE 6-12 Slow-Moving Basalt Magma
Creeping pahoehoe lava from Kilauea approaches Pahoa, Hawaii, on October 26, 2014. The black, smooth-topped basalt on the surface of the flow covers the red-hot molten basalt.

Flank eruptions from rifts that divide the volcano into three large pie slices have built three radial ridges similar to those of the Hawaiian volcanoes. Escaping gases blow cinders out of vents in these ridges, forming basalt cinder cones that also quietly produce lava flows that burst from the bases of the cones. People easily avoid the flows, but their houses cannot; ruined houses litter the flanks of the volcano **(FIGURE 6-13)**.

Etna's almost continuous—but frequently erratic—eruptions feature occasional violent episodes. On July 22, 1998, for example, one of the craters produced an eruption column 10 km high that dumped ash over a wide area to

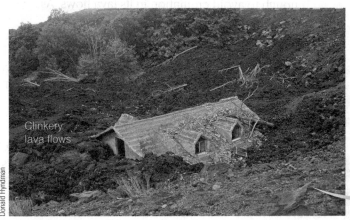

FIGURE 6-13 Mt. Etna
This is all that remains of a house that was in the path of a 1983 Mt. Etna lava flow north of Nicolosi, Sicily. Mt. Etna erupted a prominent plume of dark ash on October 30, 2002, while fires were ignited by lava pouring down its north flank. Light-colored plumes are gas emissions from a line of vents along a rift extending out from the summit. Roads and towns on the flanks of the volcano are faintly visible in the upper right and across the bottom of the photo.

the east. It has also staged at least 24 sub-Plinian eruptions in the past 13,000 years. Plinian eruptions of basalt volcanoes are rare but hazardous, especially with towns creeping ever higher on Etna's flanks. A Plinian eruption in 122 BC wreaked havoc on the Roman town of Catania 25 km south of Etna. The main Plinian deposit was 20 cm of coarse ash that started fires and collapsed roofs. As the eruption rate dropped, magma apparently boiled groundwater into steam that carried clouds of ash high into the air.

The low viscosity and low water content of basaltic magmas normally prevent Plinian eruptions that are dominated by voluminous ash. Etna's may develop when the magma rises so rapidly that gases cannot slowly escape as they normally do. Instead, bubbles of steam explode within the rapidly decompressing basalt magma.

Cinder Cones

Cinder cones are also basalt, but they are characterized by their smaller size, low viscosity, steep sides, and moderate volatile content (**FIGURE 6-14**). They erupt where rising basalt magma encounters near-surface groundwater, and escaping steam coughs cinders of bubbly molten lava out of a vent or summit crater. These cinders fall around the vent and build a loose, steep-sided pile (**FIGURE 6-15**). The smallest shreds drift downwind in a black cloud and fall as basalt ash. When the water in the ground dries up, a single basalt lava flow pushes out from the base.

Along extensional faults, such as the Great Rift at Craters of the Moon in southern Idaho, basalt also rises to feed lava flows. Where it encounters water, it flashes into steam to build cinder cones.

Most cinder cones erupt over only one short period, a few months to a few years. They typically build a pile of cinders 100 to 200 m high. The heavy lava flow emerges from

FIGURE 6-15 Cinder Cone Eruption
Red-hot cinders blasted out of the vent fall, remaining hot as they tumble down the cone of Paricutín volcano in Mexico.

the base in the same way that milk poured on top of a pile of Rice Krispies emerges from the bottom of the pile. The next eruption in the area will probably build a new cinder cone where water is present rather than reactivate an old one. Many volcanic areas, such as Haleakala on Maui in Hawaii, Lassen in northern California, and Newberry in Oregon, are liberally peppered with cinder cones, many visible within a single view.

Cinder cones provide an exciting nighttime fireworks display as glowing "bombs" and cinders arc through the air and then roll, still glowing, down flanks of the cone (**FIGURE 6-16**). The upper parts of lava flows look like glowing red streams running downslope; the lower parts darken as they cool (FIGURE 6-15).

Most people are sensible enough to evade the rain of glowing cinders from a growing cinder cone and agile enough to avoid cremation in its lava flows. As a result, these eruptions mostly endanger property.

The dark basalt ash that drifts downwind from erupting cinder cones eventually weathers into fertile soil that supports abundant and nutritious crops if water is available. Such areas generally support large agricultural communities.

Stratovolcanoes

When most people imagine a volcano, they envision a **stratovolcano**—a large, steep-sided cone over a subduction zone (**FIGURE 6-17**). Stratovolcanoes, also called *composite volcanoes*, are characterized by their moderate volume and size, moderate viscosity and slope, and moderate-to-high volatile content. Big stratovolcanoes, such as those of the High Cascades, are dominated by andesitic compositions, but darker varieties range to basaltic andesite; lighter varieties range from dacite to rhyolite. The names *stratovolcano* and *composite volcano* both convey the fact that they typically consist of layers of lava

FIGURE 6-14 Cinder Cone Volcano
Sunset Crater, near Flagstaff, Arizona, a typical cinder cone, shows a smooth-sided cone capped by a central crater.

FIGURE 6-16 An Old Pile of Cinders

Cross-section exposure of a cinder cone showing layers parallel to slope and scattered large volcanic bombs, northern California. Volcanic bombs in the cinder cone were twisted and tapered as the molten magma flew through the air. The two large bombs, broken open and filling most of the inset view, show chilled rims enclosing gas bubbles that could not escape through the rims. The red color forms later as volcanic gases percolate up through the porous pile. The red pocket knife in the lower left is 8 cm long.

flows, fragmental debris, and ash. The slopes of the flanks of stratovolcanoes depend on two factors:

1. Their magma has moderate viscosity, so lavas are not especially fluid; they flow only on moderately steep slopes before they cool and solidify.
2. Escape of dissolved volatiles through viscous magma typically causes large eruptions of ash and broken rubble that concentrate near vents and tumble to form slopes of about 30 degrees; this builds the cone higher near the vent.

Lava flows give a stratovolcano's cone enough mechanical strength to hold an internal column of magma. This enables the volcano to erupt repeatedly from a summit crater and permits it to grow into a tall cone.

Examples of stratovolcanoes include the High Cascades of the western United States and Canada, Mt. Vesuvius in Italy, and Mt. Fuji in Japan. Most stratovolcanoes grow in long chains above a slab of ocean floor that subducts into the interior of Earth. The volcanic chain, generally between

FIGURE 6-17 Stratovolcano

Mt. Hood east of Portland, Oregon, is a stratovolcano.

75 and 200 km inland, trends parallel to the oceanic trench that is swallowing the ocean floor.

Many large stratovolcanoes begin growing with a large basalt shield and then build a tall andesite cone on that base. Finally, they may erupt dacite or rhyolite, which may end in a massive explosion that destroys their andesite cone. In some of those cases, renewed activity builds a new andesite cone in the ruins of the old one. The eruptive behavior and intervals between eruptions vary widely. Mt. St. Helens has erupted at least 14 times in the last 4000 years. Mt. Lassen, in northern California, remained almost dormant for 27,000 years before a moderate eruption about 1000 years ago and another in 1914. An absence of activity for tens of thousands of years does not indicate that a volcano is no longer active.

Lava Domes

Lava domes are rhyolitic volcanoes characterized by their small to moderate size, high magma viscosity, steep flanks, and low to moderate volatile content. They generally form at the crest or on the flanks of stratovolcanoes. Rhyolite to dacite magmas sometimes erupt with little steam. They emerge slowly and quietly expand over months or years, like a giant spring mushroom, to make a small mountain. As the eruption proceeds, the lava on the outside of the dome solidifies while the still-molten, but extremely viscous, lava within continues to rise. The solid rock on the outside cracks off in pieces and tumbles down the side of the growing dome, making steep talus slopes of angular rubble. The typical result is a steep mountain so cloaked in sliding rubble that solid rock is exposed only on the summit.

A single lava dome may erupt only once, though it may be replaced by another dome as magma below continues to rise. Collapse of a big bulge growing on the flank of Mt. St. Helens in 1980 released pressure on the underlying magma. That was the final trigger for its catastrophic eruption. Sometimes magma that solidifies in the throat of a stratovolcano is slowly extruded as magma below continues to rise; that extruded magma may boil out and pour down the volcano flanks as a hot pyroclastic flow.

Giant Continental Calderas

Giant **continental calderas** are rhyolite volcanoes characterized by high-viscosity and high volatile content but gently sloping flanks due to a predominant ash content spread over large areas. Few of the millions of people who visit Yellowstone National Park realize they are on one of the world's largest volcanoes. The Yellowstone volcano is a typical giant rhyolite caldera volcano (**Case in Point:** Future Eruptions of a Giant Caldera Volcano—Yellowstone Volcano, Wyoming, p. 142).

These giant volcanoes erupt rhyolite, typically in enormous volume, most of it explosively. Their pyroclastic flows cover tens of thousands of square kilometers, and sheets of airborne ash cover millions of square kilometers. The volumes of magma involved are typically in the range of hundreds to more than a thousand cubic kilometers.

As an emptying magma chamber withdraws support from the ground above, the surface collapses to open a broad caldera (**FIGURE 6-18**). Many large eruptions from dacite magma chambers open calderas up to 25 km across. Eruptions from giant rhyolite magma chambers commonly open calderas 50 or more kilometers across. These eruptions generally fill the sinking caldera with enough rhyolite ash to ensure that only a subdued depression remains in the landscape.

As magma continues to rise beneath the filled caldera, it slowly raises a **resurgent dome** in its surface. The dome may become as large as a small mountain, and it may or may not develop into a new eruption, perhaps after hundreds of thousands of years. Some ancient rhyolite calderas still display an obvious resurgent dome. The resurgent dome of Long Valley Caldera is still rising, 0.75 m in the past 33 years; bulging between 1982 and 1999 was likely caused by magma injection. Whether that foretells a new eruption is presently unknown.

Some giant rhyolite calderas erupt several times at intervals of hundreds of thousands of years. It is fortunate that their eruptions are so infrequent, because their pyroclastic flows would likely incinerate everything in valleys for 100 or more kilometers from the caldera. They would also almost certainly inject enough ash into the upper atmosphere to cause drastic climate change over a large region for years afterward.

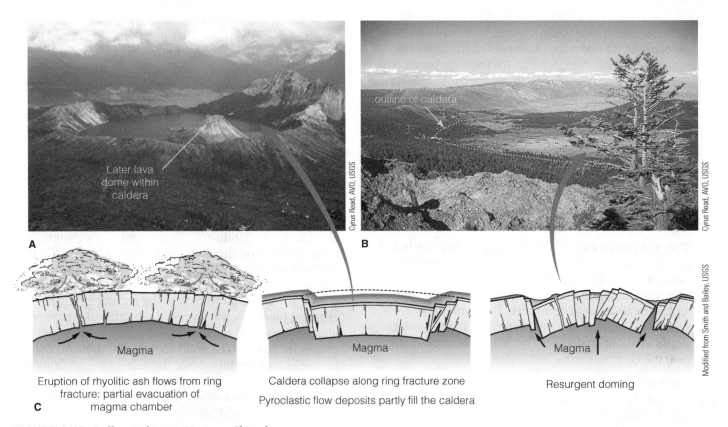

A. A small (2.5 km diameter) caldera in Kaguyak volcano on Katmai National Park, Alaska. A prominent lava dome rose in the caldera after collapse. **B.** The 32-km-diameter Long Valley Caldera of southeastern California, as seen from rim to rim. Its tree-covered resurgent dome is behind the tree on the right. **C.** The ground above an erupting rhyolite magma chamber subsides to make a caldera during the eruption of a giant pyroclastic flow; it then domes, or resurges, again as new magma refills the magma chamber.

FIGURE 6-18 Collapse into a Magma Chamber

Cases in Point

Deadly Plinian Eruption and Lahar
Mt. Pinatubo, Philippines, 1991 ▶

Mt. Pinatubo is an andesitic volcano 1745 m high on the Philippine island of Luzon, some 90 km northwest of Manila. On April 2, 1991, steam explosions suddenly piled ash on Pinatubo's upper slopes. This surprised and frightened the people who lived on the flanks of the mountain, the rice farmers on the plains below, and the 300,000 people 25 km from the crater in Angeles City. Pinatubo had not erupted in more than 400 years, and few people were aware that it was an active volcano (in the sense that it has erupted repeatedly over thousands of years and will undoubtedly do so again).

Philippine volcanologists, with the help of scientists from the U.S. Geological Survey, installed portable seismographs on and around the mountain to monitor its earthquake activity. Geologic mapping soon showed that an eruption 600 years ago had spread hot pyroclastic flows across densely populated areas south and east of the summit and over the site of Clark Air Base. Those pyroclastic flows reached 20 km east of the crater. Deep deposits of ash and widespread mudflows reached much farther.

At first, the blasts of steam and ash seemed fairly harmless to the volcanologists at the site because they contained only fragments of old altered rock, no freshly solidified ash from new magma. But then they saw the frequency of earthquakes increase and their origins migrate from deep below the north side of the volcano to shallow levels near the summit. By June 5, the numbers of small earthquakes and volumes of sulfur dioxide emissions had increased dramatically. Occasional pyroclastic flows swept down valleys. The volcanologists thought a major eruption could happen within two weeks. That led them to recommend evacuation of the area within 10 km of the summit. The volcanologists worked closely with public officials, carefully explaining the looming dangers. Then the officials went to great lengths to educate the public.

A viscous lava dome began growing, and many small earthquakes and harmonic tremors suggested that magma was moving at depth. Continuous ash eruptions accompanied expansion of the lava dome. On June 12, authorities evacuated people to a radius of 30 km from the crater.

The climactic event, a classic Plinian eruption (VEI: 6), finally began early on June 12 with a blast and a huge plume of steam and ash. The eruption became continuous by early afternoon. It climaxed on June 15 when the eruption cloud towered to a height of 35 to 40 km. Pyroclastic flows reached 16 km from the old summit.

Ash was as much as 30 cm thick at a distance of 40 km from the volcano. Then, in an unfortunate twist of fate, the climactic Plinian eruption coincided with passage of Typhoon Yunya, which brought intense rains. Heavy loads of wet air-fall

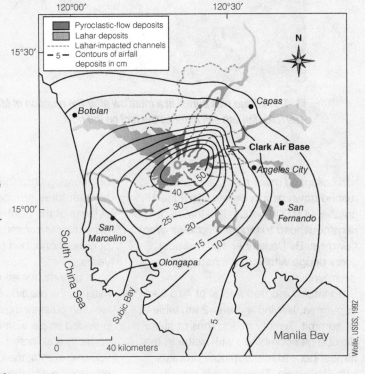

■ *The climactic eruption of Mt. Pinatubo dwarfs both the mountain and surrounding communities.*

USGS

Mountain hidden in clouds

■ *This village was buried in a mudflow after the eruption of Mt. Pinatubo in 1991. Air-fall ash still mantles the roof on the right.*

ash collapsed many roofs, and lahars rushed down nearby valleys. By June 16, another 200,000 people fled the mud. The rains continued to trigger mudflows every few days. By December 1991, almost every bridge within 30 km had been destroyed.

Pinatubo erupted a total of 4 to 5 km³ of magma, leaving a crater 2 km wide at its summit. Twenty million tons of sulfur dioxide gas combined with water in the atmosphere to make minute droplets of sulfuric acid. They hung in the air, encircling Earth and reflecting 2 to 4%

of incoming ultraviolet radiation. Mean temperatures dropped as much as 1°C in parts of the northern hemisphere. Spectacular sunsets with broad streaks of green continued for another two years.

In general, the efforts to predict and mitigate the hazards of a large eruption were an outstanding success. The volcano provided ample warning, the geologists correctly anticipated most of the major volcanic events, the local officials efficiently managed evacuations, and the local people cooperated.

■ *An air view over the Abacan River shows a bridge collapsed on August 12 by mudflows in Angeles City, the Philippines, near Clark Air Base. In the lower left, people cross the river on temporary footbridges.*

In spite of a major program to educate everyone and evacuate 58,000 people, some 350 people died, mostly when heavy wet ash collapsed buildings. Another 932 died later from disease. The large death toll notwithstanding, there is no doubt that the timely warnings and broad evacuations saved tens of thousands of lives.

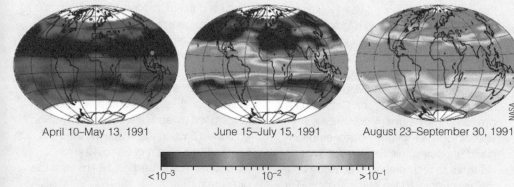

| April 10–May 13, 1991 | June 15–July 15, 1991 | August 23–September 30, 1991 |

$<10^{-3}$ 10^{-2} $>10^{-1}$

■ *Dust and SO$_2$ generated by the Pinatubo eruption encircled Earth, scattering incoming sunlight and reducing global temperature. Pinatubo is at the red dot in left view.*

CRITICAL THINKING QUESTIONS

1. Why did Pinatubo have a significant effect on climate relative to other volcanic eruptions?
2. Which volcanic hazard that is likely to kill significant numbers of people reaches farthest from the volcano and which types of locations are most at risk?

A Long History of Caldera Eruptions
Santorini, Greece ▶

Santorini is an island volcano in the eastern Mediterranean Sea, south of mainland Greece. It now appears as a ring of islands, high places along the rim of an otherwise submerged caldera 6 km across. Santorini is one of the most spectacular caldera volcanoes on Earth, similar in origin and size to Crater Lake in Oregon.

Santorini has staged 12 major explosive eruptions during the last 360,000 years, with a relatively long recurrence interval averaging one every 30,000 years. After each caldera collapse, a new andesite volcano grew within the old caldera until it sank into a new caldera during the next catastrophic eruption. Caldera eruptions happened 180,000, 70,000, 21,000, and 3600 years ago. The intervals between eruptions become progressively shorter with time, much less than half the preceding interval.

In approximately 1620 BC, a series of catastrophic Plinian eruptions of rhyolite ash and pumice erupted about 54.5 km³

of magma, evacuated the huge magma chamber, and culminated in the most recent caldera collapse. The volume was about the size of the 1815 eruption of Tambora volcano in Indonesia, the largest eruption in historic time. The remnant flanks of the volcano slope gently outward from the much steeper cliffs that face into the caldera. Thera, the main town, is on the rim and its adjacent steep cliffs. Many of the homes and tourist hotels have rooms excavated from the caldera wall into the deep ash that fell in 1620 BC.

Events during the initial stages of collapse included eruption of many meters of white rhyolite pumice. Its white lower part grades upward to dark gray andesite ash, presumably because the eruption was tapping progressively deeper levels of a differentiated magma chamber. A similarly graded pyroclastic flow exists around Crater Lake, Oregon.

Many of the rhyolite ash deposits contain inclusions of chilled basalt

magma and compositionally banded pumice, indicating that basaltic magma injected the rhyolite magma chamber from below. That would cause rapid boiling and drive intense lateral blasts and a major Plinian eruption.

At the time of the eruption in 1620 BC, the wealthy town of Akrotiri, on the lower southern flank of the volcano, had paved streets, underground sewers lined with stone, beautiful wall paintings, decorated ceramics, and attractive jewelry. The people raised sheep and pigs, farmed using surprisingly modern techniques, made barley bread and wine, gathered honey, and imported olives and nuts.

When the volcano erupted, seawater pouring into the collapsing caldera probably caused tremendous steam explosions. Some estimates suggest that the eruption raised a plume of ash 36 km into the atmosphere. It may have lasted for weeks.

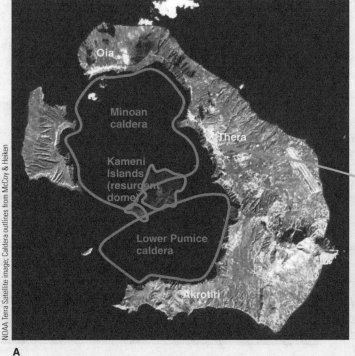

■ A. Santorini's main islands surround the caldera that collapsed during the great eruption of 1620 BC. The eruption ended the Minoan civilization. **B.** The inside wall of Santorini caldera exposes white Minoan pumice at the top right. White houses along the lower caldera rim are built mostly into the same pumice erupted in 1620 BC.

A

B

Most of the population of Akrotiri must have been sufficiently frightened to evacuate with their possessions before the main eruptions. Those who remained were buried under several meters of deposits dropped from the Plinian column, and another 56 m of pumice and ash erupted during caldera collapse. Earthquakes, ash falls, mudflows, and debris flows completed the destruction of buildings. One popular theory suggests that this eruption was the source of the Atlantis legend, in which earthquakes and floods accompanied the sudden disappearance of an island empire.

Renewed volcanic activity began in 197 BC and continued sporadically until 1950. Resurgent domes of andesite and dacite rose in the floor of the caldera. Earthquakes took a further toll in 1570 and 1672. In 1956, an earthquake of magnitude 7.8 wrecked Thera and raised a giant tsunami wave that rose along the shores to heights between 25 and 40 m. Hazard mitigation for Santorini since then involves restrictions on building on steep slopes of loose pumice in the caldera wall and monitoring of minor earthquakes and volcanic gases.

Future Eruptions of a Giant Caldera Volcano
Yellowstone Volcano, Wyoming ▶

A rhyolite giant over a continental hotspot, Yellowstone is a typical large resurgent caldera with a nearly flat summit and gently sloping sides, and it is one of the largest continental volcanoes on Earth, sometimes termed a supervolcano. The Yellowstone volcano we visit today is the relic of three monstrous eruptions that occurred 2 million, 1.3 million, and 642,000 years ago, so the recurrence interval is, crudely, 700,000 years. Smaller eruptions occurred between the

Faint layers in rhyolite ash

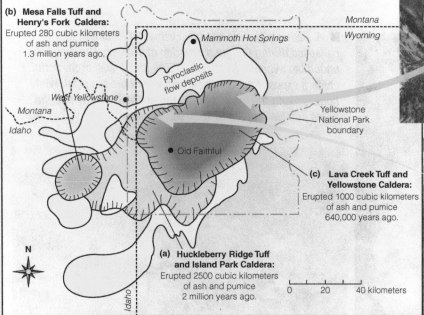

(b) Mesa Falls Tuff and Henry's Fork Caldera: Erupted 280 cubic kilometers of ash and pumice 1.3 million years ago.

Pyroclastic flow deposits

Montana
Wyoming

Mammoth Hot Springs

West Yellowstone

Montana
Idaho

Old Faithful

Yellowstone National Park boundary

(c) Lava Creek Tuff and Yellowstone Caldera: Erupted 1000 cubic kilometers of ash and pumice 640,000 years ago.

N

Idaho

(a) Huckleberry Ridge Tuff and Island Park Caldera: Erupted 2500 cubic kilometers of ash and pumice 2 million years ago.

0   20   40 kilometers

USGS

Caldera floor dropped during eruption

Robert B. Smith

Donald Hyndman

■ *The Lava Creek ash erupted 642,000 years ago during collapse of the immense Yellowstone Caldera. The northwestern rim of Yellowstone Caldera formed during that collapse. Yellowstone Canyon has eroded down through 365 m (1200 feet) of the rhyolite ash that erupted and filled the Yellowstone Caldera.*

■ *The Lava Creek ash from the Yellowstone Caldera, at the northeastern end of the Snake River Plain, covered most of the central plains of the United States. The hotspot track is marked by the white dotted line.*

large ones. Each of the three left calderas approximately 50 km across. Because the caldera collapsed into the magma chamber below, that magma body must have been at least 50 km across.

Resurgent caldera eruptions are by far the largest and presumably most destructive of all types of volcanic eruptions, often called supervolcano eruptions. No eruption even remotely comparable to the most recent eruption of the Yellowstone resurgent caldera has happened in historic time, perhaps not since the appearance of modern human beings. The first of those great eruptions produced 2500 times the volume of magma erupted by Mt. St. Helens in 1980.

Much of the ash discharged in gigantic rhyolite pyroclastic flows that reached more than 100 km down adjacent valleys, filling most of them to the brim. They were hot

enough when they finally settled to weld themselves into solid sheets of rock tens of meters thick. Airborne ash drifted east and south on the wind and settled on the High Plains. Ash from the last eruption covered the High Plains as far east as Kansas and at least as far south as the Mexican border. A large proportion of the North American wheat crop grows on soil developed in Yellowstone rhyolite ash.

If such an eruption were to happen today, it would be one of the most cataclysmic natural disasters in human history. The destruction or disruption of transportation, communication, and energy systems in the western and central United States would be major. Estimates suggest deposition of 1 m of ash over western Wyoming and southeastern Idaho and 0.3 m (1 ft) over eastern Wyoming, northwestern Colorado, and most of

Utah. About 0.03 m (1 in) would cover the western states, from San Francisco to Phoenix to much of Nebraska and Kansas. Somewhat surprisingly, the latest calculations from the USGS suggest that the thickest ash from such a giant eruption would not drift downwind, as it does in a small eruption, but would gradually thin outward in all directions from the caldera. It would form a huge umbrella cloud that would settle out without much influence from the wind. Those impacts would hardly matter relative to the serious consequences that would almost certainly follow from enormous volumes of rhyolite ash injected into the upper atmosphere. The repercussions would probably resemble those that followed the 1815 eruption of Mt. Tambora in Indonesia but on a vastly larger scale. All that ash would probably block enough of the sun's radiation to cause much colder climates within a few short weeks, leading to a worldwide agricultural disaster and probably famine.

Will Yellowstone erupt again? Absolutely—but when is less clear. All of the steam and boiling water in Yellowstone National Park is clear evidence of tremendous amounts of heat at shallow depth, but they are not indications of future eruptive hazards. The hot water originates as surface water percolates down into hot rocks below the Yellowstone Caldera. Heating of the water forms the hot springs, and intermittent boiling activates geysers. It is more ominous that a broad fringe of dead trees surrounds most of the thermal areas. This can mean either that the thermal areas are growing larger, overwhelming trees that were thriving just a few years ago, or that there is greater circulation of carbon dioxide gas up to the roots of the trees. Temperature measurements taken since the 1870s show that the thermal areas are also getting hotter.

Detailed studies of seismograph records show that the shear waves of earthquakes that pass beneath the Yellowstone resurgent caldera arrive at the seismograph later and weaker than they would in other areas. They clearly show that molten magma exists from a depth of 8 to 16 km or so under the caldera. The swarms of mini earthquakes that frequently rattle small areas in the resurgent caldera include harmonic tremors, probably an indication that the magma is moving. Two resurgent bulges in the floor of the caldera rise and fall over time, presumably because magma is rising below them. Between 1923 and 1984, it rose about 101 cm, then fell 20 cm until 1995. Between 2004 and 2007, precise GPS studies show a more rapid 21-cm rise of the caldera floor. These ups and downs are typical symptoms of a volcano getting ready to erupt. However, no one knows whether they also portend eruptions of such supervolcanoes. A 2014 study by Robert Smith and Jamie Farrell suggests that the percentage of melt in the magma chamber is only about 7 to 14%. Because shear or S waves cannot pass through liquids, researchers studied the speeds and behavior of different earthquake waves as they pass through liquid and solid parts of the magma chamber to determine the amount of melt. Unless that melt part could move to concentrate in the upper part of the magma chamber or new molten basalt magma could rise into the chamber to produce more rhyolite melt, a very large eruption of magma seems unlikely to happen any time soon.

What other warnings might we get? Because there has been no other eruption of this type or this size in historic time, we really don't know. However, smaller volcanoes generally show distinctive "harmonic" earthquakes from moving magma, earthquakes that become more frequent and closer to the surface—increasing surface temperature as the magma heats the rocks above—and broad-scale bulging of Earth's surface as the magma pushes up. This bulging stretches the crust near the surface, causing fractures likely to release water vapor, carbon dioxide, and gases richer in sulfur. These and other precursors are reviewed in Chapter 7, in the section on Eruption Warnings. As gases are released from the magma, pressure on it drops so more gases are released. The runaway reaction to these changes accelerates so that the gases dissolved in the rhyolite magma cannot diffuse fast enough. They expand within the magma violently so the magma itself froths out explosively. Because the rhyolite magma is so viscous, none of these changes are likely to happen very fast. Unfortunately, we cannot predict how fast they will happen. The enormous size of the volcano adds another unknown dimension. As the magma rises from below, the surface rocks settle into the space they leave.

An eruption equaling any of those of the past 2 million years would rise to almost 30 km, nearly three times the flight height of commercial jet aircaft. It would expel sulfur that would mix with water vapor to form sulfate aerosols in quantities that reflect sunlight and dwarf the effects of much smaller volcanoes that dramatically cooled Earth's atmosphere for more than a year and caused widespread famine. Toba, a supervolcano that erupted about 74,000 years ago on northern Sumatra, annihilated much of the human population on Earth, through such a "volcanic winter," in this case one lasting 6 to 10 years. The eruption was about three times the size of the Yellowstone eruption 642,000 years ago.

A similar-size future eruption of Yellowstone would likely cause a volcanic winter that would last for years, causing extensive crop failures and probably worldwide famine, resulting in millions of deaths. Studies indicate that rainfall worldwide would drop by 50% because of decreased evaporation from the ocean.

CRITICAL THINKING QUESTIONS

1. What evidence is there that Yellowstone Park is a volcano?
2. Discuss several different types of evidence that lead one to believe that Yellowstone will erupt again in the future. What evidence is there for how soon that could happen?
3. What signs might we expect to see if an eruption is going to begin soon?

Key Points

Introduction to Volcanoes: Generation of Magmas

■ A hot rock deep within Earth may melt by increased temperature, decreased pressure, or addition of water. **By the Numbers 6-1.**

■ The violence of a volcanic eruption depends on the magma's viscosity, volatiles, and volume.

■ The viscosity of a magma is largely controlled by the silica content, with high-silica magmas (rhyolite) having higher viscosity than low-silica magmas (basalt).

■ A volcanic eruption is likely to be more explosive for magmas with higher viscosity and larger quantities of volatiles, especially water. **TABLE 6-1.**

Tectonic Environments of Volcanoes

■ The tectonic environment dictates the volcano distribution, type, composition, and behavior.

■ Most hazardous volcanoes are near subduction zones, and most of the remainder occur at spreading centers.

■ Volcanoes that are not near plate boundaries are generally situated over hotspots.

Volcanic Eruptions and Products

■ Basaltic magma commonly produces nonexplosive eruptions, spilling out in the form of lava. Types of lava include ropy pahoehoe and rubbly aa. **FIGURE 6-7.**

■ Explosive eruptions produce pyroclastic material, solidified magma in the form of ash. Ash may rain down or be carried by the wind, or it may flow down a volcano flank in the form of a pyroclastic flow. Hot ash may combine with rain or melting snow to produce a lahar, or mudflow. **FIGURE 6-8.**

■ The size of an explosive eruption depends on the amount of magma, the magma viscosity, and its water-vapor content.

Types of Volcanoes

■ Shield volcanoes are characterized by gently sloping sides and are typically segmented into rift zones. **FIGURES 6-10** and **6-11.**

■ Cinder cones are characterized by their small size and steep sides. Erupting cinder cones produce glowing fragments of cinders that rarely cause serious injury. **FIGURES 6-14** to **6-16.**

■ Stratovolcanoes have the classic volcano shape and moderate to high volatile content. **FIGURE 6-19.**

■ As a lava dome rises and expands, lava fragments tumble down its sides while molten lava continues to rise within the dome. If the dome collapses, it may release a pyroclastic flow that can be extremely dangerous.

■ Continental calderas are formed when the roof over a giant magma chamber collapses. Their infrequent eruptions produce huge volumes of pyroclastic material and are extremely destructive. **FIGURE 6-18.**

Key Terms

aa, p. 130

andesite, p. 125

basalt, p. 124

caldera, p. 132

cinder cone, p. 136

continental caldera, p. 138

crater, p. 129

flood basalt, p. 127

hotspot volcano, p. 128

lahar, p. 131

lava, p. 123

lava dome, p. 137

lava flow, p. 129

magma, p. 123

magma chamber, p. 124

melting temperature, p. 123

pahoehoe, p. 129

peléan eruption, p. 132

phreatic eruptions, p. 131

Questions for Review

1. What can happen to heat, pressure, and water content to melt rock and create magma?

2. What factors control the violence or style of an eruption?

3. What properties of basalt magma control its eruptive behavior?

4. What properties of rhyolite magma control its eruptive behavior?

5. What drives an explosive eruption?

6. How does pahoehoe lava differ from aa lava?

7. On a huge shield volcano, such as Mauna Loa, what is the main type of eruptive site? Where on the volcano is (or are) such a site (or sites)?

8. Yellowstone Park has two huge calderas, each more than 20 km across. How do such calderas form?

9. How is a caldera different from a crater?

10. Why do shield volcanoes have such a different shape than stratovolcanoes?

11. What is the driving force behind the explosive activity of a cinder cone? Where does it come from?

12. How does dacite or rhyolite magma form in a line of volcanoes, such as the Cascades?

13. Why do the Hawaiian Islands form a chain of volcanoes?

14. On what types of plate boundaries are volcanoes typically found? Explain how each of these tectonic environments give rise to volcanoes.

Critical Thinking Questions

1. If Yellowstone volcano were to show signs of a pending eruption and you were placed in charge of evacuation, what would be the most important hazards and what would you do?

2. The Three Sisters cluster of steep-sided andesitic volcanoes, just west of Bend, Oregon, showed signs of bulging a few years ago. If that bulging were to accelerate, along with other signs of impending activity, and you were placed in charge of evacuation, what would be the most important hazards and what would you do?

3. Long Valley Caldera in southeastern California, a few years ago, showed activity generally associated with rising magma—increasing slow rise of its resurgent dome, earthquake activity, and carbon dioxide concentrations that killed areas of trees. Should the hazard zones nearby be zoned to restrict development of new subdivisions? Why?

4. Compare the nature of eruption forecasts and associated response activities to those for earthquakes. How are they different and why?

■ As Mt. St. Helens erupts, the growing bulge above the magma chamber collapses in a landslide, permitting its gases to expand explosively. A plane by chance passes close overhead—note the wing in the upper right corner of the photo.

Volcanoes: Hazards and Mitigation

7

Mt. St. Helens Erupts

May 18, 1980, was a brilliant late spring day throughout the Pacific Northwest. Early that morning, a few people were camping or logging on forestland in the restricted zone north and west of Mt. St. Helens in southwestern Washington. Some had sneaked around official safety barriers erected in response to two months of minor eruptions. Some of the interlopers wanted to watch St. Helens from nearby. They hoped to see a major eruption. A bulge high on the north flank of the volcano had been growing for weeks, showing that a large mass of magma was rising below the mountain (**Case in Point:** Volcanic Precursors—Mt. St. Helens Eruption, Washington, 1980, p. 170). At 8:32 a.m., Dr. David Johnston, the lone volcanologist from the U.S. Geological Survey (USGS) who was stationed on a high ridge facing the volcano at the time, radioed the Cascades Volcano Observatory in Vancouver, Washington: "Vancouver, Vancouver—this is it!" He died in the eruption. The night before, he had reluctantly agreed to replace someone else at that post.

Ironically, he had been especially concerned about the hazards of an impending eruption, repeatedly referring to St. Helens as "a dynamite keg with the fuse lit."

P. and C. Hickson were 17 km northeast of the crater when they saw the north side of the volcano suddenly look "fuzzy." It began to slide, the lower part faster; a dense black cloud blossomed from the summit and the north flank seemed to explode. Others nearby saw the horizontal blast and a shock wave racing ahead of the cloud. It looked like photos of nuclear explosions they had seen.

J. Downing, on the flank of Mt. Rainier 75 km to the north, saw two distinct "flows" 300 to 600 m thick—they hugged the ground, disappeared into valleys, and then "hopped" over ridges. These were pyroclastic flows, masses of steam dark with suspended ash that made them so dense they stayed on the ground.

C. McNerney was 13 km northwest of the crater, within the area doomed to imminent devastation, when he watched the north side of the volcano collapse. The leading wall of the black cloud climbed over a ridge, and a hot wind began to blow from the volcano. Two minutes after the eruption began, he started driving west at 120 km/h, but the black cloud gained on him, so he sped up to 140 km/h. The base of the black pyroclastic flow advanced "like avalanches of black chalk dust," one after another.

G. and K. Baker were 17 km northwest of the crater, also within the area destined for devastation. They saw a "big, black, inky waterfall" a few miles up the valley and began driving west on Highway 50 at 160 km/h. Even so, the black cloud almost reached them within four minutes. The cloud looked like it might be boiling oil with bubbles 2 m in diameter. It was the same pyroclastic flow that others northwest of the volcano were fleeing.

B. Cole and three others were logging 20 km northwest of the crater when the pyroclastic flow reached them. A "horrible crashing, crunching, grinding sound" came from the east. The air around them became totally dark and intensely hot. Cole and his companions gasped to breathe. Their mouths and throats burned, and they were knocked down along with the trees. Everything was covered by a foot of gray ash. The heat burned large parts of their bodies but not their clothing. Cole's three companions later died.

D. and L. Davis and A. Brooks were 19 km north of the crater. They watched the black, boiling cloud of the pyroclastic flows bear down on them. Its leading edge snatched trees out of the ground and tossed them in the air. Then the black cloud swallowed them, and everything went pitch-black and burning hot. A physician later said their burns were similar to those caused by a microwave.

Keith and Dorothy Stoffel were ready to drive home after attending a geoscience conference in Yakima, southeast of St. Helens, when they decided to hire a pilot to fly them around for a quick look at the mountain. As the plane rounded the north flank of the volcano, they saw that the snow had melted off the bulge overnight. Then they watched the bulge detach in a great landslide while an enormous cloud of steam, black with ash, spouted behind the slide and blossomed to fill the sky (see chapter opener photo). The pilot banked out of range and headed for Spokane.

Volcanic Hazards

In Chapter 6, we discussed the geologic processes that produce volcanoes and the products of eruptions. Here we describe how volcanoes impact people and what measures can be taken to mitigate volcanic hazards. Millions around the world live in risk of volcanic hazards. A history of volcanic activity gives an area rich and fertile soil, ideal for agriculture. And landscapes of soaring, majestic mountain peaks make desirable spots for homes with sweeping vistas. As a result, a number of densely populated urban areas around the world exist in the shadow of active volcanoes. The main volcanic hazard areas of the United States are shown in **FIGURE 7-1**.

Although few volcanic eruptions kill more than a few hundred people, those few can produce massive casualties. Deaths from eruptions depend heavily on the numbers of people living in close proximity to a volcano and on the eruption product. An assessment by the USGS of deadly volcanic events worldwide showed that pyroclastic flows killed about 70,800 in 19 major eruptions since 1631, 41% of them at Mt. Pelée in 1902. Lahars killed 51,300 at 11 volcanoes in the same period, 44% of them at Nevado del Ruiz in 1985. Ash falls killed relatively few, including 300 at Mt. Pinatubo in 1991. Tsunami initiated by eruptions killed about 50,900, 62% of them at Krakatau in 1883. In some cases, the high casualties were a result of misjudgments or politically motivated decisions by authorities.

FIGURE 7-1 Volcanic Hazards in the United States

Red triangles are volcanoes. Dark orange is high hazard; lighter orange is lower hazard. Dark gray area has higher ash-fall hazard; light-gray area has lower ash-fall hazard.

Lava Flows

An erupting volcano can send molten rock, or lava, flowing or spewing forth. Even well downslope from vents, where the lava exterior is no longer red, it is still hot enough to ignite wooden structures (**FIGURE 7-2**). Even if an object does not burn, it is still overwhelmed and often buried by the flow. Where basalt lava flows surround a green tree, its heat boils moisture in the trunk and chills the lava next to the tree. The woody tissue either chars or later rots away, leaving a cast of the tree trunk. Excellent examples of such

FIGURE 7-2 Fire Ignited by Lava

Dark, but still hot, pahoehoe basalt lava from Kilauea torches a home.

casts can be found on Kilauea Volcano and at Lava Cast Forest, south of Bend, Oregon.

Although lava flows are incredibly destructive to anything in their path, they move so slowly and cover such a small area that they generally pose little threat to human life (**Case in Point: Kilauea's East Rift Eruptions Continue**—Kilauea, Hawaii, 1983–2015, p. 175). Basaltic lava flows commonly advance at speeds from 1 m/s near the vent to less than one-tenth that as the lava cools. Most travel up to 25 km from the vent, but some may reach more than 50. Andesitic lava flows are more viscous and travel shorter distances. In a few cases, however, lava can flow rapidly. The especially fluid alkali-rich lava flow from Vesuvius killed 3000 people in 1631. Even more extreme was lava that erupted from Vesuvius in 1805; it traveled from the crater to the base of the mountain in four minutes, twice the speed of an Olympic sprinter. Although basaltic magma typically forms lava flows, it may occasionally contain enough gas to blast out a Plinian eruption of hot fragments. Mt. Etna in southern Italy occasionally erupts that way.

Pyroclastic Flows and Surges

A **pyroclastic flow** is a mixture of hot volcanic ash and steam that pours downslope because it is too dense to rise (**FIGURE 7-3**). A pyroclastic flow can also be referred to as an ash flow, *nuée ardente,* glowing avalanche, or ignimbrite. Many flows develop when rapidly erupting steam carries a large volume of ash in a column that rises high above the volcano. When the rush of steam slows, part of the column collapses, and the cloud of ash pours down the flank of the volcano. The main flow hugs the ground, its less dense part billowing above as loose ash is stirred into the turbulent air. Fast-moving flows tend to hug valley bottoms, but their high velocity can carry them over intervening hills and ridges.

Glowing hot pyroclastic flows can race down the flank of a volcano at speeds from 50 to more than 200 km/h, incinerating everything flammable in their path, including forests and people (**FIGURE 7-4**). People engulfed in a pyroclastic flow face certain death unless they are near its outer fringes, preferably in a building or vehicle. Even locations many kilometers from the base of a volcano may not be safe.

In some cases, high-speed ash-rich shock waves called **surges** may race across the ground ahead of a pyroclastic flow. They commonly originate as lateral blasts of ash and steam in the first stage of an ash-flow eruption. Because of their larger proportion of steam, they are generally less dense than standard pyroclastic flows. Some surges hug the ground and travel at speeds up to 600 km/h and may cover hundreds of square kilometers. They flatten forests and kill almost every living thing they meet, by heat, abrasion, and impact (**FIGURE 7-5**). Surges commonly pick rocks off the ground and carry them along, leaving in their wake deposits of ash mixed with rocks. Surges leave dunes of ash with cross beds much like those in sand dunes.

Both steam and ash in the hottest pyroclastic flows show a dull red heat, between 800°C and 850°C—hot

C. Newhall, USGS

Dr. Paul Cole

FIGURE 7-3 Pyroclastic Flows

A. Several pyroclastic flows pour down the slope of Mayon volcano in the Philippines during its 1984 eruption.

B. Pyroclastic flows race down across abandoned farms on the flank of Soufrière Hills volcano, June 25, 1997.

USGS

FIGURE 7-4 No Match for a Pyroclastic Flow

This car was singed, abraded, and crumpled by a pyroclastic flow from Mt. St. Helens, 11 km north of the crater.

Donald Hyndman

FIGURE 7-5 Evidence of a Surge

The directed blast of the May 1980 eruption of Mt. St. Helens stripped and flattened trees on the windward side of a hill. Where the trees were sheltered from the volcano on the far side, they remained standing, but their tops were snapped off.

enough to glow in the dark. The ash and pumice particles in those flows are still extremely hot and plastic. As they come to rest and begin to cool, they may fuse into a solid mass of hard rocks. The hardened flows form sheets of *welded ash* that may cover hundreds of square kilometers, all deposited during a single eruption, in a matter of hours. Not only does the ash flow kill people caught in its path but also most animals and plants. Within a few months, some animals move in from other areas and new plants begin to grow.

The travel distance of a pyroclastic flow is related to the density and the quantity of the pyroclastic materials produced in a particular eruption. Heavy ash falling from the dense part of an erupting ash column accelerates down a volcano flank due to gravity and picks up speed that can carry it over nearby hills. We can approximate which hills the pyroclastic flow will go over by imagining an *energy line*, from the top of the dense ash column outward toward the ground (**FIGURE 7-6**). For a small ash eruption, the energy-line slope may be as steep as 30°. For the largest Plinian eruptions, the

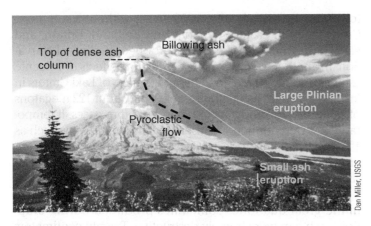

FIGURE 7-6 Energy Line

An energy line sloping from the top of the dense ash column of an eruption can estimate the height of hills that might be overridden by a pyroclastic flow from a stratovolcano eruption. The energy line has a lower slope from larger, more ash-rich eruptions, so their pyroclastic flows reach farther from the vent. The energy-line slopes are variable, depending on factors such as the density of ash in the gas-thrust zone.

slope may be as flat as 12°, so the danger zone extends much farther from the volcano. However, the actual hazard distance may be much greater if the ash is driven by a lateral blast, such as the one that felled the forest at Mt. St. Helens in 1980. Because pyroclastic flows pour downslope, the most dangerous place to be at the time of an eruption is in the bottom of a valley.

Recent research indicates that pyroclastic flows can overtop hills less than 1.5 times the thickness of the flow.

In doing so, the ash will lift into a plume above the hill, descending the lee side in a turbulent flow with back eddies. Because many flows are hundreds of meters thick, even small pyroclastic flows can go over barriers several hundred meters high.

You might imagine yourself safe from a pyroclastic flow if you were watching from the other side of a body of water, but you could be wrong! Pyroclastic flows can actually cross rivers, bays, and lakes. The lower part of the flow is the main mass. Less dense ash billows into the air above the descending flow (**FIGURE 7-7**). Although pyroclastic flows are too dense to rise into the air, only part of the main flow may have particles large and dense enough to sink into water. That part may incorporate loose sediment in its path or trigger underwater sliding. The less dense ash cloud above it may skim across bodies of water, even some that are tens of kilometers across. During the eruption of Vesuvius in AD 79, a glowing pyroclastic flow crossed some 30 km of the Bay of Naples. Similarly, 2000 people died in hot ash flows that crossed more than 40 km of open sea in southeastern Sumatra during the giant eruption of Krakatau in 1883. In 1902, pyroclastic flows raced down the slopes of Mt. Pelée in Martinique and continued offshore to capsize and burn boats anchored in the harbor. A body of water is clearly not an effective safety barrier from an erupting volcano.

Ash and Pumice Falls

Volcanic ash is composed of bits of pumice less than 2 mm across, light enough to drift some distance on the wind (see Figure 6-8B, p.130). Ash erupts into a column that can rise 6 to 20 km in the air. On occasion, a volcano may only blow

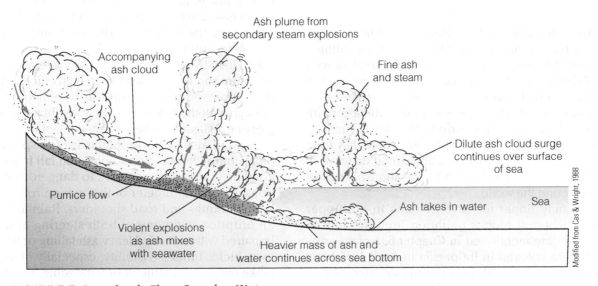

FIGURE 7-7 Pyroclastic Flow Crossing Water

One explanation for the movement of hot pyroclastic flows over water is that some of the dense part of a pumice flow may sink as the still glowing lighter part races over the water surface.

FIGURE 7-8 Dark Ash

Dark clouds of new volcanic ash blast out in the early stages of the June 12, 2009, eruption from Sarychev Peak in the Kuril Islands off eastern Asia. The white ring and cap atop the dark ash probably formed by rapid expansion, cooling, and condensation of water vapor in the eruption cloud.

Ash cloud

Water vapor condenses to form a cloud

Dark lava flows

Eruptive vent

Pyroclastic flow

NASA Astronaut

off steam that rises in a nearly white column. On the other hand, if new magma has reached the surface, signaling a new eruption, the rising eruptive column will be dark with new ash (**FIGURE 7-8**). The ash particles are suspended in a cloud of steam that condenses into water droplets as it expands and cools. Much of the water coats the ash particles, which fall like snow downwind of the vent. The largest, heavier particles fall closer to the vent, where they form dangerous projectiles. Finer particles are carried downwind; especially fine ash is carried high into the atmosphere and may spread around the world.

Some ash may linger for several years in the upper atmosphere, where it blocks radiation from the sun. In one notorious case mentioned in Chapter 6, a giant eruption of Tambora volcano in Indonesia in April 1815 blew an immense amount of ash into the upper atmosphere. For several years the ash remained suspended above the altitudes of weather, where sulfur dioxide and ash from the eruption effectively reflected the incoming solar energy, resulting in cooler temperatures around the planet. The

following year was known as "the year without a summer." Crops failed in eastern Canada, New England, Britain, and elsewhere, and tens of thousands of people died of starvation. In another example of the devastating effects of volcanic ashfalls, the June 1783 flood-basalt eruption of Laki Craters in Iceland lasted for eight months, releasing about 122 megatons of sulfur dioxide and lesser amounts of other gases. It temporarily lowered northern hemisphere temperatures as much as 1°C, caused killing frosts in May and June 1816, shortened the growing season, and contributed to widespread famine. The same rift zone had a disruptive ash eruption in 2010.

Although individual pieces of ash are small and light, in the great quantities sometimes produced during eruptions, ash can be extremely destructive. Twenty centimeters of ash is enough to collapse most roofs; less than that is required in warm regions where roofs are not designed to bear snow loads. Wet ash is much heavier than dry ash. Rain falling during an eruption is common because ash eruptions generate their own **volcanic weather**. The rising hot ash draws in, heats, and lifts the surrounding air, which expands and cools, causing its dissolved water vapor to condense and fall as rain. Heavy loading of ash on roofs, especially if saturated by rainwater, can cause collapses that injure or kill people (**FIGURE 7-9**). The hazard can be especially severe in regions where people flock to their places of worship when an eruption begins.

Heavy ash falls pose several threats to humans, including the inhalation of fine particles. At the time of an eruption, people should move out from under a dense plume of ash by heading perpendicular to the wind direction. Even outside a heavy ash-fall area, ash in the air or on the ground can cause serious health problems. Eye and throat irritation is common. Those with existing respiratory illnesses or damage—including bronchitis, emphysema, asthma, or those who are heavy smokers—face the greatest risk. Although silicosis has been a concern, the Centers for Disease Control do not believe that short-term exposure to volcanic ash presents significant risk to healthy people. Those with long-term or concentrated exposure should wear approved high-efficiency dust masks. If an approved mask is not available, people can breathe through a cloth moistened with water. In fine dust areas, goggles or eyeglasses are better than contact lenses.

Ash fall also poses problems for transportation, potentially complicating evacuations. Ash fall from an eruption, combined with rain, can lead to dangerous driving conditions. Even a millimeter of fine ash on roads can obscure lane markings and road shoulders. Rainfall at the time of an eruption is common; both steering and braking are impaired with wet ash. Heavy ash falling or being stirred up by vehicles hinders visibility, especially at night. This can make evacuation difficult or impossible. Headlights cannot penetrate falling ash; roads and familiar landmarks disappear; ash on wiper blades can scratch windshields and further obscure visibility. Rear-end collisions are common due to very poor visibility.

Ash accumulation on roof

FIGURE 7-9 Heavy Ash

Heavy ash from Rabaul Caldera, in September 1994, collapsed the roof on the left and thickly coated the roof on the right.

Ash in the air can also damage vehicles. It can clog radiators and be drawn through air filters into engines, causing them to seize up. Tiny fragments of rock grains and frothy glass can abrade the surfaces of moving engines, transmissions, and brakes. These particles clog air filters, leading to engine overheating and even failure. Except for conditions with only very light ash, vehicles should be driven only when absolutely necessary; oil and air filters may need changing every 100 to 1000 km.

ASH AND AIRCRAFT Ash can also cause serious problems for aircraft because jet engines can freeze up and stall when they enter an ash cloud. Without power, a jet aircraft flying at an altitude of 7 to 12 km can drop 5 km or more within a few minutes and potentially crash. Heavy air traffic from North America and Europe to eastern Asia—about 20,000 people per day and tens of thousands of flights per year—is of special concern because of the active subduction-zone volcanoes at the north edge of the Pacific tectonic plate.

Planes accidentally strayed into fairly dense clouds of volcanic ash more than 100 times between 1980 and 2004; 7 of those cases resulted in loss of engine power and near crashes. In June 1982, a Boeing 747 with 263 passengers and crew on a flight from Malaysia to Australia flew into an ash cloud erupting from a volcano in Java. All four engines failed. The plane fell from 11,470 m to 4030 m before the crew was able to restart the engines. On December 14, 1989, a new Boeing 747-400, heading from the Netherlands to Japan with 245 passengers and crew, flew into an ash cloud from Redoubt volcano, 160 km southwest of Anchorage, Alaska. The plane lost power in all engines

and dropped from 7500 to 3500 m in 12 minutes before pilots were able to restart the engines. Generators tripped, causing shutdown of airspeed indicators and other cockpit instruments not powered by batteries. The plane managed to reach Anchorage but repairs cost $80 million. In this and other cases, ash entered jet intakes, melted, and coated fuel injectors and turbine vanes. It abraded all forward-facing aircraft surfaces, including cockpit windows, making it difficult for pilots to see the runway during landing.

Monitoring systems installed near some active volcanoes now warn of potential danger, although most Aleutian volcanoes are still unmonitored. Nine Volcanic Ash Advisory Centers gather and convey eruption information to air traffic control centers and aircraft. However, the warning time can be short; ash from a new eruption can reach flying altitudes within five minutes, as was the case with Mt. St. Helens in 1980. Continuous satellite monitoring and pilot reports help, but clouds often obscure the view. Airplanes still fly into clouds of volcanic ash because cockpit radar cannot detect the fine particles. Warning signs for pilots include volcanic dust in the cockpit and cabin, acrid or sulfurous odors, heavy electrical static discharge around the windshield, a bright white flow in the engine exhaust, and engine surge or flameout, all in less than a minute. Pilots flying unexpectedly into a volcanic ash plume are instructed to slow the engines to lower their operating temperatures below the melting point of the ash, and then fly back out of the plume.

In mid-April 2010, an eruption along the northern Mid-Atlantic Ridge in Iceland caused a massive disruption in European air travel. Basalt lava under Eyjafjallajökull glacier in southern Iceland began erupting on March 20 outside the

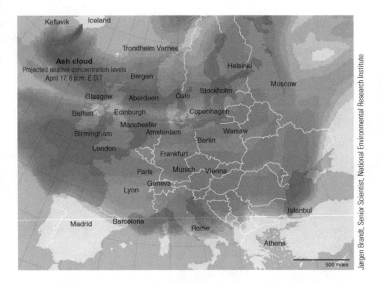

Jørgen Brandt, Senior Scientist, National Environmental Research Institute

FIGURE 7-10 Ash Cloud Over Europe
By April 17 the ash had spread over most of Europe.

T. J. Casadevall, USGS

FIGURE 7-11 Mudflows
Mud lines high on tree trunks show the depth of the Toutle River mudflow on May 18, 1980. The person in yellow on the right is almost 2 m. (6 ft.) tall.

edge of the glacier, but then expanded on April 14, along a 2-km-long north–south series of vents. It began melting ice under the glacier and rapidly chilled the magma. Lava fountains reacted with surface water and drove expanding steam eruptions that blasted fine ash into the air.

The ash cloud followed the jet stream, spreading east and southeast across Europe at about 40 km/h, rising to altitudes of 6 to 9 km (20,000 to 30,000 ft) (**FIGURE 7-10**). It severely disrupted air traffic for six days. Of Europe's 28,000 flights scheduled for Friday, April 16, about 16,000 were grounded, stranding close to 2,000,000 passengers. The disruption cost airlines more than $200 million per day in lost revenue.

Volcanic Mudflows

Volcanic **mudflows** form when ash combines with water, primarily on stratovolcanoes, and pours down their flanks at high speeds with a consistency similar to very wet concrete. Racing down valleys with the velocity of floodwater, mudflows inundate and fill lower valleys where people live (**FIGURES 7-11** and **7-12**). As the mud moves downslope, it gathers rocks of all sizes and carries them along, accelerating as it goes. Many mudflows carry rocks the size of cars. People are often killed by boulders in fast-moving flows.

Mudflows can be triggered by the eruption of a volcano covered in ice or snow that is rapidly melted or mobilized by hot ash. Many volcanologists use the term **lahar** for any volcanic mud or debris flow formed during an eruption. Others include any mudflow full of volcanic ash and rocks. Big blocks of rhyolitic magma buried within a lahar may blow steam for months before they finally cool.

Weather can also be a factor because rain on loose ash washes it downslope as a heavy slurry. Even on an otherwise

clear day, an eruption cloud may create thunderstorms and heavy rain due to the rapidly rising heat plume. Water pouring down the slope of a stratovolcano may develop into a mudflow whether or not the volcano is erupting. Thus mudflows can range from icy to boiling temperatures.

Stratovolcanoes pose an especially great mudflow risk because they are inherently unstable piles of lava and rubble. Many stratovolcanoes emit steam that quietly drifts out of rocks near the summit. In most such cases, the steam is simply meltwater from snow or rain on the high slopes that boils as it sinks into the hot rocks below. Although it probably does not foreshadow an eruption, it does tell us that the interior of

T. Pierson, USGS

FIGURE 7-12 Strong House Buried
Mudflow from Mt. Unzen in Japan completely buried the first story of many houses.

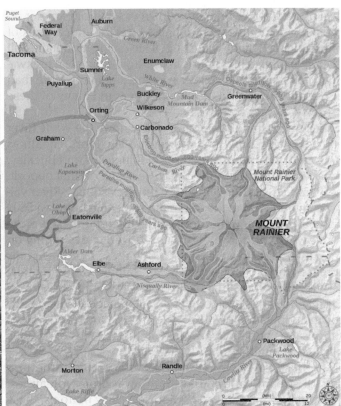

FIGURE 7-13 Development at Risk of Mudflows

A. The town of Orting, pictured here, is one of several communities built on old mudflows. The bottom of the valley—an area with significant mudflow hazard—was extensively developed between 1995 and 2015.

B. The Osceola mudflow inundated areas now occupied by hundreds of thousands of people. The Electron mudflow filled the floor of the valley that Orting now occupies.

the volcano is extremely hot. The rocks in there are stewing in the hot water and steam that fills the fractures.

In one example, Mt. Rainier, in Washington State, poses serious mudflow risks for the rapidly developing communities of the Seattle-Tacoma area (**FIGURE 7-13**). Like many large stratovolcanoes, Mt. Rainier consists of weak material. Hot volcanic gases and groundwater have degraded the edges of fractures in its rocks, and much of its andesite ash and rubble weather into soft clay that becomes even softer when wet. Enormous snow and ice deposits on the higher slopes of the mountain make it even less stable. An eruption, a season of unusually rapid snowmelt, or shaking by a major subduction-zone earthquake could easily mobilize immense volumes of ash and rubble into enormous mudflows.

The historical record of this area provides an indication of future hazards. Every 500 years, on average, large mudflows reach as far as 100 km from Mt. Rainier to cover large parts of the Puget Sound lowland. The Electron mudflow ran 48 km down the Puyallup River valley 500 years ago, then spread onto the Puget Sound lowland, including the area where Orting now stands. In places, the flow is 30 m thick, 60 km from its source on the volcano (**FIGURE 7-14**). Smaller

FIGURE 7-14 Evidence of a Huge Mudflow

This large Douglas fir stump, buried by the 5-m-thick Electron mudflow, was excavated at a new subdivision being built on the surface of the mudflow near Orting, Washington. Mt. Rainier, the source of the mudflow 60 km away, is in the background.

flows of mud and debris surge down valleys from Rainier on occasion, including one on August 14, 2001, after a week of hot weather accelerated melting of ice on the mountain. Still larger events, glacial outburst floods from subglacial meltwater have rushed down valleys from the mountain. More than 35 such floods occurred in the last century; many of these destroyed bridges and roads.

The Osceola mudflow poured 73 km down the White River 5000 years ago, burying the broad valley floor beneath as much as 20 m of mud. The towns of Enumclaw and Buckley, home to tens of thousands of people, stand on that deposit. The Paradise mudflow filled the upper Nisqually River valley south and west of the mountain sometime between 5800 and 6600 years ago. Future years will certainly see more mudflows follow those old routes, covering everything on them. Every new development along those mudflow paths increases the hazard posed by Mt. Rainier. Unfortunately, the flat valley floors that drain from Mt. Rainier are attracting large housing developments to accommodate the rapidly growing population of the Seattle-Tacoma area. Today, tens of thousands of people now live on the surfaces of geologically recent mudflows.

The best way to avoid mudflows is to avoid the bottoms of stream valleys that drain away from volcanoes. Unfortunately, most towns develop in valley bottoms; these are the very places that mudflows follow as they move downslope. Trying to outrun a mudflow almost always results in death. Climbing well up the valley side or even running back into the forest, away from the channel, is a much better idea. Broad floodplains are especially dangerous because the valley sides may be too far away. Indonesia, a country with many volcanoes, has built safety hillocks several meters high where people can hopefully climb above the mudflows. (Additional details on debris flow and mudflow processes are included in Chapters 13 and 14.)

Gas Outbursts and Poisonous Gases

Even if no eruption occurs, gases emitted by volcanoes can pose a hazard to people, animals, and trees. At depths of more than a few kilometers, gases dissolved in the magma are under tremendous pressure. As the magma rises, the pressure drops, which permits gases to *exsolve*, or come out of solution. Thus, an increasing volume of escaping gases commonly precedes—and may warn of—an impending eruption. A sudden and unexpected blast of gas and rocks from Mount Ontake, Japan, on September 27, 2014, killed 47 people. The volcano, 100 km northeast of Nagoya, is a popular hiking destination. Because it was a Saturday, just before noon, and the weather was good, more than 500 people were around the summit. There were no warning earthquakes, and the single tilt sensor on the volcano gave just seven minutes warning before the event, not enough time to evacuate. The blast of low-temperature gas, ash, and boulders up to 0.5 m in diameter lasted for five minutes. Many people were struck by large rocks, succumbed to

volcanic gases, or were buried by the heavy ash. Some survivors took cover behind large rocks or ran to mountain lodges. Searches on the ground and by helicopter continued for several days except when called off during expulsion of dangerous hydrogen sulfide gas. A previous steam blast occurred at the same site in 2007.

Volcanic gases react with sunlight, moisture, and oxygen in the air to produce aerosols—tiny particles and droplets. The volcanic gases and aerosols create an acidic volcanic smog, or **vog**, which can pose a threat to life and health (**FIGURE 7-15**).

Although carbon dioxide (CO_2) is a familiar gas in the air we breathe, it is deadly at high concentrations. Because it is heavier than air, the CO_2 expelled from some volcanoes can pour downslope, where it concentrates in depressions. Because it is colorless and odorless, it can suffocate people and other animals without warning.

A tragic case in 1986 in Cameroon, West Africa, involved CO_2 that bubbled out of Lake Nyos, a volcanic crater lake some 200 m deep. The gas had seeped into and dissolved in the deeper waters of the lake over many years. On the evening of August 21, 1986, the CO_2-saturated deep waters of the lake rumbled loudly and belched an immense volume of the gas. A rainy-season landslide into the lake, pushing some of the deep water upward, may have prompted overturn of the lake waters. A drop in pressure on the dissolved CO_2 would have rapidly released it from solution.

Heavier than air, the gas swept rapidly down through several small villages as a stream 50 m thick and 16 km long. More than 1700 people, some 3000 cattle, and countless small animals died of asphyxiation. An air mixture with 10% CO_2 can kill people; in this case, the concentration was nearly 100%.

There is risk of an even more catastrophic event at Lake Kivu, in the East African Rift Zone, at the eastern border of the Congo. Much larger than Lake Nyos, Kivu is 48 by 89 km

FIGURE 7-15 Dangerous Gases
Bluish vog drifts downwind from Kupaianaha Lava Lake in Hawaii. The lake depression is about 100 m in diameter.

(or 2700 km²); it contains 350 times as much CO_2 and methane below a depth of about 300 m, posing a potentially fatal threat to more than 2 million people living near the lake.

In another example, CO_2 seeps from within the Long Valley Caldera north of Bishop, California. Evidence of the CO_2 concentration includes dying trees in several areas around Mammoth Mountain on the rim of the caldera (**FIGURE 7-16**). On March 11, 1990, Fred Richter, a forest ranger, found refuge from a blizzard in an old cabin surrounded by snowdrifts. Entering the cabin through a hatch, he almost suffocated before fighting his way back up the ladder into fresh air. The denser-than-air CO_2 had collected and concentrated in the nearly sealed cabin. Although people walking through an area are probably not in danger, someone camping in a depression out of the wind or entering a confined space below ground could be asphyxiated. Even the trees, located near caldera faults, often succumb to the CO_2 concentrating around their roots.

In addition to carbon dioxide, another dangerous volcanic gas is sulfur dioxide, a noxious gas with a sharply acrid smell and choking effect. Some eruptions produce large quantities of sulfur dioxide (**FIGURE 7-17)**, which is harmful to animals, even in small concentrations. It reacts with oxygen in the atmosphere to make sulfur trioxide, which reacts with water vapor to make minute droplets of sulfuric acid.

FIGURE 7-17 Monitoring Volcano Emissions

Volcanologists sample gases on Mt. Baker, Washington. They are not wearing masks because they are upwind of the gases. Volcanic gases in a fissure deposit the bright yellow coatings of sulfur.

They hang suspended in the atmosphere for months, partially blocking incoming sunlight and thus cooling the climate.

Some volcanoes also expel poisonous hydrogen sulfide, various chlorine compounds, fluorine, and small quantities of other gases. Most of these emissions have little to do with events during an eruption, though fluorine breaks bonds and can make magmas less viscous. Hydrogen sulfide settles into depressions and can quickly kill people or animals. Fluorine, a neurotoxin, expelled during the immense several-months-long Laki flood basalt eruption of June 1783 in Iceland, appears to have killed about 10,000 people (more than 20% of the population) and many more in Europe. Analysis of their malformed bones suggests that fluorine contaminated food supplies and drinking water. Fluorine compounds in the volcanic haze precipitated on grass contributed to the death of about 75% of their sheep, cattle, and horses.

More than 100 million tons of sulfur dioxide was blown into the atmosphere in the eight-month-long eruption, spreading a dense, choking haze across western Europe. When breathed in, the gas mixes with moisture in the respiratory system to form sulfurous acid, which is irritating and can be deadly. Weather changes weakened monsoons in Africa and India, leading to crop failure, and produced the longest, coldest winter on record in the northeastern United States. Direct and indirect effects, including famine and starvation, killed an estimated 1 million people worldwide. Several other Iceland volcanoes including Hekla and Katla could lead to similar eruptions.

Like other kinds of smog, vog droplets are small enough to be retained in lungs, where they degrade function and compromise immune systems. Especially endangered are children and those with asthma or other respiratory problems. Knowledge of other effects—especially over the long term—is limited, but it is clear that we should avoid breathing

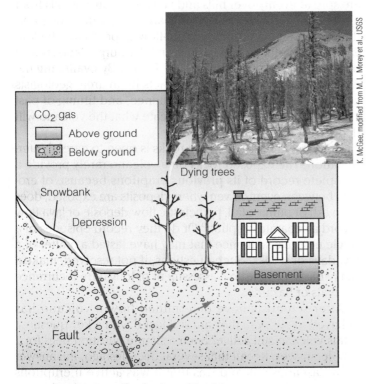

FIGURE 7-16 Deadly Carbon Dioxide

CO_2 gas rises along faults from molten magma at depth and can collect in depressions or confined areas, like the basement of a house. This gas killed these trees on the south side of Mammoth Mountain, California, shown in September 1996.

volcanic smog by staying indoors. If unexpectedly caught in a thick cloud of volcanic smog, breathing through a damp handkerchief may help.

Vog can also produce acid rain. In Hawaii, residents may unintentionally expose themselves to toxic metals when they collect rainwater from roofs for both washing and drinking. The acid rain caused by vog can leach metals from the roof that are then retained in drinking water. This means that people downwind of some volcanic vents unintentionally ingest significant amounts of metals, sometimes including lead, which can cause brain damage and other defects.

Predicting Volcanic Eruptions

As with most hazards, the best protection from volcanic hazards is to predict an occurrence well in advance in order to evacuate a population or take other precautions. When asked to anticipate future events, scientists commonly use historical records to assess the long-term prospects for volcanic activity in a certain location. They also rely on short-term indications to warn of impending eruptions.

Examining Ancient Eruptions

It is not possible in the current state of knowledge to predict exactly when a volcano will erupt—or, in many cases, whether it will erupt at all. But geologists have learned much about the history of many volcanoes. That knowledge makes it possible to assess the likely future behavior of a volcano and to mitigate the hazards it poses to life and property.

A volcano's historical record can help geoscientists understand patterns of recurrence in eruptions. The number of eruptions in the last 100 or even 1000 years may be documented in populated areas with long historical records, such as in Italy or Japan. Elsewhere, such as in the western United States, the historical record may cover only the past century or so. These records mean little if a volcano erupts once or twice in 10,000 years. In such cases, information is provided through **paleovolcanology**, which involves interpreting deposits from prehistoric eruptions and reconstructing a record using age dates on plant material charred in past eruptions or dates on the volcanic rocks themselves (**FIGURE 7-18**). The archives are not in the written record but in the rocks.

In the eruptive products, we may recognize lavas, pyroclastic flows, ash-fall deposits, and mudflows, as well as the magnitude and lateral extent of the events. Ash after deposition forms a rock called **tuff**. Ash-fall material becomes ash-fall tuff; pyroclastic-flow material becomes pyroclastic-flow tuff (also called ash-flow tuff). Deposits of ash-fall tuff can be distinguished from pyroclastic-flow tuff based on characteristics and deposition over the landscape (**FIGURE 7-19**). The most obvious distinction is that ash-fall tuff is generally distinctly layered, whereas pyroclastic-flow tuff is unlayered, at least near the vent. Either may contain lumps of pumice, the frothy rock that is typically light enough to float on water.

FIGURE 7-18 Evidence of Past Eruptions
Charcoal logs are all that remain of the forest that was overwhelmed by the Taupo ash-flow eruption in New Zealand some 2000 years ago. The pyroclastic flow must have been hot enough to burn the logs.

The distribution of deposits over hills and valleys is also an indication of what type of flow occurred. Ash-fall deposits are spread evenly over hills and valleys because it falls from above, like snow. Surge deposits are thicker in valleys and thin on hills, whereas pyroclastic-flow deposits are thick in valleys and virtually nonexistent on hills. Surge deposits also often show cross-bedding (Figure 7-19B). By evaluating the type and extent of volcanic products in an area, geologists can estimate the magnitude, sequence, and timing of past eruptions. Past behavior may indicate what the volcano will do in the future.

However, the record in the rocks is open to careful interpretation. The exposed rocks on a volcano rarely provide a complete record of its previous eruptions because of erosion between events. Even where deposits are exposed, does each ash or pumice fall, pyroclastic flow deposit, or lava flow record a separate eruption? Or do they record episodes of a single eruptive sequence that may have lasted a matter of a few days or weeks? Such questions, if not resolved, make a shaky basis for statistical assessment of long-term recurrence intervals or hazards.

After scientists have determined how a volcano has behaved in the past, they are still faced with a key question: Is this volcano still active? Such a simple question begs for a simple answer. Recall that several decades ago, the answer for a Cascades volcano was that it was active if eruptions had covered evidence of glacial activity from the last ice age about 10,000 years ago. We now know that some volcanoes lie dormant for much longer periods before erupting again. Mt. Lassen in northern California paused for some 27,000 years before again erupting in 1915. In the eastern Mediterranean Sea, Santorini rests for an average of 30,000 years

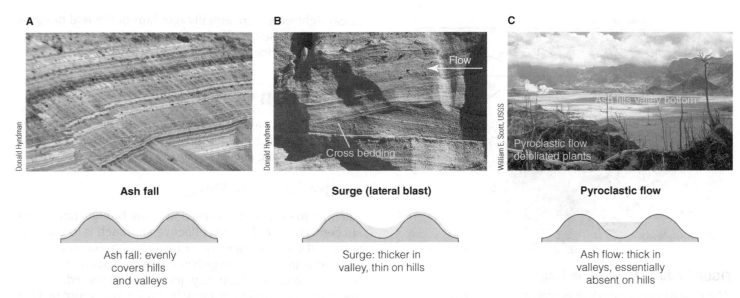

A Ash fall

Ash fall: evenly covers hills and valleys

B Surge (lateral blast)

Surge: thicker in valley, thin on hills

C Pyroclastic flow

Ash flow: thick in valleys, essentially absent on hills

FIGURE 7-19 Identifying Volcanic Deposits

Pyroclastic fall layers, surge deposits, and pyroclastic flows can be distinguished by their distribution over hills. **A.** Thin layers of air-fall ash and pumice from the Cappadocia region of central Turkey. **B.** Cross-bedded surge deposit from the 1600 BC eruption of Santorini, Greece. **C.** Pyroclastic-flow deposit from the 1991 Mt. Pinatubo eruption filled the valley bottom of Marella River. The foreground vegetation was stripped by hot ash in the upper part of the same pyroclastic flow. Ash coating surfaces in the foreground settled either from the billowy upper part of the pyroclastic flow or from the ash fall.

between major eruptions, though the intervals are becoming shorter. And Yellowstone, the gigantic volcano in northwestern Wyoming, erupts only every 600,000 to 700,000 years. Because the last of its massive eruptions occurred 640,000 years ago and seismic studies also show that molten magma lies only 5 to 16 km beneath the surface, the volcano is being closely monitored. Careful surveying within the Yellowstone Caldera shows resurgent domes that periodically rise and fall over decades or centuries, apparently in conjunction with magma movements underground. In 2002, geysers that had not erupted for a long time became quite active. Clearly, there is no simple answer for whether a volcano is active. Even if a volcano has not erupted in tens of thousands of years, scientists must be on the lookout for short-term indications of an impending eruption.

Eruption Warnings: Volcanic Precursors

Forecasting volcanic behavior for the long term is one thing. It is much more difficult to predict what a volcano may do in the next few days or weeks. Accurate eruption predictions are especially critical in areas where a large population lives close to the base of a stratovolcano. Mt. Vesuvius, near Naples, Italy, is such a case.

Seismograph records of volcanic earthquakes have been used to infer magma movement underground and to project the likelihood of an eruption. *Harmonic tremors* are the low-frequency rolling ground movements that precede many eruptions. USGS volcanologists recorded a series of minor earthquakes originating beneath Kilauea Volcano

in Hawaii in 1959. Over a period of two months, the earthquakes became more frequent as they rose from a depth of 60 km, finally reaching the surface as the volcano erupted. Similar earthquakes have been recorded below many volcanoes. These earthquakes often rapidly increase in their frequency and magnitude and, in some cases, migrate toward the summit before an eruption. That happened in the New Hebrides Islands during the 1950s and 1960s and at Mt. Pinatubo in 1991. But earthquakes are not always a reliable indicator; the frequency and magnitude of earthquakes did not change much at Mt. St. Helens in 1980 during the two months between the first activity and the climactic eruption of May 18.

Changes in the surface temperatures of volcanoes and the steam they erupt can be another indication of an impending eruption. Telescopes fitted with thermometers instead of ordinary optical eyepieces can observe temperatures of distant objects with reasonable accuracy. But their usefulness in predicting eruptions is limited because it is sometimes hard to know whether observed temperature changes are the result of volcanic activity or extraneous factors, such as rainfall cooling rocks.

Small changes in summit elevations and slope steepness associated with eruptions have been observed at some Japanese volcanoes and at Kilauea volcano in Hawaii (**FIGURE 7-20**). *Tiltmeters*, instruments that measure changes in the slope of a volcano, were first installed at Kilauea in the 1920s. Initial devices were simple levels made with water tubes 25 m long that could detect tilts as gentle as 1 mm in a kilometer. More precise modern

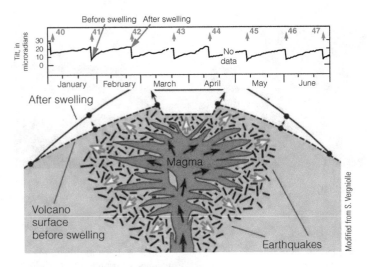

FIGURE 7-20 Volcano Flank Bulge

When rising magma gets close enough to the surface, it sometimes pushes overlying rocks to create a bulge in the flank of a volcano. The graph shows tiltmeter measurements of the changing slope on the volcano flank at Pu'u 'O'o, Hawaii in 1986. The bulge grows until the slope suddenly decreases during eruption. The process repeats each time the magma approaches the surface.

instruments use lasers to measure changes in both elevation and distance. Volcanologists also now commonly employ satellite-based Global Positioning System (GPS) devices to accurately measure changes in position. They show that volcano summits swell as magma rises into them and then deflate as they erupt.

A change in the gases emitted from a volcano is also associated with eruptions. As magma rises toward the surface, steam and other gases are released, and in some volcanoes, there are abrupt increases in sulfur or the ratio of sulfur to chlorine. Geologists collect and analyze fumarole gases to watch for these ominous signs (see again Figure 7-17).

Although none of these precursors is a sure sign of an eruption, they provide evidence of an increasing threat posed by a volcano. However, even when scientists feel reasonably sure that an eruption is coming, a prediction is useful only to the extent that authorities act on it. The USGS was reasonably successful in predicting eruption of Mt. St. Helens as a major event in 1980. But many people deeply resented the effects of the prediction upon their personal freedoms. The local loggers and timber companies fought closure of the nearby forests because they feared loss of income. The civil authorities relented and permitted their continued access. When the eruption occurred, loggers died, and logging equipment and millions of trees were destroyed. It is unfortunate that civil authorities allowed them into the area and that the loggers who pressed for such access were not aware of the extent of the geologic hazards and the consequences of ignoring them.

Local sightseers were equally ignorant of the real dangers, and, as discussed at the beginning of this chapter, some paid a heavy price.

Mitigation of Damage

Because it is not always possible to predict volcanoes, various strategies for mitigating volcano damage have been tried, with varying degrees of success.

Controlling Lava Flows

Attempts to slow or divert lava flows have brought only partial success. Perhaps the most effective approach is to cool and solidify the flow front with copious amounts of water delivered from fire hoses. A large basalt lava flow advancing on the town and harbor of Heimaey, Iceland, was cooled, slowed, and partly diverted in this way in 1973. Flows erupting from Mauna Loa were bombed in 1935 and 1942 to break the solid levee of cooling lava along the edges, thus diverting it into another path. These attempts were not successful enough, however, to inspire adoption as a standard procedure. Emergency diversion barriers were erected in nine days in 1960 to divert a flow erupting from Mauna Loa near Kapoho in westernmost Hawaii. These were partly successful in diverting the flow away from beach property and a lighthouse. Bulldozing levees of broken basalt lava into the path of a basalt lava flow proved unsuccessful, presumably because the broken rock of the levee was less dense than the flowing lava.

Warning of Mudflows

Fast-moving mudflows kill thousands of people living in valleys on the lower slopes of volcanoes. Automatic detection systems could provide enough warning for many people to evacuate from valley floors to higher ground. Warning systems high in valleys on the flanks of active volcanoes would provide warning when temperature sensors detected the passage of hot mudflows. New Zealand, uses them on ski slopes on the flank of Ruapehu, an active volcano. Vibration sensors have been installed high in major valleys on Rainier to detect the frequency of mudflow ground vibrations (between 0.6 and 20 Hz) and exclude the lower frequencies of earthquake tremors and volcanic eruptions. GPS, seismic, and tilt meters are also installed.

As with all warning systems, however, identifying an imminent threat is only the first step toward saving lives. The travel time for a large mudflow from Mt. Rainier to reach the 4000-plus residents of Orting, Washington, may be less than an hour. By the time a flow reaches the detector downslope, the event size is analyzed, likely false alarms eliminated, and the alarm sounded, residents might have only 30 minutes to flee. Successful evacuation will depend on a clear and timely warning, people's understanding of the hazard, and immediate response.

See the Volcano Survival Guide for information about protecting yourself from volcano hazards.

VOLCANO SURVIVAL GUIDE

U.S. High Hazard Area	■ Cascade Range of Washington, Oregon, northern California; eastern California (Mammoth area), Alaska, Hawaii, Yellowstone.
Preparation	■ Learn about volcanic hazards for the type of volcano nearby—air-fall ash, pyroclastic flows, lava flows, mudflows, and their extents. For example, pyroclastic flows and mudflows are most hazardous in valleys draining the volcano.
	■ Check with state and local emergency preparedness offices and schools about appropriate preparations.
	■ Prepare a survival kit appropriate to your location: 3 days of drinking water (1 gal/person/day), emergency food for a few days or weeks, dust mask and goggles, medications, spare contact lenses/glasses, first aid kit, flash light and batteries, critical papers, portable radio, cell phone and charging cord, sheet plastic and tape to seal ash from house, warm clothing and blankets, some extra cash. Shovel to remove ash from roof.
	■ Prepare with family what to do in evacuation, where to meet and who and how to contact them (in another state) in case of separation. Educate family and friends.
Warning Signs	■ Watch for "rolling" earthquakes around volcanoes, steam containing ash and other gas emissions, bulging volcano flank.
	■ Check Volcano Notification Service website (http://volcanoes.usgs.gov/vns/) for present activity.
During the Event	■ Evacuate according to plan and emergency services instructions. Don't stay to watch the eruption.
	■ For lahars (mudflows), get to higher ground away from valley bottoms, especially if you hear the roar of a mudflow; some lahars can move down-valley faster than 30 mph.
	■ For pyroclastic flows, get well away from volcano (more than 100 km), and to higher hills. For heavy ash falls, take shelter inside a sturdy building that will not collapse under heavy ash; stay upwind.
Aftermath	■ Stay inside, out of airborne dust; cover your nose and mouth with a dust mask or damp cloth if ash is in the air; drive only if necessary; protect filter air intake and front of vehicle radiator to minimize abrasive ash. Wear a long-sleeve shirt and pants. Use phone only for emergency calls.
	■ Follow instructions of state and local emergency preparedness centers. Listen to emergency radio. Evacuate dangerous areas and don't return to those areas because more eruptions and lahars may follow for months later. Shelter in place and help others in need.
	■ Lahars may fill valley bottoms, destroy bridges and roads, and disrupt shipping and air traffic for weeks or months. Plan accordingly. House roofs with a low pitch often collapse under 10 cm (4 in) of wet ash; shovel it off if necessary.

Populations at Risk

Although much has been learned about volcano behavior and warning of an impending eruption often can be provided, a false alarm could destroy the credibility of all involved. The social and economic consequences of a needless evacuation of hundreds of thousands of people would be devastating, if it were even possible. The region of Italy surrounding Mt. Vesuvius and the Cascades Range of northwestern North America provide examples of how scientists assess risk based on historical records of volcanic activity as well as volcanic precursors. They also raise important questions about disaster planning and preparedness in these regions. The same issues arise in other active volcanic regions, especially the Pacific Ring of Fire, which includes the Cascades, the Andes of South America, Japan, and the Aleutians of Alaska, and the active volcanoes of Indonesia.

Vesuvius and Its Neighbors

Large numbers of people living in close proximity to active volcanoes in Italy provide glimpses into increasing volcanic hazards both there and in North America. People do not differ much from one place to another. Their behavior during the eruption of Vesuvius in AD 79 was probably similar to the way people now living in towns near an erupting volcano react.

How could 700,000 people near Vesuvius leave on roads and rail lines that are taxed beyond their capacity on normal days? And where could so many people find convenient refuge on short notice?

The northeastern part of the African Plate descends under southern Europe along a short collision boundary that raises the Alps and drives a chain of volcanoes in Italy and the eastern Mediterranean region. These volcanoes have devastated populations and caused enormous property damage for thousands of years. Some have even completely destroyed ancient civilizations. This is one of the most dangerous volcanic environments on Earth. The modern city of Naples, with a population of 3.7 million, nestles between two volcanoes—Mt. Vesuvius, just 13 km to the east (**FIGURE 7-21**), and Campi Flegrei, centered 13 km to the west.

MT. VESUVIUS Vesuvius has a long history of major eruptions (**Table 7-1**), including one around 3580 BC and another

FIGURE 7-21 Italian Volcanoes

A chain of active volcanoes, including Vesuvius and Etna, runs down the western side of Italy. Subduction zones are shown in green lines with pointers sloping down dip. Roads and houses crowd onto the lower flanks of Mt. Vesuvius. Some new houses are built in a valley right next to historic lava flows. Campi Flegrei volcano is just west of Vesuvius—at the position of the final "s" in Vesuvius on the inset map.

Basemap: modified from NASA image; inset map: NASA; photo: Donald Hyndman

Table 7-1	Notable Eruptions of Mt. Vesuvius
DATE	EVENT
3580 BC	Pyroclastic flows cover current area of Naples.
1800 BC	Pyroclastic flows kill many people in Pompeii and cover much of the area of present-day Naples.
AD 79	Destruction of Pompeii and Herculaneum kills about 4000.
1631	Heavy pyroclastic flows and ash falls kill 4000.
1906	More than 500 are killed.
1944	Lava flows partly destroy two communities.

around 1800 BC, in which pyroclastic flows covered much of the present area of Naples and killed many people (**FIGURE 7-22**). In AD 79, during the reign of Nero, Vesuvius erupted catastrophically, burying Pompeii and Herculaneum and killing one fifth of the 20,000 people living there (**Case in Point:** Catastrophic Pyroclastic Flow: Mt. Vesuvius, Italy, p. 173).

Throughout its history, the area around Vesuvius has gone through cycles of volcanic destruction and rebuilding. The recent lack of significant activity is again leading to complacency as development continues in this high-risk area (**FIGURE 7-23**). Today, 700,000 people are living on borrowed time within the potentially fatal zone of the next major eruption of Vesuvius.

So far as anyone knows, Vesuvius could erupt anytime. If past experience is any guide, some 0.5 km³ of magma may have accumulated beneath Vesuvius after more than 70 years of inactivity. Seismic and tiltmeter studies suggest that molten magma now exists 5 to 10 km below the crater of Vesuvius. Tiltmeters detected a half-meter of expansion from 1982 to 1984, with an increase in numbers of minor earthquakes. However, in approximately half of such cases, no eruption follows. This large uncertainty leaves volcanologists wondering how to advise civil authorities.

Experts suggest that the long pause since the last major eruption means that the next may be the largest since 1631. Some volcanologists think the next Vesuvius eruption will probably resemble those of 1906 and 1944, with lava flows and heavy ash falls that could collapse as many as 20% of the roofs in Naples and stop all traffic and activity. Pyroclastic flows racing downslope at more than 100 km/h with temperatures of 1000°C would reach populated areas between five and seven minutes after the eruption column collapsed. Modern Pompeii and a dozen other towns, each with thousands of people, ring the base of Vesuvius hardly more than 6 km from the crater. Instead of several thousand deaths, the toll of a new eruption resembling the one in AD 79 could kill several hundred thousand people.

In the current state of knowledge, disaster prevention depends on early sensing of volcanic warnings such as swarms of small earthquakes or compositions of emitted gases. A special commission formulated an eruption contingency plan in 1996, which assumed that a warning could be issued at least two weeks before an eruption and 600,000 people could be evacuated within a week. Authorities now argue that a complete evacuation would take about three days. Those who have driven in Naples or used trains and buses almost anywhere in Italy might consider such a goal

Guiseppe Mastrolorenzo and PNAS, National Academy of Sciences, USA

Guiseppe Mastrolorenzo and PNAS/National Academy of Sciences

FIGURE 7-22 The Past is a Sign

A. People ran from an earlier, even larger eruption of Vesuvius 3800 years ago, leaving these footprints preserved in ash.

B. Some skeletons have been uncovered.

FIGURE 7-23 Vesuvius Looming Over Naples
Naples lies in the shadow of Mt. Vesuvius (seen in the background). Note the houses on the flanks of the volcano.

FIGURE 7-24 A Magma Chamber Refills
In the Roman market Temple of Jupiter Serapis in Pozzuoli, gray pitted parts of columns show borings from marine mollusks from when part of the columns were underwater.

hopelessly optimistic. Vocal critics argue that it is impossible to predict an eruption more than a few hours or days in advance. They suggest that even with the best available monitoring, an evacuation alarm would almost certainly come too late. The alternative of major urban planning in a region that already contains hundreds of thousands of people would require truly ruthless resettlement on a scale almost beyond imagining! If you were in charge, what would you do?

CAMPI FLEGREI Campi Flegrei is an unquestionably active resurgent caldera within the western suburbs of Naples. Two million people live on the floor of the caldera, with 400,000 residing within the most active part, making Campi Flegrei one of the most hazardous volcanic areas in the world. Giant rhyolitic pyroclastic-flow eruptions, which spread searing pumice to the southeast, were hot enough for the fragments to flatten and fuse, forming a solid mass. These events left a caldera 16 km in diameter 39,000 years ago and another caldera 21 km in diameter, at the same site, 15,000 years ago. Intense explosive activity along the faults that define the margin of the caldera has happened repeatedly during the past 12,000 years. Eruptions came at intervals of approximately 50 to 70 years from 12,000 to 9500 years ago; 8600 to 8200 years ago; and 4800 to 3800 years ago. That is an extremely high level of activity for any volcano, especially a giant rhyolitic caldera volcano.

As with the Yellowstone volcano, a pattern of swelling and subsiding magma under this area points to continued volcanic activity. Pozzuoli, a city of 83,000 people on the western outskirts of Naples, near the center of Campi Flegrei, has a history of rising and sinking three times in the last 1360 years. The Temple of Jupiter Serapis, a Roman marketplace within Pozzuoli dating from the second century BC, has spent some of its history below sea level. The lower parts of its columns show borings of

mussels that live in seawater (**FIGURE 7-24**), indicating that the market sank about 12 m after its construction and ended up below sea level. During the fifth century AD, and right before the 1538 eruption of Monte Nuovo within the caldera, the pillars showed a total of 7 m relative uplift, as magma rising under the caldera pushed it back above sea level. The marketplace generally subsided after 1538, but from 1969 to 1972 and from 1982 to 1984 it rose a total of 3.5 m. In 1983 and '84, it rose at more than 1 m per year! Rapid subsidence from 1985 to 2004 was followed by rapid 2-cm uplift in two months of late 2006. In early 2013, the ground rose at 3 cm per month. If the rise indicates that another eruption is coming, what impact would it have on this region today?

Past eruptions dumped ash to a depth of 1 m or more over much of Naples. If such an eruption were to happen now, it would kill many thousands to hundreds of thousands of people. Recall that 1 m of volcanic ash is heavy enough to collapse almost any roof, especially if rain falls on it. The potential death toll and property damage from a resurgent caldera eruption in Campi Flegrei is frightening. It is hard to imagine any workable evacuation plan.

The Cascades of Western North America

The stratovolcanoes making up the Cascades Range lie over an active subduction zone and inland from an oceanic trench, although the trench, being filled with sediment, was not at first apparent. The volcanoes erupt along a nearly continuous line inland from the subduction zone (**FIGURE 7-25**). In an effort to assess volcanic hazards, the USGS embarked on a vigorous campaign of research in the High Cascades to determine when each volcano last erupted, how

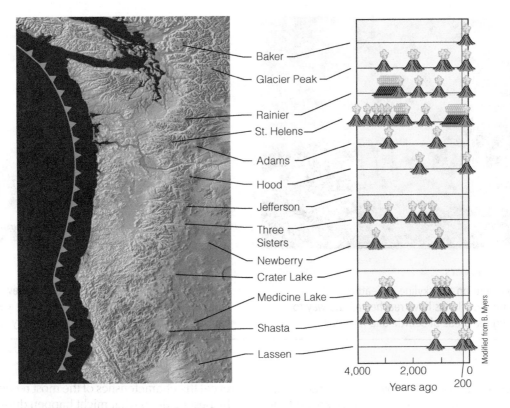

FIGURE 7-25 Active Cascades Volcanoes
This diagram shows the major Cascade Range volcanoes and their eruptions through time.
The chain of active volcanoes lies above the active subduction zone, shown as a green line
with pointers sloping down dip.

frequently, and with what types of activity. Those investigations revealed that the chain is far more active than most geologists had supposed; the Mt. St. Helens eruption in 1980 confirmed those findings. Based on USGS studies, the most dangerous volcanoes in the High Cascades are Glacier Peak, Mt. Rainier, and Mt. St. Helens, all of which are likely to produce catastrophic mudflows. Future large eruptions in the High Cascades should not surprise anyone.

Although the volcanic processes and products in the Cascades are similar to those in Italy, each region has unique hazards associated with it. The risk to human life and property in the Cascades is limited by the fact that although there are some large cities near volcanoes, they are sufficiently distant to provide some time for warning and minimize the numbers of people who need to be evacuated. With a few exceptions, the designation of volcanic areas as national parks, national monuments, national forests, or wilderness areas restricts settlement around many of the large and active volcanoes. One factor that compounds hazard in the Cascades region, however, is that these volcanoes lie in the belt of the westerly winds and close to the western coast of North America. Those winds carry heavy moisture from the Pacific Ocean, dump it on the mountains as rain and snow, and thus increase mudflow hazards. We discuss a few of the more dangerous Cascades volcanoes below.

MT. RAINIER Mt. Rainier is the largest, highest, and most spectacular volcano in the main line of the High Cascades, rising from sea level to 4393 m and dominating the skyline of a large part of western Washington (**FIGURE 7-26**). Mt. Rainier is arguably the most dangerous of the High Cascades volcanoes simply because of the large population close to its base. Rainier's status as a national park prevents most kinds of development that might attract even more people to crowd in too close.

Although Rainier last erupted in the 1840s, studies show 11 eruptions in the 10,000 years or so since the last ice age; nothing suggests that it is extinct. A substantial eruption could dump heavy ash falls on cities built on old ash falls downwind, most likely to the east and northeast. Several smaller towns lie within range of large rockfalls or pyroclastic flows.

MT. ST. HELENS In 1980, Mt. St. Helens was well known to be the most active volcano in the Cascade Range and the most likely to erupt. Its smooth, symmetrical shape showed that it must have erupted recently and filled deep valleys that mountain glaciers had carved in its flanks during the last ice age. When Mt. St. Helens erupted on May 18, 1980 (**FIGURE 7-27**), it was observed as closely and studied as exhaustively as any volcanic event to that time. St. Helens had been quiet since 1857, when it produced andesite lava flows, and 1843 when it pushed up a dacite dome (a composition roughly between

FIGURE 7-26 Rainier Hovers

A. Mt. Rainier is prominent on the skyline of Seattle, Washington. Lahars from its flanks have reached all the way to Puget Sound, including the areas of Seattle and Tacoma.

B. Orting, Washington, with spectacular views of Mt. Rainier, is built on a giant, ancient mudflow from the volcano. If mudflows happened in the past, they almost certainly will happen again.

rhyolite and andesite). Pyroclastic flows from earlier eruptions reached at least 20 km down the valleys around the volcano, and mudflows traveled at least 75 km.

Radiocarbon dates on wood charred and buried during previous eruptions were available before 1980 to show that most of the modern volcanic cone grew during the last 2500 years. That pace of activity in the geologically recent past had already persuaded most geologists to consider St. Helens the Cascades volcano most likely to erupt.

The USGS issued hazard forecasts more than two years before St. Helens erupted in 1980. The forecasts proved generally accurate, although they understated the probable size and violence of the upcoming eruption. The agency informed the public of the dangers that existed and tried to dispel imaginary ones. It did not attempt to predict specific eruption times, though many people believed it could and should have done so. The USGS provided information but did not dictate which areas should be closed because such public policy decisions were beyond its authority. The whole episode provided a learning experience in both prediction and public policy for volcanic eruptions.

MT. HOOD Mt. Hood, the spectacular volcano 75 km east of Portland, Oregon (see Chapter 6 opener, p 122), has not erupted since the 1790s. That is not a significant hiatus for a large andesite volcano. Radiocarbon dates on charcoal preserved in volcanic deposits reveal two major eruptions during the last 2000 years. Volcanic domes near Crater Rock have repeatedly collapsed over the lifetime of the volcano to shed numerous hot pyroclastic flows down the southwest flank. Some reached as far as 11 km from the peak.

There is little doubt that Mt. Hood will produce more eruptions, probably like those of the past 2000 years, but it is not currently possible to predict when. Close study of deposits that reveal the characteristics of the most recent eruptions provides the best guide to what might happen during the next one.

Another eruption would probably send pyroclastic flows as far as 10 km down the south and west flanks of the cone. Mt. Hood does not generally erupt large amounts of ash in clouds that drift downwind. Water from melting snow and ice would lift ash and rubble from the surface of the volcano and send mudflows racing down Sandy River and its tributaries (**FIGURE 7-28**), toward the eastern edge of Portland and probably north toward the town of Hood River (FIGURE 6-17). They would reach nearby towns to the west on the Sandy River in 30 minutes and the larger towns of Sandy and Troutdale in 2 hours and 3.5 hours, respectively.

In all likelihood, Mt. Hood will provide some warning of its next eruption in the form of swarms of small earthquakes and clouds of steam dark with ash. Unfortunately, its frequent small earthquakes decrease the usefulness of seismic activity in predicting an eruption. The slopes of the volcano are almost uninhabited, so property damage there would be limited. However, mud and debris flows pouring down river valleys toward Portland or Hood River would inflict immense property damage. If warning did not come several days before an eruption, mudflows could kill large numbers of people.

THREE SISTERS Among the lesser known of the Cascades volcanoes are the Three Sisters, a spectacular trio of volcanic cones just west of Bend, Oregon. Although any of the Cascades volcanoes could reawaken, the Three Sisters were not high on most people's list of the most likely mountains to erupt anytime soon. All that changed in March 2001 when USGS scientists using satellite radar interferometry discovered a low but broad bulge that had been rising since early 1997. By spring 2002, the 16-km-wide bulge had risen a total of about 16 cm—3.3 cm per year. The crest of the bulge is 5 km west of South Sister. Its shape

FIGURE 7-27 Eruption of Mt. St. Helens

In the climactic eruption of Mt. St. Helens in May 1980, hot ash-laden gas boiled out of the vent, expanding into the cooler surrounding air. The darker ash-rich upper and outer parts of the cloud are beginning to cool and descend.

suggests a magma chamber 5 or 6 km below the surface. A large eruption on the east flank of South Sister could be catastrophic for Bend, with a population of 60,000 people. Recent earthquakes add to the concern about renewed volcanic activity.

MT. MAZAMA (CRATER LAKE) Mt. Mazama was an enormous andesite volcano that hovered over a large area of southwestern Oregon until around 7700 years ago, when it destroyed itself in a gigantic rhyolitic eruption. Crater Lake now floods the caldera that opened where Mazama once stood (**FIGURE 7-29**). The geologic record of the Mazama eruption has provided much of our present understanding of such events.

The rim of Crater Lake is the stump of Mazama. It is easy to reconstruct an approximation of Mazama by projecting the outer slopes of the rim upward at the usual angle for large andesite volcanoes, then lopping something off the top for a crater. The result is a volcano fully on the scale of Mt. Shasta. Enormous amounts of rhyolite pumice and ash surround Crater Lake for tens of kilometers. Rhyolite ash

FIGURE 7-28 Signs of Debris Flow

Debris flows at the east base of Mt. Hood have raced north toward the town of Hood River on the Columbia River, reaching it in just over an hour. The larger boulders are carried along near the top of the flow.

at least 40 cm thick that spread over much of the Pacific Northwest and northern Rocky Mountains provides evidence of an extremely large and violent eruption. Radiocarbon dates on the remains of plants charred in that eruption indicate that it happened 7700 years ago. By most estimates, the Mazama eruption produced at least 35 times as much magma as the main 1980 St. Helens eruption. Mazama dumped at least 40 cm of volcanic ash across large areas of the northern Rocky Mountains. Today, an eruption that size would be a major natural disaster.

FIGURE 7-29 Crater Lake

Crater Lake is the caldera depression that formed when Mt. Mazama erupted 7,700 years ago. The event was many times the size of the 1980 eruption of Mt. St. Helens.

FIGURE 7-30 **Thick Ash Deposits**

A. A thick off-white pyroclastic flow layer from Mazama grades up to a darker cap, evidence that the eruption tapped progressively deeper into a differentiated magma chamber.

B. Pumice exposed in this quarry east of Crater Lake was deposited from the eruption of Mazama 7700 years ago. Pumice lumps vary in size. The area shown is approximately 2 m high.

Deep deposits of the culminating pyroclastic flow from the Mazama eruption are pale and grade upward into much darker rock (**FIGURE 7-30**). Most geologists interpret the gradation in color as evidence that the eruption tapped a magma mass that had separated into two components: the upper part richer in silica than the deeper part. The pale upper magma erupted first to make the lower part of the pyroclastic deposit, and the darker ash erupted last as the deeper part of the magma chamber emptied.

For many years, geologists generally agreed that Mazama had blown itself to smithereens in an enormous steam explosion that opened a gaping crater 10 km across. Then rain and groundwater filled the caldera to make Crater Lake. But more detailed study revealed an absence of old and altered volcanic rock in the debris blanket around Crater Lake. All the debris is freshly erupted volcanic ash and pumice. That discovery led geologists to conclude that Mazama's peak simply sank into the emptying magma chamber beneath the volcano. The basin that holds Crater Lake is a caldera formed by collapse, not a crater blasted out by explosion. Evidently, the eruption that destroyed Mazama's peak was not the end of volcanic activity on the site. A new version of Mazama may yet rise from the ruins of the old one. One thousand years ago, a smaller volcano grew in the floor of the caldera until it barely emerged above the surface of Crater Lake to become Wizard Island.

Crater Lake has become a picture of peace and serenity. It does not present any great hazard because the region supports few people and has enough roads to enable easy evacuation. However, we too easily forget that it was the site of an extraordinarily large and violent volcanic eruption.

MT. SHASTA French naval captain Jean-François de Lapérouse and his crew cruised north along the coast of California in 1786 shortly after failing to see the Golden Gate

and the great bay behind it, probably because of coastal fog. But they did report seeing what they thought was an erupting volcano some distance inland at the latitude of Mt. Shasta. That may have been Shasta's most recent eruption, the last of 11 in the past 3400 years.

Shasta, at 4318 m, is the second highest of the High Cascades volcanoes (**FIGURE 7-31**). It looms over a large area of northern California, a constant hovering presence in the lives of thousands of people.

Some geologists note that Shasta is approximately the same size as Mt. Mazama was before it destroyed itself and formed Crater Lake. And it may be significant that Shasta has produced small quantities of rhyolite during its most recent eruptions instead of the standard andesite that makes up most of the volcano. Finally, seismograph records of earthquake waves that pass beneath Shasta show weak S waves or none at all. This suggests that the earthquake waves pass through a liquid on their way to the seismographs, presumably molten magma in a shallow chamber. Some geologists interpret these observations as evidence that Shasta is getting ready to erupt a monstrous amount of rhyolite ash or lava. Of course, no one can tell when or even whether such an eruption will happen.

Nevertheless, Shasta is a major hazard just as it stands. It is essentially an enormous pile of loose andesite ash and rubble stacked at a steep angle to a height of 3000 m with only a few lava flows holding it together. It is likely to shed large masses of rubble with little or no provocation. Shasta did drop much of its north side in a giant rock avalanche 300,000 years ago. The debris reached 51 km northwest from the peak and covered at least 350 km^2 of Shasta Valley. Later eruptions repaired the scar on the volcano. The towns of Mt. Shasta City and Weed stand on older rock avalanche deposits directly downslope from valleys on Shasta. They also lie within easy range of large pyroclastic flows.

FIGURE 7-31 Mt. Shasta and Mt. Shastina

A. Mt. Shasta punctuates the skyline of northern California.

B. Shasta looms over the small town of Weed, on its western flank.

North America's third highest volcano, Citlaltépetl, similar to Shasta and situated at the east end of the Trans-Mexican Volcanic Belt, has seen massive flank collapse twice in its history and may do so again. Its young summit cone is highly fractured and much has been altered to clays from long contact with acidic volcanic gases. Landslide volumes of 0.5 km³ have run out to distances of 20 km in less than 12 minutes. Collapse would endanger as many as 1 million people!

MT. LASSEN Visitors to Mt. Lassen today see a quiet mountain that provides little indication of a potential eruption. Lassen is a grossly oversized lava dome of dacite that descends from a long history of volcanic violence. It rose within the wreckage of Tehama, a large andesitic volcano that sank into a caldera during a cataclysmic eruption 350,000 years ago. Brokeoff volcano is a remnant of one of its flanks. Tehama volcano had earlier risen in the caldera of an even larger volcano that similarly destroyed itself 450,000 years ago. Perhaps we should not be surprised to see more eruptions in this vicinity. Age dates show that six lava domes—called Chaos Crags—rose near Lassen 1100 years ago. That was some 27,000 years after its previous large eruption. Clearly the time since its last eruption, regardless of how long, does not indicate that the volcano will not again erupt.

Most of the people who lived in northern California in 1914 thought Mt. Lassen was thoroughly dead. Then on May 30, with no apparent warning, it suddenly produced an enormous cloud of steam with loud booming noises. Newspaper accounts tell of raw panic in the streets of Redding and Sacramento and places in between. Several lesser eruptions then overflowed the summit crater with extremely viscous pale dacite magma that broke off as blocks that tumbled downslope. On May 19, 1915, the north slope collapsed, sending a mass of hot blocks racing down over a 10-km² area at the base of the mountain. Eruptions climaxed on May 22 with a dark mushroom cloud of steam and ash that reached a height of 8 km. Melting snow sent an enormous hot mudflow down the east slope that filled 50 km of the Hat Creek valley overnight. Some of the blocks of hot lava that were buried in the mud spouted steam for months. Lassen continued to produce sporadic clouds of steam until 1921.

The area near Lassen is thinly populated, so any future eruptions will probably cause minimal loss of life or property. Redding, Red Bluff, and other towns to the west are 50 to 75 km downslope and against the prevailing wind for the most likely hazard: falling ash. Mudflows could possibly reach that far. Susanville, a similar distance to the east, could be heavily impacted by a major ash eruption.

A Look Ahead

Despite our knowledge of volcanic hazards and our ability to monitor volcanic activity, the opportunities for volcanic catastrophe to people and property are greater today than ever before. Growing populations and the fertility of volcanic soils encourage people to settle close to active volcanoes. The chance of dying in an eruption is small enough that most people ignore the hazard.

However, the dangers are real. Many times in the past, mudflows and pyroclastic flows rushed down valleys that drain the flanks of Cascades volcanoes and onto the broader expanses beyond. A 1987 analysis of specific volcanic hazards and their distribution around the High Cascades volcanoes showed them to be far more dangerous than previously expected.

Most large volcanic eruptions are both dangerous and destructive. Some are genuine catastrophes that kill tens of thousands of people and destroy billions of dollars worth of property. Although the aggregate toll of death and devastation from volcanoes pales in comparison to those of earthquakes, landslides, and floods, the potential remains for a truly catastrophic volcanic event.

Cases in Point*

Volcanic Precursors
Mt. St. Helens Eruption, Washington, 1980 ▶

The USGS had recently mapped and determined the ages of the young eruptive deposits and concluded that Mt. St. Helens could erupt again at any time. Then, in March 1980, it showed renewed signs of life—swarms of small earthquakes and blasts of steam and ash—culminating in the climactic eruption in mid-May. Preliminary activity started suddenly in the afternoon of March 23, 1980, when seismographs detected swarms of small earthquakes followed by larger ones. Eruption prediction became an issue, and several research organizations rapidly installed arrays of portable seismographs around the volcano. Then, just after noon on March 27, St. Helens produced a loud boom and a large cloud of white steam.

The steam rose 600 m above the volcano while a new crater began to open in its summit. A fissure 1500 m long opened high across the north flank of the volcano, while lesser fractures opened and closed.

Those first clouds of steam were neither hot enough nor dark enough with suspended ash to convince geologists that they were the first phase of a genuine eruption. Many volcanoes blow off clouds of steam in late spring and early summer as melting snow sends water percolating down into the hot rocks within. Starting early on March 28, St. Helens again blew off clouds of steam. Flames began to light the craters at night on March 29; their pale blue color suggested burning methane. The ejected clouds of steam became

hotter and much darker with ash. Study of the seismic records showed a longer-period wave-like pattern that came to be known as harmonic tremors. These began on April 1 and continued intermittently and with growing strength for 12 days. Although steam eruptions decreased to one per day between April 12 and April 22, they were becoming richer in ash. The situation was beginning to seem serious.

By the end of April, two craters had combined into a single oval crater 300 to 500 m across. Seismographs recorded more than 30 small earthquakes with magnitudes greater than 3 every day, the movements originating below the north flank of the volcano at depths that became shallower day by day. Microscopic study of the newly

P. W. Lipman, USGS

■ *The growth of the huge bulge on the flank of Mt. St. Helens a few weeks before the climactic eruption in May 1980. The umbrella shades the surveying instrument.*

*Visit the book's website for these additional Cases in Point: *Deadly Pyroclastic Flow*—Mt. Pelée, Martinique, 1902; *Debris Avalanche Triggered by a Small Eruption*—Nevado del Ruiz, Colombia, 1985

erupted ash showed that the eruptions were still producing only fragments of old volcanic rocks altered by steam within the volcano. There was no new magma.

A bulge slowly grew on the volcano's north flank in early May, eventually expanding to 106 m outward from the original slope and about 1067 m across. A large mass of magma was clearly rising into the north flank of the volcano. Meanwhile, the ash falling from clouds of steam was clearly derived from new magma rising from below. It had the color and composition of dacite.

At this point, it seemed abundantly clear that St. Helens was ready to erupt, but the type of eruption remained open to vigorous debate. Many geologists expected a dome

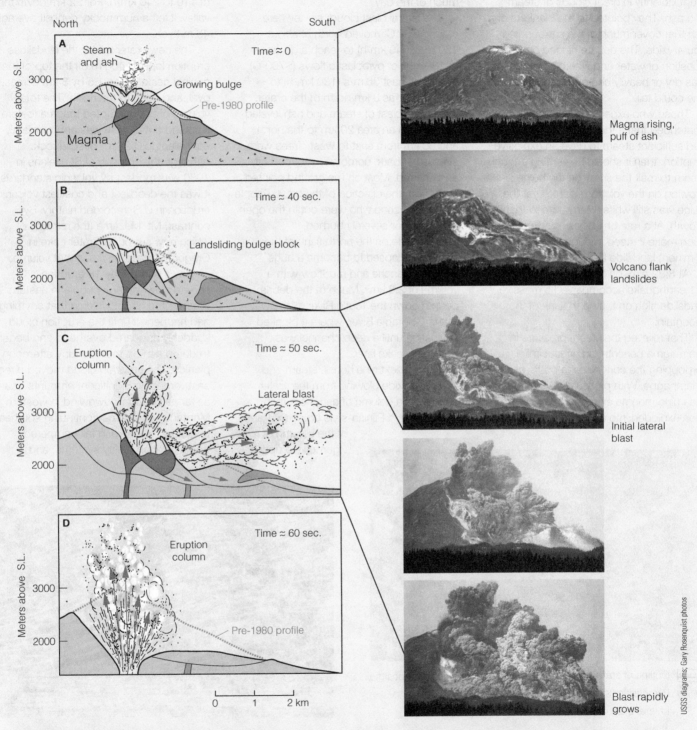

Magma rising, puff of ash

Volcano flank landslides

Initial lateral blast

Blast rapidly grows

■ *Mt. St. Helens eruptive sequence, May 18, 1980:* **A.** *Magma approaches the surface and weakens the dome at the top of the magma chamber.* **B.** *The dome collapses in a giant landslide.* **C.** *Pressure release on magma permits gases to bubble out rapidly. The lateral blast blows out to north.* **D.** *A full Plinian-type eruption ensues. S.L. = sea level.*

eruption in which the extremely viscous magma would rise quietly through the north flank of the volcano for months or a few years. They pointed to the bulge that swelled in the flank of the volcano. Others reminded them that dacite magma may accumulate water at depth and then erupt violently in great clouds of steam and ash. They pointed to the blankets of ash that cover much of the surrounding countryside. The debate hinged on the question of water and whether the magma was dry or heavily charged with steam. No one could tell.

Those who expected a dome eruption contended that if the mass of magma had sufficient steam to drive an explosive eruption, then it should be venting enough steam to melt the snow off the bulge growing on the volcano's flank—yet the bulge was still white. Many agreed that its growth, at a rate of 1 m per day, would soon make it steep enough to pose an imminent landslide threat.

At 8:32 a.m. on May 18, a magnitude 5.1 earthquake accompanied a massive landslide high on the north flank of the mountain.

That relieved the steam pressure in the magma beneath, an effect similar to popping the cork out of a bottle of champagne. With pressure relief, a mass of steaming magma expanded into pumice, then exploded into a dark cloud of steam and ash. A lateral blast shot north, as the eruption cloud soared to 24 km above the volcano. (See sequence of diagrams and photos, p. 171) The event was far larger than the USGS volcanologists expected. The immense vertical column of ash continued to blow out of the crater for much of the day.

The lateral blast cloud, somewhere around 350°C, moved north at about 150 m/s (540 km/h) to reach a distance of 13 km. Hot pyroclastic flows (>700°C) moved at least 36 m/s (130 km/h) to reach as far as 8 km north of the crater. The lateral blast of steam and ash leveled the forest in an area 20 km to the north and 30 km from east to west. Trees were stripped of bark, uprooted, snapped off, and charred. Most on the ground pointed away from the direction of the blast. People in the blast zone who were out in the open were killed or severely burned.

The bulge on the north flank of the volcano collapsed to become a huge debris avalanche and mudflow with a volume of 2.8 km^3. Much of the debris poured down the Toutle River all the way to the Columbia River, where it blocked navigation until a deep channel was dredged weeks later.

For the next nine hours, steam and carbon dioxide blowing from the crater supported a column of ash more than 20 km high: a Plinian-style eruption. Parts of the column of vigorously erupting ash collapsed at times to drop pyroclastic flows down the north flank of the volcano.

The collapsing bulge and debris avalanche displaced the water from Spirit Lake at the base of the volcano and swept over the ridge to the north. It raced down the north fork of the Toutle River at speeds of 110 to 240 km/h. For 22 km down the valley, it left a hummocky deposit averaging 45 m thick.

The new crater left by the landslides and eruption lopped 400 m off the top of the original peak; it was 1.5 by 3 km across, oval, and open to the north. The total volume of newly erupted magma (before it spread out as ash) was approximately equivalent to 0.2 km^3 of solid rock. Although the eruption of St. Helens in 1980 was modest by volcanic standards, it was the deadliest and costliest volcanic eruption in U.S. recorded history. By contrast, Mt. Mazama (the remains of which now surround Crater Lake in Oregon) erupted 35 times that volume.

News of the May 18 eruption spread slowly. Many people in eastern Washington had no inkling that anything had happened until the eruption cloud suddenly appeared overhead and began to dump ash on their Sunday afternoon picnics. The eruption cloud moved directly east, dropping significant amounts of ash as far as 800 km downwind in western Montana, with some continuing southeast to Colorado. The ash fall hampered transportation, utility systems, and

USGS

■ *A map of Mt. St. Helens after the 1980 eruption shows the crater open to the north, with pyroclastic flow, lateral blast, and mudflow—volcanic mud- or debris-flow—deposits.*

▭▭▭▭ Outline of crater
▮ Pyroclastic flow deposits
▢ Lateral blast deposits
▮ Debris avalanche deposits
▮ Mudflow deposits

Lyn Topinka, USGS

■ *This forest was flattened by the lateral blast from the 1980 Mt. St. Helens eruption. People circled in the lower right show the size of these large trees.*

A. *Huge mounds were left by the May 18 debris avalanche and mudflow that raced down the Toutle River.*
B. *Lines from the Toutle River mudflow were left high on trees after the flow continued to drain downvalley.*

Bouldery mudflow piles

Height of mudflow

Lyn Topinka, USGS

Lyn Topinka, USGS

A

B

outdoor activities until a heavy rain cleared the air four days later.

Few people in eastern Washington and the northern Rocky Mountains had experienced a volcanic eruption. Many found the ash fall frightening. As far east as Missoula, Montana, the late afternoon western sky turned a ghastly greenish black before fine light gray ash began falling like persistent snow. Under an otherwise blue sky with no wind, the finer particles hung suspended in the air for several days. People stayed indoors with the windows closed, and many wore masks for brief excursions to work or for groceries.

Water from rapidly melting snow, Spirit Lake, and the Toutle River created mudflows on May 18 that poured down the river's north fork at 16 to 40 km/h. They flushed thousands of cut logs downstream, destroyed 27 of 31 bridges, and deposited sediment in navigation channels, including the Columbia River shipping channel near Portland, Oregon. Mudflows diverted streams and raised valley floors as much as 3 m and channel beds as much as 5 m. Airborne ash blocked Interstate 90 and other highways. Airports more than 400 km downwind were closed for a week while crews cleared 6 to 10 cm of ash.

CRITICAL THINKING QUESTIONS

1. What signals from Mt. St. Helens led geologists to believe that the volcano was likely to erupt in the near future? Explain what produced each of these signals and how they related to the eruption itself.
2. Who, specifically, decided to force evacuation of the area around the volcano and was that a correct decision? What exceptions were made to the evacuation order, why were they made, and were those wise decisions?
3. Were ridgetops or valley bottoms the safest places near the volcano? Explain.
4. What was the final trigger to the eruption and how did that process work?

Catastrophic Pyroclastic Flow

Mt. Vesuvius, Italy, AD 79 ▶

During the reign of Nero, residents of the bustling Roman towns of Pompeii and Herculaneum may not have thought much of potential danger from Mt. Vesuvius, the volcano looming over them. They had forgotten during its 700 years of quiet that

Vesuvius was capable of erupting. But in the summer of AD 79, they were reminded of the volcano by earthquakes, along with reports of rising ground levels. On the morning of August 24, a series of steam explosions dropped a few centimeters

of coarse ash close to the volcano. Pliny the Younger, whose account of the event has been preserved, sat across the bay, watching the eruption unfold.

■ *Vesuvius looms over the excavated ruins of Pompeii.*

■ *These are casts of the cavities left by bodies killed in the ash, as they were found in Pompeii on top of the air-fall pumice.*

Meanwhile, his uncle, Pliny the Elder, who was commander of the Roman fleet in the Bay of Naples, dashed about in a ship equipped with oars, trying to establish some sort of order.

Vesuvius went into full eruption shortly after noon as ash and pumice rose in a scalding column of steam. High-altitude winds carried the ash plume directly over Pompeii, situated 8 km south and east of the summit. Approximately 12 to 15 cm of white pumice fell per hour until evening, a total of 1.3 m. Roofs must have begun to collapse under the weight of the ash after the first few hours.

The volatile content of the magma decreased, or the vent widened, or both, after 12 hours of continuous eruption. The proportion of ash in the erupting column of steam increased to reach a height of 33 km. The increasing ash content finally made the rising column of steam so dense that it collapsed onto the flanks of the volcano. It became a series of pyroclastic flows that killed everyone in their paths, including many who had escaped west to the shore of the Bay of Naples.

Many of the survivors walked around on the pumice during a pause in the activity. Then another big surge of scalding steam and ash swept over them and buried what remained of Pompeii, killing another 2000 people. Archaeologists found many of those last victims inside their houses. They were lying on the earlier deposits of ash and pumice, some holding cloths over their faces. Perhaps they had not left their homes because it surely seemed safer than the frightening scene outside. Ultimately, those who remained were buried under 2.5 m of ash and pumice.

In the early hours of August 25, a series of pyroclastic flows overwhelmed Herculaneum. They covered the 6 km from the summit of Vesuvius in four minutes.

Archaeologists found only six skeletons in the town of 5000 and hundreds of others in the boathouses along the bay where people had sought shelter. They had died instantly when they inhaled the hot ash and steam.

The pyroclastic-flow deposits are 20 m deep at Herculaneum. They toppled walls of some buildings and buried nearly

Modern town of Ercolano

Top of pyroclastic flow

Boat chambers

■ *Boat chambers at Herculaneum are just below the foreground. The current town of Ercolano in the background rests on top of the ash that buried Herculaneum. Vesuvius looms in the background.*

all of the others. The people who built the current town of Ercolano on top of these flows did not realize that its ancestors were entombed below. A farmer digging a well discovered part of Herculaneum in the early 1700s.

At breakfast time on August 25, a culminating sixth surge of ash and pumice swept across one town, 15 km south of Vesuvius, where it dumped 2 cm of ash. Pliny the Younger wrote about the scene, across the Bay of Naples, 28 km west of the crater. He fled with others as the dense black cloud of background hot ash and steam raced toward them across the surface of the water. The dark cloud

followed them, hugging the ground— behind it, "fire." His must have been a narrow escape.

Rapid expansion of the eruption column quickly cooled the hot steam, condensing its water vapor on particles of ash, which fell as muddy rain onto the loose debris on the flanks of Vesuvius. Much of it poured rapidly downslope as mudflows that killed more people. As usual with Plinian eruptions, no lava flows appeared. The total volume of magma erupted was 3.6 km^3 within less than 24 hours, approximately 18 times the volume of magma that erupted from Mt. St. Helens in 1980.

CRITICAL THINKING QUESTIONS

1. If you were a fisherman living along the coast at the base of Vesuvius, comment on the likelihood of survival if you put out to sea at the beginning of the main eruption. Consider available evacuation time and distance from shore. Consider the events at Mt. Pelée, 1902 (available online, see bottom of page 170), and Vesuvius in your answer.
2. If Vesuvius were raining ash on your home but it seemed to be a moderate rather than violent eruption, what important factors should you consider?
3. Which eruption type is most dangerous, lava or ash, and why?

Kilauea's East Rift Eruptions Continue
Kilauea, Hawaii, 1983–2015 ▶

Eruption of basalt lava flows along Kilauea's East Rift in January 1983 began with lava fountains that quickly spread along about 7 km of the rift. By June 1983, activity largely centered a few kilometers east of the initial eruption, where it built a cinder and spatter cone, later called PuʻuʻŌʻō. Individual eruptions lasting less than a day were separated by quiet periods of a few weeks erupting from a newly developed long, thin magma chamber that was open at the top. Additional fissures spread along the

rift for about 6 km. In 1986, activity shifted a few kilometers northeast to a lava lake from where it spread lava flows southward down the flank of the rift for the next five years. They overran Royal Gardens and Kalapana subdivisions in 1986 and remaining houses in 2010. New vents opened on the flank of PuʻuʻŌʻō, sending lava flows downslope to the ocean, except for a few months-long pauses, almost continuously until 2014.

Many of the flows were fed from vents in lava tubes on the volcano's flank. It was the longest and largest eruption from the East Rift in the past 500 years.

If you lived in the laid-back community of Pāhoa farther east near the East Rift Zone of Hawaii's Kilauea volcano, you would likely know that you lived on old lava flows. Although the East Rift has erupted almost continuously in the past 32 years, it hadn't come close to this area until recently.

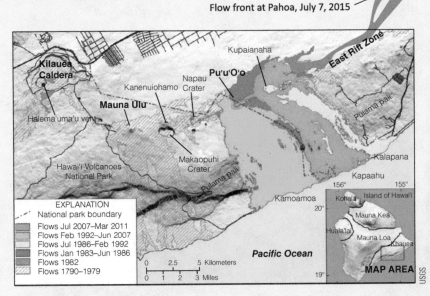

Flow front at Pahoa, July 7, 2015

EXPLANATION

⌐⌐ National park boundary

Flows Jul 2007–Mar 2011
Flows Feb 1992–Jun 2007
Flows Jul 1986–Feb 1992
Flows Jan 1983–Jun 1986
Flows 1982
Flows 1790–1979

■ A. *Kilauea's eruptions since 1983 have been on its East Rift Zone.*

■ B. *PuʻuʻŌʻō was the center of early eruptions and has been intermittently since.*

C. *Remains of Royal Gardens subdivision low on the flank of Kilauea, which was destroyed by lava flows in 1986.*

In most recent activity, most of the basalt lava flows poured down the south-facing flank of the high-standing East Rift, some invading parts of the Royal Gardens subdivision. In 2012 and again in 2013, new lava flows running northeast from the PuʻuʻŌʻō crater continued for a few months before they ceased. In June 27, 2014, basalt lava again broke out northeast of PuʻuʻŌʻō along rift-zone cracks, then flowed eastward toward the small town of Pāhoa. The thin river of lava flowed slowly in fits and starts, frequently bursting out of breaches in lava tubes. Sometimes stalling, sometimes moving at 5 to 20 m/h, they reached the edge of Pāhoa where they stalled on October 30. Breakouts formed at intervals upslope from the stalled tip for months thereafter but flows again advanced in June and July, 2015 to the edge of Pāhoa.

CRITICAL THINKING QUESTIONS

1. What precursors suggest that a new eruption has started? Explain how each works.
2. Why do you think that some eruptions at PuʻuʻŌʻō are explosive ash instead of quieter lava flows? Explain.
3. Why do most of Kilauea's eruptions occur along ridge crests? Why do initial eruptions rarely occur at much lower elevations such as along the coast?
4. How does new flowing lava appear far down the flank of the volcano when there is no lava flowing higher on the same slope?
5. Why is it dangerous to walk on a solid black lava surface that is near red-hot moving lava?

Critical View

A Ash deposits from Redoubt volcano, Alaska provide clues to the nature of its eruptions.

C. Neil, USGS

1. What type of eruption process would deposit each of the two main layers in this photo? What characteristics indicate the processes involved?

2. Where would you expect to find the thickest and thinnest parts of each layer?

C This astronaut view of the Rio Grande Valley, New Mexico, shows the top of a giant volcano, where a large depressed area is flanked by stream valleys draining downslope.

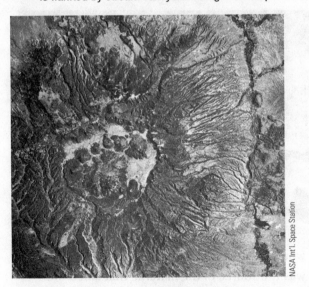

NASA Int'l. Space Station

1. What is such a depressed area called and how does it form?

2. What is the minimum diameter of the underlying magma chamber?

3. Is this a pre- or post-eruption view?

4. Is such a volcano likely to erupt again? Why?

B Hawaii lava flows provide evidence to the style of their eruptions.

Donald Hyndman

1. What are the lava flows called and what kind of rock forms them?

2. What does the distinctive surface of the lava flow tell us about how the magma erupted, how fast it likely flowed, and how viscous the lava was?

3. The nearly black flows in the background did not flow as far. What does that tell you about their viscosity compared with those in the foreground?

D A lava flow blocks a road in Hawaii.

USGS

1. Are the people in this picture in danger? Why or why not?

2. What is burning and why is it burning?

3. The slope on the horizon to the right of the brown hut is almost as flat as the surface of the ocean on the left. What built that slope and why is it almost flat?

E Only the roof is visible of this house situated at the foot of Mt. Pinatubo, in the Philippines. The inset photo shows the house before the 1991 eruption.

USGS and Raymundo Punongbayan, Philippine Inst. Volc. & Seismol

1. What material makes up the lower half of the main photo and where did it come from?
2. Did this material likely form before, during, shortly after, or months or years after the eruption? Explain the possibilities and how each would develop.

G This battered car near the foot of Mt. St. Helens, in the early 1980s is surrounded by dead trees that have no branches and are mostly lying on the ground.

J. Rosenbaum, USGS

1. What specific process or processes would have produced this kind of damage?
2. What is the light gray material on the car and on the log and specifically how did it get here?
3. Would the car driver have survived this event and if not, specifically what would have killed him or her?

F This hummocky landscape was located near the foot of Mt. St. Helens in the early 1980s.

Hyndman

1. What material makes up the lumpy hills in the foreground?
2. Where did it come from and how did it get here?
3. Why are these small hills so lumpy whereas farther down valley (in distance) the valley floor is flat?

H This forest is located near Mammoth Mountain, in southeastern California.

USGS

1. What would have killed this large patch of dead trees?
2. What was the source of the material that killed the trees?
3. Could this same material kill animals and humans in the same area?
4. Have you ever come in contact with this same deadly material? When? Under what circumstances?

Chapter Review

Key Points

Volcanic Hazards

- Hazard posed by a volcanic eruption is highly dependent on the population near the erupting volcano as well as the eruption product. **FIGURE 7-1**.

- Lava flows are destructive to property, because they ignite and overwhelm everything in their path. They do not pose a significant threat to human lives because they are slow moving. **FIGURE 7-2**.

- Pyroclastic flows consist of ash and steam that rush down volcano flanks. They may be preceded by an ash-rich shock wave called a surge. They can also cross bodies of water. The distance a flow will travel is related to the height of the main eruption column and the density of the ash in it. **FIGURES 7-3** to **7-7**.

- Ash falls pose a threat of collapsing roofs if the deposits are thick enough, and even a small amount of ash can disrupt our main forms of transport. Planes are especially susceptible if they fly through an ash cloud. Suspended ash can also block solar radiation and lead to crop failure. **FIGURES 7-8** to **7-10.**

- Mudflows can be triggered when hot ash combines with snow or rain. Hot mudflows are called lahars. Stratovolcanoes are particularly susceptible to mudflows because of their unstable makeup. Fast-moving mudflows often pick up boulders and are a major threat to communities downslope from volcanoes. **FIGURES 7-11** to **7-14**.

- Volcanoes emit poisonous gases, called vog, including carbon dioxide, sulfur dioxide, and hydrogen sulfide. High concentrations of vog can kill people and animals. Vog can also produce acid rain that may contaminate drinking water. **FIGURES 7-15** to **7-17**.

Predicting Volcanic Eruptions

- Predicting a volcanic eruption well in advance is not currently possible, but scientists can provide long-term forecasts and, once activity begins, often warn of an impending eruption.

- Understanding the hazard presented by a particular volcano depends on analysis of the type, size, rock composition, and frequency of past eruptions. **FIGURES 7-18** and **7-19**.

- A volcano that has not erupted in thousands of years may still be active—intervals between eruptions can be very long.

- Precursors to eruptions include harmonic earthquake tremors that migrate toward the earth's surface, changes in the level or tilt of the ground surface, and changes in erupted gases. **FIGURE 7-20**.

Mitigation of Damage

- Lava flows cannot generally be stopped or redirected without extraordinary measures.

- Warning of ongoing mudflows can be given with modern technology, but the time before arrival can be quite short.

Populations at Risk

- Areas with significant volcano hazards include Italy and the Cascades of western North America. **FIGURES 7-21** to **7-31**.

- Rainier's greatest threat probably lies in the paths of mudflows likely to pour down the broad valleys to the north and west at speeds of more than 100 km/h. Such mudflows could develop without a volcanic eruption.

Key Terms

lahar, p. 154
mudflow, p. 154
paleovolcanology, p. 158

pyroclastic flow, p. 149
surge, p. 149

tuff, p. 158
vog, p. 156

volcanic ash, p. 151
volcanic weather, p. 152

Questions for Review

1. What products of a volcano can kill large numbers of people long after an eruption has ceased? How?

2. Which is more dangerous, a lava flow or a pyroclastic flow? Why?

3. How can you tell whether a plume rising from a stirring volcano contains new magma that may soon erupt?

4. What product of a volcanic eruption causes a widespread drop in temperature and possible crop failures?

5. Why does it often rain during a volcanic eruption?

6. What commonly triggers mudflows?

7. Why is erupting volcanic ash dangerous to jet aircraft?

8. If you visit Mt. St. Helens, Washington, you will see thousands of trees lying on the ground, all parallel to one another. Explain how they got that way.

9. If a pyroclastic flow approaches you from across a 1-km-wide lake, are you likely to be safe? Explain why or why not.

10. What characteristics of an old ash-fall tuff will permit you to distinguish it from an old pyroclastic-flow tuff?

11. Which erupting volcanic hazard kills more people than anything else? What accounts for this danger?

12. Scientists once believed that if a volcano had not erupted in the last 10,000 to 12,000 years, it was extinct. Give at least one example that shows this is not correct.

13. What evidence do scientists use to decide whether a volcano may be getting ready to erupt?

14. What causes a big bulge to slowly grow on the flank of an active Cascades volcano?

Critical Thinking Questions

1. Mt. Vesuvius, near Naples, Italy, is considered to be an active volcano that is a hazard to nearby cities. If you were in charge of civil defence in Naples and the chief volcanologist in charge of monitoring Vesuvius advised you that Vesuvius had a 60% chance of a major eruption within two weeks, what would you do? Consider evacuation of large numbers of people, how you would do it, and when you would do it.

2. Discuss why volcanic eruptions are unpredictable in the short term. What factors likely contribute to difficulties in such predictions?

3. If you were designing a study to evaluate the risks of an eruption of Mt. Rainier, what type of data would you collect and how would you propose to use these data to minimize risk to those living in the region?

Homes in Laguna Beach, southern California, collapsed in a 2005 landslide after the slope became water saturated from El Niño winter storms.

Landslides and Other Downslope Movements

8

Unstable Hills

I n October 1978, a landslide destroyed 24 homes in the steep, forested hills above the town of Laguna Beach, south of Los Angeles. The 15- to 20-m-thick slide moved along a surface parallel to the slope, 6 m above an ancient slide surface sloping 30 to 35° on slippery clay. Heavy rains during the previous winter slowly soaked the ground. Months later this caused high water pressure within the tiny pore spaces between soil grains, which helped buoy up the overlying mass, permitting it to slide.

Surprisingly, the next year the slope was regraded and the homes were rebuilt. Then, 27 years later, in June 2005, a new slide destroyed 19 homes on the same slopes. Again water from heavy winter rains had saturated the ground a few months earlier. (See **Case in Point:** Ongoing Landslide Problems—Coastal Area of Los Angeles, p. 210.)

Forces on a Slope

As with many of the hazards discussed in previous chapters, downslope ground movement is a natural part of landscape evolution. Mountains rise up and crumble down. Rocks and other materials are constantly moving downslope. Gravity pulls a rock on a slope vertically downward. It may move slowly in a gradual process called *creep*, but unless the underlying soil or rock prevents it, the rock can slide or roll down the hill. When a large volume of material moves downslope quickly, it can be catastrophic.

The ability of a slope to resist sliding depends on the total driving force pulling it down versus the resisting force holding it back. The **driving force** consists primarily of the force of gravity working on the weight of the material, while the **resisting force** consists of the strength of the material and the friction holding it in place. Factors such as slope steepness, material weight, and moisture content all play roles that determine when a slope will fail.

Slope and Load

The relationship between **slope angle,** or steepness of a slope, and the **load,** or weight of material on the slope, is a key factor in slope failure. The steeper the slope, the greater the driving force, and the greater the likelihood that the slope will fail. The **angle of repose** is the steepest angle at which any loose material is stable. Different materials stand at different angles of repose, depending mainly on the angularity and size of their grains and their moisture content. Rounded, dry sand grains on a pile will stand no steeper than approximately 30° (measured down from horizontal). Adding more grains to the pile will cause some of them to roll down or patches of them to slide until the slope angle flattens to its angle of repose. In contrast, wet sand will stand almost vertical (**FIGURE 8-1**).

The downslope driving force is determined by the relationship between the slope steepness and the load, or weight, of materials on it. The load imposed by a rock is a factor in both the driving and resisting forces on a slope. The load can be separated into two components, the driving force that pulls down parallel to the slope, and the resisting force that holds it back. On a steeper incline, a greater proportion of the sediment or rock load is directed parallel to the slope, which increases the driving force and raises the chances of slope failure. Increasing the load on a slope will also increase the likelihood of slope failure.

Frictional Resistance and Cohesion

The driving force pulling materials down a slope is countered by the resisting forces of frictional resistance and cohesion. **Frictional resistance** depends on the slope angle (α) and the load (L) of the body. The area of contact between a mass and the underlying slope does not affect frictional resistance,

FIGURE 8-1 Angle of Repose
Dry sand slides down a slope until it reaches the angle of repose at about 30°. Wet sand (dark pile on left) can stand nearly vertical.

which means that both small and large masses of the same material will slide at the same slope. The mass will slide, or the slope will fail, when the force (F) exceeds the frictional resistance (f) (**FIGURE 8-2**). Anything that reduces the friction on the slope will increase the likelihood of slope failure.

Cohesion (C) is an important force holding soil grains together. It is generally provided by the **surface tension** of water between loose grains or cement between grains. **Cohesion** results from the static charge attraction between minute clay particles, the surface tension attraction of water between grains, or the strong chemical bonds of a cementing material. Most particles have tiny static charges on their outer surfaces. If you walk across a carpet on a cold day, you may feel the shock when you touch a doorknob and the negative-charged electrons you picked up from the carpet jump to the doorknob. Billions of extremely fine clay particles (<30 microns, or <0.03 mm) have sufficient charges to hold tightly together.

For somewhat larger grain sizes, such as sand, the thin films of water between grains hold them together through cohesion. A child playing with sand quickly learns that a little water makes sand stick together so it can be molded into sand castles. That little bit of water is sufficient to wet the surfaces of sand grains but not fill the spaces between them (**FIGURE 8-3**). For the same reason, two wet boards are hard to separate, as is a piece of glass on a wet countertop. The surface tension attraction of water in a narrow space pulls the glass to the countertop.

Cohesion is like a weak glue that can be overcome if the sliding force is large enough. When the ground is saturated with water, there is a buoyancy effect that also decreases the mass pushing against the slope. Clays that get wet also lose cohesion.

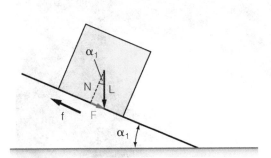

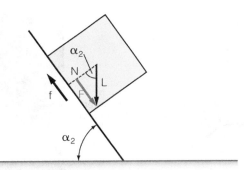

α = Slope angle
L = Load or weight
N = Force perpendicular to the slope
F = Force parallel to the slope
f = Friction holding the mass from sliding

For a moderate slope of 30°
Force parallel to the slope = Load × sin 30° = Load × 0.5 (e.g., 100 kg × 0.5 = 50 kg).

For a steep slope of 60°
Force parallel to the slope = Load × sin 60° = Load × 0.87 (e.g., 100 kg × 0.87 = 87 kg).

The driving force (F) parallel to the slope is proportional to slope angle.

FIGURE 8-2 Driving Force and Resisting Force

These two diagrams show the forces on a mass resting on a slope. A steeper slope has a larger force parallel to the slope (red arrow) and is therefore more likely to slide. Note that for the gentler slope, the friction force (f) is larger than the force pulling parallel to the slope (F), whreas for the steeper slope the opposite is true. Clearly the mass on the steeper slope is more likely to slide.

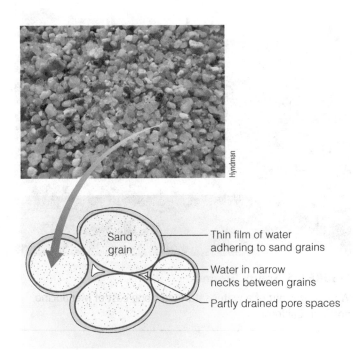

FIGURE 8-3 Surface Tension

Loose grains with water (blue) filling spaces between them. If water does not completely fill the pores, surface tension holds the grains together.

In summary, slopes fail when a driving force is large enough to overcome both the force from a resisting mass and the friction along the surface, as well as the cohesion strength of materials (**By the Numbers 8-1**: Slope Failure). As the upper surface of a slide rotates back toward a hill, the driving mass decreases and the resisting mass increases until the forces are again in balance and the slide slows or

By the Numbers 8-1

Slope Failure

Mass will slide if the:

Frictional resistance + Cohesion < Driving Force

$$f + C < F \text{ or } (N - p) \times \tan \alpha + C < F$$

resisting force < driving force

where

N = force perpendicular to the slope

$(N - p) \times \tan \alpha$ is "the force against the slope minus the pore pressure of water" times "the tangent of the slope angle"

p = pore pressure

α = slope angle

C = cohesion: soil cohesion includes the soil strength plus the root strength.

stops. As with many geological processes, slopes maintain a dynamic equilibrium to keep those forces in balance. Over time, slopes adjust to near-equilibrium values controlled by the local environment; that is, they reflect the slope material, climate, and thus the water content of the soil.

Slope Material

The material that makes up a slope—as well as its topography and moisture content—also play a role in slides. Loose material above the bedrock is inherently weak. So are loose sedimentary deposits not yet cemented into solid rocks or soft sedimentary materials, such as clay and shale. These materials are the most likely to slide.

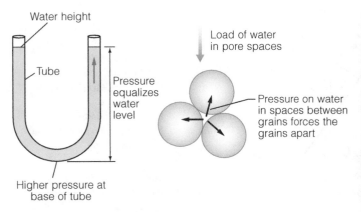

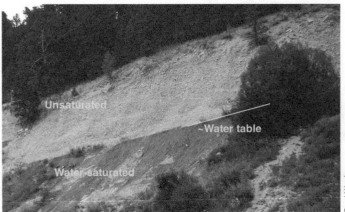

FIGURE 8-4 Water in the Ground

A. Water pressure at depth is equal to the load or weight of the overlying water.

B. Water seeps out from saturated soil, exposing the sharply defined darker water table in this road cut of Glacier National Park, Montana.

Moisture Content

As with loose grains in a child's sand castle, the amount of water between grains determines what effect the water will have on the strength of a slope. A small amount of water provides cohesion and helps hold grains together. Too much water (that is, an increase in water pressure) eliminates cohesion and pushes the grains apart.

The spaces between grains are called *pore spaces*. Loose soils have from 10 to 45% pore space. When the pore spaces are saturated with water, we can imagine them as continuous vertical columns of water within the soil. The **water pressure** at the base of each column is determined by the height of the water above it and therefore its load. More water in the slope raises the level of water in the soil, called the *water table*, and increases the pressure in the pore spaces at depth (**FIGURE 8-4A**). The weight of water above a point in the ground provides water pressure that pushes mineral grains apart. That further weakens the soil and makes it more likely to slide. In some cases, you can tell the level of water in the ground (the water table) by the fact that it is wet up to a certain height (**FIGURE 8-4B**). When the ground is fully saturated, water will ooze out of the surface, signaling an increased likelihood of slope failure (**FIGURE 8-5**).

Internal Surfaces

Most solid rocks, such as granite, basalt, and limestone, are inherently strong and unlikely to slide. However, rocks commonly contain more or less planar internal surfaces of weakness that may be tilted at any angle. These include layers in sedimentary rocks, fractures in any kind of rock, contacts between rocks of different strengths, such as between soils and underlying bedrock, faults, and slip surfaces of old landslides. If any of these zones of weakness happen to lie

FIGURE 8-5 Soggy Ground

Water oozes out of a water-saturated slope east of Mendocino, northern California.

nearly parallel to a slope, they are likely to become a slip surface. Such rocks are especially prone to slide if their zones of weakness are angled downslope.

Internal surfaces that dip at gentler angles than the slope of a hill, called **daylighted beds**, may intersect the lower slope. These daylighted beds, which are often exposed at their lower ends by a road cut or stream (**FIGURE 8-6**), make ideal zones for slippage. Because there is no resisting load holding them back, only the friction between the layers, along with cohesion, can keep the mass from sliding. The rock above the slip surface does not have to push any rock out of the way to start sliding.

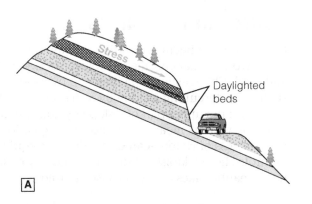

A

B

Donald Hyndman

FIGURE 8-6 Daylighted Beds

A. Rock layers dipping about parallel to a slope, with their edges coming to the surface, are said to "daylight."

B. Daylighted joints in slate, north-central Idaho.

Clays and Slope Failure

Some soils or rock materials contain clays that absorb water and expand, thereby weakening the rock and even lifting it. Feldspars, the most abundant minerals in many rocks, are grains of aluminum silicates, which also contain calcium, sodium, or potassium in various proportions. Chemical weathering of all minerals consists primarily of their reaction with water. As they weather, feldspars lose most or all of their calcium, sodium, and potassium, while their aluminum, silicon, and oxygen reorganize into clays with submicroscopic sheets of aluminum or silicon atoms that are each surrounded by four oxygen atoms. Two important clay minerals, kaolinite and smectite, have structures that can lead to landslides (these structures are on the scale of individual molecules and are too small to be seen with a microscope; see more details in Appendix 2 online).

Kaolinite flakes have no overall charge, but they do have weak positive charges on one surface (the top) and weak negative charges on the other. The weak positive and negative charges attract, holding the layers together. Individual kaolinite flakes do not absorb water and do not expand when wet. They form by weathering in warm, wet environments. However, their overall structure is soft and weak, soaks up water, and thus contributes to landslides.

Smectite flakes have an open structure between their molecular layers, which when filled with water, causes the clay to expand dramatically; these are swelling soils. Because water has virtually no strength, almost any load will cause layers to slide easily over other layers. Smectite forms readily by weathering of volcanic ash, so soils on old volcanic ash tend to be extremely slippery and prone to landslides when they are wet (**Case in Point:** Slippery Smectite Deposits Create Conditions for Landslide—Forest City Bridge, South Dakota, p. 212).

Water-saturated muds in marine bays, estuaries, and old saline lakebeds are called **quick clays** because they are especially prone to collapse and flow when disturbed. Silt

and clay grains are so fine that water cannot move through the tiny pore spaces quickly enough to escape. When flakes of clay are deposited in salty water, static negative charges on their tiny grains hold them apart; the flakes remain in random orientations so the mass has a total pore space of 50% or more. This "house of cards" with water and sea salts in between is unstable. Salt dissolved in water separates into positively charged sodium ions and negatively charged chlorine ions. Because tiny clay flakes have negative charges, the positive charges in the water hold the combination together. If this loose arrangement is disturbed by an earthquake or a heavy load above, the flakes can collapse, permitting the water to escape and "float" the flakes in a process called **liquefaction** (**FIGURE 8-7**).

Liquefaction occurs as grains settle into a closer packing arrangement with a lower porosity, expelling a fluid mix of grains and water—a quicksand or quick clay. Liquefaction can also occur on a flat surface, where it does not cause a slide but can still collapse buildings. The deposit liquefies

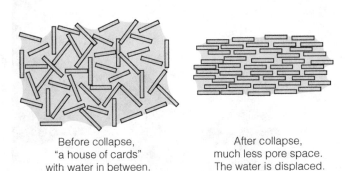

Before collapse,
"a house of cards"
with water in between.

After collapse,
much less pore space.
The water is displaced.

FIGURE 8-7 Collapse of Quick Clays

Clay grains deposited in random orientation (left) have especially large pore spaces—a "house of cards" arrangement. After collapse (right), the compacted clays take up much less space, so water in the pore spaces must escape.

and flows almost like water during the minutes that the clay flakes are moving into their new orientation. Then it again becomes solidly stable and will not liquefy again. The puddles of water that appear during the event filled the much larger volume of pore space that existed before the collapse.

Quick clays are common along northern coasts, including those of Canada, Alaska, and northern Europe. They make perilous foundations for almost any building. When tectonic movements raise a marine or salty lake sediment above the level of the surface water body, freshwater generally washes the salt from between the flakes of clay, leaving the mass even less stable. Water seeping through the deposit eventually removes positive sodium ions that attach to the negative-charged clay flakes. Then the deposit becomes a stack of randomly oriented clay flakes with nothing holding them rigidly in position. That is an unstable situation. One day, the deposit liquefies as the clay flakes collapse into their usual parallel orientation, like sheets of paper scattered across the floor. This commonly happens without warning, even with little or no triggering event.

Vibrations caused by an earthquake, pile driver, or heavy equipment can cause failure of quick clays. Even loading a surface can be enough to send a mass downslope as a muddy liquid. In Rissa, Norway, in 1978, a farmer piled soil at the edge of a lake, thereby adding a small load. It triggered a quick-clay slide covering 33 hectares (330,000 m^2) of farmland. The widespread Leda clay in the St. Lawrence River valley of Ontario and Quebec is another sensitive marine clay. On June 20, 1993, a nearly flat 2.8 million m^3 clay terrace settled and slowly flowed down a gentle slope into the South Nations River near the town of Lemieux (**FIGURE 8-8**).

Dr. G. R. Brooks, Geological Survey of Canada

FIGURE 8-8 Flowing Ground
The Lemieux flow in a horizontal terrace of the Leda clay of the St. Lawrence River valley near Ottawa, Ontario, settled and flowed into the adjacent river on June 20, 1993.

Causes of Landslides

Recall from the beginning of this chapter that slope equilibrium involves balancing the relationship between slope angle and load. Changes in slope imposed by external factors—such as undercutting a slope by a stream or road building; loading of the upper part of a slope by construction; adding water by various means; or removing vegetation to permit more access of water—can destabilize this equilibrium and promote sliding. Moist conditions, instability of slope material, or earthquakes can also cause landslides.

Oversteepening and Overloading

The balance between the forces acting on a slope can be upset by making the slope steeper or by increasing the load on the slope. The load can be increased by adding material or other load at the top, or removing material at the toe. Load is sometimes added in the form of buildings and soil fill or naturally by rain or snow. Slope angle is increased when fill is added above or when slopes are undercut below. Because steeper slopes are less stable, oversteepening of a slope increases the likelihood of slope failure.

Oversteepening can occur through natural means, such as erosion at the base of a slope. Or it can be human-induced through excavation or adding fill. Often the problem is a combination of such factors. One notorious example is Highway 1 along the coast of central California south of Monterey. Excavated through steep slopes, this highway is frequently damaged and sometimes closed due to landslides. The largest recent landslide events in the area coincided with large El Niño events in 1982–83 and 1997–98, when big waves, strong coastal erosion, and heavy rains hammered the coast. Wave action at the base of the coastal cliffs and rains soaked the fractured rocks to add internal pore pressure, providing ideal conditions for landsliding (**FIGURE 8-9**).

Aside from convenience of location, people build at the tops of steep slopes to take in magnificent views of the sea, lakes, or rivers. Unfortunately, these slopes are steep for a reason; they are being oversteepened by processes that remove material at their bases. Waves or rivers undercut cliffs, and roads undercut the slope bases. Some of these slopes are soft materials, able to stand for long periods until people dig into them to provide flat areas for houses or yards. In doing so, they undercut the slope above and overload the slope below them, making the ground more prone to sliding (**FIGURE 8-10**). This type of development often includes septic systems or swimming pools, which increase soil moisture content and can further contribute to landslide risk.

On some particularly steep slopes, such as the face of a terrace steepened by wave action, the toe of a slide may provide little or no resisting mass. In one example along the wave-eroded coast north of Newport, Oregon, expensive homes at the top of 20-m-high cliffs lost more than 6 m of property on November 6, 2006. Landslides also threatened

FIGURE 8-9 **Wave Action**

Waves undercutting coastal cliffs of Big Sur, south of Monterey, California, cause frequent slides that destroy the highway, shown here in 2009.

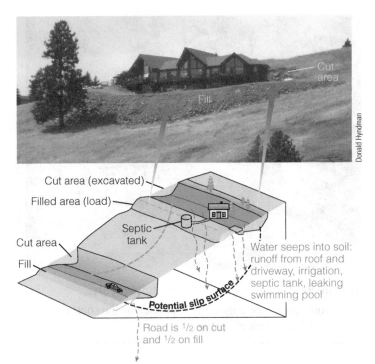

Cut area (excavated)
Filled area (load)
Septic tank
Cut area
Fill
Water seeps into soil: runoff from roof and driveway, irrigation, septic tank, leaking swimming pool
Potential slip surface
Road is ¹/₂ on cut and ¹/₂ on fill

FIGURE 8-10 **Recipe for a Landslide**

Adding weight or load (fill), slope steepening (cut and fill), building with loose material (fill), and adding water all contribute to landslide risk.

to take the homes. Unusually heavy rains triggered the slides, closing roads in northwestern Oregon; most rivers in western Washington and Oregon reached flood stage. The huge waves and heavy warm rains are a common result of an El Niño event.

In some such cases, a slope failure may occur rapidly, with tragic results. In one such case, part of the face of a steep tree- and brush-covered bluff collapsed suddenly, crushing a home in which a teacher and his young family

were sleeping, instantly killing them (**FIGURE 8-11**). The home was part of a single row of houses on the narrow strip of beach on Puget Sound near Seattle. The teacher's parents had expressed concern about the home's location

Puget sound

FIGURE 8-11 **Dangerous Hillsides**

A. A section of coastal cliff at the edge of Puget Sound, near Seattle, collapsed, crushing and burying a home and family on the narrow strip of beach.

Puget sound   Wave-undercut

B. Magnolia Bluffs, a suburb of Seattle, with spectacular views over Puget Sound, is continually plagued by landslides that threaten and destroy homes. Slides are often promoted by wave erosion at the cliff base.

FIGURE 8-12 Dangerous Cliff-Top Building Site
This house was wrecked by a 2013 landslide on the coast of
Whidbey Island, north of Seattle.

FIGURE 8-13 Saturated Slopes
The community of La Conchita, California, huddles at the base
of a soft-sediment terrace that gave way when soaked with too
much water. The 1995 landslide buried houses at the edge of the
town but no one died. In 2005, a second collapse of the earlier
landslide mass destroyed more houses to the right, killing
10 people. The irrigated avocado field is visible above the slide,
and an ancient slide scarp is visible just below the terrace, to the
left of the slide.

but had received assurances that it had been there for
70 years and there had never been a problem. In another
tragic example, at about 3:45 AM on March 27, 2013, a huge
landslide on Whidbey Island, 80 km north of Seattle, took
out 300 m of road, destroyed one house, and required the
evacuation of 33 nearby homes (**FIGURE 8-12**). The homes
were on top of a 200-foot-high bluff above Puget Sound.
The toe of the deep-seated slide lifted beach gravels as
much as 10 m.

Adding Water

Adding water to a slope makes it more likely to slide,
because water increases the load on the slope and reduces
its strength. Periods of heavy or prolonged rainfall tend to
saturate soil, increasing pore water pressure, thereby pushing apart grains and causing slides. Other things being
equal, slopes in wet climates are generally more prone to
landsliding.

Human actions also add water to slopes and increase
landslide potential. Leaking water or sewer pipes or cracked
swimming pools feed water into the soil, as do septic drainage fields and irrigation canals. Prolonged watering of lawns
or excessive crop irrigation will raise the water table and
make a slope less stable.

Water infiltration triggered deadly landslides in La Conchita, California (**FIGURE 8-13**). The near-vertical face of
the 100-m-high terrace above the town is made of soft,
weak, and porous sediments that are not cemented into
rock. An irrigated avocado orchard at the top of the terrace and heavy winter rains soaked it with water. Open
cracks visible on the slope in the summer of 1994 were
feeding water into the sediments making up the bluff, and
water seeping out along a horizontal surface suggested an

old slip plane that could easily be reactivated. A primitive
road excavated across the face of the steep slope made it
weaker and likely permitted the infiltration of even more
water. On January 10, 2005, a huge debris flow of material swept rapidly downslope and buried 15 houses under
10 m of lumpy wet sand and clay, killing 10 people. Even
after the second fatal tragedy, many residents chose to
remain, including one man whose home was destroyed
and brother was killed. On December 22, 2010, there was
a new small landslide a few blocks north of the previous
slides, emphasizing the continued instability of slopes in
the area.

Filling a reservoir behind a dam may also raise the water
table enough to cause slides around the edges of the reservoir (**FIGURE 8-14**). In some cases, the rising water fills fractures in surrounding sedimentary layers sloping toward the
reservoir, causing massive sliding into the reservoir (**Case in
Point:** Slide Triggered by Filling a Reservoir—Vaiont Landslide, Italy, 1963, p. 213).

Removal of vegetation from a slope will permit more water
infiltration because brush and trees remove water by evaporation from their leaves and take up water through their roots.
Vegetation is often removed by wildfire or intentional burning, overgrazing, or clear-cut logging. A prominent example
of the result of logging occurred in southwestern Washington State when heavy rainstorms in December 2007 triggered
730 landslides and debris flows that thoroughly dissected
areas clear-cut by logging, leaving adjacent forests
untouched (**FIGURE 8-15**).

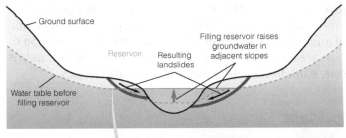

FIGURE 8-14 Rising Reservoirs

Filling a reservoir behind a dam raises groundwater in the adjacent slopes, often leading to sliding into the reservoir. Rising water in the reservoir behind the Three Gorges Dam on the Yangtze River in China triggered landslides in the adjacent slopes.

FIGURE 8-15 Landslides on Clear-Cut Slopes

Landslides and debris flows gulley clear-cut slopes along Stillman Creek in southwestern Washington, following an intense rainstorm, exposing underlying orange soils. Adjacent forest was not affected.

Overlapping Causes

Most causes and triggers of landslides, such as steep slopes, undercutting, overloading, excess water, and earthquakes, are independent of one another. However,

it is when multiple factors combine that the greatest catastrophes often occur. In one unfortunate coincidence, the devastating mudflows that accompanied the 1991 eruption of Mt. Pinatubo in the Philippines developed in large part because a major tropical typhoon hit the area just in time for the eruption. Mudflows are common on recently erupted volcanoes; heavy rains easily saturate loose ash, which flows down valleys to endanger streamside communities.

Consider a possible scenario for Mt. Rainier, the highest and largest volcano in the Cascades. It has not erupted in 2500 years but is still thought to be active. Its snow-clad flanks are steep but weak; they are heavily fractured and have altered to clays. Winters are wet in the Pacific Northwest, with especially heavy snowpacks at high elevations. If the next giant megathrust earthquake (since the last one in January 1700) were also to happen in winter, strong shaking lasting three minutes or longer could cause a large part of the mountain's flank to collapse, potentially similar to the gigantic flank collapse of Mt. Shasta 300,000 years ago (**FIGURE 8-16**). If Mt. Rainier collapsed toward communities to the northwest, the consequences would be tragic. What about the possibility that water-bearing magma could be resting within the volcano at the time of collapse? A sudden decrease in pressure by flank collapse could trigger rapid expansion of magmatic gases and eruption, as in the case of Mt. St. Helens in 1980. None of these possibilities is implausible. The results could be truly disastrous.

In March 2014, one of the most tragic landslides in recent history occurred in response to a series of overlapping circumstances (**Case in Point:** Overlapping Causes for a Landslide—The Oso Slide, Western Washington, 2014, p. 209). Overlapping circumstances included loose sand over weak clay, an undercutting river meander, heavy rains before the event, and clear-cut logging on the terrace above.

FIGURE 8-16 Hummocky Landscape

The hummocks that cover most of this photo were deposited in a massive landslide from Mt. Shasta between 300,000 and 380,000 years ago.

Landslides and Other Downslope Movements **189**

Types of Downslope Movement

Landslides and other downslope movements are generally classified on the basis of material type, movement type, and rate of movement. Materials are classified into categories of blocks of solid bedrock; debris of various sizes mostly coarser than 2 mm; and earth or soil mostly finer than 2 mm. Water plays a major role in many of these, especially those involving smaller-size particles.

Rates of downslope movements are highly variable, even for individual mechanisms of movement. Rates depend on many factors, including slope steepness, grain size, water content, thickness of the moving mass, clay mineral type, and amount of clay. **Table 8-1** provides approximate movement rates.

Styles of movement include falls from cliffs, topples, slides, lateral spreads, and flows. Note that any of them can involve rock, debris, or soil. A continuous range of characteristics exists between most of these types of materials and styles of movement, and one type often transforms into another as it moves downslope. Moving masses are typically described by names that are a combination of the type of moving material and the style of movement. Among the most common of these are rockfalls, rock slides (talus), debris slides, debris flows, earth flows, mudflows, and snow avalanches.

Rockfalls

Rockfalls develop in steep, mountainous regions marked by cliffs with nearly vertical fractures or other zones of weakness. Large masses of rock separate from a steep slope or cliff, sometimes pried loose by freezing water, to fall, break into smaller fragments, and sweep down the slopes below. Factors that favor rockfalls include cliffs or steep slopes of at least 40°. Rocks most likely to cause rockfalls are those that break easily into fragments: granite, metamorphic rocks, and sandstone. Rockfalls can be triggered by strong earthquakes, passing trains, slope undercutting, and blasting during mining (**Case in Point:** A Rockfall Triggered by Blasting—Frank Slide, Alberta, 1903, p. 214).

Table 8-1		Typical Velocities for Various Types of Downslope Movement	
VELOCITY (CM/S)	**RATES**	**TYPE OF MOVEMENT**	
$10^5 = 100{,}000$ cm/s **1km/s**			Rockfall
$10^4 = $ **100 m/s** $=$ **360 km/h** (race car speed)	Extremely rapid		
$10^3 = $ **10 m/s** $=$ **36 km/h**		Rockfalls and debris avalanches (dry)	
$10^2 = $ **1 m/s** $=$ **3.6 km/h** (walking speed)	Very rapid		
$10^1 = $ **10 cm/s**			
$10^0 = $ **1 cm/s**	Rapid	Debris flows (wet) and mudflows (water-saturated)	Debris flow
$10^{-1} = $ **6 cm/min.**			
$10^{-2} = $ **36 cm/h**	Moderate		
$10^{-3} = $ **3.6 cm/h**			
$10^{-4} = $ **8.6 cm/day**	Slow	Landslides and slumps (moderately wet)	
$10^{-5} = $ **0.86 cm/day**			Rotational landslide
$10^{-6} = $ **2.6 cm/month**	Very slow		
$10^{-7} = $ **3.1 cm/yr**		Solifluction	
$10^{-8} = $ **3.1 mm/yr**	Extremely slow	Creep and expansive soils	
$10^{-9} = $ **0.3 mm/yr**			Creep

Diagrams modified from USGS

FIGURE 8-17 Talus Slopes

These sloping accumulations—talus slopes—of volcanic-rock blocks formed as rocks fell from the cliffs and rolled downslope, coming to rest at their angles of repose along the Salmon River, Idaho.

These rocks often collect in **talus** slopes, the fan-shaped piles of rock fragments banked up against the base of a cliff (**FIGURE 8-17**). Individual rocks, especially large boulders, may bounce or roll well away from the base of the slope. These can be particularly dangerous because of their size,

speed, and distance they travel. Giant boulders can sometimes demolish a house.

Rockfall hazards are widespread wherever rocky cliffs cap steep slopes. Denver is on the edge of the Great Plains, but communities just to the west are in the foothills of the Colorado Rockies, where steeply tilted layers of sandstone make ridges and cliffs. Volcanic rocks cap some bluffs, such as North Table Mountain, which sits just north of Golden. Big boulders that tumbled from its caprock cliffs litter the steep, grassy slopes below. Houses now cover the lower parts of the slopes. The county zoned undeveloped land immediately north of it as a no-build area in response to a rockfall hazard map published by the U.S. Geological Survey (USGS). A developer, determined to build a subdivision in the area, challenged this designation less than 10 years later and staged a rock-rolling demonstration to prove his claim that the area was safe. Unfortunately for him, several boulders rolled into the area of his proposed subdivision, one bouncing high enough to take out a power-line tower at the mountain's base. His application was denied.

Even in areas of relatively subdued topography and horizontal sedimentary layers, cliffs or high road cuts can be prone to rockfalls. Strong layers of sandstone or thick beds of limestone often break on vertical fractures. When intervening soft layers, such as shale, weather back to leave an overhang, the strong but fractured layers can break off and fall. The base of a steep slope capped by vertical cliffs that have shed big boulders in the past is no place for houses (**FIGURE 8-18**). Rockfalls can be hazardous not only to houses and people but also to highways. Roads often follow the base of precipitous cliffs or are built by excavating high road cuts that

FIGURE 8-18 Rockfall Hazard

A. Some rockfall problems arise where a strong layer, like sandstone, overlies a weak layer, like shale or clay, as in this southwestern Utah photo from March 2007. The largest rocks tend to roll well out from the base of the cliff. Is this a safe place to live? The question was unfortunately answered in 2013.

B. The same location, pictured in December 2013, shortly after more huge boulders fell from the upper cliffs, destroying the house and killing its two occupants. Remnants of the house an be seen below the huge rocks in the center of the photo.

destabilize original slopes. For example, huge rockfalls repeatedly occur at the De Beque Canyon landslide along Interstate 70 in Glenwood Canyon, western Colorado. The highway in this area is crowded between a massive cliff of sandstone with shale interlayers and the river.

Colorado landslides are most common in areas of Jurassic- and Cretaceous-age marine shales interlayered with stronger sedimentary rocks. Such formations range from about 140 m for the Morrison formation to 1800 or 2100 m for the Mancos and Pierre formations. These muddy units, especially the Pierre, with its volcanic ash-derived swelling clays, are especially weak when wet. Cliffs on some mountainsides or road cuts are held at near-vertical angles by interlayered sandstones, but the shales erode back into the cliff, leaving the sandstone layers protruding. Fractures across the sandstone layers permit boulders to split off and tumble into canyons or onto highways (**FIGURE 8-19**). A relatively small rockfall on November 10, 2003, closed Washington Highway 20 in the northern Cascades for an extended period of time.

High mountain areas often have high cliffs left by receding mountain glaciers. The ground under the glacier is cold, so water from intermittent melting refreezes in rock fractures. That frozen ground, permafrost, is exposed when the glacier melts, leaving it open to warmer air in summer. As long as the water in the fractures remains frozen, it may glue the rock together. When the rock warms, especially on south-facing cliffs (north-facing in southern hemisphere), the permafrost thaws and can release rockfalls. Such south-facing cliffs in areas down-valley from receding glaciers are especially dangerous for a few decades after glacial cover disappears. Many such areas have been identified in the Alps, Himalayas, Canadian Rockies, and Alaska.

ROCKFALL RUNOUT The distance a rockfall will travel, called its **runout**, is generally related to the height from which a rock falls as well as its mass—that is, to its potential

FIGURE 8-19 Hard and Soft
Weak shales and hard, fractured sandstones of the Cretaceous-age Dakota formation pose a rockfall hazard above I-70, east of Vail, Colorado. Note car at right edge of photo.

By the Numbers 8-2

Potential Energy of a Rock on a Slope

Potential energy of a given rock on a slope depends on its mass and its height on the slope.

Potential Energy = m × g × h

where

m = mass (kg)

g = gravitational acceleration (m/s^2)

h = height (m)

When the rock falls, its potential energy becomes kinetic energy—the energy of movement. Its highest velocity is near the bottom of its fall, so its kinetic energy is greatest there.

Kinetic Energy = ½ m × v^2

where

m = mass (kg)

v = velocity (m/s)

Thus, a larger rock higher on a cliff has more potential energy, will generally accelerate to a higher velocity, and will travel farther out from the base of the slope.

energy before it accelerates downslope (**By the Numbers 8-2: Potential Energy of a Rock on a Slope**). Higher, more massive rockfalls travel farther because they have greater potential energy. The same principle guides a ski jumper or a child on a sled: they start higher on a hill to reach farther out on the flat below. Much smaller rockfalls from lesser heights run out much shorter distances. This principle can be demonstrated by a slope of boulders in which the largest rocks are concentrated at the base of the field (**FIGURE 8-20**). The biggest boulders that fall from a cliff generally reach the base of the slope for two reasons: First, large blocks high on a cliff have more potential energy than smaller rocks and thus roll faster and farther; second, larger blocks roll easily on the slope's smaller material and keep going. Little rocks tend to get caught between any larger boulders on a slope.

A formula for the relationship between height and runout can predict the distance of runout for a given rockfall (**By the Numbers 8-3**). The relationship is similar to the energy line for volcanic pyroclastic flows.

Fast-moving rockfalls can run out for distances greater than the height of an original slide scarp. This happens through conversion of the potential energy of a material's original height into the energy of movement. Very large rockfalls can run out horizontally as much as 6 to 17 times their vertical fall. Thus, a mountainside 1000 m high may collapse to run out as far as 17 km (about 10.6 miles)! In midwinter 1965, the massive Hope landslide, east of Vancouver, British Columbia, buried Highway 3 when it swept across a valley and continued well up the opposite slope (**FIGURE 8-21**).

FIGURE 8-20 Runout and Rock Mass

A big talus slope, shed from the cliff above, shows a pronounced concentration of huge boulders at its base. This photo was taken at Canyon Campground, south of Livingston, Montana.

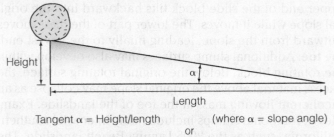

FIGURE 8-21 Highway Buried by Rockfall

The massive Hope slide, 18 km east of Hope, British Columbia, buried Highway 3 on January 9, 1965, with no warning and no apparent triggering mechanism. In 2009, when this photo was taken, the extent of the slide is still apparent.

DEBRIS AVALANCHES Rockfalls in which a material breaks into numerous small fragments that flow at high velocity as a coherent stream are called **debris avalanches.** Many of the largest and most destructive landslides in recorded history started as ordinary rockfalls that developed into debris avalanches. A classic case at Elm, Switzerland, more than a century ago, shows how a large rockfall can transform into a fast-moving catastrophic debris avalanche.

Debris avalanches can flow downslope at speeds of 100 to 300 km/h. Some contain boulders as big as houses. A catastrophic debris avalanche struck Yungay, Peru, on May 31, 1970. A magnitude 7.7 earthquake 130 km away, along the subduction zone offshore, triggered the slide on Mt. Nevados Huascarán, the highest mountain in Peru. It began with a loud boom and a cloud of dust as 50 to 100 million m³ of granite, glacial debris, and ice fell 400 to 900 m from a vertical cliff to the surface of a glacier and raced down the valley. It disintegrated and then picked up water from ice, streams, irrigation ditches, and soil, traveling downslope at speeds of 270 km/h (**FIGURE 8-22**).

Halfway down the slope, much of the debris, including huge boulders, hit a ridge of sediment and rocks deposited by a glacier and launched into the air, raining down on houses, people, and animals. Boulders weighing several tons flew as far as 4 km. Trees blew down for at least half a kilometer beyond the area of boulder impact. Mud splattered more than 1 km farther, in a blast of wind strong enough to knock people off their feet and shred bare skin. Survivors recalled that the strong wind arrived first, followed by flying rocks, then a huge wave of wet debris with a "rolling confused motion." The 1970 avalanche buried the entire city of Yungay, near the end of the flow, in 30 m of bouldery mud, killing more than 18,000 people.

FIGURE 8-22 Slide Buries a Town

The Mt. Nevados Huascarán debris avalanche in 1970 fell from the peak at the top of the photo and raced down the valley to bury the town of Yungay, which occupied the lower half of the photo.

On September 27, 1962, two American scientists reported in the local newspaper that a giant vertical slab of rock being undermined by a glacier was likely to fall and destroy Yungay. The government ordered them to retract the statement or face prison. Instead, they left the country. Unfortunately, eight years later, in 1970, their prediction came true. A mudflow on January 10, 1962, killed 4000 people in the same area.

Similar debris avalanches triggered by the eruption of volcano Nevado del Ruiz devastated the town of Armero, Columbia, in 1595, 1845, and 1985. More than 20,000 people were buried in the 1985 event. In that case, scientists had made the hazard clear, but national and provincial governments refused to order evacuations until the precise time of the event could be guaranteed. They were concerned about the cost of evacuation, the likelihood of looting, the possibility of a false alarm, and the political fallout. Officials failed to take responsibility, with devastating consequences.

Eyewitness accounts indicate that the total time from the earthquake to the flow's arrival at Yungay (14 km downslope) was approximately 3 minutes! The average velocity must have been 270 km/h, but the initial velocity on the mountain must have been faster than 750 km/h to fling rocks as far as 4 km. The victims never had a chance; they could not have seen this one coming. Equally precipitous cliffs of the avalanche scar now mark the peak, and broad fresh cracks parallel the cliffs in ice on the peak. The hazard of further collapse remains. People should avoid building on fans or in valleys that show evidence of previous debris deposits, especially after a recent large event like the 1962 flow.

A mass of rock that falls but does not disintegrate will not run far. The mechanism that permits high speeds and long runouts of debris avalanches has been a matter of considerable debate that has still not been resolved. Some authorities argue that debris avalanches ride on a cushion of compressed air trapped beneath them. Air entrained between rock fragments lubricates the mass. But major debris avalanches, such as the one at Elm, Switzerland, scoured deep furrows and excavated the ground beneath them, so some parts could not have ridden on a cushion of air.

Many characteristics of debris avalanche deposits coupled with witness descriptions suggest that debris avalanches flow as a fluid composed of rock fragments suspended in air. The mechanism, called **fluidization,** may work if air cannot readily escape the small spaces between rock fragments during the extremely brief period of movement. With tiny spaces between small particles, rockfalls composed of small grains should be more subject to fluidization than those composed of coarser particles. The air briefly supports the fragments and lubricates their flow.

Rotational Slumps

One of the most common landslide types is a **rotational slump**, also called a rotational slide (**FIGURE 8-23**). Homogeneous, cohesive, soft materials, those that lack a planar surface that guides landslide movement, commonly slide on a curving slip surface concave to the sky. The surface curves because at the top of the moving mass, gravity pulls it straight down; that vertical part of the slip surface is the **headscarp**. Farther downslope, the mass is also pushing outward, toward the open air where less load pushes down. The combination of the two forces rotates more and more outward toward the slope (**FIGURE 8-24**). The curvature of the slip surface rotates the slide mass as it moves, so the upper end of the slide block tilts backward into the original slope while it moves. The lower part of the mass moves outward from the slope, leading finally to the lowest end, the **toe**. Additional slump surfaces may also develop within the rotating block. Below the original rotating surface, the excess material above the original slope may collapse as an incoherent flowing mass at the toe of the landslide. Examples of rotational slumps include many in coastal southern California, such as the 2005 Laguna Beach landslide. The lower part of the slip surface commonly dips back into the slope. That provides some resistance so movement ultimately stops.

Engineers can estimate whether a rotational slump will move by calculating the forces on a slope. Because a rotational slump generally moves on a curving cylindrical surface (the green circular arc in Figure 8-24), engineers find the *center of rotation* by projecting perpendicular to any exposed part of that surface—such as the headscarp. They may also drill holes through a landslide to find the slip

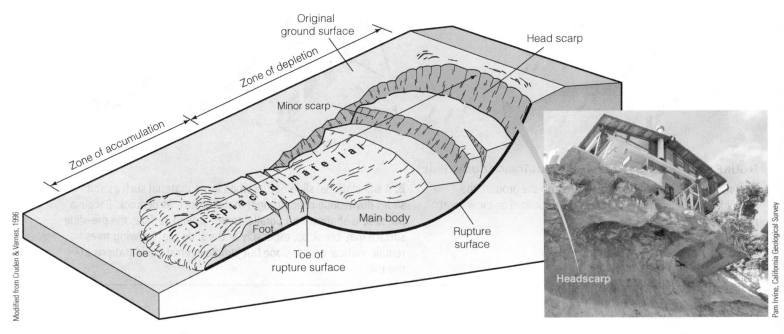

FIGURE 8-23 Rotational Slump

The main features of a rotational slump. A landslide in the hills above Laguna Beach, California, just south of Los Angeles, June 1, 2005, left this home hanging over the nearly vertical headscarp. This is the brown house in the upper left part of the chapter opener photo.

surface at depth. The slip surface may appear as a thin zone of smeared-out soil or a thin zone of thoroughly broken rock.

Having determined the shape of the moving mass, engineers calculate the total of the forces pulling the slide downslope (the driving mass) and compare that with the total forces holding it back (the resisting mass). If the driving mass is large enough to overcome the force from the

resisting mass, the friction along the potential slip surface, and cohesion, then the slide will move.

In some cases, backward rotation of the block leaves closed depressions at the top that may collect even more water or snow that can soak into the slide. In addition to load, this water increases pore pressure and may facilitate further movement.

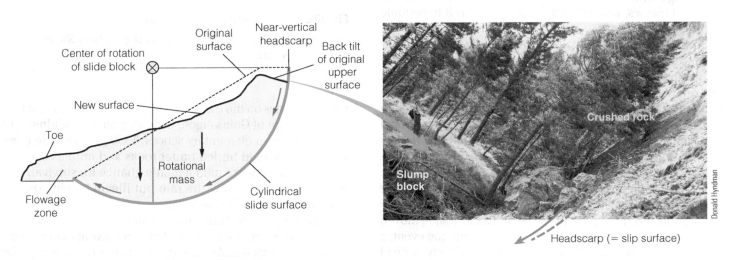

FIGURE 8-24 Anatomy of a Rotational Slump

A. This cross-section shows a rotational slump. Note that the failure surface for the driving mass slopes downward (right side of vertical line), and the failure surface for the resisting mass (left side of vertical line) slopes back into the slope.

B. Rotation of a slump block on an arcuate surface also rotates everything on the block. Trees growing on an originally horizontal upper terrace (the Blackfoot landslide in Montana) now tilt back toward the headscarp as a result of rotation. Note the person standing at the left. The headscarp is on the right.

Landslides and Other Downslope Movements **195**

 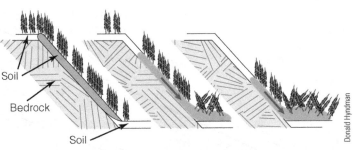

FIGURE 8-25 Rotational Slumps versus Translational Slides

A. A rotational slump penetrates deep into the ground; the pre-slide surface and growing trees tilt backward as movement progresses.

B. A translational slide occurs when loose material such as soil slides downslope over stronger material such as bedrock. Because the slide is shallow and parallel to the ground surface, the pre-slide surface may break up, especially at the slide toe. Growing trees remain vertical on the slope but jumble in erratic orientations at the toe.

Translational Slides

Translational slides move on preexisting weak surfaces that lie more or less parallel to a slope. These may be planes along inherently weak layers, such as shale, old fault or slide surfaces, or fractures. Some involve soil sliding off underlying bedrock. Compared to a rotational slump, a translational slide is shallow, which is demonstrated by the fact that trees slip down the surface and remain vertical rather than rotating with the sliding surface (**FIGURE 8-25**).

Translational slides are especially dangerous because they often move faster and farther than rotational slumps. The range of internal behavior is large. Some move as coherent masses, while others break up in transit to become debris slides.

Smaller examples that have destroyed homes—sometimes because of clearing for subdivisions or farming—include the destructive Aldercrest slide of southwestern Washington State and many small slides around Pittsburgh, Pennsylvania. In the Finger Lakes area of New York, the Tully Valley landslide in glacial lake clays is an example of an incoherent translational slide initiated by a long period of heavy rainfall and rapidly melting snow: On April 27, 1993, a large area of gently sloping farmland collapsed and flowed downslope (**FIGURE 8-26**). The gently sloping bench of soft glacial lake clays, downslope from steeper forestland, was capped by bouldery dirt left by melting glaciers more than 10,000 years ago. The slide moved as a rapid slump and earth flow. It was not a unique event; a similar slide more than 200 years old lies immediately north of the new one, and many recent but smaller slides can be found in the region. The slope of the affected farmland was only 9 to 12°, but shear strength of the clays was reduced by the high pore-water pressure.

In another example, between 9 and 10 a.m. on February 17, 2006, a steep, unstable slope collapsed following 10 days of heavy rain (totaling 63 cm) that saturated

FIGURE 8-26 Sliding of a Gentle Slope
The Tully Valley landslide affected gently sloping farmland in western New York.

mountainsides on the island of Leyte in the Philippines. It buried the village of Guinsaugon, including all 246 children and 7 teachers in an elementary school; 1350 people were missing and presumed buried under rocks and mud as deep as 30 m. The wet, dense mud left little chance for survival. Survivors blamed not only the rain but illegal logging on the slope above the 375-home village.

A variant of the translational slide type is sometimes called a lateral-spreading slide. One such event occurred on the marine terrace in Anchorage during the 1964 earthquake (**FIGURE 8-27**). If loose, water-rich sands or quick clays are present at shallow depth, then liquefaction or collapse may send the mass moving downslope. Parts of the moving mass may sink, leaving some blocks standing higher than others. Those in quick clay (described previously) can be fast and destructive when the randomly oriented clay flakes collapse and glide on thin, water-rich zones.

FIGURE 8-27 The Ground Drops

Subsidence and seaward spreading of the marine terrace in Anchorage, Alaska, during the 1964 earthquake left most wood-frame buildings intact but severed or tilted some of them. This dropped trough formed at the head of the L Street landslide.

Soil Creep

Soil creep, the slow downslope movement of surface soils and weak rock, involves near-surface movement and is not especially dangerous. It tilts fences, power poles, and walls. Rates of movement decrease at greater depth because most driving processes operate close to the surface. Alternating soil expansion and shrinkage from several processes, including wetting and drying or freezing and thawing, causes creep to accelerate. When the soil expands, it moves out perpendicular to the slope; when it shrinks, it moves more nearly straight down under the pull of gravity.

The net change is a slight movement downslope. When burrowing animals tunnel into the soil, their cavities eventually collapse. Plant roots expand the soil and then later rot, resulting in the collapse of their cavities. Even trampling by animals and people is significant over a long period of time (**FIGURE 8-28**). Trees that stand straight but with their bases curving back into a slope, so-called pistol-butt trees, have trunks that initially grow upward but become tilted downslope by creep. Even bedrock layers can bend downslope.

Solifluction is another type of near-surface downslope movement that occurs in extremely cold Arctic or alpine areas where water-saturated ground freezes to great depth. It is common downslope from snowdrifts. When near-surface layers thaw, the water cannot drain downward because the soil below is still frozen. The soggy near-surface layers slowly ooze downslope.

Snow Avalanches

In cold climates, **snow avalanches** can be deadly examples of downslope ground movements (**FIGURE 8-29**). An avalanche as little as 30 cm deep (1 ft) moves tons of snow downslope at high speed, sweeping victims off their feet and burying them. Every winter from 2000–01 to 2013–14, between 28 and 58 people died in avalanches, one-third of them in Canada. About 40% were skiers and snowboarders and 40% were on snowmobiles; most were men. Conditions for avalanche formation depend upon slope steepness,

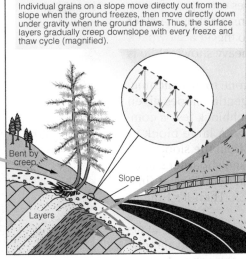

Individual grains on a slope move directly out from the slope when the ground freezes, then move directly down under gravity when the ground thaws. Thus, the surface layers gradually creep downslope with every freeze and thaw cycle (magnified).

Bent by creep

Slope

Layers

FIGURE 8-28 Creeping Ground

These trees along California's Highway 1 west of Leggett have been bent by soil creep, the slow downslope movement of the upper layers of soil or soft rocks. They originally grew upright but were tilted outward by creep. Continued upward growth produced the bending. Creep of sedimentary layers near the ground surface bends their upturned ends downslope (to the right, in the photo on the right).

FIGURE 8-29 Snow Avalanche

A powder snow avalanche races downslope.

weather, temperature, slope orientation (north or south), wind speed and direction, vegetation, and conditions within the snowpack. Although snow avalanches can occur spontaneously, they are often triggered by human actions. A skier crossing a slope can add enough load to trigger a snowpack failure. The weight and vibration of a heavy snowmobile is even more likely to trigger failure.

Steep slopes of 30 to 45° are most prone to avalanching, but avalanches can occur on much gentler slopes given the right weather conditions. The Rutschblock test is commonly used to examine slope stability. A large block of snow more than a meter across is cut from the slope, and if the block collapses when a skier or snowboarder stands on it, the slopes are unsafe. If it doesn't collapse at first, jumping up and down may do the trick. Then it's time to go home!

Particular weather circumstances cause unstable layers to form within a snowpack, leading to extreme avalanche danger. New snow is one risk factor for avalanches. A heavy snowstorm adds a load of loose snow to a slope, just the conditions favored by many skiers. Within 24 hours after a snowfall, the snow is least stable. Ten cm of new snow rarely produces dangerous conditions, but more than 30 cm is hazardous.

Temperature changes lasting hours or days can also cause unstable layers to form in the upper part of a snowpack. Higher temperatures during a spring day can melt grain surfaces enough to fill any pore spaces with water. Water trickles down through the snowpack and collects at its base. If the ground below is not frozen, that water percolates down to raise the water table. If the ground is frozen, the water may run downslope, over the ground surface but under the snow, leading to slope failure as an avalanche. If water seeps down to an internal boundary between layers of open-textured dry snow against tightly packed or frozen snow, the internal boundary between the layers provides a zone of weakness that can also lead to sliding.

Unstable layers are also formed when hoarfrost on the surface is buried by new snow. Loss of ground heat by radiation at night under a clear sky can cause formation of *hoarfrost*— loose, flaky ice crystals—on a snow surface (**FIGURE 8-30**). Buried under later snow, hoarfrost leaves a weak layer on which an overlying snowpack can slide. Recognition of such loose, flaky layers can help keep you out of trouble in the backcountry.

Wet weather in fall and early winter can leave mountains coated with wet, heavy snow that can freeze as nights get colder. Newer, drier snow that falls on the solid base may leave a weak boundary that can be prone to sliding (**FIGURE 8-31**). Winter storms loaded with moisture from the Pacific Ocean often follow warm, moist air that settles a snowpack, making it denser. An arriving storm brings colder air and new snow that collects to form oversteepened—and often overhanging—cornices high on the leeward sides of mountains. Such unstable masses of snow may break loose if triggered by animals, skiers, or even strong wind gusts.

FIGURE 8-30 Hoarfrost

Loose, flaky ice crystals can grow at night under very dry conditions. Their layers lead to very dangerous conditions in a snowpack.

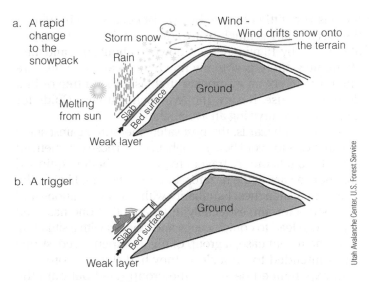

a. A rapid change to the snowpack

Storm snow

Rain

Wind - Wind drifts snow onto the terrain

Melting from sun

Ground

Slab

Bed surface

Weak layer

b. A trigger

Ground

Slab

Bed surface

Weak layer

Utah Avalanche Center, U.S. Forest Service

FIGURE 8-31 Breaking Loose

Ingredients for a snow avalanche include a weak layer in the snowpack; a rapid change in the snow from sun melting, rain, or heavy snowfall; and a trigger, such as a snowmobile or skier.

In November 2008, an early snowpack in the Pacific Northwest states and southern British Columbia was soaked by a mild rain that refroze to form a continuous sheet of ice. Light, fluffy snow fell for the next few weeks; it was extremely cold and dry, so the grains did not stick together nor freeze to the underlying ice. High winds blew some of the light snow over ridges, building thick deposits in downwind areas and gullies as heavier wind-packed snow. In the last few days of December, high temperatures packed the upper parts of the snowpack into a heavy, dense slab that loaded the underlying layer of still-cold, fluffy snow. That heavy slab was not well anchored to the ground, leading to severe avalanche conditions. Additional snowfall in some areas provided too much of a load on the underlying snow, and throughout January, avalanches swept downslope on the icy base. Elsewhere the slab stayed put until unsuspecting snowmobilers, snowboarders, and skiers added their weight, causing it to let loose.

Recognition of previous dangerous weather conditions or excavation of a section through a snowpack can sometimes reveal hazardous unstable layers. Those who monitor snow-packed slopes to warn against avalanche danger dig trenches in snow to find such weak internal zones.

The orientation of a slope is also a factor in assessing avalanche risk. Winds in the westerlies belt of the northwestern United States and western Canada blow upslope from the west, depositing snow on the leeward eastern sides of ridges. Those downwind sides are often steeper because the snow is deeper, and snow and freezing conditions endure longer in the spring. Because these avalanche-prone slopes collect more snow, many ski areas locate there. North-facing slopes that remain in shadow all day in the winter are dangerous

because they don't warm enough during the day to cause the localized melting and refreezing that helps solidify a snowpack.

Landscape features on a slope can also signal heightened risk for snow avalanches. Among the most dangerous places on a slope are under a cornice or in a "bowl," just the places that attract adventuresome skiers, snowboarders, and snowmobile riders. Cracks or breakaway zones tend to form where the snowpack thins as it goes over a rock outcropping or convex slope, at the base of a cliff, immediately downslope from a tree, or under an overhanging cornice (**FIGURE 8-32**). A fresh avalanche on an adjacent slope facing the same direction and with a similar slope is a very dangerous sign.

The breakaway for an avalanche triggered by a skier or snowmobile rider can appear as a tension crack (compression of a low-density layer at depth, causing localized bending and crack formation). Load can be added to a slope by the weight of a person, a heavy snowfall, or even wind force against the surface of a snowpack. Windblown snow in a cornice is tightly packed and heavy; it can easily break off and avalanche downslope. Standing on top of such a ridge is especially dangerous because it is almost impossible to see where the snow overlies solid ground rather than an unstable overhang. A snow ridge's highest point can sit well over an overhang.

Gullies, or the chutes left by past avalanches, are other high-risk avalanche areas. Snow blows off higher surfaces and collects in such hollows. Avalanching down steep mountainside gullies, snow tears out trees and brush, leaving an avalanche chute or "greenslide" (**FIGURE 8-33**). A forested hillside with no trees or distinctly smaller trees in a fall-line gulley often signifies an avalanche chute. The greatest danger for mountain skiers or snowmobile riders occurs

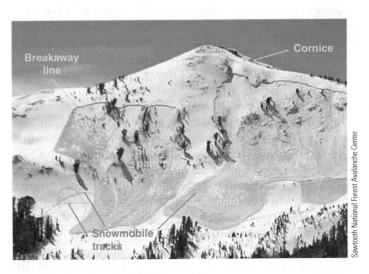

Breakaway line

Cornice

Slide path

Runout

Snowmobile tracks

Sawtooth National Forest Avalanche Center

FIGURE 8-32 Avalanche-Prone Slopes

Avalanches were triggered by snowmobile "highlining" at Baker Peak (tracks visible at left of left-most slide).

FIGURE 8-33 Avalanche Chute

A late-season dirty avalanche over an earlier clean avalanche just west of Stevens Pass, Washington. The avalanche crosses a railroad snowshed near the base. After the snow melts, avalanche chutes remain as treeless gashes through the forest.

when their weight triggers an avalanche while crossing a ridge-crest cornice or avalanche chute. These chutes can also lead to sites of dangerous debris slides.

On an active glacier, an icefall avalanche can occur at a steep drop. Huge chunks of ice break off steep faces and fall onto slopes below. Such an avalanche can be triggered by movement in the glacier; changes in temperature; external vibrations, such as earthquakes; or disturbance by hikers or ice climbers. The deaths of 15 experienced Nepalese Sherpa guides transporting supplies on Mt. Everest on April 18, 2014, emphasized the instability of icefall areas. They can happen at any temperature or time of day. Intermittent cracking or explosive sounds in the ice indicate instability.

SURVIVING AN AVALANCHE Deaths from avalanches sometimes occur by blunt trauma when a victim is rammed into a tree or rock, but more people die by suffocation under the snow. When a person's core temperature (37°C or 98.6°F) drops below about 35°C (95°F), their internal organ functions begin to slow and fail to replenish external heat loss. When core temperature drops below 68°C (82°F) hypothermia is severe, as heart rate, blood pressure, and respiration rate drop dramatically and survival is around 50%. The person becomes confused, incoherent, and loses consciousness. Once a person loses consciousness, the chances of survival dwindle rapidly.

Time is critical because if the person is dug out within 10 minutes, they have a 90% chance of surviving. By 15 minutes, the chance drops to about 60%, and after 35 minutes,

there is very little chance. In a moist coastal environment, the survivability is much worse. After 10 minutes, it drops to about 75%; by 15 minutes, it is down to about 20%, and after 20 minutes, there is virtually no chance except in the rare circumstance of an air pocket such as under a tree or in a destroyed house. Review the Avalanche Survival Guide for guidelines on surviving an avalanche.

As with all hazards, the best safety measure against snow avalanches is to avoid being caught in one. Those experienced in winter backcountry travel are generally better equipped and more aware of the hazards, but even they are likely to be caught in avalanches. Familiarity with snow conditions and steep snow terrain, especially in an area that one has used before, can lead to complacency and comfort with a situation.

In one recent case, a group of highly experienced, strong skiers intended to ski a slope they had skied before. The most experienced person of the group skied out onto the slope, trying to trigger a small slab avalanche below that would make the slope safer for them to ski. Unfortunately, a slab avalanche more than 1 m thick released suddenly above him; it rapidly swept him downslope into a narrow chute and over a cliff. He was knocked off his feet and his mouth was jammed full of snow; he lost a ski, repeatedly banged his head and body against ice and rocks, but was lucky to remain conscious. As the slide slowed, he managed to "swim" to the surface. He and his friends were amazed he didn't die. Reflecting on the accident later, he emphasized his errors. He was lulled by a false sense of security being with a group of strong skiers on a slope they had skied before. As a result, he didn't critically evaluate the snow conditions and left himself no escape route if the slope failed. He had always felt that if he got caught in an avalanche, he would just point his skis straight downslope and ski out of trouble; he no longer believes this.

URBAN AVALANCHES Avalanches can also occur in urban areas, but with different concerns. When an avalanche strikes homes in an urban area, people inside may hear rumbling moments before impact but are otherwise completely unprepared. They are not dressed for cold weather and have no chance to escape; the house can collapse on them, typically causing more severe injuries.

At about 4 p.m. on February 28, 2014, a snowboarder triggered an avalanche on a steep west-facing mountainside above a group of homes in Missoula, Montana, a town in the northern Rocky Mountains. Less than a meter thick, the slide was primed by compacted and iced upper layers created by warm weather following earlier snows. A cold front brought a blizzard that loaded the whole snowpack. A snowboarder crossing the upper slope about 400 m (1300 ft) above triggered the dense "hard-slab" mass that raced downslope that avalanche experts estimated as 175 km/hr. It destroyed two houses, burying an eight-year-old boy and a retired couple under dense, compacted snow (**FIGURE 8-34**). All three were injured but amazingly survived in cavities that provided enough breathing space. The boy, buried for an hour, was released from the

FIGURE 8-34 Unseen Hazard

A. Small basin or "bowl," which collected windblown snow, the avalanche chute, and house location.

B. Remains of demolished house; the recently added third floor lies toppled on top of the snow mass.

hospital the next day and the man, buried for two hours, was released after two days, with many broken ribs and other fractures. The woman, found alive after three hours, never regained consciousness. She died of hypothermia and internal injuries three days later. Such avalanches in urban areas are, fortunately, very uncommon.

Hazards Related to Landslides

Landslides are closely related to several other hazards. They can be triggered by storms and flooding or by earthquakes and volcanic eruptions. When a landslide blocks a waterway and then collapses after water backs up, it leads to flooding.

Earthquakes

Many eyewitness accounts tell of great clouds of dust rising from hillsides during and after earthquakes. In most cases, they rise from slides an earthquake has just shaken loose. If a slope is at all unstable, an earthquake is likely to send it downslope. Even without water, sudden shaking may trigger failure.

Earthquakes below magnitude 4 trigger few landslides. Progressively larger earthquakes trigger more and more landslides, especially closer to an earthquake's epicenter. Larger earthquakes may also start rockfalls. The 1959 magnitude 7.3 West Yellowstone earthquake, for example, triggered the massive Madison landslide and rockfall that collapsed from a 340-m-high cliff above the Madison River channel, west of West Yellowstone, Montana. At 11:37 p.m., the fast-moving slide mass spread out in both directions as it reached the valley floor. It buried 23 people in a campground in the valley bottom. The toe of the slide in the Madison Valley is 1.5 km

wide, twice as wide as the slide scar on the south wall of the canyon. A blast of air from under the falling slide mass swept away two people and tumbled one automobile. The bedrock on the mountainside, a huge mass of weathered schist in which the layering was nearly parallel to the south canyon wall, was an obvious rockfall hazard. A strong mass of dolomite marble anchoring the base of the mountain broke during the earthquake, leaving the weak schist and gneiss above it unsupported. The mass crossed the valley floor and moved 200 m up the opposite slope. The earthquake also triggered smaller slides farther away that took out the main highway (**FIGURE 8-35**).

FIGURE 8-35 Landslide Triggered by an Earthquake

The 1959 Yellowstone earthquake triggered many landslides in the vicinity. This one took out the highway and this house.

FIGURE 8-36 Landslide Over Ice

A major earthquake on the Denali Fault in 2003 sent the side of this mountain down on top of the Black Rapids Glacier in the Alaska Range.

Of all the different downslope movements an earthquake may trigger, debris avalanches and rapid soil flows make up less than 1%. Since they move at high speeds on slopes as gentle as a few degrees, they are even more deadly than rockfalls, often killing more people than the earthquake that triggers them. In some cases, they bury towns several kilometres from their starting points. Slow-moving soil and rock slumps and lateral spreads rarely kill many people, but they do collapse buildings.

Earthquakes also often cause the failure of inherently unstable slopes (**FIGURE 8-36**). Among the most susceptible are recently raised marine terraces composed of soft, wet clays and associated sediments. A prominent case involved Anchorage, Alaska, in the 1964 magnitude 9.2 earthquake. Much of Anchorage is built on a flat to gently sloping terrace as much as 22 m above sea level. Shaking lasted 72 seconds, causing liquefaction of clays in the terrace and collapse. Wood-frame houses and other buildings survived

moderately well except where they happened to straddle a slump scarp.

Many accounts of earthquakes include reports of sand spouting from the ground or surfacing in big sandboils. This activity is evidence of liquefaction, as discussed in Chapter 3. When some earthquake waves pass through soil saturated with water, the sudden shock jostles the grains, causing them to settle into a more closely packed arrangement with less pore space. Because this leaves more water than the remaining space can accommodate, water must escape. During this event, the soil grains are largely suspended in water rather than pressing tightly against one another. The soil settles and is free to flow down any available slope.

You can easily demonstrate the process by filling a glass of water with fine sand. Then repeatedly tap the side of the glass with something hard. The sand will progressively settle as the grains rearrange themselves; the displaced water rises to cover the mass of sand grains. A loosely packed sand with 45% porosity might collapse to about 30% porosity. Buildings on liquefied soils can tilt and may even fall over when their underpinnings settle or spread (**FIGURES 8-37** and **8-38**).

Failure of Landslide Dams

Any moderately fast-moving landslide can block a river or stream to create a dam (**FIGURE 8-39**). When a landslide dam fails, flooding can be catastrophic. Examples of landslide dams include those listed in **Table 8-2**. Often, water will back up to form a temporary lake before the dam eventually fails. Of those dams that failed, roughly a quarter eroded through in less than a day; half failed within 10 days; and some lasted for a long time. What controls this behavior, and what are the downstream dangers? The time before a dam fails and the size of a resulting flood depends on:

■ The size, height, and geometry of a dam.
■ The material making up a dam.
■ The rate of stream flow and how fast a lake rises.
■ The use of engineering controls, such as the excavation of artificial breaches, artificial spillways, or tunnels.

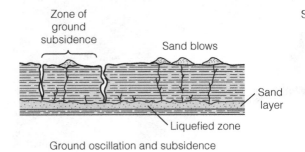

FIGURE 8-37 Liquefaction

Two typical types of ground failure occur during liquefaction. As water is expelled from between loose grains, the sediment structure settles, along with anything built above it.

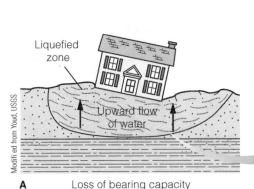

A Loss of bearing capacity

Liquefied zone

Upward flow of water

Doorway originally at street level

FIGURE 8-38 The Ground Settles

A. Tilting of a house as the soil below it liquefies. **B.** These houses on loose fill at the edge of San Francisco Bay did not collapse during the 1906 earthquake but settled as the fill liquefied. **C.** Sunken doorways were recently still visible in the Marina district of San Francisco.

Mudflows, debris flows, and earth flows create many natural dams that block rivers quickly, are not high, are composed of noncohesive material, and breach soon after formation. Other kinds of flows are often long-lived. Most landslide dams fail because the water behind them overflows and erodes a spillway that drains the lake. That may not happen if the dam consists of large rocks and is so permeable that the lake drains by seepage instead of through an overflow spillway. A small landslide dam is likely to fail

FIGURE 8-39 Landslide Dam

The 1983 Thistle landslide at Thistle, southeast of Salt Lake City, Utah, began flowing down valley in response to rising groundwater levels from heavy spring rains during the melting of a deep snowpack. Within a few weeks, the slide dammed the Spanish Fork River and took out U.S. Highway 6 and a major railroad line. Water behind the slide dam submerged the town of Thistle. With costs exceeding $474 million (2015 $), it was the most expensive landslide event in U.S. history.

soon after formation if it blocks a large stream. Permeable, easily eroded sediment is vulnerable to piping, seepage, and undermining that can lead to dam failure.

To minimize the chance of a catastrophic flood downstream if a dam fails, officials commonly try to stabilize landslide dams by constructing a channelized spillway across or around them. The U.S. Army Corps of Engineers handled the Madison rockslide this way. At the Thistle slide, the Corps used pipe and tunnel outlets.

The height and volume of impounded water, and therefore its potential energy, control the maximum height of a flood from a landslide dam failure. The higher the water level and the greater the volume of water behind the dam, the higher the flood level will be downstream. Most dam-failure flood flows decrease rapidly downstream. However, if a flood incorporates a significant amount of easily eroded sediment into the flow, then it can turn into a debris deluge many times larger than the flow at the dam itself.

Why not construct a useful dam on a landslide dam? Actually, in the twentieth century, at least 167 dams were constructed this way. Most of these were built on top of rockfalls or rock slides because those are most stable. One outstanding example of a landslide-dammed reservoir that produces hydropower is Lake Waikaremoana, the largest landslide-dammed lake in New Zealand.

Only one of these modified landslide dams failed catastrophically. In 1928, the St. Francis high-arch concrete dam, north of Los Angeles, failed while the reservoir was being filled because it was built on the toe of a large Pleistocene landslide in schist bedrock, killing 450. One seldom-addressed concern about constructing a dam atop a landslide is whether other slopes around the new reservoir are also prone to landsliding. If so, filling a reservoir would raise the water pressure in surrounding slopes, possibly leading to major slope failure that would drive water to surge over the dam.

LANDSLIDE DAM	TYPE (AND MATERIAL)	YEAR	RIVER BLOCKED	DAM HEIGHT (m)	DAM LENGTH (m)	DAM WIDTH (m)	LAKE LENGTH (km)	DAM FAILED?
Madison, Montana	Rock slide (broken rock)	1959	Madison	60–70	500	1600	10	No
Mayunmarca, Peru	Rock slide (broken rock)	1974	Mantaro	170	1000	3800	31	Yes
Gros Ventre, Wyoming	Slide (soil)	1925	Gros Ventre	70	900	2400	6.5	Yes
Slumgullion, Colorado	Earth flow (altered volcanic rock, smectite clay, soil)	1300	Gunnison	40	500	1700	3	No
Polallie Creek, Oregon	Debris flow (gravel, boulders)	1980	Hood, East Fork	11	—	230	—	Yes
Thistle, Utah (FIGURE 8-39)	Earth slide	1983	Spanish Fork	60	200	600	5	No
Usoy, Tajikistan	Landslide (broken rock)	1911	Murgab	500–700	1000	1000	60	Partial, may yet fail

Costa and Schuster, 1988, USGS

Mitigation of Damages from Landslides

Landslides are widespread throughout the United States and Canada, and their damages can be extremely costly (**FIGURE 8-40**). Few insurance policies cover them or any other type of ground movement. Landslides in the United States cost more than $2 billion and 25 to 50 deaths per year. Worldwide, landslides have caused an average of 7500 deaths per year over the last century and $20 billion per year over the period from 1980 to 2000. As with other hazards, major landslide disasters increase with the growth of world population as people settle in less suitable areas; the vast majority of deaths occur in less-developed countries. People build on landslide-prone areas because they are attracted by scenic views or because the land is more affordable.

People often build at the base of cliffs to have their homes nestled in scenic environments. In some cases, they build among large boulders that provide convenient highlights for landscaping. They tend to not think much about where the boulders came from; if they did, they would quickly conclude that they came off the cliff above. These are truly dangerous places to live (See FIGURE 8-18).

As with all hazards, however, understanding the processes that control downslope movement allows scientists to evaluate risks and implement strategies to reduce damages.

Record of Past Landslides

As with many hazards, a record of past landslides indicates whether future landslides are likely. A large slide following a long period without them does not preclude soon having another. In fact, the existence of landslides in an area indicates favorable circumstances for slides.

A hummocky hillside, sometimes obscured by vegetation, may be the remnant of an old landslide. A stretch of

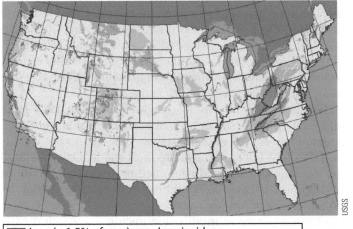

USGS

☐ Low (< 1.5% of area)   Low incidence:
☐ Moderate (1.5%–15%)   ☐ Moderate susceptibility
■ High (greater than 15%)   ☐ High susceptibility

FIGURE 8-40 Landslide Hazard Map

Landslides are widespread, not only in mountainous areas of the western United States and in the Appalachians, but also in the subdued terrain of the central United States .

FIGURE 8-41 **Evidence of Landslides**

A. This hummocky landslide terrain is near Gardena, north of Boise, Idaho.

B. Wavy road across landslide terrain near Zion National Park, Utah.

road that cracks or exhibits broad waves in pavement, often requiring repaving, is also suggestive of sliding terrain (**FIGURE 8-41**).

Building a road across, constructing a building on, or removing material from the base of such hummocky topography would be unwise. Old landslides can be reactivated by any of the processes that initiate new landslides—that is, adding water, steepening the slope, undercutting the toe (or removing toe material), loading the upper part of a slope, removing vegetation, or earthquakes. Many old landslides are reactivated by removing material from the toe because that material encroached on a road, railroad, or construction site.

What about reduction in slope following landsliding? Does that make the slope less prone to future sliding? Not necessarily; preexisting slip surfaces and surface fractures that permit further water penetration both contribute to further sliding. Building of roads or structures on an existing landslide merely aids additional slope movement. Clearly, if the conditions are appropriate for landsliding, preexisting slip surfaces can reactivate.

Landslide Hazard Maps

The best strategy is to avoid building in places prone to landslides. A Geographic Information System (GIS), which analyzes and displays the separate factors affecting a slope, can be used to build debris-flow and landslide-hazard maps. Such GIS maps can be used to prescribe restrictions in land use such as on road building, timber harvesting, or even housing subdivisions. High-risk landslide areas where development should be restricted include:

- Steep slopes and clearly mountainous areas.
- Local slopes that exceed the local angle of repose (30 to 45° on a hillside).
- Areas with abundant loose debris on a slope.

- Slopes with low permeability, as in fine-grained soils.
- Areas where large amounts of rainfall or snowmelt enter the ground.
- Locations where shallow slides commonly develop at the interface between bedrock and the broken rock and soil that covers it.
- Locations of previous shallow landslides of any size.

Shallow slides are more likely to develop on slopes with sparse vegetation and a lack of significant tree roots to hold shallow material in place. This factor is relevant only for debris flows and translational slides that are shallower than the depth of tree root penetration.

In the GIS approach, the area of concern is the area is divided into a set of polygons. Each polygon is chosen as having consistent internal attributes such as slope, concave-upward curvature, soil texture and depth, ease of slope drainage, slope-facing direction, type of vegetation, bedrock type, length of roads within the polygon, and presence of slope failures. Polygons with slopes of less than 15°, for example, may be excluded from study because they typically lack evidence of landsliding and are less likely to slide.

Unfortunately, there are sometimes political barriers to mapping landslide hazards. In June 2011, when the North Carolina legislature debated funding for mapping slope hazards in mountainous parts of the state, some argued that homeowners and homebuyers deserve to know whether a property is at risk. The group opposing hazard mapping included homebuilders and real estate agents, who are legally required to disclose that a property lies in a landslide zone. They felt that the information that a property was in a landslide hazard zone would make it hard to sell or develop, devaluing the property and leading to further regulation. Although five people died and 15 homes were destroyed in September 2004 at Peek's Creek in western North Carolina, some in this group dismissed this incident as a "random act of God." At the close of debate, the state legislature voted to cut funding for slope hazard mapping!

Engineering Solutions

Because the relationship between forces that keep slopes from sliding is established, engineers can sometimes restore the balance among forces to keep a slope stable.

For a slope overloaded at its top, we can add load to the lower part of the slide to resist movement. To stop a slope from moving, highway engineers sometimes pile heavy boulders on the toe area to increase the resisting mass (**FIGURE 8-42**). The slope angle can also be changed by removing the slope's top, adding weight to its base, or remaking the entire slope with a lower angle.

Rock cliffs or slopes can be sprayed with a cement mixture called *shotcrete*, or gunite, to restrict water access. Some are draped with heavy wire mesh to prevent falling rocks from reaching buildings or highways (**FIGURE 8-43**). Some are drilled and anchored in place by *rockbolts* (**FIGURE 8-44**).

Removing water from soil can increase its strength, making it less likely to slide. One of the most effective mechanisms for water removal involves trees and shrubs taking up water from the soil through their roots, a process called *evapotranspiration* that dries the soil. Water from roots reaches leaves, where it is transpired into the air. In addition, a portion of the rain falling on leaves or soil evaporates at the surface rather than percolating into the ground.

David Hyndman

FIGURE 8-43 Screened Rocks
Heavy cable mesh draped over crumbly road cut to protect a highway in the French Alps.

Some kinds of trees and shrubs take up water from the soil more eagerly than others. In general, those that grow prolifically adjacent to streams or lakes use great quantities of water. The most common of these are willows, cottonwoods, and aspens. Tamarisk trees may completely drain irrigation ditches in the arid Southwest. Where those same plants grow well above a floodplain, the ground probably contains excessive water and may be in danger of sliding. Planting

Donald Hyndman

FIGURE 8-42 Weight the Bottom
Heavy boulders are often piled on the lower part of a slide to resist movement. This road, cut through a landslide on U.S. Highway 101 near Garberville in northern California, has been stabilized by loading its toe area.

Hyndman

FIGURE 8-44 Drill and Bolt
Long rock bolts and concrete stabilize a fractured slope near Frigiliana, Spain.

FIGURE 8-45 Leaky Swimming Pool?

After the 1999 Dana Point landslide north of San Diego, California, black plastic was spread on the ground in the lower left to prevent water infiltration. Note that the pale gray concrete pool (right center) was amputated by the headscarp. Did a leaking swimming pool add water to the slope?

trees or shrubs that use large amounts of water can help stabilize a slope. People sometimes cover potential slide areas with plastic to prevent water penetration (**FIGURE 8-45**), although this also has the unintended consequence of shutting down evapotranspiration.

It is possible to artificially drain and thus stabilize many slopes. If pore spaces in the soil are very small or disconnected from each other, then the soil has low permeability; water does not flow easily through the small pores. One widely used method to stabilize slopes involves drilling holes inclined slightly upward into the slope and inserting *perforated pipes*. Water drains into the pipes and trickles out to the surface (**FIGURE 8-46**). A more expensive approach is to dig deep trenches in the slope with a backhoe, line the trenches with *geotextile fabric* (cloth that permits water but not sediment to flow through), then backfill them with coarse gravel. Water will trickle out through the gravel for years. If the situation is truly desperate, it may help to drill wells into the slope and pump the water out. All these methods reduce the water pressure in the soil, which makes it less likely to slide.

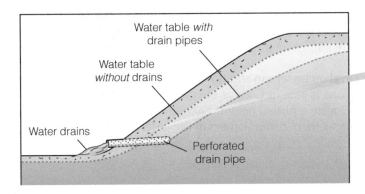

FIGURE 8-46 Drain the Water

Installation of a perforated drainpipe can lower the water table and reduce the chance of sliding. The low permeability of this shale road cut on U.S. Highway 101 north of Garberville, California necessitates many drainpipes.

Be aware that these lists don't cover everything. Use your own common sense.

High Hazard Area	▪ Mountain areas that collect a lot of snow, especially under snowdrift cornices, pronounced gullies or swales on a mountainside, especially slopes greater than about 30°. The safest areas are on high patches of rocks or in areas of large trees on the upper lateral edges of a slope.
Preparation	▪ Avoid traveling alone in avalanche country. If you must cross a dangerous slope, do so one at a time, testing the open slope carefully. Pay close attention to each skier's location in case of a problem. Even if some people safely cross a slope, the next one may trigger an avalanche. Carry a shovel, probe, and an avalanche beacon. Evaluate weather and mountain conditions critically and take appropriate precautions.
Warning Signs	▪ Watch for overhanging cornices near ridge crests, and gullies or swales that show signs of former avalanches—forested areas with treeless gullies, or broken branches and tree debris at the base of a treeless gulley.
During the Event	▪ If someone is caught in an avalanche, watch carefully to determine where you last saw the person, most likely in the debris pile of the runout zone. Look for ski poles or clothing to determine the path of the victim. ▪ If you are caught in an avalanche, and you are not at the surface, as the snow slows try to swim towards the surface; take a deep breath to expand your chest so you have room to breathe later; punch out an air space around your face. Conserve energy by not panicking and yell only when rescuers are almost on top of you; snow rapidly dampens sound waves.
Aftermath	▪ People traveling in potential avalanche terrain should carry lightweight but strong portable shovels in case someone is caught. Collapsible probes that join to form a pole at least 3 m long can be used to locate a buried person. Each member of a ski party should carry an avalanche transceiver set to "transmit." Others in the party then can hone in on a buried transceiver.

Cases in Point*

Overlapping Causes for a Landslide
The Oso Slide, Western Washington, 2014 ▶

At 10:37, the morning of March 22, 2014, a large landslide that turned into a debris flow downslope buried the small community of Oso, 90 km (55 mi) northeast of Seattle. It released from a broad terrace 200 m (650 ft) above the valley floor, spread rapidly across the Stillaguamish River, and instantly obliterated most of the homes on the floodplain of the river. Forty-three bodies, including one motorist driving on Highway 530, were recovered from a total population of 180. Except for dam collapses, volcanic eruptions, or earthquake events, it was the deadliest landslide in U.S. history. The site, commonly referred to as the Oso (or Hazel) mudslide, has a long history of movement—in 1952, 1967, 1988, and 2006; the subdivision at the toe of the old slide should never have been permitted.

The terrace formed during the last ice age as mountain streams initially deposited clay in lake beds, capped by a thick delta of loose sand and gravel formed against ice filling Puget Sound. Retreat of the ice permitted the Stillaguamish River to erode down to its present level. A meander bend of the river cuts into the base of the terrace, steepening the slope. Rain in this coastal climate soaks into the porous sediments easily, increasing internal pore pressure, percolating down to the impermeable clay at the base, and making for a nearly horizontal zone of weakness that is prone to sliding. This environment led to numerous landslides on both sides of the valley.

The sloppy nature of the 2013 debris flow, its high velocity, and long runout can be partly attributed to abundant water from heavy rains, high groundwater level, and the river water and wet ground under the toe of the slide as it moved. In the three weeks prior to slide movement, twice the normal amount of rain fell, which led to especially high internal pore pressure and greater weight of the mass. A large clear-cut area at the top of the slide may have contributed by permitting greater rainwater infiltration. The slide, 700 m wide at the scarp, swept across the river and spread out to 1500 m wide and 1700 m long, buried the subdivision by as much as 4.5 m depth, and dammed the river to cause upstream flooding. A second smaller mass pulled down more of the headwall about four minutes later. The main slide moved at about 32 km/hr (20 mi/hr) making it impossible to outrun. Most of the victims were recovered near the leading edges of the slide, which covered about 2.5 km². Search continued day and night for more than a month, though after a few days there was little chance for those buried in the watery mud mixed with debris. Heavy equipment helped dig up deeper material that needed hazardous disposal (such as refrigerators or propane tanks), or items that residents might want to claim.

There was ample reason to avoid building homes at that location. It was the site of several previous landslides including one in 2006 that also pushed the river away from the cut bank but did not reach the subdivision. The headscarp of the 2006 slide was still plainly visible as a white scar on the slope above. However, the scalloped form of many of the slopes

■ **A.** *Drenching weather on soft mud made the search for victims difficult and hazardous.*

Spc. Matthew Sissel, Wash. Nat'l Guard

■ **B.** *The Oso slide viewed from forest debris at the toe.*

From NSF-sponsored GEER Report dated 30/06/2015/GEER association

*Visit the book's Website for these additional Cases in Point: A High-Velocity Rock Avalanche Buoyed by Air—Elm, Switzerland; *Water Leaking from a Canal Triggers a Deadly Landslide*—Logan Slide, Utah, 2009.

The Oso (Hazel) landslide of March 22, 2014 reactivated an old slide detached from an old glacial-deposit terrace. Horizontal layers of sand and gravel are visible in the headscarp.

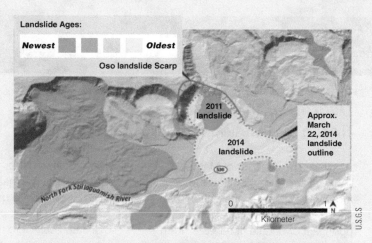

A graphic representation of widespread landslides along the Stillaguamish River valley is provided by a lidar topographic map of the area. Sharply defined scallops in the upper left, north of the river, mark headscarps from the most recent landslides; less-distinct scallops south of the river also mark older slides. The 2014 scarp and landslide deposit are outlined in red.

was not readily apparent to those not trained to see it. Amazingly, several houses were built even after the 2006 slide. Reports done for the Army Corps of Engineers, a county hydraulic engineer, and others appear to have focused on floodplain flooding and river siltation issues from undercutting the landslide toe and hillslope movement were available, but most members of the public had not seen them. The 1999 study indicated the soft, loose material in the slope could collapse in a "large catastrophic failure." However, most longtime residents of the area were aware of landsliding on the slope either through personal experience or by talking with others.

As described in this chapter, clear-cut logging (FIGURE 8-15) permits more rain or snowmelt to enter the ground (less evaporation from vegetation), making it more susceptible to landslides. The terrace above the slide had been logged around 1960 and again after being permitted in 2004. Many old landslides predate logging so the site was very susceptible, but logging added to the hazard. Since the March landslide, Washington now requires a geologic review of potential hazards before logging permits are provided, unfortunately too late for the victims.

Ongoing Landslide Problems
Coastal Area of Los Angeles ▶

Much of the material making up the coastal bluffs near Los Angeles consists of sedimentary basin silica-rich sediments, sand, and shale younger than 5 to 15 million years old. Those geologically young rocks are notorious for breaking up and landsliding, especially when soaked by unusual amounts of water, undercut by waves or road construction, overloaded by addition of fill, or disturbed during almost any kind of construction. This is especially the case in the Santa Monica Mountain areas around Malibu and in the hills of the Palos Verdes peninsula. In the latter area, more than half of the coastal slopes from Abalone Cove to Los Angeles harbor 7 km southeast have developed landslides.

In 1956, about 1 km² of the Portuguese Bend area of the Palos Verdes Hills began moving slowly downslope, apparently in response to building houses with sprinkler irrigation, septic tanks, and pools. Building Palos Verdes Drive South near the coast and loading fill by another road near the head of the slide may also

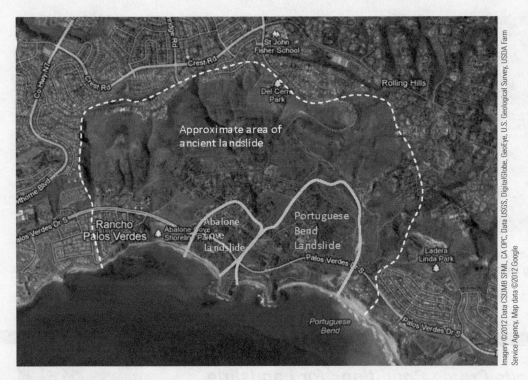

Imagery ©2012 Data CSUMB SFML, CA OPC, Data USGS, DigitalGlobe, GeoEye, U.S. Geological Survey, USDA Farm Service Agency, Map data ©2012 Google

■ *The Abalone Cove and Portuguese Bend landslides are reactivated parts of an old, much larger landslide. View is about 5 km wide.*

have contributed. Movement in clays and shales about 10 million years old reached 7 to 10 cm per day and damaged most of the 150 houses in the area. Movement continued intermittently until in 1979, and homeowners formed a group to assess the problem and decide on a plan.

They placed a moratorium on new building and hired an engineering geologist who determined that the new movements were actually reactivation of an old landslide 820 m wide and 3900 m long. Movement was a translational slide, on a smooth surface 30 m below the

ground, and dipping about 6° toward the ocean. Slippage is primarily along a volcanic ash layer altered to shale containing smectite swelling clays. Twenty-five 1.3-m diameter concrete columns installed across the rupture surface failed to stop the movement. This led to formation of a "Geological Hazard Abatement District" in which local homeowners voluntarily funded dewatering wells, storm-drain culverts, and other changes to slow ongoing movements. Slide movements are monitored by GPS. As of 2006, 17 dewatering wells were in use but some had rusted out, clogged, and needed to be redrilled. The water is discharged into the ocean.

Early problems at Point Fermin in the San Pedro area just west of Los Angeles Harbor began in 1929 when a big area, now known as Sunken City, began sliding oceanward. In 1941, a broken water main accelerated movement. Intermittent movements continued until 2010 when another large section of Sunken City collapsed 100 m onto the beach. In July 2011, small cracks appeared parallel to the 36-m-high coastal bluffs, then widened at more than 20 cm/day. During a rainstorm on November 20, 130 m of the bluff completely separated, dropped, and moved seaward as cracks on the landward side slowly widened. Just to the

Don Knabe, L.A. Co. Supervisor; USGS

■ *The San Pedro area slide, November 20, 2011.*

Donald Hyndman

■ *Homes in the Malibu area are built on steep, vegetated slopes with ongoing landslide problems.*

west, between 1969 and 1970, another large collapse, about 60,000 m² of Ocean Trails Golf Course, collapsed into the ocean.

In the Santa Monica Mountains area, farther west, homes near the coast are built in deeply gullied, heavily vegetated steep slopes with long-standing stability problems. In Malibu, Rambla Pacifico Road collapsed in February

2010, sending a section of the road sliding into the canyon and necessitating a detour for more than four months.

■ Homes on old landslide terrain near Piedra Gorda Canyon, west of Los Angeles. Coastal homes along highway are part of the same old landslide.

Donald Hyndman

CRITICAL THINKING QUESTIONS

1. When homes in the Portuguese Bend area began cracking and moving in 1956, homeowners hired an engineering geologist to suggest remedies. What were those remedies and why should each of them help slow landsliding?
2. What are all of the main characteristics of this area that favor landsliding?
3. In nearby areas what several actions by homeowners are likely to initiate or accelerate landsliding?

Slippery Smectite Deposits Create Conditions for Landslide
Forest City Bridge, South Dakota ▶

Even the subdued topography of the Great Plains can be susceptible to landslides. Problems began when the U.S. Army Corps of Engineers built Oahe Dam across the Missouri River at Pierre in the early 1950s, and the new reservoir behind the dam rose about 30 m. The steep river banks were marked by ancient landslides because smectite that formed from ancient deposits of volcanic ash in the notorious Pierre Shale swelled and became extremely slippery. The slopes stabilized when the river deposited gravel in the channel at the end of the last ice age, but water rising in the reservoir and an increase in water pressure in the slopes again made the slopes more prone to sliding. A 1.4-km-long U.S. 212 bridge built across the new reservoir in 1958 crossed those old slides; its approaches immediately began to move slowly down into the reservoir. Drilling showed the slide to be about 21 m thick. At one point, repeated repaving of the highway approaches accumulated a total thickness of 2.1 m of asphalt! Engineers helped stabilize the slopes by excavating a broad area of material upslope—which lessened the slope angle and unloaded the head of the slide—and piling heavy rock riprap to load the toe of the slide.

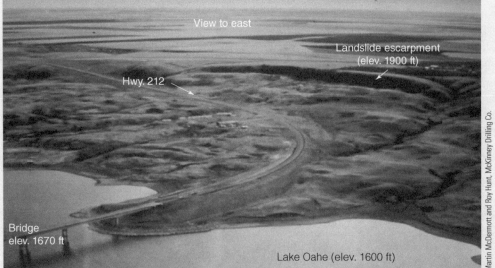

View to east

Landslide escarpment (elev. 1900 ft)

Hwy. 212

Bridge elev. 1670 ft

Lake Oahe (elev. 1600 ft)

Martin McDermott and Roy Hunt, McKinney Drilling Co.

■ A huge landslide at the U.S. 212 Forest City Bridge across the Missouri River was reactivated by raising the reservoir level behind Oahe Dam.

CRITICAL THINKING QUESTIONS

1. The Forest City Bridge landslide occurred on a very gentle slope. What three distinctive conditions caused the site to slide on such a gentle slope?
2. For each of these conditions, explain the process that makes the site more likely to slide.
3. What did engineers do to lessen the likelihood of further movement?

Slide Triggered by Filling a Reservoir

Vaiont Landslide, Italy, 1963 ▶

In a classic case in the southern Alps of northeastern Italy north of Venice, filling a reservoir behind the newly completed Vaiont Dam caused a catastrophic mountainside collapse. Engineers completed the modern, 264-m-high, thin-arch concrete dam in 1960 across a narrow, rockbound gorge. It was designed to provide both flood control and hydroelectric power. When the dam was ready for use, the engineers filled the reservoir behind it to 23 m below the spillway.

Heavy summer rains in 1963 filled the reservoir to only 12 m below the spillway of the dam. The mountainside on the south side of the reservoir consisted of limestone and shale layers parallel to the 35 to 40° slope that was known to be unstable. An ancient slide plane that was partly exposed was not recognized (or perhaps not acknowledged).*

Engineers had monitored the slope just upstream from the dam for three years, and small landslides were

expected. The slope had been creeping at 1 to 30 cm per week. By September 1963, the rate had increased to 25 cm per day; by October 8, it was up to 100 cm per day. At that point, engineers finally realized the size of the mass that was moving. They quickly began lowering the reservoir, but it was too late. Continued rain slowed reservoir draining and saturated the mountainsides. As often

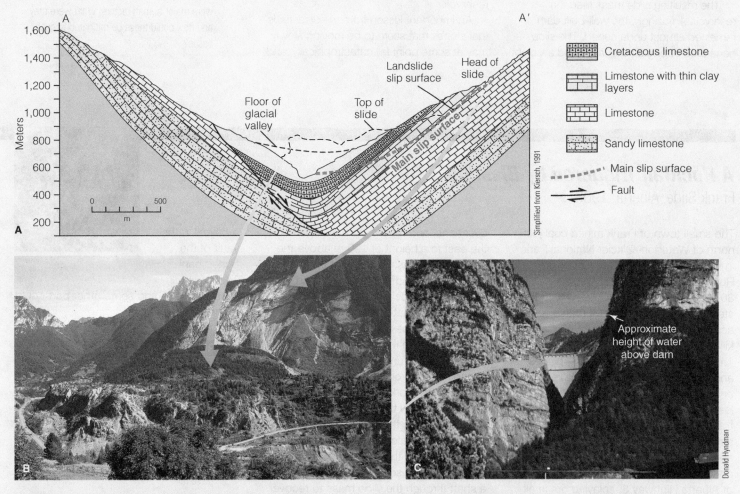

■ The 1963 Vaiont slide moved catastrophically on weak layers of shale within limestone beds parallel to the mountain face. **A.** Cross-section of the valley. **B.** The slip surface is in the upper right; the slide mass fills the center of the view, and the dam is just out of sight in the lower right. The highway looping around the toe of the slide mass in mid-view provides scale. **C.** The dam, viewed from downstream, survived even though a high-velocity wave of water 125 m high swept over it (up to the yellow line) and killed more than 2500 people downstream.

---

* Sometimes, if we want something badly enough, we ignore significant negative aspects. The deep narrow rock gorge seemed an ideal place to build a dam.

happens, grazing animals sensed danger a week before final failure and moved off that part of the slope.

At 10:41 p.m. on October 9, 238 million m³ of rock and debris collapsed catastrophically from the mountain face just upstream from the dam. It moved down at 90 km/h, filled 2 km of the downstream length of the reservoir, and moved 260 m up the far mountainside. The displaced water, in a wave 125 m high, swept over the dam and downstream, destroying Longarone and four smaller villages just 2 km below the dam. The destruction occurred two minutes after the slide began. People had no chance. Another huge wave swept into a town on the upstream end of the reservoir. The two waves killed 2,533 people.

The resulting slide mass filled the reservoir. Amazingly, the well-built dam remained almost undamaged. The slide generated strong earthquakes and a violent blast of air that shattered windows and blew the roof off a house well above the final level of the slide mass.

Engineering failures contributed significantly to the disaster. Exploratory drilling before dam construction intersected zones in which little or no drill core was recovered, a condition suggesting broken rock with much pore space. A tunnel excavated during dam building crossed a strongly sheared zone, but work continued without thorough examination of the implications. A study of the surrounding area would have shown that heavy surface runoff from higher slopes disappeared into innumerable solution fractures upslope of the slide plane. That water would dramatically increase the internal pore-water pressure in the rocks above the reservoir.

An important lesson from the disaster is that slopes that seem to be moving slowly may at some point fail catastrophically, and catastrophic landslides sometimes show precursory movement.

In addition to the lives lost, the cost of the landslide was considerable. Loss of the dam and reservoir cost $1.56 billion (2015 $) other property damage downstream came to a similar amount, and civil lawsuits for personal injury and loss of life cost even more.

CRITICAL THINKING QUESTIONS

1. Landslides in response to rising reservoir levels are well known in soft sediments. In this case one occurred in seemingly solid rocks. What were the main contributing factors here? Explain each process.
2. As with many "natural disasters," there were major human factors. What were they and how could these be mitigated in such cases?

A Rockfall Triggered by Blasting
Frank Slide, Alberta, 1903 ▶

The small town of Frank mined coal just north of Waterton–Glacier National Park in the Front Ranges of the Canadian Rockies. At 4:10 a.m. on April 29, 1903, 30 million m³ of rock fell 762 m from the steep east face of Turtle Mountain, swept across the town, buried most of it, and killed more than 90 of the 600 people living there. Moving on bedding planes and approximately parallel fractures in crystalline limestone, the event took less than 100 seconds. It began as a translational slide, which developed into a rock or debris avalanche as it gained speed. The giant pile of rock rubble continued north across the current route of Alberta Highway 3, splaying out in all directions, and moved up the slope to the east to a height of 120 m above the valley floor. Consensus is that the rubble moved as a "fluid" with compressed air, pulverized rock, and possibly water from the river that crossed its path. The immediate cause of the tragedy was mine cuts (low on the mountainside, now under the debris) that undermined tilted layers of sedimentary rocks. Although 17 miners were actively working within the mountain, they were below the slide scarp. They managed to dig themselves out; for once, a mine was a safer place to be than in the town below the mountain. Local tradition tells of several attempts to sink a shaft through the slide mass to recover a payroll in the vault of the buried bank. The mountain is still unstable and some other part could come down in future.

CRITICAL THINKING QUESTIONS

1. The Frank slide occurred from a mountain consisting of apparently very solid rock. What characteristics in that rock made the slide more likely?
2. Huge blocks of that solid rock moved far out from the base of the mountain and up the other side of the valley. What permitted it to go so far?

A. *The Frank slide peeled off the whole east face of Turtle Mountain and splayed a spectacular bouldery deposit across the valley and up the slope to the west.*

■ B. *Huge limestone boulders were carried far beyond the base of the slide.*

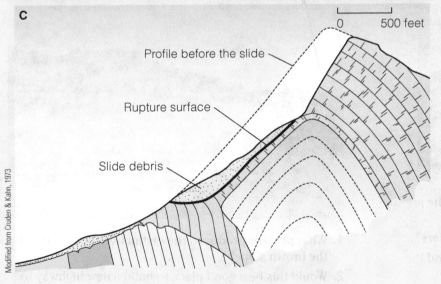

■ C. *This cross-section shows the slope before and after the Frank slide. Fracture sets in some rocks and approximately parallel bedding surfaces in other rocks provided zones of weakness.*

Landslides and Other Downslope Movements   **215**

Critical View

A A large stream lies just to the right of this photo, in southwestern Montana.

1. What would have caused slumping of the road at the right edge of the photo? Explain.
2. What would have caused slumping above the road and below the house?

C Giant boulders are strewn across this campground near Yellowstone Park.

1. Where did the boulders in the foreground of the photo come from?
2. What is the process called that moved them here?
3. What is the evenly sloping rocky surface behind the trees called?
4. Why are the largest boulders concentrated at the base of the slope?

B A treeless path runs from the cliffs at the top of the mountain to the bottom of the photo. Coquihalla (Hwy 5), southwestern British Columbia.

1. What would have caused the treeless area?
2. What processes would have promoted that movement?

D This brownish, grassy landscape lies near the north entrance of Yellowstone Park.

1. What process produced the distinctive topography of the brown slope?
2. Would this be a good place to build a new highway to reach the area at the upper left of the photo? Why or why not?

E This expensive new home was built on a flat area excavated in an evenly-sloping soil hillside (below the trees in the background) in a dry climate. The brownish area in the foreground is another flat area excavated for another house. The green area below the house has been recently planted with grass.

1. What caused the patch of gully erosion in the lower left? Where, specifically, did the water come from to erode the gully?
2. What could be done to stop the process that produced the gullying?

G The new houses at the base of barren cliffs with thick layers of sandstone are surrounded by some huge boulders that make for interesting landscaping.

1. Where did those boulders come from?
2. Is this a good place for houses? Why or why not?
3. What other occurrence described in this chapter shows the same natural hazard environment? What ultimately happened in that other case?

F The green, brush-covered slopes northwest of Los Angeles have many recent landslides.

1. Why are these slopes covered with small bushes and no trees? What is the relationship between the ground vegetation and the landslides?
2. What else likely contributed to the sliding in this area?
3. What should homeowners at the top of this slide area do to minimize the danger of a new large landslide?

H This highway runs along a very steep coastal cliff south of San Francisco, California.

1. This part of the roadway was evenly graded but now is wavy—up and down. What caused that waviness?
2. At the base of the roadcut and protruding from the concrete are many huge bolts. Why are they there—what do they do and how do they work?
3. Coarse screen covers the upper road cut. What is its purpose?

Chapter Review

Key Points

Forces on a Slope

- A driving force pulls materials downslope, while a resisting force holds them in place. When the driving force is greater than the resisting force, a landslide occurs, bringing the two forces back into equilibrium.

- The relationship between slope angle and load is a key factor in slope failure. Steeper slopes and slopes with heavier loads are more likely to overcome frictional resistance and fail. **FIGURE 8-2** and **By the Numbers 8-1**.

- Water is a key factor in slope stability. A little water coating holds grains together through cohesion. More water fills pore spaces and adds load to the slope—it pushes grains apart and makes a slope more prone to slide. **FIGURES 8-3** and **8-4**.

Slope Material

- Internal surfaces sloping in the same direction as the land surface provide weak zones that facilitate slip, especially if the surfaces daylight—that is, come back to the surface at their lower end—or if they are old landslide slip surfaces. **FIGURE 8-6**.

- Quick clays, consisting of clay flakes deposited in saltwater, can stand on edge like a house of cards. If the clays are raised, drained, and rinsed with freshwater, they are susceptible to collapse. **FIGURE 8-7**.

Causes of Landslides

- Landslides can be caused by adding soil moisture, slope material instability, or jarring by earthquakes.

- Changes in slope imposed by external factors—such as undercutting by a stream or building a road, loading of the upper part of a slope caused by construction, addition of water by various means, or removal of vegetation—can also destabilize equilibrium and promote sliding. **FIGURES 8-10** and **8-13 to 8-15**.

Types of Downslope Movement

- Rockfalls occur in areas with steep cliffs when rocks are broken loose by freezing or ground shaking. The distance a rockfall travels, its runout, is related to the mass of the rock and the distance from which it falls. Higher, more massive rocks fall farther. **By the Numbers 8-2**.

- Debris avalanches are similar to rockfalls except that their fragments disintegrate into tiny pieces that entrain air or water and flow at high velocity, much like a dense liquid. As with rockfalls, their travel distance depends primarily on their height of fall.

- Rotational slumps are common in weak, homogeneous material. Sliding masses rotate on a curving surface. **FIGURES 8-23** and **8-24**.

- Translational slides are facilitated by slip surfaces inclined nearly parallel to the ground surface. **FIGURE 8-25B**.

- Soil creep is the slow downslope movement of soil and rock, which can be accelerated by alternating periods of freezing and thawing. **FIGURE 8-28**.

- Conditions for snow avalanches include slope steepness, weather, and temperature. Unstable layers in a snowpack can be caused by new snowfall or periods of melting and refreezing. Snow avalanches can occur spontaneously or they can be triggered by human activities, such as skiing or snowmobiling. **FIGURES 8-29 to 8-34**.

Hazards Related to Landslides

- Earthquakes trigger many landslides, especially rockfalls.

- Loosely packed sandy soil saturated with water can "liquefy," settle into a smaller volume, and spread when shaken by an earthquake. **FIGURES 8-35 to 8-37**.

- Landslides can block waterways, creating a landslide dam, which can later fail and cause flooding. Failure of landslide dams depends on the size of the dam, its material, and the rate at which the lake behind the dam rises. **FIGURE 8-39**.

Mitigation of Damages from Landslides

- Landslide hazard maps can be used to restrict development in landslide-prone areas. **FIGURE 8-40**.

- Evidence of past landslides is a good indicator of future landslide risk. Hummocky landscapes and wavy roads are a sign of landslide activity. **FIGURE 8-41**.

- Engineering solutions to landslides—including water removal—are geared toward restoring the balance between the forces that act on a slope, as well as reducing the penetration of water. **FIGURES 8-42 to 8-46**.

Key Terms

angle of repose, p. 182
cohesion, p. 182
daylighted bed, p. 184
debris avalanche, p. 193
driving force, p. 182
fluidization, p. 194
frictional resistance, p. 182

headscarp, p. 194
kaolinite, p. 185
liquefaction, p. 185
load, p. 182
quick clay, p. 185
resisting force, p. 182
rockfall, p. 190

rotational slump, p. 194
runout, p. 192
slope angle, p. 182
smectite, p. 185
snow avalanche, p. 197
soil creep, p. 197
surface tension, p. 182

talus, p. 191
toe, p. 194
translational slide, p. 196
water pressure, p. 184

Questions for Review

1. Describe the forces exerted on a slope. Use a diagram to illustrate your description.

2. Why does a pile of dry sand have gently sloping sides whereas a pile of wet sand can have nearly vertical sides?

3. Why is the friction force on a gently sloping slip surface greater than that on a steeply sloping slip surface?

4. Explain why raising the water table may lead to slope failure.

5. Describe the sequence of events that leads to quick clay formation.

6. List the main factors that affect whether a slope will fail in a landslide.

7. What factors determine the distance a rockfall will travel laterally from its point of origin?

8. What is the difference between a rockfall and a debris avalanche?

9. Why does the top of a rotational slump tilt back into the slope?

10. What is the difference between a rotational slump and a translational slide?

11. What factors accelerate soil creep?

12. What conditions commonly lead to snow avalanches?

13. What steps should you take if you are caught in an avalanche?

14. What causes liquefaction of sediments? Briefly explain the process.

15. Under what circumstances is flooding a hazard related to landslides?

16. What parts of the United States are most susceptible to landslides?

17. Name several landscape features that indicate a record of past landslides.

18. Why might people want to build in landslide-prone areas?

19. List several ways in which old landslides are commonly reactivated.

20. What can be done to slow or stop the movement of a rotational slump?

21. List several distinctly different ways in which water can be removed from a wet slope that has begun to slide.

Critical Thinking Questions

1. A developer in the steep coastal area northwest of Los Angeles requests a permit from the county to subdivide a large plot of land for expensive homes. Old landslides are documented in the vicinity. Should the county provide the permit? Considerations include personal freedom for use of one's property, road building, water and sewer lines, property taxes to the county, aspects of liability, and insurance. In case of damages because of ground movement, who is liable and who should pay?

2. A moderate-size earthquake (e.g., magnitude 6) triggers a large landslide that severely damages many homes in a large 15-year-old subdivision near San Francisco. Homeowners' insurance companies refuse to pay for damages. Who should pay for the damages? Is there shared liability and if so by whom and why?

3. Landslides often occur on gentle slopes on the flanks of valleys away from mountainous terrains. What are the main factors that contribute to such landslides and what can be done to prevent or stop such movement?

■ Shrinkage and settling of ground near Apache Junction, east of Phoenix, Arizona, caused by excessive withdrawals of groundwater.

Donald Hyndman

9

Sinkholes, Land Subsidence, and Swelling Soils

Shrinking Ground

For more than 100 years, groundwater has been critical for agriculture, mining, and municipal uses in the Phoenix and Tucson areas of southern Arizona. The rapidly growing population demands ever more water. With potential evapotranspiration of more than 100 cm per year and desert-climate precipitation as low as 7 cm per year, groundwater supplies are drawn down more quickly than they are being replenished. In some areas the groundwater level has fallen more than 180 m (600 ft). As the water table lowers, the ground above sinks and sometimes collapses. Differential subsidence of an area of more than 7700 km² has led to earth fissures as much as 6 m deep and 9 m across. Widening of the fissures has damaged highways, sewer lines, buildings, and other structures. The region pumped almost all of its water from underground until the Central Arizona Project was built to take water from the Colorodo River.

Sinkholes

A Florida sinkhole opened up under the bedroom of a home in Seffner, Florida, on February 28, 2013, burying a 37-year-old man. Such sinkholes are most common at the end of the dry season, when the wet season begins, generally in February, or when strawberry farmers pump water from the ground to spray on their crops during cold weather. Ice coats the berries, preventing them from freezing. Ground movements may not be as dramatic as earthquakes and volcanoes, but they cause far more monetary damage in North America because they deform and effectively destroy roads, utility lines, homes, and other structures. **Sinkholes** develop when the overlying ground collapses into underground soil cavities over limestone.

In some places, especially in limestone terrains of the eastern United States, the ground may suddenly collapse, leaving sinkholes that are tens to hundreds of meters across. Not only can sinkholes damage houses and roads, but they can drain streams, lakes, and wetlands. In low areas, they can channel contaminants directly into underground aquifers, the main source of water used for drinking and other purposes in many parts of the country.

Groundwater

To understand sinkholes and how they form, we first need to understand water underground: its distribution and how it moves. All of the water on land—including rain, streams, lakes, and the water underground that many cities pump out to use for drinking water—ultimately came from the oceans. Evaporation from the surface of the oceans condenses in the atmosphere to form clouds, which are merely tiny droplets of water. Wherever the atmosphere becomes oversaturated with water, droplets of water form and eventually grow large enough through collisions to fall as rain. We describe how this works in more detail in Chapter 10, but for now we need to know that although some of the rain reaching the ground (or snowmelt) flows over the surface to streams and some evaporates and is taken up by vegetation, much of it seeps into the ground to become groundwater. The level to which **groundwater** rises in a well or hole in the ground is called the **water table**.

In the ground, that water fills open cracks in all kinds of rocks or the pore spaces between grains of sand, gravel, and clay (**FIGURE 9-1A**). Often there is not enough water to fill the space all the way to the surface. Above the water table, there is a small zone in which water rises and saturates the sediment, and above this there is a little water that merely wets the surface of the cracks or mineral grains. Because the groundwater under a hill commonly stands higher than that in a stream, it naturally flows down the slope of the water table into the stream under the pull of gravity (**FIGURE 9-1B**). Thus, the water table under the hill is commonly shaped about like the hill, but more subdued. Flow of water into the stream slowly feeds the stream as fast as it can travel through the cracks or pore spaces in the rocks or sediments. That is what keeps many streams flowing year-round, even long after the last rain.

Flow of groundwater is normally downslope, away from the highest point on the water table, commonly near the top of the hill. Because gravity pulls the water down, it tends to move downward. However, because the upper surface of water in the stream or lake remains nearly fixed, the water follows a curved path underground, as shown in **FIGURE 9-1C**. If a well is drilled down below the top of the water table and water is pumped out, that pulls water toward the pumping well and lowers the water table around the well as shown in the figure.

Formation of Sinkholes

Water droplets in clouds are surrounded by air, which is mostly made up of nitrogen, oxygen, and a little carbon dioxide. Lightning storms create oxidation reactions between these elements, resulting in dilute nitric acid and carbonic acid. These, combined with chemicals in agricultural fertilizer and industrial emissions, make most rain somewhat acidic. Additional chemical interactions occurring in the soil make groundwater still more acidic.

This acidic groundwater dissolves the surrounding rock surfaces. Most rocks dissolve very slowly. Some rocks, such as limestones ($CaCO_3$), dissolve a little faster. Reaction of acidic groundwater with limestone forms calcium bicarbonate, which is soluble in water (**By the Numbers 9-1**). Some common sedimentary rocks, such as salt and gypsum (often called *evaporites*), are highly soluble but not nearly as abundant as limestone.

Water that quickly percolates down through limestone bedrock tends to follow cracks. The water slowly dissolves the walls of a crack, concentrating on the weakest spots near the water table and slowly widening the cavity to form caves, caverns (**FIGURE 9-2**), and underground streams. The rate of solution is slow, on the order of a few millimeters per 1000 years. Faster flow of water through larger cavities permits it to dissolve the adjacent limestone more quickly, enlarging those cavities. Where the soluble rocks are close to the surface, the solution cavities and caverns can grow large enough for the roof to collapse, thus forming sinkholes (**FIGURE 9-3**).

By the Numbers 9-1

Formation of Calcium Bicarbonate

$$H_2O \ + \ CO_2 \ = \ H_2CO_3$$

(rainwater)   (carbon dioxide)   (carbonic acid in water)

Carbonic acid reacts with limestone to form calcium and bicarbonate:

$$H_2CO_3 \ + \ CaCO_3 \ = \ [Ca^{++} + 2HCO_3^{-}]$$

(carbonic acid)   (limestone)   (Calcium biocarbonate)

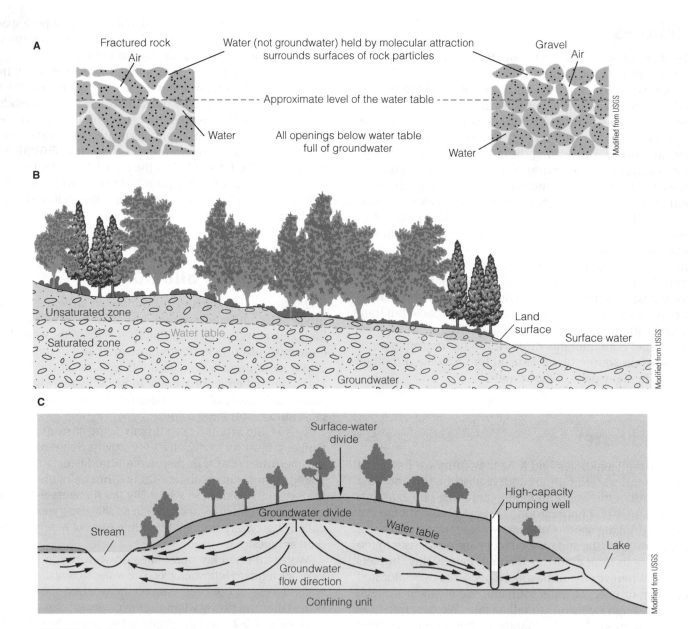

A Fractured rock
Air

Water (not groundwater) held by molecular attraction surrounds surfaces of rock particles

Gravel
Air

Approximate level of the water table

Water

All openings below water table full of groundwater

Water

Modified from USGS

B

Unsaturated zone

Water table

Saturated zone

Land surface

Surface water

Groundwater

Modified from USGS

C

Surface-water divide

Groundwater divide

High-capacity pumping well

Stream

Water table

Lake

Groundwater flow direction

Confining unit

Modified from USGS

FIGURE 9-1 The Water Table

A. The upper surface of water-filled open cracks or pore spaces in rocks or sediments is the water table.
B. The water table (dashed black line) normally lies below the ground surface, sloping downward toward connected surface-water bodies such as lakes or streams in most humid areas. In fact, the lake or stream is the exposed level of the water table. **C.** Groundwater flow in a humid region follows a curved path from the water table, through the subsurface sediments or rock toward the stream or lake. Pumping from a well pulls out groundwater, lowering the level of the water near the well.

Because the reaction between acidic water and limestone occurs more rapidly under warm, moist conditions, caverns and sinkholes are most common in tropical or subtropical climates. Dissolution is amplified by rain that is more acidic because of air pollution or heavy fertilization—acid rain. In both cases, water flows through fractures and cavities in the limestone and carries away the soluble calcium bicarbonate.

Limestone can dissolve above, at, and below the water table. Where limestone is above the water table, acidic water running down through fractures slowly widens them.

Just below the water table, cavities can rapidly widen if horizontal bedding surfaces are open enough to conduct large amounts of water. In this case, the large flow past limestone surfaces brings fresh acidic water to these reactive surfaces and carries away the calcium bicarbonate products. This is the environment in which large caverns generally form. When the water table drops below the top of a cavern, water percolating through fractures above can precipitate calcium carbonate as the water locally evaporates and loses its carbon dioxide. These formations are called *stalactites* when

FIGURE 9-2 Caverns

A limestone cavern exposed along Interstate Highway 44 near Springfield, Missouri, shows sagging and fracturing of the rocks above it.

A

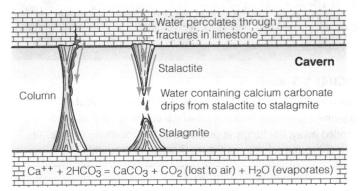

$$Ca^{++} + 2HCO_3^- = CaCO_3 + CO_2 \text{ (lost to air)} + H_2O \text{ (evaporates)}$$

B

they hang from the roof and *stalagmites* when they grow from the floor (**FIGURE 9-4**). Caverns found high on hillsides above a current water table suggest that the water table has dropped, often the result of a nearby stream eroding its valley deeper.

Limestone bedrock near the water table dissolves along fractures to create an uneven and potholed upper surface. Later erosion of the soil cover may expose that surface as a limestone landscape, called **karst** (**FIGURE 9-5**). In extreme cases, deep solution of the limestone along dominantly vertical fractures can form an extremely ragged, toothy-looking, and often picturesque landscape. Perhaps the best known of these are in the Guilin and Kunming areas of southwestern China.

FIGURE 9-4 Stalactites and Stalagmites

A. This cavern near Tulum, in eastern Mexico, shows stalactites hanging from the ceiling and stalagmites growing from the floor. They may ultimately build into continuous columns. **B.** Water containing dissolved calcium carbonate seeps down through limestone into a cavern. There the evaporating water loses carbon dioxide and again precipitates calcium carbonate as stalactites and stalagmites.

FIGURE 9-3 Sinkhole Collapse

In May 1981, a large sinkhole developed in Winter Park, near Orlando, Florida. It swallowed a house, a couple of Porsches, a camper, and half of a municipal pool.

FIGURE 9-5 Karst

This exposed karst on the west coast of Turkey is typical of dissolved limestone bedrock on which covering soil has been eroded away. The lumpy appearance of the limestone emphasizes below-ground solution as the mechanism of erosion.

Types of Sinkholes

Three types of sinkhole formation are common, with gradations between them:

1. **Dissolution.** Where the soil cover is thin and highly permeable, acidic groundwater seeps through it and dissolves the underlying limestone along fractures. The upper parts of fractures dissolve to form a lumpy or jagged karst surface. The overlying soil can slowly percolate, or ravel, down into the fractures to create a surface depression. Where the groundwater level is high or the fracture plumbing becomes clogged with sediment, the depression may fill to form a pond. These depressions are shallow and not generally dangerous.

2. **Cover subsidence.** Where tens of meters of sandy and permeable sediment exist on top of limestone bedrock, numerous sinkholes can form as the soil slowly fills expanding fractures and cavities in the limestone. Depressions generally form gradually.

3. **Cover collapse.** Where overlying sediments (called *overburden*) contain significant amounts of clay, this cover is more cohesive and less permeable. As a result, it does not easily ravel into underlying cavities in limestone. This can allow a soil cavity to grow large and unstable, leading to the sudden collapse of its thinning roof. The lack of warning makes these steep-sided sinkholes destructive and dangerous.

Cover-collapse sinkholes open with little or no warning, taking with them roads, parking lots, cars, and occasionally even houses and other buildings, as in the February 25, 2002, collapse at a road intersection in Warren County, Kentucky (**FIGURE 9-6**). In this case, storm water funneled underground at three corners of the intersection and caused further solution and instability in a limestone cavern underground. Before the road was built, karst experts who knew that the underground cave passage had an unstable roof recommended a different, safe route for the road, but their

Sediments spall into a cavity.

As spalling continues, the cohesive covering sediments form a structural arch.

The cavity migrates upward by progressive roof collapse.

The cavity eventually breaches the ground surface, creating sudden and dramatic sinkholes.

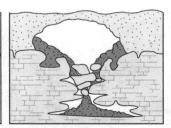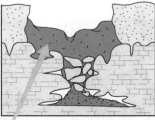

A

FIGURE 9-6 Formation of a Cover-Collapse Sinkhole

A. This sequence of events can lead to a cover-collapse sinkhole. **B.** The sinkhole on the right formed on February 25, 2002, by cover collapse over a limestone cavern in Warren County, Kentucky, where storm-drain water was funneled underground at a road intersection.

B

advice was not followed. Individual collapses such as this cost more than $1 million each to repair.

Detection of subsurface cavities that may develop into sinkholes is often possible with ground-penetrating radar (GPR), which uses high-frequency radio waves to distinguish different layers of sand, clay, and limestone. GPR can be used in conjunction with a cone penetrometer test (CPT), which helps identify potential sinkholes by determining the properties of sediment layers.

To help determine potential sinkhole locations over a broad area, NASA has been testing ways to scan the ground surface from planes and satellites. It appears that soil often shifts slowly toward a potential sinkhole site. This slow shift can be tracked by satellite imaging. On the ground, home and building owners can spot sinkhole warning signs by looking for sinking ground near newly exposed building foundations and fence posts, tilting fences, doors that don't close properly, or structural cracks in walls, floors, and ground surface (see the Sinkhole, Subsidence, and Swelling Soils Survival Guide at the end of this chapter).

Areas That Experience Sinkholes

Slowly dissolving carbonate rocks underlie more than 40% of the humid areas of the United States east of Oklahoma (**FIGURE 9-7**). Approximately 55% of Kentucky is on limestone weathering to karst, including one of the largest and most famous caverns in the United States, Mammoth Cave National Park. Slow solution of this limestone has left the landscape pockmarked with more than 60,000 sinkholes.

As recently as 20,000 years ago, during the last ice age, when much ocean water was tied up in continental ice sheets, the whole carbonate platform of Florida was above sea level. The peninsula then was roughly twice as wide as at present. The groundwater level was much lower so many caverns sat above the water table. When the ice sheets melted, sea level rose and the water table rose to fill most of the caverns. Sand and clay up to 60 m thick cover most of the cavernous limestone that is weathered to a karst surface. Sinkholes are prominent in central Florida, and many of the lakes and ponds in west-central Florida are water-filled sinkholes (**FIGURE 9-8**).

Fluctuations of the groundwater level can contribute to sinkhole formation. Periods of heavy rainfall tend to loosen soil, enlarge cavities, and promote sinkhole formation. More sinkholes tend to form during dry seasons or when excessive pumping drops groundwater levels. One circumstance that leads to unusual groundwater use and lowering of the aquifer is pumping during a prolonged spell of freezing weather—for example, when strawberry and citrus farmers spray warm groundwater on their plants to form an insulating coating of ice (**FIGURE 9-9**).

Areas with the greatest potential for sinkholes are those where surface water tends to percolate into the ground, recharging the aquifers below. This potential is greatest where the water table lies below the tops of limestone caverns, allowing for unsupported open space in cavities above the water level. Widespread limestone near Pittsburgh, Pennsylvania, is broken by numerous near-vertical fractures. Acidic rainwater percolating down through those

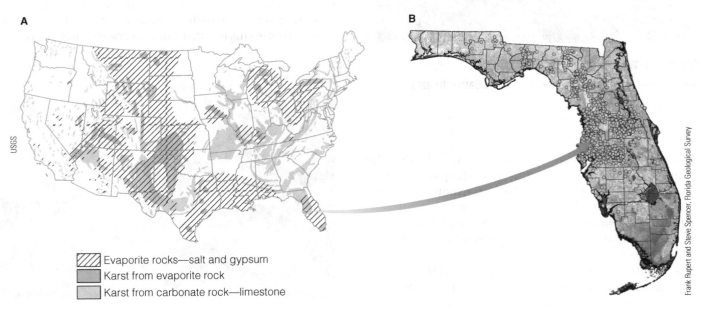

FIGURE 9-7 Sinkhole Distribution in the United States

A. This map shows the distribution of areas susceptible to sinkhole formation and other types of subsidence. **B.** Florida sinkholes are concentrated in the central part of the state (blue dots).

FIGURE 9-8 A Swarm of Water-Filled Sinkholes
Circular ponds mark water-filled sinkholes in Florida.

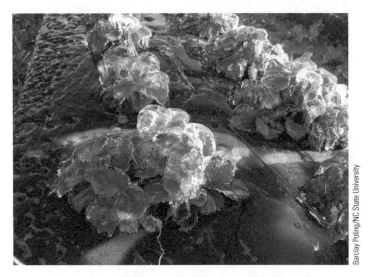

FIGURE 9-9 Winter Groundwater Use
Water sprayed on strawberry plants in cold weather freezes, protecting the plants.

fractures slowly dissolves and widens them. Soil above the limestone cracks may intermittently break apart to form a progressively enlarging cavity. Gradual expansion of that cavity eventually leads to formation of a cover-collapse sinkhole at the surface. Pennsylvania sinkholes are typically 1 to 6 m in diameter, and fractures seldom enlarge to form underground caverns.

Areas with the least potential for sinkholes are those where water is being discharged to the surface. There the water is likely to fill the cavities and helps support their roofs; the flowing groundwater is unlikely to be corrosive to carbonates.

In urban areas, large volumes of water carried by storm drains and leaking water mains can enhance the development of sinkholes. A large water flow can quickly flush soil from above fractured limestone to form large sinkholes. Roads and buildings above such sinkholes are vulnerable to severe damage (**FIGURE 9-10**). In Allentown, Pennsylvania, a nearly new eight-story office building broke up in 1994 after a pair of large sinkholes developed below major support columns. Damage required razing of the $10 million building.

On August 13, 2006, an 18-m-wide by 23-m-deep sinkhole collapsed under one end of a home in Nixa, just south of Springfield, Missouri. The two-car garage, initially cantilevered over the hole, collapsed shortly afterward, and the house continued to break up. Sinkholes are abundant in the Karst Plain in southwestern Missouri, and most appear to be cover-collapse features. The Nixa sinkhole developed at the intersection of crossing fractures, which are typical for the area. Examination of the site suggests that sinkhole collapse may have been

FIGURE 9-10 Building Damage Due to Sinkholes
A sinkhole formed under this house in Palmyra Borough, Lebanon County, Pennsylvania.

promoted by water seepage from an abandoned septic tank and its tile drain field; decayed roots of an old tree; roof downspouts draining into a gravel field on the gentle upslope side of the house; and well-drained soils that conveyed water to the fractured bedrock. Following destruction of the house, the sinkhole was filled with 1–2 m boulders and then capped with soil. Collapse apparently occurred because of drainage water percolating through soil filling a cavity above a limestone cavern. Because the soil cavity was under the concrete floor of a two-car garage, it was not noticed until it grew beyond the edge of the garage and collapse occurred. More commonly, such sinkholes appear as small holes that progressively enlarge.

Construction activities can also lead to sinkhole development through increased loads on ground surfaces, dehydration of foundation soils, and well drilling. Heavy construction equipment or the weight of a new structure can impose loads (**FIGURE 9-11**).

Sediments such as salt and gypsum also produce karst landscapes and underlie large areas of the United States. Their relative ease of solution in water can lead to cavity formation in just days. They are widespread in the southeastern states, the Appalachians, and western Texas.

In some areas, salt is mined from giant fingers of salt, or *salt domes*, which rise from deep layers of salt under coastal plains of the southeastern states. On August 3, 2012, a huge sinkhole collapsed into an underlying mined-out cavity 100 km west of New Orleans. Salt was being mined by the Texas Brine Co., and the cavity was being used to store oil and natural gas for later use. A few weeks before the collapse, nearby residents noticed shaking and bubbles rising from the bayou. On the day of the collapse, methane gas bubbles rose in the water and people could smell petroleum. A test well showed collapse of the outer wall of the salt dome. By April 2014, cypress trees that fringed the sinkhole were dying and the 300 by 350 m sinkhole was still growing. On May 8, 2008, a sinkhole 45 m (150 ft) deep and 180 by 160 m (600 by 525 ft) across developed above a huge, buried salt dome. It swallowed oil drilling equipment, storage tanks, a tractor, trees, and telephone poles. The nearby well was being used for disposal of waste saltwater from oil well production when the ground began to settle. Other sinkholes over the same salt dome formed in 1969 and 1981. It is suspected that injected water dissolved cavities in the salt, leading to collapse of the ground above.

In another example, early on the morning of March 12, 1994, a section of shale roof rock 150 by 150 m collapsed into an underground mining excavation in the Retsof Salt Mine in the Genesee Valley of New York State. The hole in the shale barrier allowed groundwater to pour into the mine, gradually filling its cavities and destroying the largest salt mine in the United States. Insurance for sinkhole collapse is not covered in many homeowner policies though it may be covered by payment of an additional premium. Florida law, however, requires authorized insurers to cover "catastrophic ground cover collapse." Subsidence not clearly associated with a sinkhole would not be covered.

Land Subsidence

Land subsidence occurs when the ground settles due to changes in fluid levels underground. Because subsidence occurs across large regions, its effects are less obvious and dramatic than those of sinkholes. Nonetheless, subsidence also causes considerable damage. Small faults cause some areas to drop more than others, damaging houses and utilities. In some areas, *earth fissures* form due to differential subsidence between adjacent areas. In coastal areas, subsidence can sink communities closer to sea level and leave them more vulnerable to flooding.

Subsidence of the ground is a serious problem throughout North America. This lowering of the ground surface is caused by a variety of human activities, including extraction of groundwater, drainage of organic or clay-rich soils, and thawing of permafrost.

Mining Groundwater and Petroleum

One of the most common causes of land subsidence is the pumping of water or petroleum out of the ground. As rainwater soaks into the ground and flows toward rivers and streams, it travels through porous rocks—*aquifers*. Many communities use these aquifers as a source of fresh water by pumping the groundwater. In areas where the amount pumped exceeds the recharge from precipitation, this is considered **groundwater mining**. A similar effect can also result from pumping of underground supplies of oil and gas.

FIGURE 9-11 Heavy Equipment Causes Collapse

A truck-and-drill rig that was drilling a new water well near Tampa, Florida, overloaded the roof over a limestone cavern, causing it to collapse. The truck and equipment eventually sank out of sight in a crater 100 m deep and 100 m wide.

Florida Geological Survey

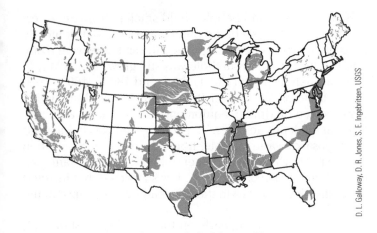

FIGURE 9-12 Groundwater Subsidence Map
Areas of ground subsidence by groundwater pumpage in the United States.

D. L. Galloway, D. R. Jones, S. E. Ingebritsen, USGS

Ground subsidence resulting from groundwater and petroleum extraction has been significant in many parts of the world, including the United States (**FIGURE 9-12**). **Table 9-1** lists a few of the more significant cases.

Many aquifers consist of thick accumulations of sand and gravel with water filling the spaces between the grains. Withdrawal of water from the pore spaces of loosely packed sand grains permits them to pack more tightly together and take up less space (recall FIGURE 8-3, p. 183). The original groundwater reservoir capacity cannot be regained by allowing water to again fill the pore spaces, because silt and clay grains cannot be pushed apart into a more open structure once their pore space has collapsed.

Large water withdrawals can cause permanent reduction of aquifer capacity and subsidence of the ground. Prominent examples of subsidence due to long-term groundwater withdrawal can be found in Silicon Valley around the southern end of San Francisco Bay; the southern half of the Central Valley of California; the Long Beach area south of Los Angeles; southern Arizona; the Texas Gulf Coast region near Houston; and Venice, Italy (**Case in Point:** Subsidence Due to Groundwater Extraction—Venice, Italy, p. 239).

In the San Jose area at the southern end of San Francisco Bay, depletion of groundwater accompanying the rapid growth in population caused about 4 m (13.1 ft) of ground subsidence from about 1920 to 1995 (**FIGURE 9-13**, p. 229). Following rapid subsidence of 80 cm in 4 years around 1960, groundwater pumping was slowed and water imported, beginning in 1963. This has nearly stopped the subsidence.

In the Houston area, oil and gas were produced early along with groundwater. More than 1 m of subsidence has been observed in parts of this area. As a consequence, coastal flooding that was formerly a problem only with the

Table 9-1	Examples of Ground Subsidence Resulting from Groundwater and Petroleum Extraction		
LOCATION	**SUBSIDENCE (m)**	**FLUID EXTRACTED**	**SOME CONSEQUENCES**
Sacramento–San Joaquin Valley, California	9 in some places	Groundwater for agricultural use	Fields in the delta-area sink below sea level
San Jose and Santa Clara Valley (Silicon Valley), California	>2–2.5	Groundwater for orchards, industrial, and municipal use	Tidal flooding of San Jose by San Francisco Bay
Phoenix to Tucson, Arizona	Up to 5	Groundwater for agricultural and municipal uses	Subsidence and earth fissures
Las Vegas, Nevada	Up to 2	Groundwater for municipal use	Subsidence, ground failure
Long Beach, California	9	Oil and gas	Pipelines and harbor facilities damaged; dikes needed to prevent seawater flooding
Houston–Galveston, Texas (a circular area 130 km across)	>2	Oil and gas, groundwater for municipal use	Frequent flooding, sometimes severe; submerged wetlands
Everglades, Florida	2–3.5	Drained for agricultural land and urban development	Wetlands now reduced by 50%; subsidence; saltwater intrusion into groundwater now slowed
Mexico City, Mexico	8–10	Groundwater for industrial and municipal use	Tilting buildings; broken water, sewer, and utility lines
Pisa, Italy	Several	Groundwater for municipal use	Danger to Leaning Tower from uneven subsidence
Venice, Italy	Several	Groundwater for industrial and municipal use	Tilting and settling of buildings below sea level, severe flooding at especially high tides
Delta of the Rhine and Meuse Rivers, the Netherlands	Large areas to 6 m, locally up to 6.7 m	Drainage of agricultural land	Dehydration and oxidation of peat-rich soils, and further ground subsidence

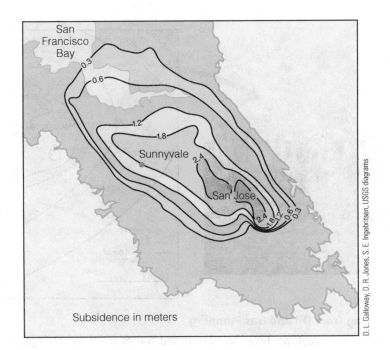

A

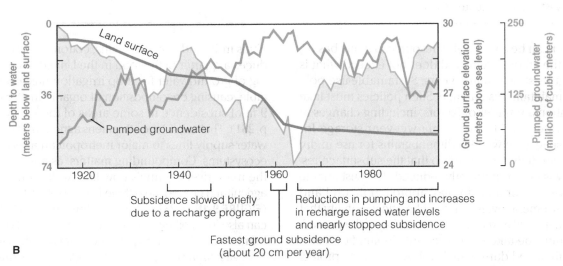

Subsidence slowed briefly
due to a recharge program

Reductions in pumping and increases
in recharge raised water levels
and nearly stopped subsidence

Fastest ground subsidence
(about 20 cm per year)

B

FIGURE 9-13 Subsidence Due to Groundwater Pumping

A. Map of subsidence in the Santa Clara Valley due to groundwater pumping. The area of greatest subsidence (dark orange) reached a total of 4 m in about 70 years from 1920 to 1990. **B.** Groundwater pumping in Santa Clara, California (shown in green), increased significantly from 1915 through 1960. As expected, the increases in pumping increased the depth to groundwater (shown in blue). It also had the unanticipated effect of dropping the land surface by subsidence (shown in brown), until water was imported beginning in 1963 to recharge the groundwater. This recharge and reduced groundwater pumping nearly stopped the subsidence.

occasional major storm is now a significant problem with many much smaller storms (**FIGURE 9-14**, p. 230).

In the coastal areas of Wilmington and Long Beach just south of Los Angeles, subsidence began in the 1940s with pumping of groundwater at the naval shipyard and later with oil extraction. By 1958, subsidence occurred over 52 km² by as much as almost 9 m, warping rail lines and pipelines and damaging buildings and shearing off oil wells.

Mitigation of ground subsidence due to groundwater or petroleum extraction initially involves stopping the activity that causes the problem. In some cases, governments institute regulations that prohibit or limit groundwater or petroleum extraction or initiate programs to reinject fluid into the ground. This can slow or eventually stop the subsidence. In 1958, when the California Subsidence Act was passed, groups of property owners in the Los Angeles area injected water underground to repressurize the pore spaces and stabilize the rate of ground subsidence. However, most of the subsidence or compaction of sediments cannot be reversed once it has occurred. Reduction in aquifer porosity and permeability is also generally permanent.

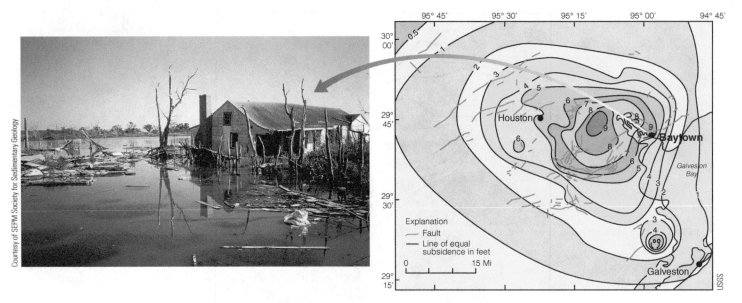

FIGURE 9-14 Subsidence Resulting from Oil and Gas Pumping
Subsidence of the Brownwood Subdivision of Baytown, Texas, has surrounded houses with brackish lagoon water. Subsidence in the Houston–Galveston region; contours in feet.

Groundwater can be a renewable resource in many basins, provided that the amount removed does not exceed what is replenished, at least over many years. Sustainable-use policies can help maintain this balance. Such policies must take into consideration a number of factors, including changes in climate and water quality. Available wet-year storage for water in pore spaces between sediment grains for use in dry years must be carefully monitored so that the subsurface reservoir capacity is not permanently reduced. The sustainable use, or "safe yield," for a basin may approach natural and artificial replenishment levels, but the climatic variability that affects both the natural and artificial restocking of the groundwater system must be taken into account. Stream flows also need to be maintained during the low-flow season to protect the health of stream ecosystems.

Drainage of Organic Soils

Subsidence can also be caused by draining organic-rich soils such as peat. In their natural state, these organic soils are saturated with anaerobic (without free oxygen) water, which allows them to maintain stability. When groundwater levels drop because of activities such as municipal and agricultural pumping, the organic soils are exposed to aerobic (oxygen-rich) water that percolates from the surface down to the water table. This allows bacteria in the sediment to oxidize the organic matter, which causes decomposition largely into carbon dioxide, a gas linked with additional natural hazards through global climate change (see Chapters 11 and 12). This decomposition occurs at a rate as much as 100 times greater than that at which such organic material accumulates.

The semi-arid Sacramento–San Joaquin Valley of central California is noted for its prolific agricultural products, from citrus fruits and wines to grains, cotton, and rice. The area produces a quarter of the food in the United States. Heavy pumping of groundwater for crop irrigation has lowered the water table, causing decomposition of organic matter and more than 9 m of subsidence in some areas of the valley (**FIGURE 9-15**, p. 231). That subsidence threatens agricultural production, large water-supply lines to major metropolitan areas, and important ecosystems. Compounding matters, as a result of subsidence, the main river channels now flow well above the level of the agricultural ground, held there behind dikes, leading to flood risk (**FIGURE 9-16**, p. 231). The agricultural ground behind the levees can also lose soil by blowing away in the wind or by agricultural burning, which destroys peat and compacts the ground.

A vast area of the Florida Everglades was drained decades ago for farming development, and flood control, causing almost 2 m of subsidence. An Everglades restoration plan is ongoing to reduce these and other impacts and take the Everglades back toward a natural state.

Many major river deltas are now sinking at an alarming rate. This is caused by reduced sediment loads related to upstream dams and placement of levees that prevent flood overwash, as well as extraction of groundwater, oil, and gas. The Mississippi River delta, for example, is sinking (relative to rising sea level) at 0.5 to 2.5 cm per year. Other imperiled deltas (with relative sinking of 0.5 to 6 cm/yr) include the Yangtze, Yellow, and Pearl Rivers in China, the Mekong River in Vietnam, the Ganges River in Bangladesh, the Po River in Italy, and the Nile River in Egypt. Most of these huge deltas lie within 2 m of sea level. Much of the area of many of these deltas is subject to flooding by storm surges.

Sinking of the Mississippi River delta area, including New Orleans, has increased because of levees that now prevent the river from adding silt to its floodplain during floods. That

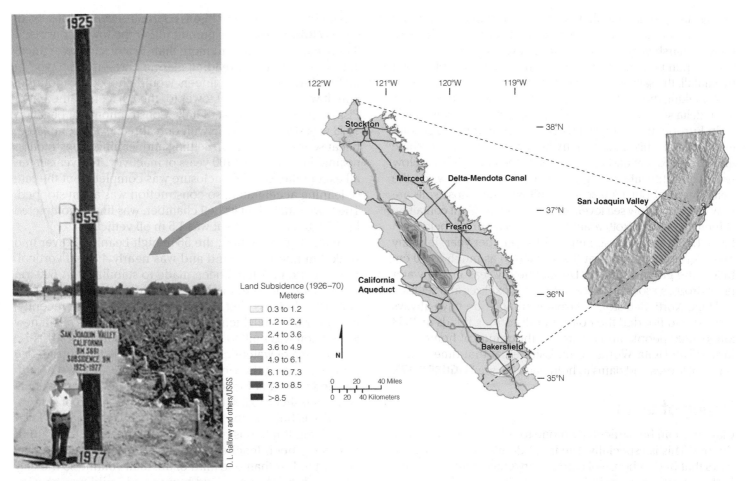

FIGURE 9-15 San Joaquin Valley Settles

Parts of the San Joaquin Valley, California, settled more than 8 m (more than 26 ft) in the 50 years between 1925 and 1977.

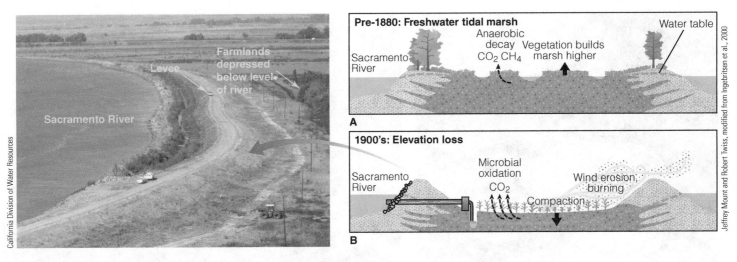

FIGURE 9-16 Subsidence of Drained Delta Marshes

A. Before 1800, the Sacramento–San Joaquin delta was a freshwater tidal marsh. **B.** Pumping water out of these marshes so they can be used for farming has lowered the water table and caused compaction, leaving them below the level of the river, which is now held back by high levees, shown in the photo.

sinking also promotes further saltwater incursion where heavier saltwater seeps in under the fresh water of an aquifer. That kills coastal marsh vegetation, including cypress forests. If fresh water is pumped out of a well faster than it can be replenished by rainfall, the salty water mixes with the water drawn from the well, making the well water unusable. An additional large factor in delta subsidence is oxidation of organic-rich sediments.

In Europe, the Netherlands is most vulnerable to subsidence because this small country lies on the delta of the Rhine and Meuse Rivers and one-quarter of the country lies below sea level. Artificial drainage of the peat-soil meadows on the delta began about 1800 BC to permit farming. Those soils are now 1 to 2 m below sea level; areas around the giant container-shipping center of Rotterdam lie as much as 6.76 m below sea level. Large areas are sinking at 1 to 8 cm per year. In many areas, groundwater lies at the surface to as much as 70 cm below the ground surface. During the dry season, the water table drops, so soils dry out and the peat oxidizes and settles.

Huge North Sea storms arriving at high tide have always eroded and flooded the coast, but disastrous floods in 1953 killed 1800 people and destroyed thousands of homes and farms. The Delta Works commissioned at that time built a series of levees and dams to hold back the sea (**FIGURE 9-17**).

Drying of Clays

Clay soils can be particularly prone to subsidence when they dry out. This is especially true in randomly oriented marine clays that have a house-of-cards arrangement with extremely high porosity (see again Figure 8-7). Collapse of that open arrangement can result in rapid land subsidence. The abundant inter-grain spaces of quick clays can collapse when jarred by an earthquake or heavy equipment, or if they are flushed with fresh water.

The Leaning Tower of Pisa is a famous example of subsidence (**FIGURE 9-18**). The ground under the tower consists of 3 m of clay-rich sand and more than 6 m of sand over thick, compressible, and spreading marine clay—not the type of material that today's engineers would choose as a foundation. It was built as a bell tower for the adjacent cathedral of Pisa, near the west coast of Italy. Construction that began in 1173 was stopped in 1185 with only three levels completed, because the tower began settling and leaning. Construction resumed in 1274 after 100 years of inactivity. Ten years later, all except the top bell enclosure was complete, but the rate of leaning accelerated, so construction was again stopped. The tower, including its bell chamber, was finally completed in 1372, but by this time it was 1.5 m off vertical.

Since its completion, the 56-m-high Leaning Tower had sunk 2 m into the ground and was nearly 4 m off vertical. Various attempts have been made to stabilize it. A 600-ton weight was added to the base of the tower on the side that had subsided less (left side of Figure 9-18). More recently, the tower was straightened by nearly a half meter, and a new foundation was built to keep it stable and safe for tourists.

One area of water-rich marine clays subject to subsidence is the Ottawa–St. Lawrence River lowland, where seawater invaded lakes that formed at the edge of the retreating continental ice sheet 10,000 to 13,000 years ago. Much of Ottawa, Ontario, is built on the notorious Leda clay, a marine layer up to 70 m thick. It is especially weak because the natural salts have been leached out. The solid grains of the clay amount to less than one-third of the total volume so the clay dries and shrinks under old homes and buildings, causing differential settling and distortion (**FIGURE 9-19**). Landslides throughout the St. Lawrence River valley and its tributaries have involved the Leda clay. In an ironic twist, the headquarters building for the Geological Survey of Canada, completed

FIGURE 9-17 Storm Surge Barriers

A. Satellite view of southern Holland with storm-surge barriers marked with red triangles.

B. Maeslant barrier gates that close to block storm surge from reaching Rotterdam.

FIGURE 9-18 Leaning Tower
Compression of marine clays by the Leaning Tower of Pisa's load caused the tower to lean. The cables that help keep it from falling are just visible between the tower and the church on the left. The tower curves slightly because builders tried to straighten the upper floors during construction.

FIGURE 9-19 Distortion by Clay Shrinkage
The diagonal zigzag crack in the brick wall is a telltale sign that this house in Ottawa, Ontario, is settling and being distorted by shrinkage of its clay substrate.

in 1911, was built on a site underlain by more than 40 m of Leda clay without consulting the geologists who worked there. The heavy outer stone walls immediately began to sink into the clay, causing cracking of the structure and separation of its tower from the main building. The tower was finally torn down in 1915 and the building and its foundations strengthened at considerable cost.

An infamous case involving Leda clay began in April 1971 at St. Jean Vianney, Quebec, where a new housing development had been built on a terrace of Leda clay 30 m thick. Heavy rains loaded the clay, and 11 days later it collapsed and flowed downslope toward the Saguenay River at 25 km/h for 3 km. The slide took 31 lives and 38 homes (Figure 8-8).

Thaw and Ground Settling

Geothermal gradient refers to the fact that the temperature at the surface of the Earth increases with depth. Rocks deep in the Earth are hotter. Near Earth's surface, the ground temperature increases by about 0.6°C per 100 m of depth (6°/km). In temperate climates where average year-round temperatures are above freezing, the ground at depth is well above freezing. Arctic climates of Canada, Alaska, northern Europe, and Asia have in-ground temperatures that remain below freezing year-round. Water in the ground remains frozen, leading to a ground condition called **permafrost**. The ice occurs as coatings on grains or it completely fills the pores between grains and supporting strands of peat moss. Ice veins and even wedges, layers, and masses of more-or-less solid ice form near the surface.

In other northern climates or high mountain areas such as the Alps and Tibet, where average temperatures remain below freezing, the ground is frozen all winter, but a meter or two of the surface, called the *active zone*, may thaw for two months in summer. Still farther north, in the Arctic, only a thin surface layer may thaw in the warmest summer weather. The distribution of permafrost around the North Pole is shown in **FIGURE 9-20**.

Local environmental factors also affect ground temperature. Large areas of water in lakes and rivers help keep the ground warmer, as does vegetation that maintains deeper snow cover. However, forest fires or removal of vegetation for construction can also lead to thawing of permafrost.

When permafrost thaws, water runs out of the frozen ground, causing the ground to settle and flow downslope (**FIGURE 9-21**). Landslides become more frequent both on slopes in soil and in bedrock as the once-solid ice turns to water between grains and in rock fractures. Most of the slides are shallow, a few meters deep, as the thawed layer slides off the remaining permafrost.

Permafrost zones

Continuous

Discontinuous

Sporadic

Asia

Europe

North
Pole

Greenland

North
America

G. D. Arendal, UNEP

FIGURE 9-20 Frozen Ground in the Arctic

Distribution of permafrost and thawing permafrost in northern
North America and Asia.

Structures such as buildings, roads, railroads, pipelines, and power poles in permafrost areas have foundations built on or in the permanently frozen ground. The
structural integrity of buildings and other structures can be
compromised if the permafrost begins to thaw (**FIGURES 9-22**
and **9-23**). Buildings that were solidly anchored on the ice
settle and twist out of shape.

A

Sarah Laxton, Yukon Geological Survey

B

Kelin Wang, Pacific Geoscience Center/Geological Survey of Canada

FIGURE 9-21 Thawing Permafrost

A. Buildings in Dawson City, Yukon, settled and leaned as the
permafrost under them thawed.
B. Permafrost near Tuktoyaktuk, Northwest Territories of Canada.

Vladimir Romanovsky

Building raised
off the ground

Alaska State Department of Transportation

FIGURE 9-22 Building Damage Due to Permafrost Thaw

A. This house near Fairbanks, Alaska, settled and
deformed as permafrost thawed.

B. Modern buildings are often raised off the ground so that (cold) air can circulate under
a heated building, minimizing the possibility of permafrost thawing. Nome, Alaska.

FIGURE 9-23 Railway Warped by Ground Thaw
The Copper River Railway in Alaska was abandoned by 1960 because thawing permafrost severely deformed its line.

colder by an ingenious device called a *thermosiphon*. The vertical posts that support the pipeline contain a gas that condenses to a liquid at cold air temperatures. The liquid settles to the bottom of the pipe within the frozen ground. Heat from the permafrost is sufficient to vaporize the liquid; that vapor rises to the top of the pipe where it again cools and condenses and sinks. This process dissipates the heat at ground level and helps keep the ground frozen so structures on it remain stable.

The pipeline is also built in a manner that permits it to slide back and forth on supports; that allows for ground movements, thermal expansion of the pipe, and earthquakes. One strategy long used in building roads and railroads is covering the permafrost with crushed rock to promote heat loss in winter and minimize heat access in summer.

Damages due to permafrost thaw are increasing as a result of global climate change. With a warming climate, ground temperatures increase, the permafrost thins, and the active zone of freeze–thaw thickens. The permafrost in some areas is now beginning to thaw to deeper levels during the summer, causing damage to more and more structures and contributing to climate change (discussed in Chapter 11).

Thawing permafrost not only wreaks havoc with the ground surface, but meltwater flowing into the Arctic Ocean promotes the melting of Arctic sea ice. Destruction and decay of surface vegetation and thawed peat can release immense amounts of methane and carbon dioxide, fostering even more global warming. These topics are discussed in more detail in Chapter 11.

Those building in permafrost areas should take into acount the possibility of thawing ground and its consequences. The Trans-Alaska Pipeline is suspended above the ground surface because oil flow causes frictional heat in the pipe; that would thaw the permafrost, which could break the pipeline (**FIGURE 9-24**). The ground is also kept

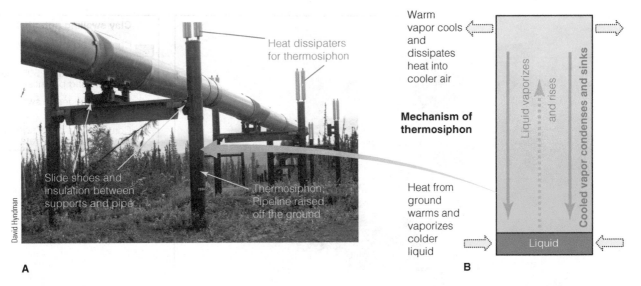

FIGURE 9-24 Mitigation of Permafrost Thaw
A. The Trans-Alaska Pipeline carries crude oil from the North Slope of Alaska to the ice-free port of Valdez. Above-ground segments of the pipeline are built in a zigzag geometry with slide shoes designed to allow the pipe to freely accommodate ground movements and the thermal expansion and contraction of the pipe. **B.** Thermosiphons dissipate heat in the air so that it will not thaw the ground.

Swelling Soils

The same process that causes subsidence in some clays when they dry out is responsible for **swelling soils** when water soaks into the interlayer spaces of the clay mineral structure. During long dry periods, the soil shrinks; in long wet spells, it expands.

Swelling soils can be just as damaging to structures as sinkholes, subsidence, and permafrost thaw. Much of the United States is affected by swelling soils, including most large valleys in the Mountain West region; all of the Great Plains; much of the broad coastal plain area of the Gulf of Mexico and the southeastern states; and much of the Ohio River valley (**FIGURE 9-25**). Even some areas not normally associated with swelling clays can have local problems.

The annual cost in the United States from swelling soils exceeds $4.4 billion; highways and streets account for approximately 50% of this total, homes and commercial buildings for 14%. When buying a home in an area subject to swelling soils, people would be well advised to have an expert examine it carefully before purchase.

Swelling soils and shales are those that contain **smectite**, a group of clay minerals that expands when they get wet. You can often spot the presence of swelling clays where the expansion and cracking of surface soils forms *popcorn clay* (**FIGURE 9-26A**). These minerals form by weathering of both aluminum silicate minerals and glass in volcanic ash (their behaviors are discussed in Chapter 8). Driving on a dirt road built on smectite-rich soils can also be hazardous when roads are wetted and the clay becomes extremely slippery.

Mud builds up on tires and shoes—the "gumbo" that is so familiar to those who live in such areas (**FIGURE 9-26B**).

The expansion and contraction of soils causes the cracking of foundations, walls, chimneys, and driveways. Damage also results when adjacent areas swell at different rates. Swelling is problematic for roads and structures when layers of clay are interspersed with layers of sedimentary rocks. Swelling of the clay layers creates differential expansion, deforming houses and roadways. Parts of the southwestern suburbs of Denver, south of U.S. Highway 285 and east of state Highway 470, were built from the 1970s through the 1990s over the upturned edges of these formations. Where horizontal, smectite-bearing layers under the High Plains tilt up to reach the surface, those layers create differential expansion under houses and streets (**FIGURE 9-27** and **Case in Point:** Differential Expansion over Layers of Smectite Clay—Denver, Colorado, p. 241). The problem appears when rainfall, snowmelt, yard watering, and, sometimes, a leaking underground pipe wets the clay. Long ridges parallel to the mountain front, and separated by troughs, grow 10 to 20 m apart. A ridge rising under a house stretches and pulls it apart. Walls buckle and crack, chimneys fracture, windows pull apart, and doors no longer close.

Homes built on smectite clays can experience problems because of different moisture conditions on separate parts of a property. For example, the area under a house remains drier, whereas the ground surrounding the house is open to moisture from rain, snow, lawn watering, house-gutter drainage, or a septic tank or drain field. Or prevailing wind-driven rain against one side of a house can cause the ground

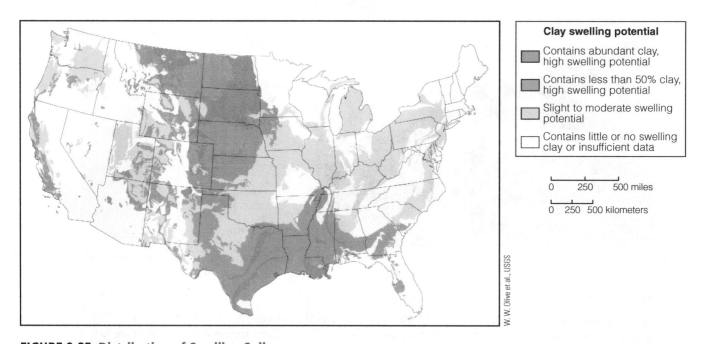

FIGURE 9-25 Distribution of Swelling Soils

Areas of swelling soils in the United States include the Mountain West region, the Great Plains, and the Gulf of Mexico states.

FIGURE 9-26 Signs of Swelling Clay

A. Popcorn clay is a telltale sign of swelling clays in the Makoshika badlands near Glendive, easternmost Montana. View is 60 cm across.

B. Gumbo that builds up on shoes and tires on wet dirt roads is a good indication of smectite-rich soil.

around that side to swell (**FIGURE 9-28**). Vegetation concentrated along one side of a house can dry out the soil on that side. Even the removal of vegetation can increase moisture in the soil by decreasing evapotranspiration. In each case, the ground receiving more moisture expands, while the ground receiving less moisture contracts, deforming and cracking buildings.

Swelling mostly occurs within several meters of the ground surface because the greater pressure at depth typically prevents water from entering a clay's mineral structure and lifting everything above it. Thus, the problem of swelling and deformation of buildings does not extend to larger, heavier structures such as large commercial buildings and large bridges.

The obvious solution to damage from swelling soils is to build elsewhere. Where this is not feasible, near-surface effects can be minimized by preventing water access to the soil, removing the expanding near-surface soils, or sinking the foundation to greater depth. The addition of hydrated lime—calcium hydroxide, $Ca(OH)_2$—to an expandable clay can reduce its expandability by exchanging calcium in molecular interlayer spaces so water cannot easily enter. However, this should not be done with quick clay soils, because that would markedly reduce soil cohesiveness, and foster soil collapse and landslides.

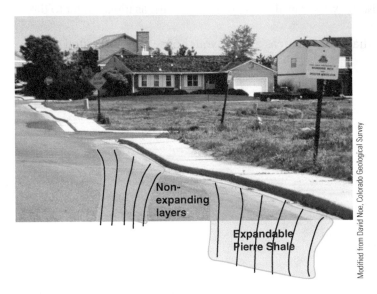

FIGURE 9-27 Swelling Near Denver Bulges Road

Houses in parts of Denver are built over inclined beds of smectite-bearing Pierre shale. Wetting of the shale beds (schematically drawn in under the road) causes swelling that bulges the ground and deforms both the road and houses. Sangre de Cristo Road in southwestern Denver has since been leveled and repaved, eliminating the evidence.

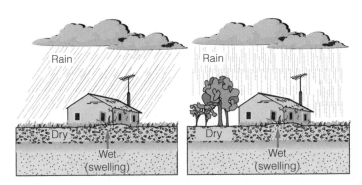

FIGURE 9-28 Differential Wetting Deforms Homes

The soil under the wet side of a house, either because of a prevailing rain direction or lack of vegetation, can swell or deform the house.

SINKHOLE, SUBSIDENCE, SWELLING SOILS SURVIVAL GUIDE

> Be aware that these lists don't cover everything. Use your own common sense.

U.S. High Hazard Area	■ **Sinkholes:** Areas of soluble bedrock in which large amounts of groundwater are pumped out at times but have high precipitation at other times. See **Figure 9-7**. ■ **Land subsidence:** Areas in which large amounts of groundwater, or oil and gas, are pumped out, generally for irrigation or drainage of organic soils; also areas of Arctic permafrost. See **Figure 9-12**. ■ **Swelling soils:** Soils containing significant amounts of ancient volcanic ash. See **Figure 9-25**.
Preparation	■ Minimize pumping of large amounts of water or other fluids from the ground. ■ Learn about sinkhole distribution and sinkhole formation in your area. ■ Techniques for finding buried sinkholes include test borings, CPTs, GPR, all done by specialists. Swelling soils can often be recognized from popcorn clays or sticky clays that build up on tires in wet weather.
Warning Signs	■ Sinking ground newly exposes building foundations and fence posts, tilting fences, doors that don't close properly; structural cracks in walls, floors, ground surface; water ponding where it hasn't before. ■ Ground settling below the tops of pipes protruding from the ground.
During the Event	■ For an abrupt sinkhole, make sure you and your family are safe; evacuate if necessary. ■ Subsidence and swelling soils tend to be slow processes.
Aftermath	■ For sinkholes, notify local authorities and your insurance company. Rope or tape off the site to prevent injury to others. ■ For subsidence from groundwater withdrawal, stop pumping; return as much water to the ground as possible. ■ For swelling soils, make sure that different areas around your house and not wetted differentially.

Cases in Point*

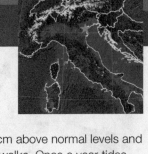

Subsidence Due to Groundwater Extraction
Venice, Italy ▶

Venice, along the northeastern coast of Italy, is built at sea level on the now-submerged delta of the Brenta River. The delta and the city have, for 1500 years, been subsiding for a variety of reasons, the most important being extraction of groundwater from the delta sediments. The load of buildings squeezes water out of the soft delta sediments, and pumping groundwater for domestic and industrial use collapses the pore spaces in the sediments. That decreases the volume of the sediment, which sinks the original surface of the delta below sea level.

The high tides that accompany winter storms compound the problem. They submerge most of the few normally dry walkways under as much as a meter of water. Maps on the water taxis advise people of the few dry routes available for walking. Pumping groundwater from under the city is now prohibited, and sinking has all but stopped. But the high winter tides that formerly flooded the city twice a year now come several times annually. Ancient drainage pipes installed to carry off rainwater now carry seawater back into the city during times of high winter tides, along with whatever effluent they normally carry.

As the delta subsides, so does Venice. Some parts have now sunk as much as several meters. The ground floor of Marco Polo's thirteenth-century house is now 2 m below present ground level. Below that are older floors from the eleventh, eighth, and sixth centuries. To accommodate the sinking, Venetians continued to raise the street levels. Streets that were above sea level when they were built are now canals, and doorways and the lower floors of homes are now partly submerged, a condition that prompted Ulysses S. Grant to remark that the place appeared to have a plumbing problem.

Tides in most of the Mediterranean Sea range less than 30 cm, but where the tidal bulge sweeps north up the Adriatic Sea near the east end of the Mediterranean, they average 1 m. About 10 times per year, tides reach 40 cm above normal levels and flood most sidewalks. Once a year tides often reach 65 to 70 cm above normal levels and flood about two-thirds of the city's buildings. When the low atmospheric pressure of a storm coincides with high tide, sea level rises even higher; coinciding with a sirocco, a warm winter wind blowing north out of Africa that pushes the tidal bulge ahead of it, the tidal rise can be especially high. On November 4, 1966, when three days of rain had already raised both river and lagoon levels to cause catastrophic flooding, tides reached about 130 cm above normal high tide.

Merely raising the sea walls around the city will not cure this problem. Climate change and the concurrent rise in sea levels also contribute. A few historical highlights illustrate the political background that makes any solution to the problem difficult:

- **1534** A major dike is built to divert the mouth of the Sile River to the east, away from Venice.

■ *Increasingly frequent high tides reverse the ancient drainage system of Venice such that polluted lagoon water backflows, bubbling up through drains, and floods St. Mark's Square. In the satellite view, the twisting green channel through the center of Venice is the "Grand Canal," part of the original Brenta River delta.*

*Visit the book's Website for this additional Case in Point: Subsidence Due to Groundwater Extraction, Mexico City, Mexico

Donald Hyndman

■ *At high tide, some streets and sidewalks are flooded, creating problems for deliveries.*

Donald Hyndman

■ *Buildings sinking into Venice's mud are flooded and deteriorating. First-floor doorways and flooded ground floors are permanently abandoned.*

- **1613** Venice, on the delta of the Brenta River, diverts the Brenta around the city. New sediment is no longer added to that part of the delta.

- **1744–1782** Seawalls are constructed along the offshore barrier islands to protect the Venice lagoon from the Adriatic Sea.

- **1925** Large-scale pumping of groundwater from under Venice begins for industrial use.

- **1966** A major storm on November 3, compounded by an especially high tide during a full moon, pushes Adriatic water into the Venice lagoon, which is normally about 60 cm deep. The water reaches nearly 2 m above normal sea level. The city floods for 15 hours, causing severe damage to buildings and works of art. Heating-oil tanks float, breaking attached pipes and leaking oil into canals and homes. Sinking permanently floods the first floors of most buildings; most families now live on their second floors.

- **1970** After 45 years of pumping and gradual sinking, the government begins to prohibit industrial pumping of groundwater from under Venice.

- **1981–1982** A technical team of the Public Works Ministry and Venice City Council approves plans for a major flood-control project to protect the shallow lagoon from flooding. The plan is to build four major mobile barriers

across the three barrier island inlets to the lagoon that can be raised at times of potential high water.

- **1984** The Italian Parliament appropriates the equivalent of billions of dollars to build the barriers and raise parts of the urban center.

Since 1908, the combination of sea-level rise and natural and human-induced subsidence has reached 24 cm. Although the Italian government agreed decades ago to spend hundreds of millions of dollars on a solution, nothing happened until recently.

The controversial Mose project, a system of giant, moveable gates across the barrier island inlets, began in 2003. The gates, like a layer of hollow boxes, lie in a concrete trough on the seafloor. With the arrival of an especially high tide, pumped air floats one edge of a gate above sea level, forming a wall to block the incoming tide. As of 2013, the project was 80% done, with an expected completion date in 2018 or later. However, given Italian politics, governments that change hands every year or two, and internal bickering, the budget for this now $6 billion project continues to be debated. There are those who argue that it is more important to spend available funds on building industrial capacity and promoting new jobs rather than on preserving Venice as one of the world's historic and artistic centers. If this sentiment wins out, development would involve a third

industrial zone at the landward edge of the lagoon in Marghera.

During floods, the impermeable marble bases of most of the old buildings are submerged, and seawater seeps into the porous bricks of their walls, which then begin to crumble. More frequent submergence of drainpipes that carry rainwater from the city into the lagoon creates another major problem. These pipes act as the city's sewers; they carry untreated sewage, including laundry suds and toilet effluent, into the canals, then into the lagoon, where the tides periodically flush it out to sea. During especially high tides, of course, the drains carry the effluent back into the canals and walkways of the city. Some of the controversy concerning the proposed movable barriers at the

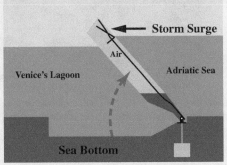

Venice Water Authority

■ *The new Mose flood barrier, pneumatically raised during storm high tides, floats a wall across the incoming tide.*

mouths of the lagoon involves restriction of the tidal action that flushes the lagoon.

One of the highest Venice floods in decades reached a peak in November 2012 but it still ranked only 6th highest, well below both the record level of 193 cm reached in 1966). Flood level is designated as 101.6 cm, and although the city erects wooden platforms that permit pedestrians to cross flooded areas around St. Mark's Square, the water rose so fast in 2008 that the platforms could not be emplaced in time. Ground-floor homes and shops flooded. Alarms sounded to warn both residents and visitors to stay indoors or use high wading boots. Such flooding has become more frequent and generally higher in recent years. Since groundwater pumping has been halted, the higher tides are blamed on climate change.

CRITICAL THINKING QUESTIONS

1. Consider other river delta areas on which people live and what their residents did to address the problems of flooding during storms. What changes did Venetians make that made their flooding worse than in other areas?
2. What detrimental effects are likely from construction of the moveable storm barriers?

Differential Expansion over Layers of Smectite Clay
Denver, Colorado ▶

Denver is built on the central High Plains, which are underlain by flat-lying 70- to 90-million year-old sedimentary rocks that curve up to nearly vertical at Earth's surface just east of the Front Range of the Rockies. Among these rocks is the Pierre shale in the Denver–Boulder area, a subtle villain 1.6 to 2.4 km thick. The shale contains smectite-rich clay layers formed by the weathering of volcanic ash, blown in from volcanoes far to the west. When wet, the clays swell upward and outward, lifting parts of houses and twisting them out of shape.

Within a few years of construction, more than 0.6 m of differential movement has created broad waves in roads and cracked and offset underground utility lines. Parts of houses rise, causing twisting, bending, and cracking of foundations, walls, and driveways. Windows fracture and chimneys separate from houses.

KLP Consulting Engineers

■ *Swelling clays in southwestern Denver deformed this house's basement foundation in spite of heavy steel pipes utilized to stabilize it.*

■ *The driveway and garage door deformed in response to the swelling clays on which this home was built in a southwestern Denver suburb. The brick wall pulled away from the garage door, leaving a gap at the bottom (see arrow).*

Donald Hyndman

Until 1990, engineers attributed the problem to swelling soils. Their mitigation plans called for drilled piers under foundations or rigid floating-slab floors. Most proved unsuccessful. More recent research by the Colorado Geological Survey and the U.S. Geological Survey demonstrated that the problem is the Pierre shale bedrock. That "bedrock" is fairly young and about as soft as the overlying soil. Recent work in the Denver area shows that excavation and homogenization of both clay- and non-clay-bearing layers to depths of less than 10 m in preparation for new subdivisions can increase uniformity of expansion and minimize problems. Regulations enacted since 1995 require new subdivisions to excavate to a depth of at least 3 m and replace this with homogenized fill before pouring foundations. Significant reduction in damage has resulted.

CRITICAL THINKING QUESTIONS

1. If people find that their house is beginning to be damaged by swelling clays, what several things can they do to minimize further damages?

2. Considering to what extent insurance companies are willing to cover damages from landslides and other ground movements in Chapters 8 and 9, do you expect them to cover damage from swelling clays? Should they?

Critical View

A The door of this garage has become skewed on this house south of Denver, Colorado.

Donald Hyndman

1. What happened to cause this type of damage?
2. What could be done to minimize further damage here?
3. What could have been done before the house was built to avoid most of this type of damage?

B These rock formations are found in Pennsylvania.

William Kochanov, Pennsylvania Geological Survey

1. What is this strange landscape called?
2. What kind of rock, visible here, develops into this type of landscape?
3. How does that landscape form?

C This huge crack in the ground formed in Arizona.

Michael C. Carpenter, USGS

1. What probably caused this crack?
2. What likely caused that process?
3. What could be done to stop further damage of this type?

D The roof of this cavern in southern Texas collapsed to leave a large opening. Large blocks in the lower right collapsed from the roof in the upper right.

Donald Hyndman

1. How would such a large cavern form and in what type of rocks would it form?
2. The small lake in the lower left is the level of the groundwater table. Explain how various levels of the groundwater might affect collapse of the roof and collapse of the slabs of rock in the lower right.

E This aerial view, west of Houston, Texas, shows a landscape dotted with ponds from tens to hundreds of meters across.

Donald Hyndman

1. How do the depressions form?
2. What type of bedrock lends itself to formation of these depressions?
3. Sinkholes may form in such areas. Are the collapse sinkholes likely to form in very wet weather or very dry weather? Why?

F Long, thin, pencil-shaped formations descend from the ceiling of this cave in Carlsbad, New Mexico.

National Park Service

1. What are these long formations called and how are they formed?
2. Sometimes you can see a drop of water dripping from the end of one of these features. Where does that water come from?
3. What is in the water that adds to the lengths of these formations?

Chapter Review

Key Points

Sinkholes

- Rainwater, naturally acidic because of its dissolved carbon dioxide, can dissolve limestone, forming large caverns underground. **FIGURES 9-2** and **9-4** and **By the Numbers 9-1.**

- Soil above limestone caverns can percolate, leaving a cavity that can collapse to form a sinkhole. **FIGURE 9-6.**

- Sinkhole collapse can be triggered by lowering the level of groundwater that previously provided support for a cavity roof; loading a roof; or an increasing water flow that causes the rapid flushing of soil into an underlying cavern.

Land Subsidence

- Removal of groundwater or petroleum from loose porous sediments can cause compaction, resulting in subsidence of the land surface. **FIGURES 9-13** to **9-16.**

- Subsidence of an area below sea level can result in flooding and saltwater invasion of freshwater aquifers.

- Land subsidence can also be caused by drainage of peat or clay-rich soils; oxygen-bearing surface water invades the peat, causing it to oxidize and decompose.

- Clays contract as they dry out, leading to subsidence.

Thaw and Ground Settling

- In cold climates, water in the ground is frozen as permafrost for all or part of the year. **FIGURE 9-21**.

- When permafrost thaws, water runs out and the ground settles, causing fractures and landslides.

- Global climate change has increased the depth of permafrost thaw in some areas and may lead to greater structural damages. **FIGURES 9-22** and **9-23**.

Swelling Soils

- Smectite soils, formed by the alteration of soils rich in volcanic ash, swell when water invades the clay mineral structure. Such soils are abundant in the midcontinent region, the Gulf Coast states, and western mountain valleys. **FIGURE 9-26**.

- Swelling soils can crack and deform foundations, driveways, and walls. Damages can occur when soils swell at different rates—for example, when one area of a property experiences more or less moisture than another due to vegetation or protection from rain. **FIGURES 9-27** and **9-28.**

Key Terms

cover collapse, p. 224
cover subsidence, p. 224
dissolution, p. 224

groundwater, p. 221
groundwater mining, p. 227
karst, p. 223

land subsidence, p. 227
permafrost, p. 233
sinkhole, p. 221

smectite, p. 236
swelling soils, p. 236
water table, p. 221

Questions for Review

1. What causes sinkhole collapse—that is, where is the cavity that it collapses into, and how does that cavity form?

2. Explain how caverns form in limestone.

3. Where in the United States are sinkholes most prevalent? Why?

4. What is karst and how does it form?

5. What are the main characteristics of soil above a limestone cavern that lead to the formation of sinkholes?

6. What weather conditions are most likely to foster the formation of sinkholes? Why?

7. Other than limestone, what other types of rocks are soluble and can form cavities that collapse?

8. What types of human behavior typically lead to widespread ground subsidence?

9. Name three areas in the United States that have seen major subsidence and in each case explain what caused it.

10. How can subsidence contribute to flooding?

11. Explain how permafrost thaw causes ground settling.

12. What are the relationships between permafrost thaw and global climate change?

13. What kind of soil swells, and in what regions of the United States is this soil common?

14. What causes swelling soils to swell?

15. How do swelling soils lead to structural damages?

16. Give an example of how a property might experience differential swelling. Make a sketch to illustrate your explanation.

Critical Thinking Questions

1. Given that lowering groundwater levels in an area of sinkholes removes support from the tops of underground caverns, what restrictions, if any, should be placed on pumping groundwater out for crop irrigation?

2. In case of collapse of a sinkhole that damages a home or commercial building, who should pay for the damages? Why?

3. If a housing subdivision is built on ground that damages homes when the soil swells, who should be held liable? Why?

4. What restrictions should be placed on building in areas subjected to swelling soils damage? Why?

5. If you find a popcorn clay surface on dried mud, what natural hazards might you be concerned with? What could you do to mitigate the hazard?

6. If, in your travels, you see an oddly tilted tall, old building that otherwise looks well built, what influences could have caused the tilt? Which of those influences are likely to be natural, and which likely were caused by human behavior? How so?

FEMA workers search for survivors in what remains of a residential area of Moore, Oklahoma, after an EF5 tornado on May 20, 2013.

FEMA, Andrea Broher

Weather, Thunderstorms, and Tornadoes

10

A Devastating Tornado

The Oklahoma City suburb of Moore seems to attract tornadoes, with four major events in 14 years—1999, 2003, 2010, and then again in 2013. The 1999 and 2013 events were the largest on the tornado scale. The collision of cold Arctic air pushing under warm humid air from the Gulf of Mexico along a northeast-trending cold front was a perfect recipe for severe supercell thunderstorms. Tornado sirens sounded at 2:40 in the afternoon. What happened in Moore in the next 39 minutes was terrifying. (See Case in Point: Lack of Shelters Despite a History of Tornadoes—Moore, Oklahoma, 2013, p. 292.)

Basic Elements of Weather

Weather refers to the conditions and activity in the atmosphere at a particular place and time—temperature, air pressure, humidity, precipitation, and air motion. Weather drives some of the most deadly and costly natural hazards, discussed in Chapters 10 through 17 of this book. Learning about weather processes provides a foundation to understand the risks we face in different parts of the world for direct weather hazards (thunderstorms, tornadoes, and hurricanes), along with hazards that result from weather (floods, coastal erosion, and wildfires).

Hydrologic Cycle

Water in the oceans covers more than 70% of Earth's surface. Unfortunately, the oceans are salty; humans need freshwater for life, and less than 3% of the liquid water on Earth is fresh. This freshwater is derived from evaporation of ocean water, some of which moves over continents and falls as rain or snow. Most of the rain that is not taken up by plants soaks in to become groundwater, which seeps back to the surface into lakes and rivers and ultimately flows back into the sea. This is called the **hydrologic cycle** (**FIGURE 10-1**).

The sun warms the ocean surface and water evaporates into the air above it, as water vapor. Under most conditions, we do not see the water vapor because it is colorless and invisible, like most gases. The amount of water vapor that air can hold, called its **relative humidity**, depends on its temperature; cold air can only hold a small amount of water vapor, whereas warm air can hold a lot. When air contains the maximum that it can hold, it is saturated with water vapor; its relative humidity is 100%. If it contains half as much as it can hold, its relative humidity is 50%.

If air cools to the temperature at which its relative humidity is 100%, the water-saturated air is still invisible. This is the *dew point*. If it cools further, it will have more water vapor than it can hold, causing some of the water to condense into tiny water droplets and form clouds or fog. If enough water droplets form, they coalesce into larger droplets and eventually fall as rain or snow.

As breezes blow across the oceans, the air picks up water vapor. Air cannot hold significant amounts of solids, so water vapor evaporates into the air, but the salt stays in the oceans. Much of the water vapor in the air drifts over the oceans with prevailing winds until it reaches a continent where some will drop out as rain or snow.

Rain or snow falling on the continents can take different paths. Some falls on leaves of trees or other vegetation; some of that evaporates directly back into air. Some of the moisture reaching the ground may run off the surface into streams or lakes, but a large part soaks into the ground to become *groundwater*—the water in spaces between soil grains or in cracks through rocks. The roots of trees and other vegetation draw on some of that moisture, taking it up to carry nutrients to the leaves. The leaves, in turn, transpire water back into the atmosphere because the air generally has less than 100% humidity. Direct evaporation and transpiration by vegetation are both included in **evapotranspiration (ET)**, which reduces the amount of moisture in slopes, and thus can reduce the chance of floods and landslides. As discussed in later chapters, when people develop an area and remove natural vegetation, this generally reduces ET and increases the risk of these hazards.

Groundwater slowly flows in the direction the water table slopes to reach streams, lakes, and wetlands; it is the main source of water for streams except in arid regions. Streams in turn carry the water downslope into larger rivers and

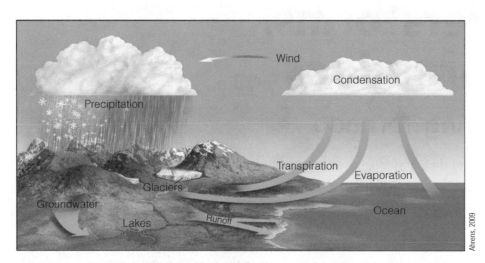

FIGURE 10-1 Hydrologic Cycle

The hydrologic cycle shows how a small but important part of water from the vast oceans evaporates to cycle through clouds and precipitation to feed lakes and streams, vegetation, and groundwater, before much of it again reaches the oceans.

ultimately back into the ocean to complete the cycle. Variations in the hydrologic cycle can lead to droughts and floods.

Adiabatic Cooling and Warming

Adiabatic cooling occurs when rising air expands without a change in heat content. Whenever an air mass expands, the available heat is distributed over a larger volume, so the air becomes cooler (**FIGURE 10-2**). The rate of cooling as an air mass rises is called the **adiabatic lapse rate**.

During condensation, the water releases the same amount of heat originally required to evaporate liquid water into water vapor. In fact, this heat of condensation is large enough to reduce the adiabatic lapse rate from 10°C per 1000 m of rise for dry air to 5°C per 1000 m for water-saturated air (**FIGURE 10-3**). In other words, very humid air cools at half the rate of dry air as it rises. When an air mass falls, its temperature rises at the dry adiabatic lapse rate because it is below saturation and can thus hold more moisture.

When this moist air reaches land and is forced to rise over a mountain range, it expands adiabatically and cools, in what is called an **orographic effect**. Because cool air can dissolve less water, the moisture separates as water droplets that form clouds and in turn rain, hail, or snow (**FIGURE 10-4**). When an air mass moves down the other side of a mountain after much of the moisture has fallen as rain or snow, the air becomes warmer and drier. This rain shadow effect is the reason deserts commonly exist on the downwind sides of mountain ranges.

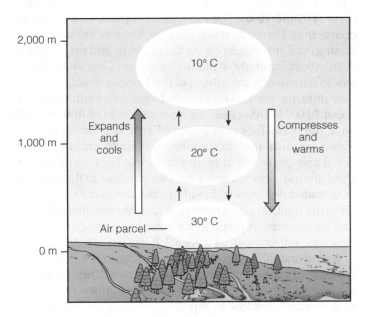

FIGURE 10-2 Adiabatic Processes

There is less atmospheric pressure at higher altitudes, so rising air expands and cools, while falling air compresses and warms. Under dry conditions, an air parcel will change temperature by 10°C per 1000 m of altitude change.

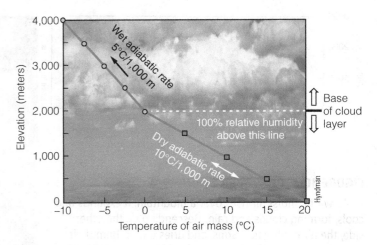

FIGURE 10-3 Adiabatic Rate

The wet adiabatic lapse rate is half the dry adiabatic lapse rate for rising air. Once a rising air mass reaches 100% relative humidity, condensation occurs; this releases heat back to the air mass. Note that the wet adiabatic rate applies only to rising air, as indicated by the upward arrow; the dry adiabatic rate can apply for rising or falling air.

Atmospheric Pressure and Weather

Air pressure is related to the weight of a column of air from the ground to the top of the atmosphere. When an air mass near the ground is heated, the air molecules vibrate faster and have more collisions, causing the air to expand, decreasing its density. The lower-density air rises to create a **low-pressure system** (a "low") (**FIGURE 10-5**). It has lower pressure because rising air exerts less downward pressure than still or falling air. Air near the ground surface is pulled in toward the low-pressure center to replace the rising air. The opposite occurs in **high-pressure systems**, where cool air in the upper atmosphere has a higher density than its surrounding warmer air and thus falls. As it moves toward the ground, it has to spread out or diverge, so air moves away from high-pressure systems near Earth's surface.

Differences in atmospheric pressure drive air movements and winds. Air flows from high- to low-pressure areas, causing winds; high pressure pushes outward and low pressure pulls inward. Where these horizontal pressure differences are large, winds are strong.

Coriolis Effect

Because Earth rotates from west to east, large masses of air and water on its surface tend to lag behind a bit. It completes a rotation around its axis through the North and South Poles once every 24 hours; thus, a point at the equator has to travel roughly 40,000 km in a day (the approximate circumference of the planet), but Earth's rotation causes no movement for

FIGURE 10-4 Orographic Effect

A. As warm, moist air rises over a mountain, it expands and cools, forming clouds and rain. Descending on the other side, the air contracts, warms, and dries out. **B.** Humid air rising from the Pacific Ocean south of Monterey, California, cools and condenses to form clouds over the Santa Lucia Range.

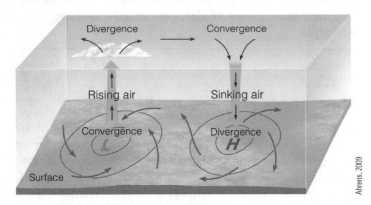

FIGURE 10-5 High- and Low-Pressure Systems

Relatively warm, humid air rises to create low atmospheric-pressure areas. Near the land surface, the air is drawn in from the sides, rises, and eventually condenses to form clouds and rain. Cold dry air descends to create high atmospheric-pressure areas and spreads out as it sinks to the land surface; skies are clear.

points on either the North or South Poles. This large rotational velocity of Earth's surface makes both oceans and air masses near the equator move from east to west because they are fluid and thus not pulled as fast as the rotating solid Earth below them. The water and air masses move above the planet, which rotates faster under them, thus they lag behind Earth's rotation. In the northern hemisphere, for water or air currents moving south, Earth rotates faster to the east below them—the lag causes them to veer off to the west. The opposite occurs as water or air currents move toward the poles. In this case, the air or water moves over areas of Earth with slower surface rotation, and the resulting lag causes a curve off to the east. This forced curvature due to Earth's rotation is called the **Coriolis effect**.

Thus, in the northern hemisphere, large currents of air and surface water tend to curve to the right; they curve to the left in the southern hemisphere. Ocean currents move westward as they flow toward the equator and eastward as they head toward the poles. This causes ocean currents to circulate clockwise in the northern hemisphere and counterclockwise in the southern hemisphere. Similarly, hurricanes, paths with tracks to the west across the Atlantic Ocean, curve to the right (North) direction as they approach North America.

One clear demonstration of the Coriolis effect is in the flight paths of airplanes as they fly above Earth, which rotates under them (**FIGURE 10-6**). For example, if a plane flew a straight course from Detroit to reach Cancún, Mexico, the west-to-east rotation of Earth would cause the plane to end up west of its destination. Similarly, a plane flying from Cancún to Detroit would experience the effects of the Coriolis effect. Before its departure, the plane is rotating eastward with Earth at a rate of about 40,000 km/day near the equator. On its northward flight path, the plane flies over parts of Earth with slower rotation speed, because the circumference of Earth around its rotational axis decreases at higher latitudes. Thus, its initial eastward ground motion carries the plane farther to the east than its intended destination. In either case, the plane's path veers off to the right, so pilots must take this into account when setting their course. A plane in the southern hemisphere is similarly affected, except that it veers left instead of right.

Rising and falling air currents also rotate (see Figure 10-5) because of converging or diverging winds. When air rises in a low-pressure system, air near Earth's surface has to converge toward this low to replace the rising air. The Coriolis effect causes these winds to shift to the right of straight in the northern hemisphere, which causes these low-pressure systems to rotate counterclockwise. Falling air diverges away from a high-pressure system as it pushes down on the planet. Again the Coriolis effect shifts these northern hemisphere winds to the right, which causes a clockwise rotation of these

FIGURE 10-6 The Coriolis Effect

In the northern hemisphere, a plane flying due south (red dotted arrow) from Detroit to Cancún would veer off to the right (yellow arrow) as a result of the Coriolis effect. The same plane, flying due north from Cancún, would also veer off to the right. In the southern hemisphere, a plane flying due south from La Paz, Bolivia, to Tierra del Fuego, off Argentina, would veer off to the left. It would also veer off to the left in its return trip north.

high-pressure systems. The opposite is true in the southern hemisphere: Low-pressure systems rotate clockwise, and high-pressure systems rotate counterclockwise. A simple way to remember this rotation is to use the right-hand rule in the northern hemisphere. That rule says that if you point your right thumb in the direction of rising or falling air, your bent fingers point in the direction of air rotation (in the southern hemisphere, use your left hand instead). For example, with low-pressure air rising, winds rotate counterclockwise (as viewed looking down on a map). The opposite rotation occurs in the southern hemisphere.

Air that rises faster also rotates faster. Because storms are localized in zones of much lower pressure, the fast-rising air there rotates rapidly counterclockwise (in the northern hemisphere). On a weather map of air pressure, the closer together the pressure contours, the higher the winds. Circulating storms such as thunderstorms are small-diameter cyclones; tornadoes have still smaller diameters with higher-velocity winds. Severe thunderstorms accompanying fronts can spawn rapid circulation and tornadoes. These smaller low-pressure systems also rotate counterclockwise (in the northern hemisphere) unless spun off by some uncommon storm effect.

Global Air Circulation

Atmospheric heating is especially prominent near the equator, while cooling occurs at the poles. Thus less-dense warm air rises near the equator, and nearby cooler surface air moves toward those low-pressure regions (**FIGURE 10-7**).

Higher in the atmosphere above the tropics, the rising air spreads both to the north and the south. At approximately 30° north and south latitude, the cooled air sinks to form subtropical high-pressure zones before returning to the equator to complete the circulation. The air warms adiabatically where it sinks; because warm air can hold more moisture, the humidity in these regions is generally low; the climate is dry or even desert like. At these latitudes lie the deserts of southern Arizona and northern Mexico, the Sahara and Arabia, central Australia, and northern Chile.

The combination of these global air movements with the Coriolis effect gives rise to the southwest-moving **trade winds** between the equator and 30° north latitude and the northeast-moving **westerly winds** between latitudes 30 and 60° north. In North America, the westerly winds bring moist air from the Pacific Ocean to make the wet coastal climate of coastal Oregon, Washington, and southern British Columbia. To the south, the trade winds bring moist air from the Atlantic Ocean to make the wet climate of the Caribbean islands, eastern Mexico, and northeastern South America.

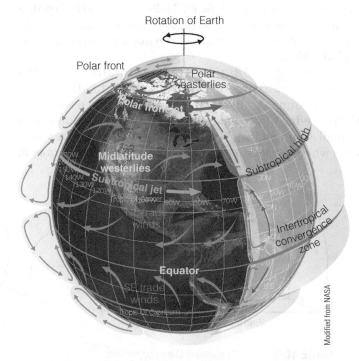

FIGURE 10-7 Global Air Circulation

Global air circulation is dominated by the prevailing westerly winds that come from the southwest or west between 30 and 60° north latitude and the trade winds that come from the northeast between the equator and 30° north latitude.

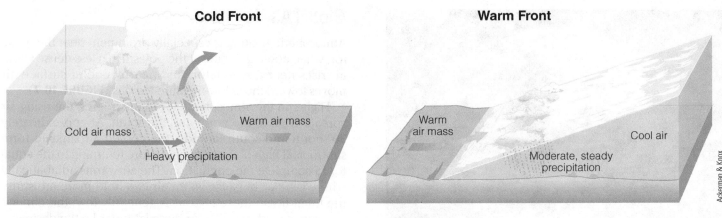

Cold Front

Warm Front

Cold air mass

Warm air mass

Heavy precipitation

Warm
air mass

Cool air

Moderate, steady
precipitation

Ackerman & Knox

FIGURE 10-8 Cold and Warm Fronts

A. A schematic diagram of a slice through a cold front shows how the steep front of a cold air mass advances and pushes the warm air mass upward.

B. A slice through a warm front shows a warm air mass advancing over a cold air mass.

Weather Fronts

Severe weather is often associated with **weather fronts**, or the boundaries between cold and warm air masses. Rainy weather develops at such fronts. Fronts accompany moist, unstable air rising, cooling, condensing, and raining. In a **cold front**, a cold air mass moves more rapidly than an adjacent warm air mass, causing rapid lifting and displacement of the warm air (**FIGURE 10-8A**). In a **warm front**, a warm air mass moves more rapidly than the adjacent cold air mass, and thus it rises over the adjacent cold air mass (**FIGURE 10-8B**).

Advancing cold fronts rapidly lift warm air to high altitudes, causing instability, condensation, and thunderstorms. Air in advancing warm fronts rises over a flatter wedge of cold air, causing widespread clouds and rainfall. Counterclockwise, midlatitude, low-pressure winds form where the jet stream dips to the south to form a *trough*.

Cold and warm fronts sometimes intersect, such as at a low-pressure center. The rising air mass in the low-pressure center cools, causing condensation and rain. When cold air moves toward the equator, it collides with warm air, which is forced to rise.

A low-pressure center can develop as warm moist air to the south moves more rapidly to the east than cold dry air to the north. That counterclockwise shear between air masses can initiate a low-pressure cell where part of the warm air begins to move to the north, such as along a flex in the jet stream (**FIGURE 10-9**). As this occurs, air is forced upward along both the cold and warm fronts, with the strongest upward motion occurring where these fronts come together. Such a system can continue to develop so that the angle between the cold and warm air masses will become tighter and tighter. The low-pressure cell may finally spin off as a cutoff low that can remain in one place, dropping rain for a long period.

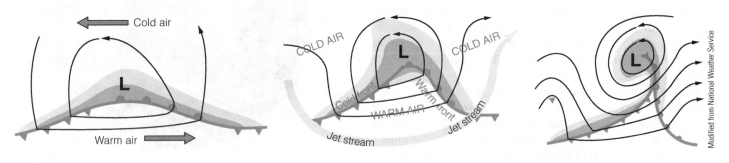

Cold air

L

Warm air

COLD AIR

L

Cold front

Warm front

WARM AIR

Jet stream

Jet stream

COLD AIR

L

Modified from National Weather Service

FIGURE 10-9 Low Pressure Development

A. A low-pressure cell can be initiated by interaction between a warm front and a cold front. Warm air moving east against the cold air causes shear rotation counterclockwise.

B. The warm air pushes northeast over the front to the east, causing development of a warm front. Cold air moving southeast under the front to the west similarly forms a cold front.

C. More rapid counterclockwise rotation strengthens the low, which can in some cases spin off from the intersection of the cold and warm fronts.

In eastern North America, coastal mountain ranges such as the Appalachians increase atmospheric convection (wind). Cold near-shore air can push warmer onshore air upward, causing expansion, cooling, and rain. Storms moving onshore tend to turn northward toward the pole. If they do not turn far offshore, they can persist for many days, dumping heavy rain along the coast and causing significant storm damage. Because such higher-latitude storms can have similar wind strengths and rainfall as tropical cyclones, they can be just as destructive.

Weather Inversions and Smog

Under normal circumstances, the temperature of the lower atmosphere decreases with elevation because the sun heats Earth and the air above it. Under certain conditions, dense cold air may be trapped under overlying warm air to produce a temperature inversion. This can occur when a less dense, warmer air mass moves over a colder one, at a warm front (see again Figure 10-8B). It can also occur under calm-air conditions during the winter when the sun provides less heat, or at night when the ground and surface air cool faster than the air above. With enough humidity in the air, fog may form. When smoke or other pollutants accumulate in this cooler layer, especially in large cities, in valleys, or next to mountains where the trapped cold air cannot escape, the polluted air—smog—may remain for days or longer. North American examples of cities susceptible to smog include Los Angeles near the San Gabriel Mountains, Denver near the Rocky Mountains, Salt Lake City near the Wasatch Range, and Mexico City, which is surrounded by mountains.

Jet Stream

Meteorologists often use weather maps that show the subtropical **jet stream** meandering across North America (**FIGURE 10-10**). This narrow, 3- to 4-km-thick ribbon of high-velocity winds blowing from west to east across North America sits at altitudes near 12 km (that of many jet aircraft). The jet stream moves east along the interface between circulating high- and low-pressure cells (**H** and **L** in Figure 10-10); the high-altitude rotation of these cells helps propel the stream. Because the jet stream tends to locate along the boundary between cold air to the north and warm air to the south, weather fronts and storms are commonly in the same areas. Low-pressure cells are dragged along with it.

When a meander in the jet stream shifts farther south, a large polar low-pressure area in the North Pacific often lies off western Canada. The counterclockwise circulation of the low carries warm moist air from the Pacific Ocean directly into California. Subtropical storms initiated in the mid-Pacific tend to follow the south edge of the polar cold front into California, an event often called the *pineapple express*. These storms carry heat and moisture to the Pacific coast, where they can dump intense rainfall for a short time. In some cases, back-to-back storms every day or two provide heavy rainfall over broad areas and cause widespread flooding. Such frequent storms saturate the soils; runoff is magnified and landslides are common. **Table 10-1** lists some extreme rainfall events.

With more warm-air production to the south in the summer, the jet stream shifts northward over Canada; in winter, it moves down over the United States. Because the jet

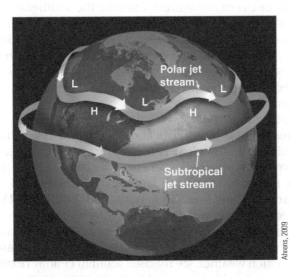

Ahrens, 2009

Modified from National Weather Service

FIGURE 10-10 Jet Stream

A. Polar and subtropical jet streams meander from west to east across North America. Shown here as continuous bands, they are actually discontinuous and variable in daily position.

B. An example weather forecast shows the meandering jet stream with a large polar low off western Canada that funnels rain into coastal California. Another Arctic low over the Great Lakes funnels rain into the Ohio River valley.

Table 10-1	Some Extreme Rainfalls*	
LOCATION	DURATION	AMOUNT
Cherrapunji, India, 1861 (mostly monsoons)	1 year	26.47 m
Cilaos, Reunion, 1952 (Indian Ocean)	1 day	1.83 m
Commerson, Reunion, February 2007	4 days	487 cm
Foc Foc, Reunion, January 1966	12 hours	114 cm
Guangdong, China, June 2010	6 hours	60 cm
Rapid City, South Dakota, 1972	6 hours	37 cm
Holt, Missouri, 1947	42 minutes	30.5 cm
Curtea de Arges, Romania, 1889	12 minutes	21 cm
Guadeloupe, West Indies, November 1970	1 minute	3.8 cm

*Compiled from various sources.

stream marks the boundary between warm, moist subtropical air and cold, drier air to the north, storms develop along weather fronts in this area.

The jet stream has wind speeds of 100 to 400 km/h. Long-distance aircraft flying east across the continents or oceans use the jet stream to their advantage to save both time and fuel. Those aircraft flying west avoid it for the same reason.

Regional Cycles or Oscillations

Several predicable patterns affect sea-surface temperatures (SSTs) and the atmospheric conditions above them, causing major effect on regional weather patterns over periods of a year to several decades. They can cause long periods of especially warm or cold weather, droughts, heavy precipitation, floods, or heavy snowfalls, although it is not yet entirely clear how they might be affected by global warming or climate change. Such lengthy adjustments in regional weather can affect people's understanding of climate changes.

Major weather cycles, spanning a year to more than a decade, help describe weather patterns in different regions. Most important to North America and Europe include:

AO = Polar vortex and Arctic Oscillation

NAO = North Atlantic Oscillation

PDO = Pacific Decadal Oscillation

El Niño and La Niña (ENSO)

The Polar Vortex and Arctic Oscillation (AO)

The **polar vortex** is a region of cold air permanently centered over both the North and South Poles. Its fastest winds are high in the atmosphere, in the stratosphere, and the jet stream marks its outer edge. Most commonly the polar vortex is strong and tightly focused, with cold temperatures, low atmospheric pressure, and a high-velocity jet stream. However, when the polar vortex weakens, with warmer temperatures and higher atmospheric pressure in the stratosphere, the jet stream slows and begins to wobble and meander. Cold Arctic air can then dip farther south. At the same time, some parts of the Arctic are warmer than usual. Westerly winds are stronger and storm tracks remain farther north.

The *polar vortex* refers to conditions higher in the atmosphere (in the stratosphere) whereas the related **Arctic oscillation (AO)** refers to northern hemisphere conditions near Earth's surface. The AO is an index of the relative intensity of a low-pressure center over the North Pole with upper-level winds of the stratosphere circulating around this center and affecting the whole northern hemisphere. When that circulation is intense (positive AO), its stronger winds (fast jet stream) lock cold air over the Arctic Ocean. When the circulation weakens (negative AO), cold air can escape southward into North America, Europe, and Asia. Since the late 1980s, winters have generally seen positive AO values.

Closely related to the AO, and often considered part of it, is the *North Atlantic Oscillation (NAO)*, discussed in more detail later in the section. The National Snow and Ice Data Center notes that the Arctic Oscillation (AO) and the North Atlantic Oscillation (NAO) "are different ways of describing the same phenomenon." North Atlantic Oscillation (NAO) phases correlate directly with AO. A positive NAO concentrates cold air over northeastern Canada; the southeastern United States is warm; and northern Europe is warm and wet. In a negative NAO, the eastern United States is colder and snowy; northern Europe is cold and dry, but southern Europe is warm and wet. Melting of ice on the Arctic Ocean also provides more open water and more moisture to the atmosphere.

Our understanding of these cycles is in a state of flux as weather researchers learn more about them and their interrelationships. What is clear is the real impact that these cycles have on the weather experienced by large population centers in the United States. In January to February 2014, a weak polar vortex and negative AO and NAO (**FIGURE 10-11**) brought extreme cold weather to the eastern United States. There was repeated snow in the northeast, and freezing rain and ice crippled Atlanta, Georgia, and other places that are not used to such weather. Ice reached south to central Texas, with deep snows in Michigan and throughout the Northeast. Ice downed trees and power lines, leaving hundreds of thousands without power; ice on roads caused gridock, stranding thousands of motorists, and canceling tens of thousands of flights. Temperatures fell to $-26°C$ ($-15°F$) in Chicago and $-30°C$ in Minneapolis.

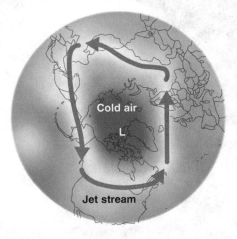

November 14–16, 2013

Polar vortex is strong
Lower atmospheric pressure
Cold stratosphere
Fast jet stream – nearly circular
Positive AO and NAO

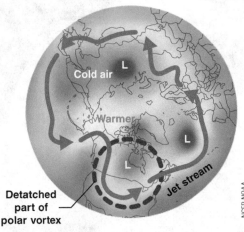

January 5, 2014

Polar vortex is weak
Higher atmospheric pressure
Warmer stratosphere
Slower, weaker jet stream – wobbles
Negative AO and NAO

FIGURE 10-11 **The Arctic Oscillation and Polar Vortex**
White marks the outer edge of the polar vortex, the jet stream, and transition between cold (blue) and warm (brown) regions. A fast jet stream traces a nearly circular path, whereas a slow jet stream wobbles with large meanders.

North Atlantic Oscillation (NAO)

As with El Niño in the Pacific Ocean, there is a recurring atmospheric pressure pattern in the northern Atlantic Ocean, although there is no clear correlation between patterns in the two oceans. In contrast to El Niño, which is recognized by variations in Pacific Ocean water conditions, the **North Atlantic oscillation (NAO)** is defined by variations in *winter atmospheric pressure* over the northern Atlantic Ocean, especially from December to March. Summer effects are much weaker. The variation in strength and position of the high- and low-pressure centers defines the NAO. There is, of course, interaction between the North Atlantic atmosphere and the ocean, but changes in the ocean lag behind those of the atmosphere by two years or more. The time between positive or negative events is highly variable, from a few years to a decade or more.

In a positive NAO, a strong high-pressure cell is generally centered west of southern Portugal, near the Azores in the eastern Atlantic Ocean, while a strong low-pressure cell is centered over the North Atlantic southeast of Greenland (**FIGURE 10-12**). Driven by the large pressure differences, the prevailing westerly winds are strong. In central Europe summers are cool and winters are wetter and warmer. In northeastern North America, southwesterly flow around the Icelandic low prevents cold Arctic air from moving south; winters are wetter and milder.

In a negative NAO, a weaker high-pressure cell forms farther south in the central Atlantic Ocean, closer to the equator, and a weaker low-pressure cell intensifies farther south, east of Newfoundland. The westerly and trade winds are both weaker. In central Europe, winter storms are fewer and weaker. Cold air moves into northern Europe and moist Atlantic air moves into the Mediterranean. Winter along the east coast of the United States is colder with more snow. Periodicity is not regular, but positive NAO comes along every 3 to 10 years.

Cold air is normally trapped in the Arctic by a strong low-pressure system in which winds circulate counterclockwise around the pole. Occasionally the barometric pressure drops farther south, permitting cold Arctic air to pour southward into the mid-continent. This change, sometimes called the Arctic oscillation (AO), is related to the NAO. Such a change occurred in January to early February 2011, leading to a massive ice and snow storm covering much of the United States and eastern Canada.

Longer-term changes in the sea-surface temperature (SST) of the northern Atlantic Ocean, such as the *Atlantic multidecadal oscillation (AMO)*, are also apparent. The SST appears to oscillate from slightly cooler to slightly warmer over several decades, typically about 65 to 70 years. The total temperature swing is only about 0.8°C (**FIGURE 10-13**), but because higher temperatures drive more frequent and stronger Atlantic storms, warmer periods may lead to wetter summers in the southeastern United States and more or stronger hurricanes.

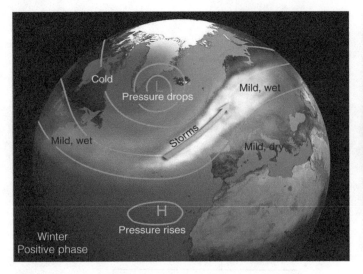

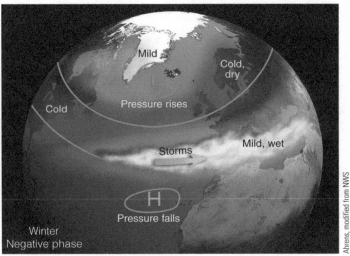

Ahrens, modified from NWS

FIGURE 10-12 North Atlantic Oscillation

A. With a positive NAO (+), strong high pressure lies over the eastern Atlantic Ocean near the Azores while strong low pressure sits near Iceland. The westerly and trade winds are strong. Warm weather develops in the western Atlantic and northwestern Europe.

B. With a negative NAO (−), both systems are weaker and lie farther to the southwest; the westerly and trade winds are weak. Darkest blue colors are unusually cold water.

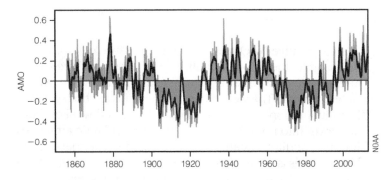

FIGURE 10-13 Atlantic Multidecadal Index

Oscillation in the sea-surface temperature in the northern Atlantic Ocean from 1850 to 2013.

Pacific Decadal Oscillation (PDO)

The Pacific Ocean shows other cyclic changes in addition to El Niño/La Niña, called the **Pacific decadal oscillation (PDO)**. From about 1925 to 1998, these occurred on a scale of about 20 to 30 years—sometimes described as a long-lived El Niño/La Niña–like pattern. Earlier changes, reconstructed from tree-ring data, were more variable. From 993 to 1300 A.D. was a distinct cool phase; then 1450 to 1620 followed a distinct warm phase. In the warm phase (PDO is positive) the eastern equatorial Pacific and along the North American coast sea-surface temperatures (SST) are warm. In the North Pacific Ocean SSTs are cool and sea-level atmospheric

pressures are low. In the cool phase everything is reversed (**FIGURE 10-14**).

In addition to the cycle length, the PDO differs from the El Niño/La Niña pattern in that with the PDO the strongest effect is in the North Pacific, whereas with El Niño/La Niña the strongest effect is in the equatorial Pacific.

A related phenomenon is the recently recognized sea-surface temperature (SST) warm blob in the northeastern Pacific Ocean. This giant area of warm surface water sits variably offshore or partly overlaps the North American continent. Its humid air currents rotate clockwise, deflecting weather systems north, then eastward before reaching the continent. The matching trough on the east brings cold Arctic weather down into the eastern United States, especially in winter.

The same high-pressure Pacific ridge steers moist Pacific storms northeastward past California, apparently contributing to the drought conditions that persisted from 2011 to 2015 (**Case in Point:** Lack of Winter Rain and Mountain Snow Leads to Severe Drought—California, 2012–15, p. 289).

El Niño/La Niña–Southern Oscillation (ENSO)

Warm shallow *ocean temperatures* of the equatorial western Pacific Ocean, known as **El Niño**, oscillate with cool temperatures in the same region, known as **La Niña**, over periods that alternate every three to five years, on average (**FIGURE 10-15**). Such conditions generally last for 6 to 18 months. Surface *air pressure* (the southern oscillation) is high with El Niño and

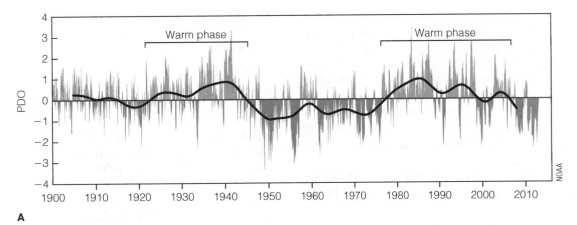

A

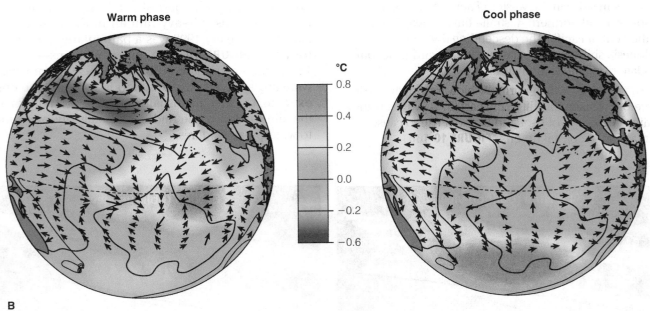

B

FIGURE 10-14 The Pacific Decadal Oscillation

A. From 1900 to 2015 the Pacific decadal oscillation alternated between warm and cool phases. **B.** Typical winter sea-surface temperatures (°C) are shown as colors. Arrows show surface winds.

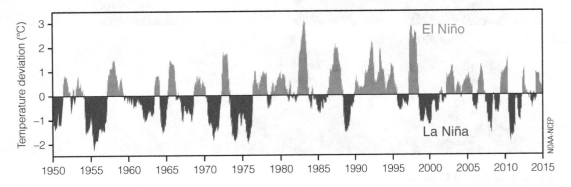

FIGURE 10-15 El Niño/La Niña Sea-Surface Temperature Deviations

Sea-surface temperature anomalies in the mid-Pacific Ocean are shown from 1950 to 2011. El Niño years show abnormally high sea-surface temperatures. Lower temperatures designate La Niña conditions. The horizontal 0 line is average.

low with La Niña. *El Niño* is Spanish for "boy child," a name given by Peruvian fisherman because the weather phenomenon typically comes every few years around Christmas.

Normal conditions over the Pacific Ocean see high pressure with dry weather over the central Pacific and low pressure with rainy weather over the warm ocean water of the western Pacific. Descending dry air over South America (**FIGURE 10-16A**) and rising moist air over the western Pacific circulate trade winds that pull warm ocean currents westward toward Indonesia.

El Niño conditions can be predicted by watching for changes in water temperatures or air pressures over broad areas as follows.

■ Air pressures rise in the western Pacific basin near Indonesia and northern Australia but fall near Tahiti in the central Pacific. The descending dry air over Indonesia shifts the regional winds eastward to accommodate rising moist air near Tahiti.

■ The trade winds, which normally blow toward the west, therefore weaken or shift to the east. Those winds rising against the Andes of Peru and Ecuador rise to cause rainfall in the coastal deserts (**FIGURE 10-16B**).

■ Those winds slowly push warm water from the western Pacific Ocean eastward to pile up against northern South America.

El Niño results in extreme weather and weather-related hazards, but it affects weather patterns differently around the world. In contrast to rains and floods in western South America, Mexico, and southern California, droughts and widespread fires affect Australia, Indonesia, and Southeast Asia. When El Niño years bring increased rainfall to the normally dry coast of Peru, a big low-pressure cell in the Pacific pulls the high-altitude, meandering winds of the subtropical jet stream southward before it loops north to approach western North America. The warm water piles up against the west coast of South America, rises, evaporates, and causes heavy rainfall in the deserts of Peru. Late 1997 to early 1998 was a particularly intense El Niño for Peru (**FIGURE 10-16C**). Incessant rains—as much as 12 cm/day—caused big problems for peasant farmers. Fields were soggy; rivers rose and flooded. During the opposite extreme, La Niña, the usually lush Amazon rain forest is in the rain shadow of the Andes, unusually dry, and subjected to widespread wildfires.

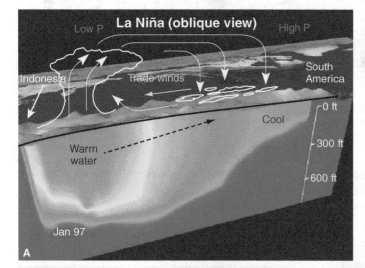
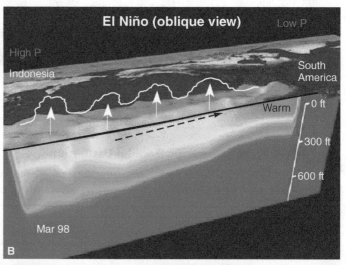

FIGURE 10-16 Warm Water of El Niño
A. During La Niña the near-surface trade winds blow toward the west, keeping the pool of warm water in the western Pacific Ocean around Indonesia. There the warm, moist air rises and cools, and rains are abundant. **B.** During El Niño, the trade winds weaken and the pool of warm water sloshes back across the Pacific Ocean to the east. **C.** A full-year time series of sea-surface temperatures during the strong El Niño of 1997–98 shows the warm water (red) sloshing slowly eastward to pile up against the west coast of South America.

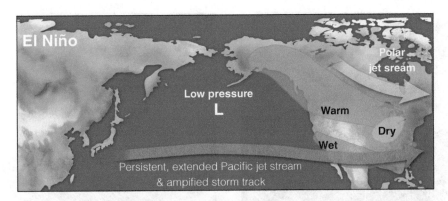

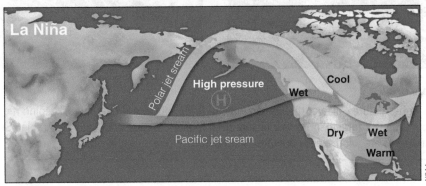

FIGURE 10-17 Weather Changes During El Niño
The maps show weather and jet stream changes from January to March for El Niño and La Niña. During El Niño, the southern states tend to be wet while the northern states are warmer than usual. During La Niña, the Pacific Northwest tends to be cool and wet, while the southern states are drier than normal.

If some areas like Peru are abnormally wet during an El Niño year, others are abnormally dry, because El Niño simply involves a large-scale redistribution of moisture rather than a change in the worldwide average. In the United States, part of the warm water sweeps north along the coast to bring warm subtropical rains to the southwestern region. This keeps cold Arctic air farther north. Eastern Pacific weather systems off Mexico repeatedly hammer southwestern North America, especially coastal parts of California and northern Mexico. El Niño years in the generally dry Southwest bring more local flooding at lower elevations. In contrast, some northern parts of the United States tend to be drier than normal.

La Niña in the south-central United States often brings drought; trade winds that normally blow westward across Mexico and the southern United States weaken or even reverse. Warm water in the equatorial Pacific has moved to the west and cool water resides off Peru (as shown in FIGURE 10-16A). The big low-pressure cell in the North Pacific during El Niño is replaced by a big high-pressure cell. The polar jet stream looping around the north side of that high pressure dips south under the Great Lakes, bringing drought conditions to the southern states (**FIGURE 10-17**).

Regional Winds

Regional winds include monsoons, as well as Santa Ana and Chinook winds. Monsoons are often beneficial but can be disastrous. They are best known from Southeast Asia in summer but also have strong effects in Arizona, New Mexico, and Nevada in mid-July.

Monsoons

The **monsoon** of southern Asia is a seasonally reversing period of mild winter breezes that comes from the northeast, and reverses in summer to bring moisture-laden winds and heavy rains from the southwest. That seasonal change in atmospheric circulation and precipitation associated with the nearly east-west intertropical convergence zone (ITCZ) across India. The zone sweeps to the north around the big low-pressure cell over the hot Indian continental mass. During June, the ITCZ is over southern India, but it moves north in July to northern India, and sometimes Pakistan. Warm, moist winds from the Arabian Sea to the southwest collide in northern India with cool winds from the Himalayas to the northeast (**FIGURE 10-18**).

The summer monsoon affects India, occasionally Pakistan, and surrounding areas from June to September. Because 80%

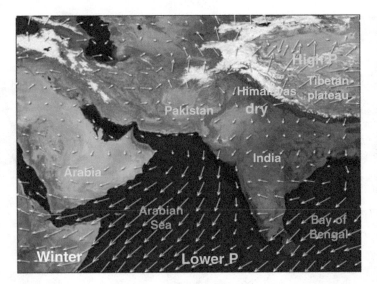
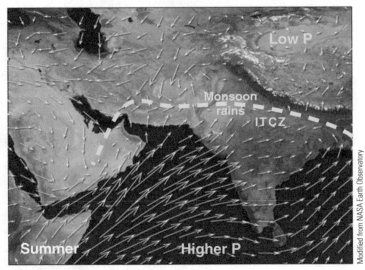

FIGURE 10-18 Indian Monsoon

In winter, cold, dry air from a high-pressure cell over the Tibetan Plateau drifts southward over India, keeping it cool and dry. In summer, the Tibetan Plateau becomes a low-pressure zone that draws moisture from higher pressure area over the Arabian Sea, bringing monsoon rains to northern India. The longest arrows are wind speeds of more than 15 m/s.

of the rain in India comes with the summer monsoons, crops receive most of their moisture from them. However, the monsoons also bring floods that can be devastating. The related winter monsoon brings dry winds from the continent on the northeast. Because the Himalayas form a barrier to cold winds from northern Asia, those winds are very warm.

Black carbon emissions produced by inefficient burning of wood and dried dung, along with forest fires and industrial pollution in India, also accumulate in the atmosphere against the base of the Himalayas in late spring, absorbing the sun's heat (**FIGURE 10-19**). The low-pressure cell over India and Tibet draws moist air from the Arabian

Sea toward the Himalayas. Those warm, moisture-laden winds are forced to rise at the wall-like barrier of the southern edge of the Himalayas, condense, and produce heavy rains.

Similar to the Indian monsoon, the East Asian monsoon moves north from Vietnam and Cambodia in May to China in June, Korea in July, returning to South China in August. The Arizona–Mexico Monsoon (North American Monsoon) also develops in summer as intense heat inland from the coast builds to develop a regional low-pressure ridge about parallel to the coast. Hot air rising from that low-pressure zone draws in cool, moist air from the Gulf of California and the eastern Pacific Ocean. That moist air rises, cools, and condenses, causing frequent thunderstorms and heavy rain. The change from hot, dry weather to frequent summer rains occurs in about a week, beginning in southern Mexico in late May and shifting northwestward to Arizona and New Mexico by early July. Localized heavy downpours produce runoff that fills normally dry washes, producing flash floods, raging torrents that flush all loose gravel and boulders downslope. Because the rainfall can be out of sight, 10 or 20 mi away, camping in a dry wash can be fatal. The summer monsoon season generally trails off in September.

Santa Ana and Chinook (Foehn) Winds

Strong winds cresting a mountain range often warm adiabatically as they descend the leeward side. Called *Foehn winds*, they include the well-known Santa Ana winds of southern California and the Chinook winds of the east slopes of the Rocky Mountains.

Santa Ana winds are very strong, dry winds moving southwestward from a high-pressure cell over the Great Basin of

FIGURE 10-19 Air Pollution

Air pollution in northern India and Bangladesh, rising against the southern Himalayas to produce rain in June, early in the Monsoon season.

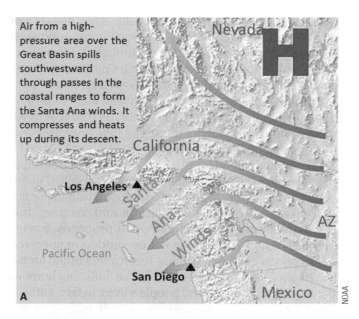

Air from a high-pressure area over the Great Basin spills southwestward through passes in the coastal ranges to form the Santa Ana winds. It compresses and heats up during its descent.

NOAA

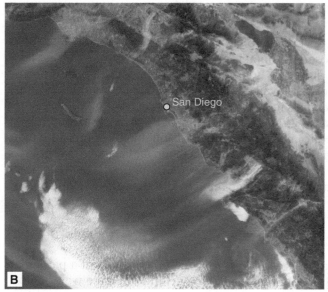

NOAA

FIGURE 10-20 Santa Ana Winds

A. Source of Santa Ana winds that descend to the coast and offshore from a strong high-pressure cell centered over the Great Basin/Basin and Range of Nevada, Utah, and Arizona.

B. Satellite view of winds near San Diego.

Nevada, Utah, and Arizona (**FIGURE 10-20**). Dense, colder air trapped between the Rockies and Sierra Nevada and circulating clockwise around the atmospheric "high" drains from the high deserts down through mountain canyons of southern California and Baja California. Compressed into a smaller volume as it descends, the air heats up adiabatically and dries out, the relative humidity often dropping to 10 to 25%. Typically coming between October and March, these adiabatic winds vary in temperature, depending on the weather in the Great Basin. During hot weather in early fall, the winds can reach speeds of 35 km/h, often fanning wildfires burning in already dehydrated vegetation. In October 2003 wildfires with 64 to 96 km/h winds burned almost 3000 km², and in October 2007, another 1700 km². On December 1, 2011, hurricane-force winds reached 164 km/h in Utah and 198 km/h northwest of Denver, Colorado. Elsewhere, winds of 156 km/h downed numerous trees, severing power lines, and blew over trucks on highways. 340,000 people lost power in the Los Angeles area.

Chinook winds similarly warm by descending the leeward side of a mountain range (**FIGURE 10-21**). During winter, strong winds from the west cool and lose their moisture as they rise across the Rocky Mountains in Alberta, Montana, Wyoming, Colorado, and New Mexico. Continuing east, they form strong downslope winds that become warmer adiabatically as the air compresses into a smaller space at lower elevations. At Boulder, Colorado, such winds have reached speeds of 225 km/h (140 mph). These warm, dry winds can rapidly melt any snowpack and cause flooding (see Chapters 13 and 14).

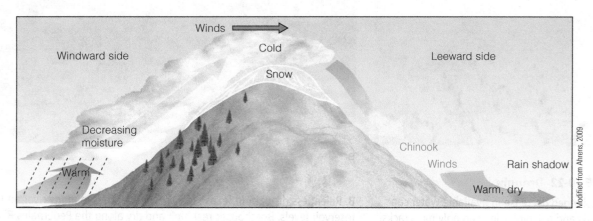

Modified from Ahrens, 2009

FIGURE 10-21 Chinook Winds

Chinook winds warm as they descend the lee side of a mountain range.

Drought, Dust, and Desertification

Water is critical to life in all of Earth's environments. Some of those environments are typically very dry; others are typically quite wet. Both plants and animals adapt to the amount of moisture available in the particular environment: cactus and camels in the semiarid plains; beavers, moose, and rain forests in wet areas. When any of these environments is subjected to months or years of distinctly lower moisture than normal (called *drought*) vegetation and animals may not survive, and the region may even turn to desert. Such dry periods are often accompanied by unusually high temperatures, sometimes for long enough periods to be called *heat waves*.

Drought

Drought is a prolonged dry climatic event in a particular region that dramatically lowers the available water below levels normally used by humans, animals, and vegetation. Drought does not require extremely dry conditions, merely a distinctly lower-than-usual amount of precipitation in a region. It can involve a significant drop in the usable water available in reservoirs, groundwater storage, or stream flow (**FIGURE 10-22**).

In 2014, a 14-year drought in the southwestern states (the longest in 100 years) led to a water crisis, necessitating a U.S. government decision to reduce by 9% the Colorado River flow downstream from Lake Powell's Glen Canyon Dam. Dropping the water level in Lake Mead, which supplies Las Vegas, has prompted drilling of a third, still lower

5-km-long intake tunnel to supply the city. Drought conditions from 2000 to 2004 were the worst in the southwestern states in 800 years, and conditions through 2014 remained as bad. Flows in the Sacramento River in California and the Rio Grande in Texas also dropped dramatically. Contributing to the prolonged drought in the Southwest was the prolonged downturn in El Niño/La Niña oscillation and loss of Arctic sea ice, which led to more transfer of heat from the ocean to the atmosphere. Finally in May 2015, return of El Niño in the Pacific accompanied long-delayed rains in Texas, even more than people hoped. Torrential downpours resulted in tens of centimeters per day in some places, along with major floods.

Unlike many other natural hazards and disasters, drought is not an abrupt or dramatic event; it proceeds slowly and progressively. Drought may not be sudden or spectacular, but its effects can be disastrous. It can affect broad areas and last for months, years, or even decades, inflicting horrendous damage to soils, crops, and people's lives. In fact, annual U.S. losses from drought average more than $6 billion, more than twice that from floods or hurricanes. Widespread drought covers more than 40% of the United States every few years (15 times from 1900 to 2010), or about once every seven years on average (**FIGURE 10-23**).

The most severe droughts in United States and Canada were in the Dust Bowl years of 1930–36. Low soil moisture was aggravated by intensive farming, deep plowing of the native prairie sod, and lack of crop rotation. Severe droughts occurred especially in the 1950s and 1980s. Losses from the 1988–89 U.S. drought totaled almost $40 billion. In 2002, widespread severe drought covered much of the western

Donald Hyndman

A

Karen Ward

B

FIGURE 10-22 Drought

A. A prolonged decrease in available water can dry ponds and reservoirs, leaving only mudcracks, such as in this former lake bottom.

B. Relentless drought in Texas led to severe declines in many river and reservoir levels. Boathouses rest high and dry along the Pedernales River near Austin in August 2009 and again in August 2011.

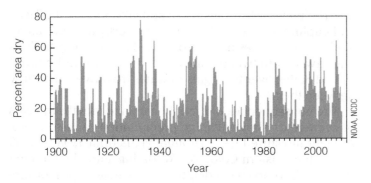

FIGURE 10-23 U.S. Drought Through Time

Percentage of the United States that was in moderate to extreme drought, January 1900 to January 2010, based on the Palmer Drought Index.

FIGURE 10-24 Disappearing Snowpack

A. March 27, 2010 ("normal"). The snowpack is broad and deep.
B. March 29, 2015 ("severe drought"). Sparse snow coverage when the mountain snowpack should be deepest.

and Midwestern states and Canadian Prairie Provinces and was accompanied by major wildfires. Extreme drought affected areas of the southeastern United States, including parts of Florida and Texas, between 2002 and 2015, and California from 2010 to 2015.

Meteorologists commonly refer to drought as an extended period with precipitation much less than average for a region. Farmers, ranchers, and foresters recognize it as an extended period of insufficient moisture for plant growth in an area. The amount of precipitation required to sustain local crops depends on the moisture needs of those crops and the agricultural techniques used. Rice and cotton, for example, require tremendous amounts of water; open range grasses used by freely roaming cattle need very little. In some regions, farmers must use irrigation water to allow intensive farming of crops that could not be sustained by rainfall alone. Drought is often aggravated by overuse of the long-term naturally available supply of water. In the Great Plains and Midwest, the widespread drought from 1987 to 1989 was made worse by depletion of groundwater in many areas through heavy pumping and farming of marginal lands.

Farmers compete for water with residential users, and drought conditions can lead to increased value for water. In a dry period during the fall and winter of 2008–09, some northern California farmers found it more profitable to sell their water rights to cities and towns that needed it rather than use it to irrigate their crops.

Timing of precipitation can also be a factor leading to drought conditions. Winter snowpack tends to melt slowly, soaking more water into the ground and keeping streams flowing year round. Rain falling in warmer seasons can rapidly run off or evaporate in the sun and wind, leaving less for groundwater and streams. During California's multiyear drought, its normal winter snowpack dwindled to very little so it did not store water for agriculture and other uses later in the season (**FIGURE 10-24**).

In North America, drought often begins with changes in the jet stream flow that normally moves westward from the warm tropical Atlantic Ocean, over the Gulf of Mexico, and then curves north into the Great Plains of North America. That northward trend, around a major high-pressure cell over the northern Atlantic Ocean, generally carries abundant moisture from the Gulf of Mexico into the Great Plains. During periods of extreme drought, such as in the Dust Bowl period of the 1930s, much of the jet stream continues west over Mexico rather than curving north to bring rain to the Great Plains. Such drought appears to correlate with the warm phase of the AMO. Drought in the south-central states often correlates with La Niña years when warm water in the equatorial Pacific moves to the west and cool water resides off Peru.

India is particularly susceptible to drought because the region is dependent on the June to September monsoons for its annual rainfall. Severe droughts in India have occurred at times of La Niña, when a low-pressure cell can pull dry air from Asia across India, causing severe drought in the mid-southern to western parts of India during periods that should have been monsoon seasons. Millions died in this region in a series of droughts and famines: 5 million in 1877 and 1899, and more than 1.5 million in the 1960s.

Humans have been plagued by droughts for thousands of years. The consequences of drought include famine, migration of large populations, social and political upheaval, war, and even the total collapse of civilizations.

Extreme droughts have caused some well-known civilization collapses, for example, in Egypt, Arabia, and Iraq, about 2200 BC, in Greece, Turkey, and the Middle East from 1300 to 700 BC, and Angkor Wat in Cambodia in early 1400s. In the Americas, a number of civilizations fell during major droughts: the Mayan civilization of Mexico and Central

America collapsed about 900 AD; the Tiwanaku of Peru about 1100 AD; and the Anasazi of the southwestern deserts of the United States at the end of the 1200s.

Southeast Asia, with its huge populations, has seen more than its share of droughts. Major droughts in China killed 9.5 million in 1877–78, 3 million in 1928–30, and 36 to 45 million in 1958–61, the last one aggravated by Mao Zedong's failed attempt to move peasants off farms and quickly transform China from an agricultural to an industrial economy.

Developing countries are especially vulnerable to famine and political unrest during times of drought. In the arid climates of northern Africa, hundreds of thousands died during the drought and desertification in Darfur (western Sudan) and Ethiopia in the early 1980s and again in 2010–11. In more-affluent areas, the consequences of drought can be severe when crops fail, causing financial distress to farm families and the businesses that depend on them. In cases of prolonged drought, farmers are forced to abandon their land and move elsewhere to look for work. In industrialized countries, droughts kill few people but are responsible for the loss of fertile cropland and necessitate the slaughter of tens of thousands of animals.

Drought is not always a result of lower-than-normal rainfall; it can also be caused by human interventions, such as river damming or increased use of water by upstream areas. Dams built to retain water for irrigation often severely affect downstream flow, leading to crop failure, thirst, and disease from decreased water quality and downstream water contamination. When a country upstream builds a dam, the drop in downstream flow can initiate violent international conflict. Downstream countries such as Iraq and Bangladesh that depend heavily on irrigation have been strongly affected by other countries' upstream dams. Former wetlands dry up, and formerly irrigated lands are left dry. The previously fertile marsh and lake areas of the Tigris and Euphrates Rivers in Iraq are now mostly dry and saline. Similar water issues have also caused intense negotiations and lawsuits between states in the western and central United States over the rights to water from rivers that flow between them.

The consequences of drought can take months to years of average or above-average precipitation to restore pre-drought moisture levels. Drought years from 1998 to 2002 in much of eastern Colorado, Wyoming, and Montana left soils dry down to 60 cm depth in some areas. Low moisture levels in the soil can contribute to related hazards such as dust bowls (discussed in the following section) and wildfires (discussed in Chapter 17) and take years of above-average precipitation to stabilize.

Drought also makes plants more susceptible to disease and infestation. Crops such as grains and corn, weakened by drought, can be overrun and destroyed by swarms of invading locusts. Weakened forests can be invaded and killed by pine-bark beetles (**FIGURE 10-25**). The risk of major wildfires is dramatically increased by all of the dead and dry pine needles. Areas of widespread infestation tend to be prime candidates for wildfires, including those in southern California in 2003, 2006, 2007, and 2009.

Are we likely to experience the kind of severe drought that devastated the American Midwest during the Dust Bowl of the 1930s? Many experts believe we could. By early summer of 2007, drought was widespread throughout the United States, from southern California, Arizona, Nevada, and western Colorado to most of the southeastern states. Many experts expect that drought in the Southwest will become the norm, an especially serious problem with the rapidly growing populations in this region. A severe drought in Texas and Oklahoma suddenly ended with

FIGURE 10-25 Pine Beetles Devastate Forests

A. Sap-filled pine-beetle bore holes. **B.** Thousands of originally green pine trees weakened by drought and killed by pine-bark beetles in central British Columbia.

massive floods in April 2007. By 2008 and 2009, however, the drought had returned and worsened. The weather pattern in the southeast was partly caused by the placement of the Bermuda High (with clockwise air circulation) in the Atlantic east of the Carolinas, which pulled moisture from the Gulf of Mexico and dumped heavy rain on Texas and the plains states. This huge high-pressure system moved far enough west to keep the southeastern states dry. Severe drought returned in 2010–11 (see **Case in Point**: Extreme Drought—Texas and Adjacent States, 2010–11, p. 289).

Dust Storms

A significant reduction of soil moisture permits dust to be carried in the air. Dust becomes a hazard when winds can pick up loose, dry dust or soil, creating a **dust storm** (**FIGURE 10-26**). Such storms not only result in soil loss from the area where they originate, they can also obscure visibility, reduce sunlight, and cause health problems in humans. If you see a dust storm approaching while driving, you should pull off the pavement as far as possible, set the emergency brake, turn *off* your lights, and take your foot off the brake, so cars will not crash into you thinking that you are still driving on the highway. On June 12, 2013, on Interstate 80 near Winnemucca, northern Nevada, a blinding sandstorm caused a 27-vehicle pileup that killed one truck driver and injured 26 others, three of whom were in critical condition.

Dust storms most commonly develop in deserts but can occur in any area where soil moisture has been reduced. A reduction of soil moisture leading to dust can occur because of drought or poor land management, such as cultivation or overgrazing of marginally arid or semiarid land. In many poor countries, this type of cultivation is almost unavoidable, as people struggle to avoid starvation, but the practice occurs in developed nations as well. Elsewhere, tropical hardwood forests on marginal soils are stripped for the value of their timber, cut for firewood and cooking, or slashed and burned for cultivation. During much of the 1950s to 1990s, China cut most of its forests to fuel its factories and clear land for crops. When much of the new farmland on marginal soils failed, frequent choking dust storms enveloped Beijing and other parts of eastern China, and forced the country to make changes. The government now pays many farmers to leave the ground fallow and embarked on a massive campaign to plant extensive swaths of trees to help retain ground moisture.

Dust storms decrease visibility on the ground and in the air, affecting transportation by automobile and by aircraft. They also carry pollutants and soil particles. Giant dust plumes in spring and summer from deserts in western China and Mongolia also carry smog, black carbon particles, industrial smoke and fumes, sulfates, and nitrates. The plumes can be 500 km wide and up to 10 km thick. The sulfates actually reflect more than 10% of the sunlight they are exposed to, but the black carbon or soot absorbs sunlight, increasing warming.

Dust can also affect human health when people breathe in particles. Recent studies show that significant quantities of bacteria and fungi travel with dust over large distances. For years, many people in the San Joaquin valley of central California have experienced flu-like symptoms known as Valley Fever. The condition is not normally serious except for those with weakened immune systems. Fungal spores from soil generally lie dormant in the dry season. In the wet season, they form a mold that can be circulated when soil is disturbed by farming or construction, which causes some residents to contract the fever. The same disease is endemic in most of the southwestern states and northern Mexico. Much of the pollution comes from small garden plots in impoverished areas where people fertilize the soil by burning garbage that now includes not only waste from plants and animals but plastics and tires.

Dust storms are a global problem because winds carry dust around the world. In 2006, a dust storm from the Gobi Desert, just north of China, dumped 270 million kg of silt and fine sand on Beijing. Dust from eastern Asia travels across the Pacific Ocean, creating haze in North America, and even out into the Atlantic Ocean. Dust plumes can cross the Pacific Ocean to reach North America in only one or two weeks. Dust from North African storms often covers Europe and even the Caribbean, affecting the health of people, coral reefs, and other organisms (**FIGURE 10-27**).

One of the most severe natural disasters in American history involved drought and dust. Extreme drought hit North America in the mid-1930s, turning the U.S. plains states into a "Dust Bowl" (**FIGURE 10-28**). Coinciding with the economic collapse of the Great Depression, huge areas of the Great

FIGURE 10-26 Dust Storms

A dust storm invades Cimarron, western Oklahoma on January, 14, 2014.

NASA

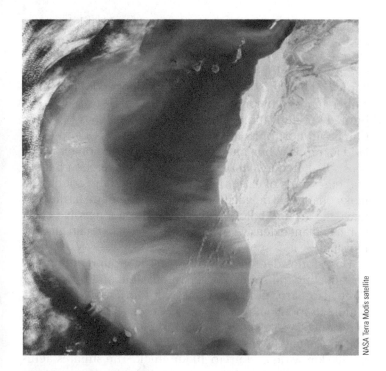

FIGURE 10-27 Dust Storm
A giant dust storm sweeps out over the Atlantic Ocean from the Sahara Desert in northwestern Africa, in March 2003.

NASA Terra Modis satellite

Plains were abandoned as people migrated west seeking work. The Dust Bowl crisis of the 1930s was brought on by a combination of extended drought and disastrous farming practices on the Great Plains. Advances in mechanical plowing permitted removal of natural grass sod to facilitate planting of wheat. Unfortunately, when the drought hit, the fertile soils dried out and were picked up by the wind and blown away to the east and northeast as immense dark clouds of

dust. At its peak in July 1934, the drought affected more than 60% of the area of the United States (**FIGURE 10-29**).

Financial hardships imposed by the Great Depression of the early 1930s forced farmers to grow more crops, which drove down prices, forcing them to increase production to pay for their equipment and land. Unable to farm and unable to pay their debts, hundreds of thousands of people left the southern plains states of Texas, Oklahoma, Arkansas, Colorado, and Kansas. Most had no skills beyond farming; most were uneducated and worked menial migratory jobs for little pay. Living in roadside tents or tin shacks, with little food, they barely survived. Many died from dust pneumonia. Autopsies of cattle that died of starvation showed their stomachs packed with dust and dirt. Lessons learned during the Dust Bowl years led to better farming practices and reduced impact from later droughts. Much of the land was reseeded with native prairie grass rather than tilled and planted with wheat. Widespread irrigation has limited the impacts of drought, but groundwater levels are rapidly declining in areas such as the Southern High Plains aquifer; a return of Dust Bowl–like conditions could develop in some areas if farmers cannot irrigate their land.

Much of the Great Plains of North America is better suited to moderate grazing than cultivation. Semiarid areas affected by drought can recover when rains return if the ground has not been misused. Even prior to European arrival in North America, there were times of extreme drought and large areas of desert. The Sand Hills of Nebraska are giant sand dunes, active less than 1000 years ago but now grown over with grass and brush. When the rains returned, vegetation sprouted, beginning in the low areas. Some of these areas are again cultivated, but the hills, with their thin, fragile soils, are not able to sustain cultivation without again reverting to open areas of sand. Grazing cattle often break the fragile cover, leaving persistent trails.

NOAA

Sloan, USDA

FIGURE 10-28 Dust Bowl
A. A huge wall of dust bears down on Stratford, Texas, on April 8, 1935.

B. Windblown dust buried farms and their machinery in Dallas, South Dakota, May 13, 1936.

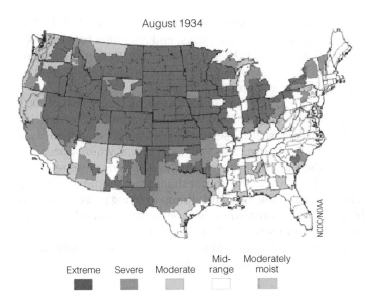

August 1934

Extreme   Severe   Moderate   Mid-range   Moderately moist

FIGURE 10-29 Dust Bowl Area

Extreme drought during the Dust Bowl in August 1934 spread over much of the central United States.

Desertification

The vegetation cover of parts of semiarid scattered-tree grasslands, such as the Sahel region along the southern edge of the Sahara Desert in Africa, has lost its cover of grass, some shrubs, and patches of trees, mostly since the late 1960s, because of increased population, intensive cultivation in marginal areas, and overgrazing. Rainfall ranges from about 20 to 60 cm (8 to 24 in) per year, equivalent to that in most of the Rocky Mountain West, east to the High

Plains of central North Dakota and down to West Texas. Decades of drought in the semiarid Sahel has prevented regrowth of grass and fostered **desertification**, the evolution of new desert environments. This is a major problem in parts of Chile, Brazil, Mexico, Morocco, Ethiopia, Pakistan, Afghanistan, Kazakhstan and other central Asian countries, Mongolia, China. In China 150,000 km² per year of cropland is damaged by drought and desertification; in 2009 it was costing $42 billion per year. The main causes were firewood collection, overgrazing, and overcultivation. Studies suggest that warmer SSTs of the Indian Ocean weaken the monsoon movement over Africa and promote drought in a belt across the continent. The United Nations Convention to Combat Desertification defines its concern as "land degradation in arid, semi-arid, and sub-humid areas resulting from factors including climatic variations and human activities." Desertification affects about 70% of Earth's dry lands, which equates to 30% of Earth's land surface (**FIGURE 10-30**).

Overgrazing, sometimes fostered by fences that prevent animal migration, breaks up surface soil, causes compaction of deeper soil, and reduces water infiltration. Artificial watering holes or watering tanks can have the same effect because animals don't migrate. Cultivation dries out the soil. If topsoil is exposed during high winds, dust storms remove its most productive parts or raindrop impacts loosen and wash it away. In Africa, the effect is severe due to overgrazing and overcultivation, especially on marginally productive lands, and worsened by climate change. Two belts across Africa, the Sahel, south of the Sahara Desert, and the other south of the equator, show growing drought. It has been estimated that one-half to two-thirds of productive nonirrigated land has been at least moderately subjected to desertification, as in the Dust Bowl of the 1930s.

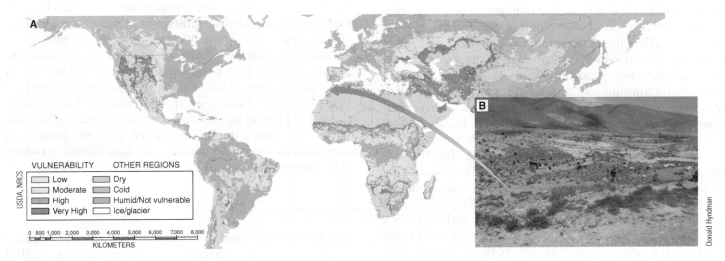

VULNERABILITY
Low
Moderate
High
Very High

OTHER REGIONS
Dry
Cold
Humid/Not vulnerable
Ice/glacier

0 500 1,000 2,000 3,000 4,000 5,000 6,000 7,000 8,000
KILOMETERS

FIGURE 10-30 Desertification

A. Map of the vulnerability of regions to desertification. **B.** Grazing of sheep and goats in Morocco, at the northwest end of the Sahara Desert, contributes to desertification. The High Atlas Mountains are in the background.

Heat Waves

In contrast to dramatic or violent weather hazards—such as floods, hurricanes, and tornadoes—the effects of excessive temperatures build more slowly but can be even more deadly, killing greater numbers of people than floods, hurricanes, and tornadoes combined. Part of a heat wave's danger is that people tend to view hot weather more as a discomfort or inconvenience than a health emergency. Deaths during heat waves can occur when a person's core body temperature, normally 37°C (98.6°F), reaches 40° to 41.1°C (104° to 106°F), resulting in heat stroke. High summer temperatures in urban areas also accelerate the chemical reactions that create ground-level ozone and smog, further increasing stress on people's health. In some cases, heat adds stress to frail, elderly people, who die of heart attacks. Adequate water intake is especially important.

Urban areas bear an increased risk of high temperatures because of the **heat-island effect**. Extreme summer temperatures in a major city can be as much as 5°C (9°F) hotter than in the adjacent countryside. Cities and airports with many buildings and vast areas of dark pavement collect and absorb heat from the sun. Exhaust from cars, trucks, air conditioners, and factories trap more heat because the dark particles absorb heat from the sun in the same way. Urban areas also cool slower than rural areas at night because buildings and pavement retain heat longer, often for three to five hours after sunset. Although in the winter warmer temperatures reduce heating needs and melt snow and ice on roads, the adverse summer heat-island effects much outweigh the winter benefits.

July of 1995 brought Chicago a period of high humidity and extraordinarily high temperatures that stayed above 37.8°C (100.2°F) for days and reached 41°C (106°F) on July 13. Some 600 to 700 people died of heat-related causes during this heat wave. Most were elderly, poor, and living in the inner city, where they were afraid to open their windows at night because of crime. Contributing to the problem was the heat-island effect, power failures, lack of warning, and inadequate ambulances and hospital facilities. In 1966, hundreds of people in St. Louis died from heat waves, and dozens more succumbed in 1980 and 1995. In July of 1993 and 1995, dozens died from heat waves in Philadelphia.

High humidity makes a considerable difference by limiting people's ability to cool by perspiration. The effect is dramatic at temperatures near or above the core temperature of the human body. For example, at 35°C (95°F) and 80% relative humidity, it feels like a very dangerous 56.1°C (133°F). Direct sunlight can add 15°F to these temperatures. In very dry climates, however, such as parts of the Mountain West, where the relative humidity can be as low as 20 to 30%, the apparent temperature can be lower than the temperature on the thermometer.

The western United States experienced an extended heat wave in July of 2006 that caused at least 164 deaths in southern California, mostly elderly people without air conditioners. Temperatures in Fresno exceeded 43.3°C (110°F) for six consecutive days, and Sacramento went over 37.8°C (100°F) for 11 days. Compounding the problem were electrical system failures that left more than a million people without power and air conditioning.

More than 35,000 died in the European head wave of 2003 (**Case in Point:** Deadly Heat Waves—Europe, 2003 and 2010, p. 291). In August 2013, severe heat waves affected the Plains states and the Midwest, as well as in China, where dozens of people died. Manhole covers in China in some cases reached 60°C (140°F) and were so hot that some people reportedly used them to cook eggs and shrimp in frying pans. Four died in Japan and Korea where temperatures reached the highest levels ever recorded. Hardest hit were large cities such as Shanghai, with 28 days above 35°C (95°F), and Hong Kong, because of the heat-island effect.

Snow, Ice Storms, and Blizzards

Winter at northern latitudes, including Canada and much of the United States, brings cold temperatures and sometimes snow and ice. We bundle up to stay warm, especially when the wind blows. The wind makes a difference because evaporation of skin moisture causes heat loss, known as **wind chill**, which makes us feel colder. Wind chills can reach −51°C (−60°F) in Midwest or Canadian Prairie blizzards.

Such bitter cold can lead to **frostbite**, the freezing of body tissue on noses, ears, or fingers. Frostbite is indicated when the affected areas turn pale or white and lose feeling. Victims need immediate medical attention; the frostbitten areas should be warmed slowly. When a person's core body temperature drops below 35°C (95°F), 2°C below normal, the person is experiencing *hypothermia*. Symptoms include uncontrollable shivering, drowsiness, disorientation, slurred speech, and exhaustion. Victims need immediate medical attention, and they should be warmed *slowly*. Outfit the victim with dry clothes and wrap in a warm blanket. Warm the person beginning with their core, ideally with warm (not hot) broth. Avoid warming arms and legs first because that drives cold blood to the heart, potentially leading to heart failure. Wearing a hat, hood, and coat in cold weather is important in order to minimize heat loss through exposed skin. To keep fingers warm, mitts are better than gloves.

Snow

Compared with land nearby, areas downwind of large lakes often have unusually large snowfalls, called **lake-effect snow**, in winter. In the westerly wind belts of North America, the prevailing winds come from the west. Thus northeastern, eastern, and southeastern sides of most of the Great Lakes can receive extraordinary amounts of snow in a single storm. This can amount to one-third to one-half of the total annual snowfall. Factors

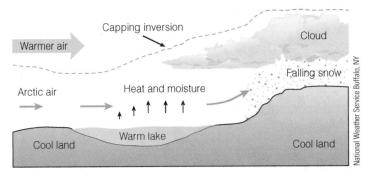

FIGURE 10-31 Lake-Effect Snow

Cold Arctic air moving across a warm lake picks up moisture evaporating from the lake, rises, cools, and condenses to form tiny ice crystals that fall as snow.

---

that affect the amount of snow (**FIGURE 10-31**) include expanse of ice-free water under the storm track (for water evaporation); duration of strong wind blowing over the water (for greater evaporation); amount of moisture in the atmosphere (to condense as snow); and topographic rise (hills) downwind of the water (to lift moisture and cause condensation and snowfall).

Cold, Arctic air moving over relatively warm water of the Great Lakes evaporates large amounts of moisture from the lakes. The cold air freezes the moisture. Reaching land, the moisture-laden air rises and cools even more, to precipitate the moisture as tiny ice crystals—snow (**FIGURE 10-32**). As long as the cold wind keeps blowing across the water, snow continues to form and fall downwind. Snowfalls as heavy as 15 cm/h have been recorded. Often the low-pressure cell, around which the wind is circulating counterclockwise, is off to the northeast. If that low-pressure cell remains stationary, the wind may continue to blow for days, bringing heavy snows to the eastern and southeastern sides of the lakes. Will global warming lead to more ice-free winters on the Great Lakes and more lake-effect snow? The concern is real because recent winters have seen less ice and more open water, leading to more lake evaporation. In early February 2007, lake-effect snows from the east end of Lake Ontario buried part of upstate New York in 3 to 4 m of snow. People shoveled off roofs to prevent collapse, leaving snow piled above their eaves.

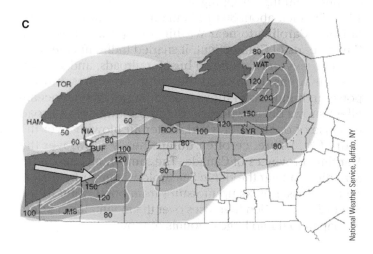

FIGURE 10-32 Lake-Effect Snow

A. A satellite view of the western Great Lakes, Superior, and Michigan, shows winds sweeping southeast, picking up moisture from the lakes, and dropping heavy snows on the southeast to east sides of the lakes.
B. Heavy lake-effect snow in western New York, November 21, 2014.
C. The average annual snowfall around Lake Erie and Lake Ontario shows three to four times as much downwind of the lakes. Contours in inches of snow.

Buffalo, at the east end of Lake Erie, is well known for its heavy winter lake-effect snows. On November 19, 2014, a storm dumped 165 cm (65 in) of snow on the city. People were stranded in cars, trucks, and buses for as long as 24 hours. Twelve people died from heart attacks while shoveling snow or pushing cars. One 46-year-old man was found dead in his car, which was buried in snow.

If the weather has been unusually cold for a long period, a lake may largely freeze over, so that little water evaporates from its surface. Under that circumstance, the cold wind remains dry, and it doesn't snow. In the winters of 2005–06, and 2009–10 the Great Lakes remained warmer and mostly free of ice. Cold winds formed ice near shore, but waves offshore broke up thin ice and stirred up warmer water from below, preventing ice from spreading. The winds picked up moisture from the water, which led to heavy snowfall downwind.

A similar phenomenon of snow falling downwind of a large body of water sometimes occurs during a nor'easter, an offshore low-pressure cell off the coast. In this case the result would be *ocean-effect snow*, such as the classic nor'easter that blew north through the eastern states on February 12, 2006. It arrived in New York with claps of thunder and lightning. Dropping snow at rates of 7 to more than 12 cm/h, it left about 30 to 76 cm on the ground from parts of North Carolina to near Washington, D.C., to New York City, to parts of Connecticut. It snarled traffic in cities and on highways and shut down buses, railroads, and airlines. In some areas, it downed electric power lines, cutting power to hundreds of thousands of homes. Similar ocean-effect snows occur wherever cold air blows over a broad expanse of open water. Cold winter air blowing over the Gulf of St. Lawrence often dumps heavy snow over the Maritime Provinces of Canada. The same thing happens in Scandinavia near the Baltic Sea. Cold winter air from Siberia and Russia picks up moisture from open water of the Sea of Japan, which then rises over the mountains of northern Japan to dump large amounts of snow.

Ice Storms

Weather events that cause **ice storms** depend on humid air and warmer temperatures above cold air near the ground (**FIGURE 10-33**). Such storms cause widespread power outages and can cripple transportation (**FIGURE 10-34**). Ice storms are especially common in New England and eastern Canada, where the humidity is relatively high. However, winter ice storms occasionally occur in southern areas such as Texas. A major ice storm in early December 2006 moved across Missouri and Illinois and into the northeast. Heavy snow and ice collapsed roofs and downed power lines, knocking out electricity for hundreds of thousands of people. At least 16 people died and 43 were hospitalized by carbon monoxide poisoning because they were heating with unventilated combustion heaters. Another ice storm in mid-January 2007 left

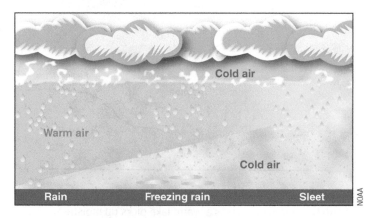

FIGURE 10-33 Ice Storm Development
Rain ahead of a front, moving north into cold Arctic air, falls through the cold air; it instantly forms ice as it hits the ground.

FIGURE 10-34 Ice Coats Everything
A classic ice storm in the northeast coats power lines.

ice coatings on everything from Texas to Maine. At least 39 died, including people in Texas, Oklahoma, Missouri, Iowa, New York, and Maine. In January 2009, an ice storm wreaked havoc in Arkansas and Kentucky (**Case in Point: A Massive Ice Storm in the Southern United States—Arkansas and Kentucky, 2009, p. 288**).

Ice on lakes in northern climates can create problems for beachfront structures in two main ways. First, ice covering a lake can crack as patches of ice shift; water filling the crack can freeze, prying the ice floes apart. Repetition of the process expands the area of ice covering the lake, forcing it up on shore, sometimes damaging structures. Second, patches of ice floating on a lake during spring thaw are often moved by frictional drag of the wind, driving them downwind to pile up onshore.

Blizzards

Winter winds can be fierce, especially on the Great Plains of the United States and the Prairie provinces of Canada, east of the Rockies. Blowing snow with winds greater than 56 km/h in the United States (35 mph) or 40 km/h in Canada can be classified as a **blizzard**. Such high winds often develop on the northwest side of a winter storm front, where the air pressure gradient is between a high to the northwest and a low to the southeast. In some cases, visibility can be very limited or even vanish, making travel extremely dangerous (**FIGURE 10-35**). Snow packed roads become even more slippery and visibility may be very limited. High winds can blow already fallen snow in a ground blizzard. Severe conditions may develop into a *whiteout* with blowing dry powdery snow, fog, or ice droplets—a condition where objects and the horizon disappear and people lose depth perception and their sense of direction. People can even lose their sense of up and down, so even though they are standing on level ground, they lose balance and fall. Extreme cold with blizzards and whiteout conditions can be very dangerous, often resulting in frostbite, hypothermia, and death to those outside. In rural areas, where communication may be limited, people caught in rapid temperature drops and poor visibility have lost their way and frozen to death on ranches or even trying to run from a rural school to a nearby house less than 100 m away.

Heavy snowfalls often create problems. In one example, snow falling in the Midwest on January 20, 2005, intensified, moved east to the Great Lakes region the next day, and continued into southern New England on January 22. The low-moisture content of the snow built to depths of 30 to 38 cm in Philadelphia and New York City and more than 90 cm in Boston and southeastern Massachusetts. Travel was severely curtailed and most schools in the northeast closed for two days.

FIGURE 10-35 Whiteout

A blowing ground blizzard or whiteout can eliminate ground-level visibility. Norton, Kansas, January 14, 2007.

Thunderstorms

Thunderstorms, as measured by the density of lightning strikes, are most common in latitudes near the equator, such as central Africa and the rain forests of Brazil (**FIGURE 10-36**). For its latitude, the United States has an unusually large number of lightning strikes and severe thunderstorms. These storms are most common from Florida and the southeastern United States through the Midwest because of the abundant moisture in the atmosphere that flows north from the warm waters of the Gulf of Mexico.

Thunderstorms form as unstable, warm, and moist air rapidly rises into colder air and condenses. As water vapor condenses, it releases heat. Because warm air is less dense than cold air, this added heat causes the rising air to continue to rise in an updraft. This eventually causes a region of falling rain in an outflow area of the storm when water droplets get large enough through collisions. If updrafts push air high enough into the atmosphere, the water droplets freeze in the tops of **cumulonimbus clouds**; these are the tall clouds that rise to high altitudes and spread to form wide, flat, anvil-shaped tops (**FIGURE 10-37**). This is where lightning and thunder commonly develop.

Cold air pushing under warm, moist air along a cold front is a common triggering mechanism for these storm systems, as the warm humid air is forced to rise rapidly. Isolated areas of rising humid air from localized afternoon heating, or warm and moist air rising against a mountain front, or pushing over cold air near the ground can have similar effects. Individual thunderstorms average 24 km across, but coherent lines of thunderstorm systems can travel for more than 1000 km. Lines of thunderstorms commonly appear in a northeast-trending belt from Texas to the Ohio River valley. Cold fronts from the northern plains states interact with warm moist air from the Gulf of Mexico along that line, so the front and its line of storms move east.

Thunderstorms produce several different hazards, including lightning, strong winds, and hail. Strong winds can down trees, power lines, and buildings. Severe thunderstorms cause numerous wildfires, and sometimes large damaging hail and tornadoes. Of the more than 100,000 thunderstorms in the United States each year, the National Weather Service classifies about 10,000 as severe. Those severe storms spawn up to 1000 tornadoes each year. The weather service classifies a storm as severe if its winds exceed reach 93 km/h, it spawns a tornado, or it drops hail larger than 2.54 cm (1 in) in diameter. Flash flooding from thunderstorms causes more than 140 fatalities per year (described in Chapter 13).

Lightning

Lightning results from a strong separation of an electric charge that builds up between the top and bottom of cumulonimbus clouds. Atmospheric scientists commonly believe that this charge separation increases when updrafts carry water droplets and ice particles toward the top of cumulonimbus clouds where they collide with downward-moving

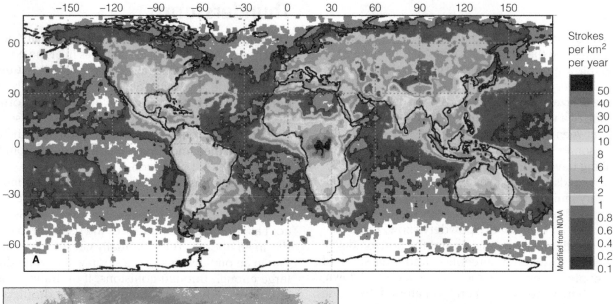

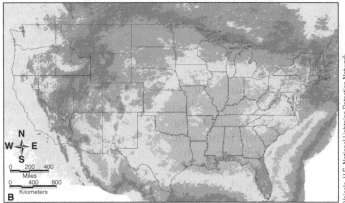

FIGURE 10-36 **Thunderstorm Hazard Areas**

A. This worldwide map shows the average density of annual lightning flashes per square kilometer.

B. North American lightning distribution from 1996 to 2005. Flash numbers per year: red = 8 to 16; orange = 4 to 8; yellow = 2 to 4; dark green = 0.5 to 1.

FIGURE 10-37 **Cumulonimbus Cloud**

This type of classic cumulonimbus anvil buildup over Utah on August 26, 2004, is where thunderstorms and lightning typically develop.

ice particles or hail. The smaller, upward-moving particles tend to acquire a positive charge, while the larger, downward-moving particles acquire a negative charge. Thus, the top of the cloud tends to carry a strong positive charge, while the lower part carries a strong negative charge (**FIGURE 10-38**, p. 273). This process is similar—though much larger—to the effects of static electricity a person can generate by dragging one's feet on carpet during dry weather—this action produces a charge that is discharged as a spark when one gets near a conductive object.

The strong negative charges near the bottom of these clouds attract positive charges toward the ground surface under the charged clouds, especially to tall objects such as buildings, trees, and radio towers. Thus, there is an enormous electrical separation, or potential, between different parts of the cloud and between the cloud and ground. This can amount to millions of volts; eventually, the electrical resistance in the air cannot keep these opposite charges apart, and the positive and negative regions join with an electrical lightning strike.

Because negative and positive charges attract one another, a negative electrical charge may jump to the positive-charged cloud top or to the positive-charged ground. Air is a poor conductor of electricity, but if the opposite charges are strong enough, they will eventually connect. Cloud-to-ground lightning is generated when charged ions in a thundercloud discharge to the best conducting location on

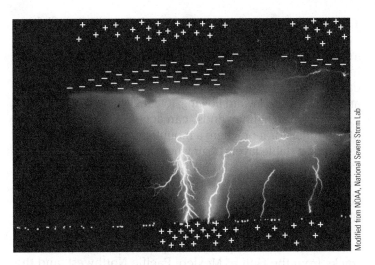

FIGURE 10-38 Opposite Charges Cause Lightning

In a thunderstorm, the lighter, positive-charged rain droplets and ice particles rise to the top of a cloud while the heavier, negative-charged particles sink to the cloud's base. The ground has a positive charge. In a lightning strike, the negative charge in the cloud base jumps to join the positive charge on the ground.

FIGURE 10-39 Forked Leaders

Lightning often follows a very irregular path between clouds and the ground, as shown by these lightning strokes in Tucson, Arizona.

FIGURE 10-40 Fulgurite

A fulgurite about 35 cm long, fused in sand at Queen Creek, near Phoenix, AZ. Top was at right.

the ground—locations such as tall trees, power poles, and people standing, for example in open areas, fields, or golf courses. Negatively charged **step leaders** angle their way toward the ground as the charge separation becomes large enough to pull electrons from atoms. When this occurs, a conductive path is created that in turn causes a chain reaction of downward-moving electrons. These leaders fork as they find different paths toward the ground (**FIGURE 10-39**). Depending on the location and strength of negative and positive charges in a cloud and in the ground, lightning may strike at a considerable distance from the storm cloud.

If the hairs on your head are ever pulled upward by what feels like a static charge during a thunderstorm, you are at high risk of being struck by lightning. When one of the pairs of leaders connects, a massive negative charge follows the conductive path of the leader stroke from the cloud to the ground. This is followed by a bright *return stroke* moving back upward to the cloud along the one established connection between the cloud and ground. The enormous power of the lightning stroke instantly heats the air in the surrounding channel to extreme temperatures approximating 28,000°C (50,000°F). The accompanying expansion of the air at supersonic speed causes the boom that we hear as **thunder**.

The lightning channel itself appears to be only 2 or 3 cm in diameter, based on holes produced in fiberglass screens and long narrow tubes (called "fulgurites") fused in loose sand (**FIGURE 10-40**). In fewer cases, lightning will strike from the ground to the base of a cloud; this can be recognized as an upwardly forking lightning stroke rather than the more common downward forks observed in cloud-to-ground strokes. Lightning also strikes from cloud to cloud to equalize its charges, although there is little hazard associated with such strokes.

Major insurance companies reported in 2007 that lightning-associated claims in the previous six years rose 77%. Costs escalated because of the growing number of electronic devices in people's homes. The voltage surge in a home's wiring from a lightning strike can destroy personal computers, HD TV sets, DVD players, game consoles, other devices, and heating systems. U.S. lightning strikes cause about $144 million in direct property damage and 6100 house fires every year. Lightning-protection systems for a whole home or a surge suppressor installed at the main circuit-breaker panel can reduce the excess voltage. Individual surge suppressors that you plug into an electrical outlet should have a Suppressed Voltage Rating of 330 or less.

Downbursts

As well as lightning, thunderstorms can also bring destructive winds. Small areas of rapidly descending air, called **down-bursts** or *microbursts*, can develop in strong thunderstorms (**FIGURE 10-41A**). Downbursts associated with thunderstorms can be wet or dry. A falling column of air driven by heavy rainfall reaches the ground and spreads out in winds of up to 240 km/h that move in straight lines radially outward from the center of the downburst. Downburst winds as fast as 200 km/h and microburst winds (a downburst with less than a 4-km radius) are caused by a descending mass of cold air, sometimes accompanied by rain.

These severe downdraft winds pose major threats to aircraft during takeoffs and landings because they cause *wind shear*, which results in a plane losing the lift from its wings and plummeting toward the ground (**FIGURE 10-41B**). A plane flying into the vertical wind shear at the edge of a downburst experiences an increase in headwind and a rapid increase in lift on its wings. If the pilots are not aware of wind shear as the cause, a common reaction is to decrease engine power and airspeed. Then beyond the plane of shear, the plane experiences downdraft and reduced lift. During take-off or landing, when the plane is close to the ground and its airspeed is low, there may not be time to increase speed enough to escape the downdraft, and the plane may crash.

Fatal microburst aircraft crashes include a July 9, 1982, Pan American flight in New Orleans; an August 2, 1985, Delta flight in Dallas; a July 2, 1994, U.S. Air flight in Charlotte, North Carolina; and a September 16, 2007, One-Two-Go flight at Phuket, Thailand. Once Dr. Tetsuya (Ted) Fujita proved this phenomenon and circulated the information to pilots and weather professionals, the likelihood of downbursts causing airplane crashes has been greatly reduced—though not eliminated.

When these descending air masses hit the ground, they cause damage people sometimes mistake as having been caused by a tornado. On close examination, downburst damage will show evidence of straight-line winds: trees and other objects will lie in straight lines pointing away from the area where a downburst hit the ground (**FIGURE 10-42**). This differs from the rotational damage observed after tornadoes, where debris lies at many angles due to the inward flowing, rotating winds.

Other "straight-line winds" can accompany the gust front of a thunderstorm or where a mountain range forces winds to rise against stable layers of air where they must squeeze into a thinner zone and speed up. Mt. Washington, in the Presidential Range of northern New Hampshire, is such a place. From November to April, two-thirds of the days are likely to have hurricane-force winds. The record there was 372 km/h, higher than a hurricane and equal to the wind gusts in a strong tornado. Such winds are partly in response to converging storm tracks from the Gulf of Mexico, Pacific Northwest, and the

FIGURE 10-42 Downburst Damage
Downburst winds in Bloomer, Wisconsin, blew these trees down on July 30, 1977.

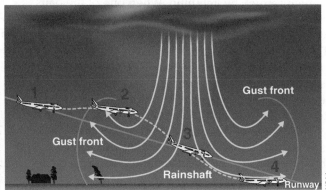

FIGURE 10-41 Downbursts

A. An invisible, dry microburst kicks up dust as the mass of descending air strikes the ground and radiates outward (downward air trajectories marked with arrows).

B. An aircraft flying into a microburst as it approaches an airport may reduce speed in response to sudden lift, possibly leading to a crash.

Atlantic Ocean. Trees blown down by such winds present a significant hazard; 11 or 12 people are killed each year in the United States. Straight-line wind gusts associated with thunderstorms in the desert southwest are frequently more than 64 km/h (40 mph) and can be more than 160 km/h (100 mph).

Hail

Hail is another weather phenomenon associated with thunderstorms. Hail causes $2.9 billion in annual damages to cars, roofs, crops, and livestock (**FIGURE 10-43**). **Hailstones** grow when warm and humid air in a thunderstorm rises rapidly into the upper atmosphere and freezes. Tiny ice crystals waft up and down in the strong updrafts, collecting more

FIGURE 10-43 Hail

A. A violent storm over Socorro, New Mexico, on October 5, 2004, unleashed hailstones, many larger than golf balls and some 7 cm in diameter. **B.** Most cars caught out in the open suffered severe denting and broken windows. In some cases, hailstones went right through car roofs and fenders.

and more ice until they are heavy enough to overcome the updrafts and fall to the ground. The largest hailstones can be larger than a baseball and are produced in the most violent storms. Hailstorms are most frequent in late spring and early summer, especially April to July, when the jet stream migrates northward across the Great Plains. The extreme temperature drop from the ground surface up into the jet stream promotes the strong updraft winds. Hailstorms are most common in the plains of northern Colorado and southeastern Wyoming but rare in coastal areas. Hail suppression using supercooled water containing silver iodide nuclei has been used successfully to reduce crop damage; the process nucleates more droplets and reduces the size of hailstones. Effectiveness of cloud seeding for increasing rainfall is still debated but new rain apparently depends on sufficient atmospheric moisture.

Safety during Thunderstorms

Thunderstorms can be deadly. Among weather-related events, only heat and floods cause more deaths. In 1940, lightning strikes in the United States killed about 400 people per year. The number dropped continuously to an average of 44 people per year from 1997 to 2006, likely due in part to increased awareness of the hazard and forecasts. Twice as many deaths occurred in Florida than in any other state. The maximum number of deaths from lightning strikes occurs at around 4 p.m., with a significant increase on Sunday—presumably because more people, such as golfers, are outside, engaging in leisure activities.

When someone is fatally struck by lightning, the immediate cause of death is heart attack, with deep burns at lightning entry and exit points. Seventy percent of lightning strike survivors have residual effects, including damage to nerves, the brain, vision, and hearing.

To reduce your risk of being struck, gauge your distance from the lightning. Lightning is visible before you hear the clap of thunder. The speed of light is almost instantaneous, whereas sound takes roughly three seconds to travel a kilometer. Thus, the time between seeing the lightning and hearing the thunder is the time it takes for the sound to get to you. Every three seconds means a kilometer between you and the lightning; if the time difference is 12 seconds, then the lightning is about 4 km away. Even if there is blue sky above and the threatening cloud is several kilometers distant, lightning can arc to the side, well away from its dark origin cloud. The National Weather Service recommends that you take cover if you hear thunder within 30 seconds of a lightning strike and stay in a safe place until you do not see lightning flash for at least 30 minutes—the "30-30 rule." Keep in mind that lightning can strike as much as 16 km (10 mi) from any rainfall.

Danger from lightning strikes can be minimized by observing the following precautions during a thunderstorm:

■ Take cover in an enclosed building; its metal plumbing and wiring will conduct the electrical charge around you to the ground. A shed or picnic shelter is not safe.

- Do not touch anything that is plugged in, including video games. Do not use a phone with a cord; cordless phones and cell phones are okay.

- Water conducts electricity. Do not take a shower or bath or wash dishes. Stay away from open water. On July 27, 2014, for example, a single lightning bolt from an isolated thunderstorm killed a 20-year-old man in the surf on Venice Beach near Los Angeles. Seven others were injured, one critically. A scuba diver was struck unconscious but revived.

- Stay away from open fields. Lightning can travel along the ground for at least 20 m and kill or burn you severely.

- Do not take refuge in a cave. People on high peaks during thunderstorms sometimes take refuge in a cave or under a deep rock overhang. Lightning can surround the inside of a cave, placing an occupant in greater danger than they would have been outside on an open slope.

- Stay away from trees, power poles, or other tall objects. Staying under low bushes well away from the trees is a better plan. About one-third of lightning deaths are people standing under or near tall trees (**FIGURE 10-44**).

- Stay away from metal objects, such as metal fences, golf carts, umbrellas, farm machinery, and the outsides of cars and trucks (**FIGURE 10-45**).

- If you are driving, pull over and stop in a safe spot. Stay inside a car with the windows rolled up and do not touch

FIGURE 10-45 Deadly Lightning
Reality can be gruesome. These cows were probably spooked by thunder and ran against the barbed wire fence, where they were electrocuted by a later lightning strike. Note that they were at the base of a hill but out in the open.

any metal, including the steering wheel, gearshift, or radio. The safety of a car is in the metal shield around you, not in any insulation from the tires. The rubber tires of a motorcycle or bicycle do not insulate you from the ground.

There are signs that you may be in imminent danger of being struck by lightning. The sky may turn a strange shade of olive gray. Storm clouds several kilometers away may indicate a lightning strike about to happen near you. People report that shortly before a strike, the surrounding air begins to hiss, buzz, and sometimes crackle, like you sometimes hear when standing under high-tension lines; strange blue sparks may leap from rocks and people's heads; hair may stand on end. Even if it strikes a few meters away, the charge through the ground may kill you, or at least do serious damage. People report a deafening explosion as the sky turns blinding white and a blast of energy exploding through their chest as they pass out.

If you are trapped in the open, and your skin tingles, or your hair stands on end, you are in immediate danger of being struck. Immediately crouch low, on the balls of your feet with your heels tightly together to minimize contact with electrified ground; much of the charge from the ground that enters your feet will pass from one heel to the other and back into the ground without passing through your body. Do not lie down because that increases your contact with the ground. Remember that the ground conducts electricity.

Most lightning-strike victims can survive with medical help such as the application of CPR. A lightning-strike victim does not carry an electrical charge or endanger others. However, rescuers should not put themselves in danger of another strike. Interestingly, men are more likely to be struck by lightning than women. Psychologists believe that for men the risk is outweighed by their desire to show others that they are not afraid.

FIGURE 10-44 Lightning Hazard
Lightning strikes a tree.

Tornadoes

Thunderstorms and other storm systems sometimes generate **tornadoes**—narrow funnels of rapidly rotating intense wind. Tornadoes are nature's most violent storms and the most significant natural hazard in much of the Midwestern United States. They often form in the right-forward quadrant of hurricanes, in areas where the wind shear is most significant. Even weak hurricanes spawn tornadoes, sometimes dozens of them.

The United States has an unusually high number of large and damaging tornadoes relative to the rest of the world, over 1000 per year on average. Canada is second, with only about 100 per year, though some are strong; a strong tornado touched down west of Winnipeg, Manitoba, in 2007. Europe experiences a moderate number.

The storms that lead to tornadoes are created through the collision of warm, humid air moving north from the Gulf of Mexico with cold air moving south from Canada. Because there is no major east–west mountain range to keep these air masses apart, they collide across the Southeastern and Midwestern United States. These collisions of contrasting air masses cause intense thunderstorms that sometimes turn into deadly tornadoes.

The average number of tornadoes is highest in Texas, followed by Oklahoma, Kansas, Florida, and Nebraska. **Tornado Alley**, covering parts of Texas, Oklahoma, Arkansas, Missouri, and Kansas, marks the belt where cold air from the north collides frequently in the spring with warm, humid air from the Gulf of Mexico, forming intense thunderstorms and tornadoes (**FIGURE 10-46**). Tornadoes are rare in the western and northeastern states. However, they are not unknown. On July 26, 2010, a tornado killed two when it leveled a farmhouse in northeastern Montana. On October 6, 2010, a tornado wrecked houses in Bellemont, just west of Flagstaff, Arizona. Then on October 23, 2014, a tornado damaged property and downed power lines in Longview downstream from Portland, Oregon, in western Washington state. An individual **tornado outbreak**—that is, a series of tornadoes spawned by a group of storms—has killed as many as several hundred people and covered as many as thirteen states (Table 10-2).

Another severe tornado outbreak occurred on April 27–30, 2014, when an EF4 tornado killed 16 in the small town of Vilonia, Arkansas (**FIGURE 10-47**), which also lost 4 people to an EF2 in 2011. A series of 84 long-track tornadoes battered a broad area, especially through Arkansas, Mississippi, Alabama, and Tennessee, killing a total of 35 people over the four day outbreak. Total damages reached more than $1 billion. Individual tornadoes changed in strength as they churned northeast.

In April 2011, a violent outbreak of tornadoes tracking from Oklahoma through Arkansas, Mississippi, Alabama, and to North Carolina killed 322 people. This **superoutbreak** caused over $11 billion in damage, making it the costliest tornado outbreak on record (**FIGURE 10-48** as indicated in the legend).

The deadliest year for tornadoes since 1950 was 2011. There are several likely reasons including growth of population

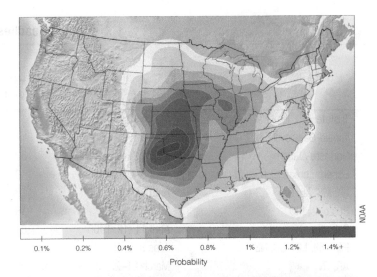

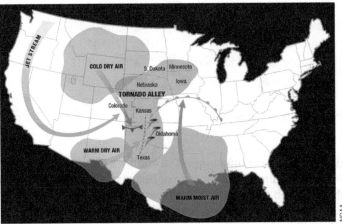

FIGURE 10-46 Tornado Risk in North America

A. The areas of greatest tornado risk include much of the eastern half of the United States (for F2 and greater; data from 1921 to 1995). The scale indicates the number of significant tornado days per year. **B.** Tornado Alley is influenced by warm, moist air from the Gulf of Mexico that collides with warm, dry air to the west and cold, dry air to the northwest.

and urban sprawl that provides a larger target for tornadoes. Although climate change may be a factor, this has not yet been confirmed. A strong La Niña weather pattern that involves a cooling of the tropical Pacific Ocean and affects weather worldwide was still in effect. Tornadoes are bred by a strong warm, moist air flow from the Gulf of Mexico, clashing with cooler, dry air from the north and west. Most vulnerable are people living in lightly built mobile homes and homes without basements because those are not well attached to the ground and provide no underground shelter. Least likely to be killed are people who are better educated and have planned where to go in advance. It may seem likely that longer warning time would be advantageous but studies show that 15 to 20 minutes is most effective. Longer times may suggest to some people that there is no rush or that after a while it must have been a false alarm. Also

Table 10-2 Deadliest U.S. Tornado Outbreaks on Record*

NAME OR LOCATION	DATE	NUMBER OF TORNADOES	NUMBER OF STATES AFFECTED	DEATHS	ESTIMATED DAMAGE IN MILLIONS (2015 $)
Tristate: MO, IL, IN	March 18, 1925	7	6	695	223
Tupelo-Gainesville (MS, GA)	April 5–6, 1936	17	5	419	275
Northern Alabama	March 21–22, 1932	33	7	334	56
Superoutbreak, E. US and Ontario	April 3–4, 1974	148	13	330	2112
LA, MS, AL, GA	April 24–25, 1908	18	5	310	60
St. Louis, MO	May 27, 1896	18	3	255	432
Palm Sunday	April 11–12, 1965	51	6	256	789
Flint: MI; MA, OH, NE	June 8–9, 1953	10	4	247	629
AR, TN	March 21–22, 1952	28	4	204	64
2011 Superoutbreak (E. US)	April 25–28, 2011	353	21	322	1144
Natchez MS	May 6, 1840	1	2	317	unknown
Joplin, MO	May 22, 2011	1	1	161	2912
2012 Outbreak (OH–KY)	March 2–3, 2012	65	4	40	1560
2014 Outbreak (OK, AR, MO, MS, AB, TN)	April 27–30, 2014	80	10	35	>$1 billion

From FEMA, NOAA, and other sources.

dangerous are "rain-wrapped" tornadoes that are not visible from some directions because of heavy rain. For comparison, the deadliest tornado on record was on March 18, 1925, before weather radar and more accurate forecasting. That event killed 695 people along a 291-mile path through Missouri, Illinois, and Indiana. Overall insured damages reached about $11 billion.

Tornado season varies, depending on location. The number of tornadoes in Alabama reaches a maximum in late April, with a secondary peak in November (**FIGURE 10-49**). An unusual tornado outbreak on February 5 and 6, 2008 killed at least 55 people in southeastern states. On December 31, 2010, at least six people were killed by tornadoes in Arkansas and Missouri. Farther north, the maximum occurs in May, and in Minnesota it is in June. These are the periods that people should be particularly vigilant for tornadoes. At these northern latitudes, tornadoes are virtually absent from November to February.

Courtesy of Southern Region Headquarters/National Weather Service/NOAA

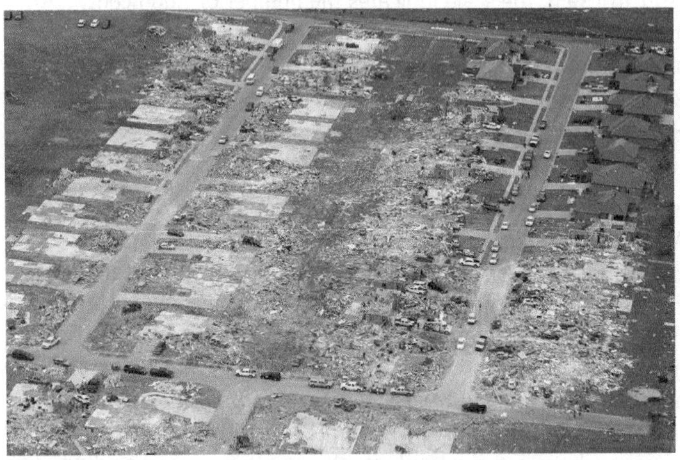

FIGURE 10-47 April 2014 Arkansas Tornado Outbreak

On April 27, 2014, an EF4 tornado selectively leveled many houses in Vilonia, Arkansas, down to their slabs. What are the chances of surviving here without a tornado shelter? Note sharp destruction cutoff.

NOAA

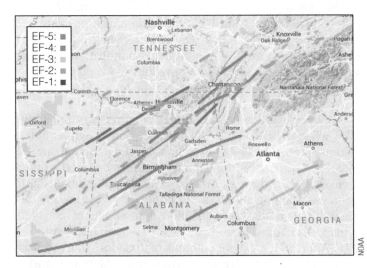

FIGURE 10-48 Superoutbreak of 2011

A total of 292 tornadoes struck 16 states on April 27, 2011, primarily in those shown. Line colors indicate tornado intensities as indicated in the legend.

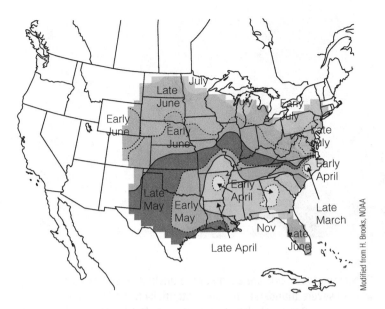

FIGURE 10-49 Tornado Season

Peak tornado season is April to May in most of the southeast, moving to June and even July farther north. Tornadoes tend to begin a month earlier and end a month later in most areas.

Even late-season tornadoes can be deadly, especially for those in mobile homes. Just after 6:30 a.m. on November 16, 2006, and without warning, a thunderstorm spawned a tornado that killed eight people in mobile homes west of Wilmington, North Carolina. On February 1, 2007, tornadoes killed 19 northwest of Orlando, Florida.

Most, though not all, tornadoes track toward the northeast. Storm chasers, individuals who are trained to gather storm data at close hand, know to approach a tornado from the south to southwest directions so they will not be in its path. They also know it is safer to chase them on the flat plains rather than along the Gulf Coast, where the lower cloud base can hide the funnel from view.

Tornado Development

Tornadoes generally form when there is a shear in wind directions, such as surface winds approaching from the southeast with winds from the west higher in the atmosphere. Such a shear can create a roll of horizontal currents in a thunderstorm as warm and humid air rises over advancing cold air (**FIGURE 10-50**). These currents, rolling on a horizontal axis, are dragged into a vertical rotation axis by warm rising air at the surface producing an updraft in the thunderstorm to form a *rotation cell* up to 10 km wide. This cell sags below the cloud base to form a distinctive slowly rotating **wall cloud**, an ominous sight that is the most obvious danger sign for the imminent formation of a tornado (**FIGURE 10-51A**).**Mammatus clouds** can be another potential danger sign, where groups of rounded pouches sag down from the cloud base (**FIGURE 10-51B**).

A *supercell*, a giant thundercloud characterized by a rapidly rising and rotating mass of warm air that spreads out at its top as a broad mushroom, can produce severe thunderstorms, torrential rain, and tornadoes (**FIGURE 10-52**). A slowly rotating, steep-sided wall cloud may descend from the nearly horizontal base of a main cloud. A smaller and more rapidly rotating funnel cloud, or *mesocyclone*, may form within the wall cloud or, less commonly, adjacent to it. After the heaviest rain and hail has passed, rotation within the mesocyclone may accelerate and tighten, typically rotating counterclockwise. If a funnel cloud descends to touch the ground, it becomes a tornado. Supercells are not common but often remain in place for extended times and produce the most violent storms.

Tornadoes derive their energy from the latent heat released when water vapor in the atmosphere condenses to form

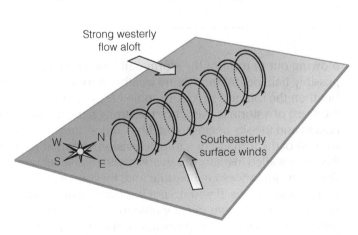

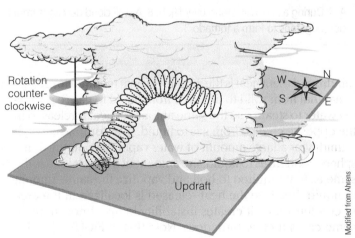

FIGURE 10-50 Wind Shear

A. Wind shear, with surface winds from the southeast and winds from the west aloft.

B. This slowly rotating vortex can be pulled up into a thunderstorm, which can result in a tornado.

FIGURE 10-51 Cloud Formations Indicating Tornado Conditions

A. A slowly rotating wall cloud descends from the base of the main cloud bank, an ominous sign for production of a tornado near Norman, Oklahoma, on June 19, 1980.

B. Mammatus clouds are a sign of the unstable weather that can lead to severe thunderstorms and potentially tornadoes. These formed over Benton Harbor, Michigan, on July 17, 2010.

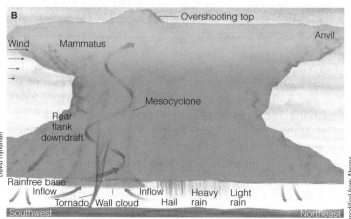

FIGURE 10-52 Supercells and Tornadoes

A. A During a supercell storm over Utah. **B.** A wall cloud descends from the main anvil cloud. The rotation of the mesocyclone at the center of the cloud tightens to form a tornado.

raindrops. Latent heat is the amount of heat required to change a material from solid to liquid or from liquid to gas (e.g., boiling water to steam), so the same amount of heat is released in the opposite change from gas to liquid. In a storm, when high humidity or a large amount of water vapor (gas) in the atmosphere condenses to liquid, the amount of heat released is the same as that required to boil and vaporize the same amount of liquid. Because the heat released is localized in the area of condensation, it creates instability in the atmosphere, in some cases fueling tornadoes. Note that in FIGURE 10-51A, the storm is moving to the right; the rainstorm passes over an area on the ground before the arrival of the tornado.

Tornadoes generally form toward the trailing end of a severe thunderstorm; this can catch people off-guard. Someone in the path of a tornado may first experience wind

blowing out in front of the storm cell along with rain, then possibly hail, before the stormy weather appears to subside. But then the tornado strikes. In some cases, people feel that the worst of a storm is over once the strong rain and hail have passed and the sky begins to brighten, unless they have been warned of the tornado by radio, television, or tornado sirens that are installed in many urban areas with significant tornado risk. Some tornadoes are invisible until they strike the ground and pick up debris. If you do not happen to have a tornado siren in your area, you may be able to hear an approaching tornado as a hissing that turns into a strong roar that many have likened to that of a loud oncoming freight train.

Conditions are favorable for tornado development when two fronts collide in a strong low-pressure center (**FIGURE 10-53**). This can often be recognized as a

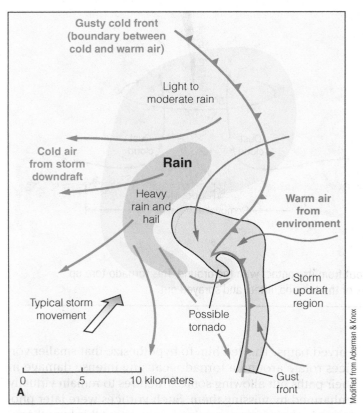

FIGURE 10-53 Colliding Cold and Warm Fronts
A. A common situation for tornado development is the collision zone between two fronts, commonly seen in a "hook echo" of a thunderstorm. A pair of curved arrows indicates horizontal rotation of wind in the lower atmosphere. **B.** Radar hook-echo image of the Lincoln County, Tennessee, tornado on April 28, 2014. The red ball at the bottom of the hook is a "debris ball" in the core of the tornado.

hook echo, or hook-shaped band of heavy rain, on weather radar. This is a sign that often causes weather experts to put storm spotters on alert to watch for tornadoes. Tornadoes in the Midwest typically develop as very warm, moist air from the Gulf of Mexico sweeps north against a cold front, against cold, dry air to the northwest (FIGURE 10-46B).

A strong tornado can form rather quickly, within a minute or so, and last for 10 minutes to more than an hour. A tornado rarely stays on the ground for more than 30 minutes, generally leaving a path less than 1 km wide and up to 30 km long. Typical tornado speeds across the ground range from 50 to 80 km/h, but internal winds can be as high as 572 km/h—the most intense winds on Earth. In a major research project, VORTEX2, on the Great Plains in May 2009, scientists were fortunate to be able to see up into the core of a tornado as it rolled over to nearly horizontal. The highest-velocity winds of the funnel spun around an open core reminiscent of the quiet eye of a hurricane.

Comparison of the winds of tornadoes with those of hurricanes (compare Figure 10-57, p. 283, with TABLE 16-1, p. 437) shows that the maximum wind velocities in tornadoes are twice those of hurricanes. Wind forces are proportional to the wind speed squared, so the forces exerted by the strongest tornado wind forces are four times those of the strongest hurricane winds. In many cases, much of the localized wind damage in hurricanes is caused by embedded tornadoes.

As a tornado matures, it becomes wider and more intense. White or clear when descending, tornadoes become dark when water vapor inside condenses in updrafts, which pull in ground debris (**FIGURE 10-54**). In its waning stages, a tornado then narrows, sometimes becoming ropelike, before finally breaking up and dissipating (**FIGURE 10-55**). At that waning stage, tightening of the funnel causes it to spin faster, so the tornado can still be extremely destructive.

Classification of Tornadoes

The severity of tornadoes is based on their internal wind speeds and the damage produced. In his research, Dr. Ted Fujita of the University of Chicago separated estimated tornado wind speeds into a six-point nonlinear scale from F0 to F5. F0 causes minimal damage and F5 blows away even strong frame homes. Early in 2007 the scale was updated as the **Enhanced Fujita (EF) scale**, based on detailed wind measurements and long-term records of damage. The new scale uses three-second wind-gust estimates at the site of damage and is considered more reliable than the old scale. The main difference between the scales is that the new EF scale shows that particular damages occur at much lower wind speeds.

In addition, Dr. Fujita compiled a damage chart and photographs corresponding to these wind speeds. Reference photographs of damage are distributed to National Weather Service offices to aid in evaluating storm intensities (**FIGURE 10-56**).

You can see from **FIGURE 10-57** that, as with other hazards, moderate tornadoes are more frequent than severe ones. Although EF5 tornadoes are the most severe and most destructive, they kill fewer people because they happen rarely. EF5 tornadoes are infrequent, but on May 4, 2007, one ripped through Greensburg in south-central Kansas and flattened 95% of the town. Although there was a 20-minute warning, winds reached 328 km/h and killed 10 people.

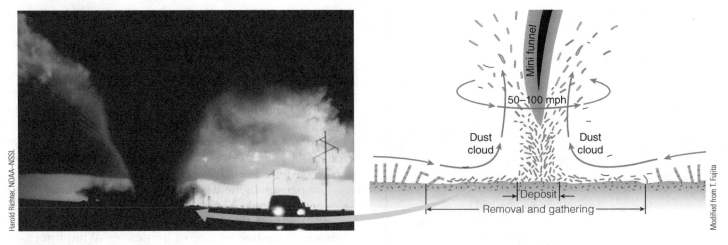

FIGURE 10-54 Tornado Picks Up Debris

A big tornado south of Dimmitt, Texas, on June 2, 1995, sprays debris out from its contact with the ground. This tornado tore up 100 m of the highway where it crossed. Debris is drawn into the vortex of the tornado, lifted, and sprayed out.

FIGURE 10-55 Dissipation of a Tornado

This thin, ropelike tornado was photographed at Cordell, Oklahoma, on May 22, 1981, just before it broke up and dissipated. A strong tornado has descended from the cloud base, picked up, and violently spewed out everything loose or breakable. As the cloud and head of the tornado moved to the left, its foot on the ground lagged behind. Finally the weakening tornado thinned to a narrow rope and lifted. Note that the foot of the tornado on the ground can sweep well away from its position at the cloud base.

Tornado Damages

In his research on tornadoes, Dr. Fujita examined damage patterns. He noticed that there were commonly swaths of severe damage adjacent to areas with only minor damage. He also examined damage patterns in urban areas and cornfields, where swaths of debris would be left in curved paths. This led him to hypothesize that smaller vortices rotate around a tornado, causing intense damage in their paths but allowing some structures to remain virtually unharmed by missing them. Such vortices were later photographed on many occasions, supporting this hypothesis. The most intense winds are within these embedded vortices, so the pattern of damage can vary greatly over short distances (**FIGURE 10-58**).

If a building is unlucky enough to be in the path of one of these destructive vortices, the resulting damage is dependent on the severity of the tornado as well as the building construction. Wind speeds and damage to be expected in buildings of differing strengths are shown in **Table 10-3**. Note that walls are likely to collapse in an EF3 tornado in even strongly built frame houses; and in an EF4, the house is likely to be blown down. Brick buildings perform better. In an EF5 tornado, even concrete walls are likely to collapse.

Damage to a home generally begins with the progressive loss of roofing material, followed by glass breakage from flying debris. Then a roof may lift off and garage doors may fail. Exterior walls collapse, then interior walls, beginning with upper floors. Small interior rooms, halls, and closets fail last. The ease of destruction depends on how well roofs are attached to walls and walls to floors. Metal "hurricane straps" connecting horizontal and vertical members make a big difference. As with many other circumstances, damage and destruction are controlled by the weakest link.

Most susceptible to damage and destruction are farm outbuildings, followed by mobile homes, apartment and condo buildings, and then family homes. Winds of 225 km/h will cause severe damage to a typical home or apartment building and endanger people; 260 km/h will effectively destroy such structures. Winds of 145 to 160 km/h will severely damage a typical mobile home and

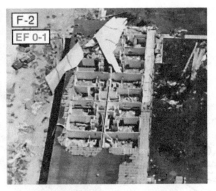

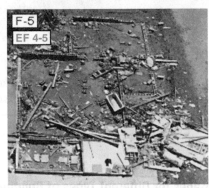

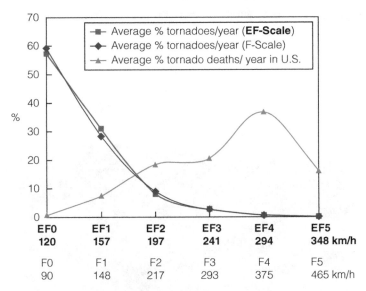

FIGURE 10-56 Fujita Scale of Tornado Damages

Dr. Ted Fujita developed the "F scale" for tornadoes by examining damage and evaluating the wind speeds that caused it. He used this set of photos as his standard for comparison. Conversion of his damage to the EF scale is approximated in the red numbers. EF5 would level virtually everything except for the strongest reinforced buildings.

FIGURE 10-57 Relationship Between Tornado Strength and Deaths

The average percentage of strong tornadoes is much less than for weaker tornadoes. However, the percentage of deaths from stronger tornadoes is much higher.

endanger the occupants; 169 km/h may destroy it, likely killing any occupants. Other than a reinforced tornado shelter, a basement with interior walls seems to be the safest place, but the variable directions of swirling winds leave no preferred location. Although the southwest corner of a basement was once thought to be safest, studies show that a house may be blown off its foundation and the house may collapse into that corner of the basement.

High-rise office buildings and hotels framed with structural steel are less susceptible to structural damage, but their exterior walls may be blown out. Downtown areas of cities with tall buildings are not immune to tornado damage. For example, instead of dissipating from turbulence among the buildings, an EF2 tornado moved into Atlanta, Georgia, at 9:38 p.m. on March 14, 2008, causing about $250 million in damages (**FIGURE 10-59**). Its strength may have been intensified by the heat-island effect of the huge city.

Although many people believe that the low pressure in a tornado vacuums up cows, cars, and people, and causes buildings to explode into the low-pressure funnel, this appears to be an exaggeration. Most experts believe that the extreme winds and flying debris cause almost all of the destruction. Photographs of debris spraying outward from the ground near the base of tornadoes suggest this (**FIGURE 10-60**). However, even large and heavy objects can be carried quite a distance. The Bossier City, Louisiana, tornado ripped six 700-lb I-beams from an elementary school and carried them from 60 to 370 m away.

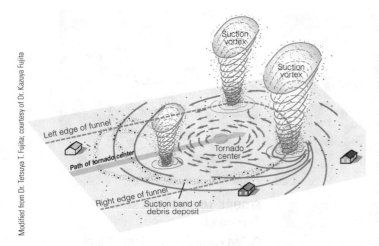

FIGURE 10-58 Damage Patterns From Tornadoes

A. Tornadoes are composed of multiple vortices that rotate around their center. **B.** The Lake County, Florida, tornado of February 3, 2007, shows how selective tornado damage can be. The hook of debris distribution, in this case clockwise, is clear in the photo.

Table 10-3	Expected Damages for Different Types of Buildings Dependent on Tornado Strength*					
TYPE OF BUILDING	EF0	EF1	EF2	EF3	EF4	EF5
Weak outbuilding	Walls collapse	Blown down	Blown away			
Strong outbuilding	Roof gone	Walls collapse	Blown down	Blown away		
Weak frame house	Minor damage	Roof gone	Walls collapse	Blown down	Blown away	
Strong frame house	Little damage	Minor damage	Roof gone	Walls collapse	Blown down	Blown away
Brick structure	No damage	Little damage	Minor damage	Roof gone	Walls collapse	Blown down
Concrete structure	No damage	No damage	Little damage	Minor damage	Roof gone	Walls collapse

*Simplified and modified from Fujita, 1992.

FIGURE 10-59 Urban Tornado Damage

Tornado damage in downtown Atlanta, March 14, 2008. Note the high-rise building on left.

FIGURE 10-60 Destructive Debris

Complete destruction from an EF4 tornado in Tuscaloosa, Alabama, April 27, 2011.

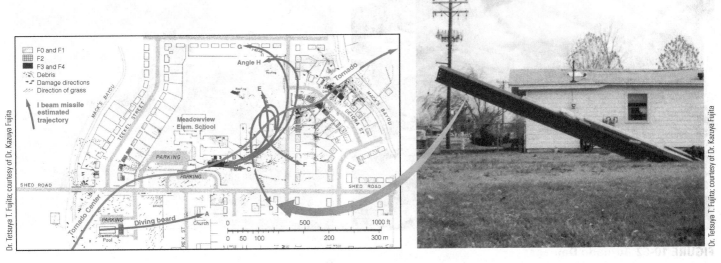

FIGURE 10-61 Tornadoes Move Heavy Objects

Six 700-pound I-beams were pulled from an elementary school in Bossier City, Louisiana, and carried by a tornado along these paths. Other objects, such as a diving board and a car, were also carried significant distances. The beam labeled "D" ended up stuck in the ground.

Another I-beam was carried to the south, where it lodged in the ground in someone's backyard (**FIGURE 10-61**). In another documented case, several empty school buses were carried up over a fence by a tornado before being slammed back to the ground.

Safety during Tornadoes

As with certain other hazards, tornadoes can be predicted with some success. Prediction and identification of tornadoes by NOAA's Storm Prediction Center in Norman, Oklahoma, uses Doppler radar, wind profilers, and automated surface observing systems. A **tornado watch** is issued when thunderstorms appear capable of producing tornadoes and telltale signs appear on radar. At this point, storm spotters often watch for severe storms. A **tornado warning** is issued when Doppler radar shows strong indications of vorticity, or rotation, or when a tornado is sighted. Warnings are broadcast on radio and television, and tornado sirens are activated, if they exist, in potential tornado paths. In contrast to the 12- to 24-hour warning often available before a hurricane's landfall, the warning time for a tornado at a particular location is 15 minutes or less in some eastern states, somewhat longer in the central states. The National Weather Service can now predict the onset of severe weather that may spawn tornadoes at least a week in advance. Then warning sirens in tornado-prone areas advise that tornadoes are imminent. However, many people disregard the warnings, probably because frequent tornadoes leave them desensitized to the danger. On May 10, 2010, for example, two people were killed and 58 injured after ignoring such warnings in and around Oklahoma City.

When a tornado warning is issued, people are advised to seek shelter underground or in specially constructed shelters whenever possible. If no such area is available, people should at least go to some interior space with strong walls and ceiling and stay away from windows. The main danger comes from flying debris. People have been saved by taking refuge in an interior closet, or even lying in a bathtub while holding a mattress or sofa cushions over them. Unfortunately, in some cases a strong tornado will completely demolish houses and everything in them (**FIGURE 10-62**).

Mobile homes are lightly built and easily ripped apart—certainly not a place to be in a tornado. Car or house windows—and even car doors—provide little protection from high-velocity flying debris, such as two-by-fours from disintegrating houses. Those in unsafe places are advised to evacuate to a strong building or storm shelter if they can get there before a storm arrives. If you cannot get to a safe building, FEMA recommends that you lie in a ditch or culvert and cover your head to provide some protection from flying debris.

What if you are in your car when you see a tornado or hear a nearby warning? Late on the afternoon of May 31, 2013, a northeast-trending line of storms west of Oklahoma City formed a supercell updraft. At 5:30 p.m., a big wall cloud descended below the main cloud base and a tornado dropped from a rain-wrapped storm a few kilometers farther south. A trio of professional researchers, including acclaimed Tim Samaras, began following the huge tornado from a few kilometers north, trying to keep at a safe distance. Twenty minutes later all three were dead, one of them still in the mutilated car. The tornado had become shrouded with rain and turned north across the researchers' path. Driving in an area during a tornado warning is perilous. The researchers

FIGURE 10-62 Tornado Damages

A. A steel-beam warehouse is not a safe place in a tornado; this warehouse in Roanoke, Illinois, was destroyed by a tornado in 2004.
B. This school bus was deposited on a roof by a tornado in Jackson, Mississippi, in January 10, 2008.

had always been very careful with tornadoes—never chasing when they were rain-wrapped, or at night, and always keeping track of an escape route. This killer tornado was the widest ever recorded, an immense 4.2 km (2.6 mi) and somewhere between an EF3 and an EF5. It tracked eastward with a forward motion reaching up to 88 km/h (55 mph).

It is as yet unclear whether vehicles provide more protection than mobile homes or lying in a ditch. Although cars are designed to protect their occupants in case of a crash, they can be rolled or thrown or penetrated by flying debris. If you are in open country, the tornado is far enough away, and you can tell what direction it is moving in, you may be able to drive to safety at right angles from the tornado's path. Recall that the path of a tornado is most commonly from southwest to northeast, so being north to east of a storm is commonly the greatest danger zone. However, some twisters, such as

the F5 Jarrell tornado, moved southwest. Remember also that the primary hazard associated with tornadoes is flying debris, and much to people's surprise, overpasses do not seem to reduce the winds associated with a tornado. Do not get out of your car under an overpass and think you are safe. In fact, an overpass can act like a wind tunnel that focuses the winds and you are higher above ground where winds are faster. Most overpasses lack hanging girders to hold onto in extreme winds. Once a few people park under an overpass, this can cause the additional problem of a traffic jam, thereby endangering others.

NOAA's weather radio and television network provides severe weather warnings. Typically, these warnings can furnish up to 10 minutes of lead time before a tornado's arrival.

Also see the Thunderstorm and Tornado Survival Guide on page 287.

THUNDERSTORM AND TORNADO SURVIVAL GUIDE

Be aware that these lists don't cover everything. Use your own common sense.

U.S. High Hazard Area	■ Tornado Alley is an area of the central to eastern United States, that is especially prone to tornadoes, including Texas, Oklahoma, Kansas, Nebraska, Iowa, and Illinois.
Preparation	■ Go to the lowest level well inside your home protected from windows, doors, and outside walls, preferably in a basement or storm shelter. If nothing else, lying in a bathtub under a mattress may help. Avoid staying in a mobile home; take shelter underground or in a strong building.
Warning Signs	■ Watch for mammatus clouds, supercell storms, especially with a slowly rotating wall cloud descending below the cloud base. Pay attention to NOAA warnings and weather radar indicating heavy rain and its direction of movement.
During the Event	■ Most supercell storms move northeastward so keep special watch to the south and west. If you spot a tornado while in your car, if possible drive perpendicular to the direction of storm movement to get out of its path. Don't park under an overpass. If your location is at all vulnerable, protect your head with a bicycle or motorcycle helmet. ■ If nothing else, get out of your vehicle and lie in a ditch below the level of flying debris.
Aftermath	■ Help others while not endangering yourself.

Cases in Point*

A Massive Ice Storm in the Southern United States
Arkansas and Kentucky, 2009 ▶

On January 26, 2009, humid air ahead of a warm weather front moving north from the Gulf of Mexico encountered cold air and light snow surrounding a large arctic high centered on Iowa. Ahead of the front, in northern Texas and Oklahoma, the moist air overriding the cold to the north turned to freezing rain and snow as it fell. In northwestern Arkansas and Kentucky, sleet, freezing rain, and snow around midnight turned to heavy freezing rain by 6 a.m. on the 27th. Freezing rain, thunder, and the cracking of tree limbs continued all day and into the evening. Many power lines can withstand only about 1.3 cm of ice buildup. Heavy ice coating up to 5 cm thick loaded, bent, and snapped tree branches, weighed down power lines, and coated roads. By evening, large tree limbs and whole trees snapped, falling on homes and power lines. Power failed locally in mid-afternoon, then totally by 8 p.m.

Approximately 1.3 million homes and businesses lost power throughout the storm region that extended northeast through Ohio and beyond. Emergency shelters were set up in most towns, though with power out at some radio stations as well, it was difficult to spread the word that shelters were available. Loss of power also shut down pumping stations and therefore water supplies. If people had running water they kept a trickle running all the time to keep pipes from freezing. FEMA provided more than 300 high-capacity electric-power generators for emergency operations and other critical facilities. Convoys of utility trucks from around the region arrived to help restore service. Five days later, thousands of Kentucky National Guard troops were still clearing branches, trees, and power lines from roads leading to remote communities and still-stranded residents. They went door-to-door to hand out food and bottled water to people still without power. A week after the storm, 260,000 remained without power.

Forty-two deaths were attributed to the storm; most were from hypothermia, traffic accidents, or carbon monoxide poisoning from inside use of charcoal grills or improperly used space heaters and generators. One police officer died when an ice-laden branch fell on him.

CRITICAL THINKING QUESTIONS

1. What weather environment led to the Arkansas–Kentucky ice storm?
2. Without electric power (which also ignites many gas furnaces and hot-water heaters) what should homeowners do to keep themselves warm and to keep their water pipes from freezing?
3. What actions by residents are likely to be particularly hazardous during a major ice storm?

■ *January 2009 ice storm in Millport, Indiana*

NWS/NOAA

■ *January 2009 ice storm in Paducah, Kentucky.*

Courtesy of Beau Dodson/Angela Compton/National Weather Service

---

*Visit the book's Website for these additional Cases in Point: Twister Demolishes Town—Greensburg, Kansas, 2007; Tornado Safety—Jarnell Tornado, Texas, 1997. *Tornado Superoutbreak*—Southeastern United States, 2011.

Extreme Drought
Texas and Adjacent States, 2010–11 ▶

Drought in Texas is not unusual; locals often joke that their weather cycles from drought to flood and back. October 2010 through August 2011 was the worst single-year drought in more than 115 years of records. The drought developed in the summer of 2010 as strong westerly winds blew across the southern plains, in contrast to the normal easterly trade winds—the pattern of a very strong La Niña in the Pacific Ocean (see FIGURE 10-16B).

It was not only drier but hotter. Wichita Falls, Texas, near the Oklahoma border, had average temperatures from June through August of 33.3°C (91.9°F) and 98 days of at least 37.7°C (100°F); Austin saw 90 days with temperatures above 100°F. For the whole of the United States it was the hottest June through August on record with an average temperature of 30.4°C (86.8°F), hotter than the average for Oklahoma in the 1934 Dust Bowl. In addition to Texas, large parts of Oklahoma, New Mexico, and Arizona were subjected to severe drought.

Surface water supplies were depleted from evaporation and heavier than normal municipal and agricultural use. Lakes and sections of major rivers dried up, such as portions of Pedernales River and Sandy Creek near Austin. By August 1, Lake Travis, from which Austin obtains much of its water, was down 8 m (26 ft) from its monthly average. Boat ramps and docks were stranded above lake level. Water suppliers instituted mandatory water restrictions. The Texas Commission on Environmental Quality suspended rights to divert water from Brazos River Basin on June 27, 2011, for those with rights dating after 1960. Groundwater levels declined as well.

Two million acres of cotton fields were abandoned. Extremely dry soils led to supplemental feeding of livestock, and ultimately to many ranchers selling off their cattle herds because there was not enough grass to feed them. By September crop and livestock losses reached $5.2 billion. Heavy use of air conditioning and stress on Texas's electrical supply forced emergency conservation measures. Agencies asked large industrial and commercial users to voluntarily go without power temporarily and drew additional power from Mexico at considerable expense. In the Dallas–Fort Worth area, 23 people died from the heat. Native trees died all around the central Texas Hill Country. As vegetation dried out, and in spite of widespread bans on burning, wildfires ignited, burning 3.8 million acres and 2742 homes by September. The worst of them, the Bastrop County fire, started on September 4, apparently by sparks from an electrical power line when 30 mph winds toppled trees onto the lines.

Karen Ward

■ *Relentless drought in Texas led to severe declines in many river and reservoir levels. Boathouses rest high and dry along the Pedernales River in August 2009 and again in August 2011.*

CRITICAL THINKING QUESTIONS

1. Because Texas appears to have recurrent droughts every few years and is heavily dependent on water from dwindling deep aquifers, what should be done about its future water dependence?
2. If severe Texas droughts can develop during strong La Niña events, how might the state mitigate associated natural hazards?

Lack of Winter Rain and Mountain Snow Leads to Severe Drought
California, 2012–15 ▶

The 2012–15 California drought was caused by decreased precipitation and higher temperatures, leading to more evapotranspiration, in combination with heavy agricultural irrigation. California has seen four other major droughts in the last century, but this was the worst. Continued dry weather conditions seem to be related to a persistent ridge of high atmospheric pressure in the northeastern Pacific Ocean.

Winters in California are typically wet, with December to February getting half of the total precipitation for the year. Winter snows build up in the Sierra Nevada Range, gradually melting during the spring and summer to provide about one-third of the water to fill reservoirs that provide water for agriculture and cities for the rest of the year.

However, recent years have brought less precipitation and a smaller snowpack. California had an average of only 18.7 cm (7.38 in) of precipitation in 2013, the driest year since 1850, when it became a state. The driest year in 119 years of record was 2014. The Sierra Nevada Range of California received 5.1 cm (2 in) of precipitation for that year (water equivalent of the mountain snowpack) compared with its normal 51 cm (20 in). As of January 31, 2014, water stored in the mountain snowpack was only 12% of average. The year 2014 was also the warmest in 119 years of record. Temperatures for January to June 2014 were about 2.7°C (4.8°F) above average. Much of the state was designated as experiencing exceptional drought, and broad water restrictions were implemented.

In response to drought conditions and water restrictions, farmers and ranchers fallowed fields and sold off their cattle herds, cities rationed water, and authorities curtailed water use where it was less critical. Disagreements arose as to how to allocate the limited supplies—what proportion to cities, to critical aspects of industry, and to fish habitat in the Sacramento-San Joaquin delta area. Some short-term solutions to the water supply problem will have long-term consequences. Some farmers were able to pump groundwater to supplement their supply; many had to drill their wells progressively deeper, effectively mining some water that cannot be replaced because much of the pore space between rock grains collapses and won't expand when water is available.

Contributing to the water shortage in California's Central Valley is the extensive use of water for agriculture. Trees that grow nuts require large amounts of water: 1 gallon per almond; 4.9 gallons per walnut. Many farmers prefer profitable almond (10% of total water used) and pistachio trees that unfortunately require year-round water. One head of broccoli needs 5 gallons; 1 large avocado, 220 gallons. One pound of cotton needs 101 gallons (1800 gallons for a cotton

T-shirt); one pound of rice needs 450 gallons of water. For beef cattle it takes 5000 gallons of water to grow 1 lb of beef. If droughts like this one persist in the future, this region may be forced to find more efficient irrigation practices or choose crops that need less water.

For an area that produces one-third to one-half of the fresh fruits and vegetables in the United States, the drought may bankrupt some farmers, force thousands of field

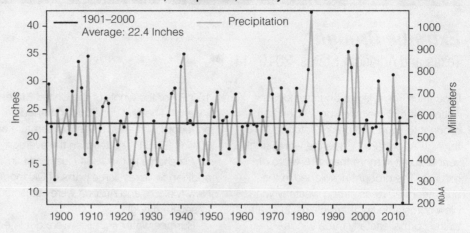

California, Precipitation, January–December

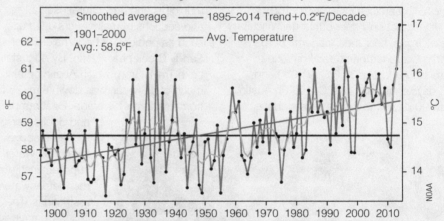

California, Average Temperature, January–August

■ **A.** *California precipitation from 1900 to 2014 shows little change overall but a significant decline since 2012* **B.** *California temperatures from 1900 to 2015 show a progressive increase, especially in the drought period of 2012 to 2015.*

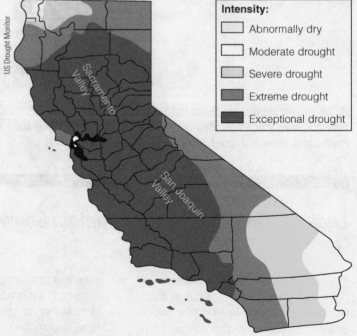

■ *California–Nevada drought map for December 9, 2014.*

A   **B**

■ *Lake Oroville, a reservoir north of Sacramento,* **A.** *at full pool, July 2011, and* **B.** *during severe drought in September 2014.*

workers out of jobs, and turn many small communities into ghost towns. The lack of moisture also brings concerns of much worse conditions for the coming fire seasons, both as a result of dehydrated vegetation and diminished water supplies to fight fires.

CRITICAL THINKING QUESTIONS

1. Under what circumstances may a series of normal annual rainfall years produce drought conditions and why? Consider both natural and human causes.
2. How do you think water supplies should be allocated among all of California's needs? What should be the priorities and why?
3. After the drought is over, should water-use restrictions continue? Why or why not?

Deadly Heat Waves
Europe, 2003 and 2010 ▶

In August 2003, western Europe sweltered in unusually high temperatures, in many places above 38°C (100°F), some 8–10°C above normal. More than 35,000 deaths in Europe were attributed to the heat; about 14,800 of those occurred in France, with thousands more in Italy and Spain. Paris, with little green vegetation, tightly packed stone buildings, and almost no air conditioning, felt like an oven with nine days in a row over 35°C (95°F). Although some parts of the world have long periods of high temperatures, people in such areas are familiar with these conditions and know to keep hydrated, stay in shady and well-ventilated areas, and use fans or air conditioners. Most of the victims in Europe were elderly, in poor health and living alone, and small children, left behind when parents or other relatives went on August vacations. The highest temperatures were in Spain, France, and Greece but the greatest deviation from normal temperatures were in France where people were less prepared for the heat and suffered most. Large areas of southern France reached 10°C (18°F) above normal. Most homes in France and other moderate climates do not have air conditioning.

■ *This color-coded map shows the temperature in Europe during the 2003 summer heat wave compared to the summer average (deviation in °C) between 1961 and 1990.*

Deviation from normal (°C): 10, 8, 6, 4, 2, 0, −2, −4, −6

Another heat wave in Europe in the summer of 2006 left southern England even drier than in 2003. Record temperatures throughout western Europe climbed above 35°C in July. In the same month, temperatures in parts of India reached 49.6°C (121.3°F), and at least 884 people died.

In July and August 2010, temperatures in Russia and elsewhere in eastern Europe reached more than 35°C. Already experiencing severe drought, the forests and peat deposits dehydrated and easily caught fire. In late July and early August, hundreds of fires became widespread. Smoldering and burning peat caused dense haze and smoke, with carbon monoxide concentrations and air particulates at four times acceptable levels. Many aircraft flights were delayed or cancelled. Deaths in Moscow, population 11 million, averaged 700 per day; total casualties for the heat wave are estimated as 55,700.

Lack of Shelters Despite a History of Tornadoes
Moore, Oklahoma, 2013 ▶

Sandra Adams and her 88-year-old mother took shelter in a bathtub as the tornado bore down on the town of Moore, Oklahoma, on May 20, 2013. It got really loud as the tornado came closer. The lights went out, debris began falling on them, and more debris was flying everywhere. Once it settled down, they could see some light through the debris and they realized their house was gone. Unable to dig themselves out, they waited until a 13-year-old boy found them and called for others to help pull remains of their house off them.

A neighbor, hearing the tornado sirens, took shelter in a ground-floor bathtub away from windows, covered himself and two small kids with a mattress, before the roaring tornado blew apart his house around him. Another heard the TV announce that an inside room was not sufficient for this tornado so she and her daughter went across the street to join a neighbor in her shelter. In nearby Plaza Towers Elementary School without a tornado shelter, children and teachers huddled in an inside hallway and covered their heads just as they had practiced. Many survived but seven died. Cars had been flipped, crushed, and wrapped around power poles, the bark stripped from trees. For some houses only the concrete foundation slab remained, swept clean of all debris.

On Monday, May 20, the jet stream dipped far south into New Mexico, northwestern Oklahoma, and into eastern Iowa, marking the southern extent of cold Arctic air. There it pushed northeastward under and lifted warm, very humid air from the Gulf of Mexico at a northeast-trending, nearly stationary cold front. Wind shear between the near-ground winds and those

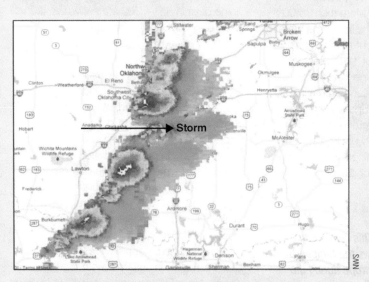

■ **A. Weather map for May 20, 2013 B. NWS storm radar, 3:00 p.m. Central Time. Note that the storm track was not perpendicular to the line of storms.**

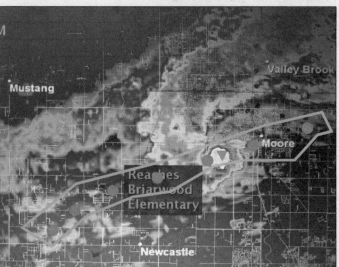

■ **Mid-afternoon of May 20, 2013, weather radar showed a pronounced hook echo of a major tornado that tracked east-northeast on a path through Moore, Oklahoma, about 15 km due south of Oklahoma City. The white ball just right of center is the radar image of debris swirled in the tornado vortex.**

■ A. *A subdivision decimated by the 2013 EF5 tornado in Moore, Oklahoma.*

B. *Rubble cleared to open a tornado shelter.*

at higher altitudes was strong. The unstable, rapidly rising warm air (28 to 30°C), as quickly developed severe supercell thunderstorms in the early afternoon. The National Weather Service issued a tornado watch at 1:10 p.m. for central and eastern Oklahoma, and as the storm strengthened, a tornado warning at 2:40 p.m. was marked by shrieking tornado sirens. With 16 minutes of warning, the tornado touched down at 2:56 p.m. west of Newcastle and tracked east-northeast toward Moore, a southern suburb of Oklahoma City. It quickly strengthened to EF4, and locally in Moore, EF5, then finally thinned to a rope tornado and dissipated; it was on the ground a total of 39 minutes. The tornado was so wide and the hail so heavy, many couldn't see it coming.

With winds topping out at 340 km/h (210 mph), the tornado leveled a path 27 km (17 mi) long and as much as 2.1 km wide, destroying houses, schools, a large hospital, and other buildings. Twenty-five

people died, including ten children, seven of them in an elementary school. Four people were killed in a destroyed 7-Eleven convenience store. About 60 people huddled in a Wal-Mart meat freezer to escape the tornado, which missed them by about 20 m.

More than 12,000 homes were damaged or destroyed by the 2013 Moore tornado, and damages were estimated as more than $2 billion. FEMA deployed Urban Search and Rescue and the Oklahoma National Guard. People searched for hours to find victims and even longer to dig any salvageable mementoes from their destroyed houses.

Unfortunately, few people had tornado shelters in Moore, in spite of three significant tornadoes in the last 14 years. On May 3, 1999, the strongest tornado ever recorded, an EF5 with winds reaching 484 kph (301 mph), followed almost the same path before swinging north to wreak havoc in Oklahoma City.

Others followed on May 8, 2003, and May 10, 2010. Neither of the two destroyed schools had built tornado shelters since the 1999 event.

Why hadn't residents and local governments taken steps to build shelters? Despite the fact that the federal government will fund 75% of the cost of school shelters, many still lack them. The state also offers rebates of up to 75% of the cost of home shelters. Calls for mandatory tornado-safe rooms in schools bogged down over whether the $1 billion cost should be paid by the state or local school districts.

CRITICAL THINKING QUESTIONS

1. Many homes in tornado-prone areas have no basement. If you don't have a tornado shelter, and the warning time is as little as 15 minutes, what should you do?
2. If you hear a tornado warning while you are in heavy traffic during your afternoon commute, what should you do?

Critical View

A A towering September cloud formation builds over the Tennessee Valley.

D. Bradford, NOAA

1. What kind of a cloud is this and what does it indicate?
2. What other weather phenomena are often associated with this type of cloud?

B This storm occurred over Norman, Oklahoma.

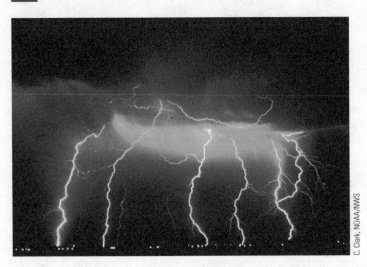

C. Clark, NOAA/NWS

1. Indicate relative positions of positive and negative static charges that lead to lightning.
2. Why is Oklahoma a likely location for thunderstorms?
3. If you were outside when this type of storm occurred, what should you do to keep safe?

C This destruction occurred in Dyersburg, Tennessee.

L. Skoogfors, FEMA

1. What would have caused this kind of damage?
2. What category of storm does this damage suggest?
3. What should you do to survive this type of event?

D These clouds formed over St. Joseph, Missouri.

NOAA

1. What is this type of cloud formation called?
2. What does this type of cloud formation suggest about developing weather?

E This damage was the result of a tornado in Paris, Missouri.

NWS, Memphis

1. What was the likely severity of the tornado that caused this damage?
2. What could the builders of this house have done to increase the chances that it would survive this type of event?
3. If you were living in such a place and a warning for this type of event were announced, what should you do to survive?

G This sheet of plastic embedded in a column, near Yazoo City, in western Mississippi on April 24, 2010.

NOAA

1. Imagine what this piece of plastic would have done to a person. How would you protect yourself from such flying debris if you were in a house lacking a tornado shelter?
2. If you were caught outside, out of range of a tornado shelter, and see a tornado coming, what would you do?

F This car was in Birmingham, Alabama.

A. Lamare, USACE

1. What would have caused this kind of damage?
2. If you are in your car and see an approaching storm of this type, what should you do?
3. If you are in open country with lots of good, straight roads, what direction should you drive to avoid being in the path of such a storm?

H This concrete slab was once under a house, in Louisville, Mississippi in April 2014.

NOAA

1. What strength of tornado (what EF level?) likely did this amount of damage?
2. How might you prepare for such a tornado, months or years in advance?
3. How might you retrofit a house on this slab so two or three people might survive inside it?

Chapter Review

Key Points

Basic Elements of Weather

■ Water continuously evaporates from oceans and other water bodies, falls as rain or snow, is transpired by plants, and flows through streams and groundwater back to the oceans, in a cycle known as the hydrologic cycle. **FIGURE 10-1**.

■ Rising air expands and cools adiabatically, that is, without loss of total heat. The rate of temperature decrease with elevation—that is, the adiabatic lapse rate—in dry air is twice the rate in humid air. **FIGURES 10-2** and **10-3**.

■ Cool air can hold less moisture; thus, as moist air rises over a mountain range and cools, it often condenses to form clouds. This is the orographic effect of mountain ranges. **FIGURE 10-4**.

■ Warm air rises, cools, and condenses to form an atmospheric low-pressure zone that circulates counterclockwise in the northern hemisphere. Cool air sinks, warms, and dries out to form a high-pressure zone that circulates clockwise in the northern hemisphere. Air moves from high to low pressure, producing winds. **FIGURE 10-5**.

■ West-to-east rotation of Earth causes air and water masses on its surface to lag behind a bit, known as the Coriolis effect. Because the rotational velocity at the equator is greater than at the poles, the lag is greater near the equator. This causes air and water masses to rotate clockwise in the northern hemisphere and counterclockwise in the southern hemisphere. **FIGURE 10-7**.

■ The prevailing winds are the westerlies north of about 30 degrees north (wind blows to the northeast) and the trade winds farther to the south (blowing to the southwest). Cells of warm air rise near the equator and descend at 30° north and south. **FIGURES 10-6** and **10-7**.

■ Warm fronts (where a warm air mass moves up over a cold air mass) and cold fronts (where a cold air mass pushes under a warm air mass) both cause thunderstorms. Such fronts often intersect at a low-pressure cell. **FIGURES 10-8** and **10-9**.

■ The subtropical jet stream meanders eastward across North America, across the interface between warm equatorial air and colder air to the north, and between high- and low-pressure cells. **FIGURE 10-10**.

Regional Cycles or Oscillations

■ The North Atlantic oscillation (NAO) is a comparable shift in winter atmospheric pressure cells that affects weather in the North Atlantic region. It also shifts every few years, but the times do not correspond to those of El Niño. **FIGURES 10-12** and **10-13**.

■ The Pacific decadal oscillation (PDO) is similar to a long-lived El Niño/La Niña pattern. In the warm phase, the eastern equatorial Pacific and North American coastal sea-surface temperatures are warm. **FIGURE 10-14**.

■ Equatorial oceanic circulation normally moves from east to west, but every few years the warm bulge in the Pacific Ocean drifts back to the east in a pattern called El Niño/La Niña-southern oscillation (ENSO), bringing winter rain to the west coast of equatorial South and North America, including southern California. **FIGURES 10-15** to **10-17**.

Regional Winds

■ Monsoon rains of India form when high pressure over the Arabian Sea pulls moisture northeastward where it rises against the Himalayas. The monsoons facilitate abundant crops but can also bring floods. **FIGURE 10-18**.

■ Santa Ana winds of southern California drain southward from a high pressure area over Nevada, heat up adiabatically as they descend toward the coast, and decrease in humidity. They come between October and March, often fanning wildfires in early fall. Chinook winds, descending the leeward side of the Rockies also warm adiabatically as they descend to the plains to the east. **FIGURES 10-20** and **10-21**.

Drought, Dust, and Desertification

- Drought involves a prolonged period of a distinctly lower-than-normal amount of precipitation in a region. It can impose severe hardship, financial stress, and even famine. **FIGURES 10-22** to **10-24**.

- Dust storms are often the result of drought, soil dehydration, and soil mismanagement such as by overgrazing. This can in turn lead to loss of fertile soils and desertification. **FIGURES 10-26** to **10-30**.

Snow, Ice Storms, and Blizzards

- Snow and blizzards bring hazards of frostbite, hypothermia, icy roads and sidewalks, and collapsed roofs. Heavy snow often forms by a "lake effect" where winds blowing across open water such as the Great Lakes, pick up moisture, then rise over land where the moisture cools, condenses, and falls as heavy snow. **FIGURES 10-31** and **10-35**.

Thunderstorms

- Storms form most commonly at a cold front when unstable warm, moist air rises rapidly into cold air and condenses to form rain and hail.

- Collisions between droplets of water carried in updrafts with downward-moving ice particles generate positive charges that rise in the clouds and negative charges that sink. Because negative and positive charges attract, a large charge separation can cause an electrical discharge—lightning—between parts of the cloud or between the cloud and the ground. **FIGURE 10-38** and **10-39**.

- You can minimize danger of being struck by lightning by taking refuge in a closed building or car, not touching water or anything metal, and staying away from high places, tall trees, and open areas. If trapped in the open, minimize contact with the ground by crouching on the balls of your feet. **FIGURES 10-44** and **10-45**.

Tornadoes

- Tornadoes are small funnels of intense wind that may descend near the trailing end of a thunderstorm; their winds move as fast as 515 km/h. They form most commonly during the collision of warm, humid air from the Gulf of Mexico with cold air to the north. They are the greatest natural hazard in much of the Midwestern United States, in a region known as Tornado Alley. **FIGURES 10-46**, **10-49**, **10-51** and **10-52**

- Tornadoes form when warm, humid air shears over cold air in a strong thunderstorm. The horizontal rolling wind flexes upward to form a rotating cell up to 10 m wide. A wall cloud sagging below the main cloud base is an obvious danger sign for the formation of a tornado. **FIGURES 10-50** to **10-52**.

- The Enhanced Fujita tornado scale ranges from EF0 to EF5, where EF2 tornadoes take roofs off some well-constructed houses and EF4 tornadoes level them. **FIGURE 10-56**, **10-58**, and **TABLE 10-3**.

- The safest place to be during a tornado is in an underground shelter or the interior room of a basement.

Key Terms

adiabatic cooling, p. 249
adiabatic lapse rate, p. 249
Arctic oscillation (AO), p. 254
Atlantic multidecadal oscillation (AMO), p. 255
blizzard, p. 271
Chinook winds, p. 261
cold front, p. 252
Coriolis effect, p. 250
cumulonimbus cloud, p. 271
desertification, p. 267
downburst, p. 274
drought, p. 262

dust storm, p. 265
El Niño, p. 256
Enhanced Fujita scale, p. 282
evapotranspiration (ET), p. 248
frostbite, p. 268
hailstone, p. 275
heat-island effect, p. 268
high-pressure system, p. 249
hydrologic cycle, p. 248
hook echo, p. 281
ice storm, p. 270
jet stream, p. 253
La Niña, p. 256

lake-effect snow, p. 268
lightning, p. 272
low-pressure system, p. 249
mammatus cloud, p. 280
monsoon, p. 259
North Atlantic oscillation (NAO), p. 255
orographic effect, p. 249
Pacific decadal oscillation (PDO), p. 256
polar vortex, p. 254
relative humidity, p. 248
Santa Ana winds, p. 261
step leader, p. 273

superoutbreak, p. 278
thunder, p. 273
thunderstorm, p. 271
tornado, p. 277
Tornado Alley, p. 277
tornado outbreak, p. 277
tornado warning, p. 285
tornado watch, p. 285
trade winds, p. 251
wall cloud, p. 279
warm front, p. 252
weather front, p. 252
westerly winds, p. 251
wind chill, p. 268

Questions for Review

1. If a humid air mass has 100% relative humidity and is 20°C at sea level, what would the temperature of this same air package be if it were pushed over a 2000-m-high mountain range before returning again to sea level? Explain your answer and show your calculations.

2. Explain the orographic effect on weather.

3. An area of low atmospheric pressure is characterized by what kind of weather?

4. Why do the oceans circulate clockwise in the northern hemisphere?

5. Explain the right-hand rule as it applies to rotation of winds around a high- or low-pressure center.

6. What is the main distinction between a cold front and a warm front?

7. What main changes occur in an El Niño weather pattern?

8. Why does heavy snow fall so frequently near the southern or eastern sides of the Great Lakes? Explain how the process works.

9. Where and when are monsoons important in Asia? What processes bring such heavy rainfall?

10. What causes formation of the Santa Ana winds and where do they occur?

11. Why are Santa Ana winds so warm and dry?

12. How can a region that gets 50 cm of rain a year be in drought whereas one that gets 15 cm per year is not in drought?

13. What factors led to the Dust Bowl of the 1930s?

14. What activities of people in a region can lead to desertification?

15. What is the "heat island effect" and how is it caused?

16. When is the main tornado season?

17. What process permits hailstones to grow to a large size?

18. Why do you see lightning before you hear thunder?

19. List the most dangerous places to be in a lightning storm.

20. What should you do to avoid being struck by lightning if caught out in the open with no place to take cover?

21. In what direction do most midcontinent tornadoes travel along the ground?

22. Why does lying in a ditch provide some safety from a tornado?

23. How do weather forecasters watching weather radar identify an area that is likely to form tornadoes?

24. What is the greatest danger (what causes the most deaths) from a tornado?

Critical Thinking Questions

1. Drought is often thought of as a weather phenomenon, over which we have no control. In what ways do humans sometimes contribute to development of drought in a region?

2. How might conditions similar to the Dust Bowl of the Great Plains happen again? What should be done to prevent that from happening again?

3. For centuries, governments in China tried to increase agricultural production in their northwestern regions by cultivating land that had not previously been used. What was involved in developing those new croplands and what were downsides of that new cultivation?

4. Explain why there are more tornadoes in the United States than in any other country. Be specific about the weather conditions that lead to tornado development and why.

5. Tornado experts say that if you encounter a tornado while in your car, you should get out and lie down in a ditch. Discuss the advantages and disadvantages of both locations in terms of your risk. Consider what about a tornado kills people.

6. How would you design a house to be safe from tornadoes, while keeping the cost to a minimum and having a design that would not look out of place in a typical residential neighborhood?

■ Thin sea ice breaks up in the wake of a NOAA research ship. A large iceberg floats in the background.

Ted Scambos, NSIDC

Climate Change: Processes and History

11

Melting Arctic Ice

The Arctic Ocean freezes over every winter, reaching its maximum extent in February or March. Ice covers everything including the Bering Sea between Alaska and Russia, Hudson Bay, around all of Canada's Arctic islands, and around much of Greenland. In 1979, the first year in which satellite images were available, late summer ice still surrounded Canada's northern Arctic islands and closed the Northwest Passage to ship traffic. By 2007, summer sea ice had thinned everywhere and had disappeared from around most of those Arctic islands, opening up the Northwest Passage, and retreating far back from the western Arctic.

Scientists who study climate and have carefully examined the data agree that human-induced global warming has already begun and is increasing. The question is no longer whether global warming is occurring, but why it is occurring, what proportion of it is human-induced, and what we should do to address it.

Climate change is occurring, is caused largely by human activities, and poses significant risks for— and in many cases is already affecting—a broad range of human and natural systems.

—Statement from 2010 U.S. National Academy of Science Report on Climate Change

Principles of Climate

Climate is the weather of a broad area averaged over a long time period. Seattle's climate, for example, is moderately humid with a broad range of sunny weather, clouds, rain, and wind, with average high temperatures from 8 to 24°C (46 to 75°F). Manaus, Brazil, in the Amazon rain forest, on the other hand, has a generally humid climate with rainy weather, except for a dry period from June to November; average high temperatures of the region range from 30.4 to 33°C (87 to 91°F; 73 to 74°F at night) (**FIGURE 11-1**). Weather and climate both have significant effects on the hazards discussed in the following chapters, which address floods, coastal erosion, hurricanes, and wildfires.

Solar Energy and Climate

Earth is warmed by solar energy coming from the sun. To understand how the sun warms Earth, we first need to understand the three mechanisms by which heat is transferred: **radiation, conduction**, and **convection** (**FIGURE 11-2**). Heat transfer is normally measured in watts per square meter (W/m²). Radiant heat, whether from the sun or a campfire, is energy transferred outward as invisible electromagnetic photons from their source; when they strike an object, that energy is converted to heat. Heat energy transferred by conduction vibrates the molecules of the solid (or liquid or gas) through which it passes, causing vibration in adjacent molecules to move the heat from a higher temperature molecule to a lower temperature molecule. In convection, the heated molecules are physically moved from a higher temperature area to one with lower temperature. Conduction of heat may burn our fingers when we grab a hot pan from the stove or when we hold a bent wire coat hanger over a fire to toast a marshmallow. Convective heat may burn an overhanging branch if heated air rises to reach it.

The amount of solar energy reaching a particular location on Earth's surface depends on the time of day, the time of year, and the position of the location relative to the equator. There are also cyclic variations in solar energy over longer periods.

Solar energy changes over the course of a day because Earth rotates once per day around its north–south axis, or poles, causing every place on Earth to face most directly toward the sun once a day at noon. At that time, the sun's energy is more intense than at any other. Daily patterns of solar energy are associated with specific weather-related hazards. For example, heat building during midday hours when the sun is high overhead can develop afternoon thunderstorms, which under some circumstances can spawn tornadoes.

Solar energy changes over the course of a year because Earth's axis is tilted with respect to its orbital plane around the sun (**FIGURE 11-3**). When the sun is perpendicular to Earth's axis on the vernal equinox (spring, usually March 20) and on the autumn equinox (fall, usually September 22), the

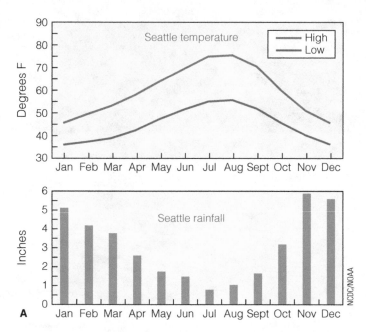

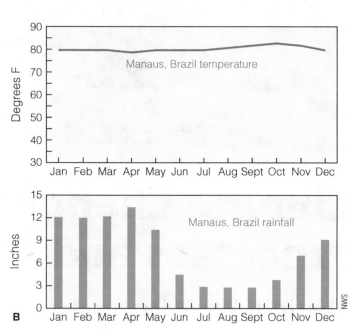

FIGURE 11-1 Climate Comparison

A. Seattle's climate is characterized by moderate temperatures and humidity, and seasonal variations that bring noticeably lower temperatures and more rain in the winter. **B.** In contrast, in the tropical climate of the Amazon region of Brazil, temperatures are hot and rainfall is high, except during a dry period from June to November.

sun is directly overhead at the equator. At these times, the sun's radiant heat strikes Earth's equatorial region directly, warming it more than at any other time. Many climate-controlled hazards are seasonal. For example, hurricanes usually occur from summer through mid-fall, while most tornadoes occur from spring through mid-summer.

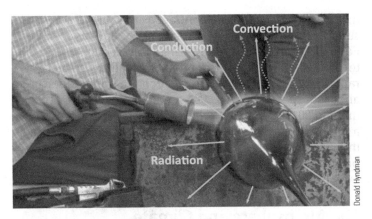

FIGURE 11-2 Transfer of Heat

A glassblower heating an ornament with a torch holds it with an attached steel rod. Heat conducted through the rod can warm his hand; he can also feel the radiant heat through the air to his hands and body. If people hold their hands above the hot ornament, they can feel warm air rising by convection from the hot glass.

Earth's axis is tilted 23.5° to the plane of its orbit around the sun; thus, when the sun is directly overhead at 23.5° north latitude on the summer solstice (usually June 21), the northern hemisphere receives its maximum solar radiation and thus the maximum heat from the sun. Although this is mid-summer in the northern hemisphere, the warmest temperatures lag the solstice by more than a month. It takes time to heat up the land and water, just as it takes time to heat up an egg in a frying pan after the stove is turned on. When the axis tilts 23.5° in the opposite direction (usually on December 21), the sun is directly above 23.5° south latitude, and it is summer in the southern hemisphere and winter in the northern hemisphere.

Because the angle of sunlight affects the intensity of solar radiation, the parts of the globe that the sun strikes more directly are warmed more than those where the sun strikes at an oblique angle. For example, the sun's energy striking straight down closer to the equator is more intense than the sun's energy striking at a glancing angle near the North Pole (**FIGURE 11-4**). Ocean and air currents then move heat energy around the globe from the equator to the poles. The warmer ocean temperatures 23.5° north of the equator in midsummer circulate ocean water clockwise in the northern hemisphere, carrying warm water to the northwestern part of the ocean and cold water moves southeastward to return the flow.

The sun's energy is also cyclic on a period of about 11 years. The sun's energy output is somewhat greater (about 0.1%) during its maximum output every 11 years, although the variation in temperature due to this effect is small compared with the overall temperature change over the last several

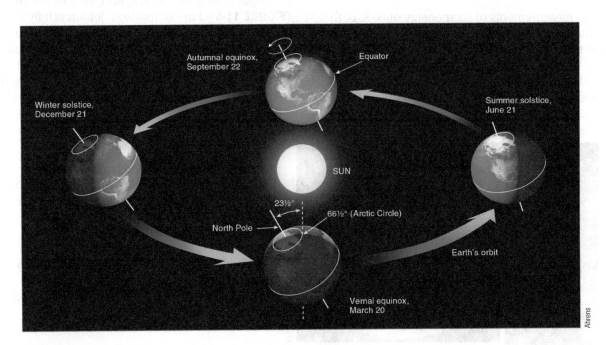

FIGURE 11-3 Seasons

Earth rotates around its north-south axis, inclined at 23.5° to Earth's orbit around the sun. Because Earth rotates around the sun once per year, the northern hemisphere is tilted toward the sun in summer, with the longest northern hemisphere day on the summer solstice (usually June 21). It then tilts away from the sun in winter, with the shortest northern hemisphere day on the winter solstice (usually December 21). The arrows from the sun show the solar energy striking the Earth from directly overhead—for example, on the northern hemisphere in summer.

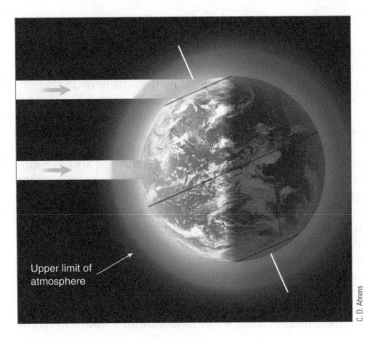

FIGURE 11-4 Angle of Sunlight

Sunlight striking straight down on Earth is intense; sunlight striking at a low angle, near Earth's poles, or in the early morning or late day, is much less intense. The atmosphere is exaggerated.

decades. Direct measurements of solar energy outside of the atmosphere have been made for only three cycles back to the late 1970s but correlation with sunspot numbers, which have a similar cycle, takes the record back to before 1750. Sunspots are darker-appearing regions of intense magnetic fields and lower than average temperatures on the surface of the sun; they expand and contract as they move across the surface of the sun. Although they appear dark as seen from Earth, the surrounding surface of the sun is brighter and hotter, so an expanding sunspot actually leads to greater solar radiation at that time. There is much we don't understand about sunspots—NASA's 2006 prediction of a sunspot maximum in 2011 did not materialize. Activity in 2010 was still at a minimum based on fading sunspots, weaker activity near the sun's poles, and a missing solar jet stream. Scientists now predict that the next sunspot cycle will be much weaker or even absent, as was the case in the late 1600s.

The Atmosphere and Climate

Radiant energy from the sun reaches Earth in forms of short-wave radiation that include visible light and heat. The amount of energy reaching the top of Earth's atmosphere affects atmospheric temperature. Once solar radiation reaches Earth, the atmosphere plays a key role in insulating the planet and protecting us from the sun's more harmful rays. The warmed surface of Earth also reflects back long-wave (infrared) radiation, some of which is contained inside the atmosphere, and some of which escapes back into space.

The atmosphere consists of broadly defined layers, including, from the ground up, the *troposphere* up to about 18 km, the *stratosphere* up to about 50 km, the *mesosphere* to about 90 km, and the *ionosphere* above that (**FIGURE 11-5**). For comparison, jet aircraft fly at altitudes of about 10 km. Excluding water vapor, the atmosphere consists of 78% nitrogen (N_2), 21% oxygen (O_2), and less than 1% of argon (Ar), carbon dioxide (CO_2), and other trace gases. Invisible water vapor (H_2O) varies widely from place to place but averages about 1 to 4% of the lower atmosphere.

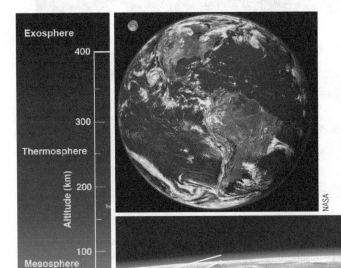

FIGURE 11-5 Earth's Atmosphere

The effluents of our cars and factories are all dumped into our thin atmosphere. A view of Earth from space emphasizes how thin that atmosphere is and therefore how vulnerable it is to pollution. In the bottom image to the west over the Pacific Ocean, Earth's atmosphere appears as a very thin, bluish line covering Earth's surface, approximately the thickness of a sheet of paper on a basketball. Without this insulating layer of gases, Earth would be uninhabitable.

Greenhouse Gases

Sunlight is short-wave radiation, most of which easily passes through Earth's atmosphere, heating the planet's surface. The heat that radiates from Earth outward is mostly long-wave radiation. Some of this radiation escapes into space, but some is absorbed by **greenhouse gases (GHGs)** in the atmosphere. In a phenomenon called the **greenhouse effect**, Earth's atmosphere traps heat similar to the way in which the glass in a greenhouse permits the sun to shine through but prevents much of the reflected long-wave radiation (heat) from escaping back through the glass. Whereas a greenhouse stays warm because the physical barrier of the glass traps warm air inside, however, the greenhouse effect warms the atmosphere because molecules of greenhouse gases absorb radiation.

Carbon dioxide (CO_2), methane (CH_4), ozone (O_3), nitrous oxide (N_2O), and hydrofluorocarbons (HFCs) are significant greenhouse gases. Non-greenhouse gases make up more than 99.9% of the dry atmosphere and do not trap long-wave radiation, though water vapor does trap a large amount. The greenhouse effect of individual gases depends on their effectiveness in blocking the outgoing long-wave radiation (heat) and their abundance in the atmosphere. Water vapor has the greatest effect as a greenhouse gas (36 to 66% under a clear sky), followed by CO_2 (9 to 26%), CH_4 (4 to 9%), and O_3 (3 to 7%). Although water vapor is the greatest contributor to atmospheric warming, it is not normally discussed as a greenhouse gas because it is completely natural and we have no control over it.

Many greenhouse gases are produced by natural processes. Most atmospheric water vapor has evaporated from the oceans, which cover about two-thirds of Earth's surface. CO_2 and CH_4 are emitted by erupting volcanoes, animals, decaying vegetation, and forest fires. For example, the extreme 2007 southern California wildfires generated about 8.7 million tons of CO_2 (this was about 20% of total emissions including those from non-wildfires). N_2O is generated from agricultural fertilizer and burning of fossil fuels. Greenhouse gases are also by-products of human activities, such as the burning of fossil fuels, which we discuss in more detail later.

The release of carbon dioxide into the atmosphere is part of the **carbon cycle** (**FIGURE 11-6**). Carbon occurs naturally

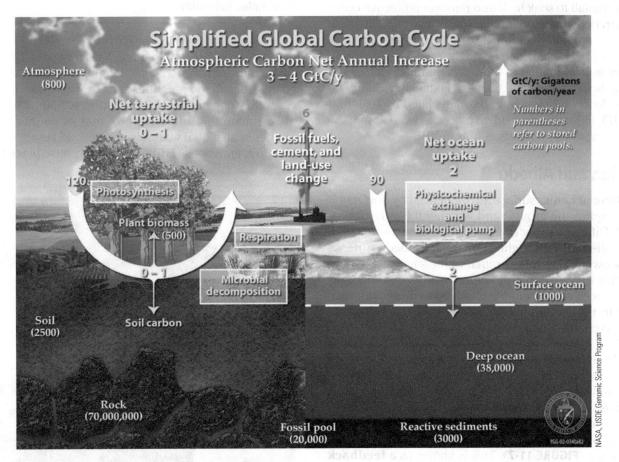

FIGURE 11-6 The Carbon Cycle

On land, carbon is held as biomass, in live plant material, and in dead material on the ground and in the soil. During photosynthesis, plants take up CO_2 from the atmosphere and release oxygen while growing, the amounts varying by season. The addition of CO_2 due to human activities is significantly larger than the plants and oceans can take up; thus, these activities add carbon into the atmosphere. Numbers are in gigatons of carbon per year.

in rocks, in soil, in the ocean, in living organisms, and in the atmosphere. In rocks it is mostly locked in place. Carbon can also precipitate from water (as calcium carbonate) to form limestone as in coral reefs or can slowly dissolve into water. CO_2 in the atmosphere also dissolves into ocean water in proportions that maintain a long-term equilibrium between them. If more CO_2 is loaded into the atmosphere, some of that dissolves into ocean water; if more were loaded into the ocean water, some of that would be released back into the atmosphere. Since the industrial revolution, human activities have been a major contributor to CO_2 in the atmosphere. Human-produced CO_2 emissions result from transportation, generation of industrial and residential energy, and land-use changes such as deforestation.

Deforestation mostly involves logging of hardwood trees for lumber, as occurs in the Amazon of Brazil, the teak forests of Myanmar (Burma), and the oil-palm forests of Indonesia and Malaysia. In each case, mature forest trees, which store large amounts of carbon, are felled, so they no longer take up CO_2. In the Amazon, that deforestation sometimes leads to progressive desertification because it leads to faster runoff rather than permitting rainfall to soak in. It also removes protective cover and leads to dehydration of undergrowth vegetation. In Indonesia, forests are felled and peat soils are burned to permit production of palm oil, which is widely used as vegetable oil because of its low content of trans fats. Burning releases the CO_2 held in both the trees and the peat. Fortunately, deforestation of the Amazon drainage in Brazil declined by 70% from 2005 to 2013. Indonesia's president placed a two-year moratorium on concessions for forest clearing in 2013.

Reflection and Albedo

Temperatures at Earth's surface are influenced by how much sunlight reaches the ground and whether it is reflected or absorbed. Different surfaces reflect more or less sunlight, a property called **albedo**. Light-colored, high-albedo surfaces such as snow and ice reflect much of the sun's energy and keep Earth's surface cooler. In contrast, dark-colored, low-albedo surfaces absorb energy and lead to more warming. Particles in the atmosphere, such as the water vapor in clouds, also contribute to changing albedo.

ALBEDO CHANGES AT EARTH'S SURFACE Changes to the albedo of Earth's surface materials affect the amount of solar energy that is reflected or absorbed. For example, white ice reflects 90% of the sun's energy back into space. Once some ice has melted, however, the dark surface of the ocean only reflects 10% of the sun's energy, leaving 90% of the original energy to heat the water and in turn melt more ice (**FIGURE 11-7**). This is known as a **feedback effect**, which is when a small disturbance in a system leads to effects that increase (positive feedback) or decrease (negative feedback) that disturbance.

The same thing happens when dark-colored carbon particles fall on a glacier or on mountain snow. Larger particles of soot that rise from incomplete burning of fossil fuels settle out

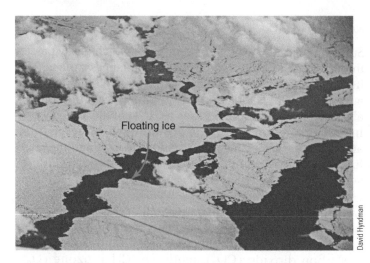

FIGURE 11-7 Albedo of Snow and Ice
The dark ocean surface absorbs more heat than white ice, so as more of the ocean surface is exposed, melting increases. Thin ice floating on the Arctic Ocean continues to break up in November 2011. The straight diagonal line in lower left is probably the path of an earlier icebreaker.

of the atmosphere within a few days or weeks. Called black carbon, they absorb all wavelengths of light and generally change the albedo of Earth's surface where they fall. In most cases they darken the surface, adding to warming. Researchers estimate that black carbon may account for about a quarter of all global warming, the second largest contributor after CO_2 emissions (**FIGURE 11-8**). Because black carbon is

FIGURE 11-8 Black Carbon
Black smoke emitted from incomplete burning of fossil fuels such as the diesel fuel of this highway truck settle out to darken surfaces and contribute to global warming.

short-lived in the atmosphere and such emissions can be curtailed with minimal expense, reducing black carbon in the atmosphere may be the fastest way to quickly slow climate change. "Brown carbon" is similar to black carbon but is primarily produced by inefficient burning of wood.

ALBEDO CHANGES IN THE ATMOSPHERE Part of the incoming short-wave radiation from the sun is reflected back into space by particles called **aerosols** suspended in the atmosphere, which reduce the amount of light that reaches Earth's surface. Water vapor, which makes up clouds, is the most abundant aerosol in our atmosphere. Thick, low clouds reflect the sun's rays away from Earth; they prevent sunlight from reaching the ground, so cloudy days stay cooler (**FIGURE 11-9A**). Because the water vapor in clouds is also a greenhouse gas, clouds contribute to both warming and cooling. Earth loses more heat on a clear night because there are no clouds to absorb outgoing radiation, allowing most of the heat to escape to outer space. A cloudy night, with its additional atmospheric water vapor, traps outgoing radiation, radiating some heat back toward Earth, keeping the lower atmosphere warmer (**FIGURE 11-9B**).

Volcanic eruptions produce large amounts of ash and sulfur dioxide (SO_2), along with other aerosols that reflect sunlight back into outer space and cool the atmosphere (**FIGURE 11-10**). The SO_2 dissolves in water droplets in clouds to form reflective sulfuric acid droplets that can remain in the atmosphere for several years until rain washes them out. Because the wind carries ash and particles around the globe, an eruption can have a significant impact on global temperatures. In 1815, the major eruption of Mt. Tambora in Indonesia injected ash into the atmosphere that blocked sunlight and reduced the average temperature in the northern hemisphere about 0.5°C (0.9°F). Over the next year, abnormally cold weather with summer frosts caused widespread crop failures and famine around the northern hemisphere.

SO_2 is also a by-product of human activities such as the burning of low-grade "bunker" fuel used in shipping. Ironically, government success at decreasing SO_2 pollution, which contributes to acid rain and respiratory diseases, will actually increase global warming. The reduction in sulfur aerosols that reflect some of the sun's energy is expected to add the equivalent of 10 years of global warming, but will reduce other environmental harm.

Volcanic aerosols such as sulfur dioxide droplets, volcanic ash, and foreign particulates such as ash, soot, and dust in the air can also increase reflection and cool the atmosphere.

Forest fires, both natural and human-caused, and dust blowing off deserts and drought-affected croplands are other major sources of particulates. Sahara dust drifting over the Atlantic has been shown to cool its equatorial waters and to hinder the development of hurricanes. Since the industrial revolution, pollution has been a major source of atmospheric particulates, especially in developing countries. Soot levels in some of the world's megacities, such as Beijing,

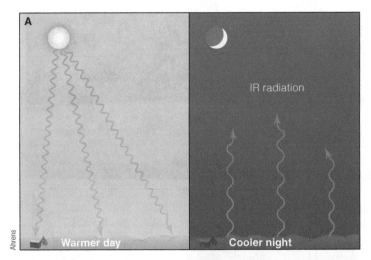

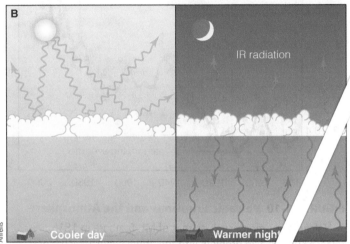

FIGURE 11-9 Effect of Clouds on Temperature
A. On a clear day, the sun warms Earth more, but at night, the absence of clouds permits that warmth to radiate heat back into space, leaving the lower atmosphere colder than it would be with clouds. **B.** On a cloudy day, clouds reflect much of the sun's energy, keeping temperatures cooler. At night they capture some of the long-wave (IR) radiation emitted from the warm Earth and radiate their heat both up into space and back down to Earth, keeping the lower atmosphere warmer than it would be without clouds.

Cairo, Mumbai, New Delhi, Seoul, Shanghai, Tehran, and Mexico City, are very high and rising, in some cases reducing sunlight by 25%. In most of these cities, the pollution comes largely from diesel-burning trucks and buses, cars, lead from leaded gasoline, road dust, coal-fired power plants, burning of fields by farmers, winter cooking, as well as campfires that burn wood, old tires, trash, and dried cow dung. Pollution drifting east from coal-fired plants in China has even shifted normal rainfall patterns, causing more rain in the south and worse droughts in the north. Although China has made progress in reducing pollutants from coal-fired power plants, the rapid increase in population, automobiles, and new factories has worsened air quality in its major cities.

A

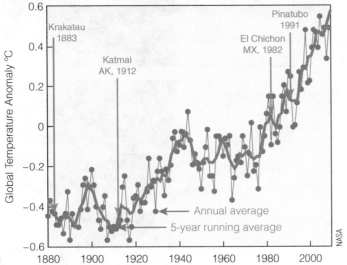

B

FIGURE 11-10 Volcanic Eruptions and the Atmosphere

A. Volcanic eruptions, such as that of Mt. Tambora in 1815, release sulfuric acid droplets into the atmosphere, which block incoming sunlight and lead to atmospheric cooling. **B.** On a graph of global temperatures, dips in temperature are apparent after major volcanic eruptions, but they are not enough to counteract the long-term warming trends of the last 100 years.

As air circulates around the globe, pollution becomes a global climate concern. Pollution emitted at a power plant or factory generally appears from a tall chimney (**FIGURE 11-11**). Commonly the more noxious the emission, the taller the chimney. Tall chimneys are built to disperse the polluted air emitted from the plant. Some of that pollution is cleared by rain; some drifts worldwide. A 3-km-thick brown cloud of smoke, soot, and dust roughly the size of the United States was recently discovered over part of the Indian Ocean. Some of that pollution is due to the huge population in India who use dried cow dung as a cheap fuel source for cooking. Pollutants can have complex effects on atmospheric conditions. Dark soot in the cloud of pollutants over the Indian Ocean absorbs heat and warms the upper atmosphere. However, it also reduces heat to the lower atmosphere and cools the surface of the ocean, which reduces evaporation. This in turn increases the intensity of regional droughts because less moisture is available to fall as rain.

FIGURE 11-11 Air Pollution

Coal-fired power plants like this one near Beijing, China, in 2011, produce much of the carbon dioxide in our atmosphere.

Earth's Energy Budget

Now you have seen that some components of the atmosphere have a cooling effect on the Earth, reflecting solar energy away, and other components have a warming effect, absorbing or trapping heat. Energy from the sun must be approximately balanced by energy lost back into space, on average, or Earth would show either major heating or cooling. Although there are regional and seasonal variations, over a year or a decade, that balance seems to hold (**FIGURE 11-12**). Of the short-wave radiation incoming from the sun, less than half reaches Earth's surface; about 23% is absorbed by water vapor and ozone in the atmosphere, about 23% is reflected back into space from clouds and the atmosphere, and 7% is reflected from the light-colored parts of the ground. The heated Earth, in turn, radiates energy back into space, in this case long-wave radiation (e.g., heat) that is absorbed by greenhouse gases, and water droplets in clouds. It cannot easily penetrate the atmosphere. The warmed atmosphere and clouds, in turn, radiate energy back downward to Earth's surface and out into space. Earth's surface also loses heat by transpiration from vegetation and evaporation from surface water, and to a lesser extent by convection, the actual movement of warm air.

Radiative forcing generally refers to the difference in the sun's radiant energy at the top of the troposphere, compared with that below it, and below which most weather occurs. It is a measure of the total incoming or outgoing radiant energy. When components of radiative forcing change, there is either a warming or cooling effect on Earth's temperatures. *Positive radiative forcing* (more incoming than outgoing radiation) warms Earth. *Negative radiative forcing* (more outgoing than incoming radiation) cools it. Some components of radiative forcing are natural but much is *anthropogenic*, or human-caused (**FIGURE 11-13**). The main *natural* radiative

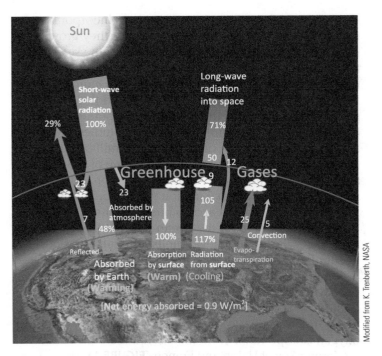

FIGURE 11-12 Earth's Energy Balance

Incoming short-wave radiation from the sun mostly passes through the atmosphere unhindered. Some is absorbed by moisture in the atmosphere, reflected off cloud tops or the surface of Earth. The warmed Earth radiates long-wave radiation back outward, much of which remains trapped by the atmosphere or absorbed by clouds, which, in turn, radiate much of that back to Earth. The warmed atmosphere also radiates energy out to space.

forcing is the sun's radiation, as described previously; it rises and falls on cycles about 11 years apart. Punctuating this are the negative (cooling) effects of aerosols ejected by volcanic eruptions.

Earth's Climate History

Earth's climate has changed many times over its history; some changes are cyclic, some are short-term results of one-time phenomena, and some changes, like the warming we have experienced since the industrial revolution, are human-induced. In this section, we discuss how Earth's temperature record is established, what it shows, and why past **climate changes** have occurred. CO_2 added to Earth's atmosphere since 1750 is the largest contributor to change in radiative forcing and increase in atmospheric temperature. It is now clear that human activities have caused more than half of the increase in global surface temperature from 1951 to 2015.

Establishing the Temperature Record

Scientists have established a long-term record of global temperatures through **proxy data**, which provide an indirect record of climate conditions. Tree rings, sedimentary deposits, and ice cores are all proxies for past climate conditions. Although these methods are approximate, they all reasonably agree throughout 800,000 years of records.

Tree-ring studies establish temperature records by using the fact that tree rings are thicker and the wood is denser when growing conditions are better (**FIGURE 11-14**).

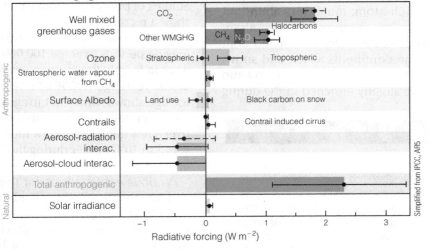

FIGURE 11-13 Radiative Forcing Components

Carbon dioxide is the primary constituent of radiative forcing in the atmosphere. The level of confidence in the data is very high for CO_2 and N_2O; high for CH_4, halocarbons, and aerosols; and medium for other short-lived gases.

FIGURE 11-14 Tree-Ring Growth

Thin tree rings mark poor growing conditions; thick ones mark good conditions.

Because tree rings also depend on moisture, scientists correlate the ring-thickness record with known historic temperatures to more reliably project temperatures back for thousands of years. Wood samples from living trees provide records of recent climate conditions, but samples from dead trees or old building materials can provide a longer record (**By the Numbers 11-1:** Radiometric Dating). Scientists also compare tree-ring results from one climatic area with those from another, and with proxy data obtained from other sources, such as from ice cores. Tree-ring patterns from desert areas are not useful for climate studies because moisture is more determinative for tree growth than temperature.

Global temperatures of the past are often estimated using oxygen-isotope ratios ($^{18}O:^{16}O$) in sedimentary materials formed at various times. Light oxygen, ^{16}O, has 8 protons and 8 neutrons in each atom; much less abundant heavy oxygen, ^{18}O, has two extra neutrons in each atom. During cooler periods, more of the water evaporated from the oceans is stored on the continents as ice and snow. Because evaporation preferentially takes the lighter ^{16}O into the air, the oceans become slightly enriched in ^{18}O during

---

By the Numbers 11-1

Radiometric Dating

The age of older samples can be established through radiometric dating. Analysis of carbon-14 (^{14}C), for example, can be used on samples as old as about 50,000 years. Carbon-14, a radioactive isotope of carbon, collects in living things such as trees when they take in carbon as they grow. After they die, the ^{14}C atoms decay gradually so that after 5730 years, half of the original amount of ^{14}C is gone; thus its half-life is 5730 years. After another 5730 years, half of the remainder is gone, leaving only one-quarter of the original, and so on—an exponential decay rate. Using measured amounts of ^{14}C, we can calculate how many years ago the tree died.

---

such periods. Because marine organisms incorporate oxygen from the seawater into their shells as they grow, their shells ($CaCO_3$) preserve the $^{18}O:^{16}O$ ratio of the seawater at the time. This permits estimates of the water temperature over long periods. Higher ^{18}O in a clamshell indicates lower northern hemisphere temperatures at the time the shell grew, because more of the water had evaporated from the oceans to form the continental glaciers. Studies show that, in general, an increase of one part per million of $^{18}O:^{16}O$ means 1.5°C higher temperature at the time of evaporation. Sediment cores from lakes or oceans can provide oxygen-isotope data up to about 5 million years ago. The snow that falls during any given year also preserves the record of the oxygen-isotope ratio for the ocean water at that time. Near the poles, where snow never melts, annual layers of snow accumulate in thick deposits of glacial ice. Scientists can drill into these deposits to obtain a cylinder of glacial ice, called an *ice core*, to determine isotope ratios over hundreds of thousands of years. Temperature changes in the Antarctic for the past 420,000 years were determined by analyzing oxygen-isotope ratios from the Vostok ice core of 1996 by the former Soviet Union and France (**FIGURE 11-15**).

Deuterium (^2H), the much less abundant heavy isotope of hydrogen, acts in a similar way to the heavy isotope of oxygen. Light hydrogen (^1H) contains one proton, whereas heavy hydrogen contains one proton and one neutron, making it twice as heavy. Like oxygen, deuterium concentration correlates with temperature, showing a similar trend to oxygen isotopes in ice cores. Deuterium concentrations collected by European Project for Ice Coring in Antarctica (EPICA) were used to create a temperature record for the past 740,000 years, reported in 2008.

Ice Ages

Studies of proxy data from Greenland and Antarctic glaciers show a record of cooler periods, or **ice ages**, alternating with warmer periods over the last two million years. These cyclic ice-age periods average 100,000 years apart but vary somewhat in length; the last one peaked approximately 18,000 to 25,000 years ago (**FIGURE 11-16**). The typical beginning of an ice age shows somewhat irregular advances to a maximum ice cover, followed by a rapid retreat to minimum ice cover over a few decades to a few thousand years. The atmosphere was 5 to 10°C cooler during the last ice age relative to modern times.

Various explanations for the cyclic nature of ice ages include changes in Earth's orbit and rotation. In the 1920s, Yugoslavian astronomer Milutin Milankovitch calculated that Earth's orbit around the sun slowly changes from nearly circular to elliptical over a period of 100,000 years, bringing Earth slightly closer to or farther from the sun during different periods. The tilt of Earth's axis also ranges from roughly 22° to 24.5° in 41,000-year cycles. A 26,000-year cycle of the planet's axis acts somewhat similar to a spinning top with a wobble in the orientation of its axis. Over this cycle, Earth's axis of rotation

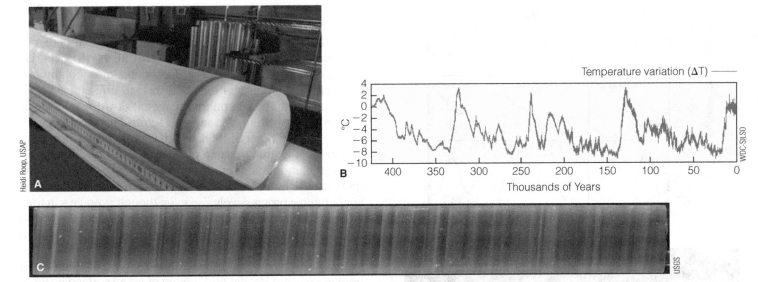

FIGURE 11-15 Ice Core Records

A. In this section of an ice core from Antarctica, the dark layer is ash from a volcanic eruption 21,000 years ago. **B.** The graph shows temperatures of lower atmosphere in °C (difference from present) determined from the Vostok ice core for the last 420,000 years. **C.** In this section from Greenland, dark layers indicate dry-season dust deposits each winter. This sample shows 38 annual layers (~16,250 years ago).

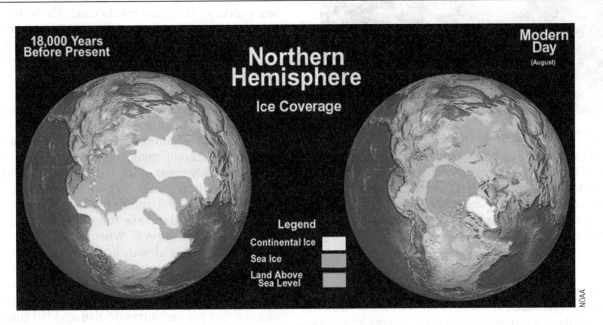

FIGURE 11-16 Extent of Ice During the Last Ice Age

The last ice age, about 18,000 to 25,000 years ago, spread continental ice over almost all of Canada and the northern United States, along with northern Europe and adjacent Asia.

points at different positions in space, which results in different amounts of solar radiation reaching different parts of Earth.

Milankovitch calculated the angle at which the sun strikes Earth at every latitude and for each of Earth's orbital variations. He inferred that climate varied in response to these orbital changes, commonly called the **Milankovitch cycles** (**FIGURE 11-17**). For example, when Earth was tilted so that the latitude of northern Europe received less sunlight in summer, less snow melted each spring; snow gradually built up year after year, and the climate cooled finally into an ice age. The Milankovitch cycles correspond well with the 20 ice ages recorded over the past two million years.

Still longer cycles of warm versus cold and wet versus dry have occurred for at least the last two billion years. These

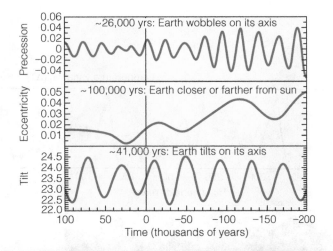

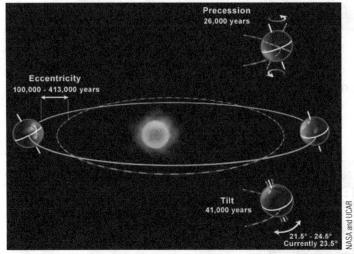

FIGURE 11-17 Milankovitch Cycles

Milankovitch charted three factors that contribute to long-term climate cycles: Earth wobbles on its axis with a cycle of about 26,000 years (top line of graph); Earth's orbit takes it closer to or farther from the sun on a cycle of about 100,000 years (middle line of graph); Earth tilts on its axis with a cycle of about 41,000 years (bottom line on graph).

cycles, generally of irregular length, can last for tens of millions of years or more. Although important in the billions of years of Earth's history, these long-term climate cycles are not important to natural hazards that affect present-day humans, nor to climate changes that affect us today, because their rate of change is vastly slower than anything that will affect human activities.

Global Warming

For more recent trends in climate, scientists have the advantage of direct measurements of temperature, along with many other types of climate data. Today, thousands of temperature measurements around the globe confirm what climate scientists have known for decades: Earth is getting warmer. Earth's average atmospheric temperature has

been rising since the industrial revolution began in the late 1700s, a phenomenon commonly called **global warming**.

The long-term global temperature record shows progressive temperature rise since about 1910 (**FIGURE 11-18**). The rise of a little more than 1°C (~1.8°F) has been almost continuous except for flatter periods from about 1945 to 1975 and between 1998 and 2014. The reason for the recent slowdown in warming has been attributed to several factors. First, some people use the starting point of 1998, which was a record warm year. Coincidentally, that began a more prolonged low in the sunspot cycle, indicating temporarily weaker solar energy. Sunspot-cycle highs and lows are related to sun energy, but overall temperatures progressively rise even though average sun temperature does not. The global economic slowdown may also have had an effect. Perhaps more importantly, there has been a strengthening of Pacific trade winds blowing warm surface water westward, which promotes the rise of cold deep water from the eastern Pacific and mixing of warm surface water to depth. As of 2015, global temperatures seem likely to resume their rapid rise. Signs of a long-delayed El Niño returning in 2015 include warm Pacific Ocean water off the West Coast and an abrupt end to the prolonged drought in Texas.

Temperatures rise slower in the oceans than on the continents because it takes more energy to heat water (called its *heat capacity*) than land and because the warmest water (in the shallowest 300 m) circulates to mix with colder, deeper ocean water. The changes are even more dramatic at high latitudes in the Arctic. In the past 40 years, the temperatures there have gone up about 2°C (> 3.6°F). The 11 warmest years on record have all been from 2002 to 2015.

These increases may not seem like very much considering day-to-day swings in temperature, but they are significant in northward migration of warmer-climate animals, plants, insects, and accompanying diseases. Glaciers are disappearing, snowfields are melting earlier in spring so dry-season stream flows drop earlier, and sea level is rising, accelerating coastal erosion. We'll discuss this more later.

Temperature data are relatively easy to obtain; the more difficult question is, What is causing the temperature increase? You have already learned that the amount of solar radiation reaching the top of Earth's atmosphere changes on 11-year cycles and that cyclical changes in Earth's orbit and rotation also have an effect on long-term climate trends. You can see that temperatures do rise and fall slightly with the sun's energy; however, over the long term the sun's energy doesn't change much but Earth's surface temperature continues to rise. (FIGURE 11-18).

If changes in incoming solar energy do not explain increased temperatures, then the temperature change must be caused by something happening after solar energy enters our atmosphere. Recall from earlier in the chapter the role that the atmosphere plays in trapping heat. In particular greenhouse gases keep long-wave radiation from escaping into space. As greenhouse gases become more abundant in Earth's atmosphere, they absorb more radiation and increase Earth's surface temperature.

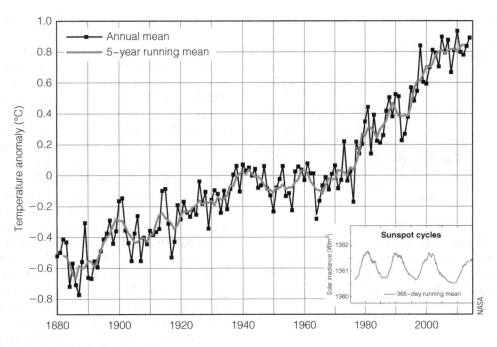

FIGURE 11-18 Surface Temperature

Global surface temperature from 1880 to 2015 (relative to the average between 1960 and 2014). Plotted temperatures are departures from average for 1951 to 1980. The sun's solar radiation cycles (shown in the inset) show no progressive change in temperature and are clearly not the cause of the significant rise in temperature over the last several decades.

Scientists have established a record of the amount of CO_2 in the atmosphere over the last 400,000 years (and less precisely over 800,000 years) by analyzing tiny amounts of air trapped in Greenland and Antarctic ice (**FIGURE 11-19**). This record reveals that some variation in greenhouse gases occurs naturally. CO_2 levels in the distant past appear to be long-term cyclic, showing variation from less than 200 to almost 300 parts per million (ppm). Initially, some people argued that the current increase in CO_2 was merely natural variation, not a result of human activity. However, the levels of CO_2 are now far higher than at any time during the past 800,000 years. The preindustrial value of 278 ppm skyrocketed to over 400 ppm by 2015, with current increases of about 2 ppm per year. The Intergovernmental

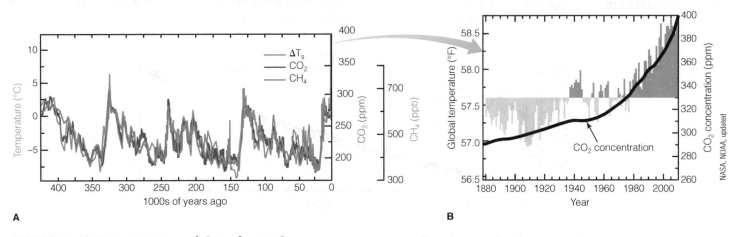

A **B**

FIGURE 11-19 Temperature and Greenhouse Gases

A. Changes in temperature (red line) correlate with changes in CO_2 (blue line) and CH_4 (green line) for at least the past 400,000 years. These data were obtained from the Vostok ice core from Antarctica. Older records to 800,000 years ago show a similar pattern, becoming slightly less distinct with greater age. **B.** The graph on the right shows rapid temperature rise in the last 130 years.

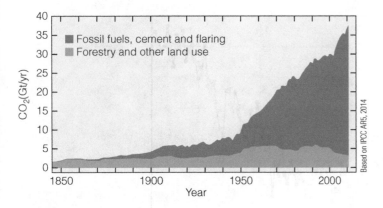

FIGURE 11-20 Global Anthropogenic CO$_2$ Emissions

Human production of CO$_2$ and other greenhouse gases worldwide rose gradually from the industrial revolution to about 1950, when it increased rapidly. As a result, the cumulate amount of CO$_2$ in the atmosphere more than doubled from about 900 to 2100 gigatons of CO$_2$.

Panel on Climate Change (IPCC) now concludes that it "is extremely likely that human influence has been the dominant cause of the observed warming since the mid-20th century." Rapid increase in human-produced atmospheric CO$_2$ began about 1950, with greater burning of fossil fuels, better roads with more vehicles, and greater production of cement for construction (**FIGURE 11-20**). Concentrations of CH$_4$ and N$_2$O have also risen far above their preindustrial values. The vast majority of scientists agree that this increase is a result of human activities. Major eruptions of volcanoes do emit large amounts of CO$_2$ and CH$_4$. However, their total contributions to the atmosphere are relatively small because large eruptions are infrequent and the source is confined to their individual eruptive vents. For comparison, USGS calculated that the very large 1991 eruption of Mt. Pinatubo produced 0.05 gigatons (Gt) of CO$_2$. That plus the average annual CO$_2$ production of 0.26 Gt from all volcanoes worldwide gives

0.31 Gt for a large-event year. This is only 6.3% of the 4.9 Gt/yr sent into the atmosphere from all human-produced sources.

The increase in atmospheric CO$_2$ levels has been established by direct measurements made since 1960, which plot an upward trend, commonly called the Keeling curve (**FIGURE 11-21**). Annual fluctuations in carbon levels (the squiggles in the line on the graph) are due to seasonal variations in plant cover. Once fall arrives to the northern hemisphere and plant uptake shuts down, the CO$_2$ concentrations rise again. Other greenhouse gases are also rising, although not so dramatically (**FIGURE 11-22**).

The dramatic rise in concentrations of greenhouse gases over the past century has occurred because these gases are a by-product of human industry, agriculture, and energy use (**FIGURE 11-23**). By far, the largest proportion of world energy use involves burning fossil fuels, which generates large amounts of CO$_2$. **Fossil fuels**, including coal, natural gas (methane, CH$_4$), and petroleum, are all hydrocarbons, meaning that they are made of various combinations of hydrogen (H), carbon (C), and oxygen (O).

Coal-fired power plants are the largest contributors of CO$_2$ into our atmosphere. A single large coal-fired power plant uses 10,000 to 15,000 tons of coal per day—the amount carried by a 100-car train. Coal burning also typically generates more of other pollutants such as sulfur that lead to smog and acid rain. Per unit of electricity generated, coal generates twice as much CO$_2$ (1020 kg/MWh) as natural gas (515 kg/MWh). In 2013, of the electricity used in the United States, coal produced 37%, followed by 30% from natural gas, 19% from nuclear power, 6% from hydroelectric power, 1.7% from biomass, and 1% from oil. Other renewable sources, including geothermal, solar, and wind, provided a combined 5.2%. Coal produces most of the energy in China, India, and many other developing countries. For comparison, 2013 per capita emissions of CO$_2$ were 16.6 metric tons (6,600 kg) in the United States. During the same time, per capita emissions were 7.4 metric tons in China and 1.7 metric tons in India.

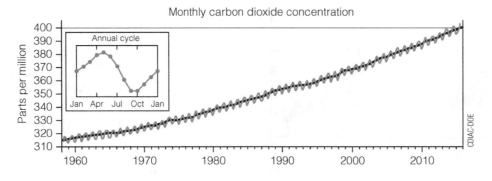

FIGURE 11-21 The Keeling Curve

Increase in carbon dioxide is shown on the Keeling curve, which shows atmospheric CO$_2$ levels from 1960 to 2015 at Mauna Loa, Hawaii. Levels are lower in the plant-growth season, higher in Northern Hemisphere winter. Note that the rate of rise is increasing.

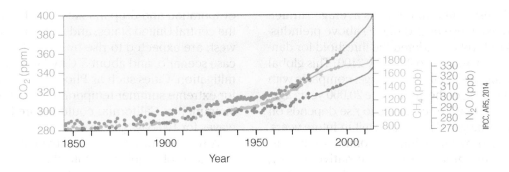

FIGURE 11-22 Concentrations of Greenhouse Gases
Global concentrations of major greenhouse gases: CO_2 in ppm, CH_4 and N_2O in ppb.

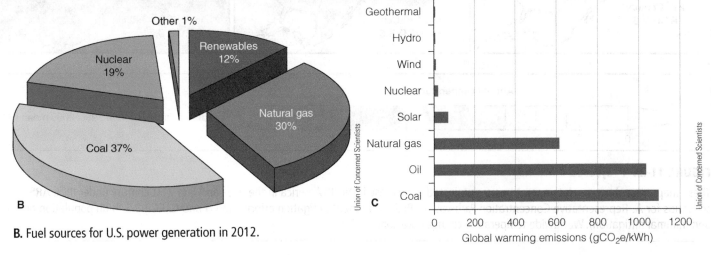

B. Fuel sources for U.S. power generation in 2012.

FIGURE 11-23 Sources of Greenhouse Gases

A. A big cement plant along the Yangtze River downstream from Chongqing, China, emits CO_2 and other pollutants. Cement plants are major contributors of CO_2; they expel it by turning limestone into the CaO in cement ($CaCO_3 \rightarrow$ $CaO + CO_2$).

C. Emissions in grams CO_2 (and equivalent) per kWh for different fuel sources of power. Estimates include all costs such as plant construction, maintenance, fuel production and combustion, and electricity transmission.

Global Climate Models

Scientists are already observing the effects of climate change around the world. Future changes can be estimated using projections of available data. Scientists use the current understanding of the relationships between CO_2 emissions and global temperatures to project future global warming. Using the basic laws of physics and chemistry, they make estimates using formulas describing expected changes in radiation, atmospheric temperature, pressure, humidity, and air motion in small parts of the atmosphere. Using these formulas, they predict future changes, compare them with the actual changes, and adjust their formulas to lead to better predictions. Because longer-term predictions depend, for example, on the amounts of CO_2, CH_4, and N_2O in the atmosphere, and on measurements and calculations by different researchers, the projected future changes vary somewhat, but all indicate accelerated warming.

The IPCC in 2014 estimated that Earth's average surface temperature would likely rise by 2°C (3.6°F) above preindustrial levels by 2050 (a level considered the threshold for dangerous climate change) and 3.7 to 4.8°C by 2100. This global warming, a significant part of climate change, compares with a rise of 4°C since the peak of the last ice age 20,000 years ago. The rate at which temperatures continue to rise depends on how much CO_2 and other GHGs humans emit in future years. This could be affected by changing economic conditions, population growth, and greater use of alternative energy, among other things. The most likely scenario involves a growth in worldwide GHG emissions as developing countries experience a boom in industry and economic growth.

Based on a stringent mitigation scenario for future fossil fuel consumption, temperatures in the Arctic are expected to rise by at least 2°C and under a minimal mitigation scenario would rise by about 5°C by mid-century (**FIGURE 11-24**). Temperatures in the Great Lakes region and eastern Canada are expected to rise by almost 3°C by mid-century and more than 5°C by late-century! They are also expected to rise 4°C in winter and 6°C in summer, likely leading to more

evaporation and drop in lake level. Temperatures in most of the central United States, and in the already hot desert Southwest, are expected to rise by 1 to 2°C, even under the best-case scenario, and about 5°C by late century under minimal mitigation. Cities such as Phoenix, Arizona, are well known for extreme summer temperatures, but major cities such as Sacramento, California, could before long experience heat waves for 100 days per year.

A recent study from the University of Hawaii, analyzed the effects of ongoing climate change, using well-established climate models, to project temperatures for the next few decades around the world. The results are disturbing. The *average* temperature in most places will be hotter in 2047 than the warmest temperature at any year from 1860 to 2005. To put this in perspective, how old will you be by 2047? How about your children? What will such temperatures mean to forests, wildfires, food crops, ocean reefs and sea life, freshwater resources, and international conflicts? At warmer latitudes, such change will come sooner: in Indonesia by about 2020; Kingston, Jamaica by about 2023; Mexico City by 2031; and Phoenix by 2043.

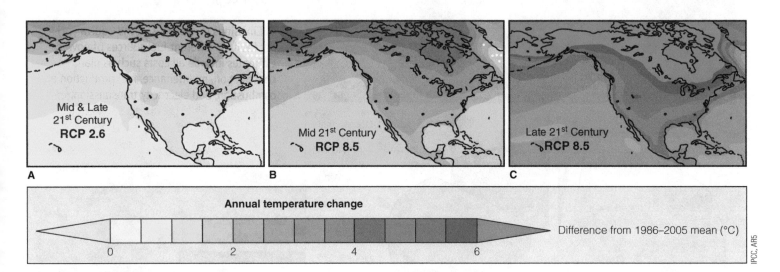

FIGURE 11-24 Projected Warming

These maps show how much temperatures are projected to increase in North America in the mid- and late 21st century under IPCC AR5 scenarios for **A.** Representative Concentration Pathway (RCP) 2.6 (stringent mitigation efforts) and **B** and **C.** RCP 8.5 (high population growth and minimal mitigation). Worldwide temperature changes are comparable.

Chapter Review

Key Points

Principles of Climate

- Climate is the weather of a broad area averaged over a long time. **FIGURE 11-1**.

- Energy is transferred from place to place by conduction, radiation, and convection. **FIGURE 11-2**.

- Temperatures on Earth are influenced by the angle of the sun's rays, which change over the course of a day and year, and vary according to the position of a location relative to the equator. **FIGURES 11-3** and **11-4**.

- Earth's atmosphere permits the sun's short-wave energy to reach Earth but reflected long-wave energy is partly trapped, keeping Earth warmer than it would be otherwise. Greenhouse gases such as carbon dioxide and methane prevent more of the sun's long-wave energy (heat) from escaping, thereby causing more warming of the atmosphere.

- Light-colored, higher-albedo surfaces reflect more energy and stay cooler; darker, lower-albedo surfaces absorb more energy and get warmer. **FIGURE 11-7**.

- Clouds reflect much of the sun's energy during the day but prevent more loss of heat at night. **FIGURE 11-9**.

Earth's Climate History

- Earth's climate record has been established by analyzing data provided by tree rings, ice cores, and other climate proxies. **FIGURES 11-14** and **11-15** and **By the Numbers 11-1**.

- Over long periods of time Earth cycles between warmer periods and cooler ice ages, in part as a result of the geometry of the Earth's axis and orbit. **FIGURES 11-16** and **11-17**.

- Earth's average surface temperature has been rising since the late 1700s in a phenomenon called global warming. This temperature rise corresponds to an increase in greenhouse gases in the atmosphere. **FIGURES 11-18** and **11-19**.

- Atmospheric carbon dioxide concentrations have increased by more than 40% since the industrial revolution and are far higher than at any time in the past 800,000 years. Most of the rise in atmospheric CO_2, greenhouse gases, and temperature appears to be human-caused, much of it as a by-product of the burning of fossil fuels. **FIGURES 11-19** to **11-23**.

- Global climate modeling allows scientists to project future warming based on different scenarios for future greenhouse gas emissions. **FIGURE 11-24**.

Key Terms

aerosol, p. 305
albedo, p. 304
carbon cycle, p. 303
climate, p. 300
climate change, p. 307

conduction, p. 300
convection, p. 300
feedback effect, p. 304
fossil fuel, p. 312
greenhouse gas (GHG), p. 303

global warming, p. 310
greenhouse effect, p. 303
ice age, p. 308
Milankovitch cycle, p. 309
proxy data, p. 307

radiation, p. 300
radiative forcing, p. 306

Questions for Review

1. What is the difference between weather and climate?
2. Describe three ways that heat energy is transferred.
3. What factors determine how much solar energy reaches Earth's surface?
4. What effects do clouds have on warming and cooling Earth at different times of day?
5. What effects do volcanic eruptions have on the atmosphere and global climate? Why?

6. How does melting snow and ice produce a feedback effect that contributes to more warming?

7. What are the most important greenhouse gases and what are their main sources?

8. How do greenhouse gases prevent some of the sun's energy from escaping Earth's atmosphere?

9. How much has the temperature of Earth's atmosphere increased in the last 1000 years? When did most of the increase begin?

10. About how much has the carbon dioxide concentration in the atmosphere increased? When did the increase begin? Why?

11. What evidence indicates that global warming is largely human-caused?

12. Other than an increase in temperature, what would be the most prominent changes in weather with global warming?

Critical Thinking Questions

1. A large proportion of carbon dioxide, blamed for much of climate change, comes from heavy industry such as burning coal for electric power and producing cement from limestone. To what extent should factories be required to incur major costs to reduce emissions, thereby increasing costs to the consumer?

2. The United States and Canada, the largest carbon emitters per capita, have been reluctant to reduce their carbon emissions at a rate greater than their rate of economic growth. Is this reasonable? Why or why not?

3. Some skeptics argue that increases in atmospheric greenhouse gases are natural rather than human caused. What evidence would you provide to show that these increases are a result of human activities?

■ Homes crowd riverbanks on Vietnam's Mekong River delta where tides reach far upstream from a rising sea level.

Hyndman

Climate Change: Impacts and Mitigation

12

Rising Seas and Life on a Sea-Level Delta

In Vietnam's Mekong delta homes are crowded along the river where residents fish and sell local farm goods, typical of delta regions in many poor countries. Even here, 100 km upstream from the ocean, tidal changes of 2 to 3 m already threaten riverfront homes. Tidal surges from major typhoons drown thousands of poor people on the world's deltas. With ongoing sea-level rise as a result of climate change, flooding hazards become even more serious, not only along river edges but across the whole delta region because water tables also rise with sea level. Despite growing threats, these residents have no means to move and nowhere else to go; such deltas are the breadbaskets for whole countries, and fish is the main protein source available.

317

The term *global warming* refers to the overall average warming aspects of climate change. However, the consequences of climate change are complex and may even bring colder weather to some areas. Abrupt changes in regional climate, either cooling or drying, would have disastrous consequences for the world's populations. Extreme weather events such as torrential rainfall are expected to become more frequent with climate change. Intense precipitation events are expected to couple with longer and more severe drought periods. In large areas, the drop in temperature and rainfall will limit agricultural production, and water supplies will be severely curtailed. Food availability will be disrupted, resulting in more malnutrition; public health will suffer from the spread of insect-borne diseases such as malaria and dengue fever to higher latitudes and elevations. Economic disruption, population migration, political upheaval, and conflicts over resources seem likely.

Mitigation of climate change involves reducing levels of greenhouse gases in the atmosphere. We can do this by consuming power more efficiently and by using cleaner, lower-carbon energy sources. We can also remove greenhouse gases from the atmosphere using either natural or artificial geoengineering processes. All of these strategies face political challenges, but whatever combination of mitigation strategies we employ, we must act quickly to protect the health and well-being of our planet and its inhabitants. An inherent problem is that we as a species are programmed to respond well to immediate, obvious hazards while postponing or not recognizing future or gradually developing hazards. Climate change falls in the latter category.

Effects on Oceans

As air temperatures increase, so do global ocean temperatures. From 1910 to 2010, the average world sea-surface temperature (SST) rose by more than 0.8°C (1.44°F). Such temperature changes have significant and complex effects on our climate, including more frequent and more energetic storms, changes in oceanic circulation and regional climate patterns, and sea-level rise. Warming oceans also have several feedback effects that increase the rate of warming.

Sea-Level Rise

Global warming leads to sea-level rise from two primary factors—partly from water added due to melting land ice but primarily (about 70%) from the heating and expansion of sea water. As water warms, water molecules vibrate faster and therefore take up more space, called thermal expansion. The actual rise in sea level from both thermal expansion and glacier melting was about 19 cm between 1870 and 2000 (0.15 cm/year), but the rate more than doubled to about 0.35 cm/year between 1994 and 2009 (**FIGURE 12-1**). The U.S. National Research Council now believes that global warming will cause sea level to rise between 0.3 and 1.2 m (about 1 to 4 ft) by 2100. Note that melting of Arctic Ocean ice does not raise the sea level because floating ice merely displaces the same volume of water as the submerged part of the ice. This is why melting ice cubes in a completely full glass of water will not make the water overflow.

Melting continental ice also contributes to sea-level rise. Global warming causes Greenland glaciers and frozen ground

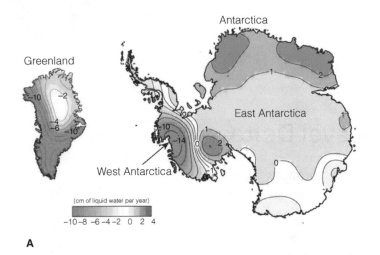

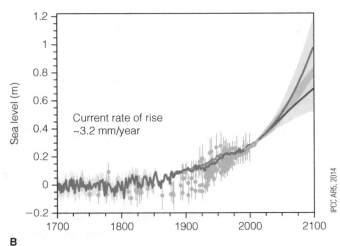

A

B

FIGURE 12-1 Global Sea Level

A. Greenland and parts of Antarctica are melting at different rates, adding to sea-level rise (data from 2003 to 2012). Red = ice loss; blue = ice gain. **B.** Sea-level rise from 1700 to 2015. Compilation of paleo sea-level data (purple), tide gauge data (blue, red, and green lines), altimeter data (light blue), and central estimates. Likely ranges for projections of global mean sea-level rise: red= RCP8.5 (little mitigation); blue= RCP2.6 (aggressive mitigation). Satellite data since 1993.

in Arctic lands to melt faster, pouring freshwater into the North Atlantic. From 2008 to 2012, the average annual rate of loss was 408 km³ of Greenland's ice melted, compared with 91.7 km³ in 1995. Antarctic glaciers are now flowing more rapidly, apparently at least in part from breaking up of the Antarctic ice shelf. Two studies in 2014 showed collapse of part of the Antarctic ice shelf by rapid disintegration of Thwaites Glacier, which had been buttressing other large glaciers. Major collapse, however, is expected to take at least 200 years. Some ice shelves are thinning because warmer water is melting their undersides. Similar collapse is occurring on Jakobshavn Glacier, Greenland, which was recently observed flowing into the ocean at a rate of 46 m/day in summer, three times as fast as it moved in 1990; the flow rate is increasing as the Arctic warms. Over the 160 years from 1850 to 2010, the toe of the glacier retreated about 40 km (only about 2.4 years' worth at the present rate). Movement accelerates because of warming ice, lubrication of the glacier's base by water (**FIGURE 12-2A**), and calving off at the glacier toe, which removes some of the resistance to flow (**FIGURE 12-2B**).

Accompanying that melting, sea level has been rising at an average of 3.3 mm per year since 1993, or 30 cm per 100 yrs (about 1 ft). In 2012, the average sea level rose to 3.5 cm above the average worldwide between 1993 and 2010; again melting glaciers contributed twice the rise of water expansion due to warming. For Greenland, half of its contribution to sea-level rise is from melting snow and ice and half to calving off from glaciers. Sea-level rates of rise are more than double at tropical latitudes compared with cold regions, partly because of expansion of water with warming. Local differences in rise rates depend on storms, tides, El Niño, and other effects. In the last 100 years, sea level has gone up 17.5 cm (almost 7 in.); on a gently sloping coast such as in the southeastern United States, which would move the coastline *landward* by about 17 m (55 ft)! Other big contributors to erosion are river dams that trap sediments from reaching the coast and dredging offshore sand for use in construction or to replenish storm-eroded coastal dunes. The rate of warming in Antarctica is likely to increase. Melting of all Antarctic ice would raise sea level about 73.5 m, while melting of all of the Greenland

FIGURE 12-2 Melting Glaciers

A. A moulin or melt hole, in Bench Glacier, southeastern Alaska, carries large amounts of water to the glacier base, increasing buoyancy and lubrication of glacier flow.

B. Jakobshavn Glacier, Greenland, flows to the right, and collapsed in April 2014 to calve off giant icebergs.

FIGURE 12-3 Rising Sea Level on a Delta

A. The Mekong River delta in Cambodia and Vietnam. **B.** Mekong River levels rise and fall with the tide at Can Tho, 120 km upstream from the South China Sea; branches of water hyacinth float in the tidal flood. **C.** Thousands living along its banks are endangered by sea-level rise.

ice cap would raise sea level by another 6.6 m. A 2014 study showed that the giant long-stable Zachariae Glacier of northeastern Greenland began shrinking rapidly in the preceding decade when warming permitted it to break through a blocking mass of sea ice—losing 10 billion tons of ice per year. The 600-km-long mass of ice, extending into the center of the ice sheet, and covering about one-sixth of Greenland, has now melted back more than 20 km. The other ice streams that drain Greenland had already become unstable. Although models suggest that complete loss of the Greenland ice sheet is not likely within 1000 years, the new results suggest that previous estimates of sea-level rise may be too low.

The consequences of a major sea-level rise would be serious for low-lying areas on which 5 billion of the world's 7 billion people now live; this includes many of the world's largest cities (**FIGURE 12-3**, and Chapter Opener photo). A rise of 3.5 to 12 m, for example, would endanger New York City, Tokyo, and Mumbai, India (formerly Bombay). Even a moderate sea-level rise of 18 to 35 cm would severely accelerate erosion of coastal areas including near Boston, Atlantic City, southern Florida, and the Gulf Coast.

A small sea-level rise along a gently sloping coast translates to a very large landward migration of the shoreline. A 1-m rise along the gently sloping southeast coast would move the beach inland by more than 100 m; at that point, many beachfront barrier-island houses would be in the ocean. Coastal barrier islands would move shoreward much faster than today, making innumerable coastal communities even more vulnerable than they are now. Large areas of coastal Louisiana, Florida, and the North Carolina Outer Banks would be submerged (**FIGURE 12-4**).

A serious consequence of sea-level rise is contamination of freshwater aquifers by salt water. Because freshwater is less dense, it floats on top of seawater (**FIGURE 12-5**). When sea level rises, as a result of land subsidence as on many river deltas, climate change, or a major storm surge, seawater can spread far inland to contaminate fresh groundwater. Heavy pumping of freshwater for domestic, industrial, or agricultural use can lower the freshwater table to below sea level, permitting contamination by salt water. Many areas in southern Florida have been affected by artificial draining of the Everglades between 1903 and 1980. Cape May in southern New Jersey has had to shut down many water-supply wells where the water table has fallen below sea level. Heavy pumping of groundwater for agriculture in the Central Valley of California has led to severe subsidence and permitted salt water intrusion into the Sacramento delta area. Groundwater levels in about half of the wells in the Central Valley dropped

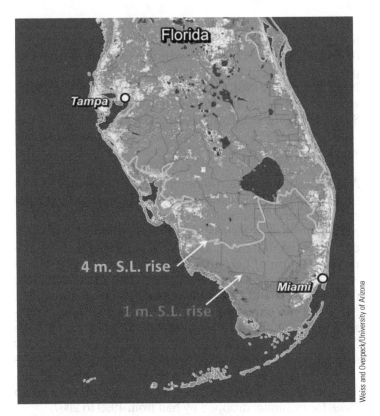

FIGURE 12-4 Potential Coastal Flooding

Whether sea level rises 1 m or 4 m, coastal areas of Florida would be underwater.

more than 3 m (10 ft) between spring 2013 and 2014. As sea level rises, coastal rivers rise as that higher sea level reaches farther inland. Raising coastal river levees to protect low-lying areas would be prohibitively expensive. Most major river deltas have many distributary channels, most of which are subject to ocean tides. Channels 100 km upstream in the Mekong delta, for example, rise and fall 2 to 3 m with each tide. Raising the levees that protect fields in the Sacramento River delta by only 15 cm (6 in.) would cost more than a billion dollars. Proposals in the Netherlands, long noted for its

levees and coastal storm barriers, suggest long-term spending of one billion euros per year because of sea-level rise accompanying global warming.

In poor countries, without the resources to build and maintain protective barriers, coastal populations will be displaced. Researchers now predict that over the next four decades, tens of millions of people will be forced to move due to climate change and rising sea level. Bangladesh, which occupies the immense near sea-level delta of the Ganges and nearby rivers, is already subjected to severe flooding during storms. Bangladesh would lose more than 17% of its land if sea level were to rise 1 m. That would have a disastrous effect on its delta agriculture and the millions of people who depend on it. Low levees along its many rivers and estuaries would be overtopped during floods. Even now, annual floods affect about 30% of the land. Deaths could run into hundreds of thousands during strong cyclones (**Case in Point:** Rising Sea Level Heightens Risk to Populations Living on a Sea-Level Delta—Bangladesh and Kolkata (Calcutta), India, p. 339).

Even if we are successful in the worldwide goal of stabilizing GHG concentrations in the atmosphere, it will take hundreds to thousands of years for temperatures to stabilize, as discussed earlier, while sea level continues to rise. It is clear that the world needs to act quickly and decisively to minimize further warming accompanied by additional rising sea levels.

Global Ocean Circulation

Changes in ocean currents resulting from changes in salinity could rapidly change climate in some areas. The large-scale circulation in the Atlantic Ocean (so-called thermohaline circulation) involves a current of warm, shallow water moving northward away from the equator. Closer to the Arctic it cools and sinks; the sinking water moves south at depth (**FIGURE 12-6**). The warm, salty Gulf Stream, which moves northeast along the east coast of North America and across to Europe, is part of this circulation. Westerly winds carry heat from the ocean to keep Europe warm in winter, in spite of its northern latitude. Changes in local ocean salinity from factors such as rapidly melting glaciers and ice sheets could

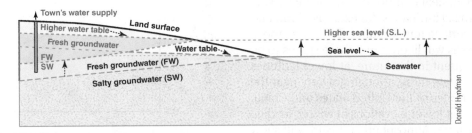

FIGURE 12-5 Saltwater Incursion With Sea-Level Rise

With present sea level, coastal towns can pump from the wedge of underground freshwater floating on top of denser salt water (black labels). As sea level rises, the wedge of lighter fresh water must float on top of the saltwater so the salt-water/freshwater boundary must also rise (blue labels). The bottom of the town's well may then be drawing from salt water.

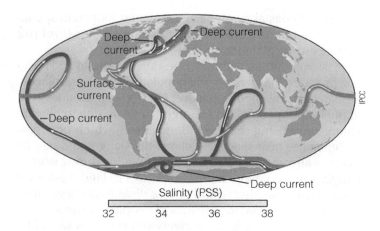

FIGURE 12-6 Global Circulation System

Red arrows represent surface warm currents and blue arrows deep cold currents.

alter the path and strength of the Gulf Stream. Lower-density freshwater cannot sink in seawater so circulation of the ocean current may slow or stop. Any reduction in the flow of the warm Gulf Stream to the north could thus cool Europe and some other northern hemisphere climates. However, unless the influx of meltwater from the Greenland ice sheet becomes much greater than expected, the thermohaline circulation is not likely to be disrupted this century.

Whether such cooling of western Europe, perhaps within the next century, would be sufficient to overcome the projected warming from climate change is not yet clear. Present models suggest that the rate of warming from climate change is likely to be strong enough to overcome any cooling from changes in thermohaline circulation.

Weather

Most of Earth's near-surface heat is concentrated in the oceans, and warmer oceans lead to more evaporation and thus greater rainfall. Satellite measurements indicate an increase of 1.3% more water vapor in the atmosphere over global oceans since 1988. More energy in the atmosphere often leads to more and/or stronger storms. Ocean heat contributes significantly to the energy that drives storms. Hurricanes, for example, form and strengthen where sea-surface temperatures (SSTs) are above 25°C (77°F). Warmer seas will likely cause these storms to be stronger and more frequent in the future, at lower latitudes where warm seas evaporate more, though not very near the equator. At middle latitudes of the United States and China, heating is concentrated at lower elevations in winter, leading to stronger storms in winter. Atmospheric pressure will show higher highs and lower lows, with associated stronger and more frequent storms. Natural disasters such as hurricanes and tornadoes will likely be more frequent and stronger. Warming sea temperatures may be responsible for the rise in frequency and strength of especially stormy weather in some areas.

Darkening of the Arctic Ocean surface as sea ice melts leads to more Arctic warming. As discussed in Chapter 10, that appears to lead to deep meanders of the jet stream that produce periods of extreme cold or warm weather at lower latitudes in both North America and Europe. Cold winter weather driven by the polar vortex of the eastern United States during the winter of 2014 is an example.

Solution of CO_2

A significant proportion (about 26%) of CO_2 expelled into the atmosphere is soaked up by dissolving into the upper ocean waters. For comparison, about 28% is taken up by plants and the rest stays in the atmosphere. Cold water can dissolve more CO_2 so cold ocean temperature promotes solution of CO_2 from the atmosphere; warm ocean water causes release of the CO_2 back into the atmosphere (**FIGURES 12-7** and **12-8**). As the ocean takes up more CO_2, the water becomes more saturated and can tolerate less additional CO_2. Because warm water can hold less CO_2, the heating of water caused by global warming dramatically slows the mixing rate of CO_2 to deeper levels, further lessening the ability of surface waters to take up additional CO_2. The rate of increase in CO_2 taken up by oceans for 1992 to 1999, for example, dropped by half from 1999 to 2007.

This phenomenon is an example of a positive feedback effect, in which a small disturbance in a system leads to effects that increase that disturbance. Another positive feedback effect is that warming increases evaporation from the oceans, increasing water vapor content in the atmosphere and thereby causing additional global warming. Recall that the feedback associated with water vapor is thought to roughly double the warming effect from CO_2 increases alone.

Warming of the oceans is a trend that cannot easily be reversed because of the high heat capacity of water (meaning it takes a lot of energy to warm it). Because water has such a high heat capacity, it soaks up more than 80% of

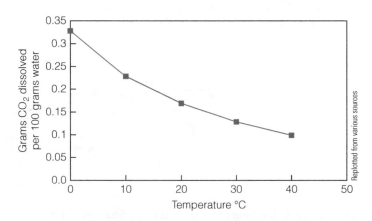

FIGURE 12-7 Water Temperature and CO_2

Increasing temperature decreases the maximum amount of CO_2 that can dissolve in water.

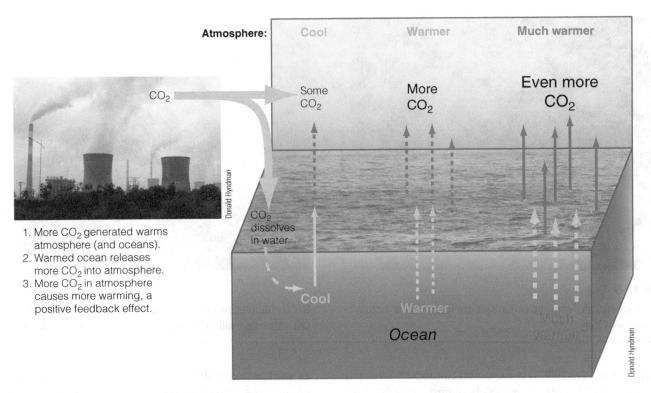

Atmosphere:

1. More CO_2 generated warms atmosphere (and oceans).
2. Warmed ocean releases more CO_2 into atmosphere.
3. More CO_2 in atmosphere causes more warming, a positive feedback effect.

FIGURE 12-8 Positive Feedback Effect
Warmer oceans absorb less CO_2, causing more warming—a positive feedback effect.

the effect of global warming. The top few hundred meters of the world oceans have warmed by about 0.3°C in the last 40 years; that heat slowly extends to deeper levels.

Because the oceans are such a huge heat reservoir, covering about two-thirds of Earth's surface, we cannot easily cool the atmosphere enough to lower ocean temperatures. Studies by National Oceanic and Atmospheric Administration (NOAA), reported in 2009, show that even if human-caused CO_2 emissions could be completely halted now (which they cannot), the amount of CO_2 in the atmosphere would decrease about 20% in the first century but then level off. Temperatures in the lower atmosphere would decrease only very slightly over hundreds of years because although human-caused CO_2 addition would cease, the huge amount already added to the oceans would slowly release back into the atmosphere. The continued return of heat from the oceans would prevent significant atmospheric temperature drop for at least 1000 years—thus, much of the human-caused warming effect to date is essentially irreversible. We instead need to focus on reducing our emissions to dramatically slow future increases in temperatures.

Ocean Acidity

Another consequence of further addition of CO_2 into the oceans is that CO_2 reacts with water to form carbonic acid (H_2CO_3), which makes the ocean more acidic (**By the Numbers 12-1:** Ocean Acidity). Normal seawater prior to the Industrial Revolution (1700s) had a pH of 8.18; by 2011 it had

By the Numbers 12-1

Ocean Acidity

Acidity of the oceans depends in large part on the amount of CO_2 dissolved: $CO_2 + H_2O = H_2CO_3$ (H_2CO_3 breaks down into charged atoms or ions that are acidic). Acidity is expressed as pH in numbers that range from 1 (extremely acid) to 14 (extremely alkaline), where 7 is neutral—neither acidic nor alkaline. Like the earthquake magnitude scale, each different whole number represents a difference of 10 times—a pH of 7 is 10 times more acidic (less alkaline) than a pH of 8.

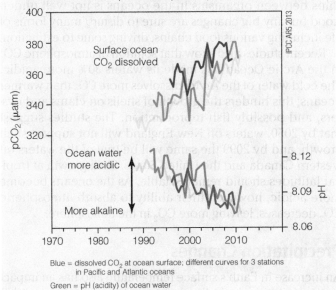

Blue = dissolved CO_2 at ocean surface; different curves for 3 stations in Pacific and Atlantic oceans

Green = pH (acidity) of ocean water

FIGURE 12-9 Ocean Acidity

A. These healthy calcifying sea organisms are located near Ischia Island, southern Italy.

B. CO_2 bubbles rising from nearby volcanic vents that acidify the seawater have killed the old reefs and prevented new growth.

dropped to 8.07. By 2050 it is expected to drop to 7.95. The increase in acidity (acidification) is occurring faster than predicted only a few years ago.

Oceans absorbed about 30% of all of the human-caused CO_2, thereby causing ocean acidification.

Changes in the acidity of oceans can be fatal to oceanic organisms such as corals, oysters, and clams that use $CaCO_3$ for their shells or skeletons. As pH falls (lower pH means higher acidity), $CaCO_3$ can no longer precipitate and may even slowly dissolve, killing the reef. Coral reefs also have a symbiotic relationship with reef algae, whereby the survival of one depends on the other. Warming of seawater tends to stress the coral and promote coral bleaching as the symbiotic algae die off (**FIGURE 12-9**). Major bleaching events have been recorded in the last few years. The variety of relationships between organisms in the oceans is not well understood but any big changes are sure to disturb many forms of life including various food chains, driving some to extinction.

Recent studies also show that solution of atmospheric CO_2 in the Arctic Ocean has made its waters 30% more acidic. The cold water of the Arctic dissolves more CO_2 than warmer oceans; this hinders the growth of shells on clams and oysters, and possibly fish reproduction. The studies suggest that by 2050, waters off New England will not support shell growth, and by 2099 the same will be true of the waters off western Canada and the United States. Shell growth at tropical latitudes should remain viable. As the oceans become more acidic, however, their ability to absorb atmospheric CO_2 decreases, leaving more CO_2 in the atmosphere.

Precipitation Changes

An increase in Earth's surface temperature also has an impact on precipitation, including greater extremes of wet and dry,

as well as more storms. Higher atmospheric temperatures cause more evaporation from the oceans and more water to dissolve into the atmosphere. Warming climates and warming sea-surface temperatures lead to greater evaporation. Thus greater moisture content in the atmosphere is likely to lead to heavier rainfall in some regions. In general, areas closer to the poles and near the equator will become wetter, and the warmer, mid-latitude regions will become drier (**FIGURE 12-10**). Thus many areas that are now dry will get drier, and many wet areas will get wetter. By 2080, much of Canada is expected to be wetter in winter, whereas Mexico is expected to be 15 to 50% drier. In summer, the precipitation across most of the United States may not change much, but the Pacific Northwest is expected to be 20 to 30% drier. Drought stress on forests in the region will likely lead to more severe wildfires and insect infestations such as by pine bark beetles. The potential for a catastrophic wildfire could affect hundreds of thousands of people living in the urban–forest interface.

Many midlatitude regions will become drier during the winter dry season because the descending dry air at about 30° north and south latitudes will expand toward the north and south poles. If atmospheric CO_2 concentrations rise from the current level of about 404 ppm to a peak of 540 to 950 ppm before the year 2100, long-term dry-season rainfalls in some areas will drop to conditions similar to those of the American Dust Bowl of the 1930s. The exceptional 2011 drought centered on Texas and the 2012–2015 California drought may be signs of some future conditions. Southern Europe and perhaps the southwestern United States would become even more arid; a long-term decrease in rainfall has already been documented in these regions.

In 2013, some farmers in the agriculturally important Central Valley of California received only 20% of their normal water allocation. San Francisco received less than

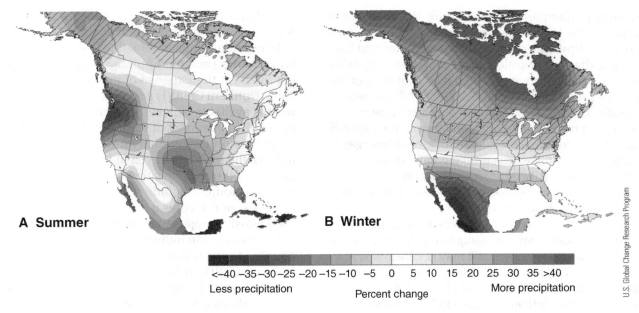

FIGURE 12-10 Projected Precipitation Changes

The relative expected change in precipitation to the end of the century, between 2010 and 2100 (higher emissions scenario) **A.** for summer (June to August) and **B.** for winter (December to February). The absolute amount of change depends on the actual emissions, but the trends are clear. Note that most of North America will become drier, especially in the summer growing season. The consequences for crop growth, water supply, wildfires, and other natural hazards are looming problems.

one-third of its normal average precipitation from July 1 to December 31. Southern California's Metropolitan Water district was told to expect only 5% of their normal water allocation for 2014, partly because reservoir levels were far below normal. Winter is normally California's wet season. Between 2013 and 2015, a stationary ridge of high pressure over the eastern Pacific Ocean was diverting winter storms farther north up the West Coast.

In contrast, higher-latitude areas, including the northern plains of the United States, Canada east of the Rockies, and eastern Asia north of China, are expected to become wetter.

Although populations in some northern climates might welcome modest increases in temperature, detrimental effects would include more droughts and floods. It might seem strange that there could be more droughts with more precipitation, but drought depends on the balance between precipitation and evaporation; higher temperatures cause more evaporation, which may have a larger effect than an increase in precipitation. Snowpacks are already melting earlier, supplying less recharge to groundwater, which is the source of most late-summer streamflows in many regions across the northern United States and Canada. As discussed in the Chapter 10 Case in Point, California Drought 2012–15, California's rains and mountain snows come primarily in winter, in time to fill its many mountain reservoirs and provide water for the long spring and summer growing season. With either drought or more rain in place of snow, more of that water would run off rather than being stored as snow for use later in the season.

Rising temperatures mean that future floods are likely to come earlier in the season with more frequency and severity. In the western United States and Canada, the spring snowpack has been melting as much as a month earlier than four decades ago. Thus snowpacks melt faster and earlier in spring, leading to earlier and higher floods. Higher floodwaters threaten levees not built for those levels. Consequences also include more severe mudflows and coastal erosion. Despite the increased flooding, these areas are also at risk for more dryness in the summer months. The western United States depends on the melting snowpack in late spring to retain moisture in groundwater for the dry summers. Higher temperatures mean that more precipitation falls as rain rather than snow, causing more to run off the surface quickly in some areas rather than slowly seeping into the ground as snow melts. Without the stored moisture from the later melting of the snowpack, summers are drier, leading to more thunderstorms and more fires.

An enterprising engineer in Kashmir and northwestern India built small artificial "glaciers" to provide later-season meltwater for irrigation. He channeled stream water into shaded canals to slow it and permit freezing. During spring thaw the built-up ice slowly melts to provide water for farms in summer.

Decreased rainfall in already dry areas could severely reduce the water supply for drinking and other municipal and industrial uses, farming, and ecosystems. Areas that already have water shortages would likely be in desperate shape during similar periods in the future. Water access in places such as the Middle East would spark even more conflict, if not war. Upstream countries, such as Turkey and Iran, have already diverted irrigation water with dams, depriving arid Iraq and Syria, which are plagued by still lower river levels. Increased shortages of potable water as climate warms could be alleviated by desalination of seawater in areas close to the coast but not without cost, increased energy expenditures, and thus additional production of greenhouse gases. Desalination can produce fresh drinking water in areas close to the sea but less easily far inland. This approach is also expensive and energy intensive, and thus does not provide a solution for large-scale agriculture.

Freshwater is necessary for every person on Earth and yet more and more people are struggling to obtain it; many others have become more aware that it may not always be readily available. The total amount of freshwater evaporated from oceans and falling on land has not changed dramatically with time, but other influences have led to shortages. The world population increase, greater industrial and agricultural use (which brings with it surface and groundwater contamination), as well as regional transfers of water supplies based on financial, political, and military issues have left some areas with adequate supplies and others with severe deficits. Variations in annual to multidecadal precipitation and long-term climate change have created ongoing concerns.

Competition for scarce water resources leads to more and more conflict—local, regional, and between countries. When an upstream country builds a dam to tap off water for their own use, the downstream country losing the water understandably takes offense, sometimes leading to war. Turkey (upstream from Iraq) and China (upstream from Vietnam and Cambodia) are recent examples. When conflict pits financial gain against basic food and water needs, issues become ethically charged. Such issues are unlikely to ever diminish.

The price of water for consumers not only depends on the abundance of water and the cost of delivering it, but also on political and other considerations. Many cities in dry environments restrict summer residential sprinkling days and charge higher rates at times to conserve water. Low prices of water delivered to some consumers in desert areas of Arizona and Nevada, however, encourage waste and further water shortages. In the 2010–2011 Texas drought and the 2012–15 California drought, authorities instituted water-use restrictions by all users and reduced water use by industry and agriculture, in many cases requiring farmers to fallow fields and ranchers to sell off herds because feed was not available. California cities were asked, in 2014, to cut water usage by 20%, but the voluntary target reached only 9%. The state then planned to require cuts of as much as 36%; some local water departments complained the rule would decrease property values, restrict swimming pool use and car washing, and raise rates. More aggressive restrictions came in 2015. Los Angeles, in the midst of the drought, still charged local residents a base rate of less than 0.2¢ for a gallon of water, and San Francisco 1.3¢/gal.; in Sacramento only about half

of the homes had water meters. Phoenix and Tucson, in the Arizona desert, charged only about 0.5¢/gal.

In northern China, alternating floods and droughts have for centuries prompted building levees, dams, and diverting rivers, at the expense of many lives and the environment. Northern China is now chronically short of water and what is available is heavily polluted by both industry and agriculture. A survey of the Yellow River and its tributaries in the north found that one-third of the water in 13,000 miles of the river is not even fit for agricultural use. Seventy percent of the groundwater under cities is not even safe for washing clothes. Including that piped in from rivers, only half the water in cities is safe to drink. China is now building a massive project to move water from the Yangtze River in the south to the industrialized water-starved north of the country. Conservation and pollution control would help. Only about 40% of water used by industry is recycled; the polluted water is dumped into rivers. China charges consumers about 0.5% of disposable income for water, compared with 2.8% in the United States. To conserve, it is now going to charge much higher rates for their heaviest users.

Arctic Thaw and Glacial Melting

The most dramatic warming is occurring in the polar regions because polar air is drier than tropical air and is thus more sensitive to the concentrations of greenhouse gases. Arctic coastal communities are now disproportionately affected by warming. The disappearance or early melting of coastal sea ice, greater wave energy, coastal erosion from longer wave fetch and larger waves, and changes in wind and storm patterns affect travel and fishing by coastal residents. Some native communities have had to relocate.

Melting Sea Ice

The surface of the Arctic Ocean has been frozen for as long as anyone can remember, preventing ships from using the Arctic as a summer pathway between the Pacific and Atlantic Oceans. In 1979, when first measurements were recorded, Arctic sea ice amounted to about 17,000 km^3 in late summer. By 2010, however, that had decreased to 4200 km^3 (**FIGURE 12-11**). Arctic sea ice, averaging 3 m thick in the 1960s, has thinned to about 1 m today. In 2012, a total of 30 ships made the passage. By the summer of 2013, 71 ships made the trip, but only 31 in 2014. Russia had icebreakers keep a sea route open along its northern coast, including escorting a Chinese container ship headed for the Netherlands; a Canadian Coast Guard icebreaker escorted the first bulk carrier loaded with coal from Vancouver through its northern Canadian coastal waters to Finland. The Arctic route saved four travel days and saved about $200,000. Although Arctic sea ice recovered somewhat from 2012, its long-term trend continued downward. There is also a feedback mechanism that accelerates melting of Arctic Ocean ice.

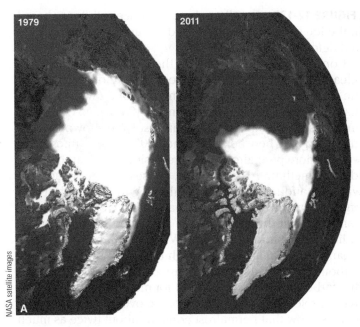

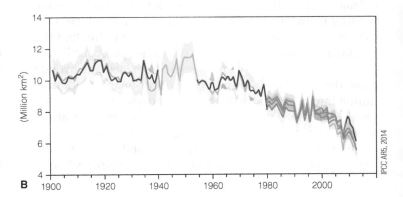

FIGURE 12-11 Melting Sea Ice

A. The minimum ice coverage on the Arctic Ocean was much less in 2011 (right) than it was in 1979 (left). In 2014, many more passages between islands were open.

B. Extent of Arctic July-August-September (summer) average sea ice. Satellite imagery since 1979.

White ice reflects solar energy, while dark sea water absorbs it, raising temperatures further. Summer Arctic sea ice is likely to disappear entirely by 2030.

Sea-floor Thaw

Sea ice melting in the shallow waters of the Arctic Ocean continental shelves appears to be promoting release of methane because the shallow sea floor is also thawing. **Methane hydrate** is a frozen methane-ice compound trapped in sediment layers at depths of 0.5 to 3 km under the seafloor of many of the world's continental shelves (**FIGURE 12-12**). That organic matter formed long ago when the sea level was lower and that surface was the vegetation-covered land.

As the Arctic Ocean is now melting and its temperature rising slowly, methane gas bubbles are rising from the thawing seafloor, especially in a huge area of the continental shelf north of Russia and Siberia. The hydrate thaws at temperatures of about 0° to 15°C. Based on projected temperature increases, huge amounts of methane could be released in coming decades. Seawater warming should destabilize some of the hydrate, leading to release of more methane. Increase in pressure from rising sea level should counter some of this effect. Preliminary calculations suggest that catastrophic release of methane should not occur this century. The impact of released methane on the greenhouse effect is compounded by the fact that methane is 20 to 25 times more effective than CO_2 in blocking outgoing thermal radiation.

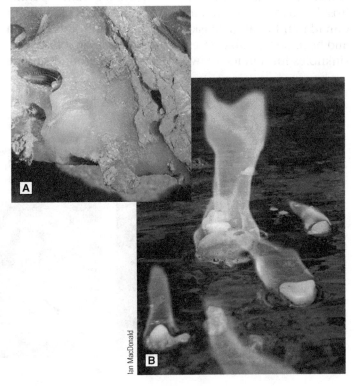

FIGURE 12-12 Methane Hydrate

A. Methane hydrate (pale orange) layer on the ocean floor.
B. Pieces of white methane hydrate decomposing and burning on the deck of a research ship.

The immense amounts of methane (natural gas) tied up in gas hydrate under the shallow ocean floor (estimated as 20,000 trillion m³) have led to extensive studies of possible commercial production for use as fuel. Because they import natural gas at a very high price, Japan, India, Korea, and Taiwan are especially eager to find alternative sources, but safe extraction from the ocean floor is still much more expensive. In March 2013, a Japanese drilling test about 600 m below the seafloor recovered about 69 m³ (4.2 million ft³) of gas. One extraction process injects CO_2 and nitrogen to fracture the methane hydrate and permit the gas to escape. A significant concern about natural gas production from methane hydrate is the danger of destabilizing the hydrate-bearing sediment during fracturing, to trigger a huge undersea landslide that would release large amounts of methane into the atmosphere. Methane is a very potent greenhouse gas, 20 times more so than CO_2. Japan is testing the feasibility of extraction of the methane, primarily on relatively flat areas of seafloor.

Permafrost Thaw

Perennially frozen ground, called **permafrost**, covers much of northern Canada, central Alaska, and northern Asia (**FIGURE 12-13A**). Permafrost around Fairbanks, in central Alaska, averages about 50 m thick, increasing to about 200 m in northern Alaska and more than 600 m near the Arctic coast. About one million km² of western Siberian permafrost has begun to thaw, along with huge areas of northern Canada and central Alaska. Roads and railroads deform and become unusable, trees tilt at odd angles in forests, and sinkholes form in ice thawed beneath the ground surface

(**FIGURE 12-13B**). Buildings that were once solidly anchored in the ice have to be abandoned because they are settling and deforming (**FIGURE 12-13C**).

Compounding the problem, the thawing of permafrost releases large amounts of additional greenhouse gases into the atmosphere. Just as occurs in seafloor thaw, permafrost thaw also releases methane gas. In a 2010 wake-up call, methane emissions from Arctic permafrost were shown to have increased 31% between 2003 and 2007. As the atmosphere warms, more permafrost thaws, releasing more carbon dioxide and methane. That, in turn, causes more atmospheric warming and more permafrost thaw. Windblown dust and organic materials found in permafrost in Siberia and Alaska contain about three-quarters as much carbon as is tied up in living vegetation on Earth. When permafrost thaws, its organic material is warm enough to decay, and some of its carbon returns to the atmosphere as CH_4 and CO_2. Much of this carbon, sealed in permafrost for more than 10,000 years, may be released into the atmosphere over the next century. Studies show that permafrost contains about twice as much carbon as is presently in our atmosphere and that more than one-half of the top three or more meters may thaw by 2050—and the rest by 2100. This would cause a huge increase in global warming, especially since CH_4 has 25 times as much of a warming effect as CO_2 for an equivalent amount of gas.

Permafrost melting also feeds more water into north-flowing rivers and into the Arctic Ocean, accelerating warming and melting there. Northern rivers are thawing several weeks earlier than in past decades. Warming and earlier ice-free Arctic waters permit greater wave heights from broader areas of open water, larger storms, and accelerated coastal erosion.

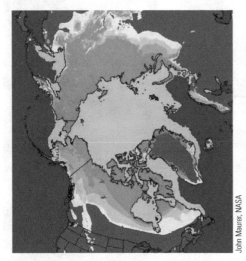

A

B

C

FIGURE 12-13 Permafrost Thaw

A. Permafrost distribution, northern hemisphere. Red and orange colors are, respectively, continuous and discontinuous permanently frozen ground. Still paler colors are sporadic occurrences of permafrost. **B.** Permafrost ice under a parking lot at the University of Alaska, Fairbanks, is thawing, leaving cavities in the ice and potholes in the ground. **C.** House settled into thawing permafrost near Fairbanks.

Average coastal retreat rates are 1 to 2 m per year. Roads across frozen ground in the Arctic of central Alaska and adjacent Canada are now useable for shorter periods because of thaw and destabilization in response to an average atmospheric temperature rise of 0.13°C (0.2°F) per decade.

Glaciers Melting

Glaciers in the world's highest mountains and on Greenland have been melting rapidly, especially since about 1930 (**FIGURE 12-14**). As glaciers melt, their smaller area is more closely surrounded by dark rocks that soak up more heat. That positive feedback effect, in turn, causes more rapid melting. Muir Glacier in Alaska, for example, is rapidly retreating, as are the remaining small glaciers in northern British Columbia, the North Cascades of Washington, and Glacier National Park, Montana. Himalayan mountain glaciers are especially important in maintaining freshwater flow for immense downstream populations during dry seasons, including those of Pakistan, northern India, Bangladesh, and China. The Ganges, Indus, Brahmaputra, Mekong, Yellow, and Yangtze Rivers support three billion people, almost half the world's population. As these people become more affluent, the demands on water for food and economic development expand. Many of the glaciers are at very high elevations in arid climates. Less snow in some recent years, black carbon in the form of soot from trucks and wood burning, and increase in greenhouse gases all contribute to the melting. That flow provides municipal water, irrigation of crops, and hydropower. Loss of Himalayan glaciers (see Chapter 2 Opener photo) threatens the water that irrigates rice and wheat crops that hundreds of millions of people depend on. Even in the southern hemisphere, in the southern Andes of Argentina and in Antarctica, huge glaciers are dramatically shrinking. In Africa, Mt. Kilimanjaro had large glaciers in 1912; by 2000, they were mostly open canyons.

Impacts on Plants, Animals, and Humans

Plants, animals, and humans adapt to their particular climate. If that climate changes, those species must either adapt or migrate to a more suitable environment.

Impacts on Plants and Animals

Many animals depend on particular plants for food; if those plants die out because of the stress of an increase in temperature or decrease in rainfall, or if the plants do not rejuvenate after a wildfire, the animals will also have to migrate or die out. The American pika, for example, a small rabbit-like mammal that lives in rocky alpine environments of western North America, stressed by warming, is migrating to cooler, higher elevations. In some mountain ranges it has disappeared completely in the last century. In a few others, it is adapting.

Some people have suggested that higher temperatures in some cool northerly climates might have some benefits such as improving agricultural productivity because plants need CO_2 to grow. For much of the coming century, that may be true for some presently cool areas with sufficient moisture and sufficient soil nitrogen. However, increased temperatures also reduce crop yields, and rising temperatures in very warm areas with marginal moisture will lead to more evaporation and drought. In many poorer countries in the Middle East, sub-Saharan Africa, and parts of Mexico, agricultural yields are more likely to decrease. In addition, for CO_2 to boost growth rates, plants need sufficient soil nitrogen; most parts of the world have a significant deficit of nitrogen. That deficiency limits rapid growth, and excess carbon dioxide can actually inhibit growth further. Plants deficient in nitrogen also provide lower protein contents and thus lower nutrition to animals and humans.

FIGURE 12-14 Glacier Melting
Muir Glacier, Alaska, August 1941 and August 2004.

Terrestrial plants and animals show their response to climate warming in their consistent migration toward the poles. Spring events are beginning earlier and in general the growing season is becoming longer. Studies of hundreds of species showed egg laying or flower blooming 2.3 to 5 days earlier per decade. In Edmonton, Alberta, for example, bud burst of aspen trees has appeared 25 days earlier than 100 years ago. Spring flowering dates for more than 150 plant species in Britain are now 15 days earlier than in the recent past. Trees appear to respond to three factors: increasing temperature and to a lesser extent increased carbon dioxide concentration and increased nitrogen deposition. Not all changes for agriculture have been negative. Black spruce at the northern limit of forest in eastern Canada grew taller in the three decades after 1970. Oranges, walnuts, and high-quality wine grapes in California have benefited from warming in the last 20 years, but avocados and cotton have suffered. By 2050, however, California's prime wine-growing area may shrink by as much as 60%. Southern Europe's main wine-growing areas may shrink by 68%, as suitable grape-growing areas shift to northern France, Germany, Poland, and parts of Russia. River-spawning Pacific salmon have showed up in Arctic rivers but are declining in warmer parts of North America.

Individual trees cannot migrate, so forest die-offs caused by fires and various pests inhibit reseeding at lower elevations and lower latitudes, while reseeding becomes successful at higher elevations and higher latitudes. This has been well documented in forest species in the Sierra Nevada range of California. Marine species show similar shifts; North Atlantic plankton have moved northward by 10° or about 1100 km in the last four decades. Coral reefs and clams also show stress from warming, though separating the effects of overfishing, ocean acidification, and water pollution is difficult. Warming lakes show similar stress.

The International Panel on Climate Change (IPCC) suggests that up to 20 to 30% of all animal species may die out by 2050 if temperatures rise 1.5 to 2.5°C (2.7 to 4.5°F). Studies of 1700 species found that about half of them show gradual migration in clear response to climate warming. Northward migration of birds and butterflies has averaged about 6 km per decade. Some animals die out in heat waves, others through lack of water in drought, scarcity of food supply because of disappearance of their primary food, invasion of other species including insects, or disease related to any of these factors. At greatest risk are birds, mammals, and amphibians, including sea turtles.

Moose, the largest members of the deer family, live primarily in cooler climates, including Canada, the northernmost United States, northern Europe, and Asia. They are dying off in record numbers, especially since the 1990s. In New Hampshire, Minnesota, and Montana, ticks and other parasites that prey on moose now survive through the warmer winters, killing many moose. In British Columbia, pine bark beetles have killed broad swaths of forest, leaving moose exposed to hunters and animal predators.

Polar bears' main food source is arctic seals that live in the ocean and on sea ice. Although they get most of their seal food between April and July, seals on sea ice are hard to catch, especially as the sea ice is breaking up. The World Wildlife Fund reports that a typical bear needs at least one fat-rich seal per eight days to maintain its body weight. Bears on land can find very little food but occasionally take reindeer, musk ox, rodents, seabirds, fish, or berries, and can scavenge other bears' kills. They are forced to fast when the ice breaks up and until it refreezes. In areas such as western Hudson Bay, ice breakup is now 3 weeks earlier than 30 years ago. Some polar bear populations in southern parts of their range declined by 22% between 1987 and 2004. The USGS, projecting a moderate rate of climate change, predicted the disappearance of two-thirds of the world's polar bears by 2050.

Effects on Humans

For humans in areas with high risk of certain hazards, those risks are increasing because of greater incidence of weather-related extreme events such as floods, droughts, hurricanes, heat waves, and wildfires. Rising temperatures lead to lower snowpacks and earlier, higher, and shorter runoff times. Dry season availability of water is declining, as is water quality, surface water and groundwater availability, and navigability of major rivers. Groundwater supplies in some areas, already stressed during droughts because of more pumping for irrigation, are likely to be severely affected in the future. In some cases, drinking water supplies have become an issue. Severe droughts in recent years in California and Texas are good examples. The Edwards aquifer of Texas and the Ogallala aquifer in the central states are also likely to suffer.

Heat waves in major European cities are increasing, as are heat-related deaths. Elsewhere in areas more accustomed to high temperatures in certain seasons, prolonged exposure to extreme temperatures can be deadly. This is especially true for those without access to air conditioning, including the inner-city poor, farm workers, and athletes. Drought-related water and food shortages are likely to be most acute in large parts of Asia. Heat waves in southern India and Pakistan in 2015 killed more than 2500 and 1100 people, respectively, a large proportion from heat stroke. Respiratory illnesses from exposure to pollution and pollen are increasing, as are malaria, dengue fever, Lyme disease, and West Nile virus. All of these are spreading northward into areas with previously lower temperatures. The geographical range of malaria and dengue will also change. Wildfires are becoming larger and more widespread not only because of decades of aggressive fire suppression but also because of earlier and more prolonged dehydration of vegetation and from invasive insects such as the mountain pine bark beetle and spruce budworm that thrive when severe winter temperatures fail to kill them off. Such insects have spread northward into areas that were previously too cold for their survival.

The U.S. military now views climate change as a significant concern in aggravating international conflicts. It views changes in climate as stressors that can impair access to food and water, thereby leading to tension between neighboring countries over limited resources, the failure of already fragile governments, and even war. They view it as a "threat multiplier" for future conflicts. Such threats include drought, floods, resulting food shortage, and mass migrations. Their concerns also address planning for natural hazards such as sea-level rise and flooding of military bases around the world.

Given that climate change is already here, we need to adapt as best we can. Even if we can quickly reduce the greenhouse gases expelled into our atmosphere, GHG concentrations will still increase for a few decades and will not decrease thereafter for many decades or centuries. Adapting to changing climate will be very expensive. However, if we decide to ignore the problem because it is already too late, the consequences for civilization will be even more severe.

Mitigation of Climate Change

The Intergovernmental Panel on Climate Change, 5th Assessment Report (AR5), finalized in 2014, reexamined climate change in great detail, including its causes, effects, and challenges. The report placed much greater emphasis on the societal effects of climate change and how mitigation efforts are likely to affect different groups of people.

The panel concluded that in order to mitigate the detrimental effects of climate change, we need to "reduce the sources and amplify the sinks of greenhouse gases." Mitigation choices should consider sustainable development, impacts on poverty, equity, justice, fairness, and the social and economic capability of groups of people to address the issues. These issues and their interrelationships are complex and involve difficult choices, many of which are beyond the scope of this discussion; nonetheless, we will outline the broad approaches to mitigation here.

Stabilization of the effects of climate change will need to involve some combination of reducing the amount of greenhouse gases emitted into the atmosphere and increasing the amount of greenhouse gases removed. Most climate scientists believe that the simplest and most effective mitigation strategies to minimize the increase in climate change include:

1. Lowering the transportation-related energy use that generates greenhouse gases: driving fewer miles, with more fuel-efficient vehicles, and flying less.

2. Using less fuel to heat our homes and buildings by adjusting thermostats and insulating ceilings, walls, and windows.

3. Replacing lightbulbs and appliances with more efficient versions such as LEDs.

4. Using energy from wind, solar, hydropower, nuclear, and to some extent natural gas where it replaces coal and oil. Capture and store underground the carbon emitted from power plants.

5. Minimizing deforestation and soil degradation and planting more trees.

Motivating world leaders to implement aggressive mitigation strategies has been challenging, however. The IPCC models consider various scenarios depending on how aggressive our mitigation is (**FIGURE 12-15**). Even in the most optimistic scenario (green line in the graph), CO_2 would continue to rise for another two decades or so. Unfortunately, even if we could immediately stop emitting all greenhouse gases from all sources, the CO_2 content and temperatures of the atmosphere would continue to rise slowly as CO_2 continued to seep back out of the oceans. Although this is a discouraging conclusion, it is still much better than unchecked emissions and the related climate effects.

Reduction of Energy Consumption

If world population increases by 10%, greenhouse gas concentrations will also increase by 10%, assuming no change in average per capita greenhouse gas production. The most direct way to slow the increase in greenhouse gases is to minimize population growth. Every person on the planet contributes to global warming, and world population is now more than 7 billion.

The rate of population growth is higher in developing nations than in the developed world, and it is in the developing world where energy demands are increasing most. The total annual CO_2 emissions from fossil fuels in developing nations have now surpassed those of developed nations, and their rate of increase is much faster (**FIGURE 12-16**). For example, the increase in U.S. energy demand is 1 to 2% per year, but in China, where a growing consumer class is newly able to own cars, it is about 10% per year. Of the total 40% projected increase in demand for fossil fuels over the next 25 years, three-quarters will likely come from developing nations, including India and China, each of which has three or four times the population of the United States.

An alternative to the challenges of minimizing population growth, to reduce greenhouse gas emissions, especially in North America and Europe, is to reduce the per capita consumption of energy through conservation and greater efficiency, including driving less; using bicycles, more fuel-efficient cars and trucks, and LED lightbulbs; and installing more insulation. Conservation is an inexpensive, simple, and easy way to save large amounts of electric power and therefore reduce carbon output used in power production. According to the International Energy Agency, energy efficiency savings have already saved more in avoided emissions of CO_2 than all of the cost of renewable energy

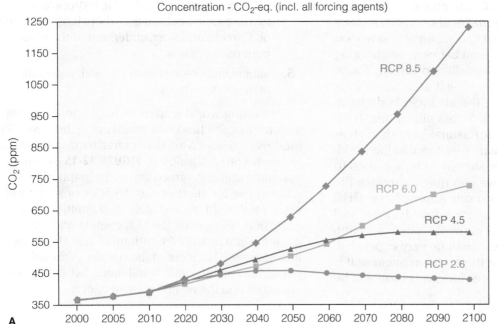

Concentration - CO₂-eq. (incl. all forcing agents)

RCP 8.5: high population growth, minimal mitigation

RCP 6.0: stabilize GHG concentrations to peak radiative forcing* of 6.0 watts/m² by the year 2100

RCP 4.5: stabilize GHG concentrations to peak radiative forcing* of 4.5 watts/m² by the year 2100

RCP 2.6: stringent mitigation, negative GHG emissions after 2070

IPCC AR5, 2014

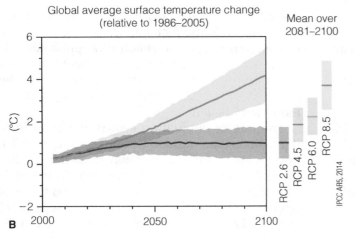

Global average surface temperature change (relative to 1986–2005)

Mean over 2081–2100

IPCC AR5, 2014

FIGURE 12-15 Scenarios for Climate Change Mitigation

A. The four scenarios for expected GHG increases through 2100 assume a mix of power generation sources: Some use of fossil fuels, some carbon capture and storage. Representative pathways are identified by radiative forcing values reached by the year 2100; various mitigation options (population, GHG emissions, etc.) lead to those final radiative forcing levels. (See *radiative forcing*, Chapter 11, p. 306–307. **B.** Projected temperature increases correlated with the four scenarios.

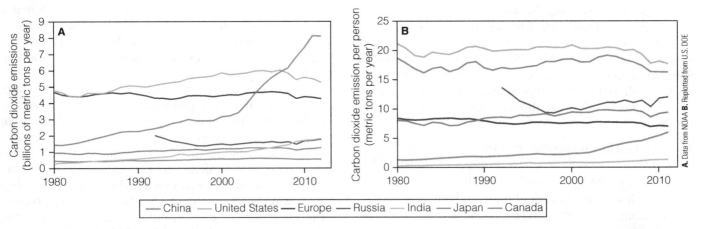

FIGURE 12-16 Emissions by Country

A. Total greenhouse gases for countries and continents that have highest total emissions. **B.** Average per person emissions in the same locations.

sources—about $300 billion worldwide in 2011. Examples include improved mileage and vehicle emissions standards in the United States and China, energy-star designations for refrigerators and other appliances, and fluorescent and LED lightbulb use.

Another approach that is gaining acceptance is energy saved during high-demand periods such as on the hottest days. Some power supply companies are paying big power users to cut back on use during high-demand times; that saves them from building more power plants that are used only during such times. Examples are to cut back on internal lighting and permitting internal temperatures to rise on warm days.

Buildings use about half of all fuels in heat, cooling, light, and ventilation. Conservation mechanisms for buildings include insulation to minimize loss of heat, make roofs light-colored or covered with vegetation, or maintain cooler environments, more-efficient lighting, and circulating and filtering air. Reductions in per capita energy consumption are possible through developing more efficient household electronics and cars, as well as changing people's energy use and driving habits, including greater use of mass transportation. Urban planning to reduce sprawl and encourage less driving would reduce miles traveled and energy use.

Cleaner Energy

Another approach to reducing carbon emissions is to get more of our energy from "cleaner" sources that produce fewer or no greenhouse gases. Additionally, researchers attempt to make some of the cheap and "dirty" traditional energy sources cleaner. As you might expect, each energy source has positive and negative trade-offs (Table 12-1).

Among the least expensive energy sources, coal and natural gas are the greatest sources of greenhouse gases and other pollutants. Energy cost values like those listed in the table do not normally reflect hidden costs such as water and air pollution and their health effects; in fact, clean energy sources that would reduce GHG emissions would also have large benefits for public health. For coal and nuclear power plants, decommissioning costs (when plants are shut down,

Table 12-1		Comparison of Energy Sources		
ENERGY SOURCE	**COST: ¢/kWh***	**FUEL & MAINTENANCE: ¢/kWh***	**PROS**	**CONS**
Coal: conventional with gasification	9.6 (8.0) 11.6	3.0	Cheaper than most other fuels	Worst CO_2 emissions, worst air and water pollution
Natural gas	6.4	4.6	Inexpensive; cleaner than coal; easily started and stopped	Emits half the GHGs of coal
Biomass	10.3	4.0	Generates GHGs but is carbon-neutral because fuel is regrown	Emits nitrogen oxides and other pollutants
Wind: on land offshore	8.0 (3.7) 20.4	Minimal	No GHGs or other pollution; 24-hour generation; costs dropping rapidly	Only when wind blows; can kill birds, though cats kill many times more
Solar: thermal photovoltaic	24.3 13.0** (7.2)	Minimal	No GHG or other pollution; costs dropping fast; photovoltaic projected to reach 6.4¢ in 2018 and continue to drop	High building cost; energy only when sun shines but thermal can store heat to generate at night
Nuclear: advanced	9.6 (10)	1.2	No GHG emissions	Difficult and very expensive waste disposal; potential nuclear accidents
Geothermal	4.8	Minimal	No GHGs or other pollution; clean; inexpensive	Suitable only in specific sites with high ground temperature
Hydro	8.4	0.6	No GHGs or other pollution	Generally seasonal; storage depends on precipitation; dams inhibit fish

*Building and operation with 30-year depreciation. Costs exclude any subsidies; figures from 2012, for newly built plants projected to come on line in 2019. Parenthetical numbers from 2015.
**Costs are for utility-scale installations.
CCS = carbon capture and storage; GHG = greenhouses gases
Source: Data from U.S. Energy Information Administration, 2014, and other sources.

such as Fukushima, Japan, and San Onofre, California) can also be very high and constitute real long-term costs of those fuels. Coal-fired plants, in particular, will need to be phased out because many are getting old, inefficient, and uneconomical.

NATURAL GAS AND FRACKING The growing popularity of **fracking**, the process of producing natural gas from underground shale, has advantages and disadvantages for climate change and the environment. The increase in oil and gas supplies in North America due to fracking and the resultant decrease in price has made coal, the most polluting and formerly least expensive fuel, much less desirable. Coal-fired power plants in the United States are rapidly being closed or refitted for natural gas. Because natural gas produces half the CO_2 compared with coal, these changes show a significant benefit in slowing global warming.

Unfortunately, fracking to produce oil and gas also has major downsides for the environment. Massive volumes of freshwater are used in the process, mostly from underground aquifers, depleting what is available for domestic use and agriculture. Contaminated wastewater may also infiltrate into nearby aquifers to endanger drinking supplies. Additionally, approximately 5% of the natural gas produced by fracking is emitted back into the atmosphere in the form of methane, a GHG 25 times more potent than CO_2 as a contributor to global warming (**FIGURE 12-17**). Small earthquakes have also been shown to correlate with some fracking operations, triggered either by injection of fracking fluids or more commonly by high-pressure underground disposal of wastewater.

Many states are now placing restrictions on fracking, such as limiting gas flaring, requiring disclosure of the chemicals used and analysis of nearby water supplies before and after drilling, and allowing residents to sue for damages. A few states have placed moratoriums on fracking or banned it entirely; others are studying the issue.

NUCLEAR ENERGY Nuclear energy, which currently produces about 20% of electrical power in the United States, is a cleaner alternative than fossil fuels, but nuclear plants are much more expensive to build and they raise safety concerns (see **Case in Point:** Hidden Costs of Nuclear Energy—Fukushima Nuclear Power Plant Failure, 2011, p. 341). The last new nuclear plant placed on line in the United States was in 1996. A nuclear plant south of Augusta, Georgia, completed two reactors in 1987; they are now planning to complete two additional reactors in 2016 or 2017. In contrast to the reactors in Japan that failed in 2011 when their electric pumps stopped working, both of the Georgia reactors use passive circulation of cooling water. However, they have encountered quality problems and cost overruns of about 10% more than originally budgeted. One nearby town seems more focused on the additional jobs and new taxes that the new reactors would bring than on any dangers of a nuclear disaster. Some older U.S. reactors have been shut down recently for economic reasons and more than 30 are considered at risk for the same reason.

A new generation of smaller, simpler nuclear reactors promises to reduce the construction costs for nuclear energy. These refrigerator-size nuclear power plants generate 25 megawatts of power, about 2.5% of the power of a traditional reactor. They can power, for example, towns with 20,000 homes, commercial ships, mining operations, or desalination plants. Also in the works is a small reactor 10 to 12 m long, mostly underground, that uses spent nuclear fuel or fuel rods (now nuclear waste), activated with a small amount of enriched uranium. The design would produce about 500 megawatts of electric power using heat exchangers to drive turbines and generate electricity.

In 2011, new nuclear power generation cost 10¢/kilowatt-hour (kWh), compared with 4¢ for natural gas. As of January 2013, some nuclear power companies were considering closing power plants because they could no longer compete with the low price of natural gas. In 2012, gas prices dropped 15% in California, 27% in New York, and 47% in Texas. Decommissioning costs of San Onofre are estimated at $4.14 billion, which would include dismantling, transporting, and disposing of radioactive equipment and fuel. Such expenses add to the hidden cost of nuclear power.

Although alternative energies are currently more expensive than fossil fuels, if the real costs of burning fossil fuel, especially coal, including the costs of health problems associated with other emissions and sequestration of CO_2, are included, wind and other power sources are similar to or lower than fossil fuels. As alternative energy technologies improve, costs should decline, making them more appealing to both industry and consumers.

WIND ENERGY For example, production of wind energy is becoming more economical and is expanding rapidly. Although wind accounted for only about 3.9% of the total U.S. use in 2013, about 43% of new electricity generated was from wind. Among the advantages of wind energy are

FIGURE 12.17 A Hidden Cost of Fracking
Natural gas being burned off at well site in North Dakota.

that it generates no CO_2, SO_X, N_2O, mercury, airborne particulates, nor waste, and uses no water. Water use by power plants becomes especially important in light of water shortages resulting from climate change. If 20% of all electric power were generated by wind, water consumption from electric power generation in the western states would be reduced by about tenfold to more than 100 billion gallons.

SOLAR ENERGY Solar still accounts for less than 2% of total energy, but because of the decreasing cost of solar power and concerns about impending restrictions on pollution from coal-fired power plants, some coal-power companies are exploring the possibility of entering the solar power market. The 392-megawatt Ivanpah solar-thermal power array in the Mojave Desert at the California–Nevada border, completed in 2013, is the largest solar power plant in the world (**FIGURE 12-18**). The plant uses 300,000 mirrors to reflect the sun's energy to boil water to supercritical steam in three huge towers to generate electricity. However, 15 months after start-up, it was producing only 40% of expected electricity. Problems include broken equipment, far more gas-powered steam for warm-up each morning, and less sun than expected, partly because of air pollution from cities upwind. Traditional photovoltaic solar panel farms have performed much better and produce electricity for much lower cost.

One challenge for solar energy is that the places with the most sunlight are not necessarily those that have the greatest energy need. For example, in Europe the sun is abundantly available in Spain and Greece, but the greatest demand for new power is in Germany, which continues to burn large amounts of heavily polluting coal.

Carbon Capture and Storage

In addition to efforts to reduce carbon emissions, some scientists and policy makers propose removing or storing greenhouse gases from the atmosphere.

STORING CARBON UNDERGROUND In a method called, **carbon sequestration**, carbon is captured at its source, for example, from fossil fuel–powered electrical generation plants (**FIGURE 12-19**). Once the CO_2 has been captured, it must be placed into long-term storage, such as underground in the pore spaces of sedimentary rocks. A few companies are now storing CO_2 in abandoned oil reservoirs, for example in Canada and one in Algeria. Others are in the works (see **Case in Point:** CO_2 Sequestration—The Weyburn Sequestration Project, p. 340). Carbon capture and storage was expected to remove 36 million tons of CO_2 emissions per ton by 2015, but the effort is falling well short of that, in large part because costs are proving to be much higher than expected. Several planned coal-fired power plants are being scrapped before being placed on line. Concerns about underground sequestration of CO_2 include fears of the gas escaping, the contamination of aquifers that have useable groundwater resources, and earthquakes caused by pumping of fluids underground at high pressure.

Instead of burning coal directly in a power plant, some new "clean coal" plants capture CO_2 expelled from coal-burning plants and permanently sequester it underground or use coal gasification to convert the coal to a flammable gas by heating with oxygen and steam; it is cleaned of CO_2 and pollutants such as sulfur and phenol and then burned to

FIGURE 12-18 Solar Thermal Power
Ivanpah solar thermal power arrays in the Mojave Desert in California.

California Energy Comm/Bright Source Energy

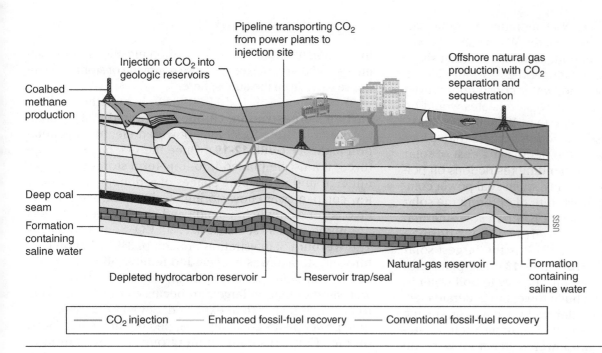

Injection of CO_2 into geologic reservoirs

Pipeline transporting CO_2 from power plants to injection site

Offshore natural gas production with CO_2 separation and sequestration

Coalbed methane production

Deep coal seam

Formation containing saline water

Depleted hydrocarbon reservoir

Reservoir trap/seal

Natural-gas reservoir

Formation containing saline water

USGS

FIGURE 12-19 **Carbon Sequestration**

Carbon sequestration involves removal of CO_2 from a fossil-fuel-burning power plant, transporting it to a disposal site, and storing it permanently underground.

— CO_2 injection   — Enhanced fossil-fuel recovery   — Conventional fossil-fuel recovery

drive the power turbines. Although cleaner, the process still generates abundant CO_2.

"Clean coal," however, is a misleading term. Although sequestration has been successful on a moderate scale to improve the recovery of petroleum from old oil fields, costs are proving higher than expected. A large power project by Mississippi Power, with expected completion by 2016, is in trouble, and power companies are pulling back because building new plants is proving cost-prohibitive. At least five large carbon-capture projects were canceled between 2012 and 2015. Concerns also include contamination of freshwater supplies and leakage back to the surface. Earthquake issues with pumping fracking fluids back underground are proving additional problems. Some states such as New York, in December 2014, have banned fracking, and the U.S. Environmental Protection Agency (EPA) has set new standards for the procedure, including disclosure of the chemicals used and standards for disposal of waste fluids. Some towns in Colorado, Oklahoma, Texas, and elsewhere have banned fracking, but in May 2015, Texas passed a law that prohibits such bans. Germany recently put a freeze on fracking until at least 2021.

Furthermore, costs of emissions and sequestration add costs that make coal uneconomical in comparison to natural gas. American Electric Power Co. and Tennessee Valley Authority, after many tries now, plan to close many of their dirtiest coal-fired plants, replacing them with natural gas or renewable resources.

Another approach is to precipitate the CO_2 as a solid compound that can be permanently stored. In 2009, in a large-scale experiment in West Virginia, American Electric Power precipitated CO_2 by reaction with ammonium carbonate solution to produce ammonium bicarbonate, a solid that could be stored. Although the process worked, carbon capture was not required by law and was not economical.

STORING CARBON IN PLANTS AND SOILS Plants are well known for taking in CO_2 and giving off oxygen. In fact, they store about 90% of the carbon in the U.S. environment and sequester 10 to 15% of emissions in the United States. How much they store in other countries depends, of course, on the amount of their plant and forest cover. Plants store carbon in their leaves, branches or stems, and roots. When the tree dies or is harvested, the carbon stays in their remains until it decomposes or burns. Carbon is also stored in rotting leaves and soil organisms; carbon stored in roots and soil can be twice as much as that stored aboveground. Forests sequester larger amounts of carbon than other plant cover. Forest and range fires release much of that carbon as does deforestation, which converts forest to ground crops. Sequestration in plants does not have the serious downsides of sequestration underground.

Geoengineering Solutions

In addition to carbon capture and storage, various means have been suggested to reduce global warming, efforts collectively called climate engineering, or **geoengineering**. There are two general approaches: removing carbon dioxide from the atmosphere and reducing the amount of sunlight that is absorbed by Earth's climate system.

In addition to the carbon capture and storage methods already discussed, some scientists propose accelerating the natural removal of atmospheric CO_2 and increasing the storage of carbon in land (e.g., forests and other vegetation). Ocean fertilization with iron would promote growth of microorganisms that would slowly sink to the bottom, carrying carbon dioxide with them. Interest has waned because of technical difficulties and potential to upset the food chain balance in the oceans. Any CO_2 removal from the atmosphere by that process would be partly offset by outgassing from the oceans and terrestrial reservoirs. Side effects of various processes include change in landscape albedo by planting forests, removal of oxygen from the ocean by fertilization, and increased N_2O emissions. Uncertainties involve the effectiveness of the method and the permanence of the storage.

The second main approach to geoengineering is **solar radiation management**, or artificially reducing the amount of sunlight absorbed by the climate system. Proposed methods include injecting aerosols such as SO_2 into the stratosphere to reflect sun energy, much as occurs naturally with very large volcanic eruptions. Although this would decrease global temperatures, it would also likely produce a small but significant decrease in global precipitation, potentially affecting freshwater supplies for drinking and agriculture, and aerosols would likely cause modest depletion of polar stratospheric ozone (i.e., increase the size of the ozone hole), which protects us from cancer-causing ultraviolet sunlight. Because the SO_2 would wash out of the atmosphere within a year or two, the injection would have to be done almost every year. Another problem is that if solar radiation management were terminated because of cost or unanticipated consequences, without other mitigation strategies surface temperatures would increase within a decade or two to what they would have been without the intervention (**FIGURE 12-20**). In addition, the management would not compensate for ocean acidification from increasing CO_2.

All of these geoengineering strategies are theoretical and few have been tested. As should be apparent throughout this chapter, our climate is a very complex system that we don't fully understand. Large-scale intervention would present tremendous risks of unintended consequences. Another criticism of geoengineering is that it would permit governments to be complacent about taking the steps needed to reduce our use of fossil fuels. Climate scientists caution that geoengineering could only supplement other mitigation strategies—it cannot be a substitute for reducing GHG emissions.

Political Solutions and Challenges

Different countries have different responses to the increasing carbon dioxide problem. As of 2015, the U.S. official position seems still focused on reducing the *rate* of *increase* in production of CO_2 by a certain amount by a given year. As of November 2014, the official plan by China's cabinet is to limit energy consumption to limit energy *growth* to 28% higher than 2013 levels by 2020 and to stop growth of CO_2 emissions by 2030. Although this may seem like progress, neither country would strive to reduce total emissions; world CO_2 levels by 2020 would continue to rise.

The important point lost on many people is that an agreement to reduce the *rate* of increase in CO_2 production or to limit growth of CO_2 emissions merely slows the *increase* of atmospheric CO_2. In fact, we need to reduce the *total amount emitted* into the atmosphere.

Many of these greenhouse gas–reduction strategies are feasible. The greatest problems are political. The public demands economic growth, jobs, and improvement of their standard of living; government leaders are reluctant to hinder economic growth. The U.S. Supreme Court ruling that the EPA can regulate carbon dioxide emissions as pollutants paves the way for use of constraints such as a carbon tax or cap and trade. Furthermore, this is a global problem and any solution will involve global cooperation. North America, Mexico, and the Canadian provinces of British Columbia and Quebec have carbon taxes. A **carbon tax** is an indirect tax that strives to account for the actual societal costs of use of a fuel, including such indirect aspects as air and water pollution, health, and climate change. There is considerable opposition to a carbon tax, especially in countries such as the United States that are heavy users of coal and are the largest per capita emitters of CO_2. Other proposals focus on taxing the production of greenhouse gases. An increase in the price of any commodity such as gasoline is an effective way to cut down on its use, regardless of whether the increase is dictated by scarcity or an imposed tax.

Another major proposal is to implement a **cap-and-trade** plan to limit carbon emissions. Such a government-controlled mechanism would provide economic incentives to reduce emissions from power plants and other factories by designating a cap on amounts of pollutants that can be emitted. The pollutants are allocated among users as "permits." If a company uses less than their allocation, they may sell their excess permits to others that need them. The combination provides an incentive to minimize emission of pollutants. If a company exceeds its cap, it is fined, thereby increasing the

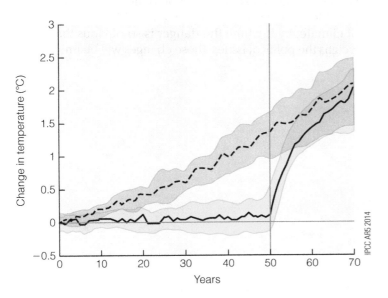

FIGURE 12-20 Solar Radiation Management

Reduction in solar radiation by adding sulfate aerosols to prevent atmospheric temperatures from rising further (yellow curve) would have to be continued indefinitely. Otherwise (e.g., after 50 years as shown here) CO_2 would quickly raise temperatures to their former trend within 10 to 20 years.

cost of its goods. The largest trading program for greenhouse gases is in the European Union, operating since 2005. Nine northeastern states operate a more limited program covering power. In 2012, California became the first U.S. state to start a cap-and-trade plan. A few times a year the state auctions off some of its carbon permits; the offsets can be a maximum of 8% of a company's carbon responsibility. Initially, the plan affects power plants and factories, with transportation fuels to follow in 2016. Households will receive a twice-yearly dividend from the money raised.

Meaningful reductions in greenhouse gas emissions can only be accomplished if all major industrial producers of CO_2 commit to the effort. Nations continue to be polarized over who is to blame and what to do about climate change. The countries that contributed most CO_2 historically, such as the United States and Canada, profited from unrestrained industrial development to become the wealthiest nations in the world. The countries that are today going through periods of rapid industrial development, such as China and India, feel that they too should be allowed to develop without environmental restraints. They argue that the industrialized nations must first reduce their emissions because they can more easily afford the associated costs. Without the cooperation of large emitters like China and India, the prospect for significant reduction of emissions is bleak. However, China, driven by choking air pollution in its giant cities, has placed major constraints on its coal-fired plants and forced many to close.

Climate-related hazards often stress the poor when they disrupt food production, prices, or security, water supply, home destruction and safety, or livelihood. Furthermore, huge populations in Southeast Asia are so poor that they cannot afford to purchase more efficient fuels. How do you tell families that have too little money for food to stop cooking with the only fuel that they can afford? The 2010 Climate Change meeting in Cancun, Mexico, committed rich countries to a $100-billion-a-year fund to help poor countries reduce their emissions but did not decide where the money would come from.

Almost all scientists are very concerned about climate change and its consequences. Public media are split between those who follow the science and present articles that convey that science to the public, and those who argue that climate change is mostly hype or not a significant concern. Individual members of the public are equally polarized, often leaning one way or the other depending on which news media they pay attention to or are in sync with, and the degree to which they are inclined to be politically liberal or conservative. Increasingly, politicians and business leaders are coming to recognize significant geopolitical impacts around the world that contribute to poverty, environmental degradation, and the further weakening of fragile governments. They note that climate change will contribute to food and water scarcity, will increase the spread of disease, and may spur mass migration. Because many climate change processes have considerable inertia and long time lags, it is mainly future generations that will have to deal with the consequences of decisions made today.

If scientists, the military, and many affected populations are so concerned about the threat of climate change, then why are national governments not taking action? The main problem seems to be the political costs of making changes with immediate financial and employment consequences, the benefits of which will not be felt until decades later. Human nature leads to putting off difficult issues as long as possible until people can no longer avoid them. In the case of climate, by the time the danger is so obvious that it outweighs the political issues, those changes will be irreversible.

Cases in Point

Rising Sea Level Heightens Risk to Populations Living on a Sea-Level Delta
Bangladesh and Kolkata (Calcutta), India ▶

The vast delta regions of Bangladesh and the northeastern coastal region of India around Kolkata (Calcutta) are low-lying and especially fertile—one of the most densely populated regions on Earth, with some 130 million people in an area not much larger than New York State. Ocean tidal range on the river delta lands at Kolkata (Calcutta) is almost 2 m. People are exceptionally poor; their livelihoods that depend on agriculture are intimately tied to weather and its associated hazards. Most people live on farms rather than in the cities, and the few rail lines connect only the larger cities. Roads are narrow and crowded with heavy trucks, buses, and rickshaws. Bridges span only a few of the smaller river channels. People get around on bicycles, small boats, and ferries. Traffic jams in the cities are frequent on normal days. Imagine

the chaos of trying to evacuate hundreds of thousands or millions of people with a couple of days' warning given such limited transportation.

Even when there is adequate satellite warning to evacuate populations in the region, most people do not leave because they believe that others will steal their belongings or because they have lived through a major cyclone and the average time between cyclones suggests to them that the next major one "will not come for many years," or that if they die it is "God's will." The situation is grim:

- In October 1737, 300,000 died in a surge that swept up the Hooghly River in Kolkata (Calcutta).

- In 1876, 200,000 died in a surge near the mouth of the Meghna River; thousands more perished in 1960 and 1965.

- In November 1970, a Category 3–equivalent cyclone with winds of 200 km/h accompanied a 12-m surge that swept across the low-lying delta of the Ganges and Brahmaputra Rivers in Bangladesh, drowning more than 500,000 people, many of them in only 20 minutes. Whole villages disappeared, along with all of their people and animals.

- On April 30, 1990, a cyclone with 233-km/h winds and a 6-m surge swept into Bangladesh, drowning 143,000 people.

- On April 29, 1991, Cyclone BOB, a Category 5 supercyclone, struck southeastern Bangladesh with winds of 250 km/h and a 6-m storm surge. It breached river levees almost everywhere, quickly spread over the surrounding lowlands, killed at least 138,000, and left 10 million homeless. The Bangladesh population of 111 million in 1990 had reached 156 million by 2013, placing millions more on the delta.

Dhaka

Ganges River

Chittagong

Bay of Bengal

NASA EarthObservatory

■ *The nearly sea-level delta area of the Ganges and Brahmaputra Rivers is laced with distributary channels and their tributaries. The area outlined in green is a preserve of mangrove swamps. The remainder is mostly farmland. Hypothetical cyclone rotation is shown in blue. View is about 200 km across.*

Staff Sgt. Val Gempis, USAF

■ *The 1991 supercyclone swept over river levees in Bangladesh, inundating low-lying fields and drowning 138,000 people. Low levees were breached almost every hundred meters.*

■ *Houses along near sea-level ground in the Mekong delta of Vietnam are just above water level even without a storm surge.*

■ *An Indian boy reaches his house near Guwahati, India, in the Brahmaputra flood of July 8, 2003.*

- On November 18, 2007, Cyclone Sidr, a Category 4 storm, made landfall in Bangladesh, killing about 15,000 people. Sustained winds reached 215 km/h. A storm surge of 3 to 5 m was widespread. Shortages of drinking water were severe because seawater reached so far inland. More than 10,300 people had typhoid fever; thousands more had pneumonia or diarrhea from contaminated water. Damages in Bangladesh and adjacent India totaled about $1.7 billion.

Heavy monsoon rains from April through October are carried by warm, moist Indian Ocean winds from the southwest and magnified by the effect of the air mass rising against the Himalayas. Torrential rains are widespread in the drainage areas of the Ganges and Brahmaputra Rivers, which come together in the broad delta region of Bangladesh. Rivers swell annually to 20 times their normal width. Upstream those drainage areas have been subjected to

widespread deforestation and plowing of the land, which causes rapid runoff, massive erosion, and heavy siltation of river channels. The decrease in channel capacity causes the rivers to overflow more frequently. Gradual compaction and subsidence of delta sediments coupled with gradual rise in sea level compounds the problem, raises the base level of the river channels, reduces their gradient, and raises the water level everywhere on the delta. Cyclone-driven storm surges can put virtually the whole delta—75% of the country—underwater in a few hours. Other deltas in Southeast Asia are subject to similar hazards.

The people have limited options. The government plans to dredge and channelize more than 1000 km of riverbeds to improve navigation and build homes for flood victims on the new levees. However, the impetus appears to be to improve food production rather than ensure safety and will do little to prevent the problem. Ironically, those same floods bring new

fertile soils to the land surface. Dredging will increase the chances of flooding downstream; and the new levees, built from those fine-grained materials dredged from the delta's river channels, will easily erode during floods. Levee breaches will lead to avulsion of channels (formation of new river pathways), widespread flooding, and destruction of fields. One useful suggestion was to construct high-ground refuges and provide flood warning systems.

CRITICAL THINKING QUESTIONS

1. In the past several decades, tropical cyclones have repeatedly killed hundreds of thousands of people living in Bangladesh. Why should sea-level rise affect the safety of people living tens of kilometers inland? Exactly how will this happen?
2. With few roads, limited public transportation, no personal vehicles, and surviving on $1 to $2 per day, how can people living on Southeast Asia coastal deltas survive a major cyclone, even if they have a few days of warning?

CO₂ Sequestration
The Weyburn Sequestration Project ▶

The world's largest-scale experiment to test the feasibility of permanent underground CO_2 storage was undertaken in southern Saskatchewan. The Weyburn

sequestration project of the International Energy Agency pumps CO_2 under pressure into depleted oil reservoirs 1450 m below, to permit recovery of any remaining oil, to permanently remove CO_2 from the atmosphere, and to test how well the CO_2 remains stored. Initial injection occurred in September 2000. Six thousand tons were injected per day, and by September 2007, 14 million tons of CO_2 had been injected from a North Dakota coal gasification plant, pumped to Weyburn by a 320-km pipeline. Apparent geological faults cut the essentially horizontal sedimentary layers, but their potential as paths for leaking fluids is unknown. Local deformation caused by injection of the high-pressure fluids complicates the situation by widening fractures or opening new ones. Review of the initial results indicates that the Weyburn site may be suitable for long-term storage of CO_2. Studies are ongoing in the Colorado Plateau of southeastern

Utah and adjacent Colorado, New Mexico, and Arizona, an area that surrounds many large coal-fired power plants that generate enormous amounts of CO_2. The area consists of horizontal sedimentary rocks with structural domes that naturally trap CO_2. Some of the natural CO_2 has reacted with rock minerals to form solid carbonate minerals, thereby trapping the CO_2 in the solid form. Preliminary data suggest that after 1000 years, 70% of the CO_2 should remain trapped underground.

A third experiment pumped CO_2 into sandstones sealed against a salt dome near the Gulf Coast of Texas. The CO_2 remained well sealed but reacted with mineral grains to mobilize iron, manganese, and toxic elements, displacing them in the salt water that originally filled the spaces

between the grains. Such toxic materials could contaminate groundwater in such regions. The amounts being sequestered at these and other sites are very modest, far from the amounts that need capturing in the future. The major concerns are how long CO_2 will remain trapped and the contamination of groundwater. Another problem is the potential triggering of earthquakes by pressure from the fluids discussed previously (see Figure 4-2, p. 65).

CRITICAL THINKING QUESTION

1. Given the additional costs in capturing CO_2 from power plants, transporting, and pumping it underground, how important is sequestration likely to be in the future?

Hidden Costs of Nuclear Energy
Fukushima Nuclear Power Plant Failure, 2011 ▶

The catastrophic failure of the Fukushima nuclear power plant following the giant earthquake and tsunami in northern Japan in March 2011 unnerved many people around the world. The 30- to 40-year-old reactors required electric pumps for backup power to circulate water to cool the fuel rods in an emergency. Those stopped working during the giant earthquake and tsunami. The latest generation of nuclear plants store water above the reactor containment vessel; in a crisis that water would flow without external power down onto the overheated system to cool it for up to three days. They also naturally circulate water using heated air in the system.

Fukushima nuclear power plant was flawed by both mechanical and operator error, with more accidents in the last several years than any other plant in Japan. A few months before the earthquake, workers on one of the plant's six reactors mistakenly used plans for the wrong reactor. A cable controlling the cooling system was removed by mistake, an error that wasn't discovered until weeks later. During the 2011 tsunami, electric power

failed at four of the six reactors, shutting down their internal cooling systems. Heat built up in the reactor cores and in pools designed to keep spent fuel rods from overheating. Two weeks later, power had not been restored to the cooling systems and to the control rooms of two of the

reactors. In May, the plant operators finally admitted that there had been a reactor meltdown and that the nuclear fuel had gone through the floor of the reactor

DigitalGlobe/Getty Images

■ *The destroyed Fukushima power plant on the Pacific coast at Sendai, Japan, March 2011.*

pressure vessel and into the floor of the containment vessel. The March 11, 2011, Fukushima nuclear disaster in Japan provided a new wake-up call concerning nuclear plant safety for plant sites in certain hazardous types of areas.

More than 100 commercial nuclear reactors are in service in the United States, mostly in the Midwest and east. A few new nuclear power plants using passive cooling systems were in advanced stages of planning but most of those are now in doubt because of the Japan disaster. A new plant costs at least $7 billion and requires approval from the Nuclear Regulatory Commission. The present uncertainty will lead to increased costs at a time when nuclear power is already borderline in cost comparison with other power sources, including natural gas. Insurance for nuclear accidents in the United States is purchased by nuclear power plant owners who are liable for damages. That insurance, from a consortium of insurance companies, is limited to $375 million for each reactor site.

Germany, a country with numerous nuclear reactors, many of them more than 30 years old, decided, after the Fukushima disaster, to abandon nuclear power entirely and to try to convert to a renewable energy power supply. The German government now proposes that their old reactors be allowed to operate until 2030. Switzerland suspended all plans for new reactors.

With the 2011 nuclear power plant failure in Japan, many people viewed nuclear power as unsafe and that its use should be discontinued. Two nuclear power plants still operating along the coast of California were San Onofre, between Los Angeles and San Diego, and Diablo Canyon, 19 km from San Luis Obispo. The active Hosgri Fault, 16 km offshore from Diablo Canyon, had a magnitude 7.1 earthquake in 1927 and the plant is capable of withstanding a magnitude 7.5 earthquake; the Shoreline Fault, only 1.6 km offshore, could have a magnitude 6.5. The plants' current

California Dept. of Forestry

■ *The San Onofre Nuclear Power Plant, on the coast of southern California.*

operation licenses expire in 2022 and 2024, respectively. Both use ocean water for cooling. Five other plants have been shut down either because of the expense of required retrofits, public referendum, or nuclear accident. Protests against both plants beginning in the 1970s continued intermittently but were renewed following the 2011 Fukushima disaster. San Onofre Nuclear Power Plant was finally closed permanently in June 2013, after a small radiation leak in January 2010 caused damage that was deemed unsafe in spite of $500 million in repairs and replacement power. Japan finally pulled all of its 50 nuclear reactors off-line in May 2012, pending safety reviews and nuclear power policy changes.

A new generation of smaller, simpler nuclear reactors promises to reduce the construction costs for nuclear energy. These refrigerator-size nuclear power plants generate 25 megawatts of power, about 2.5% of the power of a traditional

reactor. They can power, for example, towns with 20,000 homes, commercial ships, mining operations, or desalination plants. Also in the works is a small reactor 10 to 12 m long, mostly underground, that uses spent nuclear fuel or fuel rods (now nuclear waste), activated with a small amount of enriched uranium. The design would produce about 500 megawatts of electric power using heat exchangers to drive turbines and generate electricity.

CRITICAL THINKING QUESTIONS

1. Since the 2011 nuclear reactor disaster in Japan, a modern industrialized country with the most advanced earthquake detection and safety laws, and given that nuclear reactors generate no CO_2 but must dispose of their nuclear waste, what should be the future of nuclear energy?
2. If countries are choosing to avoid generating electricity by nuclear power, what should they do instead?

Chapter Review

Key Points

Effects on Oceans

■ Sea-level rise is threatening low-lying coastal areas, causing greater erosion and greater storm damages. Warming oceans may also disrupt global circulation patterns. Dissolution of more CO_2 in seawater is also making oceans more acidic, killing reefs and threatening various species of sea life. **FIGURES 12-3** to **12-9** and **By the Numbers 12-1**.

Precipitation Changes

■ Precipitation patterns are expected to change as a result of climate change, leading to more flooding and droughts. Much of North America is expected to have much drier summers, with wetter winters in northern parts **FIGURE 12-10**.

Arctic Thaw and Glacial Melting

■ Melting of Greenland and Antarctic glaciers, Arctic sea ice melting, and permafrost thaw in the Arctic is increasing. Methane is released at greater rates and provides a feedback mechanism to still greater warming. Permafrost thaw is destroying buildings, roads, and pipelines, and releasing more greenhouse gases into the atmosphere. **FIGURES 12-11** to **12-14**.

Impacts on Plants, Animals, and Humans

■ Warming will force animals to adapt or migrate to a more suitable environment. Many species will not survive; polar bears are endangered.

■ Humans will suffer from extreme weather events and water shortages. Incidences of respiratory illnesses and infectious disease will also increase.

Mitigation of Climate Change

■ Reductions in greenhouse gas emissions into the atmosphere could be accomplished by increased conservation and improved efficiency of power production and use **FIGURES 12-15** to **12-18**.

■ Switching from cheap fossil fuels with high environmental costs to cleaner sources of energy such as wind and solar power will also reduce GHG emissions (**Table 12-1**).

■ Capture and sequestration of carbon might reduce GHG levels, but also has costs and risks (**FIGURE 12-19**).

■ Geoengineering solutions, such as trapping carbon in the oceans or solar radiation management, have risks and are largely untested (**FIGURE 12-20**).

■ Political solutions such as a carbon taxes and cap-and-trade plans have been undertaken by some countries, but remain unpopular among the largest polluters.

Key Terms

cap-and-trade, p. 337

carbon sequestration, p. 335

carbon tax, p. 337

fracking, p. 334

geoengineering, p. 336

methane hydrate, p. 327

permafrost, p. 328

solar radiation management, p. 337

Questions for Review

1. Approximately how much is sea level expected to rise in the next 100 years? What country is expected to see the largest loss of life as a result of rise in sea level? Why?

2. What is the main contributor to the rise in worldwide sea level? Be specific as to what makes the level rise—not just "global warming."

Critical Thinking Questions

1. Some people advocate conservation of energy by better insulating houses, using less heat and air conditioning, and turning off lights and electric devices when not in use. How important is this and how likely are people to go along with such conservation measures?

2. Figure 12-15 outlines various scenarios for aggressive mitigation strategies to stabilize atmospheric CO_2 at present levels. How plausible is each of these stabilization measures and are they worth the economic cost? Instead of all of these mitigation efforts, should we merely adapt to climate changes as they come?

3. Because a large part of the climate change problem is overpopulation, should people be limited to having (or have incentives to have) no more than one or two children, as in China's one-child policy, to slowly reduce the world population and minimize climate change?

4. Some people argue that we should not worry about future consequences of climate change because technology will advance enough to fix the problem once it becomes dangerous. Is this possible given the current climate change situation, and if not, why?

5. Coal is a cheap and abundant fuel. Coal companies argue that we should burn clean coal by capturing the CO_2 produced and storing it underground. Is that a good idea or are there critical flaws? If so, what are they?

6. Renewable energies, like solar and wind, now produce only a few percent of energy that we use. What should we do until renewable sources become abundant?

■ No, this house was not built overhanging the cut bank of the river. When it was built, it was 10 m from the bank of the Clark Fork River near Plains, Montana. Authorities finally burned the house before it fell into the river.

Streams and Flood Processes

13

Too Close to a River

A resident of Plains, Montana, had just finished building a new house in a picturesque location on the outside of a big bend in the Clark Fork River. It had a great view of the river and the surrounding mountains. Although the site was on the floodplain, he thought it would be amply safe because it was about 10 m from the river bank.

In late May 1997, rising water in the spring runoff rapidly eroded the steep bank on the bend of the river next to the house. The fast-moving water caused progressive caving of the bank until it undercut the edge of the house. The owner hired a company to move his house, but he was too late; before they could move it, erosion had progressed so far that the moving company was not willing to risk loss of its equipment on the unstable riverbanks. Finally, local authorities decided to burn the house rather than find it floating down the river in pieces. The situation was especially embarrassing because the owner was also the local disaster relief coordinator.

Stream Flow and Sediment Transport

What the homeowner in the preceding account didn't consider is that a river is not a fixed structure like a highway but is subject to natural processes, including course change and flooding. The first step to understanding flooding is to understand the natural processes by which rivers and streams transport water and sediment.

Rivers are complex networks of interconnected channels with many small tributaries flowing to a few large streams, which in turn flow to one major river. They follow valleys that they have eroded over thousands to millions of years. Rivers respond to changes in regional climate and local weather through the amount and variability of flow and to the size and amount of sediment particles supplied to their channels.

Stream Flow

Streams and rivers collect water and carry it across the land surface to the ocean. Streams in humid regions collect most of their water from *groundwater seepage*—water percolating down through porous soils to reach the stream. Flow generally increases in the downstream direction as additional water from tributary streams and groundwater enters the channel. Streams accumulate surface water from their **watershed** (or drainage basin), the entire upstream area from which surface water will flow toward a channel (**FIGURE 13-1**).

The **discharge** of a stream, or total volume of water flowing per unit of time, is the average water velocity multiplied by the cross-sectional area of the stream (**By the Numbers 13-1**: Stream Discharge). Because there is no easy way to measure the average velocity of water in a stream, *point velocities* are measured at equal intervals and depths across a stream channel, and each point velocity is multiplied by a cross-sectional area surrounding that point (**FIGURE 13-2A**). Instruments called *acoustic doppler current profilers* have been developed to more accurately approximate stream flow by measuring water velocity at hundreds of locations based on the shift in sound frequencies due to moving particles (**FIGURE 13-2B**).

Sediment Transport and Stream Equilibrium

Rivers and streams carry sediment downstream along with water, eroding material in one place and depositing it in another. Streams change to maintain a **dynamic equilibrium** in which the inflow and outflow of sediment is in balance. A stream that is able to maintain this equilibrium is called a **graded stream**.

The cross-section of a stream adjusts to accommodate its flow, as well as the sediment volume and the grain sizes supplied to the channel. The geometry of a channel cross-section is controlled by flow velocities and the associated ability of a stream to carry sediment. Most streams are wide and shallow,

FIGURE 13-1 Watershed of a Stream
The watershed of this stream near Boise, Idaho, is outlined in blue. This area includes all of the slopes that drain water to feed the stream.

David Hyndman

By the Numbers 13-1

Stream Discharge

The discharge of water in a stream depends on the average velocity of the water times the cross-sectional area through which it flows:

$$Q = VA$$

where

Q = discharge or total flow (m^3/s)

V = average velocity (m/s)

A = cross-sectional area (m^2) = width (m) $\times$ depth (m)

---

with nearly flat bottoms, but the cross-section of a stream adjusts based on the erodibility of the bottom and banks and the nature of the transported sediment. Streams flowing through easily eroded sand and gravel at low flow generally have steep banks and broad, nearly flat bottoms. Streams flowing through bedrock or fine silt and clay tend to be narrow and deep because these materials are less easily eroded.

A stream also adjusts its gradient in response to water velocity, sediment grain size, and total sediment load in order to be able to transport its supplied sediment over time. The **gradient** of a stream, or its channel slope, is the steepness with which

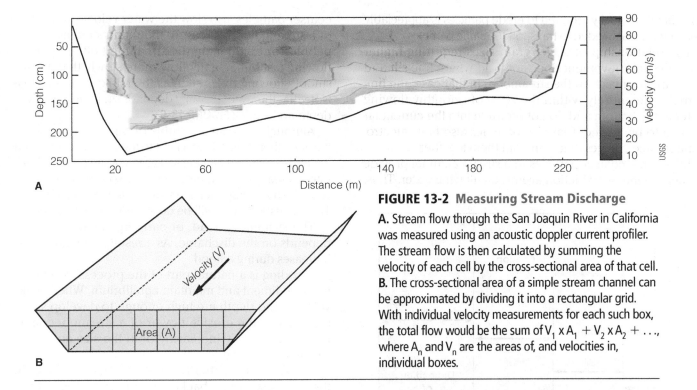

FIGURE 13-2 Measuring Stream Discharge
A. Stream flow through the San Joaquin River in California was measured using an acoustic doppler current profiler. The stream flow is then calculated by summing the velocity of each cell by the cross-sectional area of that cell. **B.** The cross-sectional area of a simple stream channel can be approximated by dividing it into a rectangular grid. With individual velocity measurements for each such box, the total flow would be the sum of $V_1 \times A_1 + V_2 \times A_2 + ...$, where A_n and V_n are the areas of, and velocities in, individual boxes.

it descends from its highest elevation to its lowest, typically expressed in meters per kilometer. Most streams begin high in their drainage basins, surrounded by steeper slopes and often by harder, less easily eroded rocks. Coarser grain sizes, such as those supplied to a stream from the steeper slopes of mountainous regions, require a steeper gradient or faster water to move the grains (**FIGURE 13-3**). There the stream moves the rocks and sediment on down the valley. The greater discharges and smaller grain sizes that exist downstream lead to gentler slopes in those channels. Thus, the gradient generally decreases downstream as sediment is worn down to smaller sizes and the larger flow there is capable of transporting the particles on a gentler slope. Ultimately, the stream will reach a lake or the ocean, a **base level** below which the stream cannot erode.

Where a tributary stream descends from a steeper gradient in mountains onto a broad valley bottom, it leaves the narrow valley that it eroded to reach a local base level of a larger valley. The rapid decrease in its gradient causes it to drop much of the sediment it was carrying (see the rocks along the river in FIGURE 13-3). Thus, where the slope decreases, the stream changes from an erosional mode, where it is picking up sediment, to a depositional mode, where it is depositing sediment. The excess sediment may spread out in a broad fan-shaped deposit called an **alluvial fan**. The term *alluvial* implies the transport of loose sediment fragments, and *fan* refers to the shape of the deposit in map view. Similarly, where a river reaches the base level of a lake or ocean, the abrupt drop in stream velocity at nearly still water causes it to deposit most of its sediment in the form of a **delta**. The delta is like an alluvial fan except that the delta sediments are deposited underwater.

FIGURE 13-3 Gradient Change
Coarse gravel brought in by the small tributary on the left creates steepening and rapids in the Wenatchee River, Washington.

Sediment Load and Grain Size

Eroding riverbanks and landslides can supply a stream with particles of any size from mud to giant boulders. The velocity and volume of the flow limit both the size and the amount of sediment that can be carried by the stream. Empirical curves can be used to estimate the maximum particle size that can be picked up or transported for a given water velocity (**FIGURE 13-4**).

Note that the velocity required to erode particles is appreciably greater than that needed to transport the same particles.

The relationship of larger grain sizes requiring higher velocities for movement does not hold where fine silt and clay particles make up the streambed. In that case, the fine particles lie entirely within the zone of smoothly flowing water at the bottom and do not protrude into the current far enough to be moved. Clay-size particles also have electrostatic surface charges that help hold them together.

In general, coarser particles in a stream channel provide greater roughness or friction against the flowing water. Thus,

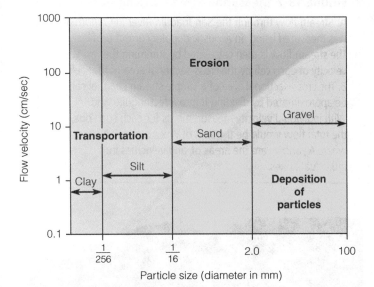

FIGURE 13-4 Stream Velocity and Erosion

This diagram shows the approximate velocity required to pick up (erode) and transport sediment particles of various sizes. Note that both axes are log scales, so the differences are much greater than it seems on the graph.

coarser particles also slow the water velocity along the base of a stream. For this reason, mountain streams with coarse pebbles or boulders in the streambed often appear to be flowing fast but actually flow more slowly than most large, smooth-flowing rivers such as the Missouri and Mississippi. Note that the water velocity depends on increases with water depth and slope (**FIGURE 13-5**).

Although a stream is capable of carrying particles of a certain size, there can still be a limit to the volume of sediment—or its **load**—a stream can carry. Large volumes of sand dumped into a stream from an easily eroded source or sediment provided by a melting glacier will overwhelm its carrying capacity. The excess sediment will be deposited in the channel.

The suspended load, or carrying capacity, of a stream depends on the discharge. As a result, the load of a stream increases during a flood.

Flooding is a natural part of the process by which rivers move sediment and maintain equilibrium. When peak flood velocity and depth are high enough to develop significant turbulence, the erosive power of a stream becomes very large, temporarily increasing both the size and volume of sediment it can carry.

At low water, virtually all of the material brought into the stream stays put, backing up the water behind it. When water rises in the stream, such as during flood, it has a higher velocity and thus can carry more sediment particles, flushing the accumulated sediment downstream.

Flooding also increases the size of sediment particles a stream can move. During floods, coarser particles are gradually broken down to smaller-sized particles in the channel and are then flushed downstream.

Similarly, if coarser material is added to the channel, such as from a steeper tributary or a landslide, it accumulates in the stream channel until a large flood with sufficient velocity occurs or the gradient of the channel increases enough to move that size of material (**FIGURE 13-6**).

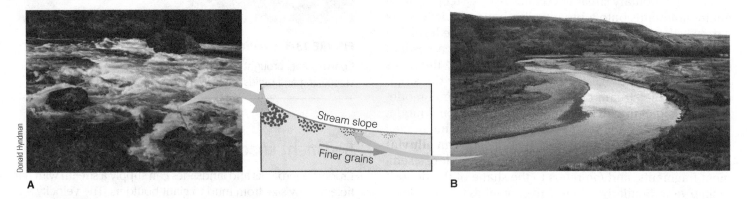

FIGURE 13-5 Stream Slope and Grain Size

Grain sizes decrease downstream as slope decreases in a stream channel. **A.** The turbulent, high-energy Lochsa River in the mountains of northeastern Idaho has a steeper gradient and carries larger grains. **B.** The Smith River south of Great Falls, Montana, has a lower gradient and carries finer particles.

FIGURE 13-6 Accumulation of Boulders in a Stream
Giant granite boulders dumped by a landslide into the Feather River in the Sierra Nevada range of California can be moved only in an extreme flood.

FIGURE 13-7 Thalweg
The thalweg of the Lochsa River in northeastern Idaho shows as whitewater rapids.

Channel Patterns

The way in which streams pick up and deposit sediment also determines the pattern of the channel and the way the channel moves over time, which in turn determines the type of flooding characteristic of each type of stream.

Meandering streams, which sweep from side to side in wide turns called meanders, are most common. Multichannel **braided streams** are much less common, and naturally straight streams are rare. Meandering streams are typical of wet climates with their finer-grained sediments, and braided streams are common in dry climates with abundant coarse sediment. The patterns of most rivers fall somewhere in between meandering and braided. Even within one river, some sections may meander and others may braid, depending on the erodibility of the banks and the amount of supplied sediment. **Bedrock streams** are fast moving, high-energy streams that occur in steep, mountainous areas where the streambed is solid rock.

Meandering Streams

Meanders in a stream are created when an irregularity such as a boulder or tree root diverts water toward one bank. As the water swings back into the main channel, it sweeps toward the opposite bank, much like a skier making slalom turns. The deep and highest-velocity part of the stream is known as the *thalweg* (**FIGURE 13-7**). The flow erodes the banks along the outside of these turns to carve out the meanders. The flow preferentially erodes the outside of meander bends because the water has forward momentum that drives it into the bank, where the higher velocity water can mobilize more sediment. As the water rounds the corner of a bend, it slows, and sediment is then deposited as a **point bar** downstream along the inside corner of the bend. The flow commonly alternates between deep pools on the outside bends, where sediment has been

eroded, and shallow riffles between there and the next outside bend, where sediment has been deposited (**FIGURE 13-8**).

Over a long time, continued erosion of the outside of meander bends causes the meanders to deepen and migrate. Meanders sometimes come closer together until floodwater breaks through the narrow neck separating them. The abandoned meander is called an **oxbow lake** (**FIGURE 13-9**). When a stream cuts across a meander bend, either naturally or by human interference, the new stretch of channel follows a shorter path for the same drop in elevation. The steepening of the channel slope increases water velocity, eroding finer particles and often leading to rapids. Engineers commonly add rock rubble, or **riprap**, along the banks of a straightened portion of a stream to diffuse the stream's energy and minimize erosion.

The size and shape of river meanders follow some general relationships (**By the Numbers 13-2:** Relative Proportions of Meandering Streams). These relative proportions are maintained regardless of stream size, whether it is a small stream, 2 m across, or the lower Mississippi River, 1000 m across. Thus, artificial attempts to change a channel by narrowing it or straightening it will be met by the river's attempts to return to a more natural equilibrium channel cross-section and meander path. If we straighten a river channel to accommodate a nearby road, for example, or if we riprap a channel bank to hinder bank erosion, the stream will adjust its path to maintain its natural proportions—often causing new problems for those who would make those changes.

Meandering streams flood in a typical way as a result of their patterns of erosion and deposition. Over the course of time, meanders erode outward and slowly migrate downstream. That process of erosion and deposition over a period of centuries gradually moves the river back and forth to erode a broad valley bottom. At high water, the flooding river spills out of its channel and over that broad area—its **floodplain** (**FIGURE 13-10**).

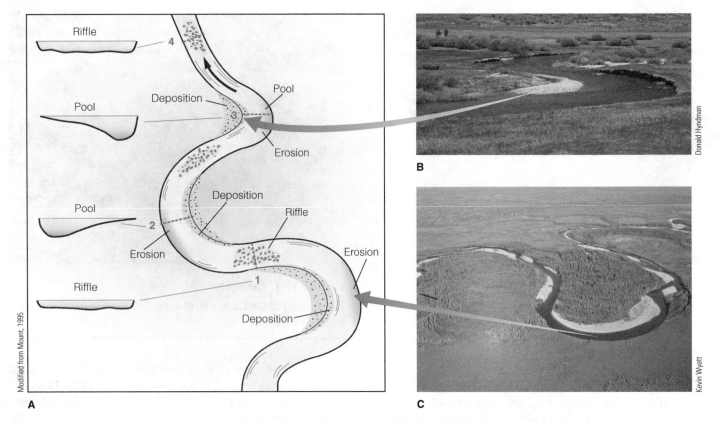

FIGURE 13-8 Patterns of Erosion and Deposition in a Meandering Stream

A. Cross-sections of a typical meandering stream channel from a riffle at 1, downstream through pools at 2 and 3, then finally through a riffle at 4. Note that the river erodes on the outside of the meander bends where it has a deep channel, and it deposits on the inside of bends where it has a shallow channel. **B.** The Carson River in Nevada illustrates the eroding cut bank on the outside of meanders and the depositional gravelly point bar on a meander's inside. Flow is toward the right. **C.** Deepwater channels on the outside of meander bends and prominent point bar deposits on the inside of meander bends along Beaver Creek, north of Fairbanks, Alaska.

FIGURE 13-9 Oxbow Lake

Meanders in this river near Houston, Texas, eroded the outsides of bends and migrated until one meander bend spilled over to the next one farther downstream, leaving an abandoned oxbow lake in the center of the photo.

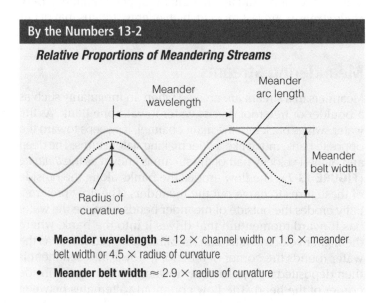

By the Numbers 13-2

Relative Proportions of Meandering Streams

- **Meander wavelength** ≈ 12 × channel width or 1.6 × meander belt width or 4.5 × radius of curvature
- **Meander belt width** ≈ 2.9 × radius of curvature

FIGURE 13-10 Floodplain

The floodplain of this central Utah river was carved out as shifting meanders of the stream and its tributary gradually widened the floodplain. A major flood would fill the floodplain wall-to-wall.

Think back to the vignette that opened this chapter, about the homeowner who built his house 10 m from a river but was surprised to find the river right at his doorstep one day. He lacked two key pieces of information about the behavior of meandering rivers when he built his home at the outside of a meander bend: He hadn't considered that meanders erode outward and move over time or that the rising water at spring runoff would speed up the erosion process.

Braided Streams

Braided streams do not meander but form broad, multichannel paths (**FIGURE 13-11**). These streams are overloaded with sediment, which they deposit in the stream channel, locally clogging it so that water must shift to one or both sides of the deposit. With further deposition, water shifts to form new channels. That behavior is promoted by dry climates in which little vegetation grows to protect slopes from erosion. Such braided streams are characterized by eroding banks and abundant stream *bedload*, which is the sediment carried along the stream bottom. With waning flow, bedload is deposited in the channel and flow diverges around it, splitting into separate channels. Depending on flow, some channels dry up, while others temporarily become dominant.

Braided channels are also characteristic of meltwater streams flowing from sediment-laden glaciers and of arid Basin and Range valleys in which heavy, intermittent rains from mountains carry abundant sediment across valley alluvial fans. When a stream channel moves from an area of high slope to one of lower slope, sediment will generally deposit in an alluvial fan because the stream can no longer carry its full load (**FIGURE 13-12**).

Alluvial fans are somewhat arched, with higher elevations along their midlines and lower elevations toward their sides, because they are built by deposition of sediment flushed out of a mountain canyon at the apex of the fan. During a flood, all of the flow from a canyon concentrates at the apex of the fan and then spreads out, decreasing in depth and intensity downslope (**FIGURE 13-13**). At different times the flow might concentrate on different parts of the fan as it finds the easiest path down the slope.

Active alluvial fans are always marked by braided streams. Like other braided environments, most alluvial fans tend to be in dry climates and lack significant vegetation. People tend to build on alluvial fans because they are inviting areas, with gentle slopes at the base of steep, often rocky, mountainsides. They typically do not realize that the fan is an active area of stream deposition because the streams rarely flow.

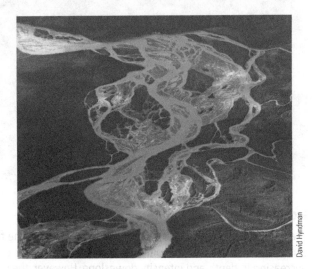

FIGURE 13-11 Braided Streams

A. This braided river channel of the Wairau River is southwest of Blenheim, New Zealand.

B. The Tanana River near Fairbanks, Alaska, has a striking braided pattern.

In fact, alluvial fans are particularly dangerous flooding areas. Torrential floods can wash out of the canyon above with little or no warning. They destroy, bury, or flush houses and cars downslope. Deaths are common, especially during rainstorms at night, when people do not hear the telltale rumble of an approaching flood.

FIGURE 13-12 Alluvial Fan
Water runs off this steep pile of sand in heavy rains despite its high permeability. Overland flow erodes gullies and carries sediment down onto depositional fans. Note that water drains into the gullies on the eroding area but spreads out over wider areas on the depositional zone.

FIGURE 13-13 Hazard Areas of Alluvial Fans
The highest hazard area is at the apex of the fan, where the flow concentrates. The hazard becomes less as the flow spreads out, decreasing in depth and intensity downslope. However, the path of the flow may change to concentrate on different parts of the fan as it finds the easiest way down the slope. Shown here is Copper Canyon, Death Valley.

Bedrock Streams

A stream is called a bedrock stream when it has eroded away all of the easily transportable sediment to reach resistant rock. Sections of bedrock streams tend to abruptly steepen. These abrupt changes in gradient from gentle to steep are called *knickpoints*. Upstream, the gentle gradient has low energy for the river and may deposit sediment. Downstream, the steeper channel provides high energy, and sediment erodes.

High-gradient bedrock channels generally have deep, narrow cross-sections that carry turbulent and highly erosive flows during floods (**FIGURE 13-14**). Their high energy and

FIGURE 13-14 Bedrock Streams
A. The turbulent, high-energy Colorado River in Grand Canyon, Arizona. B. Water rushing through this steep, bedrock channel of the Shotover River in New Zealand cleans out any loose material during every flood.

turbulence allow them to transport all of the loose material in the channel. This can include large boulders, which impact and abrade the channel sides. Bedrock is resistant to erosion, so only major floods can scour or pluck fragments from the rock.

Floods in bedrock channels have excess energy that cannot be dissipated by increasing channel roughness or greater loads of sediment transport. Extreme turbulence is typical, and large-scale vortexes or whirlpools can appear. These can be effective in swirling rocks on the bottom to drill potholes in the bedrock (**FIGURE 13-15**). Recreational rafting or boating at high water in such channels, although exciting, is especially hazardous because of the extreme turbulence and vortexes. Flotation devices can be ineffective at raising a person caught in a whirlpool.

FIGURE 13-15 Potholes

Potholes in the streambed of McDonald Creek in Glacier National Park were formed by rocks and boulders transported during a flood.

Groundwater, Precipitation, and Stream Flow

Atmospheric moisture in the form of rain is the main source of water for streams. Water stored in snow and ice is a secondary reservoir. After a rainfall, some rainwater evaporates from the ground and vegetation surfaces, some is taken up by vegetation, and some soaks into the ground. Rivers and streams collect the water that percolates down through the soil to groundwater (see also FIGURE 9-1).

The amount of precipitation on the land surface varies according to region and season, and there can also be differences year to year. Surface water interaction with groundwater is determined by these climatic factors. **Gaining streams** are fed by groundwater, whereas **losing streams** lose water into the groundwater (**FIGURE 13-16**).

In areas of moderate to high annual rainfall, groundwater levels stand higher than most streams and thus continuously feed them, ensuring a year-round flow. Stream flow in more humid regions is high during wet weather periods and low during dry weather periods. During low-water periods, the stream surface generally sits at the groundwater surface, which is the exposed water table. The rate at which water flows from groundwater into a stream depends upon both the slope of the water table and the ease of flow through the water-saturated sediments or rocks. Storm discharge into a stream includes this groundwater flow plus any water that flows over the surface. New groundwater does not need to travel all the way from its infiltration source during a storm to a river before the stream flow increases—it merely has to raise the water table at the infiltration source. That increases pressure on the groundwater and displaces "old water" farther downslope into the stream. Groundwater in such areas responds slowly to changes in rainfall, and inflow to streams from groundwater is gradual and relatively constant. Thus, wet climates tend to provide streams that flow year-round and flood in prolonged heavy rains.

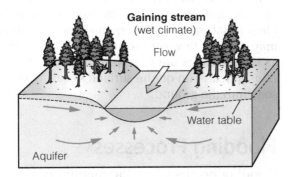

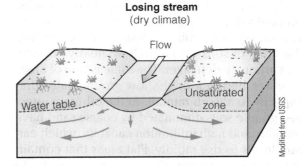

FIGURE 13-16 Gaining and Losing Streams

A. In wet climates, groundwater flows into gaining streams, ensuring year-round flow.

B. In dry climates, water from streams feeds the groundwater. These losing streams may dry up between rainstorms.

In semiarid to arid regions, however, losing streams shed water into the ground and may even dry up between storms. When they do flow, they drain water into the ground and raise the water table. In areas with low annual rainfall, little or no vegetation can grow to soften the impact of raindrops and slow their infiltration. Rain falls directly on the ground, packing it tightly, which permits less infiltration. The rain also kicks up sediment from the surface, permitting it to be carried into streams. During dry seasons, less water gets into the ground, even during the infrequent rainstorms, so less of it feeds the groundwater. Dry climates with no year-round streams can see flash floods after any major or prolonged rainfall.

Streams in deserts generally flow only during and shortly after a rainstorm but then dry up until the next storm. Because more sediment is supplied to the dwindling amount of water, sediment deposits in the gullies. This progressively chokes the flow, causing some of the water to spill over and follow another path. This gully in turn fills with sediment, and so on. The result is a braided alluvial channel that deposits sediment with time.

Precipitation and Surface Runoff

Typically, rainwater slowly soaks into the ground and feeds streams through groundwater. During torrential rainfall, however, some water may flow across the ground as **surface runoff** directly into streams. The intensity of precipitation plays a significant role in the rate of runoff to streams and, in turn, in flooding. Light precipitation can generally be absorbed into the soil without surface runoff. Heavy precipitation can overwhelm the near-surface permeability of soils, leading to rapid runoff over the surface as *overland flow*. The water that soaks into the soil raises the local groundwater level; that adds pressure to the groundwater and forces more water back downslope through the ground and into the streams.

The ability of the ground to absorb rainwater also depends on the permeability of the soil and the extent to which it is already saturated with water. Highly permeable soils absorb heavy rainfall better than impermeable soils. Rapid flood peaks during large storms are most common in areas with fine-grained soils or desert soils, especially tight clay hardpan or soils with a shallow, nearly impervious calcium carbonate-rich layer. The same is true of areas with near-surface bedrock or shallow groundwater that has little capacity to absorb rainfall. At the other extreme, decomposed granite soils dominated by coarse sand have high permeability and high infiltration capacity, which can cause stream levels to rise rapidly. Flat areas that contain many depressions can temporarily store enough water to delay runoff.

Even where the permeability of the soil is relatively high, heavy rainfall, especially over a long period or in multiple storms, can saturate near-surface sediments, thus forcing the water to flow over the surface to streams. The ground is also less able to absorb rainfall when it is frozen. In both of these cases, most of the water that reaches the ground surface runs off directly to a stream. For example, in June 1972, when Hurricane Agnes dumped 5 to 7 cm of rain in Pennsylvania on ground already saturated with water, large floods occurred. In 1993, prolonged rains oversaturated soils in a broad area of the upper Mississippi River Valley, leading to disastrous floods.

Intense precipitation often accompanies major storm systems. Such heavy rains cannot soak in quickly enough to prevent flow over the ground. Thus the water quickly reaches streams and, in turn, causes flooding. Tropical cyclones, including hurricanes, move westward in the belt of trade winds. In the Atlantic Ocean, they drift westward and then northward or eastward along the eastern fringe of the United States, where they interact with midlatitude frontal systems. Reaching the Atlantic coastal plain, hurricanes cause significant flooding, with 25 to 50 cm of rain in many areas. Heavy rainfall accompanies thunderstorms, which last for a few minutes or an hour or two. A line of thunderstorms may prolong the deluge for several hours, whereas tropical cyclones may stretch out the rains for several days. Areas such as Southeast Asia, which have seasonal monsoons, may experience extreme rains and floods, for several weeks or even months. Summer monsoon floods frequently kill hundreds of people and displace hundreds of thousands more in Bangladesh, including in 2012 and 2014.(see **Case in Point:** Monsoon Floods—Pakistan, 2010, p. 371).

At high elevations, much of the winter precipitation falls as snow. This stores the water at the surface until snowmelt in the spring. If the snowpack melts gradually, much of the water can soak into the ground. If it melts rapidly, either with prolonged high temperatures or especially with heavy warm rain, large volumes of water may flow off the surface directly into streams. When the snowpack warms to 0°C, a large proportion of meltwater remains in the pore spaces and the snowpack is "ripe." Water draining through the pack concentrates in channels at the base of the snow. Water has a large heat capacity, so dispersal of warm rainwater effectively melts snow. If the ground is already saturated with snowmelt, water from a rainstorm may entirely flow off the surface. If the ground is frozen, as is sometimes the case with minimal snowpack, heavy rains quickly run off the surface because the water cannot soak into the ground.

Flooding Processes

Flooding occurs when the amount of water entering a stream, for example, from rising groundwater or surface runoff, causes the level of the stream to surpass the capacity of its channel.

The *bankfull* level of a stream is the height at which the water reaches the highest level of its banks, which typically occurs every 1.5 to 3 years. Larger, more infrequent floods fill the channel and spill out over its floodplain. It is not simply the size of the flow that makes a flood, but how the flow compares to the normal capacity of that particular channel. Large rivers can have large flows without being above flood level. Small streams can flood with fairly small flows.

Changes in Channel Shape during Flooding

Patterns of erosion in a streambed can change dramatically during flooding. An increase in discharge during a flood involves an increase in water velocity, water depth, and sometimes width of a stream. This happens because as water depth and velocity increase during a flood, shear or drag at the bottom of the channel increases. That extra shear picks up more sediment, increasing erosion (**FIGURE 13-17**).

Channel scour, the depth of sediment eroded during floods, affects the shape of the stream channel and distribution of sediment. The grain size a stream can carry is proportional to its velocity; thus, rising water first picks up the finest grains, then coarser and coarser particles. Sediment is carried in suspension as long as grains sink more slowly than the upward velocity of turbulent eddies. Fine sediment is first picked up in eddies. At higher flows, pebbles or boulders may tumble along the bottom and even be heard as they collide with one another. This causes more erosion, deepening the channel.

As water velocity increases, the water drags much more strongly against the bottom. Increase in frictional drag on the stream bottom provides more force on particles on the streambed and thus more erosion. That friction also slows down the water.

As a flood flow wanes, the coarser sand and gravel in suspension progressively drop out, thereby raising the streambed (**FIGURE 13-18**). Thus, as water level rises, the stream begins eroding and the water gets muddy; as the level falls, the sediment in transit begins to deposit. For this reason, a cross-section of a flood deposit shows the largest grains at the bottom, grading upward to finer sediments.

As a stream spills over its floodplain, it changes from a deep, high-velocity channel to a shallow, broad, low-velocity channel. As water velocity slows at the edge of the deeper channel, sediment deposits to form a **natural levee** (**FIGURE 13-19**). These features form a nearly continuous low ridge along the edge of the channel that may keep small floods within the channel. The floodplain, with its relatively slow moving water, is part of the overall river path; it carries a significant flow during floods.

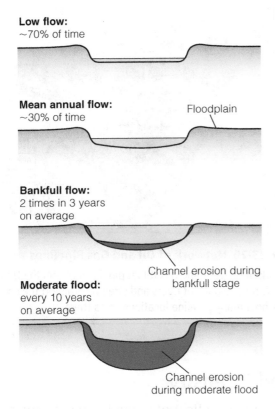

FIGURE 13-17 Stream Channel at Various Flows

The shape of a channel changes with the level of flow. The greater the flow, the more erosion occurs to deepen the channel and increase its capacity.

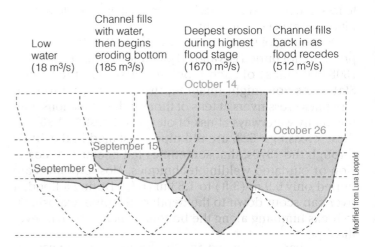

FIGURE 13-18 Channel Changes During Flooding

On September 9, 1941, the San Juan River near Bluff, Utah, was at its normal depth. Widespread rains led to early water rise (September 15); sediment from upstream deposited to raise the channel bottom. As water rose to the maximum level (October 14), the channel eroded to its deepest point. As the flood level waned (October 26), sediment again deposited to raise the channel bottom.

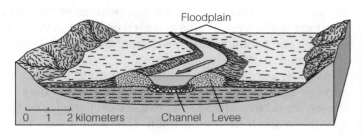

FIGURE 13-19 Natural Levees

A. The main channel of a river has the coarsest gravels at the bottom, grading to finer grains above; the natural levees are still finer grains settled out in shallower water; the floodplain consists of very fine-grained muds that settled out from almost still water during floods.

B. These houses were built on natural levees along a channel and floodplain in the Mississippi River delta.

Stream channel changes are insignificant with normal flows or even small floods, regardless of their frequency. Significant changes occur only when flow reaches a threshold level that mobilizes large volumes of material from the streambed and channel sides. Streams in humid regions adjust their channels to carry the typical annual flows that fill them. Channels in semiarid regions adjust their channels to less frequent large floods because smaller flows do not significantly affect channel shapes.

Deepening of the channel base during floods can create problems for buried gas and oil pipelines. About 492,000 km (306,000 miles) of such pipelines crisscross the United States. As you might imagine, thousands of petroleum pipelines cross rivers at tens of thousands of locations, putting many waterways at risk of oil spills (**FIGURE 13-20**). Of these crossings, 62% are old pipes, installed in the 1950s through the 1970s, that sometimes fail because of corrosion or outdated welding techniques. Most pipelines are buried only 0.9 m (3 ft) to 1.83 m (6 ft). During a flood, a river can scour down to the depth of the pipe, exposing it to rocks tumbling along the bottom, which can then sever the pipe.

In July 2010, more than 1.15 million gallons (about 4,350 m³) of heavy crude oil flowed from the Alberta tar sands into a tributary of the Kalamazoo River in western Michigan, closing 56 km of the river for 3.5 years. The 2-m break in the 0.75-m pipeline leaked for 17 hours because operators misinterpreted pressure-drop alarms to be a bubble in the pipeline. Cleanup costs were about $767 million, the highest in the history of U.S. onshore spills.

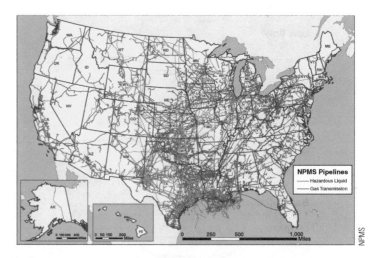

FIGURE 13-20 Network of Oil and Gas Pipelines

A network of oil (red) and gas (blue) pipelines covers the United States. With thousands of rivers and streams draining the country, imagine how many pipeline locations cross them.

Flood Intensity

The destructive effect of a flood depends primarily on its intensity. The intensity of a flood can be measured by the discharge of floodwater and the water's rate of rise. Flood intensity varies over time according to the rate of runoff, the shape of the channel, distance downstream, and the number of tributaries it has. For example, floods in small, narrowly confined drainage basins are typically much more violent than those along major rivers such as the Mississippi.

Rate of Runoff

The intensity of flooding is related to how much water flows in a stream over a period of time. Anything that increases the rate of runoff will cause more intense flooding. For example, there will be more runoff in areas where the ground is not very permeable and more where it is frozen or already saturated with water. Rapid large runoff can develop in urban settings with large areas of pavement, houses, and storm sewer systems, or in natural areas that have been deforested. Deforestation can increase the volume of storm runoff by roughly 10%.

We can depict flood intensity graphically in a **hydrograph**. A typical flood hydrograph rises steeply to the **flood crest**, where the flood reaches its peak discharge, and then falls more gently. In areas where the rainfall saturates the soil and is forced to run over the surface to rapidly feed streams, the hydrograph peaks much more quickly and rises to a greater maximum discharge (**FIGURE 13-21**).

A **flash flood**, which comes on suddenly with little warning, is a flood with a very steep hydrograph. Any type of flood can be dangerous, but flash floods are especially so because they often appear unexpectedly, and water levels rise rapidly

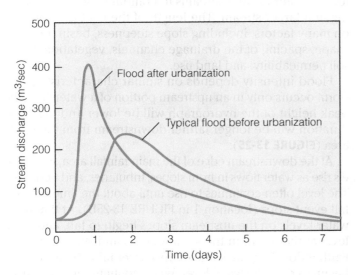

FIGURE 13-21 Urbanization and Flooding

This hydrograph is a plot of stream discharge versus time for a similar 18-hour rainfall event for the same area before and after urbanization. Note that although the flood crest is much higher after urbanization, the area under the two curves is similar, that is, both floods had approximately the same total volume of water.

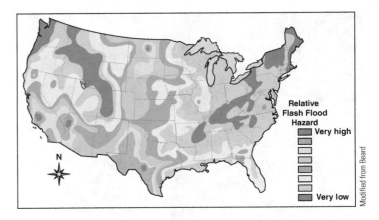

FIGURE 13-22 Flash-Flood Hazard

Regions of the United States in which red and orange areas are most susceptible to flash floods. Notice that flash floods are most severe in the driest regions of the country.

(**Case in Point:** Flash Flood in a Canyon—Colorado Front Range, 2013, p. 372). A map of flash-flood hazard tendency for the United States shows that high flash-flood danger areas are primarily located in the semiarid Southwest—Southern California to western Arizona and West Texas (**FIGURE 13-22**). Moderate flash-flood dangers exist in areas such as the eastern Rocky Mountains, the Dakotas, western Nebraska, eastern Colorado, New Mexico, and central Texas. This is not to say that other areas are not prone to flooding; rather, the floods there tend to be less extreme compared with normal stream flows.

Deaths occur often in flash floods because of the little warning they provide and their violence. Even under a clear blue sky, floodwaters may rush down a channel from a distant storm. On many occasions, people have been caught in a narrow, dry gorge because they were not aware of a storm far upstream (**FIGURE 13-23**). At night, people in their homes have been swept away.

Stream Order

The number of tributaries of a stream (its **stream order**) has a significant effect on the rate of rise of flood-waters during and following a storm. Small streams that lack tributaries are designated *first-order streams* (**FIGURE 13-24**). First-order streams join to form *second-order streams*; second-order streams join to form *third-order streams*; and so on.

Low-order streams tend to respond rapidly to storms with steep hydrographs because water has to travel only a short distance to the stream. Such streams provide less flood warning time for downstream residents. They have smaller drainage basins and carry coarser and larger amounts of sediment for a given area.

FIGURE 13-23 Canyon Dangers

Red Cave, a slot canyon in Zion National Park, Utah, that is a favorite scenic place for hikers. Rainfall far upstream can lead to a sudden, unexpected, and sometimes fatal flash flood of water flushing through the canyon.

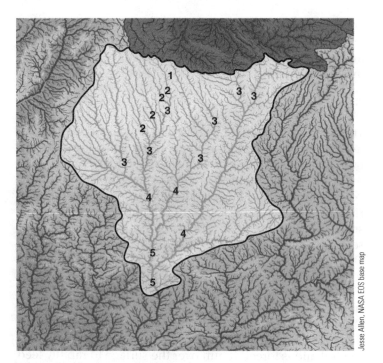

FIGURE 13-24 Stream Order

First-order streams have no tributaries but join to form a second-order stream, and so on. The watershed is outlined in black.

A storm in a headwaters area may cause flooding in several first-order streams. As the flood crest of each moves downstream, the length of time it takes to reach the second-order stream varies, so each first-order flood crest arrives at a somewhat different time. The flood crest for the second-order stream will therefore begin later and be spread over a longer period. The same goes for several second-order streams coming together in a third-order stream; its flood crest will again begin later and be spread over a still longer period. Thus, high-order streams with numerous tributaries have longer lag times between storms and downstream floods; their hydrographs are less peaked and cover longer time periods. Flood warning time for downstream residents is longer.

Downstream Flood Crest

Even with local intense rainstorms on a small drainage basin, there is a lag between a storm and the resulting flood crest. A torrential downpour may last for only 10 minutes, but it takes time for the water to saturate the surface layers of soil and to percolate down to the water table. More time is required for overland flow to collect in small gullies and for water in those gullies to flow down to a stream. In turn, it takes time

for the water in small streams to combine and cause flooding in a larger stream. The length of the lag time depends on many factors, including slope steepness, basin area and shape, spacing of the drainage channels, vegetation cover, soil permeability, and land use.

Flood intensity depends on similar characteristics. If a storm occurs only in an upstream portion of a watershed, the peak height of the hydrograph will be lower and the flood duration will be longer farther downstream from the storm area (**FIGURE 13-25**).

At the downstream edge of the main rainfall area, water levels rise as water flows in from slopes, tributaries, and upstream. The level often continues to rise until about the time the rainfall event stops (location 1 in FIGURE 13-25B). At that time, water levels on the upstream slopes begin to fall, and flood level at that point in the stream crests and starts receding. Farther downstream, some of the earlier rainfall has already begun to raise the water level. Water continues to arrive from upstream, but the size of the stream channel is larger downstream because it has adjusted to carry the flow from all upstream tributaries. Thus, the flood flow fills a wider cross-section to shallower depth (location 2). Even farther downstream, the channel is still larger and the flood crest is still lower (location 3). In cases where the precipitation occurs over the whole drainage basin, flows will increase downstream.

For flood hazards, we are normally more concerned with the maximum height of the flood crest than when the first

Christopher Magirl

FIGURE 13-25 Downstream Flooding
A. Localized afternoon rainfall over Tucson, Arizona.
B. Storm rainfall entering a stream precedes the flood crest
that it causes. The flood hydrograph nearest the rainfall area
is highest and narrowest (location 1). Farther downstream at
location 2, the flood hydrograph crests at a lower level but
lasts longer. Still farther downstream at 3, the flood crests at
an even lower level and lasts longer.

Modified from Luna Leopold

water arrives from a storm. The flood crest moves downstream more slowly than the leading water in the flood wave. Note in Figure 13-25B that the first rise in floodwater from the storm appears downstream later at locations 1, 2, and 3. Note that the flood crest, or peak discharge, also takes longer to move downstream. In general, the flood crest moves downstream at roughly half the speed of the average water velocity.

Flood Frequency and Recurrence Intervals

Flood frequency is commonly recorded as a **recurrence interval**, the average time between floods of a given size. Larger flood discharges on a given stream have longer recurrence intervals between floods.

100-Year Floods and Floodplains

A **100-year flood** is used by the U.S. Federal Emergency Management Agency (FEMA) to establish regulations for building near streams. A 100-year flood has a 1% chance of happening in any single year, although it also has a 1% chance of happening in the year following a similar-magnitude event. A 100-year floodplain is the area likely to be flooded by the largest event in 100 years—on average.

Bridges and other structures should be designed to accommodate the runoff in at least the expected life of the structure, typically between 50 and 100 years. Those who design structures such as dikes, bridges, and buildings for a given life span consider the probabilities of large floods.

As with all calculations of recurrence intervals, however, the accuracy of the 100-year floodplain is not always reliable. The 100-year floodplain is based on extrapolation from a few large events over a short and incomplete flood record. The 100-year flood level does not account for probable major changes, including later upstream alterations to the drainage basin. Although FEMA maps show the boundaries of a 100-year floodplain, this is merely an estimate based on limited data, which is subject to change and may not have been updated in decades. We know that almost all upstream human activities decrease the average number of years for floods to reach the former 100-year level and raise the height of the average 100-year event.

Complicating matters is the fact that floods due to extreme precipitation events are likely becoming more frequent as a result of climate change, upstream landscape changes, or some cyclic aspect of climate. All-time records were set, for example, in the Susquehanna River of Pennsylvania in 1972, the Santa Cruz River in Tucson in 1983, the upper Mississippi River in 1993, the Red River in North Dakota in 1997, and the American and San Joaquin Rivers of California in 1997. When an unusually large event occurs or several events occur more frequently than expected, researchers must go back and recalculate the 100-year flood flow.

Recurrence Intervals and Discharge

For a given stream, the statistical average number of years between flows of a certain discharge is the recurrence interval (T). The inverse of this (1/T) is the probability that a certain discharge or flow will be exceeded in any single year. To determine recurrence intervals for floods on a stream, the peak annual discharge is recorded for each year on record (**FIGURE 13-26**). The largest of these discharges is then ranked number 1, the next largest 2, and so on. Each discharge can then be plotted against its calculated recurrence interval using the "Weibull formula." The calculated recurrence interval of a given flood depends on the total number of years in the flood record and the rank of the flood in question. That means that any new larger flood reduces the rank of the flood in question and can dramatically reduce its recurrence interval (**By the Numbers 13-3**: Recurrence Intervals).

For example, the 1993 Mississippi flood changed the recurrence interval for flooding in St. Louis. The adjusted recurrence interval for the 1993 flood at St. Louis is 100 years or less, rather than previous estimates of up to 500 years. Any new record flood changes previous recurrence intervals because the previous largest flood is now the second largest in the recurrence interval formula. This can have major consequences for 100-year-event floodplain maps. Properties that were outside the 100-year flood hazard area are now within it; flood protection structures that appeared to provide an adequate margin of safety no longer do so.

In general, floods with large recurrence intervals are more frequent in smaller watersheds (**Table 13-1**). Such large floods are much more frequent in semiarid climates (such

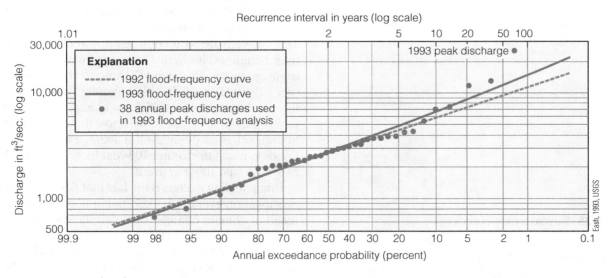

FIGURE 13-26 Flood Frequency

The flood frequency plot for Squaw Creek, a tributary of the Mississippi River at Ames, Iowa, is plotted for before and after the largest flood of historical record in 1993. The recurrence interval (plotted at top) and exceedence probability (plotted at bottom) are shown.

Table 13-1	Chance of a Flood in any Single Year
FLOOD (RECURRENCE INTERVAL)	**ANY SINGLE YEAR**
10 years	10%
50 years	2%
100 years	1%
1000 years	0.1%
10,000 years	0.01%

as the desert southwestern United States) and in monsoon climates (such as Sri Lanka) compared with tropical environments (like Congo and Guyana) because the semiarid regions have much larger variations in precipitation.

Paleoflood Analysis

A major problem in estimating the sizes and recurrence intervals of potential future catastrophic floods in North America is that we have only a short record of stream flow data; the measurement of flood magnitudes is not much more than 100 years, even less in parts of the West. The record is longer in Europe and much longer in civilizations such as China and Japan that have written records extending back a few thousand years. We can project graphical data on magnitudes and their recurrence intervals to less frequent larger events. However, as outlined previously, any record-size event can dramatically change the estimate of recurrence intervals.

In order to extend the record further into the past and include larger floods, **paleoflood analysis** uses the physical evidence of past floods—preserved in the geologic record—to reconstruct the approximate magnitude and frequency of major floods. Even where paleoflood magnitudes cannot be determined reliably from the evidence, the flood height can in some cases be fairly well determined. By itself, this can provide critical information on the minimum hazard of a past flood.

The best sections of a stream for paleoflood analysis are those with narrow canyons in bedrock, pools of still water, and areas with high concentrations of suspended sediment. Paleoflood markers include high-water marks, scarring of trees, and sediment deposits, among other things (**FIGURE 13-27**). Streams in different environments and different climates, however, are highly variable. Most usable evidence comes from meandering streams, not braided or straight streams. Studies have been conducted mainly in single regions, such as the west-central United States, where the evidence of floods is best preserved. Unfortunately, these results cannot easily be extended to other regions.

HIGH WATER MARKS The nature and magnitude of a flood is most obvious immediately after it occurs. Debris, including leaves, twigs, logs, and silt, carried in floodwaters, tends to collect at the edges of the flow, including in back eddies. These provide perhaps the best evidence for maximum flood height, though they may not be preserved long after the flood, and driftwood is likely to end up well below the maximum water level. Debris may pile up on bridges, providing a flood's minimum height. More commonly, floods leave behind drift lines marking the high-water level for a short period of time until they wash away with the next rainstorm

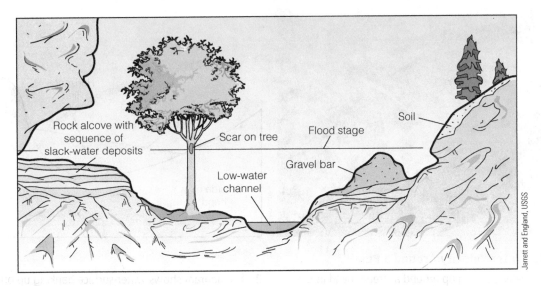

FIGURE 13-27 Paleoflood Markers

This idealized sketch shows some of the paleoflood features that are used to determine flood level. Flood heights can be estimated from the elevations of drift or silt lines, slack-water deposits, scars on trees, gravel-bar deposits, and eroded soil.

FIGURE 13-28 High-Water Marks After Floods

A. A catastrophic flood on this small stream, Shoal Creek, in Austin, Texas, on November 15, 2001, followed an intense rainstorm that dropped 7.5 cm of rain in 1.5 hours. It overtopped the bridge deck and left branches tangled in its railings, evidence of the flood height. The high water also deeply eroded the channel and undercut the supports, causing the bridge deck to sag.

B. This organic drift line marks the high-water line of the same flood.

(**FIGURE 13-28**). Stream bends can affect the interpretation of flood levels because mud lines and splashing watermarks at sharp bends will have different high water levels during a flood (**FIGURE 13-29**).

TREE RING DAMAGE Individual trees may preserve damage effects from a flood and indicate the number of years since flooding (**FIGURE 13-30**). Scars on a tree trunk or branch remain at their original heights during tree growth. The height of the damage generally indicates the minimum height of a flood, though it could be somewhat above the flood height if vegetation piles up.

The age of trees growing on a new flood-deposited sand or gravel bar indicates the minimum age since the flood that produced the bar. When all of the oldest trees on the deposit

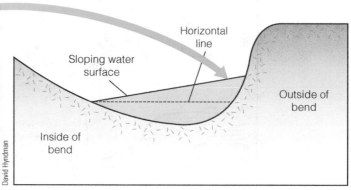

FIGURE 13-29 Water Velocity Around a Bend

A. Fast-moving water banking up around a stream bend in the Grand Canyon, Arizona. Splashing water marked by red arrows.

B. This diagram shows water-surface banking up on one side of a channel as it races around a bend. Such banking of water level can affect interpretation of flood levels because a siltline would mark the high water level at different heights along a stream bend during a flood.

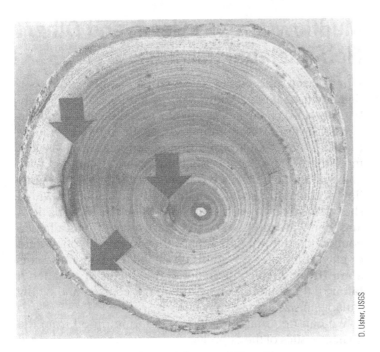

FIGURE 13-30 Tree-Ring Evidence of Flooding
The number of tree-growth rings after the point of damage (arrows) indicates the number of years since damage occurred.

are of roughly equal age, then that age is probably close to the age of the deposit.

SLACK-WATER DEPOSITS During a flood, silt and fine sand can be deposited on sheltered parts of floodplains, the mouths of minor tributaries, shallow caves in bedrock canyon walls, or downstream from major bedrock obstructions (**FIGURE 13-31**).

FIGURE 13-31 Slack-Water Deposits
These Lake Missoula slack-water sands were deposited over coarse, darker, cross-bedded flood deposits near Starbuck in southeastern Washington State.

Organic material in silt and mud layers on floodplains, and occasionally in well-preserved siltlines—stream-edge silt deposits—can be dated with radiocarbon methods to indicate the times of former floods. The elevations of these deposits provide bounds on the heights of those floods, which can be combined with data on channel geometry and slope to estimate flood flow using numerical models.

Boulders are often deposited where flood velocity decreases, such as where a channel abruptly widens or gradient decreases. These provide a minimum height for a flood.

Problems with Recurrence Intervals

Although recurrence intervals are perhaps the best estimate available of the likelihood of a flood, there are significant limitations to their accuracy.

First, for a recurrence interval to be accurate, the data on which it is based must cover a long enough time interval to be representative. (A quantitative example of how dramatic the change in recurrence interval can be with a single slightly larger record flood was shown in By the Numbers 13-3.) Second, the recurrence interval assumes that the upstream conditions that affect stream flow in the watershed must be similar through time. Although the use of paleoflood data can dramatically extend the total time interval on which a record is based, the conditions under which paleofloods occurred may have been significantly different than more recent conditions. Floods during the ice ages of the Pleistocene epoch, for example, originated in a colder, wetter climate; the conditions were different than at present, probably both for amounts and frequency of precipitation.

Human impacts, such as urbanization, channelization, the building of dikes and dams, deforestation by any process, and overgrazing, also change the conditions for flooding. Population growth clearly causes changes, especially where residential communities continue to expand the area of impermeable ground with pavement and concrete. Rapidly growing cities such as Bellevue, just east across Lake Washington from Seattle, removed vegetation and added huge areas of pavement and rooftops. The estimated 100-year recurrence interval flow for Mercer Creek, which drains the area, was estimated in 1977 as 420 ft³/s. In 1994, following 17 years of rapid growth, the 100-year flood was estimated as 950 ft³/s (**FIGURE 13-32**)! The entire character of floods here changed due to urbanization of the region.

Finally, all of the floods plotted to determine recurrence interval or exceedance probability must belong to a single group that originates from similar flood causes so that the distribution of flood sizes should be random. They all must result from an El Niño event, for example, or from storms originating at a warm front. Combinations of causes, such as a hurricane colliding with a cold front, would create aberrations that would not fall in the same group and should ideally not be plotted on the same frequency

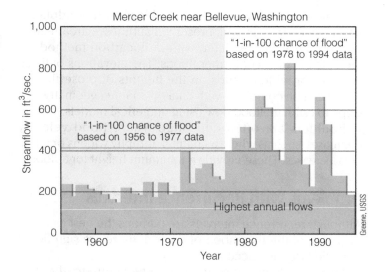

FIGURE 13-32 Effect of Urbanization on Flood Patterns
The 100-year flood for Mercer Creek, Washington, near Seattle, increased dramatically following rapid urbanization from 1977 to 1994.

distribution. Unfortunately, the data sets are so small—the period of record so short—that all floods on a river are typically lumped together, regardless of size or origin.

Although recurrence intervals are useful tools for studying floods, the complex interactions that lead to flooding mean that these intervals cannot be seen as predictions. Even worse is a simplistic perception of a 100-year flood as one that comes along every 100 years. In fact, the number is merely an average—such a flood may recur anytime, including in the year following a major flood.

Mudflows, Debris Flows, and Other Flood-Related Hazards

Where large amounts of easily eroded sediment are carried downstream by floodwaters, the nature of the sediment transport also changes. The amount of sediment transported in a flood varies widely, from less than 1% to 77% by volume (**Table 13-2**). As the proportion of sediment increases to more than 20%, a flood becomes a *hyperconcentrated flow*, and with more than 47% sediment, a **debris flow**. A debris flow is concentrated enough that you could scoop it with a shovel. In some cases, a single flow begins as a debris flow, and then with additional water downstream forms a hyperconcentrated flow, and perhaps then a water flood. If mud or clay dominates the solids choking a flowing mass, it is a **mudflow**; if volcanic material dominates, it is called a **lahar**. If the solids are more abundant, coarser, and highly variable in size, it is a debris flow.

We can determine the type of a former flood from the characteristics of the deposits:

- Water floods have little internal shear strength and are dominated by the fluid behavior. Sediment moves in suspension and by rolling and bouncing along the channel bottom.

- Hyperconcentrated flows contain higher sediment concentrations and thus have larger overall densities and viscosities. Water and grains behave separately; grains colliding with one another may not settle at all or settle at rates only one-third of those in clear water.

- Debris flows contain still higher sediment concentrations; the water and grains move as a single viscous or

Table 13-2	Characteristics of Floods with Increasing Proportions of Sediment to Water		
	WATER FLOODS	**HYPERCONCENTRATED FLOWS**	**DEBRIS FLOWS**
Sediment concentration (% by volume)	0.4–20	20–47	47–77
Bulk density (g/cm³) depends on amount of sediment in transport	1.01–1.33 Note: Water has a density of 1 g/cm³	1.33–1.80	1.80–2.3 Note: Most rock has a density of ~2.7 g/cm³
Deposits, landforms, and channel shape	Bars, fans, sheets, wide channels (large width-to-depth ratio)	Similar to floods	Coarse-grained marginal levees, terminal lobes, trapezoid to U-shaped channels
Sedimentary structures	Horizontal to inclined layers, imbrication*, cut-and-fill structures, ungraded to graded	Massive or subtle layers, subtle imbrication, normal or reverse grading**	No layers, no imbrication, reverse grading at base, normal grading at top
Sediment characteristics	Well-sorted, clast-supported, rounded particles	Poor sorting, clast-supported, open texture, mostly coarse sand	Poor sorting matrix supported, extreme range of particle sizes, ± some megaclasts

*Imbrication is the overlapping of pebbles with their flat sides sloping upstream. The current has pushed them over.
**Normal grading has coarser grains at the bottom of a deposit, finer grains at the top.

plastic mass, with water and sediment moving together at the same velocity. Internal shear strength is high, so shear is concentrated at the base and edges of a flow. As flow velocity slows, few particles settle. As water escapes from the spaces between fragments at the flow edges, the flow slows and finally stops moving.

All of these flow types are dangerous. Debris flows are most destructive because of their higher densities and the large boulders they carry; they can move at higher velocities than other floods, and their greater mass can destroy larger and stronger structures.

Mudflows and Lahars

Like landslides, mudflows are mobilized by abundant water in a slope and are often triggered by heavy or prolonged rainfall. The tiny pore spaces and low permeability of mud slurries retain water and keep the mud mobile. Boulders are rafted along near the surface of flows that have high density and resemble wet concrete. Earthquakes jar others loose.

Active volcanoes are notorious for spawning lahars, especially during eruptions, because volcanic ash supplies abundant mud-size material. However, lahars often contain rocks and boulders as well. The collapse of heavily altered ash on the flank of a volcano during heavy rains, or rapid melting of snow or ice, forms a dense slurry that collects loose volcanic debris as it races downslope (**FIGURE 13-33**). During volcanic eruptions, hot ash may mix with rain, lake, or stream water downslope to pour down nearby valleys. The rapid mixing of hot rocks and ash with ice and snow causes much more rapid melting than situations in which these hot materials simply fall on the snow surface. Even if the falling ash is too cool to melt snow or ice, large volcanic eruptions

often generate their own weather. The heat of the eruption pulls in and lifts outside air, causing it to expand and cool. That causes moisture in the air to condense; locally heavy rains fall on the newly deposited ash, flushing it downslope.

Even long after major ash-rich eruptions, the amount of ash on a volcano's flanks provides ample material for catastrophic lahars. Such lahars often have extremely long runouts, especially where they are confined to valleys. They can continue to move on slopes as low as 1%—a 1 m drop in 100 m.

Debris Flows

Debris flows are common and extremely dangerous. They are widespread, begin without warning, move quickly, and have tragic consequences for both structures and people. They are most common in steep mountainous areas such as the U.S. Southwest (**FIGURE 13-34**). They are especially common along major active faults, where tectonic movements actively build mountain belts such as the Basin and Range of Nevada, the Wasatch Front at Salt Lake City, and the Andes Mountains of South America. The steep slopes maintain a continuing supply of broken debris. Most debris flows empty onto alluvial fans, making these a hazardous place for residences (**Case in Point**: Desert Debris Flows and Housing on Alluvial Fans—Tucson, Arizona, 2006, p. 374).

As debris flows spread across a fan, they block old channels with piles of gravel and boulders. Subsequent debris flows then overflow to erode new channels. The details of the channels change, but the general style of channels on the fan remains the same. Flows continue moving on a broad alluvial fan at slopes less than 10 to 15° but rarely at slopes less than 5° or so. Because debris flows thin as they spread out on a fan, their energy dissipates, and their largest boulders drop near the head of the fan; progressively smaller ones drop downslope. Having deposited most of their bouldery load, water floods can continue to lower slopes.

Debris flows differ from stream flows in the amount of solid grains suspended within them. Because rocks are approximately 2.7 times as dense as water, a slurry of rocks with sediments in the pore spaces can be twice the density of water in a flooding stream. That permits these flows to pick up and carry huge boulders, some as large as a car or even a school bus. The high density of debris flows in steep terrain can propel them at a higher velocity than clear water.

Debris flows are characterized by internal shearing, with some parts moving faster than others. Much of their movement is by the whole mass sliding on a stream bottom, with a lesser amount of jostling between fragments. Except for all of the boulders and internal shear, movement is something like wet concrete coming down the chute of a cement truck. Slippage at the base of a flow permits it to scour material from its channel, and to entrain more material in the flow.

Debris flows tend to move in surges. When particle sizes vary, jostling moves the largest pieces to the edges and front of a flow. The same thing often happens to a human body caught in a snow avalanche. Boulders bob along a debris flow's

FIGURE 13-33 Lahar
This is the flow front of one of the many fast-moving lahars racing right to left down a valley from Mt. Pinatubo in the Philippines. Note that the surface of the flow is covered with rocks and pebbles, especially in the nose of the cresting flow.

FIGURE 13-34 Debris Flows

A. A debris-flow basin at the north edge of San Bernardino, California, filled and overflowed to surround houses in a winter event in 1980. It is being re-excavated in this photo.

B. In December 2007, a muddy debris flow swept down across Highway 30, west of Portland, Oregon, burying houses in a small community.

surface (**FIGURE 13-35A**); they are pushed to the sides and front, typically forming prominent bouldery natural levees. Although water drains easily from large spaces between boulders, movement depends on having lots of fine-grained particles and water in the pore spaces (**FIGURE 13-35B**).

Movement slows when water drains from between fragments, especially at the toe and flanks of a flow (**FIGURE 13-36A**). Fine material has small interspaces that drain water much more slowly, so movement continues in the fine-grained rear parts of a flow. The coarse, bouldery

front of a flow drains its water so it can keep moving only by riding on a core of finer-grained material (**FIGURE 13-36B**).

Many years of occasional rains tend to wash mud and assorted debris down slopes and into canyon floors. After decades or centuries, canyons contain enough accumulated debris to provide the raw material for a large mud or debris flow. That will happen as soon as one of the occasional heavy desert rains flushes the canyon floor and spreads its burden of debris out over the alluvial fan downslope. Then another long period may pass before the canyon again

FIGURE 13-35 Deposits from a Debris Flow

A. Bouldery natural levees formed in 1996 by a major debris flow on a tributary to the Columbia River, near Dodson, east of Portland, Oregon. Note the large boulders concentrated at the top of the deposit.

B. Finer-grained particles fill the spaces between the boulders and below them. It is the watery, finer-grained matrix of the flow that buoys the boulders on top. Sabino Canyon 2006 debris flow, Tucson, Arizona.

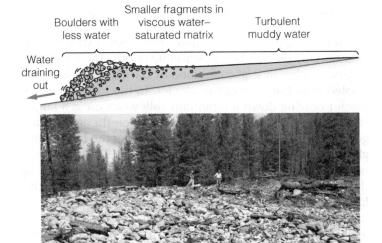

Boulders with less water | Smaller fragments in viscous water–saturated matrix | Turbulent muddy water

Water draining out

Michael Lewis, USGS

FIGURE 13-36 Distribution of Sediment

A. A schematic diagram shows the distribution of grain sizes and water in a debris flow. **B.** This debris flow, southwest of Ketchum, Idaho, formed after the Beaver Creek fire, in August 2013. To keep moving, the course, bouldery front must have ridden on a fine-grained core.

accumulates enough debris to repeat the performance. It recharges with loose debris from the sides of the canyon. Many geologists familiar with desert canyons can tell just by looking whether a canyon is sufficiently charged to produce a new debris flow. In other cases, only the more recent loose deposits in the canyon flush out as a debris flow; older deposits remain available to move in another storm. These events are most likely in watersheds that have lost their cover of brush to a fire within the previous few years. Bare ground does not absorb rain as well as it does when covered with plants, so the hills shed water like a roof, and the heavy surface run-off flushes the canyons (discussed in Chapter 17, Wildfires).

Debris flows commonly begin with a heavy rainfall or rapid snowmelt that fills pore spaces above less permeable bedrock (**Case in Point:** Intense Storms on Thick Soils—Blue Ridge Mountains Debris Flows, p. 375). This increases the pore pressure that sets a mass in motion, especially if it is disturbed by an earthquake or even a strong gust of wind on a large area of slope. Sometimes the initial movement is a landslide that begins in a swale, or trough, where soils tend to become thicker and the groundwater level is high. Sometimes they begin when water flowing downslope picks up sediment, then "bulks up" with more sediment farther downslope. Flows tend to keep moving as long as the positive

pore-water pressure is maintained. A major six-day storm in November 2006, on the flanks of Mt. Hood, Oregon, that produced 34 cm of rain, initiated debris flows, some induced by landslides. Others formed when the rain-saturated, loose pyroclastic sand material that had accumulated in upper reaches of the channels bulked up and flowed downstream. Flows in the White River backed up behind, then crested over, the Highway 35 bridge and washed out the highway in two places.

Flows tend to move in surges or waves, each one with a steep front of especially coarse boulders. The initial main surge is often followed by a series of smaller surges traveling faster than the overall flow. Individual surges may slow and stop, only to be remobilized by a subsequent surge. Although most debris flows are not witnessed, their velocities and peak discharges can be estimated from peak-flow mud lines left on channel sides and cross-channel banking angles at sharp bends (see FIGURE 13-29).

The lower ends of swales are especially risky places for homes. In coarse colluvium (loose, broken material over bedrock), landslides that develop into debris flows begin on slopes from 33 to 45°, measured down from horizontal. Where a debris flow has occurred in the past, it will occur again in the future. Evidence of a previous occurrence indicates it is likely to happen again. Disturbances such as fire, logging, housing developments, road building, and volcanic activity can change a basin to make it more prone to debris flows. The maximum amount of loose material available in catchment basin channels provides an indication of the maximum size of a future debris flow. A large debris flow that removes most of the loose material in the basin will lessen the maximum size of future flows for years until the loose material again builds up. Evidence of former debris flows include:

- A valley floor strewn with boulders that seem much too large for the modern stream to move. In exposed cross-sections, huge boulders are perched on top of finer-grained deposits that are massive and unlayered and have large, angular boulders in a finer matrix (**FIGURE 13-37**)
- Levees of coarse, angular material next to a stream (FIGURES 13-35 and 13-37)
- Deep, narrow channels cut in those deposits (FIGURES 13-35 and 13-37)
- Fan-shaped deposits forming rounded lobes with coarser material at the outer edges
- Rocks lodged against trees, deposited in tree branches, or embedded in bark (**FIGURE 13-38**)
- Bark scars high on trees on their upstream sides (FIGURE 13-38)
- Lobes of even-age vegetation younger than the surrounding vegetation
- A drainage basin with large, actively eroding areas
- Active faulting, which helps supply broken rock to a slope

FIGURE 13-37 Evidence of a Debris Flow

This debris flow in the Andes of northwestern Argentina was rich in rocks and boulders until it began losing water and slowing down. That permitted the water-rich finer-grained material (covering boulders in the middle left of the photo) to catch up and, in turn, flow over the surface of the head of the flow. Debris flows come down this channel every year, and in one case the flow eroded down to a buried natural gas pipeline, causing a massive explosion.

Glacial-Outburst Floods: Jökulhlaups

The toe of a glacier, where meltwater feeds a stream, can occasionally see sudden catastrophic floods. The water can originate either in a lake dammed by the glacier or in water pooled near the base of the glacier. In the former case, a glacier flowing down a mountain valley can cross a larger stream, damming it to form a lake. When the lake water gets high enough, the water can seep under the base of the glacier and rapidly enlarge a tunnel (**FIGURE 13-39**). If the tunnel gets sufficiently large and the roof collapses, the resulting downstream flood may be catastrophic. In other cases, the lake water may float the glacial ice dam, leading to rapid failure and flooding downstream. Camping downstream from a glacier-dammed lake can be hazardous. Such floods are called **glacial-outburst floods**, or jökulhlaups. On August 14, 2002, water dammed behind Hubbard Glacier burst through to drain the lake. In April 2010, a glacial-outburst flood during the eruption of Eyjafjalla volcano under a glacier in Iceland forced evacuation of hundreds of people, severed the coastal road, and destroyed a huge area of farmland.

Meltwater pouring down through crevasses and holes in a glacier finds its way to the glacier's base where it can lubricate glacial movement (discussed further in Chapter 11). That water continues down valley under the glacier, appearing at its toe as a meltwater stream. Sometimes large volumes of water pond under gently sloped glaciers, such as those in Antarctica. Elsewhere, glacial movement blocks a subglacial channel, leaving it vulnerable to a sudden outburst flood and potential catastrophe downstream. Active

FIGURE 13-38 Height of Debris Flows

A. Huge boulders are lodged against the upstream sides of trees and the trees are debarked (arrows) along Tumalt Creek, east of Portland, Oregon. **B.** Rocks (above and at the right of the red pocketknife) are lodged in the branches of a tree, which was debarked to about that height by a debris flow along Sleeping Child Creek, south of Hamilton, Montana, in July 2001. The area burned in a massive wildfire the year before.

FIGURE 13-39 Glacial-Outburst Flood

A. A lake formed by a tributary glacier blocking a valley drained when a tunnel eroded under it. Tulsequah area, northwestern British Columbia.

B. Advance of the Hubbard Glacier to raise Russell Lake in early May 2002 set the stage for the outburst flood on August 14.

glaciers in many parts of the world pose glacial-outburst flood hazards to campers, hikers, sightseers, and some residents. Such floods are most common in very hot or rainy weather. If a meltwater stream rises rapidly or you hear a roaring sound up-valley, move quickly up the side of the stream bank.

Many glaciers on Mt. Rainier, Washington, have experienced glacial-outburst floods, some of them triggering debris flows or lahars. Subglacial-outburst floods are also common in Iceland, where lava eruptions beneath a large ice cap may melt the ice. The meltwater flows downslope under the ice, causing bulging of the ice cap 15 km downslope. Every few years the under-ice lake periodically bursts out as a flood.

Hubbard Glacier, the largest glacier reaching the sea in North America, crossed a fjord near Yakutat, at the north end of the Alaska panhandle to fill Lake Russell. Three months later, on August 14, 2002, the ice dam failed, releasing the second largest historic flood worldwide (FIGURE 13-39B). The peak flow was about 54,000 m³/s in a river 100 m across—30 times the peak flood flow of the lower Mississippi River. The lake took 36 hours to drain. Fortunately, Yakutat, at the mouth of the bay, was not damaged. The largest glacial-lake flood on record was at the same site in 1986.

During the last ice age, huge glacial meltwater floods appear to have spread southeastward from the continental ice sheet in southeastern Alberta. Some of these giant subglacial floods had volumes of tens of thousands of cubic kilometers. Similar gigantic floods, but from glacial lakes ponded behind ice-age glaciers, have been well documented. Gigantic floods from the Altai in central Asia and multiple drainages of glacial Lake Missoula that spread across southeastern

Washington State thousands of years ago have also been well documented but pose no present danger.

Ice Dams

During winter and early spring, ice coating rivers can block stream channels; as the ice begins to melt, it breaks up and moves. A sudden warm spell that causes ice to break up can dam a channel at constriction sites, such as bridges (**FIGURE 13-40**). Large rivers flowing northward have the

FIGURE 13-40 Ice Dam

An ice jam built up at a river constriction threatens a bridge at Gorham, New Hampshire. The power shovel tries to get the river flowing again.

additional problem that ice upstream thaws before the ice downstream. As upstream meltwater flows north, it encounters ice jams that restrict downstream flow, causing floods (**Case in Point:** Spring Thaw from the South on a North-Flowing River—The Red River, North Dakota, 1997 and 2009, p. 376).

Other Hazards Related to Flooding

Different types of hazards are often interrelated, as discussed in Chapter 1, so it shouldn't surprise you to learn that flooding is related to a number of other hazards. Torrential rain can develop from thunderstorms or hurricanes (Chapters 10 and 16) or where humid, tropical air over an ocean is forced to cool and condense as it rises against a mountain range (Chapter 10). Coastal areas can flood as hurricane winds cause sea level to rise locally as it comes onshore. Heavy rains can be initiated by a volcanic eruption as hot volcanic gases pull in humid air and carry it to high elevations where it condenses (Chapter 6). Those same rains, falling on ash dumped on volcano flanks, can produce mudflows that race down nearby valleys and in turn gradually winnow into dirty floods. Wildfires often denude slopes, causing heavy rainfall to flow rapidly off their surfaces; they can cause debris flows or raise flood levels much faster than otherwise. Floods can, in turn, initiate other hazards. A flooding stream sweeping around the outside of a bend can undercut and steepen the stream bank or slope above, causing it to slide (Chapter 8).

Cases in Point*

Monsoon Floods
Pakistan, 2010 ▶

The summer of 2010 brought the usual monsoon rains to India, but normally those storms bypass Pakistan, which lies off to the northwest of the main monsoon belt. In July and August 2010, monsoon rains drowned Pakistan with the worst floods in 80 years. At least 1781 people died in the floods, mostly in the first few days.

Aggravating the torrential monsoon rains was a new La Niña following the strong El Niño in 2009. The much warmer than normal ocean evaporated more moisture into the regional atmosphere; moisture-laden summer monsoon winds from the Arabian Sea, drawn toward the atmospheric low over India, rose against the Himalayas where they condensed and dumped their moisture. The zone of convergence between warm, tropical moisture to the south against the dry continental air on the north migrated northward during the summer to move the torrential rains northward into northern India and Pakistan.

The 2010 monsoon floods in Pakistan were disastrous. Most were centered along the Indus River and its tributaries, the major river that drains the length of Pakistan. Flood levels peaked in the north at the end of July, remaining high for more than two weeks. A few days later, they peaked midway downstream, dropped somewhat, then peaked again in mid-August and remained high until the end of the month. Farther downstream, they peaked again on August 7. Two major dams upstream were not sufficient to control flood flows. Massive deforestation in some areas by militants to fund their operations decreased evapotranspiration and increased flood flows and production of landslides. Roads, bridges, and many villages were washed out in the mountainous northwest, cutting off thousands of people.

Although people were forewarned of its approach, levees failed quickly so farmland and homes were overwhelmed before residents could evacuate. Thousands were forced to wade in waist-deep water, leading a farm animal or two and the few possessions they could carry. Those on the floodplain lost everything; their land was flooded, their houses submerged under water, collapsed, or washed away. They were left without food, drinking water, or shelter. By mid-August, 36,000 cases of diarrhea were reported from people forced to drink contaminated water, and malaria cases were rising. One point two million homes were damaged or destroyed and 3.2 million hectares of crops were destroyed. The country's rice crop and much of its stored wheat were lost. Twenty million people were displaced with nowhere to go. In late August, 800,000 people remained stranded; 7 million were homeless and would need food support for the next several months. Even at the end of August, levees broke in the downstream parts of the Indus valley, displacing another 175,000 people.

Roads submerged and bridges failed, marooning thousands living on the floodplains. The only way to reach higher ground was on foot but in the north floodwaters were swift and dangerous; even downstream water in some places was chest-high or deeper. Even if they managed to reach the edge of the floodplain, there was still no food, drinkable water, medication, or shelter. Crop loss left 6 million people with too little food.

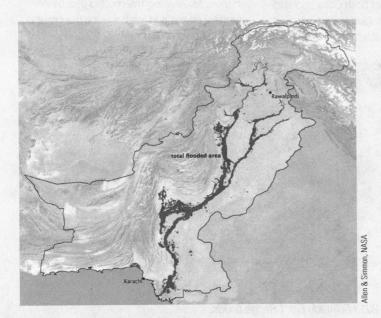

■ *Areas affected by Pakistan floods along Indus River, July–August 2010.*

Allen & Simmon, NASA

Sgt. Bryce Piper, U.S. Marine Corps

■ *In Sindh Province, Pakistan, on the lower Indus River villagers in upper left walk a submerged road toward remaining high ground, in November 2010.*

*Visit the book's Website for this additional Case in Point: Heavy Rainfall on Near-Surface Bedrock Triggers Flooding—Guadalupe River Upstream of New Braunfels, Texas, 2002.

Flood victims are stranded on this remnant of dry ground in Pannu Aqil.

Aid came slowly and in limited amounts, most of it by helicopter and military boats; flights were often grounded by torrential rains. Major roads were washed out and blocked by landslides, and bridges were washed out. Some people refused evacuation, fearing loss of their homes, livestock, and other few belongings. Damages to structures including roads, bridges, and homes was more than U.S. $4 billion. Economic losses came to 10 times that amount.

Considering the extent of the flooding and extreme number of people endangered, the amounts of international aid offered were very small, perhaps because the number of deaths was so low and the crisis built up gradually, in contrast to the large number of deaths and suddenness of the earthquake disasters in Kashmir and Haiti. Hot, humid weather fostered spread of waterborne diseases including cholera, diarrhea, and gastrointestinal diseases, along with malaria. Evacuees headed for towns that remained above flood levels. That influx aggravated already widespread unemployment and ethnic violence.

In September 2014, more than 560 people died, and tens of thousands were displaced in monsoon floods in northernmost India and Pakistan.

CRITICAL THINKING QUESTIONS

1. What caused the 2010 Pakistan flood?
2. What factors made it worse than the usual rainy-season event?
3. Why didn't more people evacuate?
4. Besides loss of food and housing, what were major secondary effects of the flood?

Flash Flood in a Canyon
Colorado Front Range, 2013 ▶

On the morning of September 9, 2013, a cold front moved eastward across the Front Range of the Colorado Rockies. At the same time, the slow-moving system pulled a deep subtropical mass of moist air from the Gulf of Mexico and Pacific monsoonal moisture from the southwest. Moist upslope winds from the southeast pushed northwestward up against the Front Range of the Rockies, where they expanded, cooled, and dumped their moisture. A high-pressure weather system to the east blocked further eastward movement of the cold front (as shown in the map). Large parts of both Colorado and New Mexico saw prolonged heavy rains and floods. Heavy rainstorms drenched the area along and north of I-70, west of Denver. This was just the beginning. Over the next week, rain fell in torrents, soaking the ground and then running off the surface because the ground couldn't absorb it all.

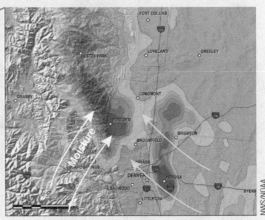

A

B

A. Inferred weather conditions leading to the September 2013 Colorado Front Range floods.
B. Intense rain areas of the September 2013 Colorado floods. Pale green areas are 10- to 50-year rainfall; dark blue areas are possible 500- to 1000-year rainfall.

■ **A.** *Remains of a streamside house in the canyon*
B. *A destroyed SUV*
C. *The boulder-strewn channel*

In the mountains west of Denver, rainfall for the week of September 9 to 16 reached more than 23 cm (9 in) in some places. In the Denver metropolitan area, it ranged from about 6 to more than 39 cm. An all-time 24-hour record of 23 cm fell in 24 hours. Farther northwest, in Boulder and to the northwest up to Estes Park, most areas received at least 26 cm (10.2 in) and as much as 45 cm (17.7 in), at a rate of 6.3 cm per hour in some places. Imagine shedding all of that water downslope to gullies and streams, with more added to that from the slopes farther down-valley. More than 1100 debris flows, most on south-facing rocky slopes with less vegetation and facing the incoming rain, began as small slides on steep slopes of 26 to 43°. As these debris flows fed into gullies, they picked up loose rocks and raced downslope.

The streams quickly became raging floods scouring everything loose from the channels, tearing out roads and bridges,

downing power lines and poles, crushing trees, houses, and everything in their path. Three dams failed, adding to the downstream damages. Floods surpassed all-time records, especially in Aurora (bordering Denver on the east), Boulder, and northwest to Estes Park, and were deemed a "point one percent annual exceedance probability" event (formerly called a 1000-year flood). On some days rain would let up a bit, only to come down in buckets again. The raging torrents of the Front Range canyons spread out on the plains to the east to inundate broad areas.

Damages were most severe around Boulder, in the mountains of Big Thompson Canyon from Estes Park downstream to Loveland, in the west, and spreading out on the plains along the South Platte River through Greeley to the east. Landslides swept away roads and filled towns with mud and debris. Upstream, giant boulders clogged the river channel. I-25 north from Denver to Wyoming was mostly closed September 13 because of flooding. Amtrak, Burlington Northern, and Union Pacific rail lines west of Denver to Grand Junction were washed out and closed for two weeks. Lyons and other mountain towns were

completely isolated when their only access roads were washed out. All told, 5958 people heeded mandatory evacuations and 8 people died. Flooding destroyed 1882 homes and 203 commercial properties, and 16,101 homes and 765 commercial properties were damaged.

Some evacuees returned to the stench of rotting food after loss of power. Others found mud everywhere, with homes beginning to smell from mold and sewage from backed-up sewers and septic tanks. To the east on the plains, oil and gas lines were severed, and crude oil spilled into the South Platte River.

In 1976, a similar intense storm ravaged Big Thompson Canyon when a cluster of thunderstorms stalled over the canyon and dumped 20 cm of rain in an hour. This led to a wall of water that flushed downstream, catching people off guard, and killing 145 when many tried to outrun the flood in their cars in the narrow canyon. On June 14 to 15, 1965, a storm dumped 30 cm of rain overnight near Fort Collins. Such storms, fed by moisture from the Gulf of Mexico, are pushed up against the Rockies; rising faster on hot summer afternoons, they dump heavy rains. Most move through quickly unless stalled against other weather systems.

The 1976 storm was similar to later events but much more widespread. As heavy rain began to fall at 6:30 p.m., police moved through the canyon telling people to leave. Unfortunately, because heavy summer thunderstorms in the area were common, many people did not believe they were in danger. By 7 p.m., a wall of water more than 6 m deep was racing down the canyon at roughly 6 m/s (~22 kph). By 8 p.m., 400 cars were on the canyon highway, and those who abandoned their cars and ran upslope survived, whereas those who tried to outrun the flood in their cars died. More than 600 people were never accounted for. An ambulance crew that drove into the canyon to render aid reported that a huge, choking dust cloud led the wall of water, picked up the ambulance and slammed it into a wedge on the canyon wall. The crew climbed out of the wrecked ambulance and up to a ledge 15 m above the highway. The water surged to their level, but they survived and were rescued by helicopter the next morning, along with hundreds of others. In the debris at the mouth of the canyon was a motel register with 23 names: none of these guests were found. The flood destroyed 400 cars, 418 houses, and 52 businesses. The flow rate was four times larger than that of any previously recorded flood in 112 years.

Desert Debris Flows and Housing on Alluvial Fans
Tucson, Arizona, 2006 ▶

Following four days of rain, July 31 saw the largest of the early morning convective thunderstorms that dumped heavy rain on the Santa Catalina and adjacent mountain ranges around Tucson, Arizona. Upper-level winds from the northwest collided with lower-level winds from the west rising against the

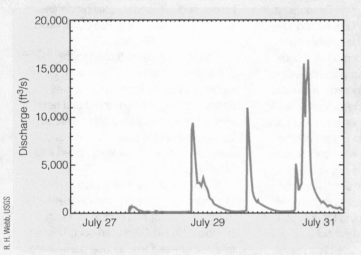

R. H. Webb, USGS

■ *Hydrograph of flood flows in Sabino Creek in July 2006.*

C. Magirl, USGS

■ *One of numerous debris flows initiated on the steep slopes of Sabino Canyon that carried huge boulders down to the main valley.*

P. Griffiths, USGS

Donald Hyndman

■ *Rest room at the end of Sabino Canyon.* Left: *Before debris flows.* Right: *After. Arrows point to paired features.*

front of the range. The rainfall intensity was not unusual for Tucson's heavy summer thunderstorms, but saturation of the thin soils by the days of preceding rain appeared to be more important than the rainfall intensity.

More than 500 slope failures occurred on the flank of the Santa Catalina Mountains, and adjacent ranges experienced 100 additional slides. Twenty to 25 cm of rain fell on Sabino Canyon, with flood flows reaching 442 m^3/s, a new 100-year record.

The flows raced downslope, coalesced, and continued down valley in all of the canyons on the west side of the Santa Catalina Mountains. Soils are only about 1 m thick over bedrock on the steep, rocky slopes. Debris flows began most commonly near the base of bedrock cliffs where excess surface water may have poured onto the already saturated soils to initiate failure. The flows quickly bulked up downslope as they collected more rocky soil material. Reaching gentler slopes at the main canyon floors, the flows rapidly thickened. In some places

they severely eroded roads; in others they clogged large culverts and buried roads and small buildings.

Intense Storms on Thick Soils
Blue Ridge Mountains Debris Flows ▶

Even relatively humid areas with higher rainfall, gentler slopes, and deeper soils can be subjected to debris flows. In August 1969, Hurricane Camille, one of only three Category 5 hurricanes to hit the United States in the twentieth century, came onshore in Mississippi, where it did severe damage. Before petering out, it stalled over the mountains of central Virginia, causing still more damage, primarily in debris flows. It dumped an amazing 71.1 cm of rain in eight hours, saturating the ground and causing shallow slides that quickly developed into debris flows.

At least 1100 slopes slid, generally where they were steeper than a grade of 17% (10° slope). Most began where groundwater is closer to the surface, near the inflexion point between convex rounded hilltops and the concave segments above main river channels. Where slopes were covered with coarsely granular soil, they were generally stripped down to bedrock. Formations that were most vulnerable to sliding contained nonresistant, well-layered, and mica-rich rocks that dipped parallel to the downslope direction.

Years later, on June 27, 1995, an intense storm stalled over the Blue Ridge Mountains in Madison County, Virginia, where it dumped more than 76 cm (2.5 ft) of rain in 16 hours. It triggered more than 1000 debris flows; one in Kinsey Run northwest of Graves Mill began as a landslide that developed into a 2.5-km-long debris flow. It raced down the slope in surges at 8 to 20 m/s, eroding and incorporating

ground material to depths up to 0.6 m. Velocities were calculated from the tilt of the flow surface where it poured around bends in the channel. Its deposits included coarse rocks with boulders up to 7 m across in a matrix of clay and sand.

Although this flow did not impact people, others living elsewhere in the Blue Ridge Mountains are vulnerable. Debris flows are common in the region, and events of this magnitude have recurrence intervals of tens of years.

A decade later, Hurricane Ivan drenched the Blue Ridge Mountains of North Carolina on September 16, 2004, with nearly 30 cm (1 ft) of rainfall. It initiated numerous landslides and debris flows, destroying many houses and killing five people.

Rick Wooten, North Carolina Geological Survey

■ *Debris flows from Hurricane Ivan destroyed this home in Starnes Cove, Buncombe County, North Carolina.*

Spring Thaw from the South on a North-Flowing River
The Red River, North Dakota—1997 and 2009 ▶

The winter of 1996–97 brought numerous heavy snowstorms to the northern plains, a snowpack two or three times the depth of the previous record. In early April, a major blizzard dumped up to 1 m of snow on parts of the area; this was the culmination of eight major blizzards that provided as much as 3 m of snow in Fargo.

The Red River that flows through Fargo and Grand Forks, North Dakota, and Winnipeg, Manitoba, freezes in winter. Come spring, it thaws first in North Dakota but remains frozen farther north, a recipe for flooding beyond the control of local residents. As the thaw progresses, ice floes drift north to pile up in ice jams that cause widespread flooding. The very flat terrain around Winnipeg, Grand Forks, and Fargo is the old flat bottom of Glacial Lake Agassiz; it covered much of southeastern Manitoba, with a prominent tongue extending south up the Red River valley. That lake was dammed on the north by the continental (Laurentide) ice sheet until about 9800 years ago. Grand

Forks, North Dakota (metropolitan area population 98,000), is entirely within a 500-year floodplain; its highest elevation lies less than 10 m above the Red River's flood level.

People in the area are used to floods, but not on the scale of 1997. On April 17, the Red River broke its 100-year record at Fargo. Discharge reached some 3400 m^3/s (compared with this river's average of 75 m^3/s). On April 19, it crested at Grand Forks and covered 90% of the city for several days. Flooding of the water and sanitation systems not only left the city without drinking water and sewage treatment but also contaminated the flood water, making the city uninhabitable. Fire broke out in downtown Grand Forks on April 19 and spread to 11 buildings, creating severe access problems for firefighters.

Sixty thousand people in the Grand Forks area were evacuated, and damage and cleanup costs exceeded $1 billion. Overland flooding from snowmelt from

farm fields on April 3 worsened the problem. To lessen damage to the larger population areas, authorities built a dike on the outskirts of town to block overland flooding. Seven people died in North Dakota and four in Minnesota. Water remained above flood level from March 26 to May 20. The federal government purchased and moved many homes in the lower parts of Grand Forks and tore them down to minimize damage from future floods. By 2006, $410 million in federal, state, and local funds had facilitated rebuilding and raising 85% of levees and flood walls, now 1 m above 1997 flood levels. The new floodway, a "greenway," provides a broader pathway for the flooding river; it includes parks, gardens and golf courses, paths for bikes and walkers, and athletic fields.

So what does the future hold? Climate predictions suggest that the average precipitation and temperature in the northern Great Plains will both increase during the next century. If this is correct, floods on the Red River may become more frequent and larger in the future.

After the 1997 floods, experts recommended building levees farther from the river in Fargo, but business owners along the river objected, and nothing was done. In March and April, 2009, flooding began farther upstream, cresting with record highs in Fargo, upstream, and Winnipeg, downstream. Intense cold and an unusually snowy winter left snow blanketing the still-frozen ground, meaning snowmelt would run off instead of soaking in. Spring rains added to the runoff, causing rapid melting. By March 27, the National Weather Service estimated a crest of 13.1 m in Fargo. To protect the city, 6000 volunteers and 1700 National Guard members filled sandbags in below-freezing temperatures to raise about 19 km of levees. The

■ *The Red River flows north along the border of North Dakota and Minnesota, through southern Manitoba, and into Lake Winnipeg. The extent of Glacial Lake Agassiz, about 9800 years ago, controlled the area of the 1997 flood. Water draining northeast on a gentle slope was dammed by the south edge of the continental ice sheet.*

■ *This hydrograph shows the flood level for the Red River at Fargo, North Dakota, March to June 1997.*

■ *Ice jams that filled the channel of the Red River below Grand Forks, North Dakota, caused the April 1997 flood.*

■ *The Red River in Manitoba also had extensive flooding during this event.*

freezing of moisture in the sand confounded these measures; one resident characterized the effort as stacking big frozen turkeys on the levee. This also

meant that the sandbags didn't sag to fill spaces between the underlying bags. But new heavy snows moved in and added to meltwater later. The threat remained of the fast-flowing water breaching weak parts of the levees; a helicopter dropped 900 kg sandbags into the river to deflect the current away from vulnerable sections. In spite of the severe damage in Grand Forks in 1997 and their nearness

to the river, many people had no flood insurance coverage for the 2009 event.

Decades earlier, farther downstream in Manitoba, a disastrous flood on May 8, 1950, caused damages of more than $790 million when eight dikes failed. In response the government built a new 47-km-long flood channel to divert the Red River around Winnipeg; it was completed in 1968. A long-proposed 56 km flood diversion channel around the west side of Fargo remained unbuilt in 2012 because of the total cost and from opposition from people whose houses would have to be removed.

CRITICAL THINKING QUESTIONS

1. What major factors influence flooding in the Grand Forks area of North Dakota?

2. Where does all the flood water in this area come from?

3. Why does flooding last so long, rather than moving downstream in a short-lived flood crest?

■ *The 1997 Red River flood at Grand Forks, North Dakota.*

■ *Downtown Grand Forks was submerged. The water level rose high enough to inundate some buildings to their rooftops as well as the bridge, which is visible near the right side of the photo.*

Streams and Flood Processes **377**

Chapter Review

Key Points

Stream Flow and Sediment Transport

- A stream's flow, or discharge, is proportional to the average water velocity and the cross-sectional area of the flowing part of the channel. **FIGURE 13-2** and **By the Numbers 13-1**.

- Stream equilibrium, or grade, is adjusted to accommodate grain size, amount of sediment in a channel, and the amount of water. Coarser grains and less water lead to steeper channel slopes. Similarly, larger particles can be moved only by higher water velocities. **FIGURE 13-4**.

- The sediment load that can be carried depends on the total flow, or discharge. Thus, floods can carry more sediment and erode more from a channel.

Channel Patterns

- Meandering streams are characterized by wide sweeping turns. This type of stream is more typical of wet climates. **FIGURE 13-8**

- Meandering streams erode on outside bends and deposit sediment in point bars on inside bends in a pattern of deep pools and shallow riffles.

- Meanders gradually migrate over a whole valley floor. The area of this migration is the area that typically floods, called the floodplain. **FIGURE 13-10**.

- Braided streams are multichannel streams that are overloaded with sediment because of either a dry climate with too little water or overabundant sediment supplied to the channel (e.g., below a melting glacier). **FIGURE 13-11**.

- Braided streams deposit sediment in alluvial fans, which can be dangerous flooding areas. **FIGURE 13-13**

- Bedrock streams are fast-moving, mountainous streams that have eroded the streambed down to bedrock. Flooding in bedrock streams is high-energy and highly erosive.

Groundwater, Precipitation, and Stream Flow

- In humid regions, most streams are gaining; thus they are fed by groundwater and flow year-round. In drier climates, there are losing streams, which lose water into groundwater. These streams flow primarily after a rain and may dry up between rainstorms. **FIGURE 13-16**.

- When water cannot be absorbed into the ground, it runs directly into a stream as surface runoff. Surface runoff is generated when rainfall is particularly intense, such as during a storm, when the ground is impermeable, frozen, or water saturated.

Flooding Processes

- Streams commonly reach bankfull level every 1.5 to 3 years on average and spill over floodplains less often. **FIGURE 13-17**.

- A flood does not have to be a large flow; rather, it is an unusually large flow for the stream.

- During floods, not only does the water surface rise higher, but the channel also erodes more deeply. Channel scour strongly affects damage. **FIGURE 13-18**.

- Patterns of erosion and deposition during flooding create natural levees along a river's banks.

Flood Intensity

- Flood intensity is a measure of stream discharge over time. Flood intensity varies with the rate of runoff, the shape of the channel, the number of tributaries, and distance downstream.

- There is a higher rate of runoff where the ground is less permeable, which can be due to soil composition, dry climate, deforestation, or urbanization. These areas are prone to sudden, severe flash floods.

- In stream order, the smallest tributaries are designated first-order; those where first-order tributaries join are second-order; and so on. Low-order streams have short lag times between rainfall and flood and are more prone to flooding. Higher-order streams with many tributaries are less prone to flooding. **FIGURE 13-24**.

- Downstream flooding is characterized by the flood peak arriving more slowly but lasting for a longer time. **FIGURE 13-25**.

Flood Frequency and Recurrence Intervals

- Flood frequency is commonly recorded as a recurrence interval, the average time between floods of a given size.

- Average flood frequency, or recurrence interval, is based on the past record for that stream. A 100-year flood has a 1% chance of occurring in any single year, regardless of the date of the last major flood. The recurrence interval for a stream is calculated as the total number of years of flood record for that stream +1, divided by the rank of the flood under consideration, where the largest flood on record has a rank of 1, the second largest has a rank of 2, and so on. **By the Numbers 13-3**.

- The 100-year floodplain is the area flooded by the largest flood in 100 years on average.

- Paleoflood analysis, the study of the magnitude and timing of past floods, includes indications of high-water marks, cross-sectional area, and meander wavelength, among other factors. **FIGURE 13-27**.

- Recurrence intervals may be poor predictors of future flooding because of limited records of past flooding, changing climate conditions, and human impacts, among other things.

Mudflows, Debris Flows, and Other Flood-Related Hazards

- With more than 47% sediment, a flood becomes a debris flow. If mud or clay dominates the solids choking a flowing mass, it is a mudflow; if volcanic material dominates, it is a lahar.

- Glacial-outburst floods occur when a glacier moving downslope crosses a stream, damming it to form a lake. When the lake water gets high enough, it may float the dam and rush downstream in a flood. **FIGURE 13-39**.

- When rivers flow north, downstream portions of the river may remain frozen after upstream portions have melted. This can lead to an ice dam that blocks water from flowing downstream, backing up water and causing flooding. **FIGURE 13-40**.

Key Terms

100-year flood, p. 359
alluvial fan, p. 347
base level, p. 347
bedrock stream, p. 349
braided stream, p. 349
channel scour, p. 355
debris flow, p. 364
delta, p. 347

discharge, p. 346
dynamic equilibrium, p. 346
flash flood, p. 357
flood crest, p. 357
floodplain, p. 349
gaining stream, p. 353
glacial-outburst flood, p. 368
graded stream, p. 346

gradient, p. 346
hydrograph, p. 357
lahar, p. 364
load, p. 348
losing stream, p. 353
meandering stream, p. 349
mudflow, p. 364
natural levee, p. 355

oxbow lake, p. 349
paleoflood analysis, p. 361
point bar, p. 349
recurrence interval, p. 359
riprap, p. 349
stream order, p. 357
surface runoff, p. 354
watershed, p. 346

Questions for Review

1. A river slope or gradient adjusts to what three factors?
2. Does water move faster in a mountain stream or in a large smooth-flowing river like the Mississippi? Why?
3. If a large amount of sediment is dumped into a stream but nothing else changes, how does the stream respond? Why?

4. Characterize the differences between flooding patterns in meandering and braided streams.
5. Explain the patterns of erosion and deposition in a meandering stream. Use a sketch to illustrate your answer.

6. How does an alluvial fan develop? Why is flooding particularly hazardous in these areas?

7. Why does a stream bottom erode more deeply when its water level rises in a flood?

8. The destructive capacity of a stream depends on what factors?

9. What aspects of weather cause a flood? (Be specific: not merely "more water.")

10. How often, on average, does a stream reach bankfull level just before spreading over its floodplain?

11. Why does removal of vegetation by any mechanism cause more surface runoff and thus more erosion?

12. In what climates are flash floods most common? (Provide an example location.)

13. Why do floods in first-order streams provide only a short warning time for people downstream but high-order streams may provide days of warning?

14. Does a flood crest move downstream at the same speed as the water flow? Explain why or why not.

15. What is the simple formula for calculating the recurrence interval for a certain size flood on a stream in 1999 if there are 69 years of record? Give a numerical example.

16. What specific evidence can be used to estimate the maximum water velocity in a prehistoric flood?

17. Why might a recurrence interval not be a good predictor of future flooding?

18. Characterize the differences between debris flows and water floods.

19. What events may lead to a lahar?

20. Describe the events that lead to a glacial-outburst flood. Use a sketch to illustrate your answer.

21. Why, in parts of Canada or the northern United States, do north-flowing rivers often flood in spring? Explain clearly.

Critical Thinking Questions involving floods are at the end of Chapter 14.

■ Houses floated down river to pile up against a railroad bridge across the Cedar River, Iowa, on June 14, 2008.

Floods and Human Interactions

14

Lessons Learned?

After the "Great Flood" of the upper Mississippi River valley in 1993, many residents chose to stay, believing strengthened levees would protect them against future flooding. Local and regional governments continued to allow construction on floodplains in the pursuit of economic development and additional tax dollars. Near St. Louis, Missouri, about 30,000 homes were built on land that was flooded in 1993. Banks no longer required federal flood insurance, people dropped their coverage, and others were permitted to build new houses and businesses behind the levees.

Mississippi River valley floods have occurred frequently, with major and widespread events in 1844, 1927, 1937, 1973, and 1993. Again in April-May 2011, the mid- to lower Mississippi valley saw 500-year flooding from heavy rain and melting snow (see Cases in Point on the Mississippi River floods of 1993 and 2011, p. 399–402, at the end of the chapter). The usual response is to raise levees higher and restrict building in the hardest-hit areas, at least temporarily.

After multiple destructive floods, have residents and government officials learned their lesson? Although many government officials are taking steps to reduce future damages in some of the affected areas, many individuals and local business owners resist.

Development Effects on Floods

Natural stream processes cause disasters only when humans place themselves in harm's way. Before Europeans settled in North America, when rivers flooded, native residents merely packed up and moved to higher ground until the water subsided. When European settlers arrived, however, they tended to establish towns along rivers because of the access that rivers provide to water and transportation. These settlers caused severe alterations of the landscape through urbanization, logging, and grazing, as well as forest fires. They changed the sediment load of rivers by adding loose gravel or excavating sand and gravel from river channels and floodplains, increasing the size and damaging nature of floods.

Modern civilizations continue to build near rivers because of the concentration of towns, jobs, and farmland along them. When modern residents build on a floodplain, they often choose a site near the river, on the outside bend of a meander, where the house stands well above the river channel to provide a great view. Unfortunately, in doing so, they locate in the area of greatest bank erosion during a flood.

When settlers first established towns near rivers, they did not realize that rivers continually strive to minimize the energy that they expend; they alter their slope and path to accommodate the amount of flow and sediment supplied to them. If we make artificial changes in a channel, the river tries to adjust to minimize those changes. Today we understand how rivers change in response to human impacts and the increased flood risk these changes can bring.

Urbanization

Increasing urbanization in many parts of the world promotes increasing numbers of flash floods and higher flood levels (**FIGURE 14-1**). Urbanization involves transforming natural landscapes through logging, paving, and building. This affects flooding because water cannot soak into pavement or rooftops, so wherever roads or buildings cover the ground, the water is forced to run off rapidly into nearby streams, causing flooding. Dense urban areas often build artificial concrete channels to rapidly move this water downstream and minimize flooding (**FIGURE 14-2**).

Floods in urban environments cause special hazards when residents try to evacuate on flooded roads. Driving through a road that is under water can be dangerous or even fatal. Even though water may appear shallow, the force of its flow against the wheels or side of a car can wash it downstream. Fast-moving water only 30 cm (1 ft) deep is extremely

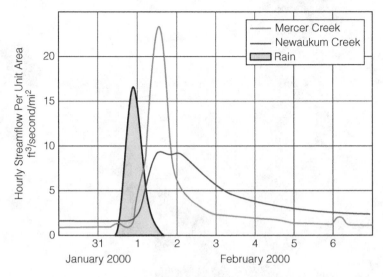

FIGURE 14-1 Flooding and Urbanization

After a period of heavy rainfall in the winter of 2000, two streams in Washington State showed different flooding patterns. Near the city of Bellevue, rainwater flowed rapidly off paved surfaces, leading to more immediate flooding and much higher flood flow in the local stream, Mercer Creek. In a nearby rural area, more rain percolated slowly down through the soil to reach Newaukum Creek, so the flooding was more gradual and flood levels were not as high.

FIGURE 14-2 Channelization

The Los Angeles River runs through a straight section of concrete channel with angled energy dissipation structures. The channel reduces flood risk to residents by collecting and moving water rapidly out of the city during periods of intense rain.

dangerous (**FIGURE 14-3**). Water 60 cm deep and above the vehicle floorboards can push a vehicle off the road and potentially drown the occupants. Compounding this danger, shallow, fast-moving water can erode a deep channel underwater that may not be visible to a driver. Driving into apparently shallow muddy water that hides a washed-out roadway can drop a vehicle into deep water and cause drowning. Even where water is still or hardly moving, the settling of part of a roadway can cause unexpected deep water.

Most cars will initially float because their weight is generally much less than the volume of water required to fill them. Eventually, however, water seeps in and they sink. If you are in a sinking car, immediately climb out through a window

or kick out the windshield. It is best to back out through a window so you are facing the car and can hang on to it. It is important to act quickly because once the car is partly submerged, it will be difficult or impossible to open the doors to escape.

Fires, Logging, and Overgrazing

Upstream human actions such as deforestation by fire, heavy logging, or overgrazing of a watershed also impact stream processes. These activities all cause excessive erosion, which dumps large amounts of sediment into channels, increasing their load. Recall from Chapter 13 the relationship between stream gradient and sediment transport. Streams choked with sediment develop a braided channel pattern and become steeper in order to transport a greater sediment load.

For example, torrential rains around Jakarta, Indonesia, in early February 2007, caused severe flooding, killing dozens and forcing the evacuation of more than 400,000 people. In about half of the city, hundreds of thousands were without clean water or electricity. Authorities blamed the severity of the flooding on deforestation of hillsides near the city, inadequate urban planning, and trash-clogged storm drains and rivers.

Deforestation can significantly impact a stream environment, including rate of runoff and sediment load. In vegetated areas, rain droplets impact leaves rather than landing directly on the ground. Rich forest soils soak up water almost like a sponge, providing a subsurface sink for rainwater that maintains a long-term moisture source for the vegetation. Fire burns away vegetation that protects the soil, permitting rain droplets to strike the ground directly and run off the surface. Intense fire also tends to seal the ground surface by sticking the soil grains together with resins developed from burning organic materials in the soil. This decrease in soil permeability reduces water infiltration. As a result, direct overland flow carves deep gullies into steep hillsides and feeds large volumes of sediment to local streams (see Chapter 17 for further discussion of the impacts of fire).

Logging by clear-cut methods sometimes involves skidding logs along the ground, which removes brush and other vegetation and leaves the ground vulnerable to erosion. In a technique called *tractor yarding*, felled trees are skidded downslope to points at which they are loaded on trucks. The skid trails focus downslope to a single point like tributaries leading to a trunk stream. Additionally, logging roads cut through the forest tend to intercept and collect downslope drainage, leading to the formation of gullies, increased erosion, and the addition of sediment to streams.

Cattle and sheep grazing on open slopes or next to stream banks similarly remove surface vegetation that formerly protected the ground. Rainfall running off the poorly protected soil erodes gullies, thereby carrying more sediment to the streams. Once gullies begin, the deeper and faster water causes rapid gully expansion (**FIGURE 14-4**).

A

Water 30 cm deep → Lateral force → **Extremely dangerous**

Buoyancy force ↑

Water 60 cm deep → Lateral force → **Fatal**

Vehicle begins to float when the water reaches its chassis, which allows the lateral forces to push it off the road.

Fatal

Muddy water hides washout

Washed-out roadway can be hidden by muddy water, allowing a vehicle to drop into unexpected deep water.

Modified from USGS

B

Don Becker, USGS

FIGURE 14-3 Hazards on Flooded Roadways

A. Even shallow floodwaters can lift a vehicle and wash it downstream. **B.** Don't venture into this type of flooded road. Cedar Rapids, Iowa, 2008.

FIGURE 14-4 Erosion Caused by Overgrazing
Heavy grazing has encouraged widespread gullying on steep slopes in the high plateau of Tibet.

FIGURE 14-5 Flooded Gravel Pit
Now flooded, these gravel pits along the South Platte River near Denver, Colorado, are separated from the river by only thin gravel barriers. The pits are flooded because they are dug down to below the level of the river.

Whether as a result of overgrazing, deforestation, or fire, vegetation removal can also increase the contribution of sediment to streams by landslides. Some water is taken up by plants in the form of *transpiration*, the natural evaporation of rainfall from leaves and removal of water from soil via roots through the leaves. When vegetation is removed, more water soaks into the ground and runs off its surface. More water penetrating a slope tends to promote landslides, which in some areas contribute as much as 85% of the sediment supplied to a stream. That upsets the balance of the stream, causing increased sediment deposition and increased flooding downstream.

Mining

Mining can also change sediment load and disrupt the equilibrium of a river. During the California Gold Rush in the late nineteenth century, floods were triggered when miners added large amounts of loose gravel to stream channels as a by-product of gold mining.

Today, large amounts of sand and gravel are used in construction materials for roads, bridges, and buildings. Much of that sand and gravel is mined from streambeds or floodplains in a process called **streambed mining**. Because the water flow in the stream is unchanged, the decrease in sediment load means the water has more energy and increased velocity, which leads to greater energy downstream. Downstream from the mining area the stream uses this excess energy to erode its channel deeper. Deepening a channel can severely damage roads and bridges. It also typically lowers the water table because more groundwater flows into the deeper stream channel; as a result, groundwater supplies decline away from the stream.

Where gravel is mined from pits on a floodplain, temporary gravel barriers are often used to channel the stream around the pits (**FIGURE 14-5**). Later, rising water may erode the barriers; water entering the pit from upstream slows and deposits sediment in the upper end of the pit, eventually filling it.

Although such mining may appear, therefore, to exploit a renewable resource, the gravel removed from flood flow by deposition is not being carried farther downstream. The increased stream energy downstream amplifies the erosion. As water nears a deepened gravel pit in the channel, its velocity increases, in many cases eroding the upstream lip of the pit and washing that gravel down into the pit. That lip migrates upstream. The deepening channel may undercut roads, bridge piers, and other structures, which can cause failure. Erosion concentrates near piers where there is increased turbulence. The cost to local governments to repair or replace such structures often exceeds the value of the mined gravel. In one prominent court case, it was shown that gravel mining from stream bars deepened the channel by as much as 3 m for many kilometers downstream from Healdsburg, California. The change threatened several bridges, destroyed fertile vineyard land, and lowered groundwater levels.

Removal of sediments from a stream often has consequences far from the site of removal, where a river deposits its sediment at the coast. Normally, much of the sediment deposited in the river delta is carried up or down the coast by longshore drift. When the sediment supply is reduced because of gravel mining or an upstream dam, the longshore movement of sediment along the coast continues but is not replenished. Beaches erode and may even disappear. Waves that normally break against the beach then break against the beach-face dunes or sea cliffs, causing severe erosion (discussed in Chapter 15).

Bridges

Bridges promote erosion in a channel when they increase water velocity by restricting the flow of water over the floodplain. The road or railroad approaches to bridges commonly cross floodplains by raising the roadway above the 100-year flood level. This typically involves bringing in fill that effectively creates a partial dam across the floodplain; the river flow is restricted to the open channel under the bridge. During a flood, when a river would usually spread over its banks and onto the floodplain, the bridge forces the water through a deeper, narrower channel (**FIGURE 14-6**). The deeper water flowing under the bridge flows faster, causing greater erosion of the channel under the bridge; that may undermine the pilings supporting the bridge (**FIGURE 14-7**), causing it to fail. Where more enlightened planners or engineers design the bridge—or where better

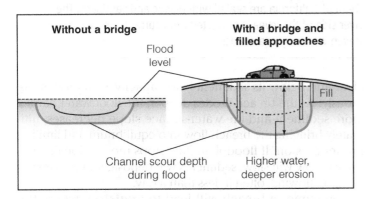

FIGURE 14-6 Bridges and Erosion

Without a bridge, the river spreads over its floodplain during a flood (shown at left). With a bridge, the river has a narrower and deeper channel, increasing channel scour (shown at right).

funding is available—the approaches may be built on pilings that permit floodwater to flow underneath the roadway.

Levees

With permanent settlement of floodplains, people feel the need to protect their structures from floods. The most common response to protecting an area from flooding is to artificially raise the riverbanks in the form of a **levee**. Individuals, municipalities, states, and the U.S. Army Corps of Engineers commonly try to protect floodplain areas from floods by building levees. **Spillways** are structures that allow for controlled release of water from a levee. Water released from a spillway may flow into a floodway or lake. Intentional release of floodwaters from levees can help policy makers and engineers control water flow and minimize damages during flooding (**Case in Point:** Managing Flood Flow through Levees—Mississippi River Flood, 2011, p. 401–402).

Recall from Chapter 13 that deposition of sediment during flooding creates natural levees along the banks of a river. Artificial levees are almost always built on top of original natural levees at the edge of the stream channel (**FIGURE 14-8A**). In the past, levees were built from locally handy materials in the floodplain, typically sand and mud from the floodplain itself or dredged from the river channel. Because these materials are often finer-grained than that carried by the river during floods, they are easily eroded and do not make good barriers to fast-moving water. Higher-quality levees built by federal agencies are often mixed with coarser gravels or faced with coarse riprap to resist erosion. Different materials have different advantages (**FIGURE 14-8B**). Compacted clay resists erosion and is nearly impermeable to floodwater but can fail by slumping; crushed rock is more permeable but less prone to slumping.

FIGURE 14-7 Erosion at Bridge Supports

A. A flood scour hole eroded next to the upstream end of one of the Alaska Highway bridge support piers in the Johnson River.

B. High-water turbulence at the upstream end of a bridge pier contributes to greater erosion. Clark Fork River, Missoula, Montana.

A

B

FIGURE 14-8 Levees

A. Levees are almost always built on old natural levees, right next to river channels. Confidence in the protection afforded by the levee has encouraged industrial and residential development behind these levees in Moorefield, West Virginia.

B. Fine materials making up this levee in the Sacramento River Delta of California are sealed with plastic and sandbags. The river side of the levee is protected from currents and wind-driven waves by coarse rocks.

Levee Failure

Although levees provide important protections in many situations, an unintended consequence of levees is that they may foster a false sense of security and even encourage development in flood hazard areas. With levees in place, more people build homes and businesses behind them, where they feel safe from floods. However, even well-built levees can fail during major floods.

The behavior of streams and rivers is governed by natural processes and fixed relationships between stream flow, channel slope, grain size, and meander shape. Human efforts to control the river are often frustrated by the river reverting to its natural course. Residents in the Nashville area realized this as a total of 0.49 m (19 in) of rain fell over a two-day period starting at the end of April 2010. The Cumberland River and nearby streams rose rapidly overnight, overtopping levees, filling the entire floodplain with muddy water (**FIGURE 14-9**), and forcing people to evacuate with little warning. At Nashville, the flood crest reached 15.8 m (51.86 ft), almost 12 ft above flood stage. Most of the damage was outside the floodplain and many houses were filled with 2 m of water. Thirty-one people died.

Although a levee may be initially of sufficient height to constrain a 100-year flood, the stream will eventually *overtop* the levee in a larger flood. Unfortunately levees are of varying quality, and, as noted in Chapter 13, major floods can come more frequently and at higher levels with time as regions become more developed. Communities can be faced with raising the level of a levee that is not high enough to contain a flooding river (**FIGURE 14-10A**).

When a flood breaks through the walls of a levee, it is called a **breach** (**FIGURE 14-10B**). When a large flood first breaches a levee, the water crossing the breach tends to be relatively

clear and below its sediment-load capacity, so it erodes vigorously. As the breach erodes deeper, the floodwater carries more sediment and the water-surface slope decreases, ultimately bringing the breach flow into equilibrium and limiting further erosion. If floodplain flow stops moving locally, the breach flow will slow, sediments will deposit, and the breach will stop flowing, often in less than a day.

Sometimes a breach will lead to **avulsion**, where the main channel of a river is redirected through the breach. If the floodplain flow is unrestricted downstream and the

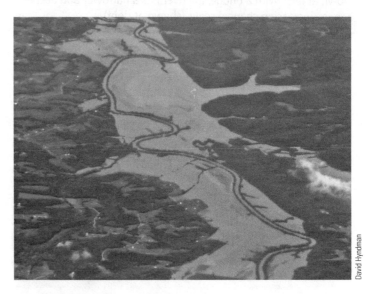

FIGURE 14-9 Submerged Floodplain

The 2010 Tennessee flood completely covered many stream floodplains. The low-water stream channel, outlined by brushy natural levees, meanders across the submerged floodplain.

FIGURE 14-10 Levee Failure

A. A barge and crane add coarse rock to shore up a levee protecting a town.

B. Breach of a major levee on the Sacramento River, California, rapidly flooded surrounding farmlands and towns in January 1997.

---

flood level remains high, the breach flow may continue to erode, ultimately diverting the main channel through the breach. The stream moves to a new lower-elevation path on the floodplain. Because the new path is lower, the stream does not return to its original path. The Mississippi River did this locally in the 1993 flood. The social and economic consequences of avulsion on such a major river can be severe (**Case in Point:** A Long History of Avulsion—Yellow River of China, p. 402–403).

Common levee failures are caused by bank erosion from river currents or waves, slumps into the channel, or *seepage* (**FIGURE 14-11**). Water seeping rapidly enough

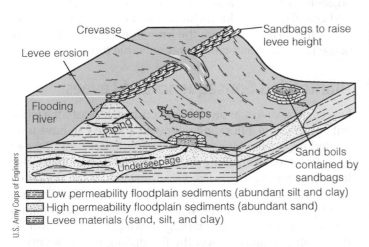

FIGURE 14-11 Types of Levee Failure

A. A levee may fail by overtopping, seeping through, or piping. Even where a levee is not overtopped, water may percolate through porous areas to cause seeps, landslides in its flanks, and ultimate collapse. Seepage from the river channel beneath the levee can rise behind the levee to form sand boils and, ultimately, flooding on the floodplain.

B. A sandbagged levee at Granite Falls, Minnesota, in danger of being overtopped.

FIGURE 14-12 Migration of Meanders

Because a river migrates laterally across its floodplain, sediment under the floodplain consists of old gravel river channels and old muddy floodplain deposits. The Mississippi River shows gradual migration of tight meanders before being cut off to form an oxbow lake. The river flows from upper right to lower left.

FIGURE 14-13 Sand Boils

A California Conservation Corps crew places sandbags around sand boils at the Sacramento River, at north Andrus Island, in the delta area southwest of Sacramento on January 4, 1997.

through parts of a levee erodes sediment at the side of the levee, which may then progress until the levee is in danger of failure. Cloudy water seeping out can indicate that soil is being washed out of the levee through a process called **piping**.

Levees can be more susceptible to these types of failures because of their composition from local material. The floodplain materials on which levees are built often are composed of old permeable sand and gravel channels surrounded by less permeable muds. The floodplain muds beyond the current channel are sediments that were left behind by the river where it spilled over a natural levee to flow on the floodplain. Recall from Chapter 13 that a river migrates across all parts of that floodplain over a period of hundreds or thousands of years. Under a mud-capped floodplain, the broad layers of sand and gravel deposited in former river channels interweave one another (**FIGURE 14-12**). These permeable layers provide avenues for transfer of high water in a river channel to lower areas behind levees on a floodplain. If a flood is prolonged, seepage beneath the levee can transmit enough water to flood surrounding areas behind the levee.

Even without overtopping or breaching the levee, water can also escape through the old gravels beneath it. Rising floodwater in a river increases water pressure in the groundwater below; this can push water to the surface on the floodplain, where it can potentially rise to nearly the water level in the river. With prolonged flooding, that water often reaches the surface as **sand boils** (**FIGURE 14-13**). The sandy water, under pressure, gurgles to the surface to build a broad pile of sand a meter or more across. Workers defending a levee generally pile sandbags around sand boils to prevent the loss of the piped sand. They leave an opening in the sandbags to let the water flow away and reduce water pressure under the levee.

Failure of artificial levees damages not only homes and businesses but also fertile cropland on floodplains. A flooding river without artificial levees spills slowly over its floodplain, first dropping the coarser particles next to the main channel to form low sand and gravel fans. Farther out on the floodplain, mud settles from the shallow, slowly moving water to coat the surface of the floodplain and replenish the topsoil. When artificial levees fail, the floodplain areas adjacent to a levee breach are commonly buried under sand and gravel from the flood channel and eroded levee. Farther away, the faster flow from the breach may gully other parts of the floodplain and carry away valuable topsoil.

Unintended Consequences of Levees

Levees do save some towns from flooding, at least for a while, but constraining a flooding river with a levee can have unintended negative consequences. Every time we build a new levee to protect something on a floodplain or to facilitate shallow-water navigation, we reduce the width of the flood-flow part of the river and raise the water level during flooding (see also By the Numbers 13-1). When a river floods under natural conditions, water flowing rapidly downstream spills over the riverbank, spreads out over the floodplain, and slows down. Although that shallow water eventually continues downstream, its temporary storage on the floodplain lessens flood levels downstream. Constructed levees along stream banks eliminate that storage, causing higher flood levels downstream.

Additionally, all of the floodwater that should have spread over the floodplain is confined between the levees, causing the flood flow to be much deeper and faster

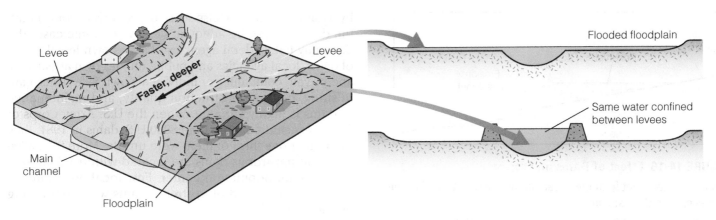

FIGURE 14-14 Effect of Levees on Flow

A. Levees that constrict the flow of a stream cause water to flow faster and deeper past levees. This causes flooding both upstream and downstream.

B. Levees confine at least the same amount of water between them as could spread out over the floodplain.

(**FIGURE 14-14**). Upstream of a levee section, water levels rise because the flow is constricted, causing flooding. Immediately downstream, the water level is higher because it is deeper within the constricted area. Flooding will thus commonly occur both upstream and downstream where it would not have occurred before the levees were built. For this reason, levees are sometimes intentionally breached during flooding to reduce flooding elsewhere (**Case in Point:** Repeated Flooding in Spite of Levees—Mississippi River Basin Flood, 1993, p. 399–401).

A study conducted for the U.S. Congress in 1995 showed that if levees upriver had been raised to confine the 1993 Mississippi River basin flood, the water level in the middle Mississippi would have been 2 m higher than it was. In fact, the increase in flood heights, for constant river flow, has been 2 to 4 m in the last century in the parts of the Mississippi River that have levees. This rise in flood level is mostly the result of levee building. The upper Missouri and Meramec Rivers do not have levees and show no increase in flood levels.

Wing Dams

In addition to levees, navigational dikes or **wing dams** constrict river channels in areas such as St. Louis to increase river depth for barge traffic during low flow (**FIGURE 14-15**). Even at high flow, when the river tops the wing dams, these structures increase resistance to flow near the riverbanks, which slows the velocity and raises the water level. As a result, for the same flood discharge, the water level or height increases. This artificial increase in river level clearly affects the inferred recurrence interval for any huge flood, such as that on the Mississippi in 1993. The 1993 peak level lies well above the recurrence interval curve adjusted for river flow without the wing dams.

FIGURE 14-15 Wing Dams

Wing dams on the Mississippi River provide a deepwater channel for shipping at lower water but narrow and slow the higher flows and thus increase flood levels.

Dams and Stream Equilibrium

Another major human intervention in rivers is the construction of dams across river channels. Dams are built for a variety of reasons, including flood control, hydroelectric power generation, water storage for irrigation, and recreation. These benefits must be weighed against significant cost to taxpayers as well as environmental impacts and increased flooding hazards. For example, one benefit, such as water storage for irrigation, may lead to high reservoir levels behind a dam and too little remaining water storage in the reservoir to prevent downstream flooding at times of high rainfall or snowmelt.

Flood-control dams are built to stop downstream flooding. These dams are built high enough to contain a certain magnitude of expected flood, perhaps a 100-year flood.

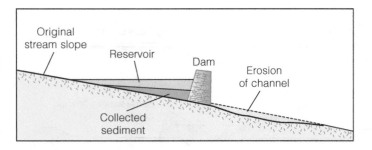

FIGURE 14-16 Effect of Dams on Erosion
Trapping of sediment from the stream in the reservoir behind a dam causes erosion downstream.

The ability of a dam to resist such a flood depends on a reservoir having a low-enough water level so it can fill up during a flood and avoid flooding downstream of the dam. This water can then be slowly released throughout drier parts of the year. Flood protection capacity will decrease as a reservoir fills with sediment; the rate of sediment infill increases with upstream deforestation or urbanization.

As with levees, dams built for flood protection can actually amplify the danger of flooding. Dams across rivers remove sediment from streams because the velocity in reservoirs behind the dams drops virtually to zero. Downstream of a dam, a stream carries little or no sediment, so it erodes its channel more deeply during flooding (**FIGURE 14-16**).

The unintended impact of dams on river equilibrium is illustrated by the 200-m-high Three Gorges Dam on the Yangtze River, completed in May 2006. The Chinese government relocated 1.4 million residents of at least 13 major cities, flooded by the dam, which was designed to provide hydroelectric power and prevent catastrophic floods. Since then the dam has led to a variety of major problems. Landslides developed along the reservoir upstream because rising groundwater levels destabilized adjacent slopes. Faster-moving, lower-silt-content water downstream of the dam is eroding riverbanks and levees. In late 2007, the government persuaded 4 million people to resettle because they were endangered by riverbank collapses. It appears that the giant port of Chongqing, at the upper end of the reservoir, will be silted up and closed by about 2021 because the river deposits its 150 million tons of sediment per year where the water slows and no longer carries the sediment downstream. This estimate takes into account the annual, high-water, summer flushing of silt through sluice gates at the base of the dam. Plans for more dams to trap sediment upstream would delay but not fix the siltation problem.

Floods Caused by Failure of Human-Made Dams

The intent of dams is not to allow people to build on floodplains, but that is often the effect. People may feel protected by a dam, but that is a false sense of security. At some point, the dam may not be adequate, and in extreme cases the dam may fail. Federal agencies or states own less than 8% of dams, local agencies and public utilities own about 19%, and private companies or individuals own 59%. Thus, care in siting, design, construction quality, and maintenance of dams is highly variable. When the U.S. Army Corps of Engineers studied more than 8000 U.S. dams in 1981, they found that one-third of them were unsafe. More than 3300 high and hazardous dams are located within 1.6 km of a downstream population center. Few local governments consider the hazard of upstream dams when permitting development.

Dams can fail for a variety of reasons, including overtopping, erosion of the dam foundation, poor design, low-quality construction or maintenance, or natural disasters such as earthquakes. Like levees, dams can fail when water overtops or seeps through them. In the 1972 flood in Rapid City, South Dakota, the reservoir was overtopped and the dam failed after a prolonged rainfall. The 11.5-m-high Delhi Dam on the Maquoketa River in northeastern Iowa was overtopped and failed on July 24, 2010, after 25 cm of heavy rain in 12 hours.

The most disastrous dam failure in recent decades was of the 24.5-m-high Banqiao Dam in central China. Built to withstand a 1000-year flood, it was hit by a 2000-year flood in 1975, accompanying a supertyphoon. Rainfall of more than 1 m in one day caused a flood that overtopped the dam and a cascading failure of 62 smaller dams along the river. A giant wave 3 to 7 m high and 10 km wide raced out onto the plains downstream, killing about 26,000 people. Subsequent epidemics and famine killed an additional 145,000. Although most dams could be built higher, the cost increases rapidly with height, in large part because a dam's length also increases rapidly with its height. Seepage of water under a dam or subsurface erosion along faults or other weak zones in the rock below it can lead to erosion of its foundation, resulting in catastrophic failure. The Teton Dam in eastern Idaho, which failed on June 5, 1976, took 11 lives and caused more than $3.2 billion in damage (**Case in Point:** Flooding Due to Dam Failure—Teton Dam, Idaho, 1976, p. 405). The 10-m-high Hope Mills Dam in North Carolina failed in May 2003 when it was overtopped after heavy rains. Rebuilt in 2008, it again failed in June 2010 when part of its foundation collapsed, apparently into a sinkhole.

Poor design and engineering standards can lead to dam failure. Improper maintenance, including failure to remove trees, repair internal seepage, or properly maintain gates and valves, and negligent operation. The 1.7-km-long, 78-m-high Wolf Creek Dam on the Cumberland River in Kentucky that impounds 265 km² Lake Cumberland is an earth-fill, concrete-gravity structure on porous limestone that was completed in 1952. Long plagued by seepage and sinkholes, the U.S. Army Corps of Engineers has designated the dam at high risk of failure. It requires ongoing longer-term

FIGURE 14-17 Damages Caused by Dam Failure

A. Damages from the June 10, 1972, flood at Rapid City, South Dakota, were extensive. Cars were mangled, and the only thing left of a nearby house is the tangle of boards in the lower right.

B. This house was carried from its foundation onto the road by the flood.

modifications that are expected to cost $584 million. The 2009 National Inventory of Dams places 1819 of them in the high-hazard category.

Other disasters can also trigger dam failure. Landslides into reservoirs can cause surges and dam overtopping, as happened when the Vaiont Dam in northeastern Italy was overtopped in 1963, killing 2600 people. Filling the reservoir behind the dam increased pore-water pressure in sedimentary rocks sloping toward the reservoir. A catastrophic landslide into the reservoir displaced most of the water, which drowned more than 2500 people downstream (see Chapter 8, Case in Point: The Vaiont Landslide, p. 213). Earthquakes can weaken earth-fill dams or cause cracks in their foundations. The Van Norman Dam, owned by the city of Los Angeles and situated less than 10 km from the epicenter of the 1971 San Fernando Valley earthquake, is immediately upstream from thousands of homes. It is an earthen structure that was 30 years old when the earthquake struck. The quake caused a large landslide in its upstream face and so drastically thinned the dam that it seemed likely to fail. Authorities evacuated 80,000 people from the area downstream until operators were able to lower the water to a safe level and avert dam failure.

The hazard potential of a failing dam to people living downstream depends on the volume of water released, the height of the dam, the valley topography, and the distance downstream. In a broad open valley, the speed and energy of the water from the failed dam would spread out and quickly dissipate, whereas in a steep narrow valley the water could maintain high velocity for a great distance. Calculations show that a dam-failure flow rate in a broad open valley would likely drop to half of its original rate in 60 km or so, but in a steep narrow valley half the original flow rate could be maintained for 130 km downstream.

Flooding on Rapid Creek in the Black Hills of South Dakota in June 1972 provided dramatic evidence of why it is dangerous to live downstream from a dam (**FIGURE 14-17**). In just six hours, 37 cm of rain fell over the Rapid Creek drainage basin. Southeast winds carrying warm, moist air from the Gulf of Mexico banked up against the Black Hills, where it encountered a cold front from the northwest.

The creek's typical flow of a few cubic meters per second became a torrent of 1400 m^3/s within a few hours. With rising water, authorities began ordering evacuation of the low-lying area close to the creek at 10:10 p.m., and the mayor urged evacuation of all low-lying areas at 10:30 p.m. The spillway of a dam just upstream from Rapid City became plugged with cars and house debris, raising the reservoir level by 3.6 m. At 10:45 p.m. the dam failed, releasing a torrential wall of water into Rapid Creek, which flows through Rapid City. The dam had been built just 20 years earlier for irrigation and flood control after an earlier flood. Building of this and other dams made people feel secure, so they built homes along the creek downstream. The flood just after midnight killed 238 people, destroyed 1335 homes and 5000 vehicles, and caused $690 million in damages. More than 2800 other homes suffered major damage.

In this case, the lesson was learned—at least for the time being. The city used $207 million in federal disaster aid to buy all of the floodplain property and turned it into a greenway, a park system, a golf course, soccer and baseball fields, jogging and bike paths, and picnic areas. Since then, building on the floodplain has been prohibited. However, decades later, pressure increased to build shopping centers and other structures in the greenbelt. The decision rested with a politically divided city council locked in the usual struggle between developers and the promise of additional jobs versus long-term costs, esthetics, and safety.

Reducing Flood Damage

Increased development of floodplains has brought greater casualties and costs related to floods. Floods are among the deadliest weather-related hazards in the United States, with 84 deaths per year over a recent 10-year period. Floods are also one of the costliest hazards, causing one-quarter to one-third of the annual dollar losses from geologic hazards, an average of $4 billion a year. Between 1929 and 2003, the long-term average annual costs of flood damages has gone up by a factor of 10, from $400 million to about $4 billion. Individual annual losses vary widely. A significant part of this is because more and more people are living in dangerous areas.

Land Use on Floodplains

If people build homes on floodplains or developers succeed in obtaining changes in local ordinances to allow building, a change to years of heavier rainfall can lead to an unpleasant surprise. Rivers in some areas, such as north of the Gulf of Mexico, can remain well within their banks for many years and then flood frequently during wet years. Such a weather shift can occur rapidly (**FIGURE 14-18**).

People have various reasons for settling on floodplains, including cheap land, fertile soil, or scenic views. As a result, many floodplains in the United States are heavily settled. Many people do not realize that a floodplain is part of the natural pathway of a stream or river. People often build their houses at the outside of a meander bend because the view of the river is best there or the trees are larger. Unfortunately, if you follow this strategy, you may find the stream running through your living room during the next flood. In mid-November 2006, major floods on the Skykomish River in northwestern Washington severely eroded riverbanks and destroyed homes along its banks. Many of these were carefully located to take advantage of the spectacular mountain and river scenery but were built too close to the river (**FIGURE 14-19**). At Mt. Rainier National Park, about 45 cm of rain fell in 36 hours, washing out roads and campgrounds and destroying a major highway near Mt. Hood, Oregon. A hunter who ignored road-closure signs was drowned when his pickup truck was swept into the flooding Cowlitz River.

After severe flooding in 1999 in the southern Mexican state of Tabasco killed more than 900 people, the government initiated a project to strengthen the levees that protect the more than half a million people of Villahermosa, the capital city. Unfortunately the work was never done. According to local reporting, flood-protection monies were used for other purposes and developers were allowed to build in high-risk areas. At the end of October and the beginning of November 2007, a cold front brought days of heavy rain. The low-lying state occupies the coastal ends of several Gulf Coast rivers that breached levees to affect 70% of the state and displaced about half of the state's two million people. For days, thousands of people remained stranded on rooftops or upper floors of houses surrounded by putrid, brown, debris-laden water. Food distribution, clean drinking water supplies, electric power, and public transportation came to a halt. Authorities were concerned about outbreaks of cholera and other waterborne diseases in the region's tropical climate. Flooding affected at least 50 of the state's medical centers. At least 30 people died.

People quickly forget most major floods that have affected an area and believe that similar or larger floods are

FIGURE 14-18 Changing Water Levels

A. Seasonal water levels in central Texas can change rapidly. Lake Travis, Austin, Texas, December 31, 2006, is so low that the lake is almost invisible in the distance. **B.** The same view six months later on June 28, 2007. Note the same "Public Notice" sign almost submerged at left. The same thing happened here in 2009–10 and in 2014–15.

FIGURE 14-19 Stream Bank Erosion

A 2006 flood on the Skykomish River, northeast of Seattle, Washington, undercut the stream bank underlying this homeowner's new house, which had spectacular views of the mountains and river.

unlikely to occur anytime soon. Similarly, if people have not seen their homes or businesses flooded in their lifetimes—or those of their parents—they also believe that such an event is not likely to happen to them. In both cases, people significantly underestimate their risk.

One way to reduce damages from flooding is to restrict development on floodplains. Natural rivers spill over their floodplains every couple of years. Because they often flood, these areas should be reserved for agriculture and related uses, not housing and industry. Parks, playing fields, and golf courses are reasonable uses in urban areas. Unfortunately, past government policy has failed to discourage development in flood-prone areas. The federal government pays millions to build and maintain levees to protect such developments and to provide disaster relief following floods. The cycle continues when governments allow relief funds to be used to rebuild in the same unsuitable places.

Federal government policies began to change after the disastrous 1993 upper Mississippi River flood. The Federal Emergency Management Agency (FEMA) began buying up floodplain land to prevent people from rebuilding there and being flooded again. Disaster relief funds were provided only if people moved out of the floodplain. The government purchased many homes on floodplains with the requirement that building new structures there was prohibited. Unfortunately, some exceptions have been made. At St. Louis, for example, billion-dollar developments have been placed on land flooded in 1993.

Some people feel that regulations on floodplain building infringe on their right to use their property as they choose. For example, there is intense public debate about development in flood-prone areas in some parts of California. However, an individual's choice to build on a floodplain often infringes on many other individuals. Huge sums of public tax dollars are spent every year to fight floods, build flood-control structures, and provide relief from flood damage for structures that should not have been built on the floodplain. Streams and rivers generally pass through many people's property. How one person or one town affects, restricts, or controls a stream commonly influences stream impacts to others both upstream and downstream. Developers and builders can make profits by building on floodplains, leaving homeowners and governments to pay the price of poor or insufficient planning. A coordinated approach is necessary to protect everyone. In many places, floodplains are not adequately zoned to minimize damages. Newly developed structures should generally be prohibited, and in some cases entire towns should be moved. Expensive as this may be, it is less expensive in the long run.

Flood Insurance

Flood insurance is one way to mitigate costs of flood damage to individuals and also influence behavior to reduce future flood damages. The **National Flood Insurance Program (NFIP)** for the United States was established by the National Flood Insurance Act of 1968 and the Flood Disaster Protection Act of 1973. The NFIP made insurance available to those living on designated floodplains at modest cost. Insurance for floods is provided by the federal government but purchased through private insurance companies. Ratings and insurance premiums were intended to be actuarial, that is, they were based on flood risks and the existence of certain mitigation measures.

Guidelines for federal flood insurance stipulate several definitions. The 100-year floodplain is formally separated into a floodway and a flood fringe (**FIGURE 14-20**). The **floodway** includes the stream channel and its banks. During flooding, this zone carries deeper water at higher velocities. Most new construction is prohibited in this area, including homes and commercial buildings. Also prohibited are structures, fills, and excavations that will significantly alter flood flows or increase 100-year flood levels. The **flood fringe** zone includes the stream channel and banks but is farther from the stream channel and still below the 100-year flood level. It is mostly floodplain, so during flooding it may be underwater; water there is generally shallower and flows more slowly. Wherever any part of the flood fringe is raised by buildings or other features, that is considered an encroachment. The available area for flood flow is lessened so the flood height will be higher than before.

Rates for this insurance depend on the likelihood and severity of flooding and are designated in mapped flood zones. Flood-hazard boundary maps and flood insurance rate maps (FIRMs) are available for communities under the regular FEMA program (**FIGURE 14-21**). To be eligible for flood insurance, a community must complete the

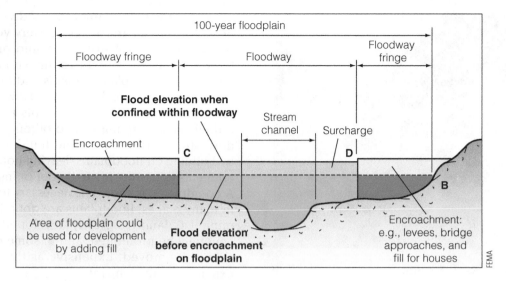

FIGURE 14-20 **Flood Insurance Definitions**

This schematic river cross-section shows FEMA definitions for flood insurance purposes. Line AB is the flood elevation before any encroachment. Line CD is the flood elevation after encroachment. The surcharge is the rise in flood-water level near the channel as a result of artificial narrowing of the floodplain. The surcharge is not to exceed 0.3 m (Federal Insurance Administration requirement) or a lesser amount specified by some states.

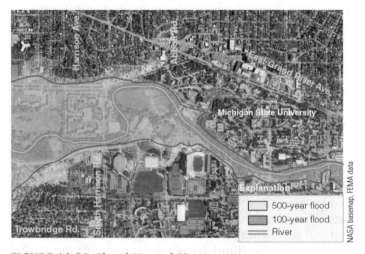

FIGURE 14-21 **Flood Hazard Map**

This example of a National Flood Insurance Program flood-hazard map is for part of East Lansing, Michigan.

required studies to designate floodplain zones and enforce its regulations. Larger amounts of insurance are available at actuarial—that is, "true risk"—rates.

Although flood insurance is clearly a good deal for those in flood-prone areas, it is not a good deal for U.S. taxpayers, who foot the cost of any losses. By 2006, the NFIP was $20 billion in debt, and insurance premiums are not high enough to pay it off. The long-term prognosis is worse; the NFIP insures $870 billion in homes and businesses in areas of high risk

for flooding. Some policyholders have filed claims again and again. Losses will continue to mount, with most of the costs ultimately paid by taxpayers across the United States.

For people's behavior with respect to a river to be appropriate, flood insurance premiums for a given location should be proportional to the risk of damage caused by flooding. Premiums are currently set too low to cover the actual cost of flood insurance. In 2003, participants in the NFIP were paying only 38% of actuarial risk rates. Clearly, premiums need to be raised to actuarial levels.

Another problem with flood insurance is convincing individuals to purchase it. Although insurance can be purchased up until 30 days before flood damage occurs, few people whose properties are damaged by flooding have purchased such insurance. Before the 1993 Mississippi River flood, only 5.2% of households in the flood hazard area had purchased flood insurance.

Some people may not purchase flood insurance because they assume that flood damage will be covered by their normal homeowner's insurance. In fact, homeowner's insurance only covers flooding events when the water source is inside the home, from a burst pipe, for example. Homeowners without separate flood insurance are liable for flood damages, such as water damage to walls, floors and furnishings; mud deposits; as well as the growth of mold that often occurs when warm water from outside stands in a home for more than a few days.

People often don't purchase insurance because they don't realize they are at risk. A survey of Missouri residents conducted seven months after the end of the catastrophic

FIGURE 14-22 Repeated Flood Damages
Davenport, Iowa, on the Mississippi River floodplain, was again underwater during the 2001 event.

flood in 1993 found that approximately 70% of floodplain residents did not know they lived on a floodplain. Some communities flooded by the Mississippi River in 1993 were again flooded in 2001 (**FIGURE 14-22**). One method of making people aware of their risk is by requiring flood insurance in order to sell a property or permits to modify or develop its land or buildings. This is often when the property loan must be secured by a flood insurance policy if it is within a floodplain. Most experts believe that real estate agents should be required to disclose flood risks when properties are offered for sale. Only since 2002 have virtually all banks and mortgage companies required that a residence be surveyed to see if it is on a floodplain. If it is, it needs to be covered by flood insurance before the mortgage company or bank will provide a mortgage on a property.

Flood insurance also provides an opportunity to encourage those making claims to relocate or rebuild homes with better flood protection. Recent, stricter controls require that where flood insurance funds are used for reclamation or rebuilding, the work has to conform to NFIP standards. For insured buildings, the rebuilt lowest floor must be on compacted fill and at least 2 ft above the 100-year flood level. After two floods in less than a decade, some 10,000 homes and businesses were approved for removal or non-rebuilding along more than 400 km² of floodplain. As of 2003, flood insurance policyholders with a record of repeated losses receive funds only for the purpose of relocating outside the flood zone, elevating their homes above flood level, or doing flood proofing or demolition. Those who reject the mitigation offer are charged insurance rates based on standard actuarial costs for properties with severe repetitive losses.

A bill passed by the U.S. Senate in 2008 requires that people buy flood insurance if a levee breach will result in their homes being flooded. In spite of the hazard, many towns and their residents resist such regulations because they claim they curtail economic development and the cost of insurance premiums is prohibitive. Flood insurance premiums in 2008 for Illinois floodplain homes, for example, average about $400 per year.

On October 1, 2013, the U.S. Congress ordered that insurance rates be increased to approximately reflect actuarial costs. That caused a steep upturn in insurance rates and anger from property owners and politicians of East Coast and Gulf Coast states. However, because the National Flood Insurance Program is $24 billion in the red (as of January 1, 2015), funds have to come from somewhere, either from those along the coasts or from federal taxes. After homeowners complained, the bill was modified to limit annual increases of most insurance premiums to 15% and to permit home sellers to pass on their subsidized rates to buyers. Changes took effect on April 1, 2015.

Environmental Protection

Human alterations to the landscape can have significant effects on the magnitude of future floods. By protecting rivers and the watersheds that feed them, governments can also reduce the intensity of future floods. Current restrictions dictate that building or encroaching on a floodway must not raise the level of a 100-year flood by more than one foot (30 cm). Some states set more stringent restrictions of no more than a half-foot rise (15 cm). Bridge approaches and levees, however, commonly encroach on a channel. Waste disposal, storage of hazardous materials, and soil-absorption sewage systems, including septic tank drain fields, are prohibited in both floodways and flood fringes.

Changing government policy toward artificial river barriers also reflects a better understanding of natural river processes and may reduce future flood damages. A 1994 committee of federal experts recommended that levees along the lower Missouri River be moved back from the river by 600 m to give the river room to meander and spill over its floodplain during high water. An ongoing federally funded wildlife-habitat project restored about 675 km² of floodplain accessible to the river. It was a patchwork dependent on volunteer sales that continued into the early 2000s. However, they also authorized raising other levees.

Reducing Damage from Debris Flows

Debris flows are among the most dangerous of downslope movements because of their sudden onsets and high velocities (**FIGURE 14-23**). Debris flows can be highly destructive, even on slopes of less than 30° that have thick brush or forest.

A broad alluvial fan spreading from the mouth of a desert canyon with picturesque boulders littering its surface is not a

FIGURE 14-23 Dangerous Flows

A. In May 1998, a muddy debris flow from the steep hillside to the right of this house in Siano, Italy, east of Mt. Vesuvius, had sufficient momentum to blow right through the walls of the house and out the other side. Debris on both balconies (arrows at left) provides an indication of the height of flow.

B. It may not be obvious, but the car this geologist is standing on is at the roof level of a house buried by a bouldery debris flow (note the roof vent pipe above the car's roof).

safe setting for residential development (**FIGURE 14-24**). The dangers include not only burial in heavy debris but also huge impact forces from fast-moving boulders. Hundreds of thousands of people live on gravelly alluvial fans in Los Angeles, Palm Springs, Phoenix, Tucson, Salt Lake City, Denver, and elsewhere. Most of these fans were built up from fast-moving slurries of sand, gravel, and boulders—debris flows. Even arid regions can have periods of intense or prolonged rainfall. People who live on such fans are at considerable risk from debris flows.

The best solution to minimize the impact of debris flows is to avoid building in vulnerable areas (**FIGURE 14-25**). Especially hazardous areas should be zoned as open space such as parks,

FIGURE 14-24 Debris Flow Hazard

People living in the modern housing subdivisions on the surface of alluvial fans in the Palm Springs area of California find that boulders make for great landscaping but seem unaware of how the boulders got there.

Debris-flow fan

FIGURE 14-25 Obvious Hazards

Some debris-flow hazards are really obvious. New homes on a debris-flow fan in Georgetown, Colorado, west of Denver. Huge boulders among houses document previous debris flows.

golf courses, and agriculture. Building on the debris fan should be prohibited, and in some cases existing development should be bought out and removed to open pathways for future flows. Where development is necessary, buildings and streets should be oriented with their lengths parallel to the downslope direction of flow to limit building exposure to flows.

Although insurance cannot be purchased for landslides or other ground movement, debris flows and mudflows may provide some exceptions. The distinction for insurance purposes is generally that if the flow is too watery to be shoveled, then damages can be claimed under flood insurance. Thus, some fast-moving, watery debris flows may be covered.

Early Warning Systems

Where buildings predate recognition of a hazard, education and early warnings can minimize problems (see the Flood Hazards Survival Guide at the end of this chapter). Conditions can be monitored and alerts provided when values reach known thresholds for triggering an event. Warnings of increased hazard, during prolonged or heavy rainfall, for example, can help, but the steep terrain in which most debris flows occur leaves little time for evacuation once a flow begins moving. This is especially true at night in a heavy rainstorm. Storm sounds can drown out that of an approaching debris flow.

Detection devices can help warn people of an already moving debris flow or mudflow. The best are acoustic flow monitors that detect the distinctive rumble frequency of ground shaking caused by debris flows. The sensed motion is telemetered automatically to downstream sirens. Once alerted, people should immediately run to higher ground off to the sides of a debris flow or mudflow path. Such sensors are used, for example, at Mt. Rainier and in Alaska, Ecuador, and the Philippines. Mudflows are most frequent around active volcanoes because of the abundance of volcanic ash. Prominent examples of recent volcanic mudflows include Mt. Hood, Oregon, in November 2006; Mt. St. Helens, Washington, in 1980; Mt. Pinatubo, the Philippines, in 1991; and Nevado del Ruiz, Colombia, in 1985.

Where likely sources of debris flows are a short distance upstream and the channel gradient is high, the warning time may be too short for evacuation. A trip wire installed in the canyon 3 km upstream from heavily populated areas on the Caraballeda fan in Venezuela after the December 1999 disaster would have provided only five minutes' warning for the large population there, almost certainly not enough for most people to move out of danger (**Case in Point:** Flood Hazard in Alluvial Fans—Venezuela Flash Flood and Debris Flow, 1999, p. 404). Inexpensive trip wire sensors are in use in many areas, but falling trees, animals, or vandalism can trigger false warnings.

Trapping Debris Flows

Damage from debris flows can be limited through construction of barriers (**FIGURE 14-26**). Walls can be built to deflect large-volume flows to a part of a fan that has little development. Debris flows can also be channeled into a debris basin large enough to contain all loose material in the channel upstream. These must be cleaned out after each flow.

Structures used to trap debris flows in canyons upstream from alluvial fans include permeable dams that stop boulders but permit water to drain, that is, grid dams consisting of cross-linked steel pipes, horizontal beams, vertical steel pipes, or reinforced columns. Widely used in Canada, Europe, Japan, China, Indonesia, and the United States, they abruptly slow the progress of debris flows by draining the water. The grid is generally spaced to permit people, animals, and fish to easily travel through the structure.

FIGURE 14-26 Debris Flow Protections

A. A modern debris-flow collection basin was built in Rubio Canyon, north of Pasadena, California. Downstream is to the upper left.

B. Owners of houses immediately below a debris-flow dam at the northern edge of Pasadena apparently trust the dam to block all debris flows and floods that may come down the canyon.

FLOOD HAZARDS SURVIVAL GUIDE

| | Be aware that these lists don't cover everything. Use your own common sense. |

U.S. High Hazard Area	■ Coastal areas, especially the East Coast and Gulf Coast ■ Along the floodplains of any stream or river ■ Along major rivers with large levees that restrict the width of channel flow
Preparation	■ Assess the risk of flood hazards to you or your home. Examine National Flood Insurance Program maps and purchase flood insurance where appropriate. ■ Flood-proof your home. ■ Determine the best location for your evacuation. Prepare a survivors' kit, including medications, important papers, food, and water.
Warning Signs	■ Watch for a persistent weather pattern that dumps large amounts of rain on upstream areas; or rising water in wet season, heavy snowpack beginning to melt, or heavy rain on large snowpack. ■ Stay tuned to radio and TV weather for government flood watches and warnings. ■ Listen for rumbling or roaring sound from upslope indicating an approaching debris flow or flash flood.
During the Event	■ Evacuate before your circumstance becomes dangerous. If you need to evacuate, take with you your survivors' kit, identification, and other important valuables. Let friends and family know where you are going and jointly arrange for an out-of-state person that you can all contact.
Aftermath	■ Be careful of contaminated water and other disease agents; avoid electric wires and appliances that are in contact with water; and discard most refrigerated food that has thawed. Dry wetted areas as quickly as possible. Mold grows rapidly in warm temperatures, including within walls and wood floors and is very difficult to remove.

Cases in Point*

Repeated Flooding in Spite of Levees
Mississippi River Basin Flood, 1993 ▶

It was a wet winter and spring in the northern Great Plains, so most of the ground was saturated with water from rain and melting snow, and the rivers were rising. Unfortunately, heavy rainstorms continued as flood waters arrived from upstream. The storms kept moving through in the same area; it rained and rained, literally for months. Highways, roads, and railroads in the upper Mississippi River basin were submerged for weeks on end, along with homes, businesses, hospitals, water-treatment plants, and factories.

Almost 100,000 km², much of it productive farmland, lay underwater. Des Moines, Iowa, went without drinking water and electric power for almost two weeks. Grafton, Illinois, at the confluence of the Illinois River with the Mississippi, is not protected by levees. When the water rises, people merely move out, then clean up afterward. But after 6 big floods in 20 years, some people began to think about moving the town. Small towns such as Cedar City and Rhineland, Missouri, accepted a governmental buyout of their flood-damaged homes and moved off the floodplain. The homes were bulldozed to create public parkland.

In spite of extensive high levees, more than 70,000 homes in the Mississippi River basin flooded, predominantly those of poor people living where the land is less expensive. After the lengthy flood, putrid gray mud coated floors and walls, and plaster was moldy to the ceiling because it and the insulation wicked it up.

Fifty people died, and damages exceeded $26.5 billion (in 2010 dollars), the worst flood disaster in North American history until the Hurricane Katrina–driven flooding of New Orleans in 2005.

The first half of 1993 had twice the average precipitation. By late June, flood-control reservoirs in the upper Mississippi River basin were nearly full. The storms kept forming in the same area. Between April and August most of the area was drenched with 60 cm of rain; regions in central Iowa, Kansas, and northern Missouri received more than 1 m.

Weather systems generally move east, but in late June and July of 1993 that part of the jet stream swept northeastward from Colorado toward northern Wisconsin. At the same time, warm, moist air pulled into the central United States from the Gulf of Mexico collided with the cool, dry air to the north to produce persistent low-pressure cells and northeast-trending lines of thunderstorms centered over Iowa and the surrounding states. Unfortunately, a persistent high-pressure system stalled over the southeastern coast and kept the storms from moving east as they would normally have done.

At times, a stationary weather front extended from northern Missouri to southeastern Wisconsin. A series of cold fronts rotated counterclockwise around a low and collided with the warm, moist air over Iowa. This collision lifted the warm air, causing condensation and thunderstorms.

While it rained, the flood wave moved downstream at approximately 2 km/h, so it was relatively simple to predict when the maximum flood height would occur at any location. Tributaries added to the flow, and broad areas of floodplain that were not blocked by levees removed flow and then gradually released it to the river. Tributaries backed up and even flowed upstream. Most places experienced multiple flood crests. In St. Louis the river crested at nearly 6 m above flood level; it was above flood level from late June through August, then again from September 13

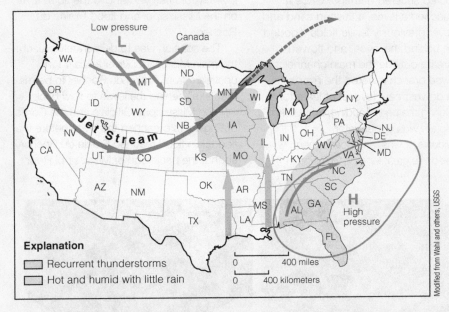

■ *The dominant weather pattern for June and July 1993 that created the Mississippi River floods included a stationary low over western Canada and a persistent high (red oval) off the Southeast coast. The main flooding was in the blue outlined area.*

---

* Visit the book's Website for these additional Cases in Point: Addition of Sediment Triggers Flooding—Hydraulic Placer Mining, California Gold Rush, 1860s; Streambed Mining Causes Erosion and Damage—Healdsburg, California; The Potential for Catastrophic Avulsion—New Orleans; Flood Severity Increases in a Flood-Prone Region—Upper Mississippi River Valley Floods, 2008; Catastrophic Floods of a Long-Established City—Arno River Flood, Florence, Italy, 1966.

U.S. Army Corps of Engineers

■ *This 1993 Missouri River flood near Jefferson City, Missouri, left many homes deeply submerged in floodwaters for an extended period of time.*

to October 5. The U.S. Army Corps of Engineers halted all river-barge traffic on the Mississippi north of Cairo, Illinois, in late June because it could no longer operate the locks and dams along the river.

In most places in this drainage, the recurrence interval of the flood was 30 to 80 years. In the lower Missouri River drainage basin in Nebraska, Iowa, and Missouri, the peak discharge was greater than a 100-year event.

As floodwaters rose, locals, National Guard personnel, and others dumped loads of crushed rock and filled sandbags to raise the height of critical levees across the region. As the higher levees raised river levels, they became saturated, causing slumping. Crushed rock in the levees minimized that. People inspecting a levee would sound an alarm if wave erosion became significant or if they found a leak.

Cloudy water indicated that soil was being washed out of a levee through a process called piping.

Rivers across the basin overtopped or breached numerous levees as the flood crest reached them. There were 1576 levees on the upper Mississippi River. Over 75% of those built by local or state agencies were damaged, while less than 20% of the 214 that were federally constructed suffered damage. Once a river breached a levee, it flushed sand and gravel over previously fertile fields, flooded the area behind the levee, and flowed downstream outside the main channel. Each levee breach lowered the river level, sparing downstream levees, at least for a while. During the peak flooding, 25,000 km^2 of floodplain was underwater.

Individuals, towns, and government agencies go to great lengths to maintain

and raise levees to protect towns from floods. So why would levee district officials and the U.S. Army Corps of Engineers intentionally sever a Mississippi River levee? They did just that in southwestern Illinois on August 1, 1993, to save the historic town of Prairie du Rocher. Fifteen kilometers north of Valmeyer, Illinois, a levee failed, permitting Mississippi River floodwaters to flow onto the floodplain on the east side of the river. There it began moving south, soon overtopping a pair of levees on a tributary stream and continuing south behind the main levee to flood Valmeyer.

Twenty-seven kilometers farther south, another tributary stream flanked by levees of its own would stop the floodplain water and protect the town of Prairie du Rocher unless they also were overtopped. However, water in part of a floodplain enclosed by levees will gradually rise to the level of the inflow breach upstream, because it behaves like a lake. The Mississippi River, of course, decreases in elevation downstream; levees protecting Prairie du Rocher were high enough to keep out the advancing flood but not as high as the Mississippi River at the breach 42 km upstream. That means that the "lake" behind the main Mississippi River levee would rise well above the flood level on the Mississippi and flood Prairie du Rocher.

The solution was to deliberately breach the main Mississippi River levee 1 km upstream from Prairie du Rocher to permit the "lake" behind the levee to flow back into the Mississippi. Officials and the Corps of Engineers breached the levee before the floodplain flood reached Prairie du Rocher so that the backflood of Mississippi River

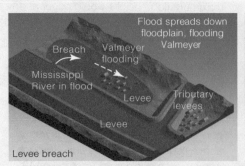
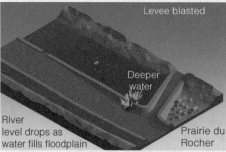
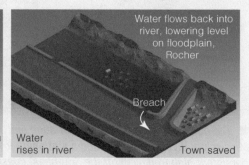

■ *Failure of a Mississippi River levee upstream from Valmeyer, Illinois, raised water in the floodplain behind the levee. To protect the historic town of Prairie du Rocher, downstream, the Corps of Engineers dynamited the levee just upstream to permit water to flow back into the main river.*

water would cushion the oncoming wall of water on the floodplain. When the "lake" behind the floodplain rose higher than the Mississippi, it again flowed back into the river.

In 2008, heavy rains again caused flooding of the upper Mississippi River valley. FEMA contracts with outside engineers certified that most of the levees can hold back a 100-year flood.

Unfortunately, the floodplain maps on which they depend were 20 years old and outdated—river conditions had changed, and levee quality varies. The continued building of homes, shopping centers, and parking lots encroached on the floodplain and caused more rapid runoff and higher flood levels. The flood at Gulfport, Illinois, was estimated to be roughly a "500-year event."

Managing Flood Flow through Levees
Mississippi River Flood, 2011

During the major Mississippi River flood of May 2011, rising water threatened to overtop many levees and flood towns on the floodplains. To lower high flood levels in the river at Cairo, Illinois, for example, the U.S. Army Corps of Engineers deliberately breached a 3-km gap in a levee to drain water into the Birds point Floodway (last activated in 1937) that follows the floodplain on the west side of the river. Water slowly moving down the floodplain was effectively "stored" until the flood crest had passed downstream. During the flood, water backed up into tributaries, causing large floods away from the main river. Most major levees

built by the Army Corps of Engineers held, though locally near the river, raw sewage seeped out of backed-up sewers.

In the following weeks, as the flood crest moved down the Mississippi River, taking on additional flow from the Arkansas River and other tributaries, it continued to threaten its levees and cities on its floodplain. Below Natchez, Mississippi, and upstream from Baton Rouge, Louisiana, the Army Corps of Engineers has built several flood relief structures. In that area, the Mississippi could carry an estimated 77,000 m³/s (2,720,000 ft³/s) in a 500-year flood,

with potentially disastrous effects downstream if a levee should fail. Hundreds of thousands of people, along with oil refineries and chemical plants, are endangered by potential levee breaches. The northernmost relief valve near the confluence of the Red River is the Old River Control Structure, which almost failed in the 1973 flood but was later repaired; it can release up to 17,500 m³/s of the river to the west to drain down the Atchafalaya River and then into the Gulf of Mexico. Small communities along the

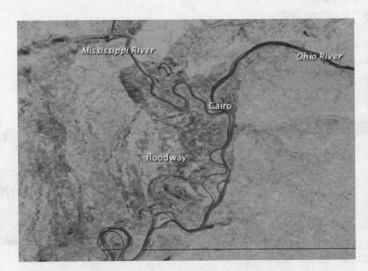

■ *Prior to the major May 2011 flood on the Mississippi River, water in both the Mississippi and Ohio Rivers was largely confined between levees.*

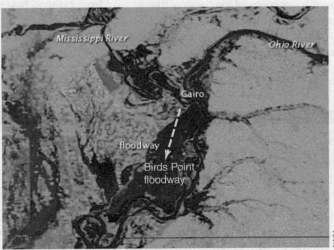

■ *A few days later, as the flood crest threatened many levees, breach of a levee near Cairo drained water into the western floodplain, thereby lowering river flood levels.*

■ *The Morganza Floodway takes water from the Mississippi River to the Atchafalaya River and the Gulf of Mexico.*

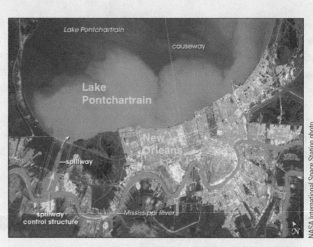

■ *The Bonnet Carré spillway carries floodwater from the Mississippi River down to Lake Pontchartrain.*

Atchafalaya were warned to evacuate the oncoming floodwaters. Flooded homes and businesses in low-lying areas remained inundated for weeks, some up to their attics. Government inspections were needed to determine whether they could be gutted and repaired or if they had to be torn down. A little farther downstream, the Morganza Floodway (last used in 1973) can take as much as 17,000 m³/s west into the Atchafalaya. This leaves 42,500 m³/s in the river. Below Baton Rouge, the Bonnet Carré Floodway can take another 7100 m³/s east into Lake Pontchartrain to further protect New Orleans.

In May 2011, the Army Corps of Engineers opened the Old River Structure to 1% beyond its flow for the maximum project flood, many of the gates of the Morganza Spillway, and the Bonnet Carré Spillway to lessen the flood risks for Baton Rouge and New Orleans. Without the

breaches, the flood at New Orleans would have been almost 0.5 m above the tops of its levees.

The U.S. Coast Guard halted tugboat and barge traffic on the lower Mississippi to minimize wave erosion of endangered levees. Because of very fast, erratic river currents, mostly submerged or floating trees and other debris, riverboats had great difficulty navigating. Two barges sank near Baton Rouge. The Mississippi is a main transportation corridor for corn and other crops from the Midwest, so transport interruption costs the U.S. economy hundreds of millions of dollars per day.

In recent years the Army Corps of Engineers has gradually moved away from building levees higher and higher with each successive flood. The new policy is to allow more flooding, while working with local and state governments

to restrict development on floodplains to reduce economic damage. In Illinois the Nature Conservancy has purchased floodplain land, converting it into natural wetlands that absorb and slow water flow. Louisiana seeks federal permission to temporarily open old natural river branches that once carried Mississippi River flow to the Gulf of Mexico.

CRITICAL THINKING QUESTIONS

1. What major changes has the U.S. Army Corps of Engineers made in recent years to protect large towns from flooding?
2. What are significant downsides of such a plan? Who should pay for such downsides? Why or why not?
3. If the Old River Control Structure were to fail completely, what would be the major consequences for New Orleans? Why?

A Long History of Avulsion
Yellow River of China ▶

The 4845-km-long Yellow River (Huang Ho) drains most of northern China, an area approximately 945,000 km². That is a similar length but less than one-third the drainage area of the Mississippi River. The upper reach of the river, flowing generally

east from the Tibetan highlands, carries relatively clear water through mountains and grasslands for more than half the river's length. The middle reach south from Baotou and the deserts and plains of Inner Mongolia drains a broad region of

yellowish wind-deposited silt, or loess, originally blown from the Gobi Desert in Mongolia to the northwest.

Sediment supply to the river from the loess plateau is vigorous because of vast arid-to-semiarid hilly areas of easily eroded silt. Before heavy agricultural use of the loess plateau began in 200 BC, it was mostly forested, and the sediment load fed to the river would have been one-tenth the current amount. After the tenth century AD, agriculture had largely destroyed the natural vegetation. Silt supplied by sheet-wash and gully erosion is so abundant that floods carry hyperconcentrated loads of yellow-colored sediment that gives the river its name. Average sediment load during a flood is generally greater than 500 kg/m^3, which is 20% by volume or 50% by weight.

The lower reach of the Yellow River, downstream from Zhengzhou, flows across a densely populated and cultivated alluvial plain, one affected repeatedly by flooding for more than 4000 years. As the river gradient decreases and spreads out over a width of several kilometers, sediment deposits progressively raise the channel bottom; this in turn requires regular raising of the levees. Near Zhengzhou, sediment deposition accumulated on the river bottom by an average of 6 to 10 cm per year, aggravated flooding, to rapidly fill the reservoirs behind dams.

After a disastrous flood in AD 1344, people used a combination of river diversion, river dredging, and dam construction. After each flood, they plugged breached levees and raised existing ones. With the channel elevated by deposition, some breaches drained the old channel and followed an entirely new path to the sea—that is, by river avulsion. Unfortunately, as with most rivers, levees here are built from the same easily eroded silt that fills the channel; the river erodes the levees just as it does the loess plateau. Downstream from a breach, the river bed between levees is left dry.

In 1887, the river topped 20-m-high levees and followed lower elevations to the south to reach the East China Sea at the delta of the Yangtze River at Shanghai. The flood and resultant famine killed more than one million people. Because the bed of the river is now as much as 10 m higher than the adjacent floodplain, the lower 600 km of the river receives no water from either surface runoff or groundwater.

In 1960, the Chinese completed the San-men Gorge Dam, 122 m high and more than 1 km wide. Its 3100-km^2 reservoir, designed for flood control and electric power generation, is now filled with sediment and is useless for both of its designed purposes. At the same time, a major effort was launched to plant trees and irrigate huge areas of the silt plateau to reduce the amount of silt reaching the Yellow River. Unfortunately, most of the trees died.

The lower 800 km of the Yellow River has repeatedly shifted its course laterally by hundreds of kilometers. China has lived with the Yellow River and its floods for thousands of years and has tried to control the floods with levees just as we do. It has not worked. Levees repeatedly failed and killed thousands of people. Avulsion dramatically changed the river course several times. With each flood, the people built levees higher and they still failed. There should be a message here; our levees fail just about as frequently.

■ *The largest remaining area of loess tableland at Dongzhiyuan in Gansu, China, is being rapidly eroded. The size of the view can be inferred from the road around the end of the deep canyon.*

George Leung

State of CA

■ *Levees of the Yellow River stand high above the surrounding floodplain, much as this river does near Sacramento, CA, but after hundreds of years of raising the Yellow River levees, they are many times higher.*

■ *Aggradation of the channel of the Yellow River, between levees, has raised it some 10 m. Even with 30 m levees, breaches often lead to avulsion and abandonment of the main channel.*

Flood Hazard in Alluvial Fans
Venezuela Flash Flood and Debris Flow, 1999 ▶

The rainy season in coastal Venezuela is normally from May through October, so a storm in the first two weeks of December 1999 was unusual. A moist southwesterly flow from the Pacific Ocean collided with a cold front to bring 29 cm of rain. Then on December 14 through 16, when soils were already soggy, torrential rains dumped 91 cm near sea level in 52 hours! Imagine almost a meter of water on the landscape in a little more than two days, all headed downslope. Higher elevations received twice as much rainfall as areas along the coast. This was the area's greatest storm in more than half a century.

Flash floods and debris flows inundated the coastal towns of Maiquetia and La Guaira, north of Caracas. Most homes and buildings in this area crowd large alluvial fans and narrow valley bottoms at the base of incredibly steep, unstable mountainsides. Floods began after 8 p.m. on December 15. Eyewitnesses who

fled the river and watched the flood from nearby rooftops reported crashing rocks and debris flows. Massive mudslides and floods killed an estimated 30,000 people and left more than 400,000 homeless. Exact numbers are difficult to determine because muddy slides buried many people or carried them out to sea. Mud and debris either swept away or buried shantytowns of tin-and-cinderblock shacks that covered extremely steep mountainsides.

Losses totaled $2.31 billion (in 2010 dollars). Because the only non-mountainside land along this coastal part of Venezuela is on alluvial fans, the large fan at Caraballeda was intensively developed with large multistory houses and many high-rise apartment buildings. On reaching the fan, the flood separated into several streams and spread debris throughout the city, up to 6 m thick in some places. Boulders more than 1 m in diameter collapsed the ends of several buildings.

Eight similar events are found in the historic record between 1798 and 1951. Examination of older debris-flow deposits shows that some flows were larger than the 1999 flow—thicker and with larger boulders.

Food, water, clothing, and antibiotics were in short supply. Evacuation and disaster recovery was especially difficult because many areas were hard to reach. The single highway along the coast was extensively blocked and cut by debris flows and flood channels.

CRITICAL THINKING QUESTIONS

1. Building on Venezuela's big alluvial fans proved to be hazardous. Where have people built on alluvial fans in North America? Are those places also dangerous and why?
2. What protections might people build to minimize such hazards?

■ *The debris flow and flash flood in Venezuela destroyed most of the homes on the low-lying Caraballeda fan at the mouth of the canyon.*

■ *The end of this apartment building collapsed when debris-flow boulders crushed key support columns. The largest boulder is more than 2 m high.*

Matt Larsen, USGS

Flooding Due to Dam Failure
Teton Dam, Idaho, 1976 ▶

The Teton Dam, near Rexburg in eastern Idaho, was built by the U.S. Bureau of Reclamation to provide not only irrigation water and hydroelectric power to east-central Idaho, but also recreation and flood control. After dismissal of several lawsuits by environmental groups, construction began in February 1972, and filling of the reservoir behind the completed dam commenced in October 1975. The dam was an earth-fill design, 93 m high and 945 m wide. It had a thin "grout curtain," or concrete core wall to prevent seepage of water through the dam. The intensely fractured rhyolite bedrock below contained large gas-vent openings; the largest openings were filled with concrete slurry along only a single line because the amount of concrete needed was more than expected. The earth-fill material deep in the dam was poorly compacted.

On June 3, 1976, workers discovered two small springs just downstream from the dam. On June 5 at 7:30 a.m., a worker discovered muddy water flowing from the right abutment (viewed downstream). Although mud in the water indicated it was carrying sediment, project engineers did not believe there was a problem. By 9:30 a.m., a wet spot appeared on the downstream face of the dam and quickly began washing out the embankment material. At 11:15 a.m., project officials told the county sheriff's office to evacuate the area downstream. At 11:55, the crest of the dam collapsed. The flood obliterated two small towns and spread to a width of 13 km over Rexburg, with a population of 14,000.

The flood killed 11 people and 13,000 head of livestock and cost the federal government almost $1.3 billion (2015 dollars) in claims and other costs. It was the highest dam that ever failed and marked the end of large dam building in the United States.

CRITICAL THINKING QUESTIONS

1. Man-made dams fail for a variety of reasons. What were the most important problems at the Teton Dam?
2. Which of these problems could have been anticipated?

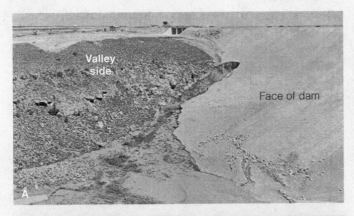

U.S. Army Corps of Engineers and U.S. Bureau of Reclamation

■ *Progressive failure of Teton Dam, eastern Idaho, June 5, 1976:* **A.** *At 11:20 a.m., muddy water pours through the right abutment of the dam.* **B.** *At 11:55 a.m., the right abutment begins to collapse.* **C.** *Just after noon, the dam fails and the reservoir floods through it.*

Critical View

A This is a dry-climate area in western Montana. A modern house and garage are built in the bottom of a canyon next to a tiny seasonal stream.

Donald Hyndman

1. What natural hazard should have been considered before building?
2. Where would have been a better location for the house in the area visible in this photo?
3. What other natural hazard is apparent? Consider the subjects of preceding chapters.

C A bouldery deposit severely damaged this house in western Colorado.

S. Cannon, USGS

1. What process likely deposited the material?
2. A widespread event affected the forest about a year before deposition of the material. What kind of event or events would lead to that movement and deposition of the boulders?
3. Why are the larger boulders sitting at the top of the deposit, rather than sinking to the bottom?

B This vehicle camped overnight in a sheltered gravelly area in Death Valley, California.

D. Steensen, National Park Service

1. What process destroyed the vehicle?
2. Describe the likely size and nature of the destructive process.
3. What clues are there in the photo that the dry channel could be prone to floods?

D The Mississippi River submerged a highway near Jefferson City, Missouri, in 1993.

Missouri Highway and Transportation Department

1. What is the area filled with muddy water called and why are the boundaries of this flooded area so well defined?
2. List as many as possible of the consequences to flooding of the small town near the freeway interchange.

E A foot-and-horse bridge across MacDonald Creek in Glacier National Park was destroyed in this flood in November 2006.

Shaun Bessinger, National Park Service

1. What could the builders of the bridge have done to ensure it survived a flood like this?
2. What could lead to such an unusual flood well after the spring wet season and after a dry summer?

G This new, unfinished home with a great view was undercut and severely damaged along a river in northwestern Washington State.

M. Nauman, FEMA

1. What normal stream process led to the failure of the riverbank under this home?
2. What conditions might have accelerated the collapse of this bank?
3. What would have been a better choice for a building site near the river?

F This flooded home is located on the Salt River floodplain near Lebanon, Kentucky.

K. Crawford, USACE

1. Would normal homeowner's insurance cover damages of this type? Why or why not?
2. Could the homeowner have purchased flood insurance to pay for damages on this home?
3. What type of damages would be expected in this environment?
4. Note the For Sale sign out front. Why would this home be a poor investment, even if offered at a bargain price?

H This drainage ditch in New Hampshire was built lined with concrete to prevent erosion.

New Hampshire office of Emergency Management

1. What would be the difference in flood height and duration before and after building such a ditch?
2. What other features in and near a city would affect flood characteristics and what would be those differences?

Chapter Review

Key Points

Development Effects on Floods

- Urbanization aggravates the possibility of flash floods because it hastens surface runoff to streams. **FIGURE 14-1**.

- Cars driven into a flooded roadway with water above their floorboards are often pushed off the road, which can cause their occupants to drown. Most vehicles will float until they fill with water and sink. **FIGURE 14-3**.

- Deforestation and overgrazing increases erosion and adds sediment to streambeds, making streams steeper and more prone to flooding.

- Mining of stream gravel removes sediment so that the excess stream energy causes erosion downstream.

- Bridges with in-filled approaches prevent water from flowing over floodplains, causing a deeper, faster flow of water under the bridge. Increased erosion around bridge pilings puts bridges at risk of failure. **FIGURE 14-6**

Levees

- Although people feel safe behind them, levees fail from overtopping or breaching, bank erosion, slumps, piping, or seepage through old river gravels below the levee. **FIGURE 14-11**.

- Avulsion occurs when a breach flow does not return to the river but follows a new path down a valley, flooding cities in its way.

- Because levees confine streams to main channels rather than permitting floodwaters to spread over floodplains, they dramatically raise water levels during a flood and increase flooding upstream and downstream from the levee. **FIGURE 14-14**.

Dams and Stream Equilibrium

- Dams are built on rivers to provide electric power, flood control, water for irrigation, and recreation. They can also cause a flood hazard to those living downstream.

- Dams slow water at their reservoirs, which accumulate sediment that would typically be transported downstream. Without that sediment, flow is more erosive downstream of a dam. **FIGURE 14-16**.

- Dams can fail during floods due to seepage and erosion under them, poor design and construction, and by major landslides into reservoirs upstream.

- Floods caused by the failure of human-made dams are worst in steep, narrow valleys.

Reducing Flood Damage

- Floods in the United States cause one-quarter to one-third of the monetary losses and nearly 90% of deaths by natural hazards.

- Flood damages could be reduced by restricting development on floodplains.

- Many people living on floodplains are eligible for national flood insurance but do not purchase it because they are not aware that they live on a floodplain or do not believe their risk is great.

Reducing Damage from Debris Flows

- Damage from debris flows can be reduced through land-use planning, early warning systems, or structures to trap or divert flows. **FIGURES 10-24 to 10-26**.

Key Terms

Questions for Review

1. What nonnatural changes imposed on a stream cause more flooding and more erosion?

2. How would a hydrograph for a drainage basin change if major urban growth were to occur upstream? Draw a sketch to illustrate your answer.

3. What are the negative effects of mining sand or gravel from a streambed?

4. Roughly what depth of flowing stream is dangerous to drive through?

5. What should you do if your car is floating in water?

6. What negative physical effect do most bridges have on the streams they cross? Draw a sketch to illustrate your answer.

7. Aside from protecting the adjacent stream bank, what effects do levees have on a stream?

8. What process can lead to the failure of a river levee?

9. What process can lead to flooding of the floodplain behind a levee (of a flooding river) even if the levee does not fail?

10. What is a common sign of seepage under a levee?

11. Under what circumstance (or for what purpose) might a levee be deliberately breached?

12. What negative physical effects do dams have?

13. Name three possible causes of dam failure.

14. What is the difference in the danger of a dam failure in a wide open valley versus a narrow closed valley?

Critical Thinking Questions

1. People should not build homes on floodplains because of the danger of flooding. What are better uses for floodplains?

2. How should flood assistance be offered in situations where an area has had damaging floods multiple times in a few decades? Who should pay for such assistance (affected individuals, local, regional, national governments)?

3. Why are engineered structures such as levees and dams much more popular as a mitigation measure than moving people out of affected areas?

■ Beach erosion and cliff collapse endanger homes in Pacifica in March 1998. Most of these are now gone.

15

Waves, Beaches, and Coastal Erosion

Coastal Cliff Collapse

In Pacifica, California, a small coastal community just south of San Francisco, modest homes line picturesque sea cliffs, with stunning views of the ocean. In recent decades, cliff erosion and collapse has threatened such homes. Dams built for flood control, irrigation, and water supply trap sediments that were once carried to the coast and distributed by coastal waves to maintain sand on the beaches. Sediment that once deposited in San Francisco Bay was removed by dredging, so tides no longer wash it out to the Pacific.

Heavy rock boulders were placed along the beach at Pacifica to protect the 20-m-high cliff intermittently since the early 1970s. Especially after the 1982–83 and 1997–98 El Niño storms, cliff collapse removed as much as several meters of the coastal terrace at a time. By 1972 the block of houses facing the ocean was gone and by March 1998, only 2 of the 12 houses remained standing. One was sold to a new buyer in 2004! Those two still remained in March 2010, but their back yards had disappeared. How long before they are gone?

Living on Dangerous Coasts

People have always lived along the shores of inlets and bays and fished in nearby streams and lagoons, but their structures along the open coast were once temporary shelters that could be moved or low-value ramshackle summer cabins that could be replaced after storms. In the early years after European settlement of North America, coastal tourism was not important because of the difficult access through local brush and wetlands and the incidence of malaria. By the 1700s, people began building more costly and permanent structures along the coasts.

By the 1850s, reduction in working hours, formation of an urban middle class with money, and expanded transportation via railroads and steamships led to the expansion of coastal tourism and resorts, especially after the late 1940s. As populations grew in numbers and affluence, more people moved to the coasts, not only to live but also for recreation. People installed utilities, paved roads, and built bridges to the islands, along with more expensive permanent homes, hotels, and resorts along the same beaches. More recently, second homes for summer use have become popular, some used for only a few weeks per year. Others have become year-round dwellings for urban retirees. By 2004, approximately 42% of the population of the continental United States lived in coastal counties. Coastal populations continue to grow. For example, the population of Florida increased by approximately 50% from 1990 to 2013.

Beaches and sea cliffs constantly change with the seasons and progressively with time. When people build permanent structures at the beach, coastal processes do not stop; the processes interact with and are affected by those new structures. Instead of living with the sea and its changing coastline, people try to hold it back and prevent natural changes to the beaches and sea cliffs.

Hurricanes and their dramatic aftermaths are often viewed as abnormal or "nature on a rampage." In fact, they are normal for a constantly evolving landscape. What is abnormal is how human actions and structures cause natural processes to impose unwanted damage. To be aware of coastal hazards, we need to understand wave processes and the formation of beaches and sea cliffs. We also need to understand how human activities affect wave action, beach response, and sea-cliff collapse.

Waves and Sediment Transport

Winds blowing across the sea push the water surface into waves because of friction between air and water. Wind-driven waves are described in terms of **wave height**, **wavelength**, and **period** (**FIGURE 15-1**). Gentle winds form small ripples. As the wind speed increases, ripples grow into waves that grow higher as they catch even more of the wind energy, somewhat like a sail. Two other factors that increase wave height are **fetch**, which is the length of water surface over

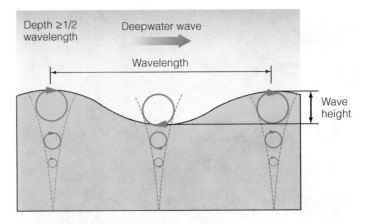

FIGURE 15-1 Wave Features
Wavelength is the distance between successive wave crests; wave height is the height from crest to trough; and wave period is the time it takes for two crests to pass a point.

which the wind blows, and the amount of time the wind blows across the water surface. Ocean waves are generally much larger than those on lakes, and prolonged storms often build huge, damaging waves.

Big waves generated by storms far offshore can travel for hundreds or thousands of kilometers, so they can appear at a beach even when there is little or no wind. The Los Angeles area recently experienced waves as high as 7 m in August 2014 due to Hurricane Marie, which was 1500 km away. Because waves can move outward from a storm center at different times as the storm moves, waves generated can be different sizes and move at different speeds. When faster waves overtake slower ones, they can interfere to increase or decrease the size of the combination (compare Figure 1-9). Such combination waves can be very large and even lead to occasional giant "rogue" waves.

Waves up to 35 m high (higher than an 11-story building) occur many times each year in all oceans. Most commonly they develop where wave travel direction is opposite to a strong current direction such as off the tip of South Africa. A number of ships have been sunk by such waves. In September 1995, the *Queen Elizabeth* 2 cruise ship encountered a 29-m wave during Hurricane Luis in the North Atlantic. In March 2001, two small cruise ships were struck by a 30-m wave between South America and Antarctica. One lost navigation and power. Giant waves also build occasionally during storms in Lake Superior.

The timing and size of waves approaching a shoreline vary by the location and size of offshore storms. Waves move out from major storm centers, becoming broad, rolling swells with large wavelengths (**FIGURE 15-2**). A constant, mild onshore wind will produce waves with smaller and shorter wavelengths.

Offshore, water does not move shoreward with the wave; it merely has a circular motion within the wave, otherwise staying in place (see again Figure 15-1). You can see that motion

FIGURE 15-2 Storm Waves

In addition to shore damage, huge storm waves can topple boats.

by watching a stick or a seagull floating on the water surface. It moves up and down, back and forth, not approaching the shore unless blown by the wind or caught in shallow water where the waves break. Waves in this circular motion are not damaging because the mass of water moves only slightly forward, then back.

When waves approach shore, they are said to "feel bottom." Waves begin to feel bottom when the water depth is less than approximately half the wavelength (**FIGURE 15-3**). Because the size of the circular motion is controlled by wave size, the depth at which wave action fades out downward is controlled by wavelength. In shallower water, the crest of the wave moves forward and rises as the base drags on the bottom and slows. The wave rises in height, leans forward, and breaks. (**By the Numbers 15-1**: Wave Velocity). The momentum of the upper mass of water carries it forward to erode the coast.

Big waves are more energetic and cause more erosion, because **wave energy** is proportional to the mass of moving water. This can be approximated by multiplying the density of water by the volume of water in a wave, which is roughly the wave height squared times the wavelength. The fact that the height term is squared tells us that waves that are twice as high have 4 times the energy; those that are 4 times as high have 16 times the energy.

Wave Refraction and Longshore Drift

Wavecrests often approach shore at an angle. The part of each wave in shallower water near shore begins to drag on the bottom first and thus slows down; the part of the wave still in deeper water moves faster, so the crest of the wave curves around toward the shore (**FIGURE 15-4**). This is called **wave refraction** because waves bend, or refract, toward shore.

When wave crests approach a **beach** at an angle, the breaking wave pushes the sand grains up the beach slope at an angle to the shore. As the wave then drains back into the sea, the water moves directly down the beach slope perpendicular to the water's edge. Thus, grains of sand follow a looping path up the beach and back toward the sea. With each looping motion, sand grains move a little farther along the shore. The angled waves thus create a **longshore drift** that essentially pushes a river of sand along the shore near the beach. Over the period of a year or so, with high and low tides, large and small waves, and storms, most of the sand on the active beach moves farther along shore.

Along both the west and east coasts of the United States, longshore drift is dominantly toward the south, but locally or seasonally it can be the reverse. Longshore drift in parts of coastal California averages a phenomenal 750,000 m³ of sand past a given point per year, more than 20,000 m³ per day. If a large dump truck carried 10 m³, this would be equivalent to 2000 dump-truck loads per day. Along the East Coast, drift rates are much less but still average 75,000 m³ per year, more than 2000 m³ per day.

Breaking wave moves forward at crest

Stirred sand is carried into incoming wave

Previous wave moves down beach slope, drags bottom of incoming wave

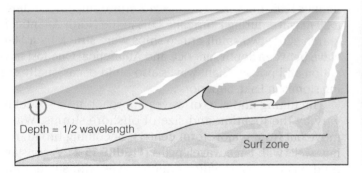

Depth = 1/2 wavelength

Surf zone

FIGURE 15-3 Breaking Waves

At depths less than one-half wavelength, motion of particles in a wave are flattened. As waves approach shore, they drag on the bottom and their crests lean forward to break, as seen in this photo taken along the West Coast. The water is brown from sand stirred from the bottom.

Wave Velocity

Wave velocity in deep water is proportional to the square root of wavelength:

$$v = \sqrt{\frac{gL}{2\pi}} = 1.25\sqrt{L}$$

where

v = velocity (m/sec)

L = wavelength (m)

g = acceleration of gravity (9.8 m/sec²)

π = 3.1416

Thus, waves with a wavelength of 5 m move at 1.25 × 2.24 = 2.8 m/sec. or 10.1 km/h. Waves with a wavelength of 100 m move at 1.25 × 10 = 12.5 m/sec, or 45 km/h. Note that if we measure the period (the time between wave crests), we can determine both the wavelength and wave velocity using the graph below.

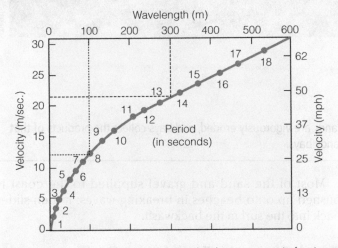

In shallow water, wave velocity is proportional to square root of water depth:

$$V = \sqrt{gD} = 3.1\sqrt{D}$$

where

V = velocity (m/sec)

D = water depth (m)

Waves on Irregular Coastlines

Waves approaching a steep coast, such as those along much of the Pacific coast of North America or the coast of New England or eastern Canada, encounter rocky points called **headlands**, separated by shallower sandy bays. Waves bend or refract toward the rocky points, causing the energy of the waves to focus against the headlands (**FIGURE 15-5**). Thus, wave refraction dissipates much of the wave energy that would otherwise have impacted sandy bays. Sand pounded off a rocky point migrates along the sides of the point due to wave refraction and is dumped along beaches in bays. Currents on both sides carry sand into adjacent bays. Over

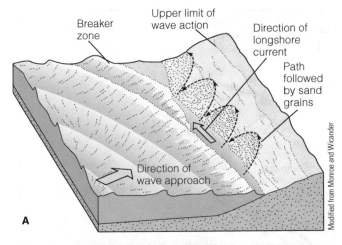

FIGURE 15-4 Longshore Drift

A. Sand grains are pushed up onto a beach in the direction of wave travel. Gravity pulls them back directly down the slope of the beach. The combination is a loop that moves each sand grain along the shore with each incoming wave.
B. Waves in shallow water near Oceanside, California, drag on the bottom and slow so their crests curve to approach nearly parallel to the shore. Direction of longshore drift shown by red arrow.

time, the activity of waves thus tends to erode the points and straighten coastlines.

Rip Currents

As waves pile up water onshore, the water may stream back offshore to create a **rip current**. Some high-energy coasts show a prominent scalloped beach with cusps 5 to 10 m apart, a suspicious sign of rip-current danger (**FIGURE 15-6**). Such rip currents can persist for long periods. Permanent rip currents can develop at groins, jetties, or rock outcrops, where water flowing against those structures piles up and is forced to flow offshore along the structure. A scalloped beach may form in an area of rip currents because each current erodes sand as it flows offshore, leaving the beach slightly indented at the point of offshore flow. Elsewhere, rip currents may appear unpredictably for a short period of less than 10 minutes after a large swell from a distant storm pushes water up onto a beach.

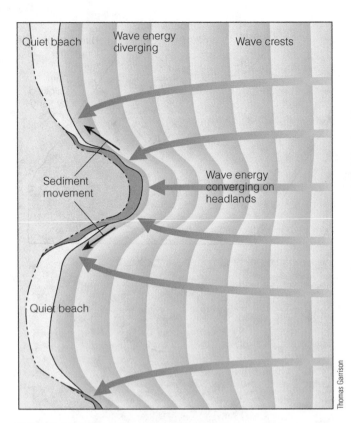

FIGURE 15-5 Erosion of Headlands

A. Wave crests bend to attack the shoreline more directly. Thus, rocky headlands are vigorously eroded, and bays collect the products of that erosion. **B.** This coastline in western Mexico shows eroded headlands and sandy bays.

These streams of muddy-looking water can be dangerous to people not familiar with them because the currents are too fast to permit even strong swimmers from swimming directly back to shore. Eighty percent of water rescues at beaches are from rip currents. Many people drown at the coast when caught in rip currents, which are especially powerful and dangerous during strong onshore winds and big waves. Rip currents are sometimes called undertows, but this is a misnomer because these currents would not drag someone under the surface. The drowning danger comes from swimmers becoming overtired while fighting the current. You should escape a rip current by swimming parallel to shore and then back to the beach (Figure 15-6C).

Beaches and Sand Supply

Beaches are accumulations of sand or gravel supplied by sea-cliff erosion and river transport of sediments to the coast. The size and number of particles provided by sea-cliff erosion depend on the energy of wave attack, the resistance to erosion of the material making up the cliff, and the particle size into which the cliff material breaks. Waves often undercut a cliff that collapses into the surf (**FIGURE 15-7**) and then break it into smaller particles. The size and amount of material supplied by rivers depends similarly on the rate of river flow and the particle size supplied to its channel.

Most of the sand and gravel supplied to the coast is pushed up onto beaches in breaking waves; it then slides back into the surf in the backwash.

Beach Slope: An Equilibrium Profile

Beaches are constantly changing as a result of the deposition of sediment and erosion by waves. The so-called active beach extends from the high-water mark to some 10 m below sea level. Grain size strongly controls the slope of a beach, called its **shore profile**. Just as in a stream, fine sand can be moved on a gentle slope—coarse sand or pebbles only on a much steeper slope. Whether or not the sediment moves shoreward depends on the balance (equilibrium) between shoreward bottom drag by the waves, size of bottom grains, and downslope pull by gravity. This balance is called the *equilibrium profile*. Thus, the slope of the bottom is controlled by the energy required to move the grains, which is related to water depth, wave height, and grain size (**FIGURE 15-8**). Shallower water, smaller waves, and coarser grains promote steeper slopes offshore, just as in rivers. In the breaker zone offshore, waves can easily move the sand, and the beach surface is gently sloping. As breakers sweep up onto the shore, the water is shallower, their available energy decreases, and the *shore profile* steepens. Sand there can move back and forth only on such a steeper slope.

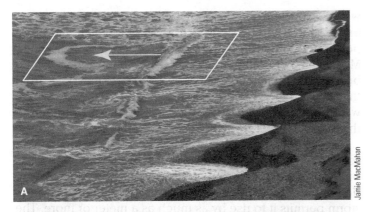

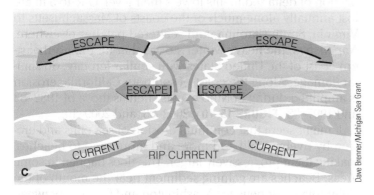

FIGURE 15-6 Rip Currents

A. Prominent scallops in beach sand are a sign of dangerous rip-current action (Lima, Peru); box outlines rip current moving offshore. **B.** A prominent rip current carries muddy water offshore in Monterey Bay, California. **C.** To escape a rip current, swim parallel to shore. Don't try to swim directly toward the beach against the current.

As waves travel into shallow water and begin to touch bottom, they shift sediment on the bottom, stirring it into motion and moving it toward the shore. Most sediment moves at water depths of less than 10 m. Long-wavelength storm waves, however, with periods of up to 20 seconds, reach deeper; they touch bottom and move sediments at depths as great as 300 m on the continental shelf. Those large waves have the energy to spread the grains into a gentler slope; they erode the grains above water level and deposit them offshore below sea level.

FIGURE 15-7 Waves Undercut a Cliff

This cliff on Kauai, Hawaii, has been undercut by waves. Large chunks of rock break off and are pounded into sand by the waves.

FIGURE 15-8 Grain Size and Beach Slope

The beach slope steepens shoreward where the breaking waves reach their upper limit. The result is a ridge, or berm, as in this case at Positano, Italy.

FIGURE 15-9 Rocky Winter Beach

A sandy summer beach left by small summer waves covers the lower part of a bouldery upper beach left by big winter waves north of Newport, Oregon.

Big waves during winter storms carry sand offshore into deeper water, leaving only the larger gravels and boulders that they cannot move. The gentle breezes and smaller waves of summer slowly move the sand back onto the beach. As a result, some cliff-bound beaches show sand near the water's edge with gravel or boulders upslope. In such areas, the beaches are commonly sand in summer but more steeply sloping gravel or boulders in winter (**FIGURE 15-9**).

On shallow, gently sloping coastlines, such as those in much of the southeastern United States, the beach both onshore and offshore becomes steeper landward. This is because the waves use energy stirring sand on the sea bottom, so that they slow as they ride up onto the beach. Most of the wave energy is used in waves breaking and moving water and sand upslope. The active beach slope is controlled by the grain size being moved and the amount of water carrying the grains. To reiterate, larger grains or less water requires a steeper slope to move the grains.

On beaches with coarser sand or gravel, much of the water soaks quickly into the ground rather than flowing back offshore over the surface; thus, less sand or gravel is eroded from the beach. Finer grained sand on beaches does not allow rapid downward seepage of water from waves; thus, most of the water flows back over the surface of the beach, carrying finer sand with it. This effect is most significant with large storm waves that still have most of their water available to flow offshore and carry sand with them. Therefore, small waves tend to leave more of their sand onshore. During low tide, winds pick up drying sand on the beach and blow it landward into dunes. The next strong storm may erode both the beach and the dune face and carry the sand back offshore.

Loss of Sand from the Beach

Sand in the surf zone moves with the waves; however, where it goes changes with the tides and with wave heights. Larger waves, especially those during a high tide or a major storm, erode sand from the shallow portion of a beach and transport much of it just offshore into less-stirred deeper water. Much of the eroded material comes from the surface of the beach, which flattens the beach's profile. More comes from the seaward face of dunes at the head of the beach, where waves either break directly against the dunes or undercut their face.

Storms bring not only higher waves but also a local rise in sea level—called a *storm surge*. High atmospheric pressure on the sea surface during clear weather holds the surface down, but low atmospheric pressure in the eye of a major storm permits it to rise by as much as a meter or more. The stronger winds of a storm also push the water ahead of the storm into a broad mound several kilometers across. The giant waves of a hurricane and the higher water level of storm surges take these effects to the extreme. They cause massive erosion and decrease the beach slope (see Chapter 16 for further discussion).

Along the Gulf Coast and southeastern coast of the United States, storms and heavy erosion are most likely to occur in the hurricane season of late summer to early fall. The most severe area of coastal erosion in the United States, with more than 5 m of loss per year, is along the Mississippi River delta, where dams upstream trap sediment, and the compaction of delta sediments lowers their level. One to 5 m are lost annually along much of the coasts of Massachusetts to Virginia, South Carolina, and scattered patches elsewhere. *Nor'easters*, the heavy winter storms that hit the northeastern coast of the United States with similar ferocity, do similar damage. Along the coast of southern California, erosion is amplified during El Niño events such as those of 1982–83 and 1997–98 (**FIGURE 15-10**), 2002–03, and 2009–10.

It might seem that continual erosion of sea cliffs and erosion by rivers would add more and more sand to beaches, making them progressively larger. This does happen in some areas. Beaches are gaining sediment near the mouth of the Columbia River between Washington and Oregon, between Los Angeles and San Diego, along much of the Georgia and North Carolina coasts, and along scattered patches in Florida and elsewhere on the Gulf of Mexico coast.

However, many coastlines are continuing to erode. What happens to the sand? Nature gets rid of some of it. Some is blown inland to form sand dunes, especially in areas without coastal cliffs. Other processes can permanently remove sand from the system. Some is beaten down to finer grains in the surf and then washed out to deeper water. Rip currents form when waves carry more water onshore than returns in the swash. That current flows back offshore in an intermittent stream that carries some of the beach sand back into deeper water. Huge storms such as hurricanes carry large amounts of sand far offshore.

Some sand drifts into inlets that cross barrier islands, where dredges remove it to keep the inlets open for boat traffic. Some migrates along coasts for a few hundred kilometers until it encounters the deeper water of a *submarine*

FIGURE 15-10 Erosion Increases During El Niño

This beach north of Point Reyes, California, was eroded by waves during the 1997 El Niño event, shown at left. By April 1998, shown at right, the beach had been naturally rebuilt.

canyon that may extend offshore from an onshore valley (**FIGURE 15-11**). Another prominent example of a submarine canyon extends offshore from the Monterey area of central California. The canyon extends across the continental shelf to where the sediment intermittently slides down the continental slope as turbidity flows onto the deep ocean floor. Thus, much of the longshore drifting sand of beaches is permanently lost to the beach environment.

Sand Supply

With European settlement of North America over the past 400 years, attempts to control nature have upset the natural sediment balance. The advent of steam locomotion in the early 1800s, followed by railroads and a large increase in population in the continental interior, also led to deforestation, land cultivation, and overgrazing on a large scale. Invention of the internal combustion engine continued the trend. This removal of protective cover from the land led to heavy surface erosion and large volumes of sediment delivered to the coasts. Along the steep Pacific coast, longshore drift of these sediments caused widespread enlargement of beaches.

The supply of river sediment to the coasts now has been severely reduced from building dams that trap sediment, and by sand and gravel mining from streams. With less sand and gravel, beaches shrink and the waves break closer to, and more frequently against, coastal cliffs. Waves undercut the cliffs, which collapse into the surf.

Anything that hinders sand supply to a beach from "upstream" on a coast or removes sand from the moving longshore current along a beach results in less sand to an area of coast and erosion of the beach. Dams on rivers trap sand, keeping it from reaching the coast, and mining sand from river channels or from beaches for construction has a similar effect. Although sand mining in many places is now permitted only offshore at depths greater than 18 to 25 m, monitoring such activity is sometimes lacking (**FIGURE 15-12**). And some communities tacitly condone mining by purchasing beach sand for use on roads. In many industrial countries, major dam-building periods on rivers began in the 1940s to generate electricity, store water for irrigation, and provide flood control. That and better land-use practices caused dramatic reductions in the amount of sediment carried by rivers and supplied to beaches. Beach erosion again accelerated. The resulting erosion of beaches and coastal cliffs is clear in California. Worldwide, shoreline recession of 5 to 10 m per year is common, but in some places, for example, at the mouth of the Nile River in Egypt, is as much as 200 m per year.

FIGURE 15-11 Longshore Drift and Sediment Loss

La Jolla Canyon just north of San Diego, California, carries sediment from the continental shelf down to the deep ocean basin. This is a submarine canyon.

Waves, Beaches, and Coastal Erosion **417**

FIGURE 15-12 Beach Sand Mining

A. This work crew is mining beach sand for construction near Puerto Vallarta, Mexico. **B.** During Hurricane Ike in 2008, almost 2 m of sand were eroded from this beach southwest of Galveston. The concrete platform at author's head level was a concrete carport floor poured onto beach sand and now suspended after beach erosion.

Building the massive Aswan Dam upstream increased the rate of Nile delta erosion by about four times its previous rate to about 50 km² per year.

Erosion of Gently Sloping Coasts and Barrier Islands

The Atlantic and Gulf of Mexico coastal plains are gently sloping, with a sandy bottom offshore that is steeper than the coastal plain, so it tops off landward in a ridge, called a **barrier island**, or barrier bar (**FIGURES 15-13** and **15-14**). Landward of the barrier island, the area below sea level is a coastal lagoon. Although barrier islands form primarily along gently sloping coastlines of the East and Gulf Coasts of the United States, they also form across the mouths of some shallow bays and estuaries along the West Coast. The sea-level rise after the last ice age drowned the mouths of these river valleys.

Offshore barrier islands are a part of the active beach, built up by the waves and constantly shifting by wave and

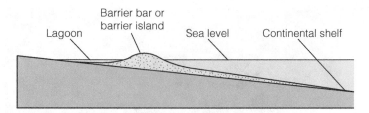

FIGURE 15-13 Barrier Island

This cross-section shows an offshore barrier bar with a sheltered lagoon behind it. The barrier bar-lagoon combination is generally hundreds of meters wide. The waves create a steeper profile for the sand than the overall slope of the coastline.

storm action. Offshore barrier islands are typically 0.4 to 4 km wide and stand less than 3 m above sea level. Winds picking up dry beach sands may locally pile dunes as high as 15 m above sea level.

High tides or storms carry the sea through low areas, inlets in the barrier bars; the water returns from lagoons to the sea at low tide through the same inlets, eroding them and keeping the channels open (**FIGURE 15-14**). Over time, inlets through barrier islands naturally shift position; some close and others open. Longshore drift of sand at times closes some gaps, and storms open others, so they intermittently change locations.

Because the equilibrium profile of a beach and the position of a barrier island are linked to wave size and water depth, a beach and barrier island shift landward as the water level rises (**FIGURE 15-15**). Current rates of sea-level

FIGURE 15-14 Inlets between Barrier Islands

A new breach from Hurricane Isabel, 2003, develops into a new inlet through a barrier bar. Here the tide is going out.

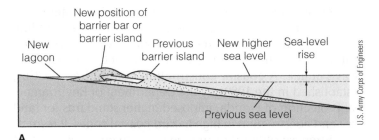

A

B

FIGURE 15-15 Barrier Island Migration
A. The barrier island migrates landward as sea level rises.
B. Storm waves reduced the level of sand by almost 4 m during one storm at Westhampton, New York. The barrier island migrated landward, leaving these houses stranded offshore.

rise are approximately 30 cm per century. On especially gently sloping coasts, like those of the southeastern United States, that 30 cm rise can move the beach and barrier island inland by 100 to 150 m or more over the course of a century. Barrier island migration happens over decades, but all of the significant movement occurs during hurricanes and other major storms. Oysters grow in the quiet waters of lagoons on the coastal side of barrier islands. If you find oyster shells on the seaward-side beach of a barrier island, one possibility is that the island gradually migrated landward, over lagoon mud. Beach waves winnow out the fine mud of the lagoon, leaving the heavier oyster shells embedded in the beach sand. Migrating barrier islands also overwhelm trees growing along lagoons; their stumps reappear later along a beach's upper edge as the bar gradually moves landward after a storm.

Development on Barrier Islands

By the 1700s, people began building more costly and permanent structures on the protected landward side of some barrier islands. Those who understood beach processes built homes on stilts on the bay side of barrier islands, with only temporary structures at the beach. Today many barrier islands are so crowded with buildings that they bear little resemblance to their natural states.

Although barrier islands help protect low-lying coastal areas from storm waves and flood damages, the islands themselves are hazardous places to live. Offshore barrier islands and the lagoons behind them are products of dynamic coastline processes: erosion, deposition, longshore drift, and wind transport. Barrier island communities live within this constantly changing environment. Most distinctively, former broad beaches erode rapidly in front of buildings, especially during hurricanes (**FIGURE 15-16**).

There are a few barrier islands at the mouths of drowned river estuaries along the West Coast; people build on those

FIGURE 15-16 Beach Migration
Several hurricanes shifted the beach shoreward at North Topsail Beach, North Carolina, destroying beachside homes one by one.
A. This photo was taken after Hurricane Bertha on July 16, 1996. Hurricane Fran on September 7, 1996, subsequently destroyed several more of the oceanfront buildings. **B.** After Hurricane Bonnie on August 28, 1998, only two of those remained.

just as they do on barrier islands on the East Coast. On gently sloping areas of some western beaches, as in parts of southern California, people build right on the beach. Perhaps they purchase homes or build them in good weather, not realizing that big winter waves come right up to the house. Some homeowners pile heavy boulders on the beach in front of their homes, hoping to protect them. The protection is temporary because big waves reflect off the boulders, washing away the beach sand in front and steepening the remaining beach. Eventually, the boulders will slide into the wave-eroded trough, leaving the homes even more vulnerable. But then the beach is gone.

Shifting sand closing an existing inlet commonly hinders access to marinas and boating in protected lagoons behind bars. It also hampers sea access to any coastal industrial sites on the mainland. Existing inlets are thus often kept open by dredging and building jetties along inlet edges. Where significant populations or large industrial sites are affected, the U.S. Army Corps of Engineers will often construct or maintain an inlet. Where significant settlement has occurred on or behind a barrier island, the maintenance of inlets severely hinders natural evolution of the island and beach.

When a storm overwashes and severs a beach-parallel road or cuts a new inlet across a barrier island, some homes and businesses are isolated on part of the bar (**FIGURE 15-17**). Generally, road engineers fill the new inlet and rebuild the road. Unless they do so immediately, the inlet typically widens over the following weeks or months as tidal currents shift sand into the lagoon and back out. Therefore, the scale of the repair project can quickly get out of hand. For example, a winter storm in 1992 at Westhampton, New York, opened a breach inlet 30.5 m wide. Within eight months, the inlet widened to 1.5 km.

Through past experience with hurricanes and other big storms, people living on barrier islands learn to build

FIGURE 15-17 Barrier-Island Breaches

This breach severed the only road to the west from Galveston to the mainland during Hurricane Ike. The yellow line marks the center line of the road. The ocean is to the left.

homes on posts, raising them above the higher water levels and bigger waves of some storms. In many areas, building codes require such construction. Codes also require preservation of beachfront dunes to help protect buildings from wave attack. A coastal construction control line (CCCL), established in the 1980s by the Department of Environmental Protection of Florida, imposed higher standards for land use and construction in high-hazard coastal zones. It restricts the purchase of flood insurance to those communities that adopt and enforce National Flood Insurance Program (NFIP) construction requirements governed by the standard building code. On the coastal side of the CCCL, requirements are more stringent for foundations, building elevations, and resistance to wind loads.

Dunes

At low tide, winds pick up drying sand and blow it landward to form **sand dunes**, which help maintain the barrier islands. Major storms or hurricanes wash sand both landward into lagoons and seaward from barrier islands. Sand moves back up onto the higher beach in milder weather, and wind blows some dry sand into the dunes again.

Unfortunately, most dunes along the coasts of the southeastern and Gulf states have disappeared as a result of development. People level the dunes to build roads, parking areas, and houses, or to improve views of the sea. They remove vegetation deliberately—to provide vistas—or inadvertently—by trampling it underfoot to reach beaches or by using off-road vehicles. They move sand onto lagoon-margin marsh areas for housing and marinas. They remove sand from dunes for construction or for reclaiming beaches damaged by erosion elsewhere.

Adding structures to a beach can also affect the formation of dunes. Sand still blows off the beach, but it drifts around houses instead of forming dunes on back beaches. It piles against houses and covers sidewalks and streets (**FIGURE 15-18**). After storms, the sand is routinely removed from roads by either private citizens or municipal employees. Shoveling snow in some northern areas can be hard work, but imagine having to shovel sand, which weighs much more and will not melt. The sand is sometimes placed back on an upper beach, but often it is pushed onto vacant lots, where it is lost to beach surfaces. In areas of lower sand supply, the wind may funnel between houses to scour local areas.

When people modify dunes, they upset the natural equilibrium of sand movement. True, they improve views, but with each storm a beach gets closer until the desired view gets too good (**FIGURE 15-19**). Houses are the next to go. Bigger storm waves will impact such structures because any protective dunes have long since disappeared.

As communities have become more aware of the importance of sand dunes in maintaining beaches, many have introduced measures to protect them. When buildings are elevated 4 m or so on pilings, as is commonly the case across the southeastern and Gulf coasts, the effect on windblown sand

FIGURE 15-18 Dunes and Development

These huge houses are right on the beach, along the former line of dunes, at Oxnard Shores, in southern California. Sand blown off the beach piles as drifts around houses and covers roads, sidewalks, and driveways. Shoveling snow in northern climates is heavy work, but this is ridiculous. "For Sale" signs are often numerous along this beach-parallel street.

FIGURE 15-19 Severe Erosion in the Absence of Dunes

With no line of dunes to protect it, the beach under this walkway at Holden Beach, North Carolina, was severely eroded by Hurricane Floyd, and the beachfront dune was completely washed away. The homes behind are next.

is much reduced. In many places, walking on dunes or otherwise damaging dune vegetation is prohibited. Homeowners often build elevated walkways to cross dunes and reach a beach; visitors get to beaches using roads and raised paths over dunes (**FIGURE 15-20**). Without walkways, such paths to beaches can become deeply eroded. Unfortunately, where dunes are managed at local levels with strong input from landowners, the dunes are often low and narrow to permit easy

access to the water and direct views of the ocean. Such low dunes provide minimal protection and storms easily remove them, exposing homes to direct wave attack (**FIGURE 15-21**).

Sand can be trapped by placing sand fences across the wind direction to slow blowing and promote sand deposition on a fence's downwind side. Fences can help keep sand on an upper beach or in dunes; they can also be used to prevent drifts from forming where they create problems for roads and driveways. The sand scraped from a beach or from overwash sediments may be used to partly rebuild artificial dunes or close gaps opened through existing dunes.

Where vegetation has been lost, it can be replanted to help stabilize the sand, though new vegetation may be difficult to establish if the sand is salty or mobile. Straw or branches from local coastal shrubs can be strewn on sand surfaces to slow movement and help establish the growth of dune vegetation. Dune vegetation that is diverse and native to an area is best for its likelihood of survival. Dune grass, or "sea oats," are often planted for that purpose, but it takes about three years without severe damage or root dehydration for the grass to establish roots in the sand and survive.

Natural revegetation of dunes may be accomplished using cuttings taken from nearby dunes, but this can be a slow process. European beach grass introduced along the Pacific coast of the United States has been even more successful in trapping sand than has native vegetation, but it creates dunes that are less natural and higher than the originals. It now dominates dunes in coastal Washington State and parts of Oregon. Residents often plant exotic vegetation because of appearance, but such plants often require artificial watering. The resulting rise in the water table can lead to the formation of gullies and increase the chance that coastal cliffs will have landslides.

A

B

FIGURE 15-20 Paths through Dunes

A. This walkway at Sunset Beach, North Carolina, once reached over a protective beachfront dune, which has since been removed by hurricane waves.

B. A vehicle and foot path cut through beachfront dunes on the coast south of Atlantic City, New Jersey. Such paths dramatically increase erosion during storms.

Donald Hyndman

FIGURE 15-21 Dune Management

Low beach dunes are partly stabilized by sand fences and beach grass south of Atlantic City, New Jersey.

Sea-Cliff Erosion

In contrast to the gently sloping coasts of the East and Gulf Coast states, much of the West Coast is characterized by steep sea cliffs. Where these cliffs consist of soft sediments, they are commonly undermined, causing landslides and cliff collapse. Disintegration of the collapsed material supplies sand to beaches down the coast. That part is good for the beach, but if your home is perched at the top of that cliff, where there is a magnificent view of the ocean, it may not be so good. As the cliff erodes, it gets closer to your house, and the view gets better, but your house eventually collapses with the cliff. Severe

erosion of sea cliffs at Martha's Vineyard in Massachusetts in 2007 prompted a group of affluent homeowners to try to protect their cliff-top property by moving some homes and importing large amounts of sand to replenish the beach. Unfortunately, the effort will succeed only temporarily.

Hard, erosionally resistant rocks—granites, basalts, metamorphic rocks, and well-cemented sedimentary rocks—mark some steep coastlines. Such coastlines often consist of rocky headlands separated by small pocket coves or beaches. Much of the coast of New England, parts of northern California, Oregon, Washington, and western Canada are of this type. The rocky headlands are subjected to intense battering by waves and drop into deep water. Sands or gravels pounded from the headlands are swept into the adjacent coves, where they form small beaches (see Figure 15-5).

Raised marine terraces held up by soft muddy or sandy sediments less than 15 to 20 million years old mark other coastlines, including much of the Oregon and central California coasts. These terraces, standing a few to tens of meters above the surf, are soft and easily eroded. They were themselves beach and near-shore sediments not so long ago. Some terraces rose during earthquakes as the ocean floor was stuffed into the oceanic trench at an offshore subduction zone. Elsewhere they may rise by movements associated with California's San Andreas Fault and related faults.

Along these coasts, cliffs line the heads of beaches, except in low areas at the mouths of coastal valleys. The beaches consist of sand, partly derived from erosion of soft cliff materials and partly brought in by longshore drift. Waves strike a balance between erosion and deposition of beach sands. Larger waves flatten a beach by taking sand farther offshore, and smaller waves steepen it. Where streams bring in little sand or the cliffs are especially resistant, a beach may be narrow. Where rivers supply much sand or where the coastal cliffs are easily eroded, a beach may be wide. A broad, sandy beach hinders cliff erosion because most of the wave energy is expended stirring up and moving around sand.

In addition to rock strength, the main factors affecting cliff erosion are wave height, sea level, and precipitation. All three factors are heavily influenced by intermittent events, from El Niño to hurricanes. When storms come in from the ocean, their frequency and magnitude strongly affect the rate of erosion. Farther north along the coasts of Washington and Oregon, storms are more common during the weakest El Niño years.

Unfortunately, especially in the West, people also choose to build houses on cliff-tops for sea views, typically not realizing how hazardous the sites are. Developers promote building in such locations, charging a premium for view land that may be gone in only a few years. Vertical cliffs made of soft, porous sediments are a recipe for landsliding and cliff collapse. Homeowners themselves exacerbate the problem by clearing beaches of driftwood that help reduce wave force. They unwittingly become agents of erosion by making paths down steep slopes, cutting steps, and excavating for foundations next to the cliffs. They irrigate vegetation and drain water into the ground from rooftops, driveways, household drains, and septic drain fields. Adding water to the ground further weakens it and promotes landslides.

Human Intervention and Mitigation of Coastal Change

People with beachfront or cliff-top homes are commonly affected by storms that cause beach erosion or threaten the destruction of their property. Until something bad happens to their beaches, sea cliffs, or homes, however, many do not really think about the constant motion of sand along the beach from waves and offshore sand movement in storms. Unless tragedy strikes near home, they do not realize that soft sediment beach cliffs gradually retreat landward as they erode at their base and collapse or slide into the ocean.

The typical response to this threat to their property is to build structures to protect communities, a process called **beach hardening**. An extreme example of beach hardening is evident along the New Jersey coast (**Case in Point: Extreme Beach Hardening—New Jersey Coast, p. 429**). Unfortunately, the structures put in place to protect communities commonly contribute to further erosion and, in the long run, make those communities more vulnerable. More recent developments in beach management policies include beach nourishment and zoning.

Engineered Beach Protection Structures

Before large-scale tourism, coastal residences and even small communities fell to the waves or moved inland as the beach gradually migrated landward. Once longer-term investments were made in shore properties, however, individuals and governments were motivated to protect those properties

FIGURE 15-22 Beach Hardening
These huge interlocking concrete pieces are designed to reduce wave energy and protect the road at Muscat, Oman.

rather than move them. Shore-protection projects followed the catastrophic Galveston, Texas, hurricane of 1900. These included building massive **seawalls** or large blocks of concrete (**FIGURE 15-22**). Individuals during this period tried to stop the erosion threat with riprap or walls built of either timber or concrete at the back of the beach in front of their property.

Construction of seawalls accelerated until the 1960s, when scientists and governments began to recognize that much of these activities had long-term disadvantages. Although a structure may slow the direct wave erosion of a beach cliff for awhile, the waves reflect back off the barrier, stir sand to deeper levels, and carry the adjacent beach sand farther offshore (**FIGURE 15-23**). A wave sweeps up onto a gently sloping beach, and the return swash moves back on the same gentle slope. A wave striking a seawall, however, is forced abruptly upward, so the swash comes down much more steeply and with greater force, eroding the sand in front of the seawall. The beach narrows and becomes steeper, the water in front of the barrier deepens, and the waves reach closer to shore before they break. Thus, instead of a protected beach, bigger waves approach closer to shore. The result often hastens erosion and removal of the beach. When the water in front of a barrier becomes sufficiently deep, the beach is totally removed, and the waves may undermine the barrier, which then topples into the surf (**FIGURE 15-24**). If the beach lies at the base of a sea cliff, the supposedly protected cliff succumbs more rapidly.

When cliff-top dwellers see their properties disappearing and recognize that part of the problem is waves undercutting the cliff, the typical response is to dump coarse rocks—**riprap**—at the base of the cliff (**FIGURE 15-25**) or build a wood, steel, or concrete wall there. Waves reaching such a resistant barrier tend to break against it, churn up adjacent sand, and sweep it offshore. The new deeper water next to the

A

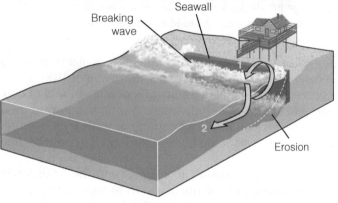

B

FIGURE 15-23 Seawalls Increase Erosion

A. Waves crash against this seawall in Puerto Vallarta, Mexico. The downward force of the return wave undermines the sand at the base of the wall.
B. This diagram shows how waves break against a seawall, causing erosion that may result in the collapse of the wall.

FIGURE 15-24 Eroded Beach in Front of a Seawall

The massive Galveston, Texas, seawall built after the 1900 hurricane ultimately accelerated erosion of the beach. Sand repeatedly dumped on the beach became scarce; heavy riprap was added later.

FIGURE 15-25 Riprap

Collapsing sea cliffs destroy homes. North of San Diego, 1998. Huge boulders attempt to arrest the erosion.

barrier undercuts the barrier, which then collapses into the deep water. The barrier has provided short-term protection to the cliff, but after a few years that "cure" has done more harm than good. A bigger slab of the cliff collapses into the deeper water in the next big storm. Some people spray *shotcrete*, a cement coating, on the cliff surface to minimize erosion (**FIGURE 15-26**). This might be effective for a few years, until the waves undercut the coating.

In a few places, people even build at the base of cliffs, on the beach itself (**FIGURE 15-27**). What do you think happens to such homes during a major storm?

For those who understand that sand grains on a beach tend to migrate along the coast, another approach has been to try to keep the sand from migrating. **Groins**, the barriers built out into the surf to trap sand from migrating down

a beach, do a good job of that. They collect sand on their upstream sides (**FIGURE 15-28**).

Riprap walls, or **jetties**, are sometimes used to maintain navigation channels for boat access into bays, lagoons, and marinas. Jetties that border such channels extend out through the beach but typically require intermittent dredging to keep a channel open (**FIGURE 15-29**). They also block sand migration along the beach. **Breakwaters**, built offshore and parallel to the shore, have a similar effect, causing deposition in the protected area behind the barrier (**FIGURE 15-30**).

Although all of these measures can trap sand, they have the unintended effect of reducing the sand migration along a beach, which causes beach erosion on the down-current

FIGURE 15-26 Shotcrete on Sea Cliffs

These large homes sit atop a sea cliff without a beach in Pismo Beach, California. Note that a new house is being built on the right, even though established houses have lost most of their yards and have spray-cemented shotcrete on their cliffs to slow cliff loss.

FIGURE 15-27 Beach Development

These homes were built on pilings on the beach west of Malibu, California. The coast highway is at the left edge of the blue sign. Would you build on the "For Sale" site in the foreground?

FIGURE 15-28 Groins

Groins in Lake Michigan near Chicago. Longshore drift is from upper left to lower right.

side of a barrier. Effectively, they displace erosion sites to adjacent areas.

Without continued supply, these beaches are starved for sand and thus erode away. So once someone builds a groin or jetty, people down-current see more beach erosion and are inclined to build groins to protect their sections. New Jersey shows these effects to the extreme. Except where replenished, its once sandy beaches are now narrow or non-existent and lined with groins and seawalls.

Most knowledgeable people view groins as a bad choice, and in many cases they are. The state of Florida now requires removal of groins that adversely impact beaches (or are simply nonfunctioning). Groins should not stop sand migration permanently—this should be a remedy only until sand is fully deposited on the upstream side.

However, in some cases, groins can work where the drift of sand farther down the coast is undesirable. For example, to keep an inlet open, dredges must remove sand moving down the coast into an inlet. In such cases, well-engineered groins can be effective. They take a wide variety of forms and sizes that depend on the specific purpose.

Beach Replenishment

Are there better ways to protect a beach—or to repair the damage after its loss? Beginning in the 1950s, replacing sand

FIGURE 15-29 Jetty

The beach on the north side of Shinnecock Inlet, Long Island, has been eroded by longshore drift of sand from top to lower right.

FIGURE 15-30 Breakwaters

Breakwaters at Chesapeake Bay, Norfolk County, Virginia, in June 2002, shelter the beach behind them, causing deposition.

on beaches became popular in the United States, in a process called **beach replenishment**, or beach nourishment (**Case in Point:** Repeated Beach Nourishment—Long Island, New York, p. 430). In 56 large federal beach projects in the United States between 1950 and 1993, the U.S. Army Corps of Engineers placed 144 million m³ of sand on 364 km of coast. Enormous volumes have been placed in some relatively small areas, such as the 24 million m³ on the shoreline of Santa Monica Bay, California.

When the federal government agrees to foot a large part of the bill for a major beach replenishment project, the U.S. Army Corps of Engineers becomes the responsible agency. Engineers, geologists, and hydrologists with expertise in beach processes design a replenishment project and oversee the private

contractors who actually do the work. Common sand sources include shore areas in which sand shows net accumulation or sources well offshore and below wave base. Sometimes sand is dredged off the bottom and transported to the beach area on large barges. Elsewhere sand is suction-pumped from the source and pumped through huge pipes as a slurry of sand and water. From there, the sand is spread across the beach using heavy earthmoving equipment (**FIGURE 15-31**).

In some cases, however, engineers choose to minimize costs by adding sand up the coast and permitting longshore drift to spread it into the area needing nourishment. Dumping sand at a cape, for example, can lead to sand migration into adjacent bays where it is needed.

Along cliff-bound coasts, sand, sediment, or easily disintegrated fill material is sometimes merely dumped over an eroding bluff to permit breakup by waves. Instead of rapidly eroding the cliff base, the waves gradually break up the dumped sediment to form new beach material.

One problem with beach replenishment is that natural coastal processes involve the migration of sand away from beaches. Unfortunately for replenishment projects, sand added to a beach in one place will quickly migrate elsewhere. In many cases the replenished beach is eroded away in the next big storm or several smaller storms—often after only a year or two. Because subsequent storms remove sand from the beach, part of the cost of a sand renourishment project involves continued small-scale nourishments at intervals of two to six years. Small trucking operations may need to be repeated every year, especially if the added sand is finer grained than that on the original beach. Sometimes, because of scarcity, sand is excavated from the lagoon side of a barrier bar where the grain size is finer than the natural sand on most beaches. Sand finer than 0.5 mm is too fine; small waves will typically wash it offshore. At Carolina Beach, North Carolina, for example, the beach, lined with private homes, was replenished eight times in the 26 years between 1965 and 1991.

Another concern is that sand placement can change a shore profile and actually increase erosion. Most people want to see the sand they pay for placed on the upper, dry part of a beach rather than offshore. It is also easier to calculate the volume of sand added to the upper beach and therefore the appropriate cost. But the active beach actually extends well offshore into shallow water. If that part of the beach is not also raised, waves will move much of the onshore sand offshore to even out the overall slope of the beach. The next storm will carry much of the new sand offshore, where people view it as lost to the beach and think the replenishment was a waste of money. If half of the sand added is offshore in shallow water where it provides a more natural equilibrium profile, people have a hard time understanding that it is not being wasted. Either scenario leads to criticism of beach replenishment as a viable solution to beach erosion. As a result of such problems, some beach experts suggest using the expression "shore nourishment" rather than "beach nourishment" to emphasize that the shallow, underwater part of the beach is equally important.

FIGURE 15-31 Beach Replenishment

A. To replenish a beach, the U.S. Army Corps of Engineers pumped a high-volume sand-and-water slurry from up the coast in 30-inch pipes. Heavy earthmoving equipment spread more than a meter of sand across the new beach at Ocean City, New Jersey, in April 2010.

B. As part of the massive beach replenishment at East Rockaway, New York, in March 1999, the groin at left minimizes the loss of the replenished sand by longshore drift.

Identifying the source for the sand is another hurdle for beach replenishment projects. For obvious reasons, mining sand from other beaches is generally not permitted. In some areas, significant sand is obtained from maintenance dredging of sand from navigation channels. Sand drifting into inlets is often lost to the beach system if inlet dredges dump it far offshore. Some areas, such as Florida, now require that dredged sand be dumped next to the down-drift side of the inlet. The operation thus permits the sand to bypass the inlet.

Other bypass operations use fixed or movable jet pumps, most commonly in conjunction with conventional dredges (**FIGURE 15-32**).

Mining sand from privately owned sand dunes well back from a beach or dredging sand from a lagoon or other site behind a barrier island is sometimes possible, though expensive. A significant drawback to such sources is that dune and lagoon sand is generally finer grained than that eroded from the beach. Because storm waves were able to move the

FIGURE 15-32 Piping in Dredged Sand

A. Jetty bordering an estuary in Pompano Beach, Florida, blocks southward drift of beach sand, starving the beach to the south and leading to its erosion. Note that the beach south of the estuary is much recessed from the straight coastline. **B.** As sand is dredged from a river outlet between the jetties, it is piped to the down-drift side of the jetties to replenish the beach farther down-drift.

coarser-grained sand from the beach, slightly finer-grained sand can be removed by even smaller waves.

Another solution used in many areas, especially along the southeastern coasts, is to dredge sand from well offshore and spread it on a beach. Because much larger equipment is required, regional or federal governments, often under the direction of the Corps of Engineers, normally undertake such projects. Taking sand from near shore deepens water and makes the beach steeper. However, waves shift sand into an equilibrium slope, a process that depends especially on the size of both sand grains and waves. An artificially steeper beach will erode down to the equilibrium slope. If the sand is taken from well offshore, the bigger waves of storms erode the beach down to a lower slope farther offshore.

Sources of usable sand in many areas are being rapidly depleted. Florida's sources of economically recoverable sand, for example, were almost exhausted by 1995. By 2014, the beach-replenishment situation had gotten so bad that Miami had used up its offshore sources and is trucking sand for small projects from central Florida. It and nearby counties are also considering shipping sand from the Bahamas and using crushed bottle glass in place of beach sand.

The scale and cost of beach nourishment projects are huge. A large dump truck carries roughly 10 m^3 of sand. If a person's lot is 30 m wide and the beach extends 65 m from the house to the water's edge at mid-tide (roughly 2000 m^2), one dump-truck load would cover that part of the beach to a depth of only half a centimeter or so; it would take 200 loads to add about a meter of sand to the beach in front of one house. Thus, if sand costs $30/m^3, it might take 200 truckloads in front of every home to replace the sand removed in one moderate storm—for a cost of roughly $60,000 for each home. More than 28,000 km of coast continue to erode.

As always, the role of government subsidising the cost of such projects is a matter for debate. Sometimes, groups of residents, towns, or counties on barrier islands lobby the local, state, and federal governments to replenish the sand on a severely eroded beach. Because large sand replacement projects typically run into millions of dollars and homeowners do not want their taxes to increase, local governments lobby state and federal governments to foot most or all of the bill. Politicians want to be reelected and bring as much money back to their communities as possible, so they lobby hard for state and federal funding. What this means, of course, is that the cost of replenishing sand to benefit a few dozen beachfront homeowners is spread statewide—or more commonly countrywide—among those who generally derive no benefit from the work. For example, the federal government paid approximately 50% of the cost of beach replenishment projects in Broward County, Florida, from 1970 to 1991, while local communities paid only 4%. To add insult to injury,

beachfront communities often try to inhibit access by numerous mainlanders who flock to beaches on warm summer days. Although beach areas below high tide are legally public, access routes are often poorly marked or illegally posted for no access. Some communities also make beach approaches difficult by severely restricting parking along nearby roads.

Since the 1990s, federal and state funds for replenishment have dwindled. Environmentalists, inland taxpayers, and government officials often argue against spending millions of public dollars for major projects to benefit a few beachfront homes when the sand is likely to wash away in the next significant storm. Lacking state or federal support, local counties sometimes try to fund their own beach projects, but these often fail because even those living on the beach sometimes argue that they cannot afford the annual assessments. Because many of those who own beachfront homes live elsewhere, local voters often reject tax increases. Even where local groups successfully fund multimillion-dollar projects, they often find that the new sand erodes in a few years.

Zoning for Appropriate Coastal Land Uses

Many coastal experts advocate a more permanent alternative to beach hardening or beach replenishment. They suggest moving buildings and roads on coasts back landward to safer locations after major damaging storms. Sand dunes behind a beach can be stabilized with vegetation in order to provide further protection for areas behind them. Riprapping and other shore hardening to prevent cliff erosion is temporary, very expensive, and accelerates erosion over longer times. Now some communities partner with various state and federal agencies to allow for the coast to erode landward unimpeded in a program called "managed retreat strategy." Endangered structures are either demolished or relocated. Costs are shared among those involved. The cost of moving buildings may be high, but it is less than the continuing long-term cost of maintaining beach-hardening structures that tend to destroy their beaches or of continually bringing in thousands of tons of sand that get removed by following storms. And it does provide a way of living with the active beach environment rather than forever trying to fight it.

On cliff-bound coasts, buildings should not be placed close to cliff edges. Those that are too close should be moved well back, and foot traffic and other activities should be restricted to areas away from cliff tops and faces. Beaches should not be cleared of natural debris such as driftwood. Mining sand and gravel from beaches and streams should be prohibited. Additional dams should not be built on rivers that discharge in coastal regions with erosion problems; removal of old dams would eventually bring more sediment back to beaches and help protect cliffs.

Cases in Point

Extreme Beach Hardening
New Jersey Coast ▶

One of the most extreme cases of beach hardening and severe erosion is along the coast of New Jersey. Early development was promoted by extending a railroad line along more than half of the New Jersey shoreline by the mid-1880s. Buildings first clustered around railroad stations, dunes were flattened for construction, natural dense scrub vegetation was removed, and marsh areas were filled. The arrival of private automobiles accelerated development. Marsh areas along the back edge of the offshore bars were both filled and dredged in the early 1900s to provide boat channels. Inlets across barrier islands were artificially closed, and jetties built after 1911 stabilized others. People did not understand then that barrier islands were part of the constantly evolving beaches and that nature would resist human attempts to control it.

The large population nearby, and the demand for recreation, stimulated the building of high-rise hotels and condominiums in Atlantic City; it is now so packed with large buildings next to the beach that they strongly affect wind flow and sand transport. Large-scale replenishment of sand on beaches is widespread in some areas. In other areas, narrow and low artificial dunes are built in front of separate detached houses, as are barriers to provide backup protection from erosion. Building the "protective" erosion barriers, riprap, and seawalls merely hastened erosion of the beach sand in front of them, leading to disappearance of the beach. Groins are numerous. A prominent seawall and continuous groins bear no resemblance either to the original barrier island environment or to the beaches that attracted people in the first place.

The beachfront of the barrier island has completely eroded away. The options now are either for continuous expensive beach replenishment for the benefit of those living at those beaches or moving the buildings back from the beach. The latter would be equally expensive and politically difficult.

CRITICAL THINKING QUESTIONS

1. By what processes does beach hardening increase beach erosion?
2. If beach hardening increases beach erosion, what alternatives are there for:
   a. building homes on barrier islands?
   b. protecting existing homes threatened by beach erosion?
3. Should sand on beaches be replenished? Who should pay for these efforts?
4. Why is it difficult to find sand of the same characteristics of an original beach?

■ *New Jersey has permitted condos and large houses to be built on the artificially replenished beach just south of Atlantic City. By 2006 the dunes were gone; their only protection is riprap and steel walls.*

■ *This is the sad ending for a mishandled beach north of Cape May, New Jersey. Building the coastal road led to protection with riprap, which hastened erosion of the beach. As of 2006, a concrete and rock seawall remains in place of the beach.*

■ *Major coastal hardening—rock groins, a seawall, and a major boardwalk— at Atlantic City failed to prevent beach destruction in a moderate storm, Hurricane Sandy, 2012.*

Repeated Beach Nourishment

Long Island, New York ▶

One of the most vulnerable coastlines in the northeastern United States is the south coast of Long Island, New York, which protrudes east into the paths of some strong Atlantic storms. As in any low-lying coast facing big waves, its southern portion is marked by a nearly continuous line of sand bars and barrier islands. The unincorporated community of Westhampton Dunes, about halfway east along the barrier island, has become a "poster child" for problems of beach renourishment and attempts to maintain a constantly shifting barrier island. A chronology of events illustrates evolution of the problem:

1950s: Erosion of the beach threatens the town of Westhampton Beach to the north. The Army Corps of Engineers proposes renourishment and construction of 23 groins "if necessary."

1962: A storm breaches the barrier island and, following political and legal pressure, 11 hastily constructed groins immediately restrict east-to-west longshore drift of sand. Beach erosion to the west, at the 4-km-long community of Westhampton Dunes, abruptly increases from 0.5 to 4.5 m per year.

1970: Four more groins are built to protect large homes and a condominium. In the late 1970s more houses are lost to erosion in severe nor'easters.

1973 and 1984: Residents with eroding beaches file lawsuits against the government because the groins accelerated erosion, but they are not compensated. Various efforts to complete the groin field stall because of local, state, or federal concerns with cost sharing to specific state requirements for shore protection projects.

1988: New York's Sea Grant Program convenes a group of coastal experts who consider available options and predict formation of a breach at one site just west of the westernmost groin, but nothing is done because lawyers are reluctant to move forward in the midst of a lawsuit.

1991 and late November 1992, nor'easter: Storms destroy many houses and breach the barrier island in several places; one breach widens to almost 1 km, destroying dozens of houses. A total of 190 out of 246 homes have now been lost.

1993: The breach at the predicted site is filled in 1993 for $8.8 million and the groins are shortened to permit more sand to drift along the beach.

1994: Residents, now as a municipality, again file a lawsuit, this time successfully. The Corps of Engineers, state, and county agree to replenish the beach periodically until 2027.

1997: The $53 million (in 2007 dollars) renourishment project is completed with 3.4 million m^3 of sand.

Most of the cumulative cost was paid by taxpayers of the state of New York. In return, they are provided with limited parking for 200 cars at two locations along 3 km of road length and entry to the beach via seven public access points. Unfortunately the homeowners were permitted to rebuild in the same locations as their destroyed homes. The history of this short stretch of barrier bar emphasizes the hazards and pitfalls of trying to control the natural processes that shape and move a beach on a gently sloping coastline. A better solution would have been to neither build groins nor replenish the beach but to abandon or move the endangered modest houses and permit the barrier bar to migrate shoreward after each storm.

CRITICAL THINKING QUESTIONS

1. Why do beaches again erode after being replenished?
2. What factors control whether sand on a replenished beach remains there or is easily eroded?
3. Where should sand be placed on a beach to have the best chance of remaining there?
4. Who typically pays for beach replenishment and who should pay for it? Why? What are the trade-offs for those who pay and those who benefit?

■ *The late-1992 storm breached the barrier bar at Westhampton Dunes and left houses stranded in the surf. The same houses are shown in Figure 15-15.*

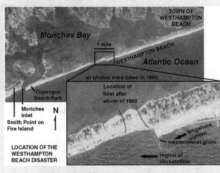

■ *The main breach was just down drift from the westernmost groin.*

■ *This December 1992 breach downstream of the westernmost groin was later filled in 1993 and the beach was replenished.*

A This beach is on a large lake in the northern Rockies.

1. Why is the upper edge of the beach so steep?
2. Why is it steeper at its top than down at the water's edge?
3. Why are pebbles larger at the top of the beach front rather than the bottom?

C These houses sit right on the beach along the California coast.

1. What hazards can you identify in this photo?
2. What measures might these beachfront homeowners take to better protect their properties?
3. What is the direction of longshore drift? How can you tell?

B This beach runs along the central California coast next to a highway.

1. Why are huge boulders piled at the back of a nice sandy beach?
2. What are the consequences to the beach from piling these boulders here? Why?

D This fence runs between these houses and the beach on Long Beach Island, north of Atlantic City, New Jersey.

1. Why is there a steep cliff in the beach sand in the foreground?
2. What effect does the slat fence have on the distribution of sand on and around the beach?
3. How does this fence reduce the hazard to these houses during a storm?

E These houses are near the New Jersey coast, south of Atlantic City. The ocean is 100 m toward the lower left corner of the photo.

Hyndman

1. Why are these houses built up on posts?
2. How do the posts help protect the home?

F This beach is along the central California coast.

Donald Hyndman

1. What dangerous swimming conditions are present?
2. What evidence of those conditions is apparent in the pattern of wet sand on the upper part of the beach?
3. What evidence is there of those dangerous swimming conditions in the water just offshore?
4. If you were swimming in the water here, what action should you take to get to safety?

G This sea cliff is in northern Monterey Bay, California.

Cheryl Hapke, USGS

1. What hazards are apparent in the photo?
2. What circumstances would lead to more imminent danger for houses at the bottom of the cliff? What about for houses at the top of the cliff?
3. What might you watch for to provide some lead time before a disaster?

Chapter Review

Key Points

Living on Dangerous Coasts

■ Beaches and sea cliffs constantly change with the seasons and progressively with time. When people build permanent structures at the beach, coastal processes do not stop; the processes interact with and are affected by those new structures.

Waves and Sediment Transport

■ Most waves are caused by wind blowing across water. The height of the waves is dictated by the strength, time, and distance of the wind blowing over the water.

■ Water in a wave moves in a circular motion rather than in the direction the wave travels. That circular motion decreases downward to disappear at an approximate depth of half the wavelength. Waves begin to feel bottom near shore in water less than that depth. **FIGURES 15-1** and **15-3**.

■ Wave energy is proportional to the wavelength multiplied by wave height squared, so doubling the wave height quadruples the wave energy.

■ Waves approaching the beach at an angle push sand parallel to shore as longshore drift. **FIGURE 15-4**.

■ Waves refract toward rocky headlands, eroding them more aggressively. Sediment eroded from headlands is deposited in sandy bays. **FIGURE 15-5**.

Beaches and Sand Supply

■ The grain size and amount of sand on a beach depend on wave energy, erodibility of sea cliffs and the size of particles they produce, and the size and amount of material brought in by rivers. Larger winter waves commonly leave a coarser-grained, steeper beach.

■ A beach at its equilibrium profile steepens toward shore. Larger waves erode the beach and spread sand on a gentler slope. **FIGURE 15-8**.

■ Sand is lost from a beach through erosion, particularly during storms. Anything that removes sand supply from a beach, for example, damming rivers upstream, results in greater erosion.

Erosion of Gently Sloping Coasts and Barrier Islands

■ Offshore barrier islands develop where the landward-steepening beach profile is steeper than the general slope of the coast. Wind blows sand shoreward into dunes. **FIGURE 15-13**.

■ Tidal currents keep open estuaries and inlets that lead into lagoons behind bars. Storm surges and waves open, close, and shift the locations of such inlets.

■ Barrier islands maintain their equilibrium profile by migrating shoreward with rising sea level. Because many barrier islands are now covered with buildings, islands cannot migrate landward but are progressively eroded. **FIGURE 15-15**.

■ Dune sand spreads over a beach, helping to protect it from erosion during storms, but most dunes have disappeared by excavation, trampling underfoot, or wave action.

Sea-Cliff Erosion

■ Sea-cliff erosion is determined by the strength of the rock, wave height, sea level, and precipitation. As a result, erosion is amplified during storms.

■ Homeowners can attempt to protect their sea-cliff property from erosion by placing riprap at the base of a cliff or covering it with concrete.

Human Intervention and Mitigation of Coastal Change

■ Beach hardening to prevent erosion or stop the longshore drift of sand includes seawalls, groins, jetties, and breakwaters. All have negative consequences either after a period of time or elsewhere along the coast. In some cases, they result in the complete loss of a beach. **FIGURES 15-22** to **15-30**.

■ Beach replenishment or nourishment involves replacing sand on a beach, an expensive proposition that needs repetition at intervals. **FIGURES 15-31** and **15-32**.

■ A less-expensive and more permanent solution to the effects of beach erosion is to move threatened structures back from a beach and implement environmental protection strategies.

Key Terms

barrier island, p. 418

beach, p. 412

beach hardening, p. 423

beach replenishment, p. 426

breakwaters, p. 424

fetch, p. 411

groin, p. 424

headlands, p. 413

jetty, p. 424

longshore drift, p. 412

period, p. 411

rip current, p. 413

riprap, p. 423

sand dune, p. 420

seawall, p. 423

shore profile, p. 414

wave energy, p. 412

wave height, p. 411

wave refraction p. 412

wavelength, p. 411

Questions for Review

1. Sketch the motion of a stick (or a water molecule) in a deepwater wave.

2. What force creates most waves?

3. How can there be big waves at the coast when there is little or no wind?

4. What factors cause growth of wind waves?

5. Why are ocean waves generally larger than those on lakes?

6. Where does the sand go that is eroded from a beach during a storm?

7. Where would a curved, sandy peninsula form on a barrier island relative to the direction of longshore drift? Draw a sketch to illustrate your answer.

8. Draw the shape of a barrier island before and after a significant rise in sea level.

9. How do sand dunes affect stability of a beach?

10. Which side of a beach groin collects sand? Why is this the case?

11. What happens to wave energy and erosion when riprap or seawalls are installed?

12. What can be done to prevent building in hazardous regions in coastal areas?

13. What can be done to minimize erosion in a coastal area, and what are some positive and negative aspects of those methods?

Critical Thinking Questions

1. Who should pay for the costs of beach nourishment projects (local residents, state taxpayers, national taxpayers)? Should beach-user fees play a role in financing these projects?

2. If you owned a house near the edge of an eroding sea cliff on the Oregon or California coast, what would you do? Keep in mind that you are morally and legally obligated to inform a buyer of any damages or imminent hazards that may affect the house.

3. If you were a Federal Emergency Management Agency (FEMA) administrator in charge of government aid to people living on a barrier island and whose homes are now cut off by hurricane breach of their barrier island, what would you do and why?

4. Where should beach-replenishment sand come from? What are the consequences of taking sand from that location?

■ After Hurricane Sandy, little was left of the Atlantic City boardwalk but its concrete pilings. Nearby coastal buildings were reduced to slabs of concrete and parts of brick walls.

Hurricanes and Nor'easters

16

Late on October 29, 2012 the eye of Category 1 Hurricane Sandy came ashore near Atlantic City, New Jersey, destroying most of the north-facing section of the Atlantic City boardwalk and many of the older waterfront buildings. Coming ashore, the high winds and surge of its right-front quadrant funneled directly against northern New Jersey, into the Lower Bay, between northern New Jersey and Long Island, and into the low-lying Staten Island section of New York City. Compared with other hurricanes, Sandy was not well organized, but its giant diameter meant that it stayed in impacted areas for a long time. The huge surge washed across the barrier island at Mantoloking 80 km north of Atlantic City, washing out the east end of its bridge access and destroying large homes nearby. Many homes floated off their foundations to come to rest in strange locations. Because the surge was more damaging than the wind, most windows remained unbroken. See the **Case in Point:** A Damaging Late-Season, Low-Category Storm—Hurricane Sandy, Atlantic Coast, October 2012, p. 470.

Hurricane Formation and Movement

A **tropical cyclone** is a storm system consisting of a low-pressure center surrounded by strong rotating winds. These storms have wind speeds of more than 120 km/h and can exceed 260 km/h. They can also bring torrential rain, thunderstorms, tornadoes, and a high surge of water from the sea along the coast.

When a large tropical cyclone occurs in the North Atlantic and eastern Pacific, it is called a **hurricane**, a term derived from a Caribbean native Indian word meaning *big wind*. The same type of storm is called a **typhoon** in the western Pacific, Japan, and Southeast Asia. In the Indian Ocean these storms are simply called cyclones. In this chapter, we will refer to all such storms as hurricanes.

Historically, hurricanes create some of the world's deadliest disasters. The worst have occurred in heavily populated poor countries in Southeast Asia, where single storms can result in hundreds of thousands of deaths.

Formation of Hurricanes

Hurricanes begin to develop over warm seawater of at least 25°C (77°F), commonly between 5 and 20° north latitude. In these tropical latitudes, temperatures are high and air-pressure gradients are weak. Air rises by localized heating, which causes condensation that can build into towering convective "chimneys" with frequent thunderstorms. Strong winds can shear off the tops of the convective chimneys and cause them to dissipate.

In the absence of winds, convection may strengthen when the air rises to high elevations. The warm, moist air over the ocean rises and spreads out at the top of the chimney. That warm air expands, cools, and releases latent heat; the air rises faster as the storm strengthens.

At the center of the storm is a low-pressure zone called the **eye**, generally 20 to 50 km in diameter and clearly visible as a small, dark area in many satellite views (**FIGURE 16-1**). The eye of a hurricane is as much as 10–11°C warmer than the surrounding air. Rise of air near the eye pulls more moist air into the eye from low elevations at the periphery of the storm (**FIGURE 16-2**). Coriolis forces initiate rotation in the rising air, the highest winds and lowest air pressures focusing toward the core of the storm. The winds rotate counterclockwise in the northern hemisphere and clockwise in the southern hemisphere, with the highest wind speeds along the edge of the eye wall. Inside the eye, wind speeds drop abruptly from, for example, 220 to 15 km/h. Air pressure in the eye drops from normal atmospheric pressure of approximately 1000 millibars (1 bar) to 960 to 970 millibars. The air in the sharply bounded eye sinks, causing skies to clear as the dry air from above warms and can hold more moisture.

The whole storm may be from 160 to more than 800 km in diameter. Once formed, storms move across the ocean with the prevailing trade winds, their forward motion averaging 25 km/h.

Classification of Hurricanes

The **Saffir–Simpson Hurricane Wind Scale** divides hurricanes into five categories based on average wind speed, with Category 1 storms being the weakest and Category 5 the strongest. The higher the wind speed, the stronger the hurricane (**Table 16-1**). A hurricane typically changes category as it develops, either intensifying or weakening. For example, Hurricane Katrina was a Category 5 storm at sea, but its winds slowed as it approached land, weakening to a Category 3 storm.

Although the Saffir–Simpson Scale is designed to indicate the intensity of a hurricane, it can be misleading as an indicator of the level of expected damage. Though people often focus their attention on the strongest hurricanes, Categories 4 and 5, lower-category storms can sometimes do almost as much damage and in some cases cost even more lives. In addition to the wind speed, the amount of damage from a hurricane is heavily dependent on the height of water rise

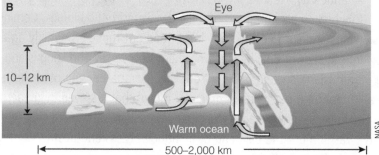

FIGURE 16-1 Formation of a Hurricane

A. An oblique astronaut view of Hurricane Katrina looking north before landfall. The storm circulates counterclockwise, with its trailing winds spreading farther out (lower right).

B. An idealized cutaway view of a hurricane as seen from the side. The top of the diagram corresponds to the astronaut view on the left. Air drawn in at sea level from outside the storm rises near the eye and spreads out; cold, dry air sinks in the eye.

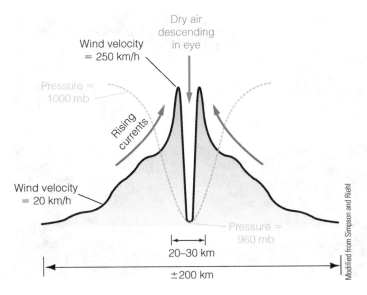

FIGURE 16-2 Wind Speed and Atmospheric Pressure
This plot of properties through a typical hurricane shows what atmospheric pressure (tan dashed line), wind velocity (heavy black line), and air motions would be for measurements taken from a plane flying through the center of a hurricane.

from the storm surge, how large an area is covered, the duration of high water and high winds, and how recently another storm has affected the area (discussed in greater detail in the next section). Because a Category 5 has much stronger winds than a Category 2 storm, it rotates faster and often has a smaller diameter. Even with the same forward speed, the smaller diameter of the Category 5 can lead to a shorter time period for passage over a given area and sometimes causes less overall damage.

Hurricane Sandy in October 2012 emphasized how destructive a Category 1 storm can be because of its immense diameter and slow passage over an area. Except for Katrina in 2005, it produced the most expensive damages of any hurricane on record.

Movement of Hurricanes and Areas at Risk

Hurricanes rotate counterclockwise but track clockwise in northern hemisphere ocean basins, the same direction as the ocean currents (**FIGURE 16-3**). Hurricanes rotate clockwise and track counterclockwise in the southern hemisphere. Northern hemisphere tropical storms begin in warm waters off the west coast of Africa, then move westward across the Atlantic Ocean with the trade winds (**FIGURE 16-4**). They

Table 16-1		The Saffir-Simpson Hurricane Wind Scale	
		WIND SPEED	
CATEGORY**	**EXAMPLE**	**KM/H** (MPH)	**DAMAGES****
Normal (no storm)		**0–61** (0–38)	
Tropical storm		**62–118** (39–74)	
1	Danny, 1997; Sandy, 2012	**119–153** (75–95)	Minor to trees and unanchored mobile homes. Sandy, only a Category 1 storm at landfall, had the largest diameter of any recorded U.S. hurricane. Its total costs were estimated as $68 billion, the second costliest hurricane in U.S. history.
2	Bertha, 1996; Isabel, 2003; Ike, 2008	**154–177** (96–110)	Moderate to major damage to trees and mobile homes, windows, doors, some roofing. Low coastal roads flooded 2 to 4 hours before arrival of hurricane eye.
3	Alicia, 1983; Fran, 1996; Katrina, 2005	**178–209** (111–130)	Major damage: Large trees down, small buildings damaged, mobile homes destroyed. Low-lying escape routes flooded 3 to 5 hours before arrival of hurricane eye. Land below 1.5 m above mean sea level flooded to 13 km inland.
4	Hugo, 1989	**210–249** (131–155)	Extreme damage: Major damage to windows, doors, roofs, coastal buildings. Flooding many kilometers inland. Land below 3 m above mean sea level flooded as far as 10 km inland.
5	Camille, 1969; Gilbert, 1988; Andrew, 1992; Mitch, 1998	**>249** (>155)	Catastrophic damage: Major damage to all buildings less than 4.5 m above sea level and 500 m from shore. All trees and signs blown down. Low-lying escape routes flooded 3 to 5 hours before arrival of hurricane.

*Standard atmospheric pressure at sea level = 29.92 inches of mercury = 1000 millibars (mb) = 1 bar or 1 atmosphere pressure.
**Wind is the primary control used to categorize hurricanes—pressure and storm surge height were formerly used, but are now just reference.
***These are highly variable and depend on many factors as discussed in the text.

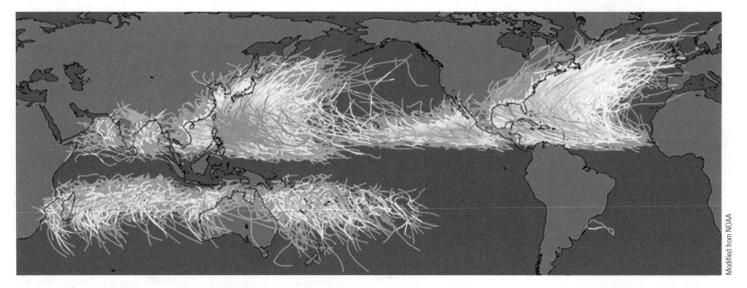

FIGURE 16-3 Hurricane Tracks

Colored lines indicate the paths of all hurricanes and tropical storms. Storm intensity is indicated by color, from weaker storms in blue to Category 5 hurricanes in red. Note that hurricanes do not form within several degrees of the equator.

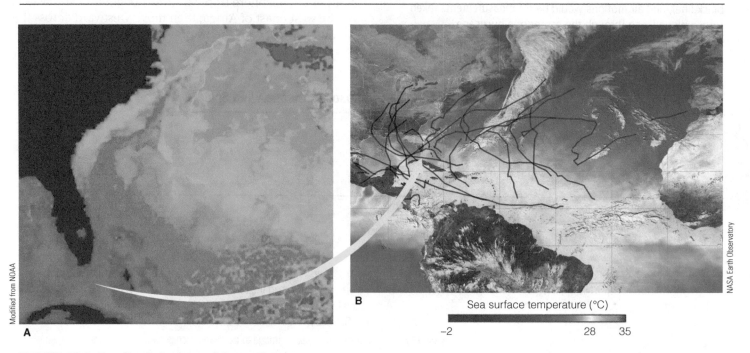

FIGURE 16-4 Hurricane Paths and Ocean Temperature

A. Warmer colors emphasize warmer water with red outlining the Gulf Stream current that sweeps up the eastern seaboard. Hurricanes passing over this warm water become more energized as they approach land. **B.** Sea-surface temperatures in the Atlantic Ocean in 2005 show warm water (orange colors) off the west coast of Africa, drifting westward with the trade winds to energize hurricanes impacting the southeastern United States (black paths indicate 2005 hurricanes). Streaky white clouds are also visible.

warm, pick up wind speed and energy, and often develop into hurricanes before they reach the Americas. They generally track west, northwest, and then north, either off the southeastern United States or sometimes into the continent. Typhoons in the western Pacific and cyclones in the north Indian Ocean have similar tracks.

Some 80 to 90 tropical storms and 45 hurricanes occur worldwide every year (130,000 such storms per thousand years). An average of six named hurricanes form annually in the Atlantic Ocean and Gulf of Mexico. Thus, in a short period of geologic time, cyclones are likely to have a major effect on erosion, deposition, and overall landscape modification, especially in exposed areas such as barrier islands.

For the United States, subtropical cyclones or hurricanes are a major concern along the Gulf of Mexico and southern Atlantic coasts. More than 44 million people live in coastal counties susceptible to such storms, roughly 15% of the total population of the United States. Because of the rapid growth of these areas, a large proportion of the population has never directly experienced a hurricane and is poorly informed about the risks. These residents, and equally inexperienced developers and builders, choose to live and work in hazardous sites such as beachfront dunes and offshore barrier bars. Rising property values amplify the monetary damage when a major storm does hit.

Hurricanes generally curve northward as they approach the southeastern coast of the United States because of the Coriolis effect. Some track fairly straight; others take erratic paths. They commonly strengthen as long as they remain over the warm water of the Gulf Stream that flows northeasterly along the east coast of Florida and continues up the coast past eastern Canada. Westward-moving storms therefore tend to track in arcs to the northwest and then north as they approach the coast. If they move over especially warm ocean eddies, they may strengthen dramatically, as in the case of Hurricane Katrina in the Gulf of Mexico in 2005 (**FIGURE 16-5**). As they move over cooler waters or continue over land, they gradually lose energy and dissipate.

Hurricane damages can be amplified by weather and climate conditions. Higher ocean temperatures or warm-water eddies add energy. Storms can be stalled by other weather systems, thereby dropping more rain in a region. Two or more hurricanes hitting the same region one after another can also amplify damages.

The number and severity of hurricanes appears to be increasing. An average of five hurricanes develop in the Atlantic Ocean every year, two of them major (Category 3 or greater). The number naturally varies, but the correlation with the Atlantic multidecadal oscillation (discussed in Chapter 10) and with annual sea-surface temperatures (SSTs) suggests that the next few decades will see not only more hurricanes but more severe hurricanes. Although 2004 was a record year for hurricanes, 2005 provided an even larger number, including Katrina, the costliest hurricane of all time, and Wilma, the second strongest ever measured. In more than 150 years of records, only 1960, 1961, 2005, and 2007 had more than one Category 5 storm. Both Dean, which struck the Yucatan Peninsula of Mexico on August 21, 2007, and Felix, which hit eastern Honduras on September 4, 2007, were still Category 5 hurricanes at landfall. The strongest hurricane ever measured was Supertyphoon Haiyan in 2013 (**Case in Point:** Catastrophic Typhoon in the Western Pacific—Supertyphoon Haiyan, the Philippines, 2013, p. 472).

Hurricane strength appears to be increasing as a result of the increase in ocean temperatures due to climate change (**FIGURE 16-6**). A warming atmosphere can hold more

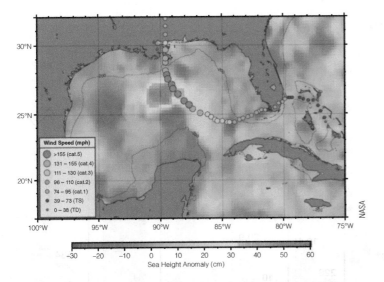

FIGURE 16-5 Hurricane Strength and Ocean Temperature

Sea-surface temperatures for the Gulf of Mexico in August 2005 show an eddy of warm water (measured by sea-surface height, indicating thermal expansion of the water). The strength of Hurricane Katrina, as indicated by wind intensity, increased dramatically where it passed over the eddy of warm water (red).

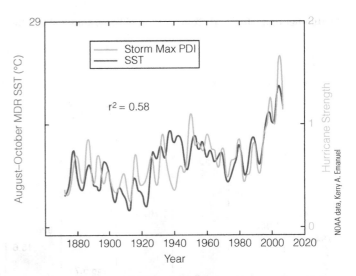

FIGURE 16-6 Correlation of Hurricane Strength and Ocean Temperature

Hurricane strength in any given year appears to correlate quite well with Atlantic Ocean sea-surface temperature. Both are rising with time, suggesting that global warming will increase the strength of hurricanes as ocean water warms. PDI = Power Dissipation Index.

moisture and thus an spawn stronger storms with higher rainfall. Over the east coast of the United States, the sea-surface temperature at the time of Hurricane Sandy was an average of 3° C higher than normal, permitting it to hold about 35% more moisture. Climate change experts estimate that global warming contributed about 20% of that temperature rise.

We need to be better prepared for more frequent and stronger events in the future. A 2012 study concluded that so-called 100-year storm events are now more likely to occur every 3 to 20 years. Compounding this is the continuing movement of more people to eroding coastal areas where rising sea level and stronger storms place them in greater danger.

Although most U.S. hurricanes affect lower-latitude areas, many of those that remain off the coast with northward trajectories reach as far north as New England. Only a few of these have intensities greater than Category 3, but those few can be highly destructive, especially where Long Island, Rhode Island, and eastern Massachusetts protrude into the Atlantic Ocean and across the hurricane paths. Based on historic accounts and prehistoric overwash deposits of sand covering layers of peat, at least 27 hurricanes of Categories 1 and 2 have hit the same area in the last 400 years, an average of one every 15 years.

Large hurricanes of Categories 3 to 5 strike most locations from Florida through the Texas Gulf Coast once every 15 to 35 years on average. Especially vulnerable are southern Florida, the Carolinas, and the Gulf Coast (**FIGURE 16-7**). Although hurricane season spans from late June or July to November, most hurricanes develop in August or September, because the sun spends all summer warming ocean

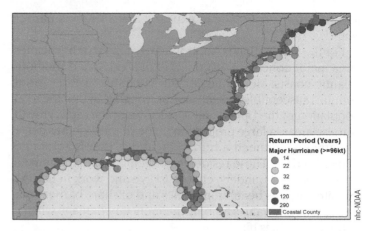

FIGURE 16-7 **Areas at Risk for Hurricanes**

Potential number of hurricane strikes (and floods from hurricanes) per hundred years based on historical data. Red = more than 60; dark blue = 40–60; light blue = 20–40.

surface temperatures. On average, nine hurricanes develop in the eastern Pacific basin each year, four of them major.

Storm Damages

Over the past century, the costs related to hurricanes in the United States have dramatically increased, while the number of deaths has decreased (**FIGURE 16-8**). In fact, 6 of the 10 most costly hurricanes occurred in 2004 and 2005 (**Table 16-2**).

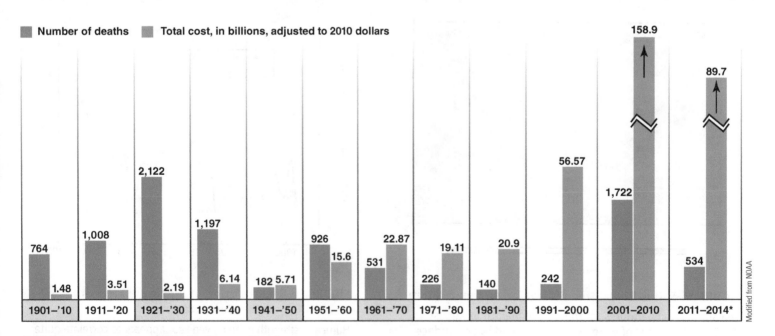

FIGURE 16-8 **Costs and Deaths Due to Hurricanes**

The death toll due to U.S. hurricanes has dropped from its peak in the 1920s, except for a recent rise due to Hurricane Katrina. In contrast, the damage costs from hurricanes have risen rapidly over this same period. *Note that 2011–2014 is for four years instead of ten years for other decade graphs. In just four years, the damage cost is already nearly two-thirds of the total for the previous decade.

Table 16-2	Some of the Costliest Hurricanes in the United States, 1900–2011					
RANK	HURRICANE	DATE	MAIN LOCATIONS	CATEGORY	TOTAL ESTIMATED LOSS*
1	Katrina	2005, Aug.	New Orleans, Gulf Coast	3	99.4
2	Sandy	2012, Oct.	East Coast: Florida to Maine especially New Jersey, New York at landfall	1	66.1
3	Andrew	1992, Aug.	Florida, Louisiana	5	56.4
4	Ike	2008, Oct.	Galveston, Texas Gulf Coast	2	31.7
5	Wilma	2005, Oct.	Southern Florida	3	25.2
6	Charley	2004, Aug.	Florida, Carolinas	4	19.1
7	Ivan	2004, Sept.	Alabama, Florida	3	18.1
8	Irene	2011, Aug.	Carolinas to New England	3	16.7
9	Hugo	1989, Sept.	Georgia, Carolinas, Puerto Rico, Virgin Islands	4	15.9
10	Agnes	1972, June	Florida, northeast United States	1	14.6
11	Rita	2005, Sept.	Louisiana, Texas, Florida	3	13.9
12	Camille	1969, Aug.	Mississippi, Louisiana	5	11.5
13	Frances	2004, Sept.	Florida, Georgia, Carolinas, New York, Virginia	2	11.4

*In billions of 2015 dollars. Figures from NOAA; Swiss Re Insurance Co., respectively.

Increased costs reflect the rapidly growing populations along the coasts, more construction in unsuitable locations, and more expensive buildings. The number of deaths from hurricanes decreased considerably from its peak in the 1920s until a recent rise with Hurricane Katrina. The earlier decline is at least partly due to the improved ability of scientists to predict the locations of landfall and the coordinated ability to evacuate populations at risk.

Deaths and damages from hurricanes can be caused by storm surge, high winds with flying debris, waves, torrential rainfall, and flooding. Lower-category storms, however, can be extremely destructive. Hurricane Sandy was only a Category 1 storm at landfall. Its immense diameter and broad surge caused flooding, which led to total damages that were the second highest on record (**FIGURE 16-9**).

Storm Surges

Coastal areas experience **storm surges**, also called storm tides, as the sea rises along with an incoming hurricane, as a result of low atmospheric pressure and high winds. Under normal conditions, atmospheric pressure pushes down on the water surface, keeping the sea at its usual height. When a low-pressure system moves over an area of ocean, the height of the storm surge rises with lower atmospheric pressure.

Prolonged high winds contribute to a storm surge by pushing seawater into huge mounds as high as 7.3 m and 80 to 160 km wide. The crest of a surge wave moves more or less at the speed of the storm because it is pushed ahead of the storm (**FIGURE 16-10**). The surge mound piles higher with greater

FIGURE 16-9 Hurricane Sandy

My car! Sandy submerged much of the low-lying areas of Staten Island, across from Manhattan, New York. Its broad surge and immense size caused catastrophic flooding. Wind was a lesser problem.

wind speed and *fetch*, the distance the wind travels over open water, and with shallower water (**By the Numbers 16-1: The Relationship among Surge Height, Wind Speed, Fetch, and Water Depth**).

The mound of water ahead of a storm slows down and piles up as it enters shallow water at a coast. Shallower water of the continental shelf forces the offshore volume of water into a smaller space, causing it to rise. A bay, inlet, harbor, or

FIGURE 16-10 Storm Surges

A. A 5-m storm surge during Hurricane Eloise attacks the west end of the Florida panhandle in September 1975. *Note buildings in lower right.* **B.** Katrina's storm surge at the Michoud-Entergy power plant at the I-510 bridge.

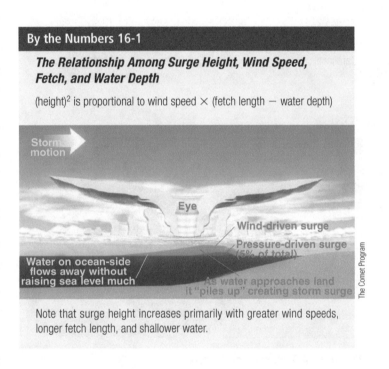

By the Numbers 16-1

The Relationship Among Surge Height, Wind Speed, Fetch, and Water Depth

(height)^2 is proportional to wind speed $\times$ (fetch length $-$ water depth)

Storm motion

Eye

Wind-driven surge

Pressure-driven surge (5% of total)

Water on ocean-side flows away without raising sea level much

As water approaches land it "piles up" creating storm surge

Note that surge height increases primarily with greater wind speeds, longer fetch length, and shallower water.

river channel that funnels the flow of water into a narrower width also causes the surge mound to rise. Even the Great Lakes are large enough to build storm surges 1 to 2 m high.

The level of surge hazard depends on a variety of factors. Because sea level rises rapidly during an incoming storm surge, low-lying coastal areas are flooded and people drown. In fact, about 90% of all deaths in tropical cyclones result from storm-surge flooding.

Contrary to many people's assumptions, the highest surge levels are not at the center of the hurricane but in its north to northeast quadrant (**FIGURE 16-11**). Because hurricanes rotate counterclockwise, winds in that "right-front" quadrant point most directly at the shore and cause the greatest surge and wave effect there. The forward movement of the storm

enhances these winds because they blow in nearly the same direction that the storm is moving. Winds south of the eye of the hurricane are moving offshore and have no onshore surge or wave effects.

The path of a hurricane compared with shore orientation also has an effect, as does the shape of a shoreline. A hurricane arriving perpendicular to the coast can lead to a higher storm surge because the whole surge mound affects the shortest length of coast. One arriving at a low angle to the coast is spread out over a greater length of shoreline but will remain along the coast for longer. Although people may feel more protected by living along the side of an inlet rather than along the open coast, the surge height in such a location may actually be higher (Figure 16-11B). For a Category 4 hurricane striking Boston Bay, Massachusetts, at high tide, the National Hurricane Center estimates that it could put the outer coastal areas under 6 m (20 ft) of water. As the surge sweeps up the inlets, depths could reach 9 m in the center of Boston.

The forward speed of a storm center can have mixed effects. A faster storm movement pushes the storm surge into a higher mound that submerges the coast in deeper water, but slow-moving hurricanes often inflict more overall damage because they remain longer over a region, dumping greater rainfall and causing more landward flooding. Similarly, a larger-diameter (often lower category) hurricane typically drops more rain because it impacts a larger area. It can also build higher waves because of longer fetch and form a wider surge mound because it pushes the water over a larger area.

The inland reach of storm-surge waters depends on a variety of factors. Most coastal areas of the southeastern United States and Gulf of Mexico stand close to sea level, so there is little to prevent surge flooding inland. However, inland progress of storm-surge waters is slowed dramatically by the presence of vegetation and dunes. Because surge waters flow inland like a broad river, they are slowed by the height of

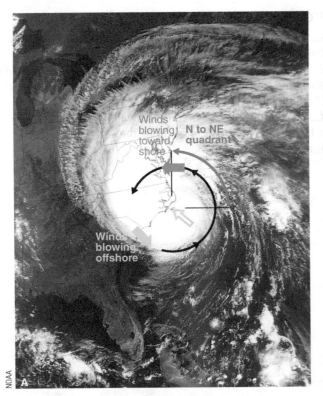

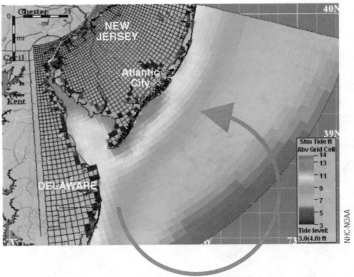

FIGURE 16-11 Wind Direction and Storm Surge

A. Winds in the northeastern quadrant of the storm are directed toward the shoreline and inflict the greatest damage; southwestern-quadrant winds are directed offshore and inflict least damage. Hurricane Isabel, 2003, is shown here at the North Carolina coast, as an example. State boundaries are superimposed on this natural color image taken at 11:50 a.m. EDT on September 18. **B.** The predicted storm surge heights for a Category 2 hurricane making landfall at Atlantic City are shown.

coastal dunes, especially those covered with brush, and by near-shore mangroves or forest.

Areas where dunes are low or absent or where vegetation has been removed permit surge waters to penetrate well inland. Thus, people who lower dunes or remove vegetation to improve their view or ease of access to the sea invite equally easy access from any significant storm surge. In many cases, pedestrian paths across dunes foster sites of erosion and overwash. A storm surge, finding a low area through dunes or an area of little or no vegetation, will flow faster through the gap, eroding it more deeply and causing more damage. This includes the destruction of buildings, roads, bridges, and piers, along with contamination of groundwater supplies with saltwater, agricultural and industrial chemicals, and sewage. Saltwater can invade aquifers and corrode buried copper electrical lines.

Buildings, bridges, and piers can be washed away by surge currents and waves. They can also float away if not well anchored to foundations or if the foundation is undermined by waves. Because larger waves feel the bottom at greater depth, they stir bottom sand and erode more deeply, undermining pilings and foundations. For this reason, most low-lying coastal homes are raised high on posts above the most frequent storm-surge heights, typically determined by local building codes (**FIGURES 16-12**). Coastal houses are often built 4 m off the ground on posts to raise them above

FIGURE 16-12 Unstable Support Posts

This house on the beach northeast of Galveston, Texas, lost its support posts during Hurricane Ike.

storm surge and wave levels. Unfortunately, storm surges can rise higher than usual, and unless set deeply into the sand, wave erosion may undermine the posts and topple houses.

Because a storm surge locally raises sea level, moves inland as a swiftly flowing current, and raises the height of wave attack, it amplifies all of the erosional aspects of a storm,

U.S. Army Corps of Engineers

Orrin Pilkey

Hyndman

FIGURE 16-13 Heavy Objects Moved by Storm Surge

A. After Hurricane Hugo, pleasure boats moored in the lagoon behind the barrier island lay stacked like fish in a basket.
B. This house, at Carolina Beach, North Carolina, was not well anchored to its slab. The surge accompanying Hurricane Fran (Category 3) in 1996 carried it off its slab and onto a road.
C. A roof, not well anchored to the walls, was lifted off a house in Crystal Beach, Texas, by the storm surge from Hurricane Ike in September 2008.

moving not only sand and sediment but also boats, cars, and houses (**FIGURE 16-13**). Anything already floating in the water will be readily moved in the direction of a surge and deposited in bays or lagoons. Houses are essentially big boxes full of air, so whole houses can be floated off their foundations and transported inland, often breaking up in the process. The nature and quality of construction are important in minimizing damage. A building's foundation should be well anchored to the ground, through deeply embedded strong piles. Floors and walls should be well anchored to the foundation and the roof securely attached to the walls with "hurricane clips."

When a surge comes at high tide, the resulting sea level rises still higher (**FIGURE 16-14**), with correspondingly greater reach inland and greater damage (**FIGURE 16-15**). Although the circumstance may be unusual, the additional load of a large surge mound on Earth's crust may be sufficient to trigger an earthquake in an area already under considerable strain—as in the Tokyo earthquake of 1923, in which 143,000 people died, mainly in the cyclone-fanned fire that followed the rupture of gas lines.

SURGES AND BARRIER ISLANDS Barrier islands are particularly vulnerable to storm surges during hurricanes. As discussed in Chapter 15, heavy erosion during storms is part of the natural

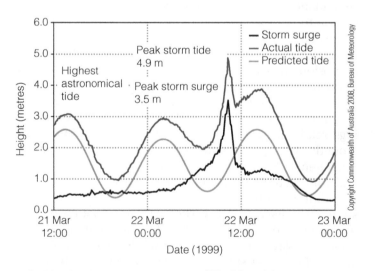

Copyright Commonwealth of Australia 2008, Bureau of Meteorology

FIGURE 16-14 Storm Surge Amplified by Tide

During Cyclone Vance, which struck Australia in March 1999, the storm surge arrived just before high tide. As a result, the surge was higher and lasted longer.

FIGURE 16-15 Surge at High Tide

Hurricane Katrina came ashore at high tide, raising the storm surge level by 0.6 m and worsening the flooding with polluted water in New Orleans.

FIGURE 16-16 Barrier Island Cut Off from Mainland

Hurricane Isabel severed Highway 12 along Hatteras Island, North Carolina. The continuation of the highway is visible directly across the breach to the right of the power poles.

process through which barrier islands are built and gradually migrate. When a storm surge moves inland over a low-lying coast, it raises water levels in lagoons and adjacent land areas, a huge volume of water that must return to the ocean after the storm passes. Much of that water reaches inland either as overwash of eroded dunes or through tidal inlets between segments of barrier islands. That overwash carries sand across an island, helping build island height, thereby making it less susceptible to erosion from later storms. Sand carried landward also builds the landward edge of the island as the storm removes sand from the seaward edge—so the island moves landward.

When sea level drops after the storm, that higher water level rushes back offshore across the already eroded beaches and through the widened inlets and breaches. New breaches can form in low or narrow areas of a barrier island depending on the storm surge height, wave size and direction, forward storm speed, and other factors. Other existing breaches can close.

When barrier islands are developed, buildings get in the way of the natural cycle of erosion and deposition that governs them. Breaches caused by storm surges can destroy bridges and access roads, leaving no road access to the mainland and isolating homes and towns (**FIGURE 16-16**).

Where a new breach severs the only road, government agencies are pressured to close the breach. However, closing such breaches decreases the number of tidal breaches along a barrier island, upsetting the natural equilibrium of the barrier island–lagoon system. Closing a breach can decrease the tidal circulation between lagoon and ocean, with a variety of negative consequences. Much of the erosion that we see after a hurricane occurs during that surge outwash, or "ebb tide." Sand carried back offshore can end up in deeper water below wave-base depth where it is lost to the beach system.

During Hurricane Katrina, barrier islands were severely eroded. For example, almost all of the homes on the beach side of the road running along Dauphin Island, Alabama, east of the hurricane's eye, disappeared, leaving only posts (**FIGURE 16-17**). In some instances, even the posts were snapped off. A large volume of sand eroded from the beach swept across the island during the storm surge.

Dauphin Island is an example of a barren offshore sand bar sprinkled with expensive homes repeatedly protected with federal taxpayer dollars; the fix is often temporary. An artificial sand dune destroyed by Hurricane Georges in 1998 was rebuilt in 2000 for $1 million, only to be washed away by Tropical Storm Isidore two years later. Again rebuilt, Hurricane Katrina in 2005 washed away not only the berm but also many of the houses. A new sand berm 3 m high and 5.7 km long was completed in May 2007 at a cost of $4 million; it was severed two weeks later by waves during a high tide.

Waves and Wave Damage

The higher waves generated by a hurricane impact the coast with much more energy. They are able to move sand on a lower slope than smaller waves. They also stir up sand to greater water depths both offshore and onshore, eroding beaches and moving sand farther offshore to form more gently sloping beaches (**FIGURE 16-18**). Because dunes dissipate

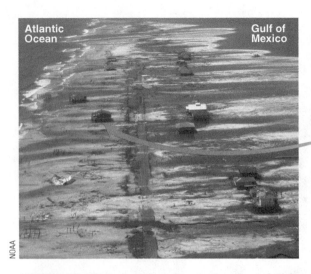

FIGURE 16-17 Severe Erosion to Barrier Islands

During Hurricane Katrina, Dauphin Island, Alabama, lost most of the three rows of homes on its beach; all that remains are some of the posts on which they once stood. A portion of the sand eroded from the island was carried to the right into Mississippi Sound.

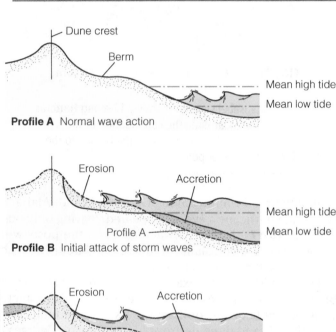

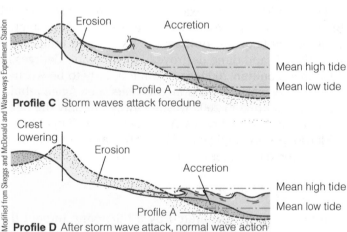

FIGURE 16-18 Beach Erosion During a Storm

This sequence of diagrams shows the effects of a storm wave attack on a beach and dune. Note the progressive erosion of original shore Profile A, in part because of waves on higher sea level of storm surge.

FIGURE 16-19 Homes on Posts

Hurricane Dennis on September 1, 1999, eroded the beach out from under these beachfront homes in Kitty Hawk, North Carolina.

wave energy and protect homes placed behind them, erosion of the sand from beaches and dunes frequently undermines any structures there (**FIGURE 16-19**). Loose debris such as boards, branches, logs, and propane tanks carried by waves act as battering rams against buildings, amplifying damage.

The level of damage at a particular site can be dramatic, as shown in photographs taken before and after individual hurricanes (**FIGURE 16-20**). Many beachfront homes are completely destroyed when the sand beneath them is washed offshore. Others nearby sustain major wind and wave damage (**FIGURE 16-21**).

The width and slope of the continental shelf has a significant effect on wave damage. On a wide, gently sloping

FIGURE 16-20 Before and after a Hurricane

These photos show part of North Topsail Beach, North Carolina, **A.** after Hurricane Bertha on July 16, 1996, and **B.** after Hurricane Fran on September 7, 1996. The area is hardly recognizable. Compare individual houses in the second photo. House 1 is pushed off its foundation and houses to the right are demolished. House 2 is pushed off its foundation into a new inlet to its left. House 3 is now gone.

continental shelf offshore, the waves drag on the bottom and stir up sand; this uses up more wave energy before the waves reach land, decreasing the damage. Large amounts of moving sand on a shallow bottom offshore reduce the wave energy available for coastal erosion. On a narrow continental shelf, as in the Outer Banks of North Carolina, or a steeper slope offshore, larger waves maintain their energy as they approach closer to shore, thereby causing more damage (**Case in Point:** Landward Migration of a Barrier Island Coast—North Carolina Outer Banks, p. 469).

It might seem that low-lying, subtropical islands such as Grand Cayman in the Caribbean and Guam in the western Pacific would be especially vulnerable to storm surge and wave damage. However, fringing coral reefs force storm waves to break well offshore, minimizing those effects. Water depths offshore of the reefs tend to be deep, and absence of a wide continental shelf prevents the storm surge from rising as high. However, those characteristics do not make these islands safe. Because many of them are low-lying, they are vulnerable to severe wind damage and storm surge.

FIGURE 16-21 Wave Damage to Homes

A. Hurricane Alicia (Category 3) in October 1983 removed the beach sand from under the concrete parking area below this house on the Texas Gulf Coast, leaving it suspended on the pilings that raised the house above surge level. It also peeled off much of the roof and the front half of the house.

B. Hurricane Ike (high Category 2) in September 2008 destroyed most houses near this beach along the Texas Gulf Coast. The house at top, just left of center, survived well because its roof slopes in four directions, it has few windows, and it sits on tall, sturdy posts.

Reefs damaged in storms often regrow, but those near coastal cities or major resort areas are vulnerable to pollution and erosion, which is exacerbated by beach construction, seawalls, and sand mining. Coral bleaching from higher water temperatures and increasing seawater acidity due to global warming also take a toll. Damaged reefs no longer provide the vital, natural coastal protection they once did.

Winds and Wind Damage

Although flooding compounded by water rise from storm surge can be catastrophic, wind damage is often ten times as great. Wind velocity has a major effect on wave height. Wind can also wreak weak buildings and blow down trees, power lines, and signs. It can fan fires and destroy crops; blow in windows, doors, and walls; and lift the roofs off houses. Wind damage is greatly magnified by flying debris (**FIGURE 16-22**). Significant harm also results from rain entering a wind-damaged structure.

Wind in some hurricanes does far more damage to buildings than flooding. To understand the force of wind, the lowest-level hurricane winds at 119 km/h apply approximately 73 kg/m^2 of pressure on the wall of a building. Thus, a force of 1360 kg (3000 lbs) would press against a wall 2.4 m high and 7.6 m long. If the wind speed doubled, the forces would be four times as strong; so in a Category 4 hurricane at 240 km/h, the force on the same wall would be 5400 kg. Reduced pressure caused by the same winds on the downwind side of a building adds to the problem because the winds pull against the wall (**FIGURE 16-23**).

Roofs often fail before walls because of additional factors. The largest wind forces are caused by suction.

Where a roof slopes toward the wind, the air is forced up and over the roof, lifting it in the same way that air flowing over an arched airplane wing lifts the plane. A steeper-sloped roof actually performs better in high wind than one sloped more like an airplane wing. The lifting forces under overhanging eaves tend to cause failure there first. Roofs that slope to all sides without overhangs deflect the wind best. Larger roof spans are more vulnerable.

Roof material also makes a difference; shingles fare poorly because they can easily blow off during high winds. Metal, slate, or tile roofs are good; a single-membrane roof is better; and a flat concrete-tile roof is much better. Concrete or steel beams supporting roofs are better than wood. An extra nail or two in each shingle can save a roof.

Because of pressure differences inside and outside a building, sidewalls, windows, and roofs are commonly sucked out rather than blown in. When a house is breached during a storm by flying debris, for example, pressure increases inside the building, exerting outward force on windows, walls, and the roof. If a garage door fails, the wind can inflate a house and blow its roof off and windows out.

Shutters or plywood well anchored over windows and exterior doors helps protect the integrity of a house more than anything else because those are the vulnerable points of entry. Impact resistance is also important for windows and doors. Tempered and laminated glass windows are significantly better than ordinary window glass. When windows or exterior doors make up more than 50% of a wall, it is especially vulnerable. Skylights are especially susceptible to penetration. Double-wide garage doors, particularly if the doors open overhead, are unusually susceptible because tracks holding doors may fail or the wind may force a door out of its track.

Evelyn Shanahan, NOAA

FEMA

FIGURE 16-22 Wind-Borne Debris

A. Yes, hurricane winds can be strong! The danger of wind-blown debris is vividly illustrated by a palm tree near Miami, which was impaled by a sheet of plywood during Hurricane Andrew in August 1992. This may have been caused by an embedded tornado.

B. A 14-cm-diameter branch impaled the side of a house in Punta Gorda Isles during Hurricane Charley, August 2004.

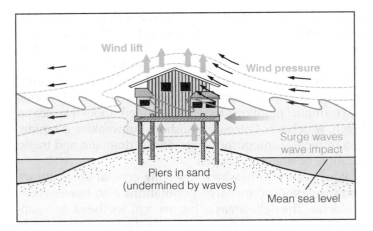

FIGURE 16-23 Wind Forces on a Building

Wind pressure exerts force on the windward side of a building, while simultaneously pulling upward on the roof. At the same time, reduced pressure on the sheltered side also pulls on the house. Wave and surge effects are also shown.

Much of the damage in homes from hurricanes, tornadoes, and earthquakes arises from the repeated flexing of wood structures nailed together. Nails pull out, or their heads pull through walls or roof sheathing, or they shear off. Recent changes in nail design have specifically addressed those failings: barbed rings around the nail shaft, larger nail heads, and higher-strength carbon-steel alloy shafts. Such nails can double resistance to high-wind damage and add 50% to resistance to hurricane damage.

Unreinforced brick chimneys fail in large numbers, even when situated against walls, because of poor attachment to the walls. Brick, stone, or reinforced-concrete block walls, however, are much stronger than wood-sheathed walls. Because much damage comes from flying debris, gravel or roof-mounted objects on nearby buildings are dangerous.

Rainfall and Flooding

Hurricanes cause significant flooding because of storm surge as well as heavy rainfall. Heavy rain and flooding during a hurricane can wash out structures, drown people, contaminate water supplies, and trigger landslides.

The characteristics of a particular storm, as well as the local topography, determine the severity of flooding. As mentioned earlier, Category 1 or 2 hurricanes or even smaller tropical storms can actually cause more flooding damage than larger storms. Lower-category, large-diameter, or slow-moving hurricanes spend much more time over an area and typically drop large amounts of rain on large parts of a drainage basin, and for a longer period, so they cause more extensive and more prolonged flooding. Hurricane Agnes, only a tropical storm when it came back on land in Pennsylvania and New York in 1972, spread over a diameter of 1600 km and provided the largest rainfall on record in that area.

In another example, late on June 5, 2001, Tropical Storm Allison drifted slowly inland to 200 km north of Houston and, over the next 36 hours, dropped heavy rain on southeastern Texas and adjacent Louisiana. Strengthening of a high pressure system over New Mexico caused Allison to loop east, then southwest to Houston a second time. In total, the storm dumped almost 1 m of rain on the Port of Houston in less than one week, in an area that has notoriously poor drainage (**FIGURE 16-24**). Damages in the Houston vicinity alone reached $4.88 billion. The Texas Medical Center was especially hard hit. Its belowground floors flooded. Backup generators were aboveground, but unfortunately switches between the two systems were belowground and destroyed by the flooding.

More recently, Hurricane Ike in 2008, a strong Category 2 storm, inflicted $29.2 billion in damages, making it the third costliest storm ever to affect the United States (**Case in Point: Extreme Effect of a Medium-Strength Hurricane on a Built-Up Barrier Island—Hurrican Ike, Galveston, Texas, 2008, p. 466**).

FIGURE 16-24 Flooding Damage
A. Tropical Storm Allison flooding in Houston.

B. Water flowing into a below-street mall.

Local topography also influences flooding severity. Several centimeters of rain dumped over a few hours collect rapidly and run off slowly because of the gentle slopes of coastal plains. Draining downslope, the water accumulates to even greater depths to cause catastrophic floods in low areas. A good example is Hurricane Floyd in 1999, following on the heels of Hurricane Dennis, which had already saturated the ground.

Deaths

Nearly 60% of the people who die in hurricanes drown because of river floods near the coast. One-quarter of hurricane deaths are of people who drown in their cars or while trying to abandon them during floods. The heaviest rainfalls are from slower-moving storms and those with larger diameters. The National Hurricane Center predicts the total rainfall in inches by dividing 100 by the forward speed of the storm in miles per hour. For example, a storm approaching at 20 mph would be expected to have 5 in of rain (100/20 = 5 in).

Deaths in hurricanes depend not only on the strength of a hurricane, through the wind velocity and surge height, but also on patterns of development and the size and education of the local population. Obvious problems include buildings too close to the coast, buildings too close to sea level, and buildings, such as mobile homes, that are too weakly constructed. Less obvious is people's lack of awareness of the wide range of hazards involved in such events. Finally, a large population in the path of a storm cannot evacuate quickly or efficiently, especially if an accident blocks a heavily used highway.

Hurricanes have much higher rates of deaths in poor countries, such as in the Caribbean and Central America. Poverty, culture, and disastrous land use practices put such populations at a much higher risk of hurricane-related damages. Much of Central America is mountainous, with fertile valley bottoms that are mostly controlled by large corporate farms. Many people have big families that survive on little food and marginal shelter; unable to afford land in the valley bottoms, they decimate the forests for building materials and fuel for cooking, leaving the slopes vulnerable to frequent landslides. Others migrate in search of work to towns in valley bottoms where they crowd into marginal conditions on floodplains close to rivers subject to flash floods and torrential mudflows fed from denuded slopes. Hurricane Mitch, which was Category 5 offshore but weakened rapidly to a tropical storm onshore, provides a dramatic and tragic example of what can happen.

Poor countries in which low-lying coastal areas attract and provide food for large populations also have severe problems. The delta areas of big rivers in Southeast Asia support large numbers of people because the land is kept fertile by the frequently flooding rivers. As those populations grow, heavy land use in the drainage basins of the rivers leads to more frequent and higher floods from both rivers and tropical storm surges (**Case in Point:** Floods, Rejection of Foreign Help, and a Tragic Death Toll in an Extremely Poor Country—Myanmar [Burma] Cyclone, 2008, p. 474).

The deadliest Atlantic hurricanes have affected the Caribbean islands with 64% of total storm losses. Before Mitch in 1998, Mexico and Central America accounted for 15% of total deaths due to storms, with the U.S. mainland at 20 or 21%. Major catastrophes in Central America include large death tolls:

- About 10,000 in Nicaragua and Honduras from Hurricane Mitch from October 22 to November 5, 1998

- Approximately 8000 in Honduras from Hurricane Fifi, September 15 to 19, 1974

- 7192 in Haiti and Cuba from Hurricane Flora, September 30 to October 8, 1963

- More than 6000 at Pointe-a-Pitre Bay, Haiti, September 6, 1776

The largest numbers of deaths in U.S. hurricanes from 1900 to 2009 are shown in **Table 16-3**. Note that Katrina is

Table 16-3	Hurricane Deaths in the United States, 1900–2011			
RANK	HURRICANE LOCATION	YEAR	CATEGORY	DEATHS
1	Galveston, Texas	1900	4	8000–12000
2	SE Florida	1928	4	3000
3	New Orleans, Gulf Coast (Katrina)	2005	3	1836
4	Louisiana	1893	4	2000
5	South Carolina, Georgia	1893	3	1000–2000
6	South Carolina, Georgia	1881	2	700
7	SW Louisiana, N Texas (Audrey)	1957	4	416
8	Florida Keys	1935	5	408
9	Louisiana	1856	4	400
10	NE United States	1944	3	390

the only recent hurricane to make the list, due to improved forecasting and warning systems, better transportation facilities, and the lower incidence of hurricane activity between 1970 and 1995.

Social and Economic Impacts

Because hurricane damage can spread over large regions, their social and economic impact can be significant. Disruptions from Katrina shut down 95% of petroleum output from the Gulf of Mexico, the biggest domestic source. Pipelines inland from the area shut down for lack of power. Several major petroleum refineries in Katrina's path were closed, almost all workers from offshore drilling platforms were evacuated, and 29 of the platforms were destroyed (**FIGURE 16-25**). When the giant anchors that held some platforms in place broke loose, they dragged, twisted, and sometimes severed seafloor pipelines that carried the crude oil to the mainland. The same thing happened with Hurricane Ivan. As a result, gasoline prices rose dramatically throughout the United States.

Dramatic disruptions in an unexpected evacuation sometimes separate immediate family members who have not made contingency plans to find each other in such a circumstance. Family members in other states can find themselves with no means to contact their kin in stricken areas because communication via telephone, email, and even snail mail may be impossible due to storm destruction.

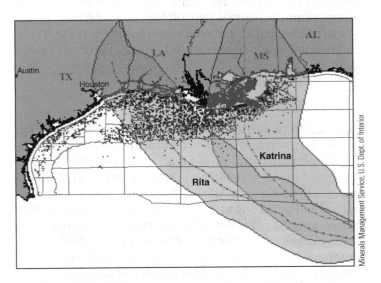

FIGURE 16-25 Oil Platforms in the Paths of Hurricanes
There are 6357 oil platforms located in the Gulf of Mexico within 230 km of the U.S. coast. The 2005 hurricane paths are shown as pink and orange lines. The colored shading indicates areas that experienced hurricane-force winds from Katrina (brown) and Rita (pink). Note that the hurricanes veered and tracked clockwise.

Specific plans, such as having a contact relative in another state, should be made before a major storm's arrival. Otherwise, local authorities or national services, such as those provided through the American Red Cross, may not be able to help people reach one another.

Many people in hurricane- and related flood-prone areas lack flood insurance. For a large proportion, the financial impact of a major event is long-term, if not permanent. Job losses are widespread. Companies cannot reopen because utilities are damaged for weeks or months. Without business, they have no income and cannot afford to pay their regular workers. Those workers need the income not only for day-to-day expenditures but also for ongoing expenses such as health insurance and housing. Many move to other areas for work and may never return.

By September 1, 2006, one year after Katrina, $44 billion of federal money had been used to pay flood-insurance claims and accommodate immediate needs after the storm—paying federal workers and providing temporary housing for victims and operating funds to local governments. Only a small amount was used to repair water and sewer lines and electric power grids. This resulted in a major dilemma—the government won't spend on infrastructure until it is sure people are going to move back, but people cannot move back until there is usable water, sewer, and power.

One year after the storm, thousands of New Orleans residents said that they would rebuild their damaged homes, even though their property remains in flood zones. Thousands of others were either uncertain or decided to not return. By late 2009, the city's population had slowly climbed back to 72% of its original 462,000. A significant percentage of the present population resides in the 20% of the city that was not flooded. In one of the poorest and hardest-hit neighborhoods, the Lower 9th Ward, however, fewer than 2000 out of more than 13,000 people have returned. On September 1, 2006, the New Orleans City Council ruled that any house being rebuilt had to be raised 3 ft (almost 1 m) aboveground level; unfortunately, many areas were flooded with 5 to 6 m of water. A $10 billion upgrade of the more than 500 km of levees around New Orleans was essentially complete by 2011, but some temporary flood gates and pumps still remained to be placed. Even now the levee system does not provide much more protection than in 2005. The effect of Hurricane Katrina on New Orleans provides many lessons in this regard (**Case in Point:** City Drowns in Spite of Levees—Hurricane Katrina, 2005, p. 460).

Hurricane Prediction and Planning

One reason that there have been fewer deaths from hurricanes over time is the great success in predicting and planning for hurricanes.

Hurricane Watches and Warnings

Storms are monitored by weather satellites and "hurricane hunter" aircraft that make daily flights into storms to collect data on winds and atmospheric pressures. Within two to four days of expected landfall, they drop dropsonde sensing instruments into a storm from 9- to 12-km altitudes, their fall slowed by small parachutes. These transmit wind, temperature, pressure, and humidity information. The path of the storm is commonly controlled by nearby high- and low-pressure systems.

Hurricane predictions include time of arrival, location, and magnitude of the event. The National Hurricane Warning Center tries to give 12 hours of warning of the hurricane path. Alerts come in two stages. A **hurricane watch** indicates that "A hurricane is possible in the watch area within 36 hours." A **hurricane warning** is provided when "A hurricane is expected in the warning area within 24 hours." On average, 640 km (400 mi) of coastline is warned of hurricane landfall within 24 hours. Of that, 200 km (125 mi) may actually be strongly affected by a storm. Thus, some $275 million in costs are borne by people ultimately not in the storm's path. Clearly, more accurate forecasting could save lives and significant evacuation costs.

One future prospect for predicting hurricane behavior is to monitor thermal anomalies in the ocean with satellite sensors. We know that a hurricane's energy is drawn from the heat in tropical ocean water. A newly recognized effect on hurricane strength is the presence of large eddies of warm water 100 or more kilometers across, which can apparently spin off larger oceanic currents and boost the energy of hurricanes passing over them, sometimes dramatically (compare FIGURE 16-5).

Uncertainty in Hurricane Prediction

Hurricane prediction, like any weather prediction, involves significant uncertainty. Hurricanes can quickly change paths, or increase or decrease in intensity (**FIGURE 16-26**). Although forecasters can provide some indication of a hurricane's time of arrival, the exact landfall location and storm strength is less predictable. The uncertainty in storm strength was emphasized in August 2011 when many in the news media predicted dire consequences of extreme winds and flooding from Hurricane Irene. Widespread states of emergency and mandatory evacuations were declared but winds were much weaker than expected. The storm maintained a nearly straight northeastward track up the east coast, with landfall as a Category 1 hurricane in North Carolina, then weakened to a tropical storm through New Jersey, New England, and Quebec.

The National Hurricane Center's current 24-hour lateral forecast error for the path of a hurricane is 80 km. Even a small shift in a hurricane's path could make a dramatic difference to its impact on coastal communities.

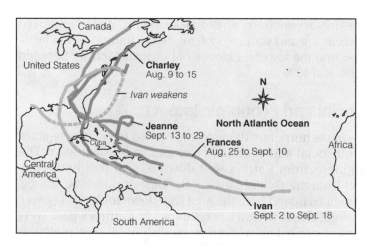

FIGURE 16-26 Unpredictable Behavior of Hurricanes
The tracks of four major hurricanes that decimated Florida in August and September 2004 show the difficulty of predicting the paths of hurricanes.

A shift of as little as 80 km in the Galveston hurricane in 1900 could have resulted in far fewer deaths. A shift of 1992's Hurricane Andrew just 32 km to the north near Miami would have caused two or three times as much damage. Much less monetary loss would have resulted from a strike 64 km to the south in the minimally populated Florida Keys.

Smaller storms can also quickly gain energy and become powerful hurricanes. Hurricane Andrew developed as a thunderstorm in warm waters off the coast of Africa, gradually strengthening as it moved westward in the trade winds, and becoming one of only a few Category 5 storms to make landfall.

Planning for Hurricanes

Planning for a hurricane should be among the first things people do when they move to a hurricane-prone coast. They should prepare their house for all of the potential hazards discussed in this chapter, and do so in anticipation of having little or no warning. Many of these preparations take days and should be an ongoing effort. If any cannot be done long before, specific plans should be in place to do them quickly. Don't assume that stores or workers will be available after a hurricane watch or warning has been announced (see the Hurricane Survival Guide at the end of this chapter).

Evacuation

Early warning in the United States, using weather satellites, allows people to either evacuate by road or take refuge in reinforced high-rise buildings, such as those in Miami Beach. Because forecasters refer to the statistical likelihood that a storm will strike a given length of coast and there is significant uncertainty in the direction of the storm track, many

FIGURE 16-27 Evacuation

A. One-way traffic evacuation near Beaumont, Texas, before Gustav. Surges quickly cover escape routes (inset).

B. Some traffic still tried to evacuate after Tropical Storm Francis in 1999 began to move in.

more people are warned to evacuate than those in the final path of the storm. The costs of evacuations are also large. The usual estimate is roughly $1 million per coastal mile evacuated—exclusive of any damages from the hurricane. This is an unfortunate but necessary cost for the safety of coastal populations.

One problem is convincing people to evacuate promptly. With recent heavy development in coastal areas, most of the 40 million people living in hurricane-prone localities have never been involved in an evacuation. Large seasonal tourist populations make matters even worse. Municipalities view hurricane publicity as bad for tourism and property investment, so they sometimes delay warnings or minimize the danger.

Some people who live through one hurricane do not leave, believing that if they have stayed before, they can do it again. Many delay their evacuation because of inconvenience and cost or because they spend time purchasing and installing materials to help protect their homes from a storm. Others stay because they feel their homes are strong enough to survive. Still others delay evacuation until the last minute, thinking it will take them only an hour or two to drive inland, or to minimize lodging costs, or hoping the storm will miss their part of the coast.

Most people believe it will take less than one day to evacuate, but studies show that because of traffic jams and related problems, most evacuations take up to 30 hours. Only single two-lane bridges to the mainland serve most barrier islands. Those bridges and roads become snarled with traffic, and accidents can cause further dangerous delays.

The huge increase in population along the coasts in recent years has outgrown highway capacity. The roads are not able to cope with evacuating populations. Changing freeways to single-direction traffic away from a coast merely moves bottlenecks inland. It also creates severe safety problems at off-ramps, where people traveling against the normal direction of traffic try to exit for lodging or fuel (**FIGURE 16-27A**). If people fail to evacuate before a storm arrives, high surge levels can flood roads and freeways under meters of water, making escape impossible (**FIGURE 16-27B**). Fallen trees, power lines, and other debris can block roads, or a surge may cover roadways, making escape impossible and rescue difficult.

In part to combat these problems with evacuation, the latest approach is to move people 32 km (20 mi) from the coast, beyond the limit of surge, to temporary protection, not hundreds of kilometers from the coast, as was done in the past. Even once residents successfully evacuate, however, lodging is completely inadequate to cope with an entire coastal population.

The most difficult area to evacuate in the United States is the Florida peninsula, which is only a few meters above sea level and less than 200 km across. A hurricane can easily cross the whole state without losing much strength. Population growth has been rapid, especially along the coasts, and it is difficult to predict the precise path of a hurricane.

When you hear a Hurricane Warning, consider the American Red Cross recommendations, included in the **Hurricane Survival Guide** at the end of the chapter.

Managing Future Damages

Governments, communities, and insurance companies are finally reacting to damages and costs from hurricanes. They are beginning to push for "disaster-resistant communities" that are less vulnerable and incur reduced costs from coastal hazards. Policies used include land-use planning, building codes, incentives, taxation, and insurance. However, because tourism is the largest source of income for most coastal areas, most governing bodies are reluctant either to publicize their vulnerability or to place many restrictions on anything related to tourist income.

Natural Protections

Beachfront sand dunes absorb the energy of waves and advancing storm surges and can reduce damages to coastal communities. Cypress forests or thickets of mangroves limit the shoreward advance of waves and dramatically slow a surge's landward advance (**FIGURE 16-28**). Unfortunately, many of these areas have been modified or damaged over the years through development. Some communities have now recognized the importance of these areas and are working to protect or restore them.

Human alterations of dunes and mangrove stands that would increase potential flood damage are prohibited. Even walking on dunes is generally prohibited because it disturbs the sand and any vegetation covering it, thereby making the dune more susceptible to erosion by both wind and waves. To protect the dunes, people build elevated walkways over dunes to beaches (see Figure 15-20A).

The damage inflicted on New Orleans by Hurricane Katrina was likely increased because decades of dredging channels for shipping and the emplacement of onshore oil-drilling sites introduced saltwater and killed off nearby cypress forests and marshes, which had previously provided natural protection from storm surges and waves. At the same time, the delta has been sinking because the levees of the Mississippi River and its tributaries have deprived it of sediment that was once replenished annually. A four-year, $500 million federal grant to restore coastal wetlands that was awarded in July 2005 should help somewhat in the future.

Building Codes

After major hurricanes obliterate a stretch of coast, you would think that people would have second thoughts about rebuilding in the same vulnerable locations. Unfortunately, the opposite seems to be true (**FIGURE 16-29**). Developers immediately descend on devastated areas and snap up coastal building sites, often paying as much or more than the value of the property prior to its destruction. Many want to build large hotels or condominium complexes. Some have been waiting to build their dream beachfront property; others are merely real estate speculators betting on a rapid comeback. Waterfront lots in the Florida panhandle, for example, go for over a million dollars; those in formerly depressed areas several blocks back from the beach go for more than $100,000.

Most people agree that buildings in hurricane-prone areas should be built stronger, and in some areas legislation has been passed to regulate such development. Studies by the research center of the National Association of Home Builders indicate that homes can be made much more resistant to hurricane damage at a cost of only a 1.8 to 3.7% increase in the total sales price of a property.

In the 1980s, the state of Florida designated a coastal construction control line (CCCL), seaward of which habitable structures were permitted only with adherence to certain standards of land use and building construction. The CCCL designates the zone that is subject to flooding, erosion, and related impacts during a 100-year storm. Hurricane Opal on October 4, 1995, provided a Category 3 test of the system. None of the 576 major habitable structures built to CCCL standards and seaward of the CCCL suffered substantial damage. Of the 1366 preexisting structures in that zone, 768 (56%) were substantially damaged.

Unfortunately, developers, builders, local governments, and many members of the public often oppose such increased standards. Their argument is that these regulations unnecessarily increase housing costs and limit economic development. Often increased standards fail to pass at all levels of government until a major disaster and huge losses make the need obvious to everyone. Even then, the issue of

FIGURE 16-28 Coastal Forests Provide Storm Protection

A. Cypress forests near New Orleans, and **B.** mangroves along southeastern coasts present a formidable barrier to an advancing storm surge.

FIGURE 16-29 Coastal Development

A. A few months after Hurricane Ike completely annihilated most of the homes on the Bolivar Peninsula, northeast of Galveston, Texas, new houses were being built on the same beach with no protective dunes.

B. Many of the hotels along Cancún's main beach sustained major damage after Hurricane Wilma struck in 2005, but the entire area has since been rebuilt.

whether the state or counties are required to pay for enforcement of new regulations often thwarts new laws. When adequate building standards are not enacted and enforced, the general public is eventually forced to pay for the unnecessary level of damage. People's federal and local taxes could be lower if it were not for such unnecessary costs.

With no national building code, state and local governments are responsible for enforcing their own codes. Some southeastern states do not have universal building codes. When Hurricane Hugo hit South Carolina in 1989, half of the area of the state, including parts of the coast, had no building codes or enforcement at all. This lack of building regulations results in much more widespread and severe destruction of property.

Flood Insurance

Flood insurance is one way to reduce monetary damages from hurricanes and potentially influence people's behavior to keep them from settling in inappropriate areas. Significant destruction of homes and other structures in historic hurricanes has prompted some states such as North Carolina to dictate a setback line based on the probability of coastal flooding. Seaward of this line, insurance companies will not insure a building against wave damage. In spite of many challenges, the courts have so far upheld the building prohibition.

The National Flood Insurance Program (NFIP) requires federal mapping of areas subject to both river and coastal floods. The purpose is to implement a Flood Insurance Rate Map (FIRM) provided by the Federal Emergency Management Agency (FEMA). The designated special flood-hazard areas have a 1% chance of being flooded in any given year, that is, they are comparable to the 100-year floodplain of streams.

Coastal communities wishing to participate in the program must use maps of these flood-hazard areas when making development decisions. Coastal areas subject to significant wave action in addition to flooding are more vulnerable to damage. Significant waves are considered those higher than one meter. Therefore, construction standards are more stringent in such coastal zones. National flood insurance premiums are higher for those who live in more vulnerable areas.

Flood insurance costs in 2007 for a single-family dwelling built after 1981, with no basement, were generally $1520 per year for a $250,000 replacement value. Covered damage to basements often includes only heating and other equipment that serves the living area. For a home in a high-risk coastal flood-risk zone, the rates rise to $3275. Contents coverage adds an additional $2000, and insurance is capped at $250,000 for a building. Land loss is not covered. As with stream flooding, there is a 30-day waiting period after purchase before the insurance takes effect. Given that hurricanes are not predictable for a specific location for more than a day or two in advance, it is prudent to maintain flood insurance if you live in an area of possible storm surge or coastal river flooding.

Insurance can include coverage for lost business, but some business owners do not file claims for fear their rates will go up. Small businesses often lack insurance and may be forced to close if they lack the resources to remain open and face weeks or months of lost income. For others, such as building supply companies, business typically increases because of a storm.

Some undeveloped areas were protected by the Coastal Barrier Resources Act of 1982, which prohibited federal incentives to development and prohibited the issuance of new flood insurance coverage. However, in areas not covered by that act, flood insurance remained available for elevated structures located as far seaward as the mean high-water line, regardless of local erosion rates.

Rebuilding after a flood may require higher building standards, including moving structures to higher elevations. Coastal building standards for the NFIP require the following:

- All new construction must be landward of mean high tide.

- All new construction and major improvements must be elevated on piles so that the lowest floor is above the base flood elevation for a 1% chance of flooding in any year.

- Areas below the lowest floor must be open or have breakaway walls. Fill for structural support is prohibited.

Raising the lowest floor above a 100-year flood level does nothing, of course, to prevent erosion. Piers may be undercut, and an eroding beach will eventually move landward from under a structure, causing its collapse.

With global warming and rising sea levels, coastal landforms including beaches and barrier islands will continue to migrate landward; it would seem wise to move coastal homes, roads, and railroads back farther from the coast. The alternative is to suffer increased damage and destruction along the coasts or require extremely expensive beach replenishment with disappearing supplies of replenishment sand.

In the last several years, Congress has attempted to reform FEMA's flood insurance program, which is now $23 billion in debt, but they face significant political challenges. In many cases, rebuilding costs have been paid on repetitively damaged buildings. In the last 32 years, the program has paid $12 billion on repetitive-loss properties. In most of those cases, politicians have undercut efforts to curtail such waste.

In 2012, Congress moved to phase out federally subsidized flood insurance rates, which would have led to flood insurance premiums increasing by 20 to 25% year for those in high-risk zones, especially along the coast of New Jersey, Long Island, and nearby areas. However, after heavy criticism and lobbying by realtors and builders, rate hikes were delayed for a few years.

Homeowners Insurance

After the four disastrous hurricanes in 2004, some small insurance companies left Florida entirely; several large companies stopped writing new policies or dropped certain policyholders. Almost all insurance companies significantly increased the cost of coverage after the storms. Recent major hurricanes including Ivan, Katrina, Rita, and Ike have led insurance companies to dramatically increase premiums for people living in susceptible areas, impose deductibles as high as $20,000 (before a company pays anything), and cap replacement and rebuilding costs. Many now exclude wind damages from policies and require additional premiums for wind coverage. In some areas of south Florida, insurance

rates were three to four times higher in 2007 than in 2005; in South Carolina, they were seven times higher.

In the Mid-Atlantic States and New England, insurance companies have canceled about one million homeowners' policies since 2004. Although most people have found other coverage, it comes with higher rates and larger deductibles. Some companies no longer provide insurance on Long Island and in New York City because of hurricane risk. Long Island protrudes directly into common hurricane paths. Exclusive summer homes on the southeast corner of Nantucket Island, Massachusetts, with an average 2007 price of $1.8 million, line a sea cliff that is rapidly eroding, endangering the homes. To slow the erosion, a group of wealthy homeowners put up $23 million of their own money to replenish the beach using sand dredged from 2.5 km offshore.

A few years ago, governments in states such as Florida, Mississippi, and Louisiana created state-run insurance programs to cover homeowners who are unable to get insurance from private companies; private insurers in the state are billed part of the cost, which is then passed on to policyholders. Now, given the severe damage from the hurricanes of 2004 and 2005, the states feel the need to dramatically raise their premiums—to the forceful complaints of residents. Surprisingly, Florida's state-run insurance pool still provides coverage to people building expensive coastal homes on sites destroyed in recent hurricanes. A recent actuarial (actual cost) analysis indicated that state-run rates should be increased by an average of 80%—and in some areas more than double that. Politicians, however, resist raising premiums on state-run policies to avoid alienating policyholder voters. Thus, if the state has insufficient funds to cover losses, either the policyholder isn't paid for a loss or the state puts pressure on Congress to step in and provide the remainder from federal funds—that is, from taxes paid by everyone in the country. The overall problem is that too many people live in dangerous coastal areas.

Costs paid by homeowners insurance commonly include physical damage to homes caused directly by winds, flying debris, falling trees, and rain penetration after wind damage. In fringe areas of a storm surge, insurance companies often argue that water came from the surge, which is not covered, rather than from wind followed by rain.

Flood insurance is not included as part of homeowners insurance, but it is available through the NFIP. In low-lying areas, flood insurance is expensive and capped at $250,000 per home. In New Orleans and its vicinity, more than half of the eligible homes were not insured for floods prior to Hurricane Katrina, either because people could not afford the additional expense or didn't believe they would be flooded. In Mississippi, less than 20% of eligible homes had flood insurance. Since a large part of the damage in low-lying areas was from storm-surge flooding, homeowners insurance did not cover the damage. Flooded homes as far as 3 km inland lacked flood insurance. Many were not in the floodplain zone designated after Hurricane Camille in 1969, which came ashore in Mississippi as a Category 5 storm, with a 7-m surge.

Following Katrina, arson fires sprang up in a number of places, leading to the suspicion that some people may have set fire to their own flood-damaged homes, hoping that fire insurance would pay for rebuilding.

Extratropical Cyclones and Nor'easters

Extratropical cyclones are cold-weather storms that behave much like hurricanes and can cause as much damage. **Nor'easters** are extratropical cyclones that strike the northeastern coast, especially New England and Nova Scotia. They differ from hurricanes in several ways:

- Named for the direction from which their winds come, nor'easters bring heavy rain and often heavy snowfall. Benjamin Franklin noted that precipitation begins in the south and spreads northward along the coast.

- They are most common from October through April, especially February (rather than late summer for hurricanes).

- They build at fronts where the horizontal temperature gradient is large. They often form as low-pressure extratropical cyclones on the east slopes of the Rocky Mountains, such as in Colorado or Alberta, when the jet stream shifts south during the winter. Similar storms can arise in the Gulf of Mexico; Cape Hatteras, North Carolina; or between the Bahamas and the east coast of Florida.

- They lack distinct, calm eyes but can spread over much of the northeastern United States and eastern Canada. As recognized in the late 1700s by Benjamin Franklin, smaller, counterclockwise-rotating cyclonic weather systems are embedded in the broader overall flow.

- They are cold-core systems that do not lose energy with height. If jet stream winds move an air mass away from the center of a storm, this drops surface pressure and increases storm strength.

- Damage is concentrated along the coast, whereas much of the damage from hurricanes occurs farther inland. With strong winds from the northeast, they typically batter northeast-facing shorelines.

Nor'easters can build when prevailing westerly winds carry these storms over the Atlantic Ocean and if the jet stream is situated to allow the storms to intensify. The annual number of nor'easters ranges from 20 to 40, of which only one or two are typically strong or extreme.

In addition to direct damage from high winds, a nor'easter also generates high waves and pushes huge volumes of water across shallow continental shelves to build up against the coast as a storm surge. Low barometric pressure in these storms also permits the water surface to rise, creating higher storm surges. Especially high tides or surge movements into bays can amplify flood heights. Like those that accompany hurricanes, these surges flood low-lying coastal plains and overwash beaches, barrier islands, and dunes. As with hurricanes, the greatest damages occur when a major storm moves slowly at the coast or hits a coast already damaged by a previous storm.

A classification scale for nor'easters by Davis and Dolan approximately parallels that of the five-category Saffir–Simpson Hurricane Wind Scale, except that the emphasis is on beach and dune effects rather than wind speeds. It infers a "storm power index" based on the maximum deepwater significant wave height (average of the highest one-third of the waves) squared multiplied by storm duration (Table 16-4).

Nor'easter wave heights are commonly 1.5 to 10 m, with energy expended on a coast being proportional to the square of their height. Thus, a 4-m wave expends four times the energy of a 2-m wave. Wave height depends on fetch, or the distance the wind travels over open water. Waves with a long fetch and constant wind direction in a slow-moving storm can therefore be much more destructive than those in a stronger, fast-moving storm with variable wind directions. Where a storm center is well offshore, the highest waves may reach the coast after the clouds and rain have passed. When high waves are stacked on top of a storm surge, the effects are magnified. A severe nor'easter can remain in place for several days and through several tide cycles. Examples of Class V storms include the Ash Wednesday storm of 1962 and the Halloween storm of 1991.

The Ash Wednesday Storm of March 7, 1962, was a high-latitude nor'easter along the Atlantic coast of the

Table 16-4	Dolan-Davis (1993) Nor'easter Scale		
CLASS	MAXIMUM DEEPWATER SIGNIFICANT WAVE HEIGHT (M)	AVERAGE DURATION (HRS)	EROSION AND DAMAGE
I (Weak)	2	10	Minor
II	2.5	20	Minor
III	3	35	Lower beach structures destroyed
IV	5	60	Severe on beach and dunes; numerous structures destroyed
V (Extreme)	6.5	95	Extreme beach and dune erosion; regional scale structure damage

United States; it stayed offshore approximately 100 km, paralleling the coast for four days. The storm began east of South Carolina and migrated slowly north and parallel to the coast before moving farther offshore at New Jersey. Its slow northward progress was blocked by a strong high-pressure system (clockwise rotation) over southeastern Canada. It affected 1000 km of coast and caused more than $2.2 billion in damages (in 2010 dollars). Sustained winds over the open ocean blew 72 to 125 km/h and produced waves as high as 10 m. Storm waves of 4 m on top of a 1- to 2-m storm surge washed over barrier sandbars that had been built up for years. A series of five high tides amplified the height of the surge. Beaches and dunes were extensively eroded and dozens of new tidal inlets formed. Almost all of heavily urbanized Fenwick Island, Delaware, was repeatedly washed over by the waves. The coastline moved inland by 10 to 100 m.

The huge Halloween nor'easter of October 1991, also called "The Perfect Storm," had 10.7-m-high deepwater waves and lasted for almost five days. Its wave crests were especially far apart, with intervals ranging from 10 to 18 seconds between crests, so they moved much faster than most storm waves. Accompanied by a major storm surge, it caused heavy damage from southern Florida to Maine, especially in New England (**FIGURE 16-30**).

A huge nor'easter ravaged the eastern states and Canadian maritime provinces in mid-April 2007. It produced six tornadoes across northern Texas, then others farther east to South Carolina. It strengthened as it moved up the East Coast. Several people died in Texas, Kansas, and South Carolina. Wind gusts of 100 km/h were common from New Jersey to Maine and reached 250 km/h at Mount Washington, New Hampshire, an area notorious for high winds. The storm

FIGURE 16-30 Nor'Easter
Storm waves batter this seawall at Sea Bright, New Jersey, during the 1991 Halloween nor'easter.

dumped up to 22 cm of rain and wet snow, which produced flash floods in the Carolinas, West Virginia, New Hampshire, and Nova Scotia. Winds pushed high waves that eroded beaches through New York, New Jersey, and New England.

Snow depth is one measure of the severity of a winter storm because of snow loading and accompanying building collapse. Snowfall occasionally exceeding a meter occurs in Maine, northern Michigan, Wisconsin, and western New York. Heavy snow can even fall in southern states such as Alabama and Georgia.

HURRICANE SURVIVAL GUIDE

Be aware that these lists don't cover everything. Use your own common sense.

U.S. High Hazard Area	■ Coastal areas along the Gulf of Mexico and Atlantic. High surge flooding and river rise can extend to many kilometers inland.
Preparation	■ Sign up for National Flood Insurance well in advance of hurricane season. Stockpile necessary building materials months in advance. Make sure there are strong attachments from your roof to the walls, the walls to the floor, and the floor to the foundation.
	■ After a hurricane watch is announced, board up windows and reinforce garage doors, remove damaged limbs and outside antennas, store as much clean drinking water as possible, and fill your vehicle's gas tank.
	■ Plan to leave if you live on the coast, an offshore island, a floodplain, or in a mobile home. Tell a relative outside the storm area where you are going.
	■ If you evacuate, take important items: identification, important papers (e.g., passports, insurance papers), prescription medicines, blankets, cash, flashlights, a battery-operated radio and extra batteries, a first aid kit, cell phones, a few days of water and nonperishable food, baby food and diapers, and any other items needed during a week or two in a shelter. Lock your house. Turn the refrigerator to the coldest setting. The power may be out for a long time. Turn off other appliances (the power surge while electricity is restored may damage them). Turn off a main natural gas line or propane tanks to your home, and anchor them securely.
Warning Signs	■ The National Hurricane Center keeps track of tropical storms during the summer and fall and issues warnings.
During the Event	■ Listen to the radio or TV for updates. Use telephones only for emergency calls.
	■ Do not drive through floodwaters. Roads may be washed out, and two feet of water can carry away most cars. Stay away from downed or dangling power lines; report them to authorities.
	■ If trapped at home before you can safely evacuate, be aware that flying debris can be deadly and hurricanes may contain tornadoes or spawn them. Stay in a small, windowless interior room; stay away from windows even if they are shuttered. Use flashlights for emergency light; open-flame sources cause many fires. Be prepared to move to a higher level. If you are in the water, rope people together; use life jackets or float tubes.
	■ If the eye of the hurricane passes over you, the storm is not over; high winds will soon begin blowing from the opposite direction, often destroying trees and buildings that were damaged in the earlier winds.
Aftermath	■ Be wary of downed power lines that may be live, gas lines that may be open (don't light matches or cause sparks), sewage lines and septic tanks that may contaminate everything; watch for broken glass and other sharp objects, and branches and trees that may fall. Don't run generators, heaters, or charcoal grills inside the home because of carbon monoxide poisoning.

City Drowns in Spite of Levees
Hurricane Katrina, 2005 ▶

Hurricane Katrina set a new standard for damage levels and for questions raised about the potential effects of hurricane flood surge, coastal development patterns, and the need for far better disaster preparedness. The results underscore the expression "Whatever can go wrong will go wrong."

As Hurricane Katrina bore down on the Louisiana and Mississippi coasts, most people complied with evacuation orders, but thousands of residents in the very poor, predominantly African American, eastern parts of New Orleans had no cars or other means of transportation and no money for travel even if transportation had been available. Many were tired of the time and expense of evacuating only to find that a storm did not strike their area. Many of them felt that New Orleans and the surrounding communities, especially areas not right on the coast, had survived hurricanes before and would do so again. Others felt that the brunt of the storm would miss them.

However, many of these people were not there in 1965 when Hurricane Betsy, a Category 3 storm, left almost half of New Orleans under water, up to 7 m deep. Category 5 Hurricanes Andrew in August 1992, and Camille in August 1969 just missed New Orleans to the west and to the east.

Pre-Katrina planning by FEMA was badly flawed. It predicted that at least 100,000 residents would not have transportation but few buses were sent to shuttle them out of the area. Hundreds of school buses were left unused because the city could not find drivers and because FEMA asked that school buses not be used because they were not air-conditioned, and evacuees might suffer heatstroke. FEMA said that it was providing suitable buses but failed to tell the governor that they would come from out of state and not be immediately available.

Many residents who sought shelter found their own way to the giant Louisiana Superdome stadium and to the New Orleans Convention Center; others were brought there by rescuers. By the time the storm arrived, 9000 residents and 550 National Guard troops

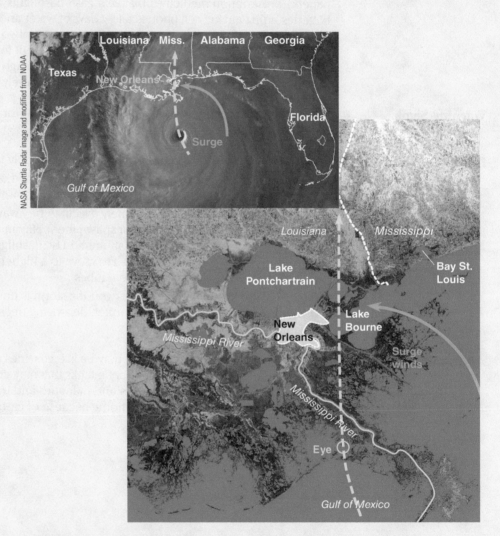

■ *Hurricane Katrina with its well-developed eye moved north from the Gulf of Mexico into Louisiana and Mississippi. Southeastern Louisiana and the Mississippi River delta with the path of Hurricane Katrina and direction of the surge and winds.*

---

*Visit the book's Website for these additional Cases in Point: Déjà Vu?—Hurricane Gustav, Gulf Coast, 2008; Unpredictable Behavior of Hurricanes—Florida Hurricanes of 2004; *Back-to-Back Hurricanes Amplify Flooding:* Hurricanes Dennis and Floyd, 1999; *Floods, Landslides, and a Huge Death Toll in Poor Countries*—Hurricane Mitch, Nicaragua and Honduras, 1998.

were housed in the Superdome. FEMA did arrange for medical disaster teams, search and rescue teams, medical supplies, and equipment, but relief for most people did not come until four days after the storm.

Ten thousand National Guard troops were finally ordered to the area days after the storm to help with rescue, public safety, and cleanup. They brought water, ice, tarps, and millions of ready-to-eat meals, but it was not enough. Winds blew off part of the roof lining of the Superdome, causing it to leak; the power went out, so the air-conditioning failed. Then water pressure dwindled, so toilets backed up. Some people were sick or went without their prescription medicine; some suffered heatstroke, and several died. Ultimately more than 20,000 people ended up crammed into the Superdome.

Approach and Landfall

The National Weather Service correctly predicted that the hurricane would cause catastrophic damage and "human suffering incredible by modern standards." Katrina weakened to a Category 3 in shallow waters before landfall in coastal Mississippi on Monday, August 29, at 7 a.m., where it obliterated nearly everything within hundreds of meters of the beach.

The Wind, Storm Surge, and Flood

With its counterclockwise rotation and landfall near the Louisiana–Mississippi line, the high near-shore winds on Katrina's north flank were directed westward toward New Orleans. Trees and power lines fell; the winds blew out windows, including those in hospitals, office buildings, and hotels; as with other hurricanes, tornadoes did some of the most severe wind damage. Much of the city was under water, so rescuers had to use small boats and helicopters to reach people.

Most of the damage to New Orleans was caused by the storm surge that raised the water level in Lake Pontchartrain and the large canals that typically drain water from the city into the lake. Pushed by fierce winds, the surge moved northwest through the deeply dredged Mississippi River Gulf Outlet into

■ *By August 31, the Superdome was surrounded by water, and sections of its roof covering had blown off. View is northeast; the Mississippi River is to the right.*

Lake Pontchartrain, which lies meters above the lower parts of New Orleans. The eastern suburbs of New Orleans were rapidly inundated, with water rising as much as a meter every three minutes. Residents described a "river" with 2-m waves rushing down streets.

Surges came over many levees on the eastern fringes of the city but did not breach them. However, concrete canal walls failed in several places, causing flooding of New Orleans that began about 18 hours after landfall. The canal walls generally consist of a ridge of dirt and rock topped by a concrete and steel floodwall 30 cm thick and 5 m tall. Homes had been built just behind the levee and floodwall, where people felt protected from the water high above them. The water poured down into the city, drowning it with up to 4 m of water. Only the homes on the high-standing natural levee along the Mississippi River were largely spared. It took two weeks to fill the breaches, using helicopters to drop giant sandbags. Levees along the canals were supposed to be over 5 m high, enough to withstand a moderate (Category 3) hurricane, but some had settled; in some cases, water surging over the walls likely undermined their bases.

Cars, houses, debris, and marsh grass floated with the surge and sometimes houses ended up on top of cars or in streets. Even brick houses well

anchored to 40-cm-thick concrete slabs floated and were swept down streets (see photo on page 463).

The Pumps Fail

With most of the city under several meters of water, the pumps failed. Electricity soon failed throughout New Orleans and east along the coast through Mississippi and Alabama to the Florida panhandle. Telephones and cell phones wouldn't work and the lack of communication greatly hampered rescue efforts. Broadcasts from reporters in nearby cities and communication over the Internet became important.

The water came in so fast that within minutes it was over people's knees, forcing them up to second floors and attics or onto roofs. A few thought to grab an axe or a saw so they could break through the roof if the water continued to rise. They waited there, sometimes for days, but had no means to contact authorities or rescuers. Searchers rescued at least 1500 people from rooftops and heard others beating on roofs from inside attics.

Contamination, Disease, and Mold

The floodwaters were littered with marsh grass, pieces of houses, old tires, garbage cans, all manner of trash, sewage, coatings of oil and gasoline from ruptured tanks, and even bodies. Many of those rescued had problems related to the polluted

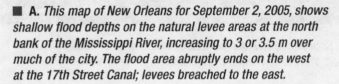

- ● Pumping station   ⇨ Breach direction   △ Main water plant

■ **A.** *This map of New Orleans for September 2, 2005, shows shallow flood depths on the natural levee areas at the north bank of the Mississippi River, increasing to 3 or 3.5 m over much of the city. The flood area abruptly ends on the west at the 17th Street Canal; levees breached to the east.*

■ **B.** *Almost the entire city lies below the average annual high-water level of the river and below the level of Lake Pontchartrain, so these giant pumps are used to drain the city.*

water, especially gastrointestinal illnesses, dehydration, and skin infections. Aircraft sprayed pesticides to kill mosquitoes because of the danger of malaria, West Nile virus, and St. Louis encephalitis.

Mold grew in most buildings in contact with the warm, contaminated water. Despite being structurally sound, many buildings were so affected that they had to be bulldozed. As the water receded,

dark gray muck coated everything. Wallboard, insulation, rugs, bedding, and almost anything else that had gotten wet had to be discarded. Even the bare studs of walls needed to be sanded, disinfected with bleach, and then dried with fans. Any wood frame structure standing in water for more than two or three weeks had to be demolished because mold was impossible to remove from deep in

the wood. The effects of mold on people with allergies, asthma, or weak immune systems can be serious.

Relief Came Slowly, Many Victims Died
Emergency response to the catastrophe was a dismal failure. Day after day, FEMA officials promised National Guard troops, supplies, and buses for evacuation—help that rarely materialized.

■ *The 17th Street Canal crosses much of New Orleans south from Lake Pontchartrain (bottom edge of photo). The breach and flooded homes are visible on the far side of the canal to the right of the bridge. The Corps of Engineers are driving sheet pilings at the bridge to block water flowing down the canal from the lake while they also drop giant sandbags into the breach.*

■ *This canal wall in the Gentilly area of New Orleans collapsed outward from the canal. The Corps of Engineers erected the corrugated steel wall as a temporary barrier against further flooding.*

Donald Hyndman

■ *Modest houses in the Lower 9th Ward were floated off their cinder-block posts and deposited on cars, other houses, or streets.*

Donald Hyndman

■ *This brick home on a 40- to 50-cm-thick concrete slab in the Chalmette area at the east edge of New Orleans floated in the surge and was carried five blocks and deposited in the middle of a residential street.*

Distribution of aid for disasters is logistically complex, but giant building materials and grocery companies have become very efficient at distribution on a global scale. Since 2003, aid organizations such as the International Red Cross have used those techniques to distribute aid quickly to victims of major disasters. FEMA presumably could have used those companies or such procedures.

Confirmed deaths totaled 1836, including 238 in Mississippi. Thirty-nine percent of the people who died were more than 75 years old. As water rapidly rose, people were swept away by surge flow and drowned because they couldn't swim. Or they drowned in their houses after retreating to a higher floor or an attic and becoming trapped. Five people died from cholera.

Some patients in hospitals died when electricity necessary to power equipment such as respirators and dialysis machines failed and backup generators ran out of fuel. There was no running water or ventilation; seriously ill patients died in the 41°C (106°F) heat. Thirty-four nursing-home patients died in the flood. Some hospitals were so damaged by flooding and mold that they did not reopen. These should be prominent lessons for any future potential disaster.

Impacts Farther South and East

News media focused on the storm damage in New Orleans because its flooding and destruction was so dramatic and catastrophic. However, elsewhere downriver and along the coast to the east, the storm's effects were no less disastrous. Downriver, near the main dredged Mississippi shipping channel, the high winds, huge surge, and waves floated houses like matchboxes, dropping them on roads or other houses. The surge lifted shrimp boats of all sizes and dumped them onto nearby levees and roads. Nests of poisonous water moccasins and other snakes swept in from the bayous added to the dangers. Northeast of New Orleans, the high surge and waves from Katrina lifted segments of Interstate 10 and U.S. 90 causeways and dropped them into Lake Pontchartrain, St. Louis Bay, and the east end of the Back Bay of Biloxi, Mississippi, in a similar series of tilted road panels.

Marvin Nauman, FEMA

■ *Some neighborhoods were still flooded weeks after the storm. Many already-damaged homes floated off their foundations and collided with other homes. Fetid water was everywhere.*

■ *A pair of 100-ft-long oil service vessels from the Mississippi River shipping channel ended up on Highway 23. The U.S. 90 bridge across St. Louis Bay, near the western end of the Mississippi coast, collapsed in the massive surge and giant waves as Katrina arrived. The surge and waves must have lifted the bridge deck segments and then dropped them either onto their supports or into the bay.*

Katrina's eye tracked almost due north, making landfall on August 29 at 7 a.m. near the border of Louisiana and Mississippi, causing collapse of an apartment complex and killing dozens of people. Some people survived the fast-rising surge by climbing into treetops. Given its counterclockwise rotation, the strongest onshore winds, as high as 224 km/h at the deadly eastern edge of the eyewall, hammered the Mississippi coast near Bay St. Louis, where the surge reached its highest level of about 9 m, the height of a three-story building!

Buildings as far as 2 km from the beach, from the border of Louisiana, through Mississippi, from Bay St. Louis, Gulfport, and Biloxi to Pascagoula, were leveled, leaving only tall posts and concrete pads to show where they once stood.

In Gulfport, the surge rose 3 m in a half hour; fierce winds tore the roofs off eight schools that were being used as shelters, and a hospital was heavily damaged. In Biloxi, seven giant casinos, floating just offshore, were wrecked. The state's largest was carried more than 1 km inland.

The storm surge at Gulfport stacked the lumber that used to be houses against the remaining heavily damaged houses.

Buildings slammed into nearby buildings are clear evidence of storm surge damage, as are huge piles of building debris banked up against one side of buildings with none on down-current sides.

In Alabama, a huge oil-drilling platform moored at a shipyard slammed into a suspension bridge across the Mobile River. Although many of the high-rise hotels survived the storm, some of the beaches that attracted them and on which they were built disappeared, as did most homes behind them. Following damage from each previous major storm in coastal Mississippi and Alabama, developers took advantage of the destruction to build larger and more expensive structures right to the edge of the beach.

Predictions, Preparation, and Response

In 2004 FEMA prepared a simulation of a major flood in New Orleans. The results, although unfinished, were unnervingly accurate: The simulation left much of the city under 3.5 m of water and showed that transportation would be a major problem.

Batteries in emergency radios used by the mayor's staff, police, and firefighters would quickly drain and could not be

recharged because the power was out. These were unlike radios used by teams fighting wildfires, which can be powered by ordinary disposable batteries.

A big contributor to the poor response to this disaster was lack of coordination among government groups with different responsibilities. Clear-cut lines of authority and communication were not in place. In some cases, the head of an agency said to proceed with a plan, but lower-level employees wanted signed papers to protect themselves from later criticism. Some did nothing because it was "not part of their jurisdiction."

FEMA, charged with handling response to disasters, proved tragically unprepared and inept. With 9500 people sheltered by the Red Cross in the Convention Center, FEMA said it had no "factual knowledge" of its use as a shelter until September 1, 4 days later.

Arranging temporary housing for 300,000 displaced people was an immediate and enormous task. By September 4, 220,000 were sheltered in Houston, San Antonio, Dallas, and other cities across the country. Outside New Orleans, FEMA provided army-style wood-frame tents and some travel trailers and mobile homes but insisted that

■ *A nearly new subdivision in a coastal area of Gulfport, Mississippi, was leveled by the surge, leaving its remains piled up against the battered houses farther inland.*

before the tens of thousands of trailers could be moved, the sites that would receive them had to have water, sewer, and electricity hooked up, services not available in many areas for months. In most of the city, even a month after the hurricane, power lines still dangled, tree branches and other debris still clogged the streets, and no stores or gas stations had reopened.

The Future of New Orleans?

An important question is whether New Orleans should be rebuilt in essentially its previous form. Years later, little had been accomplished in some areas of the city. Should people be permitted to rebuild in a huge sinking depression several meters below sea level and below the Mississippi River, or should aid for reconstruction come with the requirement that any new homes be situated above sea level and outside the floodplain? The latter was one of FEMA's main requirements for people and companies seeking funds for rebuilding structures along rivers since the 1993 upper Mississippi River flood.

Certainly the higher-elevation areas of New Orleans, those on the natural levees of the Mississippi River, should

be restored. These areas provide the shipping and industrial facilities that serve not only the Mississippi River basin but much of the rest of the country. Some port facilities could be moved upriver about 150 km to Baton Rouge, which is also a dredged deepwater port.

New Orleans' natural-levee areas, including the lightly damaged, famous French Quarter, also make up the cultural and historical part of the city frequented by tourists, who provide a large portion of the city's income. Even in areas that should be rebuilt, where do you start? Where will people live, get groceries and gas, or find schools for their children? Without people in the area, there are no jobs; without jobs, people cannot return.

Katrina was the costliest natural disaster to strike North America to date. Insured costs reached $45.1 billion. The NFIP paid out $16.1 billion, and the NFIP paid out an additional $16.1 billion, total costs are estimated at $150 billion. Nine years after the storm, the population was estimated to be about 384,000, still down 71,000 from before the storm.

The Corps of Engineers plans to repair 60 km of the 480-km levee system to withstand a Category 3 storm. Improved sections will be 5.2 m high rather than the

previous 3.8 m high. Rebuilding the system to withstand a Category 5 storm would cost more than $32 billion—that is $66,000 for each of the 485,000 original residents, $264,000 per family of four, much more than that for the many fewer who are expected to return!

A $10 billion project to rebuild and strengthen New Orleans' levees left them able to withstand a Category 3 hurricane—a 100-year flood event. However, further studies by the Corps show that a slowly rising sea level and a storm event slightly larger than Katrina would overtop the new levees to cause significant flooding.

Other severe dangers to New Orleans were recently pointed to by Weather Underground and NOAA, respectively: 1. An early-summer moderate-strength hurricane could strike New Orleans while the Mississippi River is in flood. Storm surges often move up the Mississippi River so that even a Category 1 hurricane surge could overwhelm the city's flood barriers and submerge it under several meters of water; and 2. a Category 5 storm arriving at high tide could cover everything from the Gulf of Mexico to New Orleans with 6 m of water.

Unfortunately, the city on the floodplain continues to slowly sink, as groundwater is withdrawn for municipal and industrial uses, and buildings continue to compress the underlying peat.

CRITICAL THINKING QUESTIONS

1. Katrina was a Category 3 storm at landfall. What major factors led to such a catastrophic outcome for New Orleans?
2. Pair off into two small groups: one to argue that New Orleans should be protected, even with increasing flood protection costs, the other to argue that the part of the city below sea level should be relocated upstream near Baton Rouge.
3. If you were in charge of disaster preparation for FEMA, what specific plans would you make?

Extreme Effect of a Medium-Strength Hurricane on a Built-Up Barrier Island

Hurricane Ike, Galveston, Texas, 2008 ▶

Hurricane Ike provided a vivid reminder of the damage that can be wreaked by a moderate-category hurricane on a developed barrier island. A Category 4 storm in the Atlantic Ocean, Ike weakened to Category 1 in the Gulf of Mexico, then strengthened to Category 4, before easing to a strong Category 2 at landfall at Galveston, Texas. Hurricane Rita in August 2005, a Category 5 offshore, weakened to Category 3 at landfall, near the Louisiana border east of Galveston, but the ordered full evacuation of Galveston and much of Houston did almost as much damage as the hurricane. Again before Ike, authorities ordered complete evacuation; about one million complied, but the lower wind speeds encouraged several thousand people to ride out the storm.

Following the tragic hurricane of 1900 that killed about 8000 people,

Galveston raised the level of the city and built a high seawall to protect it from further hurricanes. However, there was a downside. Aggressive waves in front of the seawall stirred up and removed sand where there was once a nice beach, the main reason most people moved there in the first place. For many years, they dumped truckloads of sand—removed from dunes elsewhere along the coast—over the wall to provide a narrow beach for recreation and protect the base of the seawall from being undermined. Continued erosion and shortages of readily available sand, however, made it necessary to "pave" the remaining sand with huge blocks of granite to minimize further erosion and undermining of the wall. During Ike, the wall did in fact protect the city from the worst of the incoming storm surge and waves, but

areas beyond the ends of the wall were not protected.

As in the majority of coastal areas, the most vulnerable homes sit on offshore barrier islands, especially those on beachfronts with no protective dunes. Natural dunes, originally present on the barrier islands, both northeast and southwest of Galveston, gradually blew landward during past hurricanes and other storms. Coastal houses built on the islands, however, stayed put, leaving them standing in the surf as the sand moved inland. Unfortunately, with so many houses built at the beach and with little zoning to control building activity, protective coastal dunes were also removed for construction or to improve views or beach access.

Ike's low atmospheric pressure and winds of about 160 km/h (about 100 mph)

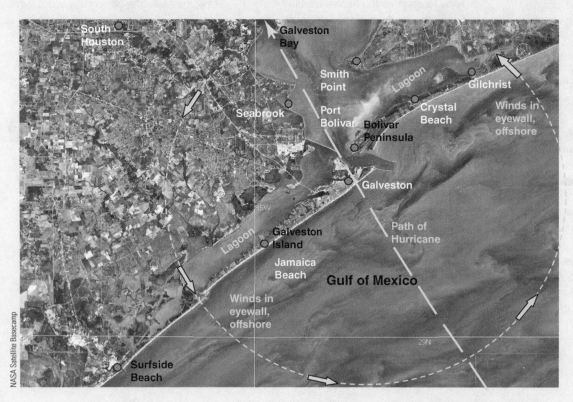

■ At landfall, the high winds of Hurricane Ike's eyewall tracked northwest, directly over Galveston and just east of Houston. Northeast of Galveston, the winds and their accompanying waves and storm surge blew directly onshore, while they blew offshore southwest of the town. As the hurricane approached from the southeast (dashed yellow path), the storm surge was approaching a broad area of the Texas coast.

NOAA

■ *Most of the houses on the west end of Bolivar Peninsula were completely obliterated, leaving only remnants of their ground-floor slabs.*

drove a 3.6-m-high (12-ft) surge onshore northeast of Galveston, over the Bolivar Peninsula, an offshore barrier island heavily populated with homes. Most were built 3 to 4 m above the sand, on top of sturdy posts, to keep them above storm waves, as required by coastal flood insurance and coastal ordinances. Although low artificial dunes were piled on the beach in some places, by the time Ike arrived, the front rows of houses were on the beach itself, essentially unprotected. Ike had a huge diameter, with hurricane-force winds extending

out from the eye for 193 km (120 mi) and tropical-storm winds extending out to 442 km. This made it the largest-diameter hurricane ever measured, a formidable storm with more kinetic energy (energy of motion) than any other Atlantic hurricane recorded by modern instruments. Winds and waves pounded the Texas coast for hours before the fierce winds of the eyewall arrived. The storm finally came ashore at Galveston with 176 km/h winds.

Because of the counterclockwise rotation of the storm, winds and

waves ahead of it blew to the west, so southwest of Galveston, winds angled toward the shore carried the high surge, or storm-tide, in over the lagoon landward of Galveston Island. Thus, although destruction near the beach northeast of the eye was almost complete, damage southwest of the eye was also severe.

At Galveston the surge raised tide levels almost 3 m 24 hours before landfall. It swept inland about 50 km, reaching about 3.6 m (almost 12 ft) along the west side of Galveston Bay and on the Louisiana border to the east.

Destruction at and near the barrier island beaches was nearly complete, with only a few especially strongly built houses remaining. The winds took a heavy toll. Winds of 176 km/h lifted off shingles and some complete roofs, blew over mobile homes, shredded and flattened signs, and carried all kinds of debris that acted as missiles, impacting other structures. Houses on elevated posts that were not sunk deeply enough into the ground were blown over and shredded in the surf. Support posts that were weak or partly rotted, snapped. High winds and waves wrecked oil platforms offshore, damaged storage tanks, and ruptured fuel pipelines; oil and gas spills were widespread.

Southwest of Galveston, as the wind turned south and subsided, the massive mound of water, standing several meters above normal sea level, swept back across the barrier island and offshore, further deeply eroded the beach, carrying as much as 2 m or more of sand offshore and undermining the posts of many remaining homes. Weak construction contributed to the destruction of houses. In many cases post tops were not well anchored to the main floors of houses, walls were not well anchored to floors, or roofs were not well anchored to walls.

Because many barrier island homes are not connected to municipal sewer lines, the sewage typically goes into septic tanks. Because liquid outflow from the tanks flows into a drain field, in this case beach sand, the outflow must be above high-tide level for drainage to occur. Thus the septic tanks for homes

Hyndman

■ *Support posts failed under this house so it collapsed and broke up. Crystal Beach, northeast of Galveston.*

■ *Well away from the beach, the storm surge floated this house off its foundation and dumped it into nearby brush, along with debris from other houses. Northeast of Crystal Beach, Bolivar Peninsula.*

■ *One and a half to two meters of sand were eroded from this beach southwest of Galveston.*

on the beach are only shallowly buried. During Ike, numerous septic tanks were uncovered, upended, their plumbing severed, lids torn off, and their contents spilled.

Thirteen thousand people in the coastal area did not evacuate, in spite of dire warnings. Some thought the storm would miss them or that they would be safe in their homes on tall stilts. Others distrust government and refuse to be told what to do. With a new Texas law, police can now arrest people that don't leave under a mandatory order. In addition, they can be forced to pay for any rescue during or after the storm.

In the United States, 112 people died, mostly in Texas and eight in Louisiana; 3500 people needed rescue. Coastal damages reached $29.6 billion and only Hurricanes Andrew in 1992 ($51.9 billion) and Katrina in 2005 ($91.4 billion) cost more (in 2010 dollars). In the Houston area, nine people died while cleaning up after the storm, from carbon monoxide poisoning from generators or from house fires.

FEMA handed out ice, millions of liters of water, and millions of meals at a dozen distribution centers in Houston and provided about 11,000 motel rooms for temporary evacuees. In contrast to

the debacle after Hurricane Katrina three years before, they received praise for their efforts after Ike. With most stores, businesses, and restaurants closed, people couldn't work or purchase food and supplies. A week after the storm, most roads were still closed because of debris or washouts, only a single gas pump was operational in Galveston, and mosquitoes were rampant because of standing water.

■ *Severe beach erosion lowered the beach level, uprooting and emptying septic tanks near Gilchrist, on the Bolivar Peninsula northeast of Galveston.*

CRITICAL THINKING QUESTIONS

1. Ike was only a strong Category 2 storm at landfall. What factors led to it causing such widespread and severe destruction?

2. The governor of Texas and others suggested building a giant seawall along much of the Texas coast to protect coastal homes. What are the positive and negative consequences of such a wall?

3. Considering the recreational values of the coastal area, as well as people's ownership of land and homes, what should be done with that area? Who should pay for the costs?

4. What should be done with sewage from coastal homes? If there are additional costs, who should pay?

Landward Migration of a Barrier Island Coast
North Carolina Outer Banks ▶

North Carolina's Outer Banks form an endless stretch of white sandy beach that draws millions of visitors each summer. The beach is actually an offshore barrier sandbar resting on a shallow-water area of especially gently sloping continental shelf that protrudes east into the Atlantic Ocean. That form places it farther into the path of many Atlantic hurricanes than anywhere else on the East Coast.

A map of some of the hurricanes that have ravaged the Outer Banks makes clear that anyone who lives in the vicinity needs to be aware that this is a dangerous place to live. The winds, waves, and surge of many hurricanes would extend to more than half the width of the map. Because the storm surge impact is concentrated in the right-front quadrant of a storm, most of the impact from offshore would lie northeast of the track of the eye.

With every significant storm, beaches are eroded and sand swept offshore; in major storms where the storm surge and waves overtop barrier bars, sand is also swept across barrier bars into back-bar lagoons, and the barrier islands literally migrate landward. Just north of Cape Hatteras on the North Carolina Outer Banks, an area with few houses and little beach hardening, the island has migrated landward about 800 m to the west during

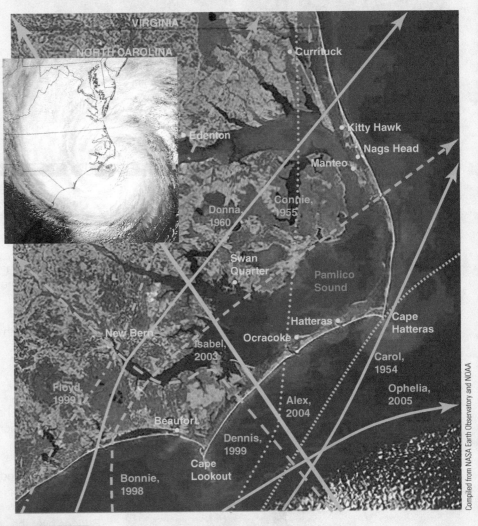

■ Some of the many hurricanes that have battered the North Carolina Outer Banks. Total width of the view is about 210 km. Inset shows Hurricane Isabel for comparison; its size covered all of the area of the main map—compare shape of coastline to map.

Compiled from NASA Earth Observatory and NOAA

Mark Wolfe, FEMA

■ *This beachfront home on stilts at Nags Head, North Carolina, was left stranded, then toppled on the beach after Hurricane Isabel in 2003.*

the past 150 years. A little farther north at South Nags Head, which has beachfront houses, shows beach migration that has progressively destroyed rows of houses during major storms. The slow but progressive rise of sea level, now about 42 cm per century in this area, heightens the problem of barrier island migration. One hundred years ago it was less than half of that.

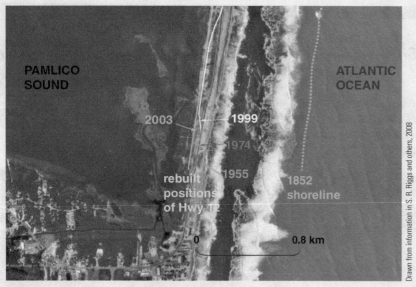

PAMLICO
SOUND

ATLANTIC
OCEAN

2003   1999

1974

rebuilt   1955
positions
of Hwy 12   1852
shoreline

0   0.8 km

Drawn from information in S. R. Riggs and others, 2008

■ *Landward migration of part of the North Carolina Outer Banks, just north of Cape Hatteras. In the last 50 years, Highway 12 has had to be moved landward several times. This 1993 USGS aerial photograph shows when the beach was approximately along the 1955 position of Highway 12 (green line). The average erosion rate in the last 150 years is 5 m per year!*

CRITICAL THINKING QUESTIONS

1. The North Carolina Outer Banks are partially protected from development as a National Seashore. Given what you have learned here, what should be the future of homes on the developed parts of the coastal areas of the Outer Banks? Who should pay for any changes? Why?

2. The coastal highway has been repaired and relocated many times. Given the costs and benefits, should that continue and if so, who should pay?

A Damaging Late-Season, Low-Category Storm
Hurricane Sandy, Atlantic Coast, 2012 ▶

In late October 2012, Hurricane Sandy, often referred to as Superstorm Sandy, moved up the East Coast of the United States a few hundred kilometers offshore. Although it was a late-season storm ranging in strength from a tropical storm to a Category 1 hurricane, its storm winds extended to an incredible 1850-km diameter before landfall. Its barometric pressure, fueled by a warm area (27° C) of the Gulf Stream offshore, dropped to 948 millibars at landfall, the lowest pressure ever recorded for the United States. On October 29, with its northeastward path blocked by an Arctic high-pressure ridge, it turned northwestward toward the New Jersey coast. On the previous day, the New Jersey governor had ordered evacuation of all residents of the barrier islands and closed almost all schools and universities. Bridges over the Chesapeake Bay were also closed because of high winds.

Although Sandy was relatively low intensity, it wreaked havoc for several reasons. As it moved north off the East Coast, its large diameter pushed up a large storm surge. This was amplified by the high tide accompanying a full moon (there is increased gravitational pull on the ocean surface when the moon and sun align with Earth). Moving north offshore, it ran into a major blocking high-pressure system over Greenland so that instead of turning east with the prevailing westerly winds, as is normally expected, it was diverted northwestward directly toward the coast. As a result, it ran ashore in more vulnerable areas not commonly impacted by major hurricanes. Sea-level rise of almost 30 cm over the last century contributed to the severe damage in low-lying areas.

Due to the counterclockwise rotation of the storm, the coast saw strong onshore winds reaching 130 km/h (80 mph), which coincided with a high tide that pushed a strong surge tide into northern New Jersey, Long Island, and New York City, raising the sea level 4.23 m (13.88 ft). Storm waves piled on top of that.

Flooding was the biggest problem. Rainfall was significant up and down the coast, with more than 17.8 cm in much

■ *Hurricane Sandy moved north off the East Coast of the United States until it turned abruptly northwest into Maryland and southern New Jersey. Hurricane-force winds are shown in red; very broad tropical storm winds (in brown) caused widespread damage. Onshore, its path turned north (dotted yellow line).*

Continued path of the storm on land

National Hurricane Center

■ *Beachfront homes along the coast of New Jersey sustained major damage—many were completely destroyed. Can you spot some houses, or their upper floors, that floated inland to land at odd angles between others?*

of Delaware and eastern Maryland, and more than 12.7 cm in southern New Jersey. In New York City, out-of-date flood zoning for a 500-year flood was for waves 3.3 m high. In places, Sandy pushed it to 4.3 m. The East River rose almost 4.3 m where it backed up against the high surge tide off Lower Manhattan. Seven subway tunnels under the East River filled with water.

Sandy killed at least 253 people, including 131 in the United States (especially New York and New Jersey). In spite of mandatory evacuations and provided shelters, many people chose to stay rather than evacuate. In Belmar, in northern New Jersey, 200 families, including many with elderly members, remained. Some had stayed through other major storms, some felt safe in their homes on raised posts, and some had no transportation. By the time they recognized they should leave, it was too late; their evacuation route was cut off by fast-rising surge water. On Staten Island, across from Manhattan, the surge came in so fast in the coastal zone it filled basements, drowning people who had taken shelter there.

In the aftermath of the storm, the New York–New Jersey region was nearly paralyzed. Electric power failed for 8.6 million people. A week later, 1.4 million homes still lacked power as overnight temperatures dropped to freezing.

Dozens of people were admitted to hospitals for carbon monoxide poisoning after they ran generators and charcoal grills indoors or in closed garages. New York City airports closed, and more than 16,000 flights were canceled in and out of the region, wreaking havoc on air transportation throughout the northeast and around the world. Downed trees and power lines closed most roads. Shortages of gasoline were caused by delivery problems and a lack of electricity needed to power gas pumps, leading to rationing. Economic losses in New York alone were estimated at more than $18 billion. The total U.S. costs of Sandy have been estimated at $66 billion.

In response to the storm, FEMA sent dozens of teams to damaged areas of southeastern New York State. Evacuation shelters were opened throughout New York City. Water and prepared meals were distributed to homebound and elderly residents and made available to everyone. Removal of piles of damaged belongings was

a monumental task. New York City, for example, trucked 246 metric tons of storm damage to create a mountain of trash 10 m high and a km long.

Following Sandy, discussions about how to protect New York City centered on hard solutions such as seawalls and levees or building a storm-surge barrier across the harbor, versus soft solutions such as restoring wetlands and reefs. Intermediate steps included moving generators to upper floors and strengthening sewage systems to handle storm surges.

What is the future for people whose decades-old modest beachfront homes were severely damaged or destroyed

■ *A ship grounded on Staten Island after the storm surge.*

■ **A.** *Many coastal homes at Ortley Beach, New Jersey, were completely wrecked. The incoming storm surge lifted many homes off of their pilings and big sheets of asphalt pavement off of the sand. Note the white parking place lines.*
B. *The lower floors of many homes were structurally weak so the surge lifted the upper floors and pushed them over; the house on the left flopped over to the right.*

by Sandy? Many do not have money to rebuild or even repair. Towns, including many in coastal New Jersey, have lost much of the tax base that pays for police, fire, water, and other services that are still necessary. Many towns desperate to survive, rebuild, and revitalize look to eminent domain, under which whole neighborhoods can be condemned and sold to large-project developers. However, many longtime residents do not want to leave.

They point to past abuse that created upscale residential and commercial developments, including big-box retail stores. When these developers came in, low-income residents were forced to leave the coast for more affordable communities. In February 2013, New York's governor announced a $648 million floodplain buyout program to low-lying homes and businesses with parks, wetlands, and dunes that would protect homes farther inland.

CRITICAL THINKING QUESTIONS

1. Sandy was a low-intensity storm well offshore for most of its time before landfall. Why might you have been concerned when it turned shoreward toward your coastal area?
2. At landfall, why did Sandy do so much damage in spite of being a low-intensity storm?
3. What major proposals have been put forward to minimize future damage? What are the main advantages and disadvantages for each proposal?

Catastrophic Typhoon in the Western Pacific:
Supertyphoon Haiyan, the Philippines, 2013 ▶

The worst natural disaster ever to hit the Philippines struck November 8, 2013, with a huge Category 5-equivalent supertyphoon, named Haiyan (or Yolanda), making landfall in the central Philippines. Its sustained winds in some areas were 236 km/h, equivalent to those in an EF2 tornado; winds lasting a full minute reached 315 km/h (196 mph),

as in an EF4 tornado. This made it the strongest-ever cyclone at landfall. Storm surge waves reached 5 to 6 m and more, partly because of its fast-forward speed of about 40 km/h. Due to the storm surge, 75% of Leyte province was considered destroyed, and deaths reached at least 6340. Low-lying areas were completely flattened with almost no

buildings left standing. After the storm, 1000 people remained missing and more than 670,000 were left homeless. Damages reached more than $2.8 billion.

The city of Tacloban, with a population of about 220,000 was especially hard hit, partly because it sits at the head of

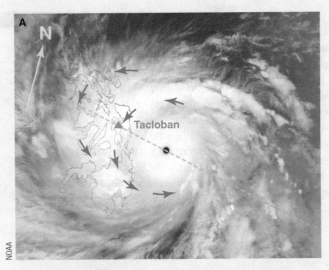

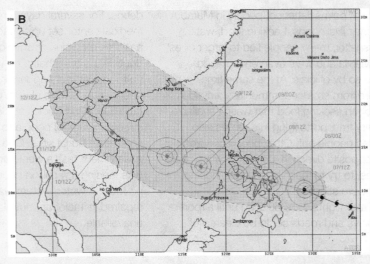

A. *Supertyphoon Haiyan, a Category 5 (equivalent) storm with extremely high wind speeds circulating around a very well-defined eye in this NOAA satellite view. It tracked west-northwest through the central Philippines. Yellow arrows show general circulation.*

B. *This regional map shows the eye at 12-hour intervals, striking Tacloban on November 8.*

a large bay that funneled and raised the surge height. Flooding reached 1 km inland. Rescue efforts were hampered by debris-covered roads and bridges out. Many people, especially those in rural areas and on outlying islands, saw no relief for six days, hampered by severed communication and blocked roads, along with lack of fuel and power. Medical facilities were destroyed, along with fishing boats and rice paddies and cornfields.

Devastation was amplified by low-lying terrain, aggravated by settling because of groundwater extraction and rising sea level, as well as by a tripling of the population since the 1970s, poverty, and shoddy construction. Bodies were strewn in the streets and hidden under debris. Three months later, people in Tacloban were still finding five bodies a day buried in the rubble. Stunned survivors wandered around searching for water, food, and shelter. Some people dug up water lines,

desperately smashing them to get water. Looting was rampant; some aid convoys were held up by armed attackers. Many people died when a stack of rice bags in a warehouse collapsed as a huge crowd raced in to carry off 100,000 bags of rice. Much of the population evacuated after being given orders to do so and following warnings of the severity of the storm. However, because people had experienced many typhoons before, many did not leave. Survivors said they were surprised by its severity and the fact that their preparations proved insufficient.

In Marabut, across the bay from Tacloban, all 15,900 homes were destroyed, though only about 20 people died. With 80% of the population earning a living from coconut products and tens of thousands of

A. *Homes in low-lying areas near the coast were floated by the storm surge and swept against one another to fill streets. Note the diagonal street from the lower right corner of photo.*

B. *Survivors in Marabut, Philippines, find minimal shelter in remains of their homes.*

coconut palms flattened, people in Marabut lost their livelihoods. Lacking food, water, and shelter, many people fled to larger cities like Manila and were even encouraged to do so by officials. At the same time, others from smaller communities arrived in Tacloban seeking food that wasn't available. Military transports flying in supplies also carried people out. Other people traveled the opposite direction, generally by regional ferries, to try to reach loved ones, hoping they were still alive.

Relief flights were delayed until airport runways and roads could be cleared of debris. For several days, an improvised medical center set up at Tacloban Airport had little medicine or antibiotics, few doctors, no anesthetic, no water, and no food. Without bandages, patients' open wounds were closed with masking tape. Decaying bodies still lay in the streets for lack of people to move them and locate burial sites. One monumental government task a week after the storm was feeding 1.4 million people who had no food. A water system was finally repaired in Tacloban, but electricity was still unavailable.

CRITICAL THINKING QUESTIONS

1. What cultural or social issues increased the damages or difficulties experienced by people in this storm?
2. What likely differences would there have been if this had been a more modern and affluent country?
3. If you were living in a small, very poor coastal community in the Philippines as a Peace Corps, UNICEF, or other aid person, as Haiyan approached, what would you do to help people prepare for the storm and respond afterward?

Floods, Rejection of Foreign Help, and a Tragic Death Toll in an Extremely Poor Country
Myanmar (Burma) Cyclone, 2008 ▶

In the Pacific, hurricanes (cyclones) reach land in Southeast Asia—the Philippines, Vietnam, Taiwan, southern Japan, and southeastern China, or occasionally Cambodia, Thailand, Myanmar (formerly Burma). On April 27 a cyclone grew in the northeastern Indian Ocean and began moving northwestward away from Myanmar, then northeastward and strengthened. By May 2, 2008,

Cyclone Nargis was a Category 4 storm with peak winds of 215 km/h, making landfall in the south coast of Myanmar. On land, it turned northeast near of Yangon (Rangoon), the largest city in Myanmar, where its winds slowed to 130 km/h. Although Indian meteorologists warned the Myanmar government more than two days before landfall, the time available was much less than the five to

seven days necessary for most people to evacuate.

The southern, hardest-hit part of the country occupies the immense delta of the Irrawaddy River, an area about 200 km across. Because of the agricultural fertility of the delta, this is also the area with the greatest population. In addition to the fierce winds, torrential rains and up to a

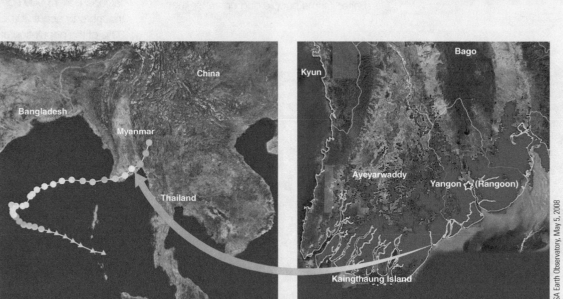

NASA

NASA Earth Observatory, May 5, 2008

■ **A.** *Cyclone Nargis path (warmer colors, higher-intensity storm) and* **B.** *flooded area (blue).*

■ *A Myanmar woman in her destroyed house.*

5-m-high (13-ft) storm surge flooded most of the delta, with 2-m storm waves on top of that. The surge moved as much as 50 km inland, causing severe destruction in this desperately poor country. Sewage mains broke, contaminating water supplies and destroying more than half of the rice crop. Most businesses and markets were closed. More than 138,000 died, making this the deadliest storm-related event since 1970; about 1.5 million were left homeless.

Damages and deaths were made worse by loss of mangroves that were removed for firewood and to make room for agriculture. People had not experienced such a storm in their lifetimes, so there was no evacuation plan. Because all of the land is almost at sea level and transportation is minimal, people had no escape.

Almost immediately, help in the form of food, medicines, water purification systems, and shelter was offered from many countries, but Myanmar's military rulers, suspicious of outside influence, especially from the United States, initially refused aid. Finally on May 7 and 8, India was permitted to bring in more than 136 tons of relief supplies, including tents, blankets, and medicine, but Myanmar denied India's search and rescue teams and media people access to the hardest-hit areas. A few days later, many other countries, especially the United States, the United Kingdom, and Australia, sent large amounts of relief supplies. Of the people left homeless—in danger of disease or in need of food—after a month only 50% had received aid. Diarrhea was widespread. Food that was available in delta towns was often provided by private businessmen and volunteers, without knowledge of the military authorities. The persistent rain was a mixed blessing; it provided the only source of clean drinking water.

The price of rice increased by 50% in Yangon, the largest city. The price of roofing tiles went up by a factor of 10.

Much of the foreign aid was being confiscated and relabeled as being provided by the Myanmar government. Some of it was sold on the black market and then appeared in local markets at extravagant prices. Access was difficult because of problems with roads and too few trucks and boats. Much of the military government's resistance to foreign access to the delta stemmed from the fact that many inhabitants were ethnic minorities who had aggressively fought the army.

With easing of government resistance to foreign distribution, the U.N. World Food Program and other aid organizations used lessons from past national disasters to organize distribution hubs, much like those of FedEx and UPS. They used local staff to set up improvised distribution centers in rented buildings in the affected area and utilized radio contact to Yangon, and the Internet to organize distribution of foreign aid from nearby Bangkok, Thailand. Locals in Myanmar would provide distribution within the country.

Damage estimates one year later were about $10 billion. Although the totals were not high by western standards, they were extreme for such a poor country.

CRITICAL THINKING QUESTIONS

1. As with many other major disasters, there are often cultural and/or political issues that compound people's problems. What were they in this case? Where else might there be such problems?

2. Water contamination and disease are major concerns after some disasters. Why was that such a concern here? In what other disasters, described in this chapter, would that have been a lesser problem? Why?

Critical View

A This building was damaged during Hurricane Charley in 2004.

FEMA

1. What process or processes led to destruction of this building?
2. How can you tell that a storm surge was not the primary cause?
3. What can you see that made the building especially vulnerable to this type of destruction?

B This damage occurred on the Texas Gulf Coast from Hurricane Ike.

Hyndman

1. What process led to removal of the sand from under this concrete foundation?
2. What process or processes led to destruction of the house?
3. What would have made the house less susceptible to destruction?

C This building was damaged during Hurricane Charley in 2004, in Punta Gorda, Florida.

Andrea Booher, FEMA

1. What happened to this large beachfront building?
2. What could have prevented the damage?

D This home in New Orleans was damaged during Hurricane Katrina in 2005.

Andrea Booher, FEMA

1. Why is beach grass and other debris piled on the edge of this roof? What does it indicate?
2. What indicates that it was not carried there by the wind?

E This photo of Nags Head, North Carolina, was taken soon after Hurricane Isabel in 2003.

U.S. Army Corps of Engineers

1. Why was this walkway and stairs elevated above the beach?
2. What happened to leave the walkway where it is?
3. What does the current state of this beach mean for future hazards to these homes?

G This photo was taken on the Texas Gulf Coast after Hurricane Ike.

Hyndman

1. Why are all of these boards left here in a long pile?
2. The large gray surface is the roof of a house. What is it doing there?
3. What does it indicate about the house?

F This house is located tens of kilometers south of New Orleans, just after Hurricane Katrina.

Donald Hyndman

1. What are the main damages to this house and from what cause?
2. This house appears much less damaged than most of those in New Orleans, even though it is much closer to the Gulf of Mexico. Explain why that makes sense.

H This building in Punta Gorda, Florida, was damaged by Hurricane Charley in 2004.

U.S Army Corps of Engineers

1. What happened here? Was this house built out on this sandy beach? How can you tell?
2. If it wasn't built on its present sand level, what changed and by how much? What likely caused this change?
3. Why are there posts sticking out of the sand in the foreground?

Chapter Review

Key Points

Hurricanes

■ Hurricanes, typhoons, and cyclones are all major storms with winds greater than 120 km/h rotating around clear, calm air in the low-pressure eye. **FIGURES 16-1** and **16-2**.

■ The strongest hurricanes, Category 5 on the Saffir–Simpson Hurricane Wind Scale, have highest wind speeds (more than 249 km/h). **Table 16-1**.

■ Hurricanes that affect the southeastern United States form as atmospheric lows over warm subtropical water. They grow off the west coast of Africa, then move westward with the trade winds. Most hurricanes occur between August and October because it takes until late summer to warm the ocean sufficiently. They strengthen over warmer water and weaken over cool water or land. **FIGURES 16-4** and **16-5**.

Storm Damages

■ Damages from a hurricane depend on its path compared with shore orientation, presence of bays, forward speed of the hurricane, and the height of dunes and coastal vegetation.

■ Storm surges, as much as 7.3 m high and 160 km wide, result from a combination of low atmospheric pressure that permits the rise of sea level and prolonged winds pushing the sea into a broad mound. The surge and associated winds are concentrated in the northeast forward quadrant of a hurricane because of wind directions. **FIGURE 16-11** and **By the Numbers 16-1**.

■ High waves have much more energy and erode the beaches and dunes lower and to a flatter profile, especially if the waves are not slowed by shallow water offshore. **FIGURE 16-18**.

■ Wind damages include blown-in windows, doors, and walls; lifted-off roofs; blown-down trees and power lines; and flying debris. **FIGURES 16-22** and **16-23**.

■ Rain and flooding from hurricanes can be greater with large-diameter, slower-moving storms.

■ Deaths from hurricanes are often related to the poverty, size, and education of populations. Poor, mountain-dwelling populations such as in Central America experience many fatalities as a result of landslides, while poor populations along deltas and rivers can be killed in storm surges and flooding.

Hurricane Prediction and Planning

■ Hurricane predictions and warnings include the time of arrival, location, and event magnitude. The National Hurricane Warning Center tries to provide a 12-hour warning, but evacuations can take as many as 30 hours.

■ Many people think they can evacuate quickly and can leave too late. Storm surges and downed trees and power lines often close roads.

Managing Future Damages

■ The National Flood Insurance Program requires that buildings in low-elevation coastal areas be landward of mean high tide and raised above heights that could be impacted by 100-year floods, including those imposed by storm surges.

■ The nature and quality of building construction have a major effect on damages. Floors, walls, and roofs need to be well anchored to one another, and buildings should be well attached to deeply anchored stilts.

Extratropical Cyclones and Nor'easters

■ Nor'easters and other extratropical cyclones are similar to hurricanes except that they form in winter, lack a distinct eye, are not circular, and spread out over a large area. Like hurricanes, they are characterized by high winds, waves, and storm surges.

Key Terms

extratropical cyclone, p. 457

eye, p. 436

hurricane, p. 436

hurricane warning, p. 452

hurricane watch, p. 452

nor'easter, p. 457

Saffir–Simpson Hurricane
Wind Scale, p. 436

storm surge, p. 441

tropical cyclone, p. 436

typhoon, p. 436

Questions for Review

1. What causes a hurricane? Where does a hurricane get its energy?

2. Where do hurricanes that strike North America originate? Why there? Why do they track toward North America?

3. Why are coastal populations so vulnerable to excessive damage (other than the fact that they live on the coast)?

4. Where in a hurricane is the atmospheric pressure lowest, and approximately how low might that be?

5. When is hurricane season (which months)? Why then?

6. What two main factors cause increased height of a storm surge? Which is more important?

7. What effects does the wind have on buildings during a hurricane?

8. How might you distinguish between storm surge damage and wind damage from a hurricane?

9. What effects do the higher waves of hurricanes have on a coast?

10. If the forward speed of a hurricane is greater, what negative effect does that have? What positive effect does it have?

11. If the eye of a west-moving hurricane were to go right over Charleston, South Carolina, where would the greatest damage be?

12. What is the difference in damage if a hurricane closely follows another hurricane—for example, a week later? Why?

13. Why is there more coastal damage during a hurricane if sand dunes are lower?

14. What shape of roof is most susceptible to being lifted off by a hurricane? Why?

15. Why is it so important to cover windows and doors with plywood or shutters?

16. What factors contribute to uncertainty in hurricane predictions?

17. What steps should you take to prepare for a hurricane?

18. Characterize the differences between a hurricane and a nor'easter.

Critical Thinking Questions

1. How is climate change expected to affect hurricane damages? How should this play a role in the discussion of costs to reduce carbon emissions?

2. Examine Figure 16-4B and 16-7 and explain why the Atlantic coast of northern Florida and Georgia has fewer hurricane strikes than coastal areas farther north and south.

3. Why do developers, builders, local governments, and many members of the public oppose higher standards for stronger houses?

4. In 2011 Hurricane Irene swept up the East Coast of the United States and into eastern Canada. State governors, mayors of towns, and emergency managers declared states of emergency and evacuations. Were these wise choices given the outcome? What would you have done?

■ The Bastrop Fire that burned near Austin in the fall of 2011 was the worst wildfire in Texas history.

17 Wildfires

A Fast-Spreading Wildfire

On a windy Sunday afternoon, September 4, 2011, fire broke out near Bastrop State Park, about 50 km east of Austin, Texas. Sparks from electric power lines ignited dry grass left by the worst drought since the 1950s and highest summer temperatures on record for any U.S. state. A century of fire suppression had left abundant dry leaves, pine needles, and branches on the ground, providing fast-burning ground fuel. Fanned by sustained winds of 43 km/h and gusts up to 50 km/h from Tropical Storm Lee, the firestorm quickly spread to produce 63 new fires in the following two days. Winds carried burning embers past firebreaks, major highways, and across rivers. The fire quickly spread through dry pine trees, overwhelming 400 homes and requiring evacuation of thousands of people. Pumper trucks and bulldozers were being used but a converted DC-10 air tanker from California could not be used because they lacked tanks and pipes to load it and a qualified available pilot. By September 8, 1386 homes had burned. The fire was finally completely contained on October 10 after burning 138 km^2, the worst wildfire in Texas history.

Fire Process and Behavior

Wildfires are unplanned and uncontrolled fires in minimally developed areas. They are commonly caused by lightning, downed or sagging power lines, or human carelessness. Wildfires are a natural part of forest evolution. They benefit ecosystems by thinning forests, reducing understory fuel, and permitting the growth of different species and age groups of trees. Wildfires become a hazard only when humans place themselves and their properties in environments that are susceptible to wildfires.

Factors that affect wildfire behavior include fuel, weather, and topography. In some cases, the fuel and conditions foster a firestorm that is virtually impossible to suppress without a change in the conditions.

The Fire Triangle

For a fire to burn, it requires three components—fuel, heat, and oxygen (**FIGURE 17-1**). Understanding each of these requirements can provide clues to means of both preventing and fighting wildfires. For homeowners or others in areas susceptible to wildfire, clearing ground *fuel* such as dry grass, leaves, needles, brush, and other fast-burning fuels from around a house or other valuable property can reduce the ease with which an encroaching fire can reach the building. Similarly, trimming dead branches from trees up to a level above the reach of flames from ground fuels can minimize the ease of igniting standing trees. For larger areas, fuel in the path of a potential fire can be removed by bulldozing a several-meter-wide path or "fire line" down to mineral soil. For an ongoing fire, firefighters commonly remove ground fuels between the active fire and unburned fuels using hand tools.

The second component of the fire triangle, *heat*, is needed to raise the temperature of the fuel—for wood this is about 300°C (572°F). That heat is often provided by a lightning strike or spark from an electric power line or from a railroad car on which metal parts rubbing together can overheat. Such ignition can also occur through a discarded match, cigarette butt, or welding equipment. In addition, fire can be ignited deliberately as in controlled burns to clear forestland for crops, to reduce ground fuels to minimize future fires, to eliminate fuel for an approaching fire by setting a back burn near a building to burn toward the main fire, or by arson. Heat can transfer from an advancing fire to ignite a building by sparks or burning embers carried ahead of the fire, by hot flames actually touching flammable material on the building, or by radiant heat heating the building. Firefighting often involves heat reduction by adding water either from the ground or from aircraft.

The third component of the fire triangle, *oxygen* in the air, is needed in the reaction of oxygen with carbon in the woody material of fuel to cause it to burn and produce carbon dioxide. Because the oxygen in the air is used up in the reaction, more must be provided by more air, either from wind or by air drawn in to replace air rising as it heats during combustion. A chemical fire retardant is sometimes designed to either coat the fuel to reduce the likelihood of its ignition or to release water to cool the fuel. Other retardants produce carbon dioxide, thereby reducing the amount of oxygen available.

Fuel

The type of fuel available to a fire, its distribution, and its moisture content determine how quickly a fire ignites and spreads, as well as how much energy it releases.

Fuel loading refers to the amount of burnable material available to a fire. Trees and dry vegetation are the primary sources of fuel for a wildfire. They burn at high temperature by reaction with oxygen in the air. The main combustible part of wood is cellulose, a compound of carbon, hydrogen, and oxygen. When it burns, cellulose breaks down to carbon dioxide, water, and heat. Shrubs and trees also contain natural oils or saps that add to the combustibles.

Surface fires spread along the ground fueled by grasses and low shrubs. Vegetation with large relative surface areas accelerates ignition and burning because heat that promotes ignition begins at the outer surface of the fuel, then works its way inward. The large relative surface area of dry grasses, tree needles, and to a lesser extent shrubs, allows them to burn more easily than trees. Fires in dry grass spread rapidly. With a small total mass of burnable matter on the ground—that is, grass, dry leaves, shrubs, and litter—a surface fire may move through an area fast enough to consume ground cover and understory brush but not ignite trees. Heavy fuels such as tree trunks have a small surface area compared with their total mass; they ignite with difficulty and burn slowly. Some tree species such as ponderosa pine have evolved with time to survive fires that burn ground vegetation.

Low brush and branches are known as **ladder fuels** because they ignite first and then allow the fire to climb into higher treetops (**FIGURE 17-2**). The ability of fire on the

FIGURE 17-1 The Fire Triangle

A fire requires fuel, oxygen, and heat. Without any one of these, a fire cannot burn.

FIGURE 17-2 Ladder Fuels

A. Dense ladder fuels in Glacier National Park, Montana.

B. Ladder fuels permit wildfire to climb into the trees; Yellowstone Park fire, 1988.

ground to easily reach tree crowns has a major effect on the rate of spread and intensity of a wildfire. Once flames reach into trees and then treetops, the smaller branches and fine needles of many trees easily ignite to form a **crown fire** as much as 10 or more meters high. A large crown fire can burn out at a single location, such as a cluster of trees, in less than a minute.

The intensity of a wildfire depends heavily on different types of vegetation. Lightweight flammable materials, such as dry grass and leaves, ignite easily but burn up quickly without generating very much heat. At the other extreme, heavy materials, such as tree trunks, are more difficult to ignite but burn much longer and generate much more heat. These include heavy coniferous (softwood) forests. By contrast, deciduous (hardwood) forests burn at somewhat lower intensities. Chaparral shrubs, conifers, and dead junipers contain volatile oils that permit them to ignite more easily and burn hotter.

Ignition and Spreading

Fires can be naturally started by lightning strikes, intentionally set for beneficial purposes, or set accidentally or by arson. Different regions of North America vary in their wildfire causes. For example, lightning-caused fires are more frequent in the Pacific southwest, Pacific coast, and northern Rockies. Human-caused fires are more frequent in the southeastern states, northeast, and California, especially near roads, campgrounds, and urban areas. In California, between 2001 and 2009, causes included lightning (5%), arson (7%), trash burning (10%), vehicles (14%), campfires and power lines (3% each), smoking (2%), and unsafe use of power equipment (about 30%). Various unintentional causes make up the remainder.

A major heat wave in southern Europe in the summer of 2007 fostered catastrophic wildfires. Some fires in Greece and Italy were ignited by arsonists in protected forests to create new areas for construction. In 2000, Italy passed a law that bans construction for 10 years after a wildfire, but the rules are not always enforced.

Investigators searching for clues to a fire's origin look for evidence left behind, including remnants of a campfire, matches, cigarettes, and evidence of accelerants such as gasoline. They also search for any glassy residue or elongated or straw-like groups of fused mineral grains (fulgurite) from a lightning strike. They can reconstruct the path of a fire by taking into account wind direction when the fire began and indications of the direction in which the fire moved. Fire-source indicators include postfire aerial photo and satellite image mapping of the burn distribution; weather records and wind patterns during a fire; topography, such as slopes and canyons; more-severe scorching and wood charring on the sides of trees facing a fire source; and lower height and degree of defoliation toward a source (**FIGURE 17-3**). Effects at the back of a fire nearer the source tend to be less intense. Unburned grass stalks sometimes fall backward toward a fire's source; burned stalks commonly fall forward in the direction of fire movement. Large noncombustibles, such as metal signs and rocks, can leave unburned patches in the direction of fire movement.

Fire ignites and progresses primarily by three mechanisms: radiation, convection, and burning embers. Radiant energy decreases with the square of the distance from a heat

By the Numbers 17-1

Radiant Heat and Distance

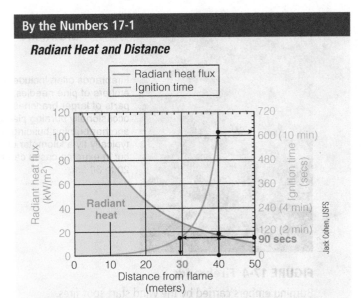

Radiant heat (red line) decreases rapidly away from a flame that is 20 m high and 50 m wide. The time it takes for flammable material, such as wood, to ignite (blue line) increases farther from the flame.

source, that is, at twice the distance the energy on a surface drops by four times (**By the Numbers 17-1:** Radiant Heat and Distance). With only radiant heating—the energy transferred by thermal radiation—without air movement or contact with flames, 50 seconds of intense crown-fire burn is sufficient to ignite dry fuels at a distance of 20 to 25 m. Convective transfer of heat is more direct and efficient; it involves direct movement of heated air or flames to ignite fuel. Fire also spreads when burning embers, called **firebrands**, are carried with the wind and then deposited, igniting **spot fires**, which burn ahead of the main fire but can quickly spread (**FIGURE 17-4**).

Firebrands are especially important because they often spread the fire far beyond the main fire front where homeowners are not expecting a problem. They can ignite dry grasses, twigs, or needles on the ground, or even standing trees. At houses they ignite flammable roof materials such as wood shakes, and they enter house vents, crawl spaces, and under stairways or decks.

Topography

Local topography, especially canyons, can funnel air, accelerate a fire, and cause more rapid spreading. In a canyon, as sometimes initiated by an untended campfire, fire can race up the valley because of this chimney-like funneling effect. Flames rise because their heat expands air, making it less dense. Thus, fires generally move rapidly upslope but slowly downslope (**FIGURE 17-5**). On a hillside, especially a steep one, flames and sparks rise with their heat, promoting the upslope movement of fires into new fuel. Although flaming branches and trees can fall downslope, advance in that direction tends to be slow. To be

FIGURE 17-3 Determining the Path of Fire

The more severely burned white side of this tree indicates the direction from which the fire came; in this case, the fire moved upslope to the right.

Crown fire

Firebrands often include glowing embers of pine needles, twigs, parts of larger branches, or occasionally burning pieces from another burning building. They typically fly a kilometer or more, but in extreme cases can travel 20 to 30 km.

USFS

Hyndman

Spot fire far downwind

FIGURE 17-4 Firebrands

Burning embers carried by the wind start spot fires.

John Newman, USFS

National Park Service

A

B

Heat

FIGURE 17-5 Fire Moves Upslope

A. Heat rising from a fire moves upslope. The 2007 Zaca Fire burns in the mountains north of Santa Barbara, California. Reaching the ridge crest, its progress slowed because of lack of fuel upslope. **B.** Close view of flames attacking new fuels above them, Yellowstone Park Fire, 1988.

upslope from a fire is to be in serious danger—as were the firefighters in the Storm King Fire, in Colorado in 1994.

At a ridgetop, fire often slows dramatically because the rising flames run out of fuel. A fire reaching a ridge crest can create an updraft of air from the far side that fans the flames, but the same updraft can reduce the chance that the fire will move down the other side.

Weather and Climate Conditions

Weather conditions can both ignite a fire and determine the rapidity of its spread. Lightning wildfires can be ignited by thunderstorms triggered at weather fronts, especially after a

period of dry weather (**FIGURE 17-6**). Lightning strikes in an area of heavy timber, at lower elevations or during high winds, can ignite a fire that progresses rapidly. Lightning strikes on rocky mountain peaks often have limited effect because fire progresses slowly downslope, and fuels there are often limited. Some climates such as the tropics are seasonally dry, leading to heightened wildfire hazards at those times. In the northern hemisphere, the tropics range from the equator to the Tropic of Cancer near 23°N and a similar distance south. Wet seasons follow high temperatures during the year so north of the equator the wet season is in the summer months and the winter dry season is subject to wildfires. Example winter fire areas include the Mediterranean climate parts of southernmost Europe and southern California, and the savanna grasslands of southern Mexico, central Brazil, and central Africa.

Mark S. Moak/USFS

FIGURE 17-6 Lightning Ignites Fires

This fire was started by lightning in the Selway-Bitterroot Wilderness area of the northern Rockies on August 9, 2005.

Fires start more easily and spread rapidly during dry weather because of low **fuel moisture**. High humidity makes it more difficult for fires to start and continue burning, whereas high temperatures and winds progressively reduce moisture over time and increase fire hazards. A few years of less-than-normal moisture dehydrates soil and lowers the water table, providing less moisture for the growth of trees and other vegetation. It may take several wet seasons to replenish the water shortfall below the surface. Thus, surface vegetation during a drought may green up in the rainy season but dehydrate quickly as dry weather returns.

Wind conditions dramatically affect the spread and direction of fires. An extreme example is the Santa Ana winds of southern California, the strong regional winds that blow southwest from Nevada in the fall. The wind descending through California's mountain passes compresses and heats up, causing the humidity to drop, and providing prime conditions for fire. Winds accelerate fires both by directing flames at new fuels and bringing in new oxygen in the air that facilitates burning. Once fires reach high into trees and crown, firebrands—burning embers carried downwind from flaming treetops—can ignite fuel as far as a kilometer away. High winds or large amounts of dry fuel can initiate **firestorms** intense enough to generate their own winds from a **convective updraft** of heat, which draws in new air from all sides and fans flames. Some types of vegetation, such as eucalyptus, junipers, and conifers, contain oils that produce tall, intense flames that in turn can spawn firestorms.

With continuing effects of climate change, more frequent and larger wildfires emit greater amounts of carbon into the atmosphere—primarily carbon monoxide (CO) and carbon dioxide (CO_2). Longer, drier summers contribute much larger amounts of burnable vegetation and more fires. Earlier snowmelt in the spring leads to a reduction in soil moisture, earlier and more complete dehydration of vegetation, and hotter fires. Weakened forests also become more susceptible to insect attack (pine-bark beetle infestations, for example) and tree deaths, so that forests are even more susceptible to fire.

Deliberate deforestation by fire also contributes. The largest areas of such deforestation in recent years have been in places like Indonesia, where forests are burned to make space for growing commercial crops, such as palm oil. The fires burn not only forest trees and shrubs but also peat that has stored carbon in the ground for hundreds of years. Studies estimate that 13 to 40% of fossil fuel emissions come from such fires. Lighting frequent small fires deliberately when weather is cooler and more humid can reduce fuels and slow the advance of hot fires in the dry season. Australia has had success with this procedure in the hot grasslands of its Northern Territory.

Secondary Effects of Wildfires

A major fire can lead to other hazards, such as floods and landslides, which are sometimes more disastrous than the fire itself. Major fires also produce air pollution.

Erosion Following Fire

Major fires often result in severe slope erosion for several years following the fire. In normal, unburned areas, vegetation has a pronounced effect in reducing the runoff after rainfall and snowmelt. Tree needles, leaves, twigs, decayed organic material, and other litter on the ground soak up water and permit its slow infiltration rather than turning it into surface runoff. Evapotranspiration by trees and other vegetation also decreases the amount of water soaking deep into the soil.

Intense fires burn all of the vegetation, including litter on the forest floor (**FIGURE 17-7A**). The organic material burns into a **hydrocarbon residue** that soaks into the ground. This material fills most of the tiny spaces between fine soil grains in the top few centimeters of soil and sticks them together (**FIGURE 17-7B**). In some cases, water can no longer infiltrate the surface soils, thus most of it runs off. Such impervious soils are called **hydrophobic soils**. The hydrophobic nature of soils is often attributed to intense burning, but the relationship to fires is not entirely clear. What is clear is that water more readily runs off the surface of charred soil.

Big rain droplets beat directly on unprotected ground. When rainfall is heavy and vegetation and ground litter have been removed by fire, the water cannot infiltrate fast enough to keep up with the rainfall. Raindrops splash fine grains off the surface, and the runoff carries them downslope as sheetwash, or overland flow, which easily carves out gullies and channels in the unprotected soils (**FIGURE 17-8**). All of this surface runoff quickly collects downslope in larger gullies and small valleys. Water collecting in a valley with a large-enough watershed quickly turns into a flash flood or debris flow. These destructive and dangerous floods rip out years of accumulated vegetation, soil, and loose rock, carrying everything down valley in a water-rich torrent.

Rapid accumulation of water from overland flow following a fire can lead to flash floods that wash out bridges, roads, and buildings. After a region has burned, it is unsafe to drive or walk in small valley bottoms during intense rainstorms because swift runoff funnels most of the precipitation into adjacent gullies and then quickly into sequentially higher-order streams. Even when storms are several kilometers away in the headwaters of a burned drainage, a flash flood downstream can appear under a clear blue sky. Homes that may have survived a fire in usually dry valley bottoms are vulnerable to debris flows from all of the new sediment washed downslope after a fire.

Mitigation of Erosion

Erosion following a fire cannot be prevented, but it can be minimized. Federal and state agencies and individuals can plant vegetation, grass, shrubs, and trees that prevent the direct impact of raindrops on bare soil. Over large areas, slopes are often seeded soon after a fire. Straw can be packed into tubes or laid out in bales to provide a barrier to overland

Surface of soil sealed by resins formed during fire

Thomas Spittler, California Department of Forestry

Hyndman

FIGURE 17-7 Soil Conditions After a Fire

A. This steep slope on the O'Brien Creek drainage at the western edge of Missoula, Montana, had been ravaged by fire one month earlier, leaving it vulnerable to erosion.

B. In hydrophobic soil such as this produced by the Banning fire in southern California in 1993, the dry layer lies beneath the dark gray hydrophobic layer, which is composed of soil and ash. The view is about 20 cm across at the base.

Hyndman

M. Rieger, FEMA

FIGURE 17-8 Flooding after a Fire

A. These gullies were eroded by a short-lived rainstorm one year after the 2000 fires in the southern Bitterroot Valley, Montana.

B. This mudslide followed the Hayman, Colorado, fire on July 5, 2002.

flow, thus reducing slope erosion. Felling dead trees across a slope can have a similar effect. Drains can direct water flow laterally to valleys to minimize surface gullying. On small steep patches of particularly vulnerable soil, sheets of plastic can be spread to prevent water from reaching the soil.

Air Pollution

Major wildfires can spread particulates in the form of smoke, soot, and ash over areas more than 1000 km away and, in some cases, around the planet (**FIGURE 17-9**). Pollutants

FIGURE 17-9 Air Pollution

A 2011 wildfire in West Texas produced smoky air pollution.

include carbon monoxide, hydrocarbons, and nitrogen oxides. Sunlight can react with these pollutants to produce ozone at low elevations. In the stratosphere, ozone helps protect us from ultraviolet radiation, but close to the ground it can destroy living tissue and cause respiratory problems in some people. Studies after the Southern California wildfires showed a large increase in hospital admissions from asthma, bronchitis, and related ailments, especially in young and elderly people. Health experts recommend avoiding outside activity and in some cases taking medication. Southern California wildfires have repeatedly increased ground-level ozone to unhealthy levels.

Wildfire Management and Mitigation

Wildfires are necessary to the evolution of forest ecosystems, but out-of-control fires pose hazards to humans. Wildfires occur in most regions of the United States, where they affect wildlands and suburban or built-up areas that encroach on fire hazard areas (**FIGURE 17-10**). From 1985 to 1994, an average of 73,000 fires per year burned on federal lands, consuming more than 121,000 km^2 and costing $411 million for fire suppression. From 2000 to 2007, annual firefighting costs exceeded $1.3 billion in four of the seven years. Additional millions were lost in timber values, and indirect costs resulting from landslides and floods are unknown.

Hazards to the public increase as more people move to forest, range, and other wildlands. Over 10 days in October 2003, catastrophic fires destroyed 3452 homes in southern California (**Case in Point:** Firestorms Threaten Major Cities—Southern California Firestorms, 2003 to 2009, p. 499). From October 25 to November 3, 1993, Santa Ana winds fanned 21 major fires in southern California, burning 76,500 acres (309.6 km^2) and 1171 buildings. Three people died.

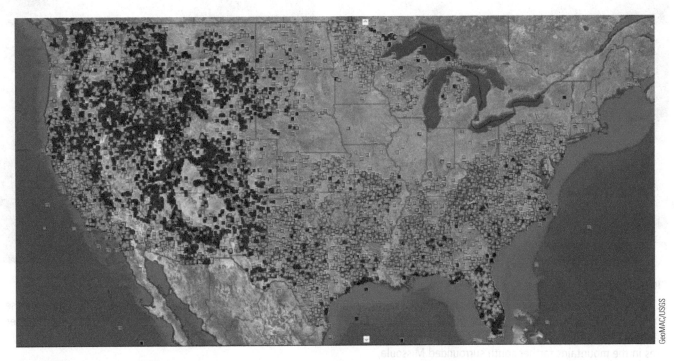

FIGURE 17-10 U.S. Wildfires

The map shows large wildfire locations from 2002 through August 2015. Black spots indicate lightning-caused fires while red spots are human caused. Pink (pale) spots indicate fires with unknown causes.

As with all hazards, a clear understanding of the natural processes underlying fire behavior allows governments and individuals to take steps to protect lives and property.

Forest Management Policy

The U.S. Forest Service was formed in response to a catastrophic wildfire that burned 12,140 km^2 of northern Idaho and western Montana in the summer of 1910. For most of the twentieth century, their policy was to aggressively fight wildfires. Until a few years ago, the Forest Service and its Smokey Bear mascot maintained that fire was bad and all fires should be extinguished as quickly as possible.

Because the United States spends upward of $1 billion per year on fighting fires, you might expect some significant long-term benefit. However, many experts argue that we get no long-term benefit. In fact, preventing fires leads to buildup of wildland fuels, which ultimately leads to worse fires. Forest tree densities and fuel loads have risen to critical levels, creating ideal conditions for fires. Many open, park-like forests became choked with closely spaced smaller trees and dense underbrush, leading to disastrous wildfires, as in the case of the Bitterroot Valley fires of westernmost Montana in 2000 (**FIGURE 17-11**). Decades of fire suppression have led to a buildup of wildland fuels. Very large wildfires in recent years include the 1036 km^2 Carlton Complex fire in Washington in 2014, the 1601 km^2 Buzzard Complex fire in central Oregon in 2014, the 1171 km^2 Whitewater-Baldy fire in New Mexico in 2012, the 2177 km^2 Wallow fire in Arizona and New Mexico in July 2011, the 2643 km^2 Murphy Complex fire in Idaho and Nevada in 2007, and the 3213 km^2 Yellowstone fires in 1988.

Federal agencies now recognize that fire is a natural part of wildland evolution, a process necessary for the health of rangelands and forests. They recently shifted to mitigation and management of fires, including setting controlled fires intentionally in **prescribed burns**. The Forest Service now permits many wildfires to burn and consume excess fuel in wilderness areas as long as they don't endanger buildings or important resources.

Fighting Wildfires

When wildfires approach human developments, several approaches are used for fire suppression. Highly trained firefighters, called **smokejumpers**, parachute into remote areas of lightning-caused spot fires to exterminate them before they spread out of control. Other firefighters hike in from the nearest roads to hand-dig meter-wide paths, **firelines**, down to bare mineral soil to slow or stop a fire's spread. Since many low-intensity wildfires mostly burn along the ground, even a narrow firebreak can provide a defensible line. Bulldozers are used to cut firebreaks, both to slow the spread of active fires and as preemptive measures in fire-prone areas long before fires break out. Often they are located along ridge crests where wildfire activity

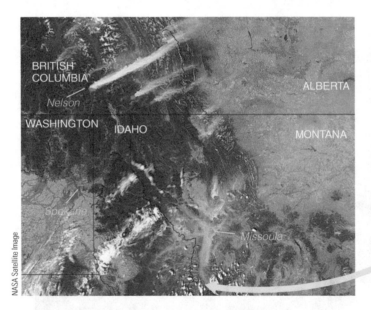 

FIGURE 17-11 **Bitterroot Valley Fires**

A. Forest fires spread throughout the northern Rockies on July 15, 2003, as shown by the red areas on the image. Fires in Glacier National Park burned just south of the Alberta-Montana border. Fires in the mountains farther south surrounded Missoula, Montana. Smoke plumes, clearly visible as bluish-gray areas, are quite different from the white clouds across the region.

B. Fires in western Montana on August 6, 2000, chased these two elk into the relative safety of the Bitterroot River's upper drainage.

FIGURE 17-12 **Firefighting**

A. A large slurry tanker (MD-87) drops retardant to constrain the edge of the Two Bulls fire northwest of Bend, Oregon, on June 7, 2014. Bright orange dye is added to show pilots coverage from previous drops.

B. A helicopter fills up at a local pond in Sonoma Valley, California, before dropping its load on a fire nearby.

naturally weakens because of the terrain. Aircraft drop giant loads of water or fire retardant along the lateral and advancing edges of fires to help direct and contain them (**FIGURE 17-12A**). Retardants include ammonium sulfate, ammonium phosphate, borate, or swelling clay. Because of their toxicity to some plants, use of retardants on federal forestland is not permitted within 300 feet (~100 m) of water bodies or in areas designed to protect endangered plants. Exceptions are made to protect people in immediate danger from flames. The restrictions do not apply to California firefighters. Helicopters scoop huge buckets of water from nearby streams or lakes and dump them on the edges of fires (**FIGURE 17-12B**).

For the largest fires that threaten towns or critical facilities, firefighters sometimes deliberately set burnouts (**FIGURE 17-13**). In a **burnout**, a large area in the path of a big fire is burned under controlled conditions and suitable weather, generally beginning at a road or river; the burnout fire burns back toward the first fire, eliminating fuel and stopping the progress of the main fire. Before lighting a burnout, the forest or range between the burnout and the area needing protecting is thoroughly drenched using planes and helicopters. The technicalities of a burnout are complex and dangerous, but they can work remarkably well. In July 2003, for example, firefighters set a large burnout that saved the town of West Glacier at Glacier National Park from a huge fire bearing down on the area.

A **back burn** is a fire deliberately set close to an advancing fire to create a firebreak by using the updraft of the wildfire to draw the back burn fire back toward the main fire. A back burn is typically set by using a *drip torch* to drip flaming droplets on the ground fuel (**FIGURE 17-14**). In 2013,

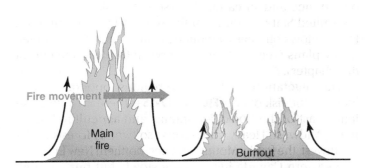

FIGURE 17-13 **Burnout**

In a burnout, firefighters intentionally set a fire that is drawn in by the updraft of the main fire. This causes the intentional fire to burn back toward the main fire, which consumes fuels and hinders the wildfire's progress.

firefighters used burnouts at California's Rim fire to protect 4500 homes, Hetch Hetchy Reservoir (San Francisco's water supply), and Yosemite National Park.

Where there is no escape route, firefighters resort to taking cover under specially designed tent shelters made of aluminum foil, woven silica, and fiberglass that can provide some protection from the heat and flames. The shelters can provide limited protection for less than 10 minutes in moderate fire at 315°C (600°F) conditions, but much less protection in a hot fire that may reach 871°C (1600°F). Firefighters using the shelters can still suffocate from smoke inhalation, carbon monoxide poisoning, or lack of oxygen, and suffer from burns.

FIGURE 17-14 Drip Torch

A drip torch used to ignite a back burn.

Risk Assessments and Warnings

Several government agencies study patterns in wildfire occurrence and spreading to assess the risk for regions of the United States at different times. Closely monitoring risk levels allows officials to reduce hazards and prepare emergency plans (see the Wildfire Survival Guide at the end of this chapter).

The vegetation, and in some regions topography, can increase the risk of fire (**FIGURE 17-15**). These include grasslands east of the Rocky Mountains and agricultural lands in the Midwest. Heavy coniferous (softwood) forests in the Northwest, the Rocky Mountain states, northern New England, and eastern Canada ignite more slowly but burn at high intensities. By contrast, deciduous (hardwood) forests of the southeastern states burn at somewhat lower intensities.

Typical weather patterns affecting wind direction or moisture for a region also influence fire risk. Fires tend to advance in the direction of prevailing winds, so fires in the northern United States and southern Canada tend to burn toward the northeast with westerly winds. In the trade-winds belt of Mexico, fires tend to burn toward the southwest. Thus, homes in the prevailing downwind direction from fire-prone or high-fuel areas are more at risk than those upwind from such areas. The desert regions in the Southwest, from West Texas to southern California and northward to southern Wyoming are at higher risk because of their low moisture levels. Areas with more moisture are less at risk. Along the West Coast, prevailing westerly winds carry moisture off the Pacific Ocean, and in the southeastern states, storms bring abundant moisture north off the Gulf of Mexico.

Agencies also monitor shorter-term weather conditions and alert the public of rising risk levels. The favorability of weather conditions for wildfire is called its fire weather potential (see Figure 17-15B). **Fire weather potential** is posted on fire danger-level signs along roads by the U.S. Forest Service and Bureau of Land Management. For a local area, it depends primarily on the number of high-temperature days, the air's relative humidity, moisture in available fuels, and wind speed. For large regions, it depends on similar factors over periods from weeks to months.

Drought maps provide a record of longer-term moisture deficits. A high-resolution infrared sensor is used to measure greenness of vegetation and inter-fuel moisture content. This greenness is compared with the typical value for each map area. Such estimates are used to infer fire danger and inform the public through news media and roadside signs.

Fire-detection methods have evolved from the fire lookout towers located on mountaintops decades ago, but some manned towers are still used, especially in remote areas. Ongoing detection tactics include the use of weather radar to detect suspect areas after lightning storms; light aircraft

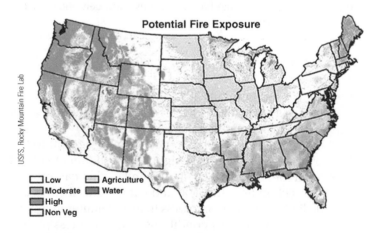

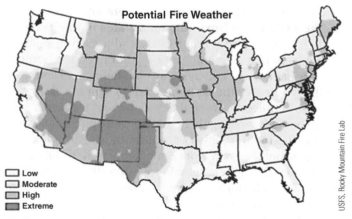

FIGURE 17-15 Fire Hazard Areas

A. Potential fire exposure depends upon different vegetation types, how quickly fire spreads, and with what intensity.

B. Extreme fire weather potential in the continental United States averaged over 16 years.

are then sent in to spot smoke. Aircraft are also used to detect new fire starts off of large fires caused by wind-carried firebrands. Another approach involves automated smoke-detection systems, mounted on tall towers, with multiple cameras powered by solar panels that communicate with central offices via microwave or satellite. False alarms from automated systems remain a problem.

Mounted in light aircraft, high-resolution infrared heat-detection cameras combined with visible-light cameras with GPS location equipment can now detect fires as small as 30 cm (~1 ft) across, even in bright sunlight. NOAA uses automated imaging by Earth-observing satellites, but results are dependent on hot fire temperatures, large fire extent, and subjective interpretation by fire analysts.

Protecting Homes from Fire

Many people who live in the woods do so to escape urban congestion and experience the beauty of nature (**FIGURE 17-16**). They like to be surrounded by trees and shrubs, but for a wildfire, trees and shrubs are fuel. Not only do forest dwellers put themselves and their property in danger, but they also endanger firefighters and compel fire-management officials to expend limited resources protecting individual properties rather than managing an overall fire.

The risk to flammable structures, especially people's houses, depends on the fire weather potential, the regional type of vegetation—how fast fire will spread and how hot it will burn—and the housing density. At one extreme, fire in a wilderness area or unpopulated desert endangers no houses. At the other extreme, a firestorm in a metropolitan area can become a catastrophe because it affects numerous homes and people.

FIGURE 17-16 Dense Trees Increase Risk to Homes
Fire consumes homes surrounded by heavy forest northwest of Payson, Arizona, in the Dude fire, June 25, 1990.

FIGURE 17-17 Roof Fires
These fires were ignited by firebrands landing on roofs. The homes might otherwise have been saved from low-intensity ground fires nearby.

Although individual homes scattered throughout a forest might constitute a low-risk area overall, the solitary homes could be in severe danger, especially if they are built from flammable materials and surrounded by ground fuels. Even those who live in an urban interface near a forest are in danger, particularly in dry, hot weather, where high winds can drive fires.

Firebrands carried from treetop crown fires can easily ignite flammable rooftops and decks (**FIGURE 17-17**). Wood-shingle roofs, which are favored by many people for their natural appearance, are often ignited by falling firebrands, even if a wildfire is not particularly close and there is no nearby vegetation. Once started, fire spreads rapidly across the roof and into the house. Wooden walls will not ignite from this process because the burning embers do not remain in contact long enough to ignite them. Composition shingles are better than wood shakes because, although they may burn if they get hot enough, the fire will not easily spread to other areas.

Those who live in the forest can minimize danger by building with flame-resistant materials. With a nonflammable roof and deck and no fuel within 30 m, a building has a good chance of surviving, even with falling firebrands. Burning embers landing on a metal roof will not ignite it, although if the roof gets sufficiently hot the wood framework supporting it will ignite. Metal roofs, though not as natural, are flame resistant.

Trees or ground fuels around a home can also be a danger, either from the spread of fire along the ground or from falling firebrands. Low-intensity fires often burn forest homes, even though surrounding trees and fences may survive, because low-intensity fires burn ground

FIGURE 17-18 Surface Fires

A. Missionary Ridge fire, Durango, Colorado, June 2002. Surface fire spread through ground needles and ignited the house but did not burn part of a wooden fence, ponderosa pines, and some low vegetation.

B. In the Cerro Grande fire at Los Alamos, New Mexico, in May 2000, houses on one side of a street in a forested community burned to the ground, whereas those on the other side remained unscathed. The ground fire was stopped by the street, indicating that it spread through ground fuels.

fuel, such as needles, leaves, and twigs, without reaching higher into trees. In the aftermath of many wildfires, investigators frequently find homes burned to the ground but the surrounding trees still mostly green (**FIGURE 17-18**). The best protection for homes in forests is fuel reduction before fires start. Homeowners should remove underbrush, low-hanging ladder fuels, and especially dry ground fuels (**FIGURE 17-19**).

FIGURE 17-19 Removal of Ladder Fuels

Ladder fuels have been removed from this slope to prevent fire from easily reaching the tree branches, but dry grass could still provide a path to the house.

Even once ground fuels are cleared, radiant heat from nearby burning trees can ignite a home. Because radiant heat decreases rapidly away from a fire, clearing fuels out to about 30 m from a home will generally protect it from ignition by that mechanism. In fact, dry wood structures may not ignite even when fire is close enough to burn people severely (**FIGURE 17-20**). Exposure to heat that will produce a second-degree burn on skin in 5 seconds will take more than 27 minutes to ignite wood!

For small, high-value wood structures, a building may be wrapped in aluminum fabric (**FIGURE 17-21**). This reflects much of the radiant heat away from the building but provides only a small amount of protection from direct flame.

In many cases, government regulations on building materials can mitigate damages from future fires. On October 20, 1991, a firestorm in the hills upslope from Oakland, California, destroyed or damaged 3354 single-family homes and 456 apartments (**Case in Point:** Firestorm in the Urban Fringe of a Major City, Oakland–Berkeley Hills, California Fire, 1991, p. 501). It killed 25 people, and damage amounted to $2.2 billion. Eighty percent of the homes were rebuilt in the same risky hillside locations with their views of San Francisco Bay and the Golden Gate Bridge. However, new restrictions dictate that roofs be fire resistant, decks and sheds be built with heavier materials that don't burn as quickly, and landscaping use fire-resistant plants.

Evacuation before a Wildfire

Preparation for evacuation in a fire-prone area is important because too little time is available when a fire approaches. Forest Service guidelines say to evacuate all family members

FIGURE 17-20 Wood Structure Withstands a Fire

A wall of flame at the edge of a forest clearing provides enough radiant heat to severely burn skin but does not burn either wood posts or the siding on this home during the 2003 Wedge Canyon fire along the North Fork, Flathead River, Montana. During the fire (left) and after the fire (right).

FIGURE 17-21 Radiant Heat Protection

Firefighters wrap wood walls and the roof of a historic Gila Forest cabin in aluminum fabric for radiant heat protection from wildfire.

and animals who are not essential to preparing the property well in advance, designating a contact person and safe meeting place beforehand. Wear cotton or wool clothes that do not burn easily, long pants, gloves, and a kerchief to cover your face. Tune into a portable radio to hear instructions. Place your vehicle in a garage with the garage door closed but an electric garage door opener disconnected so the door can be opened manually if power fails. As with other imminent disasters, put essential items in the car, including important documents. Crucial items might include bank and insurance

records, birth certificates, credit cards, medications, drivers licenses and passports, computer backup files, house photographs and an inventory of its contents, cell phone and charger, and family heirlooms. Close all interior doors, remove non-fire-resistant coverings from windows, turn off pilot lights, leave a light on in each room, and move stuffed furniture to the center of the room. Shut off gas lines, close exterior windows and vents, prop a ladder against the roof for firefighters, unlock exterior doors, connect garden hoses to faucets, and leave trash cans and buckets full of water for firefighters. If there is time, cover windows and vents with plywood at least one-half inch thick and wet down wood roofs.

Forced Evacuation

Under circumstances in which local authorities believe residents' lives are threatened, they may dictate mandatory evacuation of an area. In general, federal policy is to defer to the laws of individual states; local officials work with states to enforce those laws. Federal Emergency Management Agency (FEMA) and the American Red Cross may facilitate the evacuation and shelter of evacuees. California law authorizes officers to restrict access to hazardous areas, but because no court has upheld a law requiring people to evacuate their own residence, "mandatory evacuation" apparently means that authorities believe strongly that people should evacuate for their own safety.

What to Do if You Are Trapped by a Fire

Of course, no one ever wants to be trapped by a wildfire, but if it ever should happen, keep in mind how fire behaves, and act accordingly. Light fuels such as grass and dry brush burn fastest, especially uphill and in the direction of the wind. Fuels burn uphill faster, so try to get upwind and downslope of a fire. You are not likely to outrun a wildfire up a slope—it may advance at more than a meter per second. If you can see the extent of a fire below, you may be able to move laterally, then downslope to avoid the advancing fire front. Even on flat areas it may move at 100 to 150 m per minute. Because convective updrafts concentrate in swales or gullies on slopes, creating a chimney effect, avoid those areas—stay on outward-rounded (convex) parts of a hillside.

Try to reach flat, moist, grazed areas, ponds, or streams where vegetation is short, wetter, and not subject to updrafts. If trapped by a closely advancing fire, lie in a ditch, cover yourself with noncombustible material—such as wet clothes or wet dirt—and let the fire burn past you. If you can reach a car, park it over bare ground. Roll up the windows—covering them with opaque material to minimize access of radiant heat, if possible—and lie low. Even if the car catches fire, try to let the intense wall of fire pass before getting out. Except in movies, gasoline tanks rarely explode. Don't try to outdrive a closely approaching fire and don't drive through dense smoke—your chances of making it are slim. The safest place in an uncontrolled fire is in an area already burned. A burning line of short fuels such as grass may have a gap—a trail or road—that may provide access to already burned areas. Studies have shown that even running through a flame front 3 m high (about 10 ft) and 20 m (65 ft) deep may be survivable if you can maintain a speed of 4 m/s (about 15 mph). You would also need to hold your breath, because inhalation of hot gases causes severe and potentially fatal respiratory system damage.

Public Policy and Fires

As with all hazards, the best way to mitigate damages from fire is to keep people from building in high-risk areas. Unfortunately, with the fiercely independent attitudes of many rural residents, zoning restrictions are considered unacceptable infringements on their freedoms. They insist on their right to do with their property as they wish. Some even build in so-called **indefensible locations**, such as narrow canyons that are too dangerous for firefighters.

The most dangerous part of the day for firefighting is late afternoon when weather is hottest, and winds often develop to blow the fire in unpredictable directions. Firefighter deaths in the Storm King South Mountain, Colorado, fire in 1994, Esperanza, California, fire in 2006, and the Yarnell, Arizona, fire in 2013, all happened at about 4 to 5 p.m.

Who pays for the costs of fighting fires and trying to save the homes of those who choose to live in fire-danger zones? Insurance companies generally set premiums at levels based on risk and replacement costs. But fire insurance rates, though rising, pay only a small part of the cost because federal and state governments pay most of the fire-fighting and cleanup expenses and generally provide a portion of rebuilding costs. When governmental agencies spend millions of dollars fighting fires, and most of those efforts are expended in protecting homes, much of the real cost is borne by the general public rather than the people who choose to live in wooded areas.

Although fire insurance for homes in wildland–urban interfaces is not as high as might be expected given the greater hazard, some insurance companies are now requiring wildland homeowners to clear ground duff and brush, and cut down trees to create a "defensible space" or install fireproof roofs. They are also dramatically increasing insurance premiums. People should view their houses in the forest as a nice dry stack of kindling surrounded by green trees (see Figure 17-17). The house will likely burn; the trees may not. The country's second-largest insurer announced in May 2007 that it would no longer provide new homeowner policies in California because of risks from wildfires and earthquakes.

One needs to ask whether public funds should be used to help people rebuild in places where wildfires—and associated floods and erosion—are ever-present. Our conscience tells us that we should help those in need, but such help merely encourages them and others to live in vulnerable areas. Perhaps, as FEMA has finally learned, we should provide more help to people who are willing to relocate to less-vulnerable places.

Should people accept more responsibility for their own safety and property? Until 2009, in Australia, a country having equal problems with major disastrous wildfires, responsibility was largely left to the individual. The Australian government did not focus on protecting private wildland property, but it educated people on how and when to evacuate on their own, before fires reached them, and provided programs that teach ways to make property less vulnerable to fires. Government policy mandated either evacuating early or staying and defending your house from fire. Most deaths occurred during last-minute evacuations.

Australian summers are dry and hot, especially when heat waves push temperatures into the mid-40s°C (110°F) and humidity drops to about 5%. On February 7, 2009, a blistering summer day, those temperatures accompanied 100-km/h winds that fanned bush fires in areas near Melbourne, Victoria. One of several huge, fast-moving firestorms incinerated whole towns, killing 120 of the 173 people that died on that single day. Most of the fires were ignited by fallen power lines, others by lightning, cigarettes, power tool sparks, and arson. The fires followed a decade-long drought and gradual warming that has been associated with human-caused climate change. Following that catastrophe, the Australian government changed its stance and planned to create a nationwide fire alert system involving telephone and text messaging.

An alternative to zoning is to stipulate that if people do build in a vulnerable location, they must not expect public help in times of crisis. Such people should not count on public assistance for firefighting, stabilizing streams or hillsides before or after floods, or rebuilding after catastrophes. If people still insist on living in fire-prone areas, they could create special fire-prevention districts that would tax their members—for example, a few thousand dollars per year—to create a pool of funds to pay the costs of future fire protection for their homes.

WILDFIRE SURVIVAL GUIDE

U.S. High Hazard Area

- Many are caused by lightning in western mountains and by humans nationwide. Any area downwind of a large fire is at risk, especially during high winds.
- Worst for fire weather potential is the desert Southwest, due to high temperatures and low humidity.

Preparation

- Home protection: Clear a 30-m defensible space around the house, including removing leaves, needles, woodpiles, and other flammables. Keep any grass mowed short and green. Clear brush and ladder fuels away. Minimize use of flammable roofing and easily ignited deck and stair material. Block in overhanging eves.
- Prepare for evacuation when necessary, beginning early with most vulnerable family members and animals. Decide on a contact person and place. Take essential items. Connect garden hoses, leave a ladder propped against the roof, turn off gas lines, and cover windows and vents with plywood. Leave light on in each room and outside doors unlocked for firefighter entry.

Warning Signs

- Abundant ground vegetation after a long period with little rain. Government warnings and forecasts of hot weather, strong electrical storms, and high winds. Be wary if you see or smell smoke.

During the Event

- Wear cotton or wool clothes that do not burn easily, long pants, gloves, and face kerchief. If trapped by fire, try to stay upwind and downslope of the fire. Safest areas include flat, wet areas, lakes and streams, burned-over areas, roads, and other clear ground.

Aftermath

- Hazards in burned areas include falling snags and branches, especially with winds. Rain following fires often initiates dangerous mudflows and debris flows, which are especially hazardous near and below gullies.

Cases in Point

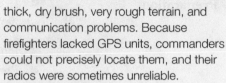

Unexpected and Deadly Change in Fire Behavior
Yarnell Hill Fire, Arizona, 2013 ▶

Late June 2013 in central Arizona was unusually hot, almost 10°C above normal, and with no precipitation for the whole month. From June 28 to 30, temperatures around Prescott and Yarnell were 38 to 40°C, (101 to 104°F). The hilly region was under severe drought so the chaparral vegetation was drier than normal. Phoenix, 167 km (85 mi) southeast, reached 46 to 48°C (115 to 119°F) those days.

The fire, about 110 km northwest of Phoenix, was ignited by dry lightning on June 28; it grew the next day from 0.8 km² (200 acres) to 34 km² with shifting winds that gusted to 65 to 80 km/h and blew in wildly different directions. Aerial tankers had been grounded because of severe shifting winds and building thunderstorms.

The 20-person hotshot crew sent to battle the blaze appears to have followed the safety guidelines. Following the tragic 1994 Storm King Mountain fire that killed 14 firefighters, government standards were revised to emphasize safety measures. These measures included having escape routes and safe zones, along with posting lookouts and paying close attention to changing weather. Firefighters wear fire-resistant clothing and carry about 20 kg (40 to 50 lbs) of gear, including an emergency shelter, an axe-like tool, food and water, and sometimes a chain saw and fuel.

Despite these increased safety measures, 19 of the firefighters died on the afternoon of June 30th, when caught in a *burn over*, in which shifting winds blew flames back over them. It was the largest loss of wildland firefighter lives in 80 years. Contributing to the tragedy were unstable winds, thick, dry brush, very rough terrain, and communication problems. Because firefighters lacked GPS units, commanders could not precisely locate them, and their radios were sometimes unreliable.

The old mining community of Yarnell, population 700, was evacuated before the fire. Ultimately, the fire destroyed 200 buildings, including 100 homes in the town. Unfortunately, only 63 of the 569 buildings in Yarnell had cleared proper buffer zones of brush and trees to protect them from wildfire; 60 of the 63 survived the fire, demonstrating the effectiveness of this prevention measure. A research report following the fire has questioned whether the firefighting team should have been deployed in an area that did not do their part to protect homes.

■ *The Yarnell Hill fire, Arizona, June 30, 2013.*

inciweb.nwcg.gov

CRITICAL THINKING QUESTIONS

1. What are the priorities for firefighters when deciding what to protect? What are the likely consequences of each decision?
2. In this fire, what went wrong that was unexpected for those involved? How would you avoid such problems in the future?
3. What factors made this fire especially dangerous for firefighters, and what changes would you have made if you had been in charge?

Wildland–Urban Fringe Fires
Waldo Canyon and Black Forest Fires, Colorado Springs, 2012 and 2013 ▶

The Waldo Canyon fire, a wildfire that began early the morning of June 23, 2012, 6.4 km northwest of Colorado Springs, and 1.6 km from a residential neighborhood, was determined to have been human-caused, but it was unclear whether it was deliberate or accidental.

The fire spread quickly in erratic winds, and voluntary evacuations were ordered. Within the first day, the fire spread to 8 km² and required the active use of hundreds of firefighters, dozens of water-pumping fire trucks, and aircraft dropping fire retardant.

On June 26, the fire jumped a containment line, fed by thunderstorms and gusty winds from the west reaching 105 km/h. It then topped a canyon and raced downslope into western Colorado Springs, burning 346 homes. Afternoon

Modified from inciweb.nwcg.gov

US Army

Adam Drake, inciweb.nwcg.gov

■ *Waldo Canyon and Black Forest wildfires near Colorado Springs.*

temperatures reached a record 38°C. By mid-afternoon, city officials asked for help from all nearby counties, including Denver. The rapidly changing firefight was hindered by missteps in managing the fire, internal miscommunication, and a lack of resources including maps for out-of-town firefighting groups. There were 26,000 evacuees jamming the roads trying to escape. One couple in their 70s died from heat and smoke inhalation while preparing to evacuate.

By the time it was 100% contained a week later, the fire had charred 74 km², destroyed 347 homes, and caused the evacuation of 32,000 people. A year later, $454 million in fire insurance claims had been filed. Scientists and fire officials remained concerned that burned areas could lead to widespread and destructive mudslides and debris flows because of lack of vegetation to slow infiltration of later rains. These could impact homes, culverts and drainage pipes, and highways, downing trees and damming storm drains. Nearly a year later, on July 2, 2013 and again on August 9, heavy rains falling on the burned area flushed through Manitou Springs, washing away cars and houses and closing Highway 24 with mud and boulders.

Black Forest Fire, 2013

Almost a year later, near midday on June 11, a fire ignited in Black Forest, about 24 km north of Colorado Springs. The prolonged, severe drought was lessened by some spring moisture, but vegetation remained dry. On June 11, temperatures in Colorado Springs reached 36°C (97°F) and in Denver 38°C (100°F). Sustained winds of 46.6 km/h gusted to 58 km/h. Humidity was an extremely low at 4%.

Strong winds fanned the flames, which traveled mostly in needles, leaves, and branches on the ground. It was the most destructive fire in Colorado history. Even worse than the Waldo Canyon fire, the Black Forest fire burned 57.5 km², destroyed 511 homes, damaged an additional 28, and forced the evacuation of 38,000 people. Two people died of smoke inhalation and heat while preparing to evacuate. The couple that died had waited for automated calls ordering them to evacuate, whereas their neighbors left when they saw approaching smoke. Many homes were built at road ends with no other route of escape and on winding mountain roads with slow travel times. Most people built with wood and neglected to clear flammable trees and dry materials from areas near their homes. Homes in some areas survived

because people had removed dead trees, ground vegetation, and forest-floor duff, and spread rocks and gravel to prevent fire from creeping up to their homes. An elementary school in the midst of the fire survived because of a large area of treeless parking lots surrounding it.

The Black Forest fire evacuation zone included 7000 homes, half of them built in the past decade. Since the 1990s, about 16 million new homes across the country have been built in and adjacent to fire-prone forest areas. Eight of the nine worst fire seasons in the United States occurred between 2000 and 2012. This combination of more people living in and near forested areas and the effects of global warming have complicated agencies' efforts to fight wildfires.

CRITICAL THINKING QUESTIONS

1. Why are homes in forested areas more vulnerable to wildfires?
2. Do people living in such areas endanger themselves and others; if so, who and what dangers are there?
3. Should people be prohibited from building in forested areas? Why or why not?
4. Should those who own homes in forests pay for the eventual cost of firefighting? If so, how?

Heat from an Erratic Wildfire

Lolo Creek Complex, Western Montana, 2013 ▶

Fires were initiated from lightning strikes on August 18, 2013, in the rugged hills west of Lolo, Montana. Multiple fires spread rapidly by high winds, through dry grass and timber on steep slopes, an area that has become part of the wildland–urban interface, and quickly became the nation's top firefighting priority. U.S. Highway 12 was closed when two fires merged, and the fire blew across the highway as if it wasn't there. Single trees torched like roman candles. Numerous power lines and phone lines burned, cutting power to the area. Smoke-obscuring visibility, falling snags, and burning debris falling or rolling down onto the highway created severe hazards. A long line of cars and semis backed up on Highway 12 to the west at the Idaho border. Traffic was escorted through the area on days when safety permitted. Temperatures hovered around 32°C (90°F), with relative humidity of only 15%. Fire low on hillsides raced upslope into unburned fuel; at night, cooler downdrafts pushed it down the next ridge. Erratic and unpredictable winds up to 80 km/h created hazardous conditions for highway travelers, homeowners, and firefighters. Those same winds left an erratic patchwork of burned areas.

An undamaged wood building stood next to a house burned to the ground; a tool shed with charred machinery rested next to a wood house that remained untouched.

At the height of the fire, with 10,902 acres burned, 882 personnel fought the fire—including 18 crews, seven helicopters, fixed-wing aircraft, 47 fire engines, 6 bulldozers, and 20 water tenders. Montana National Guard and hotshot crews from Michigan, North Carolina, and the Ontario Ministry of Natural Resources assisted. Driven by high winds in the heavy pine forest, the fire burned hot enough to melt not only window glass but also the aluminum frame of a utility trailer. Interestingly, a large propane tank was scorched but didn't explode. Its relief valve must have operated as designed.

Melted aluminum

Don Hyndman

■ *The intense fire that leveled one residence along U.S. Highway 12 was hot enough to melt window glass that dripped onto the house foundation (A), and to melt a four-wheeler off-road vehicle's aluminum that trickled down the driveway (B).*

U.S. Forest Service

■ *The intense crown fire collapsed on itself and then raced across U.S. Highway 12. Burning embers spotted the highway far from the main fire front.*

CRITICAL THINKING QUESTIONS

1. What could homeowners have done better to protect their homes from wildfire?
2. What indications were there that the heat from the fire was extreme, and what might have contributed to that heat?
3. What factors made this fire particularly dangerous for firefighters, and how might they protect themselves while fighting such a fire?

Firestorms Threaten Major Cities
Southern California Firestorms, 2003 to 2009 ▶

The first of 11 major fires of the season began in southern California on October 21, 2003. Hot, dry Santa Ana winds that blew westward out of the high deserts to the east gusted to 100 km/h. The air compressed and heated up as it descended toward the Pacific, rapidly dehydrating soils and vegetation. Sparks from downed power lines, careless campfire handling, barbeques, cigarettes, and even arson ignited the fires. Ten days later, the rapidly spreading fires had burned more than 300,000 hectares, or 3000 km^2, about the area of Rhode Island. They leveled 3600 homes and killed 24 people. Nine hundred of the homes were in the San Bernardino Mountains east of Los Angeles, caught in a fire likely started by arson on October 25. The largest, the Cedar fire in San Diego County, burned 2207 homes and caused 15 deaths. A lost hunter trying to signal rescuers started that fire.

In May 2009, someone clearing brush with a power tool sparked a wildfire that burned 33 km^2 and destroyed 77 homes at the base of the Santa Ynez Mountains behind Santa Barbara. In that case, the fires were fanned by local winds, called "sundowners," that sweep downslope as evening air cools high in the mountains, becomes denser, and flows downward. Thirty thousand people required evacuation.

Many of those who died in 2003 had ignored evacuation orders, waiting until the last minute to leave. By then, the fast-moving flames overtook their lone evacuation road, cutting off all escape. The worst fires were in San Diego County, where the high cost of available land encouraged developers and individuals to build in areas vulnerable to brush fires. Some neighborhoods consisted of closely spaced wooden houses surrounded by pine trees, with some trees wedged against flammable roofs and wooden decks. California state law now requires strict standards for building with fireproof materials and clearing brush around homes.

Four years of drought in the San Bernardino Mountains northeast of Los Angeles left the trees vulnerable to bark-beetle infestation that killed large numbers of them. Although there are tight controls on fire-resistant building materials and brush removal, the cost of tree removal can be as high as $850 per tree, so few people complied. Such beetle-killed trees make for running crown fires that are almost impossible to stop. Fifteen thousand firefighters fought the fires, along with water from helicopters and air tankers. Insured losses amounted to

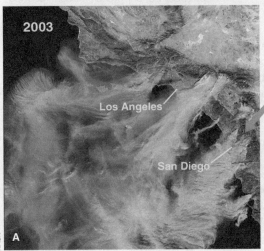

■ **A.** *The southern California fires of October 2003 and 2007 can be seen in these satellite photos. The largest area of fire at the right edge of both images is on the eastern outskirts of San Diego. The largest area of fires in the north-central part of both images is on the outskirts of Los Angeles. The Santa Ana winds carried the fires and smoke toward the southwest.* **B.** *Fires bear down on San Diego in 2003.*

2007

2003

Los Angeles

San Diego

NASA

A

B

John Gibbins/Zuma Press/Newscom

■ Many homes were closely surrounded by trees and brush, which made it impossible for firefighters to save them from the Witch fire near San Diego.

■ Debris flows following the Station fire severely damaged homes in part of Los Angeles County.

more than $1.25 billion. A change in weather on November 1 calmed the Santa Ana winds and brought moisture from the Pacific. Two or three centimeters of snow fell in the Big Bear Lake resort area near San Bernardino, slowing the fires and permitting firefighters to construct fire lines. The rain and snow caused mudslides that closed highways.

On December 25, torrential rains that fell on the unprotected soils of the burned area unleashed heavy mudslides that raced down canyons and through a trailer camp. Fifteen people died in mud as much as 4 to 5 m deep.

Costs of firefighting and cleanup were estimated as another $2 billion. Fire insurance is still available; however, insurance companies build the heightened risk into their premiums. In some areas, insurers are becoming less willing to write coverage for fire. In others, homeowners are told to replace roofs and clear brush if they want to be covered.

Some homeowners' associations in scenic areas banned property owners from cutting trees, as in the same area in the San Bernardino Mountains. They lifted the ban less than a year before the fire, but by then it was too late; 350 homes burned.

In recent years, as the population in the Sierra Nevada nearly doubled, so did the number of fires and acres burned, including large fires in 2007. Property damage went up 5000%.

In this region's Mediterranean climate, more than 90% of the rainfall comes during the winter and early spring. Chaparral or scrub brush that covers many southern California hillsides has resins that fuel fires. This vegetation burns in wildfires every 35 to 65 years, primarily during late summer or early fall, when hot, dry Santa Ana winds blow seaward off the deserts to the east. The hydrophobic soil layer produced by fires (see Figure 16-7B), along with the lack of vegetation, lead to greater overland flow and high stream flows.

Four years later, on October 21, 2007, the hot, dry, Santa Ana winds picked up and fanned catastrophic fires in nearly the same areas as in 2003. A wet winter in 2004–05 that permitted abundant growth of trees and brush was followed by a dry 2006 and 2007. Some fires started when high winds downed power lines that sparked them and arson appears to be responsible for a few others. High, shifting winds reaching gusts of more than 110 km/h, and essentially no humidity, fanned the fast-moving flames. Under intense conditions even airdrops of water and fire retardant are ineffective because they evaporate before reaching the ground. When flames are longer than about 3 m, a fire is considered unstoppable, and firefighters must retreat until conditions ease.

Four days later when the winds died down, 2000 km² had burned, more than 800,000 people were evacuated, and more than 2300 homes and businesses were destroyed. Damages totaled more than $1 billion in San Diego County. Only 7 people died in the fires—a fact partly credited to aggressive evacuation and to a "reverse 9-1-1" calling system that warned people of a dangerous fire's approach. Seven other major southern California conflagrations may have been sparked by high winds that caused arcing between high-voltage power lines. Many of those fires started on wooden decks.

Farther north in the Santa Monica Mountains west of Los Angeles, the Canyon fire burned into the center of Malibu. Erratic winds carried firebrands downwind for more than half a kilometer, igniting new fires. This, along with heavy smoke, made conditions extremely difficult for both firefighters on the ground and for firefighting aircraft.

On August 26, 2009, an arsonist ignited the enormous Station fire in the San Gabriel Mountains above Burbank and Pasadena, California. Before being contained on October 16, the fire in the Angeles National Forest had blackened 650 km² (250 mi²). Two firefighters died. By September 23, suppression costs reached $97 million (2015 dollars). In mid-January 2010, mandatory evacuations were ordered for the area, this time because heavy rains increased the danger of mud- and debris-flows from the burned areas. A few months later, on February 6, 2010, a rainstorm over the burned area generated debris flows that wreaked havoc in the La Canada-Flintridge area of Los Angeles County.

Compounding the problem in regions like southern California are two factors: Thousands of people live back in the wooded hills and rugged canyons on narrow, winding roads and federal policies call for firefighters to protect homes, even when they are built in high-risk, poorly accessible areas. If safety were the operative priority, areas would be left to burn. Yet in spite of major wildfires in 1970, 1993, 2003, 2007, and 2009, even more people rebuilt in the burned areas and with still larger homes. Something needs to change!

CRITICAL THINKING QUESTIONS

1. In what months are wildfires in southern California most common and widespread? Why?
2. What factors cause southern California wildfires to be so dangerous to fight?
3. What do rural southern California homeowners do that makes their homesites so dangerous?
4. Although many of the wildfires in this region start in mountains to the east, the fires often spread to lower elevations to the southwest. Why?

Firestorm in the Urban Fringe of a Major City
Oakland–Berkeley Hills, California Fire, 1991 ▶

The steep hills behind Oakland and Berkeley, California, provide spectacular views west over San Francisco Bay to the city and its Golden Gate Bridge. Land so close to the cities with such views and with short commutes commands premium prices and promotes building of closely spaced, expensive homes. Steepness of the terrain dictated that streets are narrow and winding. In this Mediterranean-type climate, summer and early fall weather is hot and dry with breezes blowing offshore. Homes are shaded by big Monterey pine and dense groves of eucalyptus trees; privacy is enhanced by lots of undergrowth. Most people built with wood, including untreated wood shake roofs to enhance the feeling of living in harmony with nature. Unfortunately, all of these characteristics make for a prime wildfire environment.

Overlapping this hazardous environment was a five-year drought that left the plants thoroughly dry and a freeze in December that killed and further dehydrated a lot of the vegetation.

On Saturday and Sunday, October 19 and 20, temperatures were well above 33°C (90°F), as they tend to be in October, with low humidity. Firefighters were doing final mop-up from a small grass fire (about 100 × 200 m) that had burned on Saturday but was beaten down and completely soaked by that night. It was also monitored during the night. The next morning, equipment and hoses were still in place for about two hours as firefighters searched for any hot spots and began picking up equipment.

Then just before 11 a.m., winds picked up an ember and blew it into a tree just outside the burn area; it exploded into flame

Steep hills east of this line

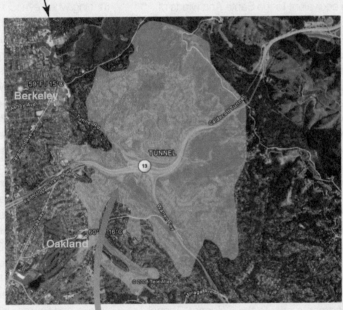

■ *In its initial stages, the fire burned rapidly, pushed down-valley and over ridges by winds from the east. It moved downslope more than 1500 m in 10 minutes.* **B.** *Area of the 1991 Oakland–Berkeley Hills fire.*

Todd Stout, San Bernardino Co. ISD

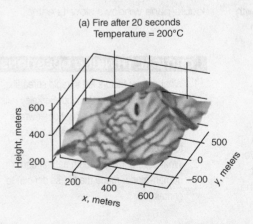

(a) Fire after 20 seconds
Temperature = 200°C

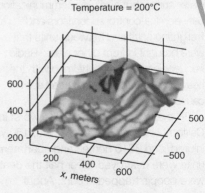

(b) Fire after 5 minutes
Temperature = 200°C

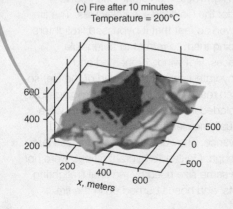

(c) Fire after 10 minutes
Temperature = 200°C

Lawrence-Livermore National Lab

■ *Little remained of expensive houses on the steep slopes of the 1991 Oakland–Berkeley Hills fire but their chimneys and foundations, along with shells of burned-out cars.*

and embers blew all over. Winds in October often come from the east. Locally called Diablo winds because they come from the direction of Mt. Diablo to the east, they are equivalent to the Santa Ana winds of southern California. These winds gusted to 80 to 100 km/h. Within a few minutes, the wildfire spread to surrounding vegetation and homes. Within the first hour it burned 790 homes, and firefighters were completely overwhelmed. The wind blew the fire downslope to the west and across ridges but the heat from the fire created its own winds, the definition of a firestorm. Heat from the fire also created cyclonic swirls that spread burning embers and carried the fire upslope.

Fire officials immediately called for reinforcements and airdrops from as far away as the Oregon and Nevada borders and Bakersfield in southern California. More than 370 fire engines and 1500 firefighters fought the fire. Within half an hour, the fire had reached the Parkwoods Apartments next to the Caldecott Tunnel on Highway 24, an eight-lane freeway under the Oakland–Berkeley Hills. The fire moved so fast that it bypassed firefighters, forcing them to retreat to defensible spaces. Hot winds preheated and dehydrated everything ahead of the fire, so when reached by the flames, buildings just exploded into flame. Ground crews were virtually helpless to slow the fire's rapid advance. They could not move hoses fast enough, Oakland's hydrant fittings were not the same size used by regional firefighting units, and hoses burned up in the fire.

At the fire's height, supply tanks and reservoirs ran out of water due to fire suppression efforts, residents hosing down roofs and shrubs, and leaving sprinklers running when evacuating. In addition, the fire destroyed power lines to 17 water pumping stations used to lift water from Oakland. Department of Forestry pilots complained that as soon as they released their huge buckets of water, the winds just blew it away into a mist. Large air tankers dropped hundreds of loads of fire retardant. By mid-afternoon, the wind slowed and came from the west. Finally that evening, the wind died down to 8 km/h, permitting firefighters to stop and ultimately contain the fire's spread.

Thirty-degree slopes and narrow, winding streets created serious problems with panicked residents trying to evacuate and firefighters rushing in. Both were hampered by sightseers. Trapped people abandoned cars and fled on foot. More than 10,000 people were evacuated, many by car through smoke and flowing debris on the steep, narrow streets. Problems arose, not only with water supply but also with communication between fire-control supervisors and firefighting units and between units that were brought in from other areas. Radio frequencies were overwhelmed with the heavy use and there were "dead spots" caused by the rugged terrain.

To that date, the fire was the worst in California history: 25 people died, 150 were injured, and 3810 homes and apartment units were destroyed. Over half the deaths were people trapped in cars. About

610 hectares (6.1 km²) burned within an overall area of about 13.6 km². Total physical damages reached about $2.5 billion (2015 dollars).

In the aftermath, fire safety codes were strengthened and fire hydrant couplings are now standardized. Water cisterns were added and radio communications improved and standardized. The Oakland Fire Department now requires brush clearing 9 to 30 m from buildings, and chimneys must have a spark arrestor and no vegetation within 3 m. Highly flammable building materials, including untreated wood walls, wood roofs, small-dimension lumber for decks, and firewood stacked next to houses, are not permitted. Ground duff, such as dry pine needles and small-dimension lumber, are more like kindling; they ignite quickly. Because flames generally move upslope, fire-resistant materials should cover deck undersides and overhanging eves. Fire entering a house will burn it from the inside, regardless of external fireproofing; double-pane windows slow its entry.

CRITICAL THINKING QUESTIONS

1. What natural factors led to rapid spreading of this wildfire in the Oakland–Berkeley Hills?
2. What are several things that homeowners living in those hills did that caused the fires to spread so fast?
3. What several factors (both natural and human-made) made the fire so difficult to contain?

Critical View

A This 2006 wildfire burned in the Kaibab National Forest, Arizona.

Jackie Denk, USFS

1. Why is this fire only on the ground?
2. What could lead to a more intense fire in which the trees burn?

C This wildfire is burning on a hillside.

National Park Service

1. Why is there so much flame in this fire?
2. In what direction is the fire moving, and how can you tell?
3. How does the topography of this location affect the direction the fire is moving?

B This community in the Los Alamos National Forest in New Mexico was destroyed by fire.

Andrea Booher, FEMA

1. Describe what burned and what was most susceptible to burning.
2. What would have helped the fire move from one burn area to the next?
3. How would the fire have crossed the paved roads?

D This fire occurred in the Las Padres National Forest, California, in July 2006.

USFS

1. This fire appears to be burning green vegetation. How can that happen?
2. Why does the fire appear to be intense? Note that yellow flames are hotter and red are cooler.

Wildfires   **503**

Chapter Review

Key Points

Fire Process and Behavior

- Fire requires fuel, oxygen, and heat. Lacking any one of these, fire cannot burn. **FIGURE 17-1**.

- Ground fires are fueled by low-growing vegetation such as grass and low shrubs. The higher surface area of dry grass and needles makes them burn faster and more easily. Ladder fuels such as the low branches of trees help fire climb into treetops where it burns as a crown fire. **FIGURE 17-2**.

- Fires can be naturally started by lightning strikes, intentionally set for beneficial purposes, or set accidentally or maliciously.

- Fires spread through radiation, convection, or wind-borne embers called firebrands. **FIGURE 17-4**.

- The spread of fire is affected by local topography. Fire can be accelerated by air funneled through a canyon. A fire tends to accelerate upslope as heat rises.

- Fires start more easily and spread more quickly during dry seasons or droughts, when fuel moisture is low. Winds accelerate fires by directing the flames to fuels and by supplying more oxygen.

Secondary Effects of Wildfires

- Hydrocarbons formed in a fire seal soils, making them hydrophobic, and force water to run off the ground surface rather than soak in; this can lead to flash floods and mudflows. **FIGURES 17-7** and **17-8**.

- Reduction in evapotranspiration from vegetation after a fire can increase the amount of water in the ground and thereby promote debris flows and landslides.

- Wildfires contribute to air pollution by spreading smoke, soot, and ash over large areas, causing respiratory problems and other health effects.

Wildfire Management and Mitigation

- U.S. Forest Management policy has changed over the years from stopping all fires to letting fires burn—as long as they are under control—in order to consume dry fuels.

- Techniques for fire suppression include cutting firebreaks down to bare ground, having helicopters dump large buckets of water, and having air tankers dump huge loads of fire retardant. In out-of-control fires that threaten critical facilities, firefighters sometimes set burnouts to burn back toward an advancing fire and thus deprive it of fuel. **FIGURES 17-12** and **17-13**.

- People who live in the woods are surrounded by fuel for fire. Many build their homes out of wood, which provides more fuel for any wildfire.

- Prolonged dry weather reduces fuel moisture and increases fire danger. Thus, satellite imaging for greenness compared with normal conditions provides an indication of regional fire danger. **FIGURE 17-15**.

- Much or most of the cost of protecting a few who choose to live in dangerous places is borne by the vast majority of the public who do not. Zoning restriction or other policies may be needed to discourage building in fire-prone areas. **FIGURES 17-16** and **17-18**.

Key Terms

back burn, p. 489
burnout, p. 489
convective updraft, p. 485
crown fire, p. 482
fire weather potential, p. 490

firebrand, p. 483
fireline, p. 488
firestorm, p. 485
fuel loading, p. 481
fuel moisture, p. 485

hydrocarbon residue, p. 485
hydrophobic soil, p. 485
indefensible location, p. 494
ladder fuel, p. 481

prescribed burn, p. 488
smokejumper, p. 488
spot fire, p. 483
surface fire, p. 481
wildfire, p. 481

Questions for Review

1. Wildfires are beneficial to forests in what two ways?

2. What are the two main causes of wildfires?

3. What two conditions lead to more fire-prone forests?

4. Why do winds accelerate fires? Give two specific reasons.

5. Why do fires typically advance faster upslope than downslope?

6. What natural conditions and processes of a fire lead to new fires well beyond the burning area of a large fire?

7. Why is hill-slope erosion more prevalent after severe wildfires? Be specific.

8. What main techniques are used to minimize post-fire erosion? Name two specific and quite different techniques.

9. Why does flash flood hazard increase after a large fire?

10. In addition to pouring water or retardant on fires, what techniques are used to fight fires? Name two specific techniques.

11. On which side of a forest is a home at greater risk to fire, and why?

12. What can people living in the forest do to minimize fire danger to their houses?

Critical Thinking Questions

1. Most wildfires cost far more in damages and suppression than any individual could afford. Homeowners' insurance premiums are generally much lower than actuarial costs because state and federal firefighters are paid by their respective governments. Who should pay the overall costs of fire suppression and damages? Why?

2. Many people live in wildlands such as brushy areas and forests that are highly susceptible to wildfire. Should these areas be zoned as off-limits to homes? Would such restrictive zoning tread on individual rights? Would that be necessary or appropriate?

3. Firefighters are often required to first defend private residences from wildfire, even when the occupants have evacuated, while neglecting the overall progress of the fire. Is this appropriate? Should homeowners who choose to live in such fire-prone environments be responsible for defense of those homes or is it the responsibility of public agencies to defend those homes?

4. Public agencies such as the U.S. Forest Service spend funds to thin forest areas in the wildland–urban fringe to minimize fire danger. Should public or private entities do such work, and who should pay for it?

■ Simulated impact of a giant asteroid on Earth.

18 Asteroid and Comet Impacts

The Ultimate Catastrophe?

An asteroid 10 to 15 km in diameter struck the Yucatán Peninsula of eastern Mexico 65 million years ago, opening a crater about 80 to 110 km in diameter. The energy released from such an impact would have been equivalent to that of 100 trillion tons of TNT or a million 1980 eruptions of Mt. St. Helens. The impact would have had dramatic consequences—a planet-shaking earthquake, giant tsunami that swept over the continental margin, widespread acid rain, and wildfires. Prolonged darkness from all the atmospheric dust would have triggered abrupt cooling of Earth for many years, resulting in extinctions of plant and animal life. It appears that the impact of this asteroid killed the dinosaurs and the majority of other species on Earth at that time. What would an impact of this size do to Earth today, and what are the chances of such an impact occurring in our lifetimes?

Projectiles from Space

Space objects that cross Earth's path include asteroids and comets, and the smaller pieces of rock that cause meteors.

Asteroids

Asteroids are chunks of space rock orbiting the sun just like Earth (**FIGURE 18-1A**). Astronomers believe that asteroids are remnants of material that did not coalesce into planets when the other planets formed around our sun. This theory is founded on the fact that most asteroids are found in the **asteroid belt** between Mars and Jupiter, where you would expect another planet to be orbiting (**FIGURE 18-1B**). Most asteroids stay in the asteroid belt, where they pose no danger to Earth.

The gravitational influence of planets perturbs the orbits of some asteroids, causing them to migrate toward the inner solar system, potentially crossing Earth's path. The dangerous few that may strike Earth within a few days of discovery are difficult to spot because their trajectories toward Earth leave them nearly stationary in the night sky. They are recognized on sequential telescope images as changing position and their approximate paths are then calculated. The majority of these asteroids are less than 2 km in diameter, and most are less than 100 m to 1 km in diameter. Smaller bodies, called *meteoroids*, are more abundant.

Comets

Comets are similar to asteroids but consist of ice and some rock; they are essentially giant, dirty snowballs (**FIGURE 18-2**).

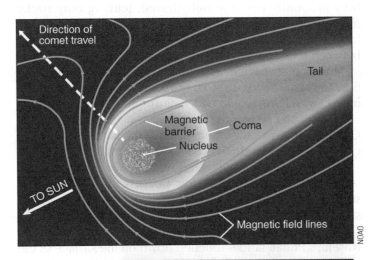

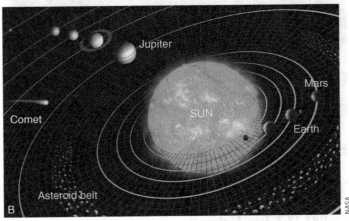

FIGURE 18-1 Asteriods

A. This photo compilation shows Asteroid Ida from four different sides. **B.** The asteroids lie in a belt between Earth and Jupiter, in the same orbital plane around the sun as all of the planets.

FIGURE 18-2 Structure of a Comet

A comet consists of a solid nucleus of a rock and ice mixture surrounded by a coma of dust and gas. The tail is a mixture of water, other volatiles, and dust that the solar wind sprays away from the direction of the sun. This close-view 2004 image of Comet Wild 2 shows its heated surface spraying incandescent dust and gas into space.

They do not come from the asteroid belt but range far beyond the planetary part of our solar system, where they make up the Oort cloud. The Oort cloud forms a vast spherical region around the sun that extends to more than 100,000 times the distance between Earth and the sun. It contains billions of comets. An inner doughnut-shaped zone of millions of comets, the Kuiper belt, lies in the plane of the solar system and extends to 55 times the distance between Earth and the sun.

When they come within the influence of the sun's solar wind, comets spray off water, other volatiles, and dust that form their glowing tails. The visible tail of a comet does not indicate its direction of travel; rather, the tail points away from the sun. As comets spray off virtually all of their water, they gradually become dehydrated, leaving only rocky material. At that point, they are not easily distinguished from asteroids. In fact, there may be a continuous gradation between rocky comets and icy asteroids.

Most comets have heads less than 15 km in diameter, but they travel at speeds up to 60 or 70 km/s. At those velocities, impact with Earth would be a catastrophe. Some comets traverse our solar system as frequently as once every 10 years. Of all space projectiles, comets have the highest chance of coming close to Earth, or even colliding with it.

Comets may be able to tell us how our own solar system was formed, and we now possess the capability to study them up close. After 10 years of travel through our solar system, the Rosetta orbiter reached Comet 67P/Churyumov-Gerasimenko, 510 million km from Earth. Rosetta's Philae lander touched down on the surface of the comet on November 12, 2014. Unfortunately, it ended up in the shade of a cliff after bouncing, which limited recharge of its solar panels and shortened its life. Before landing, the Philae detected water vapor, carbon dioxide, carbon monoxide, and traces of organic compounds. Philae was also to transmit data that let scientists know that the comet's core appears to be solid ice or ice and dust.

Meteors and Meteorites

When small pieces of rock from space enter the atmosphere at high speeds, friction with the air molecules heats the surrounding air to white-hot incandescence, forming **meteors**, or what we sometimes call *shooting stars*. Most space rocks that enter Earth's atmosphere originate in the asteroid belt between Mars and Jupiter (see again FIGURE 18-1B).

Earth's atmosphere shields us from the impact of most of these space projectiles, because small ones burn up in the upper atmosphere. When a large rock enters Earth's atmosphere, it heats up due to friction with the air and forms a fireball that glows for a period of time before it either disintegrates or survives to strike Earth. When an object is large enough to survive its passage through the atmosphere without burning up and colliding with Earth, it is called a **meteorite**.

Upon entering Earth's atmosphere, the air around a large fragment of rock heats to become incandescent, but its core typically does not get especially hot. Meteorites are cool enough when they reach the ground that they fall on buildings or dry grass without starting fires. A meteorite that fell in Colby, Wisconsin, on July 14, 1917, was cold enough to condense moisture from the air and become coated with frost.

Because meteorites travel much faster than the speed of sound, we hear only those that are relatively close. If we hear no sound, the meteorite is probably more than 100 km away.

Unless a meteorite is witnessed falling from the sky, it can be difficult to distinguish one from any other rock on Earth. Meteorite fragments are hard to collect, because relatively few meteorite falls are witnessed; fewer than 1000 have ever been seen in the United States. Only 20 or 30 witnessed falls lead to meteorite finds worldwide each year. Sometimes, large rocks break up in Earth's atmosphere and fall as a **strewn field**, spread out around the main impact site. The Allende meteorite that fell in Chihuahua, Mexico, in 1968 scattered fragments over an oval-shaped area 50 km long and up to 10 km wide.

Identification of Meteorites

Meteorites come in several types, all of which are somewhat similar to rocks thought to make up the deeper interior of the Earth. **Iron meteorites** make up 6% of all meteorites (**FIGURE 18-3**). Because metallic meteorites consist mostly of a nickel-iron alloy, they are attracted to

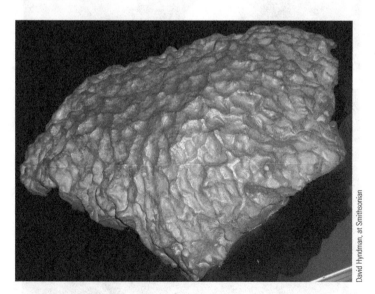

FIGURE 18-3 Iron Meteorite

The Henbury Iron Meteorite, found in northern Australia, has characteristic small depressions on its surface due to partial melting upon passage through Earth's atmosphere approximately 10,000 years ago.

magnets. Even stony meteorites often contain some iron, so they are often magnetic. Most meteorites have a *fusion coating*, a very thin layer of dark glass, formed when friction against Earth's atmosphere heats it above its melting temperature. A meteorite's coating is sharply bounded and quite different than its interior; in contrast, the weathering rind on an Earth rock is generally not sharply bounded. Weathering of the meteorite may later result in rusting that turns the coating reddish brown. Metallic meteorites probably crystallized slowly in the deep interior of a large, solid body in our solar system. Collisions between such bodies, and their breakups, lead to some collisions with Earth. Iron meteorites are extraordinarily heavy, with densities of 7.7 to 8 g/cm^3. By comparison, Earth's surface rocks have densities of 2.6 to 3 g/cm^3, and water has a density of 1 g/cm^3.

Stony-iron meteorites make up less than 1% of all meteorites (**FIGURE 18-4**). They consist of nearly equal amounts of magnesium and iron-rich silicate minerals, such as olivine and pyroxene, in a nickel-iron matrix. They probably come from a zone between the deeper, iron-rich parts of a large asteroid and the outer stony parts.

Chondrites are stony meteorites that make up 93% of all meteorites. They consist primarily of olivine and pyroxene, magnesium- and iron-rich minerals, along with a little feldspar and glass, similar to the overall composition of Earth's mantle. They have densities of approximately 3.3 g/cm^3. Millimeter-scale silicate spheres called *chondrules* enclose nickel-iron inclusions within or surrounding the chondrule. **Achondrites** are stony meteorites that are similar to basalt, a common rock on Earth. They consist of variable amounts of olivine, pyroxene, and plagioclase feldspar.

Iron meteorites are distinctive. They differ from other continental rocks and are generally black unless oxidation over many years has turned their surfaces brown. Iron meteorites are extremely hard, virtually impossible to break with a hammer. Over time, when exposed to the weather, they rust to iron oxides. Pieces of manufactured iron are more abundant and may look similar, but meteorites can be distinguished by polishing a sawn surface. Iron meteorites show intersecting sets of parallel lines marking the internal structure of the nickel-iron minerals. When these lines are accentuated by acid etching, they show the distinctive patterns characteristic of iron meteorites.

Even stony meteorites are distinctively heavy. They are made of peridotite that is 15 to 20% more dense than most common rocks. Most contain enough metallic iron to be still heavier. Stony meteorites may be broken, exposing the meteorite's fresh interior. If the broken surface is unaltered, you may see small inclusions of metallic, silver-gray-colored nickel-iron that strongly suggest a meteorite. One large meteor broke up in the atmosphere over central California in April 2012. Dark, angular fragments of the meteorite recovered near Coloma, east of Sacramento, were an unusual type of chondrite that contains organic compounds.

Some meteorites are rounded by their passage through the atmosphere. Iron meteorites are commonly more angular and sometimes twisted-looking. Some show rounded thumb-sized depressions. Others have an orientation related to their direction of travel, with a smooth leading end and a pitted rear end.

Evidence of Past Impacts

Asteroid impacts have been recognized worldwide (**FIGURE 18-5**). The largest proportion presumably fell into the oceans that cover two-thirds of Earth's surface. Most of those have remained undetected because details of the ocean floor are not well known. Because older parts of the ocean floors have subducted into oceanic trenches, evidence of early impacts is no longer available. On continents, impact sites are broadly distributed but have been found concentrated in areas of greater population or well-exposed, vegetation-poor regions. As with all natural hazards, recognizing and analyzing evidence of past impacts helps scientists determine how often meteorites strike and what consequences they have.

Impact Energy

The record we have of past collisions with space objects is a result of the tremendous energy of these impacts. Because most asteroids travel at velocities of 15 to 25 km/s, they have incredible energy. Recall that the energy of a moving object is equal to its mass times the square of its velocity (**By the Numbers 18-1**: Energy, Mass, and Velocity). Thus, the energy doubles for every doubling of the mass of the asteroid but quadruples for every doubling of the velocity. Comets typically have less mass than asteroids but greater

FIGURE 18-4 Stony-Iron Meteorite
This meteorite that crashed through the roof of a home in Auckland, New Zealand, in 2004 is approximately 13 cm long.

Meteorite Magazine/Grant Christie

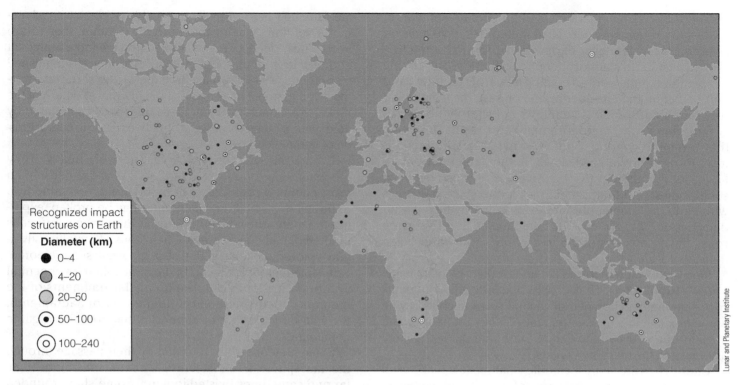

FIGURE 18-5 Recognized Impacts Worldwide

Meteorites of various sizes have peppered Earth for billions of years. The uneven distribution of recognized impacts is related mainly to how easily they are identified because of population density and land cover.

Recognized impact structures on Earth
Diameter (km)
- 0–4
- 4–20
- 20–50
- 50–100
- 100–240

Lunar and Planetary Institute

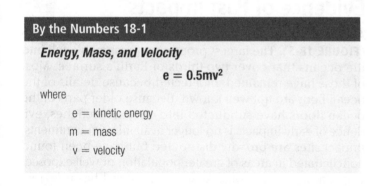

By the Numbers 18-1

Energy, Mass, and Velocity

$$e = 0.5mv^2$$

where

e = kinetic energy

m = mass

v = velocity

velocity. Because comets are mostly ice, with a density of 0.9 g/cm³, their overall densities, including their rock component, tend to be similar to that of water, 1.0 g/cm³. However, comets tend to travel at much higher velocities than asteroids—for example, 60 to 70 km/s. Because the energy quadruples for every doubling of the velocity, comets can have extremely high energy in spite of their lower densities.

On impact, the kinetic energy of the incoming object is converted to heat and vaporization of the asteroid and the target materials. This melts more rock, excavates a crater, and blasts out rock and droplets of molten glass. The result is a huge fireball that heats and melts rock and burns everything combustible.

Impact Craters

One might imagine that an asteroid striking Earth would create a big hole in the ground that has a shape dependent on the incoming angle of the object. However, the incredible velocity, and therefore the energy, of the impactor requires that it explode violently on impact. The effect is more like a missile being fired into a sand surface. It blasts a nearly round hole regardless of the impact angle (**FIGURE 18-6** and **Case in Point:** A Round Hole in the Desert—Meteor Crater, Arizona, p. 521). The largest identified open crater is the 100 km-diameter Popigai crater of Siberia.

If the impactor is large enough, the explosion violently compresses material in the bottom of the crater, accelerating it to speeds of a few kilometers per second and ejecting material outward at hypervelocity. The center of the crater rebounds rapidly to form a central cone; that cone and the outer rim almost immediately collapse inward to form a wider but shallower final crater (**FIGURE 18-7**). The older Manicouagan crater of eastern Canada is preserved as a striking ring of lakes in the basement rocks of the Canadian Shield (**FIGURE 18-8**, p. 512).

Impact craters provide evidence about the size and date of past impacts and also help us speculate about the damage those impacts would have caused. For example, the age and size of the Chicxulub crater basin in Mexico, mentioned at the beginning of this chapter, makes that impact the most

FIGURE 18-6 Open Impact Crater

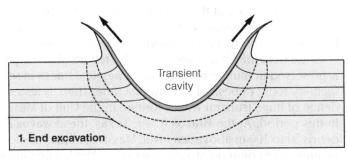

1. End excavation

Transient cavity

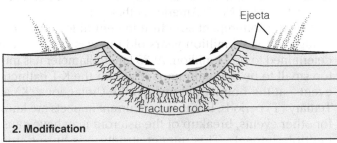

2. Modification

Ejecta

Fractured rock

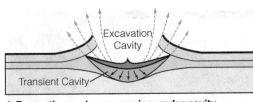

3. Final crater

Uplifted rim | Melt-rich material | Breccia lens | Ejecta blanket

Fractured rock

Bevan French, Lunar and Planetary Institute

First (1), a transient crater* is excavated, compressed, and fractured; the base of the cavity melts; and the rim is raised. Then (2) the ejecta blanket* spreads around the cavity, and the part of the rim slumps back into the cavity. Finally (3), fallback material* partly fills the cavity, along with some melt-rich material. Bacolor crater on Mars, shown in the photo, 20 km across, shows its raised margins and central uplift. Image from Mars Odyssey Orbiter.

NASA/JPL-Caltech/ASU

*Transient crater = Temporary crater formed on impact before blown-out material partly fills it in. Ejecta blanket = Material blown out and deposited around the crater. Fallback material = Material blown out and deposited back into the crater. Breccia = broken rock fragments.

FIGURE 18-7 Large Impact Crater

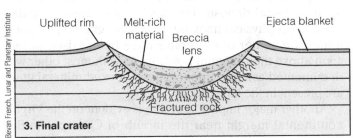

Excavation Cavity

Transient Cavity

1. Excavation and compression under cavity

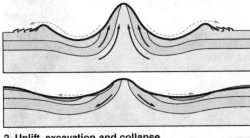

2. Uplift, excavation and collapse

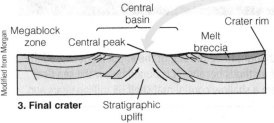

Central basin | Crater rim
Megablock zone | Central peak | Melt breccia

3. Final crater Stratigraphic uplift

Modified from Morgan

First (1), a transient crater is excavated and compressed, and the base of the cavity melts. Second (2), the base of the transient cavity rebounds as excavation continues. Then (3), the raised rim of the transient crater and central uplift both collapse to form a larger and shallower crater basin partly filled with inward-facing scarps, large blocks, smaller fragments, and melt rocks. The final crater is much broader and shallower than the initial transient crater. Some large craters on the moon have this central uplift (see photo).

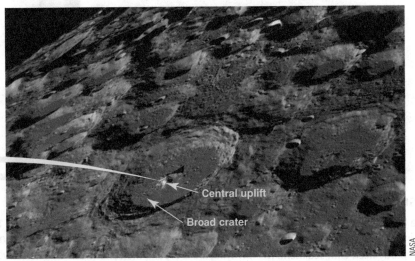

Central uplift

Broad crater

NASA

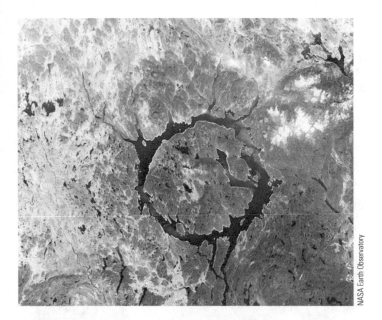

FIGURE 18-8 Large Impact Crater

The giant Manicouagan crater of the eastern Canadian Shield is one of the largest impact craters observable on the surface of the Earth. The crater shows as a dark ring of lakes 65 km in diameter in this space shuttle image.

likely cause of the mass extinction that killed the dinosaurs. That crater is 195 km in diameter, with a pronounced central uplift (**FIGURE 18-9**, p. 513). An impact large enough to have such a devastating effect should theoretically create an initial crater 200 km in diameter—with a much larger final diameter—but Chicxulub is the largest crater of the right

age found so far. The crater was slowly buried later by the quiet accumulation of limestones on the continental shelf, so it is not exposed at the surface. However, this burial also preserved some features that tend to erode with time. It has been studied through drilling and geophysical methods.

Aside from the final crater shape, the record of the impact is preserved in minute particles and chemical traces in sedimentary rocks deposited around that time. There is also evidence of huge tsunami waves formed in the Gulf of Mexico in this period. At the Brazos River, Texas, these waves left debris 50 to 100 m above sea level. Massive submarine slope failures were common around the Gulf of Mexico and along the East Coast of North America at that time.

The Manson impact structure in central Iowa was once thought to be 65 million years old but was more recently determined to be 74 million. At 35 km in diameter it is much too small to be the main impact site for the K-T extinction event—at the boundary between the Cretaceous (K) and Tertiary (T) periods. Because concurrent impacts are known for other events, breakup of the asteroid may have caused multiple impacts. This impact structure is buried under ice age glacial deposits but has been studied using drilling and geophysical imaging. Basement granite and gneiss under the center of the crater have been raised at least 4 km above their original position. As with most other well-documented asteroid impact sites, shocked mineral grains are present in the target rocks.

Another major impact structure is evident on the Virginia continental margin near the mouth of Chesapeake Bay. From its crater shape as well as other evidence, we know that a major asteroid or comet about 3 km in diameter and traveling about 110,000 km/h impacted this site about

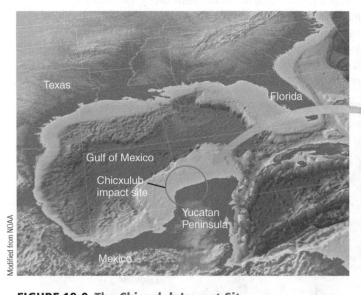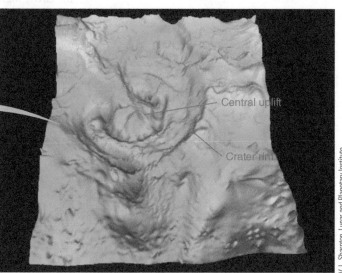

FIGURE 18-9 The Chicxulub Impact Site

The Chicxulub crater at the northern edge of the Yucatán Peninsula of eastern Mexico is thought to be the site of the impact that caused the extinction of the dinosaurs 65 million years ago. The crater was imaged using geophysical methods because it is buried below the sea floor and partly filled with sediment. The low-lying areas around the current Gulf of Mexico were a shallow continental shelf 65 million years ago.

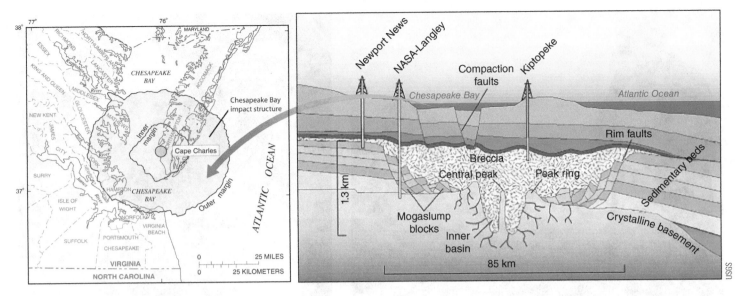

FIGURE 18-10 Chesapeake Bay Impact

The Chesapeake Bay impact site, from 35 million years ago, is now covered by a few thousand feet of sediments and water of the bay.

35 million years ago (**FIGURE 18-10**). It initially blew open a crater 35 km in diameter, followed by collapse of its near-vertical walls to form a wider, shallower crater about 85 km in diameter, with its floor 2 km below sea level. Millions of tons of fragments and dust would have been ejected tens of kilometers into the upper atmosphere. Pieces of those fragments, with shreds of impact melt that fell back into the crater, were found by drilling at the site between 2004 and 2006. The impact would have had dramatic consequences—giant tsunami that swept over the continental margin, widespread acid rain, and wildfires. Prolonged darkness from all the atmospheric dust would have triggered abrupt cooling of Earth for many years, resulting in the extinction of many plants and animals.

Shatter Cones and Impact Melt

The same energy that blasts out a crater also has an effect on rocks in the area. Rocks on the receiving end of an impact show distinctive characteristics, especially **shatter cones**. These cone-shaped features, with rough striations radiating downward and outward from the shock effect, range from 10 cm to more than 1 m long and are considered proof of asteroid impact (**FIGURE 18-11**). They form most readily in fine-grained massive rocks. Apexes of shatter cones point upward toward the shock source. Cones directly under the impactor should be vertical; those off to the sides flair down and outward from the source in "horsetail" fashion. Thus, the distribution of shatter cone orientations provides evidence for the location of the center of an impact site. Individual cones are often initiated at a point of imperfection, sometimes a tiny pebble, as in Figure 18-11B. The Sudbury structure north of Toronto, Ontario, has some of the best-known shatter cones (see the online **Case in Point**: A Nickel Mine at an

Impact Site—The Sudbury Complex, Ontario). They are distributed over an area 50 km by 70 km around the intrusion.

Sometimes melting of the asteroid produces the glass; sometimes the glass forms by melting the target material. Molten glass droplets sprayed out during impact form tiny, hollow *spherules* (**FIGURE 18-12**). These droplets of glass, generally 1 or 2 mm in diameter, and often altered to green clay, are typically the most obvious signatures of an impact feature found in sediments. In silicate target rocks, the impact melt may also form sheets, dikes (igneous rock filling a crack), and ejecta fragments, and may be disseminated in breccias. The melt develops under extremely high impact pressures. Impact-melt compositions are a mixture of the compositions of target rocks shocked above their melting temperatures. Because the impacting meteorite is often melted as well, impact melt may contain small but extraordinary amounts of nickel, iridium, platinum, and other metals that are abundant in iron meteorites.

Fallout of Meteoric Dust

Although the impact site itself is the best source for clues about past impacts, other evidence can be found far from such sites. Fragments and dust sprayed out from a large site can drift around Earth. The enormous impact at the Chicxulub site blew out enough material to deposit a thin, dark layer called Cretaceous-Tertiary boundary (**K-T boundary**) clay worldwide (**FIGURE 18-13**). The dust and elemental-carbon soot layer from fires ignited by the impact fireball 65 million years ago is generally only a centimeter or so thick and chocolate brown to almost black in color. The large amount of soot is related to worldwide fires that burned much of the vegetation on the planet.

FIGURE 18-11 Shatter Cones

A. Robert Hargraves, who discovered the Beaverhead impact site in Medicine Lodge Valley southwest of Dillon, Montana, points to a shatter cone cluster in an outcrop. Note that the apex of each shatter cone points upward.

B. Note the tiny pebble at the apex of the cone (arrow) in this close view of a shatter cone at the Beaverhead impact site. These are thought to be the flaws that trigger the radiating striations.

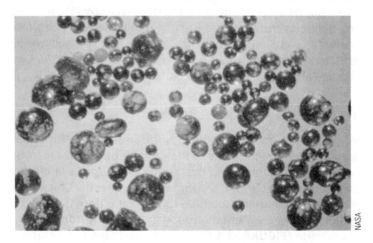

FIGURE 18-12 Spherules

Green glass spherules collected from the surface of the moon by Apollo 15 astronauts. Those found on Earth are often altered and dull.

The clay in this boundary layer contains grains of shocked quartz and other minerals, as well as spherules. The abundance of quartz in the boundary clay strongly suggests that the dinosaur-killing impact occurred in quartz-rich continental rocks, probably in an area rich in granite, gneiss, or sandstone.

The 15.1-million-year-old, 24-km-diameter Ries crater in Germany formed in limestone, shale, and sandstone over crystalline basement rocks. The impact ejected a blanket of sedimentary rock fragments that were cold when they landed 260 to 400 km to the east. These fragments contain droplets of frozen melt, or *tektites*, derived from the explosive melting of underlying crystalline basement rocks. Passage through the atmosphere aerodynamically shaped the glass, but it landed cold. Shock features are widespread. The interpreted sequence of events begins with high-speed shock waves and vapor blown out at shallow angles to Earth's surface, quickly followed by high-speed ejection of target material. The tektites formed as melted target material probably ejected above the atmosphere before falling back to Earth.

FIGURE 18-13 K-T boundary

The K-T boundary layer is exposed at Bug Creek in northeastern Montana. The inconspicuous lowest dark layer (arrow) marks this important boundary.

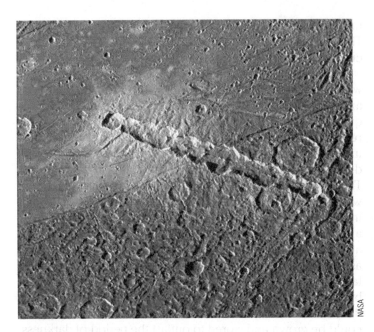

FIGURE 18-14 Multiple Impacts

A chain of impacts on Jupiter's moon Ganymede from fragments of the Shoemaker-Levy 9 comet that broke up on its approach to Jupiter.

The crater is circular, but the eastward fallout pattern of the tektites suggests that the impactor arrived from the west.

Also found in the boundary clay of the 65-million-year-old K-T event are anomalous amounts of iridium and other platinum-group elements, an occurrence called the **iridium anomaly**. The anomalous amounts are tiny, some 0.5 to 10 parts per billion, but those elements are essentially absent in most rocks except for meteorites and dark, dense rocks like peridotite and others derived directly from Earth's mantle. Some iridium is vaporized during the eruption of large basaltic volcanoes, such as those in Hawaii, but the amounts are too small to account for the quantities found in the K-T boundary clay. The boundary clay from the initial find contains 30 times as much iridium as would be expected from the normal fallout of meteoric dust that rains down on Earth daily. In Denmark, the clay contains up to 340 times as much. Vaporization of the impacting asteroid is thought to have spread the iridium worldwide, where it has now been found at roughly 100 sites. That boundary in sedimentary rocks marks the demise of the dinosaurs and 60 to 75% of all other species.

Multiple Impacts

An asteroid would be likely to break up in the atmosphere, so we should expect multiple impacts in a sequence. Most impacts arrive obliquely to Earth's surface, a significant proportion at 5 to 15° from horizontal. Such asteroids ricochet and often break up in the atmosphere into five to ten fragments that still have about 50% of the original velocity.

A dramatic example of multiple impacts occurred on Jupiter. In 1992, the Shoemaker-Levy 9 comet broke up into 21 fragments during a close approach to Jupiter as it was pulled apart by the huge planet's intense gravitational field. Then, in July 1994, the fragments, all less than 1 km in diameter, impacted one after another in an arc across Jupiter, over a period of six days (**FIGURE 18-14**). Comets typically travel at higher velocities than asteroids; this one was traveling at roughly 60 km/s. Impacts that appear to have been simultaneous at different sites on Earth may be multiple fragments of larger asteroids or comets.

Consequences of Impacts with Earth

Impact of a modest-sized asteroid 1.5 to 2 km in diameter is thought to be enough to kill perhaps one-quarter of the people on Earth and threaten civilization as we know it. The consequences of such an impact would be disastrous for life. All other hazards and disasters pale in comparison.

Immediate Impact Effects

The fireball or ejecta from the impact would ignite fires within hundreds of kilometers of the impact site. A heavy plume of smoke would linger for years in the atmosphere. Sulfate aerosols and water would be added to the atmosphere as well. A large portion of the ozone layer, which protects us from the sun's ultraviolet rays, probably would be destroyed.

If the asteroid broke into fragments, numerous smaller masses still traveling at hypervelocities would provide the heat and energy to cause widespread reaction between nitrogen and oxygen in the atmosphere. That would form nitrates that would combine with water to form nitric acid. The resulting acidic rain would damage buildings, as well as crops and natural vegetation.

At about the same time, dust blown into the stratosphere would block sunlight almost worldwide to the equivalent of an especially cloudy day. Large particles would settle out quickly, but dust particles smaller than 1/1000 mm would remain in the stratosphere for months. Any temporary increase in temperature from widespread fires would quickly decrease due to less solar radiation reaching Earth's surface. The dust would be distributed worldwide because much of it would be blown out of the atmosphere before settling back into it. Agriculture would probably be wiped out for a year, and summertime freezes would threaten most agriculture after that. Many specialists view global-scale wars over food as inevitable. Such desperate conflicts would likely kill a large percentage of the world's population. Others argue that humans are more resourceful than that. If there were, say, a decade of warning before the event, sufficient food supplies could be grown and stored to outlast the period of darkness. However, that is a pretty big "if"!

Impacts as Triggers for Other Hazards

Impact of a large asteroid would trigger many other hazards. Earthquakes would occur within hundreds of kilometers of the impact site. Asteroid ocean impacts would form tsunami as water sloshed into and out of the crater. Some of these tsunami could be extremely high. The waves would inundate coastal areas for tens of kilometers inland, areas inhabited by a large proportion of the world's population. As an example, the Chicxulub collision is calculated to have produced a 200-m-high wave. The run-up likely averaged more than 150 m in height, with a maximum of 300 m.

An impact might also cause volcanic activity. The large, essentially circular features that cover most of the moon's surface that can be seen on a dark night are known as lunar maria ("lunar seas"). Most are filled with basalt. These are recognized as the products of major impacts, mostly during the early evolution of the moon and Earth. Many are enormous, much larger than any known impact sites on Earth. Why have no such giant sites been found on Earth, especially given that the much greater gravitational attraction of Earth should pull in many more asteroids than the moon?

Flood basalts on Earth have a curious habit of forming at about the time of major extinctions of life, so a relationship between them seems possible. The immense Deccan basalts of western India, for example, erupted at about the time of the dinosaurs' demise (along with a large proportion of other species) 65.7 million years ago. If there is a cause-and-effect relationship, what is it? Most of the other handful of major flood basalt fields on Earth erupted at times of major extinctions of animal species. One imaginative but controversial suggestion is that the impacts of giant asteroids not only caused major extinction events but also triggered mantle melting and the eruption of flood basalts through a sudden

relief of pressure in Earth's mantle. One problem with this proposal is that a layer containing shocked quartz grains has been found recently—not at the base of the 65-million-year-old Deccan flood basalts in India, but between two major periods of basalt eruption.

Mass Extinctions

The impact of an asteroid 10 to 15 km in diameter 65 million years ago was associated with the demise of the dinosaurs and the death of between 40 and 70% of all species (see chapter introduction). If an asteroid of that size were to impact Earth today, it would annihilate virtually everyone on the planet and a large proportion of other species. Almost immediate flash incineration near ground zero would accompany strong shock waves. The 10-km-diameter impactor of 65 million years ago most likely produced sufficient nitric acid rain to be the primary cause of the mass extinction.

For those species lucky enough to be far away from the impact, extinction would follow the impact from various indirect causes. Acid rain would kill vegetation and sea life around the planet. Dust, soot from fires, and nitrogen dioxide would blot out the sun, so animals that were not incinerated would freeze and starve to death. Plants would also die because of the drop in temperature and sunlight. With the impact of a 10-km asteroid, land temperatures would, depending on assumptions, likely drop worldwide to freezing levels within a week to two months. Because of the large heat capacity of water, sea-surface temperatures would drop only slightly. Widespread fires would ignite from lightning strikes after much of the vegetation had died from either freezing or toxic atmospheric effects.

The vaporization of a 10-km-diameter chondritic asteroid on impact with Earth would probably not only generate strong acid rain but also spread nickel concentrations of between 130 and 1300 parts per million, even when diluted with 10 to 100 times the amount of target Earth material. A nickel-iron impactor would generate even more. This is many times the toxic level for chlorophyll production in plants. Seeds and roots would likely not recover for a long time.

Other mass extinctions befell Earth millions of years ago, including extreme events that marked the end of Triassic time, the end of Permian time, late in Devonian time, and the end of Ordovician time (see Appendix 1 online). The end-of-Permian event, 250 million years ago, wiped out approximately 90% of all species. The time of that extinction event correlates with the time of a gigantic outpouring of flood basalt lavas in Siberia, which erupted in the geologically short period of about one million years.

It has been suggested that the extinctions could have been caused by huge amounts of CO_2, methane, and other gases expelled by the basalt eruptions. Such flood basalts do not appear related to normal types of plate tectonics but rather to the rise of a deep mantle plume and hotspots, as described in Chapter 2. Another hypothesis, though controversial, is that both the extinctions and the flood basalt eruptions were caused by the impact of a massive asteroid. Although such a giant collision would almost certainly kill us all, the chances of it happening are exceedingly small.

Evaluating Impact Risks

Projectiles from space are not significant natural hazards on most people's horizon, although Hollywood's fictional dramatizations of such catastrophes have heightened awareness for some people. In reality, however, no other physical hazard has such a dire potential. Although the odds of a huge asteroid colliding with Earth are tiny, the consequences of such an impact would be truly catastrophic. Even without impact, grazing contact with Earth can be tremendously destructive (**Case in Point:** A Close Grazing Encounter— Tunguska, Siberia, p. 521). A large impact could wipe out civilization on Earth.

As with other hazards, small impacts are quite common, while giant events are rare (**FIGURE 18-15**). On average, a 6-m-diameter asteroid collides with Earth every year; one 200 m in diameter strikes the planet every 10,000 years on average. Because land covers only one-third of Earth's surface, compared to two-thirds for oceans, impact intervals on land would be about three times longer than those for the whole Earth. Only 1500 or so asteroids larger than 1 km in diameter are known to be in Earth-crossing orbits (those that pass through Earth's orbit around our sun). The largest is 41 km in diameter. Most of these cross Earth's orbit only at long intervals, so the chances of a collision are fortunately extremely small. The orbits of some asteroids vary from time to time because of the gravitational pull of various planets. Innumerable smaller asteroids also cross Earth's orbit. We live in a cosmic shooting gallery with no way to predict when one will hit the bull's-eye. We can only estimate the odds.

Your Personal Chance of Being Hit by a Meteorite

Meteorites fall from the sky daily, but has anyone ever been struck? The only well-documented case occurred on November 30, 1954, in Sylacauga, Alabama, when a 3.8-kg meteorite crashed through the roof of a house, bounced off a radio, and hit Mrs. Hulitt Hodge, who was sleeping on a sofa. She was badly bruised but otherwise okay.

There have been some close calls. On the night of April 8, 1971, a 0.3-kg meteorite came through the roof into the living room of a house in Wethersfield, Connecticut. No one was injured. In 1982 a 2.7-kg meteorite struck a house just 3 km away. In another unusual case, on October 9, 1992, a bright fireball seen by thousands of people from Kentucky to New Jersey came down in Peekskill, New York, where it mangled the trunk of Michelle Knapp's car. Hearing the crash, she went outside to find the 11.8-kg meteorite in a shallow pit under the car. It was still warm and smelled of sulfur. On June 8, 1997, a 24-kg meteorite crashed into a garden 90 km northeast of Moscow, Russia, where it excavated a 1-m deep crater. On the night of March 26, 2003, a 10-cm stony chondrite meteorite crashed through the ceiling of a house in Park Forest, near Chicago. It was one of many, most of them small, that damaged six houses and three cars over an area of about 3 by 10 km. On Saturday, June 12, 2004, at 9:30 a.m., a 1.3-kg stony meteorite crashed through the roof of a home in Auckland, New Zealand. The rock was 7 by 13 cm, gray

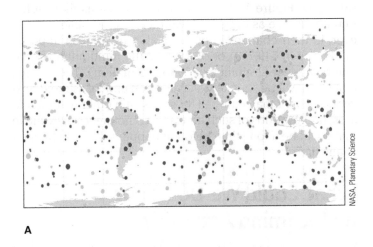

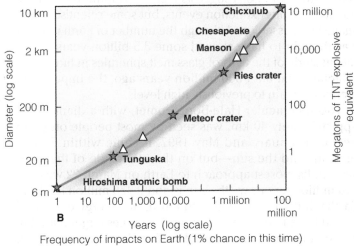

Frequency of impacts on Earth (1% chance in this time)

FIGURE 18-15 Risk of Asteroids and Comets

A. A 20-year sample (1994 to 2013) of small asteroids or comets, 1 to 20 m in diameter. The largest size shown, equivalent to 240 kilotons of TNT explosive or 16 times the size of the nuclear bomb dropped on Hiroshima in 1945. Larger impactors are fewer but may have similar distribution. Orange = daytime impacts; blue = night **B.** This log graph plots the approximate impact chances for asteroids of given sizes hitting Earth.

with a black rind, and had rounded edges (see Figure 18-4). It was hot when it landed on the Archer family's sofa in the living room, but no one was hurt.

On November 20, 2008, a 13-kg meteorite fell on a farmer's land south of Lloydminster, Saskatchewan. More than 1000 fragments of the meteor have been found, but three larger pieces tracked on eyewitness videos have yet to be located. Given their estimated sizes of 45 to 200 kg they are expected to have created holes in the ground. More recently, on January 18, 2010, a fist-sized 0.3-kg chondrite meteorite fell through the roof of a doctor's office in Lorton, Virginia, a suburb of Washington, D.C.; it broke into several pieces on impact.

In 1860, a meteorite falling on a field near New Concord, Ohio, killed a colt, and in 1911, another killed a dog in Nakhla, Egypt. There are no records of a person having been killed—at least so far! Clearly, you need not stay awake at night worrying about the possibility of being hit by a meteorite.

Chances of a Significant Impact on Earth

Earth's gravitational field is constantly sweeping up stray asteroids, so their number should be decreasing with time. However, collisions in the asteroid belt create new asteroids that leave the belt; some of those fall into Earth-crossing orbits.

Also dangerous are comets that have orbits outside our solar system but become visible when they pass close to the sun or Earth. About 10% of impacts on Earth and the moon are caused by comets. The periodicity of major impacts, based on the ages of impact craters and on theoretical aspects of interacting orbits and other movements within our galaxy, suggests an average interval of 33 million years. Study of fossils suggests a 26- to 31-million-year periodicity of genus-extinction events, but some scientists argue against that assertion. Although the number of Earth impacts was thought to have slowed some 3.5 billion years ago, a recent study of the ages of glass-melt spherules in lunar soils indicates that after 500 million years ago, the impact rate increased again to previously high levels.

The spectacular Hale-Bopp comet, with a diameter of approximately 40 km, was seen by most people on Earth between January and May 1997. It came within Earth's orbit around the sun—but on the other side of the solar system. Its closest approach to Earth on March 22 was still 320 million km away (the sun is about 150 million km from Earth). If it had collided with Earth, the energy expended would have been tens to hundreds of times larger than that of the dinosaur-killing asteroid of 65 million years ago. It is a long-period comet that spends most of its orbit far beyond our solar system before blowing through it again after thousands of years. Whether one of these objects annihilates us is a matter of where Earth is in this shooting gallery when one of these objects passes through.

On December 6, 1997, astronomers of the University of Arizona Spacewatch program spotted a huge chunk of space rock, an asteroid 1.5 km in diameter, apparently on a near-collision course with Earth. Asteroids spotted in telescopes are tracked over time until astronomers have enough information to determine how close their trajectories will take them to Earth. In March 1998, the astronomer who spotted this particular rock, Asteroid 1997 XF11, was able to plot its path and determined it would come dangerously close to Earth; he predicted the nearest approach to Earth to be on October 26, 2028. However, in his excitement to announce the event to the press, he neglected to check earlier measurements on the same object. The corrected results indicate that the asteroid will come no closer than 2.5 times the distance to the moon.

If a mass the size of Asteroid 1997 XF11 ever did strike Earth, the energy expended would be that of two million Hiroshima-size atomic bombs. According to Jack G. Hills of the Los Alamos National Laboratory, if it struck the Atlantic Ocean, it would create a tsunami more than 100 m high that would obliterate most of the coastal cities around that ocean. If it hit on land, the crater formed would be 30 km across and darken the sky for weeks or months with dust and vapor. For comparison, the infamous asteroid that exterminated the dinosaurs and as much as 75% of all other species on Earth 65 million years ago was 10 to 15 km in diameter.

Asteroid 2004 MN4 (Apophis) was expected to pass very close to Earth on Friday, April 13, 2029, with collision odds initially calculated as 1 chance in 37. NASA's further measurements and calculations now suggest that it will miss Earth by about 30,000 km (about 2.8 times Earth's diameter), and the odds of a 2029 Earth impact are less than one in 100,000.

The chance of that smaller-sized asteroid striking Earth in any one year is estimated at one in several hundred thousand (see Figure 18-15). This probably seems like such a minute chance as to be irrelevant. People do believe, however, in their chances of being dealt a royal flush in poker (1 chance in 649,739) or of winning the multimillion-dollar lottery jackpot (1 chance in 10 to 100 million). The chance that Earth will be struck by a civilization-ending asteroid next year is greater than either of those. As with other natural hazards, the low odds of such an event do not necessarily mean that it will be a long time before it happens. It could happen at any time.

What Could We Do about an Incoming Asteroid?

What if astronomers were to discover a very large asteroid or comet, at least as large as the one that did in the dinosaurs, and they determined it was on a collision course with Earth? What would we see without a telescope, and when would we first see it? By the time it reached inside the moon's orbit, it would be three hours from impact. It would first appear like a bright star, becoming noticeably brighter every few minutes. An hour from impact, it would be as bright as Venus.

Fifteen minutes from impact, it would appear as an irregular mass, rapidly growing in size. Three seconds from impact, it would enter Earth's atmosphere with a blinding flash of light, traveling at perhaps 30 km/s. Then instant annihilation! After that, the impact would produce the same effects recorded in the extinction event 65 million years ago.

The Torino Scale of risk and consequences for an asteroid or comet striking Earth is shown in **Table 18-1**.

Clearly, we could not survive the scenario outlined. So what, if anything, can we do about it—that is, other than bury our heads in the sand and wait for the inevitable? Suggestions include blasting the asteroid into pieces with a nuclear bomb or attaching a rocket engine to it to deflect it away from Earth. For an asteroid discovered within days or months of impact, blasting it into smaller pieces might unfortunately just pepper a large part of Earth with thousands of smaller pieces—not a comforting scenario. For an asteroid discovered years or centuries before impact, blasting it into thousands or millions of pieces with a nuclear weapon would spread the fragments out over time so that most or all could miss us. Conventional explosives detonated on one side of an asteroid or striking it with a heavy rocket might deflect it rather than shatter it. Another possibility involves deflecting an asteroid by changing the amount of heat radiated from one side; for example, we could coat part of it with white paint. The effect is weak, but over a long time, it could shift its orbit enough to narrowly miss Earth.

Unfortunately, we don't yet have a spacecraft capable of deflecting an asteroid in deep space and it would take several years to design, build, and send it out to hit its target.

NEOShield is a joint research project of the United States, Russia, and several European countries, to study three main ways to divert an asteroid headed for Earth:

1. Use of a *kinetic impactor*, a heavy, fast-moving spacecraft designed to hit the asteroid hard enough to divert it slightly.

2. Use of a *gravity tractor*, a spacecraft stationed near the asteroid so that its tiny mass can slowly pull the asteroid off its trajectory.

3. For a very large asteroid, we could try to blast it into a new orbit using nuclear weapons.

If all else fails and the predicted impact site is populated, many people could be evacuated.

NASA catalogs **near-Earth objects** (**NEOs**) larger than 100 m in diameter and that come within 74.9 million km of Earth. The Jet Propulsion Laboratory in Pasadena, California, can detect objects as small as 10 to 20 m or so in diameter. Most are tracked by three telescopes in Arizona, New Mexico, and Hawaii, and by the NEOWISE satellite. Smaller ones are not important; most flame out before they reach Earth's surface. However, as noted, the impact of a large one could be catastrophic. To be classified as potentially hazardous, an object must be larger than 140 m in diameter and come within 0.05 times the 150,000,000-km distance of Earth from the sun—750,000 km. Sometimes scientists tracking asteroids do not get much warning before one comes close. On December 26, 2001, scientists spotted an object 0.3 km across. Twelve days later, it came within 800,000 km of Earth, roughly twice the distance to the moon, a frighteningly small distance. If it had collided with us, it would likely have destroyed an area the size of Texas or all of the northeastern states and southern Ontario. Another object 60 m across came within 460,000 km of Earth on March 12, 2002. Astronomers did not detect it until four days after it passed because it came from the direction of the sun and thus could not be seen. Another 100-m-diameter object was also first spotted in June 2002, three days after it missed us by only 120,000 km.

Government satellites are designed to look down; now some are being programmed to look up to spot bright flashes that suggest a meteor burning up in the upper atmosphere.

Table 18-1	The Torino Scale of Asteroid/Comet Risk
NO HAZARD	**0:** Likelihood of collision near 0; rarely cause damage.
NORMAL	**1:** Asteroid/comet collision unlikely; no public concern.
MERITS ASTRONOMERS' ATTENTION	**2:** Somewhat close pass but collision very unlikely. **3:** Close encounter capable of localized destruction. Notify public if encounter within a decade. **4:** Same as **3** but capable of regional destruction.
THREATENING	**5:** Close encounter capable of regional destruction. Critical attention from astronomers to determine if collision likely. Government contingency planning may be warranted if collision within a decade. **6:** Same as **5** but capable of global catastrophe. **7:** Same as **6** but very close encounter. If collision expected this century, international contingency planning is warranted.
CERTAIN COLLISION	**8:** Localized destruction. Approximately every 50 to several thousand years. **9:** Same as **8** but regional destruction. Approximately every 10,000 to 100,000 years. **10:** Same as **8** but global catastrophe that may threaten world civilization. Approximately every 100,000 to several million years.

Modified from D. Morrison et al. NASA

Prompted by the grazing near-miss of the Chelyabinsk meteor (**Case in Point:** A Near Miss—The Chelyabinsk Meteor, Mongolia, 2013, p. 522) and recognizing the special hazard of NEOs larger than 140 m across and large enough to destroy a small country, NASA asked in 2013 for a budget to find 90% of them. The NEO program estimates that, currently, asteroids smaller than 100 m in diameter and on an Earth-collision course can only be detected a few weeks before impact.

The impacts of asteroids and comets with Earth only rarely affect individuals and have not killed groups of people in historic time, at least none that have been recorded. A long time without an event, and in this case an especially long time compared with human life spans, leads to the widespread belief that it will never happen, at least not to us. However, we now have enough information on objects in Earth-crossing orbits that we have a pretty good idea of the odds of a significant impact with Earth. Scientists have now discovered 95% of the 1-km and larger objects but less than 40% of those larger than 140 m diameter that are potentially hazardous asteroids, and we find more every year. The odds of an undiscovered one striking Earth in the next 200 years are very small. But it is quite clear that it will eventually happen—we just do not know when.

Are we prepared for that inevitable event? The short answer is, clearly no. There is no point in staying awake at night worrying about being hit by a stray space rock. Certainly, the odds of a person being hit in a human lifetime are extremely small. In addition, worrying about the possibility would do no good. We can neither predict the impact of a small but deadly rock nor see it coming before being struck. That conclusion holds for even larger impactors that could kill thousands of people. For still larger asteroids, the difficulty of spotting incoming objects heading directly for Earth creates a major predicament. For a larger doomsday object somehow spotted as it grows larger on Earth approach, we have neither decided on a formal plan of action nationally or internationally nor set up the mechanism for implementation of that action. We just do not want to think about the possibility.

If you were in a position of power or influence, what would you do?

Cases in Point

A Round Hole in the Desert
Meteor Crater, Arizona ▶

Meteor Crater is the classic open-crater impact site, 65 km east of Flagstaff, that is so well known to the general public. It is small as impact craters go, only 1.2 km across and 180 m deep, but it is nicely circular and has distinctly raised rims (see Figure18-7). Being only 50,000 years old, it is also well preserved. The projectile was an iron meteorite with a diameter of 60 m. It came in at 15 km per second and exploded with the energy of 20 million tons of TNT, equivalent to that of the largest nuclear devices. The target rock, Coconino sandstone, shows good evidence of shock features, including shocked quartz and lechatelierite, a fused silica glass. A shepherd found a piece of iron in 1886, and a prospector found many more in 1891. One piece found its way to a mineral dealer in Philadelphia, who recognized it as an iron meteorite. The dealer visited the site and found numerous fragments of iron meteorites. Unfortunately, no craters were then known to be formed by meteorite impacts, and even the great G. K. Gilbert, chief geologist of the U.S. Geological Survey, misinterpreted the crater as being of volcanic origin or a limestone sinkhole around which the meteorite fragments had fallen by coincidence.

Daniel Barringer, a mining engineer interested in the iron as an ore, filed claim to the site in 1903 and began intensive exploration of it with many drill holes. He found meteorites under rim debris and under boulders thrown out from deep in the crater. Barringer presented scientific papers in which he concluded that meteorite impact had formed the crater, but few scientists were convinced. A few years later, two astronomers separately visiting the crater concluded it had been caused by a meteorite, but that given its size and velocity, the meteorite would have vaporized or disintegrated completely. The scientific community remained unconvinced until 1960, when Eugene Shoemaker, then a graduate student, studied the crater and its materials in detail, finding shock-melted glass containing meteoritic droplets and the extremely high-pressure minerals coesite and stishovite. Stishovite also requires temperatures greater than 750°C.

■ *Meteor Crater, Arizona, is the best-known impact crater on Earth.*

David Brody, USGS

CRITICAL THINKING QUESTION

1. If you found a large, well-exposed circular crater, what things would you look for to distinguish one formed by impact instead of a volcanic blast?

A Close Grazing Encounter
Tunguska, Siberia ▶

An asteroid estimated to be 50 m in diameter blew down and charred some 1000 km^2 of forest on June 30, 1908, but failed to create a crater. It must have been a grazing encounter, approximately 8 km above the ground and traveling some 15 km/s before it exploded and disintegrated. Colliding with even a thin atmosphere at that altitude, its high velocity abruptly encountered severe resistance and caused instant disintegration. Its energy was equivalent to 1000 Hiroshima-sized atomic bombs. Seismographs around the world recorded seismic waves suggesting the equivalent of a magnitude 5 earthquake. Many people saw the huge fireball as it traveled north over remote villages in central Siberia. At the time of "impact," 670 km northeast of Krasnoyarsk, people saw a bright flash and heard loud bangs. The ground shook, windows shattered, many were knocked off their feet, and people felt blasts of hot air. One man 58 km southeast of the explosion felt

■ *Trees blown down by the Tunguska asteroid event in 1908.*

searing heat and was knocked more than 6 m off his porch by the blast. Some of the reindeer and dogs of native people near the site were killed by the explosion.

The first expedition into the remote area, in 1921, discovered that scorched trees, over a 50-km diameter, were blown radially outward from the site of the explosion.

Several intensive expeditions that included trenching and drilling failed to find any meteorite fragments, but microscope examination of the soil in 1957 showed tiny spheres of iron oxide meteoritic dust. That suggested that the object was a stony meteorite.

It remains controversial whether the object was an icy comet or a small rocky asteroid. Different pieces of evidence suggest one versus the other.

CRITICAL THINKING QUESTION

1. Why are the trees at Tunguska oriented in a radial pattern even though the object apparently came in at a low angle?

A Near Miss
The Chelyabinsk Meteor, Russia, 2013 ▶

At 9:20 a.m. on February 15, 2013, a small asteroid, DA14, entered Earth's upper atmosphere over western Mongolia, just north of Kazakhstan. It tracked west-northwest for 32 seconds with a low-angle trajectory to an elevation of about 15 to 20 km where it exploded in a brilliant fireball over the southern Ural Mountains area of Russia. Called the Chelyabinsk meteor, it traveled at 18.6 km/s, weighed about 11,000 metric tons (11 million kg), and had an estimated diameter of about 17 to 20 m. People on the ground felt intense heat and a shock wave that blew out windows in thousands of buildings and injured about

1500 people, mostly by flying glass from the shock wave. The outside temperature of −15°C (5°F) forced residents to scramble to cover the open window spaces in their apartments, schools, and factories. The shower of meteorite fragments spread over the snow in a *strewn field* near Chelyabinsk, Russia. Recovered fragments showed it to be a silica-rich ordinary chondrite, a stony meteorite. The largest recovered fragment, about 1.5 m wide, was found at the bottom of a frozen lake.

This was the largest such event since the 1908 Tunguska event. It now seems that there were about 10 other large fireballs

within a few days of February 15. Calculations suggest that all are part of a cluster of fragments; because of their low-entry angle, some bounced off the atmosphere and reorbited Earth to return again.

Asteroid-detection telescopes based both on the ground and in space did not detect the object because they were blinded by the low 15° angle of view to the sun. Prompted by Chelyabinsk, concern has recently shifted from massive NEOs (capable of destroying a small country) to much more numerous bodies 10 to 50 m in

■ *Chelyabinsk meteor burns as it enters Earth's atmosphere.*

diameter that would trigger explosions the size of nuclear weapons.

Another meteor, Apophis, is many times larger than Chelyabinsk, and is expected to pass within 20,000 km of Earth on April 13, 2029. Instead of it being a single mass, if it has a cloud of fragments like the Chelyabinsk meteor, Earth could be hammered with many large impacts.

1. If you find a dark-colored rock, how can you tell whether or not it was part of a meteorite (what characteristics such as density or heft, color, mineral color, coating, ease of breakage, nature of surrounding rocks, and so on, would help you determine this)?
2. If a large asteroid or comet is detected headed for Earth, what option do governments have to stave off catastrophe? What are the pros and cons of each option?
3. With all of the scientists and equipment focused on spotting incoming space objects, why was the Chelyabinsk meteor not seen?

Chapter Review

Key Points

Projectiles from Space

- Asteroids are pieces of space rock orbiting the sun like Earth does. Comets are ice with some rock that occasionally loop through our solar system. **FIGURES 18-1** and **18-2**.

- When space rocks enter Earth's atmosphere, they create meteors, light-producing objects streaking through the sky. Meteorites are the pieces of rock that actually collide with Earth. **FIGURES 18-3** and **18-4**.

- Meteorites include those consisting of iron, chondrites composed of the dark minerals olivine and pyroxene, and achondrites that are similar to the basalt found on Earth. Proportions of the three types are similar to such rocks in the interior of Earth. **FIGURES 18-3** and **18-4**.

Evidence of Past Impacts

- The energy of a moving object is related to its mass and its velocity. The energy of an incoming asteroid doubles for every doubling of its mass and quadruples for every doubling of its velocity. **By the Numbers 18-1**.

- Impacts of projectiles from space blast a deep, round crater that compresses and melts the rock below, spraying it outward in all directions. Moderate-sized impacts have open, rimmed craters. Larger impacts collapse to form a broader, shallower basin with a central peak. **FIGURES 18-6** and **18-7**.

- The asteroid that annihilated the dinosaurs and most other life forms 65 million years ago was likely 10 to 15 km in diameter and struck in the Yucatán Peninsula of eastern Mexico. **FIGURE 18-9**.

- Distinctive features at and around the site of a major impact include shatter cones of rock radiating downward and outward, droplets of molten glass, quartz grains showing shock features, and, for giant impacts, a layer of carbon-rich clay containing the metal iridium. **FIGURES 18-11** to **18-13**.

Consequences of Impacts with Earth

- An asteroid impact could wipe out civilization on Earth.

- Hazards from a large impact include tsunami, firestorms, soot that would block the sun and cause prolonged freezing and plant death, strong acid rain, and nickel poisoning of plants.

Evaluating the Risk of Impact

- Numerous small meteorites, far fewer large ones, and only rarely a giant one strike Earth.

- On average, a 200-m-diameter asteroid impacts Earth once in 10,000 years. Impact of a 10-km-diameter object that could wipe out most or all of civilization is expected to occur once every 100 million years on average. **FIGURE 18-15**.

What Could We Do about an Incoming Asteroid?

- NASA catalogs and tracks near-Earth objects that have the potential of colliding with Earth. Unfortunately, such objects are often undetectable.

- If an asteroid were headed for Earth, some propose that it could be blasted into smaller pieces or deflected.

Key Terms

achondrite, p. 509
asteroid, p. 507
asteroid belt, p. 507
chondrite, p. 509

comet, p. 507
iridium anomaly, p. 515
iron meteorite, p. 508
K-T boundary, p. 513

meteor, p. 508
meteorite, p. 508
near-Earth object, p. 519
shatter cone, p. 513

stony-iron meteorite, p. 509
strewn field, p. 508

Questions for Review

1. Where do asteroids originate in space?
2. What is the path of comets around the sun?
3. How is the tail of a comet oriented? (Which way does it point?)
4. Name two general kinds of meteorites and describe their composition.
5. How fast do asteroids travel in space?
6. Why do asteroids create a more or less semicircular hole in the ground, regardless of whether they come in perpendicular to Earth's surface or with a glancing blow?
7. What is the sequence of events for the impact of a large asteroid immediately after it blows out of the crater?
8. What evidence is there that some large comets or asteroids break up on close encounter with a planet?
9. List five quite different direct physical or environmental effects of impacts of a large asteroid in addition to the excavation of a large crater.
10. Roughly what proportion of Earth's human population would likely be killed by impact of an asteroid 1.5 to 2 km in diameter?
11. Sixty-five million years ago, a very large asteroid struck Earth. Where did it apparently happen?
12. What is the relationship between the size of meteorites and the number of a given size?
13. Why would astronomers have difficulty recognizing a large incoming asteroid headed for direct impact on Earth?

Critical Thinking Questions

1. Why should scientists look for dangerous asteroids?
2. If a 15-km-diameter asteroid were on collision course with Earth and you were the head of NASA, what would you do if the president of the United States gave you complete control over actions to deal with it? Why?
3. The total energy of an asteroid dictates the crater size. What major factors dictate the amount of energy an asteroid has?

Conversion Factors

Length

1 millimeter (mm)	= 0.03937	inches
1 centimeter (cm)	= 0.3937	inches
1 meter (m)	= 3.281	feet
1 kilometer (km)	= 0.6214	miles (~5/8 mile)
1 inch (in)	*= 2.540*	*cm*
1 foot (ft)	*= 30.48*	*cm*
1 mile (mi)	*= 1609*	*m = 1.609 km*

Area

1 square centimeter (cm^2)	= 0.1550	square inches (in^2)
1 square meter (m^2)	= 10.76	square feet (ft^2)
1 square kilometer (km^2)	= 0.3861	square miles
1 hectare (510,000 m^2)	= 2.471	acres
1 in^2	*= 6.452*	*cm^2*
1 ft^2	*= 929.0*	*cm^2*
1 mi^2	*= 2.590*	*km^2*

Volume

1 cubic centimeter (cm^3)	= 0.06102	cubic inches (in^3)
1 cubic meter (m^3)	= 35.31	cubic feet (ft^3) = 1.308 cubic yards (yd^3)
1 liter	= 61.02	in^3 = 1.057 qt. = 0.2642 gallons
1 ft^3	*= 0.02832*	*m^3*
1 yd^3	*= 0.7646*	*m^3*
1 gallon U.S.	*= 3.785*	*liter = 0.1337 ft^3*

Mass

1 gram (gm)	= 0.03527	oz = 0.002205 pounds (lb)
1 kg	= 2.205	lb
1 oz	*= 28.35*	*gm*
1 pound (= 16 oz.)	*= 453.6*	*gm = 0.4536 kg*

Velocity

1 meter per second (m/sec.)	= 3.281	feet per second = 3.6 km/hr
1 mile per hour	*= 1.609*	*kilometer per hour (km/hr)*

Temperature

°C	= 5/9 (°F−32)
°F	*= 9/5 °C + 32*

Prefixes for SI units

Giga	= 10^9	5 1,000,000,000	times
Mega	= 10^6	5 1,000,000	times
Kilo	= 10^3	5 1000	times

100-year flood A flood magnitude that comes along once every 100 years on average.

aa Blocky basalt flow with a ragged, clinkery surface.

acceleration The rate of increase in velocity with time. During an earthquake, the ground accelerates from being stationary to a maximum velocity before slowing and reversing its movement.

achondrite A stony meteorite that is similar to basalt in composition.

adiabatic cooling Change in volume and temperature, for example in the atmosphere, without change in total heat content.

adiabatic lapse rate (dry and wet) The rate of change of temperature as an air mass changes elevation. As an air mass rises, for example, it cools.

aerosol Tiny particles or droplets suspended in the atmosphere that reflect short-wave radiation from the sun and reduce the amount of light reaching Earth's surface.

aftershocks Smaller earthquakes after a major earthquake that occur on or near the same fault. They may occur for weeks or months after the main shock.

albedo The fraction of the sun's energy reflected away from Earth's surface. Snow has a high albedo.

alluvial Related to sand and gravel deposited by running water.

alluvial fan A fan-shaped deposit of sand and gravel at the mouth of a mountain canyon, where the stream gradient becomes gentler or flattens at a main valley floor.

amplitude The size of the back-and-forth motion of earthquake shaking on a seismograph.

andesite An intermediate-colored, fine-grained volcanic rock. It has a silica content of approximately 60%, has intermediate viscosity, and forms slow-moving lavas, fragments, and volcanic ash.

angle of repose The maximum stable slope for loose material; for dry sand it is 30 to 35°.

aquifer A permeable underground formation saturated with enough water to supply a spring or well; a water-bearing area of rock that will provide usable amounts of water.

Arctic oscillation (AO) An oscillating pattern of sea-level atmospheric pressure anomalies, one in the Arctic and one of the opposite sign centered near 37 to 45°N latitude. With surface pressure low in the Arctic, the AO is positive, keeping cold air near the pole; the midlatitude jet stream flows rapidly from west to east. With surface pressure high, AO is negative and the jet stream slows and wobbles.

ash Loose particles of volcanic dust or small fragments; formally it refers to particles less than 2 mm in diameter but informally almost any small-size particles.

ash fall Volcanic ash that falls through the air and collects loosely on the ground.

ash flow A mixture of hot volcanic ash and steam that pours at high velocity down the flank of a volcano. Also called a pyroclastic flow or nuée ardente.

asteroid A chunk of rock orbiting the sun in the same way that Earth orbits the sun.

asteroid belt The distribution of asteroids in a disk-shaped ring orbiting the sun.

asthenosphere Part of Earth's mantle, below the lithosphere, that behaves in a plastic manner. The rigid and brittle lithosphere moves slowly over it.

Atlantic multidecadal oscillation (AMO) The oscillation of the sea-surface temperature of the North Atlantic Ocean by about 0.8°C over several decades, typically about 70 years.

atom The smallest particle of matter that takes part in an ordinary chemical reaction.

average The usual or ordinary amount of something. Specifically, the sum of a group of values divided by the number of values used. Same as "mean."

avulsion The permanent change in a stream channel, generally during a flood, when a stream breaches its levee to send most of its flow outside of its channel.

back burn Fire deliberately set close to an advancing fire to create a firebreak by using the updraft of the wildfire to draw the back burn fire back toward the main fire.

bankfull capacity The maximum capacity of a stream before it overflows its banks.

barrier bar A natural ridge of sand built by waves a few meters above sea level that is parallel to the shoreline and just offshore. The term is also commonly used for a sandbar across the mouth of a bay or inlet.

barrier island A near-shore, coast-parallel island of sand built up by waves and commonly capped by windblown sand dunes. Also called a barrier bar.

basalt A black or dark-colored, fine-grained volcanic rock. It has a silica content of approximately 50% and low viscosity, and can flow downslope rapidly.

base isolation A mechanism to isolate a structure from earthquake shaking in the ground; often flexible pads between a building and the ground.

base level The level below which a stream cannot erode, typically at a lake or ocean.

base surge The high-energy blast of steam and ash that blows laterally from the base of an eruptive column during the initial stages of a strong volcanic eruption.

beach The deposit of loose sand, pebbles, or rock fragments along the shore. It includes sand seaward of either cliffs or dunes and extends offshore to a depth of a few to perhaps 10 m.

beach hardening The placement of beach "protection" structures: seawalls, riprap, groins, and related features.

beach replenishment Addition of sand to a beach to replace that lost to the waves. *Also called beach nourishment.*

bedload Heavier sediment in a stream that moves along the streambed rather than in suspension.

bedrock stream A stream that flows over bedrock.

berm The uppermost accumulation of beach sand and other sediment left by waves.

blind thrust A thrust fault that does not reach Earth's surface. It is not evident at the surface.

blizzard Blowing snow with high winds stronger than 56 km/h in the United States or 40 km/h and with visibility reduced to 400 m or less in Canada.

block A mass of solid rock ejected from a volcano and larger than 25 cm across.

body wave A seismic wave that travels through Earth's interior. P and S waves are body waves. Surface waves travel only near Earth's surface and are not body waves.

bomb A 6- to 25-cm-diameter mass of solid rock that is ejected from a volcano.

brackish Water intermediate in salinity between sea water and fresh water. Often true of the water in lagoons connected to the ocean.

braided stream A stream characterized by interlacing channels that separate and come together at different places. The stream has more sediment than it can carry, so it frequently deposits some of it.

breach Failure of part of a river's levee, leading to some flow outside the main channel.

breakwater An artificial offshore barrier to waves constructed to create calm water on the shoreward side for a beach or for anchoring boats.

burnout A burnout fire is deliberately set to burn back toward the main fire, eliminating fuel and stopping the progress of the main fire.

buyer beware The old saying that warned a buyer to be wary that what he or she was purchasing might not be as good as it appears. The buyer pays the consequences.

caldera A large depression, generally more than 1 km across, in the summit of a volcano and formed by collapse into the underlying magma chamber.

cap and trade A government policy to limit carbon emissions by permitting companies to buy and sell pollutants to remain within their designated cap amount.

carbon cycle The movement of different forms of carbon through various Earth forms or processes, for example vegetation, coal, and carbon dioxide.

carbon sequestration Removal and long-term storage of greenhouse gases.

carbon tax An indirect tax on the societal costs of using carbon-based fuel, costs such as air and water pollution, health, and climate change.

catastrophe Disaster or calamity.

cavern A large, natural underground cave or tunnel, most commonly in limestone. Also used for a soil cavern developed over a limestone cavern.

Celsius (°C) The temperature scale based on 0 for the freezing point of water and 100 for the boiling point of water at Earth's surface and under normal conditions. Used in science and in most countries. Degrees Celsius times 5/9 equals the *size* of degrees Fahrenheit, the scale used by the public in the United States.

chance The likelihood of an event. For example, if the chance of an event is 1%, then the event will occur, on average, one time in 100 tries.

channel scour The depth of sediment eroded during stream or river floods.

channelization Straightening and confining stream flow within artificial barriers such as riprap or concrete along the sides of the stream.

Chinook winds Winds that warm by adiabatic compression as they descend from high elevations of a mountain range to lower elevations on the plains to the east.

chondrite A stony meteorite consisting primarily of the minerals olivine and pyroxene; the most common kind of meteorite.

cinder cone A small steep-sided volcano consisting of basaltic cinders.

cinders Small fragments of basalt full of gas holes that were blown out of a cinder cone.

clay Particles of sediment smaller than 0.004 mm. *See also* clay minerals.

clay minerals Exceptionally fine-grained (often clay-sized), soft, hydrous aluminum-rich silicate minerals with layered molecular structures.

climate The weather of an area averaged over a long period of time.

climate change Future long-term changes in worldwide climate; human-induced changes are often blamed.

coastal bulge The region above a subduction zone in which the overlying continental plate flexes upward before slip on the subduction zone causes a major earthquake.

cohesion The attraction between small soil particles that is provided by the surface tension of water between the particles.

cold front The line of boundary between a regional mass of cold air advancing under an adjacent large mass of warm air.

collision zone The zone of convergence between two lithospheric plates.

combustible Material that can be set on fire or burned.

comet A mass of ice and some rock material traveling at high velocity in the gravitational field of the sun but traveling outside our solar system before passing through it on occasion.

compaction The reduction in volume of clay, soil, or other fine-grained sediments in response to an overlying load.

composite volcano A large, steep-sided stratovolcano consisting of layers of ash, lava, and assorted volcanic rubble.

conduction Vibration energy transferred from one molecule to the next, causing transfer of heat from higher temperature to lower temperature areas.

continental caldera Large rhyolite volcanoes characterized by their high viscosity and high volatile content but having gently sloping flanks.

continental crust The upper 30 to 60 km of the continental lithosphere that has an average density of 2.7 g/cm^3 and an average composition similar to that of granite.

continental drift The gradual movement of continents as their positions change on the Earth's surface.

continental margin The boundary between continental lithosphere and oceanic lithosphere.

continental collision zones Collision between two continental plates.

convection Physical movement, of mases of heated air, water, or deep Earth from higher temperature to lower temperature areas.

convective updraft The rapid rise of a heated, expanded air mass driven by its lower density. The removal of air from below draws outside air into the rising "chimney."

convergent boundary A tectonic plate boundary along which two plates come together by either subduction or continent–continent collision.

core Innermost part of Earth consisting of nickel and iron. The inner core is solid; the outer core is liquid.

Coriolis effect Rotation of Earth from west to east under a fluid such as the atmosphere or oceans permits that fluid to lag behind Earth's rotation. Compared to the direction of travel, fluid flows shift to the right in the northern hemisphere and to the left in the southern hemisphere as a result of this effect. Thus a southward moving fluid appears to curve off to the west in the northern hemisphere.

cover collapse Collapse of the roof material over an underground cavity, often a soil cavern over a limestone cavern.

cover subsidence Gradual depression of the roof material over an underground cavity.

crater A depression created as ash blasts out of a volcano.

creep (1) Surface layers of soil on a slope gradually move downslope more rapidly than subsurface layers. (2) Slow, more-or-less continuous movement on a fault. Creeping faults either lack earthquakes or have only small ones.

crown fire A fire that burns into the treetops.

crust The thin outer layer of Earth, overlying the mantle. About 7 km thick under oceans; about 40 km thick on continents.

cumulonimbus cloud A thundercloud with a flat anvil-shaped top; a sign of a thunderstorm.

current The continuous movement of a mass of fluid in a definite direction.

cutoff When adjacent meanders in a stream come close together, the stream may cut through the narrow neck between them to bypass and abandon the intervening meander loop. The cut through the narrow neck is a cutoff.

cyclic events Events that come at more-or-less evenly spaced times, such as every so many hours or years.

cyclone A large low-pressure weather system that circulates counterclockwise in the northern hemisphere and clockwise in the southern hemisphere. It is equivalent to a hurricane or a typhoon.

dacite A light-colored volcanic rock intermediate in composition and appearance between andesite and rhyolite.

daylighted beds Layers inclined less steeply than the slope so that their lower ends are exposed at the surface such as on a steep slope or road cut.

debris avalanche A fast-moving avalanche or rockfall of loose debris that flows out a considerable distance from its source; it is often very destructive.

debris flow A slurry of rocks, sand, and water flowing down a valley. Water generally makes up less than half the flow's volume.

deepwater waves A wave in water deep enough that its water motion does not touch bottom—that is, a wave in water that is deeper than half the wave's length.

deformation The change in shape of an object.

delta The accumulation of sediment where a river reaches the base level of a lake or ocean.

density The mass (e.g., grams) of a material per volume (e.g., cubic centimeter). Water's density is 1 g/cm^3; granite's is 2.7 g/cm^3.

deposition The accumulation of sediment previously carried by moving water or air.

desert A region with a particularly dry climate in which evaporation exceeds precipitation. An area with less than 25 cm of precipitation annually. A desert does not need to be hot.

desertification The growth of new desert environments in areas where drought prevents regrowth of grass and shrubs stripped away by increased population, intensive cultivation in marginal areas, or overgrazing.

destructive interference Where independent wave highs overlap lows and lows overlap highs; the lows tend to cancel out the effects of the highs.

developed countries Industrial countries that are wealthy enough to have a large middle class.

dip The angle of inclination of a plane as measured down from the horizontal.

discharge Measured volume of water flowing past a cross section of a river in a given amount of time. Usually expressed in m^3/s or ft^3/s.

displacement The distance between an initial position and the position after movement on a fault.

dissolution Minerals or rocks dissolving in water.

divergent boundary A spreading plate boundary such as a mid-oceanic ridge.

dome A bulge on a volcano that is pushed up by molten magma below.

downburst A localized, violent, downward-directed wind below a strong thunderstorm.

drainage basin The area of slopes upslope from a point in a valley that drains water to that point.

dredging Scooping up sediment blocking a channel or loose sediment such as gravel from a stream or harbor.

driving force Any force on an object that causes it to move or keep movement. Gravity provides a downward pull on a mass that can start it moving or keep it moving.

drought A prolonged dry-climatic event in a particular region that dramatically lowers the available water below that normally used by humans, animals, and vegetation.

dry climate Sometimes described as an area with evaporation greater than precipitation, sometimes as an area with too little precipitation to support significant plant life.

dune A ridge or mound of sand deposited by the wind, most commonly at the head of a beach or in an arid environment.

dust storm A large area of airborne dust or fine sand carried in a dense cloud with a distinct front, and close to Earth's surface (up to a few km thick). It occurs mainly in arid areas or during drought where bare ground is exposed. It is often initiated at a cold front or as outflow from a strong thunderstorm.

dynamic equilibrium The condition of a system in which the inflow and outflow of materials is balanced.

earthquake The ground shaking that accompanies sudden movement on a fault, movement of magma underground, or a fast-moving landslide.

earthquake early warning (EEW) systems An advance warning system to alert people or equipment that an earthquake has occurred and approximately when shaking will arrive at their location.

effluent stream = gaining stream One which gains water from groundwater at the water table.

elastic A type of deformation in which a rock deforms without breaking. If the deforming stress is relaxed, the rock returns to its original shape. Contrast with plastic.

elastic rebound theory The theory applied to most earthquakes in which movement on two sides of a fault leads to bending of the rocks until they slip during an earthquake to release the bending strain.

element One of ninety-two naturally occurring materials such as iron, calcium, and oxygen that cannot be further subdivided chemically. Elements combine to make minerals or other compounds.

El Niño Elevated sea-surface temperatures off the west coast of South America that lead to dramatic changes in weather in some areas every few years—for example, rains in coastal Peru, western Mexico, and southwestern California. *Compare with* La Niña.

emissions Gases, generally undesirable, given off by power plants, factories, other human activities, or by volcanoes.

energy The capacity for doing work. Energy from the sun heats the surface, for example, to evaporate water and melt ice. The kinetic energy of a large wave can move huge boulders. The heat energy of a rising mantle plume may melt the overlying mantle or crust.

Enhanced Fujita scale Tornado strength based on internal wind speeds and long-term records of damage. A revision of the original Fujita scale.

ENSO; El Niño-southern oscillation The alternation between surface air pressure during El Niño and its opposite extreme La Niña.

epicenter The point on Earth's surface directly above the earthquake focus; surface location above the initial rupture point on a fault.

equilibrium profile (1) The longitudinal profile or slope of an equilibrium stream (see *graded stream*); (2) the seaward slope of a beach that is adjusted so that the amount of sediment that waves bring onto the beach is adjusted to the amount of sediment that the waves carry back offshore.

erosion The wearing down and transport of loose material at Earth's surface, including removal of material by streams, waves, and landslides.

eruption The processes by which a volcano expels ash, steam and other gases, and magma.

eruption column The mass of ash and gases forcefully blown upward from a volcanic vent during an eruption.

eruptive rift A crack in the ground from which lava erupts.

eruptive vent The site of eruption of material from a volcano.

estuary The submerged mouth of a river valley where fresh water mixes with seawater.

evaporation The change of water or some other substance from a liquid to a vapor.

evapotranspiration Water returned to the atmosphere by a combination of evaporation of water from the soil and leaves, and transpiration of moisture from the leaves.

expanding soil Clay-rich soil that expands when it absorbs water. Most expanding soil contains smectite, a swelling clay.

exsolve The process by which gas or other material separates from water or a magma.

extratropical cyclone A cyclone formed outside the tropics and commonly associated with weather fronts. Nor'easters are an example.

eye The small-diameter core of a hurricane; characterized by lack of clouds and little or no wind.

fault An Earth fracture along which rocks on one side move relative to those on the other side.

fault creep The slow, more-or-less continuous movement on a fault, in contrast to the sudden movement of a section of fault during an earthquake.

feedback effect Where a small disturbance in a system leads to effects that increase that disturbance (positive feedback) or decrease it (negative feedback).

fetch The distance over which wind blows over a body of water. The longer the fetch, the larger the waves that will be developed.

firebrands Burning embers carried by the wind, which have potential to ignite new fires.

fireline A narrow firebreak dug or cleared in soil to stop the advance of a fire.

firestorm A large, extremely hot fire generating a large updraft that pulls in vast amounts of air from the sides to fan the flames.

fire weather potential The favorability of weather conditions for wildfire.

flash flood A short-lived flood that appears suddenly, generally in a dry climate, in response to an upstream storm.

flood basalt A broad expanse of basalt lava that cooled to fill in low-lying areas of the landscape.

flood crest The point where the flood reaches its peak discharge.

flood fringe The fringe or backwater areas of a flood that store nearly still water during a flood and gradually release it downstream after the flood.

floodplain Relatively flat lowland that borders a river, usually dry but subject to flooding every few years.

floodway Area of floodplain regulated by Federal Emergency Management Agency (FEMA) for federal flood insurance.

fluidization The process of changing a soil saturated with water to a fluid mass that flows downslope.

focus The initial rupture point to produce an earthquake on a fault. The epicenter is the point on Earth's surface directly above the focus. Focus is also called hypocenter.

foehn wind A strong wind that warms adiabatically as it descends the leeward side of a mountain range. In specific areas they are called a Chinook wind or Santa Ana wind.

forecast The statement that a future event will occur in a certain area in a given span of time (often decades) with a particular probability. *Compare with* prediction.

foreshocks Smaller earthquakes that precede some large earthquakes on or near the same fault. They are inferred to mark the initial relief of stresses before final failure.

fossil fuel Coal, petroleum, and natural gas; hydrocarbons formed by burial and decomposition of organic materials.

fracking A method of producing oil and natural gas from source rocks by injecting high pressure water, chemicals, and sand to open up fractures in the rock.

fractal Features that look basically the same regardless of size; for example, the coast of Norway looks serrated on the scale of a map of the world or on a map of only part of that coast.

fracture Any crack in a rock.

frequency The number of events in a given time, such as the number of back-and-forth motions of an earthquake per second.

frictional resistance The resistance to downslope movement of a flow, landslide, or earthquake fault.

front (weather) The boundary between one air mass and another of different temperature and moisture content.

frostbite The freezing of body tissue as a result of exposure to extreme cold temperature and winds.

fuel load The amount of burnable material, such as trees and dry vegetation, available for a wildfire.

fuel moisture Amount of moisture in a fuel. The lower the moisture content, the more readily the fuel will burn and at a higher temperature.

Fujita scale The scale of tornado wind speed and damage devised by Dr. Tetsuya Fujita.

fumarole Vent that emits steam or other gases.

funnel cloud Narrow, rapidly spinning funnel-shaped "cloud" that descends from a violent storm cloud. *See also* tornado.

gaining stream A stream, typically in a climate with abundant rainfall, that lies below the water table and gains water from groundwater as it flows downstream.

geoengineering The process of artificially manipulating Earth systems to counteract climate change effects.

geotextile fabric Strong permeable cloth that permits water to pass through it but not soil and rock particles.

geyser Intermittent eruption of hot and boiling water from a vent in an area of volcanic rocks.

glacial-outburst flood Flooding of glacial meltwater when ice dams float or collapse, or ice tunnels collapse. Also called jökulhlaups.

global warming Warming of Earth's surface temperatures in the atmosphere, especially over the past 100 years. Much of the recent warming has been attributed to increases in human-generated greenhouse gases.

graded stream Stream in equilibrium with its environment; its slope is adjusted to accommodate the amount of water and sediment amounts and grain sizes provided to it.

gradient Slope along the channel of a streambed, typically expressed in m/km or ft/mi.

granite Light-colored, grainy, plutonic igneous rock consisting of the minerals quartz, the two feldspars, orthoclase and plagioclase, and sometimes a dark mineral, either mica or amphibole.

greenhouse effect Increased atmospheric temperatures caused when atmospheric gases such as carbon dioxide and methane trap heat in Earth's atmosphere.

greenhouse gases Atmospheric gases such as carbon dioxide and methane that retain heat in Earth's atmosphere; solar energy can get in, but the heat cannot easily escape.

groin Wall of boulders, concrete, or wood built out into the surf from the water's edge to trap sand that moves in longshore drift. The intent is to hinder loss of sand from the beach or to keep it from blocking inlets.

groundwater Water in the saturated zone below the ground surface.

groundwater mining The amount of water pumped from the ground exceeds recharge from precipitation.

Gulf Stream Warm oceanic current that moves north from the Gulf of Mexico, past the East Coast of North America, and across the North Atlantic to maintain mild temperatures in northern Europe.

gunite = shotcrete A fluid cement-type material that is sprayed on a slope to prevent water penetration.

hailstone Single pellet of hail formed when water droplets freeze and grow before falling to the ground.

"hard" solutions Solutions to a problem that involve building protective structures such as riprap and levees or continuous pumping of water from a landslide-prone slope.

harmonic tremor Rhythmic shaking of the ground that often accompanies magma movement under a volcano.

Hawaiian-type lava Fluid basalt lava of the type that erupts from Hawaiian volcanoes.

hazard An environment that could lead to a disaster if it affects people.

headland An erosionally resistant point of high land jutting out into the sea (or a large lake).

headscarp The exposure of the slip surface at the top of a landslide.

heat Energy produced by the collisions between vibrating atoms or molecules in a gas, liquid, or solid. The faster the vibration, the greater the heat and the higher the temperature.

heat capacity The capacity to hold heat. Metals such as iron heat and cool easily; they have low heat capacity. Water has a high heat capacity.

heat-island effect Increased temperatures in urban areas due to buildings and paved areas absorbing more solar radiation, while exhaust from cars and factories traps heat.

Hertz The number of cycles of earthquake shaking per second, a measurement of frequency of shaking.

high-pressure system An area characterized by descending cooler dry air in the atmosphere and clear skies. Winds rotate clockwise around a high-pressure system (in the northern hemisphere). *See also* right-hand rule.

hook echo A curved, hook-shaped pattern on weather radar maps that is indicative of a supercell thunderstorm and potential tornado.

hotspot volcano An isolated volcano, typically not on a lithospheric plate boundary, but lying above a plume or hot column of rock in Earth's mantle.

hurricane A large tropical cyclone in the North Atlantic or east Pacific ocean, with winds greater than 118 km/h. Similar storms in the western Pacific and elsewhere are called typhoons or cyclones.

hurricane warning A hurricane watch is upgraded to a warning when dangerous conditions of a hurricane are likely to strike a particular area within 24 hours.

hurricane watch Indicates that a hurricane may affect a region within 36 hours.

hydrocarbon residue Remaining organic material that soaks into shallow ground after a wildfire.

hydrograph A graph that shows changes in discharge or river stage with time.

hydrologic cycle The gradual circulation of water from ocean evaporation to rain or snow falling on the continents, where some of it soaks in to feed vegetation, groundwater, and streams. From there, some water evaporates while a portion flows off the surface in streams and returns to the oceans.

hydrophobic soil Soil sealed by hydrocarbon resins. These soils restrict water penetration.

hypothermia When a person's core body temperature drops below 35°C/95°F. Symptoms include uncontrollable shivering, drowsiness, disorientation, slurred speech, and exhaustion.

hypothesis A proposal to explain a set of data or information, which may be confirmed or disproved by further study.

ice age A period of low temperature lasting thousands of years and marked by widespread ice sheets covering much of the northern hemisphere. Two to four ice ages are recorded in North America, and as many as 20 in deep-sea sediments of the last 2 million years.

ice storm A winter storm in which moisture falls through a warm air layer, then refreezes on passing through a cold air layer near the ground.

igneous rocks Rocks that crystallized from molten magma, either within Earth as plutonic rocks (e.g., granite and gabbro) or at Earth's surface as volcanic rocks (e.g., rhyolite, andesite, and basalt).

indefensible location A home or building location that is especially vulnerable to wildfire. Particularly important are dry vegetation and other fuels near the building and a location high on a hill, one that fire can easily reach upslope.

influent stream = losing stream One which loses water to the groundwater at the water table.

insurance A service by which people pay a premium, usually annually, to protect themselves from a major financial loss they cannot afford. The insurance company that collects the premiums pays for any covered losses.

intensity scale The severity of an earthquake in terms of the damage that it inflicts on structures and people. It is normally written as a Roman numeral on a scale of I to XII.

ion An atom with a positive or negative charge.

iridium anomaly A thin clay-bearing layer of sediment 65 million years old that has an abnormally high content of the platinum-like trace-metal iridium and suggesting asteroid impact.

iron meteorite A meteorite consisting of a nickel and iron alloy.

isostacy = isostatic equilibrium Lower-density crust floats in Earth's higher-density mantle.

jet stream The high-speed stream of air current traveling, generally in meanders, from west to east across North America at an approximate altitude of 10 to 12 km.

jetties Walls of boulders, concrete, or wood built perpendicular to the coast at the edges of a harbor, estuary, or river mouth. The intent is to prevent sediment from blocking a shipping channel.

joint One of a group of fractures with similar orientation.

kaolinite A common clay mineral formed by weathering and which may contribute to landsliding; it does not expand when wet.

karst The ragged top of limestone exposed at the surface, resulting from dissolution by acidic rainfall and groundwater. The ground is often marked by sinkholes and caverns.

K-T boundary The boundary between the Cretaceous and Tertiary geological time periods, approximately 65 million years ago. The boundary is marked by a thin layer of sooty clay that contains an anomalous amount of iridium.

ladder fuels Forest fuels of different heights that permit fire to climb progressively from burning ground material to brush, low branches, and finally to the tops of trees.

lagoon A narrow, shallow body of seawater or brackish water that parallels the coast just inland from a barrier island.

lahar A volcanic mudflow.

lake-effect snow Increased snowfall in humid-atmosphere areas downwind from large lakes or other water bodies.

land subsidence Settling of the ground in response to extraction of water or oil in subsurface soil and sediments, drying of peat, or formation of sinkholes.

land-use planning Restriction of development according to practical and ethical considerations, including the risk of natural hazards.

landslide Downslope movement of soil or rock; it can be slow or fast.

La Niña The opposite of El Nino, which is characterized by unusually warm ocean temperatures in the equatorial region of the Pacific Ocean.

lapilli A particle of volcanic ash between 2 and 6 mm.

latent heat The heat added to a gram of a solid to cause melting or to a liquid to cause evaporation. Also the heat produced by the reverse processes.

latitude Imaginary lines on Earth's surface that parallel the equator. The equator is 0° latitude; Earth's poles are 90° north and south latitude.

lava Magma that flows out onto the ground.

lava dome A large bulge or protrusion, either extremely viscous or solid on the outside, and pushed up by magma.

lava flow Lava that spills out of a volcano and flows down its sides.

left-lateral One sense of offset on a strike-slip fault; if you stand on one side of the fault, the opposite side has moved to the left.

levee An artificial embankment of material placed at the edge of a channel to prevent floodwater from spreading over a floodplain.

lightning The visible electrical discharge that marks the joining of positive and negative electric charges between clouds or between clouds and the ground.

liquefaction A process in which water-saturated sands jostled by an earthquake rearrange themselves into a closer packing arrangement; the ground may liquefy and flow. The expelled water spouts to form a sand boil.

lithosphere Rigid outer layer of Earth approximately 60 to 100 or so km thick; it includes Earth's crust and uppermost mantle and forms the lithospheric plates.

lithospheric plates A dozen or so segments of the lithosphere that cover Earth's outer part. Above the asthenosphere.

load (1) Related to landslides, the weight of material on a slope. (2) Related to flood processes, the volume of sediment a stream can carry.

longitude Imaginary lines on Earth's surface oriented north–south and perpendicular to the equator.

longshore current A current parallel to the shore caused by refraction of waves striking the beach at an angle.

longshore drift The gradual migration of sand or gravel along the shoreline resulting from waves repeatedly carrying the sediment grains obliquely up onto shore and then back down to the water's edge.

losing stream A stream, typically in a dry climate, that lies above the water table and loses water to the groundwater.

low-pressure system An area of low atmospheric pressure that is characterized by rising warmer and humid air and cloudy skies. Winds rotate counterclockwise around a low-pressure cell (in the northern hemisphere). *See also* right-hand rule.

low-velocity zone Zone of low seismic velocity that marks the boundary zone between the lithosphere and asthenosphere. Presumed to contain a small percentage of partial melt.

maar A broad, shallow crater with low rims, often on nearly flat topography, and formed by gas-rich volcanic eruptions.

mafic A dark-colored igneous rock.

magma Molten rock. When it flows out on the ground, it is called lava.

magma chamber Large masses of molten magma that rise through Earth's crust, often erupting at the surface to build a volcano.

magnetic field The region around a magnet in which magnetism is felt.

magnetic stripes Stripes of rocks on the ocean floor, parallel to an oceanic ridge, that alternate between weak and strong magnetism parallel to Earth's current magnetic field.

magnitude The relative size of an earthquake, recorded as the amplitude of shaking on a seismograph. The number (open-ended but always less than 10) is recorded as a logarithm of the amplitude. Amplitudes differ for different seismic waves: M_L = Richter or local magnitude; M_s = surface wave magnitude; M_b = body wave (P or S wave) magnitude; M_w = moment magnitude (most important for extremely large earthquakes).

mammatus clouds Downward rounded pouches protruding from the base of an overhanging anvil of a thundercloud.

mangrove Large leafy bushes or trees that form dense thickets in shallow brackish waters of coastal lagoons and estuaries.

mantle The thick layer of Earth below the thin crust and above the core. Its upper part is mostly peridotite in composition. Its density approximates 3.2 to 3.3 g/cm^3 in the upper part and 4.5 g/cm^3 in the lower part.

mass A measure of the quantity of matter, regardless of the acceleration of gravity. Not the same as weight.

meandering stream Streams that sweep from side to side in wide turns called meanders.

mean sea level The average height of the sea, averaged over a long time, specifically 19 years. Inferred to be midway between average high and low tides.

melting temperature The temperature at which a rock melts (actually a range of temperatures). Granite commonly melts between 700° and 900°C. Basalt commonly melts near 1200 to 1400°C.

Mercalli Intensity Scale The severity of an earthquake in terms of the damage that it inflicts on structures and people. It is normally written as a Roman numeral on a scale of I to XII.

metamorphic rocks Rocks of any kind that have been changed by heat, pressure, and often deformation; sometimes changes occur in chemical composition.

meteor A piece of space rock that heats to a white hot incandescence when it streaks through Earth's atmosphere.

meteorite An asteroid that passes through the atmosphere to reach Earth.

methane A gaseous hydrocarbon formed by decay of plant or animal matter after burial. Also called natural gas, it can be burned to generate heat. If released into the atmosphere, it is a potent greenhouse gas. Also emitted during volcanic eruptions.

methane hydrate Frozen methane-ice "compound," trapped in layers deeper than 1 km in continental permafrost and at shallow depths under the seafloor of many of the world's continental slopes. A potential source of greenhouse gas.

microearthquakes Minute underground tremors that may reveal a previously unknown fault system and may someday precede an earthquake.

mid-oceanic ridge A high-standing rift or spreading zone in an ocean—for example, the Mid-Atlantic Ridge or the East Pacific Rise.

migrating earthquakes Earthquakes that occur, over time, in a sequential manner along a fault.

Milankovitch cycles Cycles of sun's energy calculated from variations in Earth's orbit around the sun and measured in tens of thousands of years.

mineral A naturally occurring inorganic compound with its atoms arranged in a regular crystalline structure. Rocks are groups of minerals.

mining groundwater Removal of water from the ground without equal replacement.

mitigation Changes in an environment to minimize loss from a future disaster.

Modified Mercalli Intensity Scale Scale measuring the severity of an earthquake in terms of the damage that it inflicts on structures and people. It is normally written as a Roman numeral on a scale of I to XII.

Mohorovičic discontinuity (moho) Boundary between Earth's crust and mantle. It is detected from the contrast between the slower seismic velocities of the crust (5 to 7 km/s.) and the upper mantle (approximately 8 km/s.).

molecule The smallest combination of two or more atoms held together by balanced atomic charges.

moment magnitude (M_w) The magnitude of an earthquake based on its seismic moment which depends on the rock strength, area of rock broken, and amount of offset across the fault.

monsoon A rainy season of atmospheric circulation that depends on changes in heating of land and sea. In India, it is a wind that blows from the southeast for half of the year bringing warm, moist air and heavy rains from the Indian Ocean onto the south Asian continent.

mudflow A flow of mud, rocks, and water dominated by clay or mud-sized particles.

multipurpose dam A dam justified by perceived multi-purpose benefits—for example, flood control, hydroelectric power, water supply, and recreation.

National Flood Insurance Program The flood insurance program guaranteed and partly funded by the U.S. government.

natural disaster A natural event that causes significant damage to life or property.

natural hazards Hazards in nature that endanger us and our property.

natural levee Natural embankment of sediment at the edge of a stream, where coarser-grained sediment is deposited as flood waters slow and spread over an adjacent floodplain.

nature's rampages The perception of many people that a natural catastrophe involves nature going on an abnormal rampage. Actually, nature is not doing anything that it has not done for millions of years; we humans have merely put ourselves in harm's way.

near-Earth object A space object such as an asteroid or comet that narrowly misses Earth; closer than 1.3 times the distance from Earth to the sun.

Nor'easter A strong extratropical winter storm that moves up the East Coast of North America with high winds and high waves. They can be as damaging as hurricanes.

normal fault A fault (generally, steeply inclined) on which the upper block of rock moved down compared with the lower block. Found in extensional environments.

North Atlantic oscillation (NAO) The winter atmospheric pressure pattern over the North Atlantic Ocean that brings major storms every few years. A positive NAO correlates with a positive Arctic Oscillation and a strong Polar vortex.

obsidian Volcanic glass of rhyolite composition. It is generally a water-poor magma that solidified before it could nucleate and crystallize.

oceanic crust The upper part of Earth's lithosphere under the oceans. It consists of basalt and gabbro and is approximately 7 km thick.

offset The distance of movement across a fault during an earthquake.

orographic effect The effect created when moisture-bearing winds rise against a mountain range: They condense to form clouds and rain or snow.

overburden Soil or other material above the bedrock. Often used in reference to waste material that must be removed to get at economically valuable material below.

overland flow Surface runoff of water in excess of that which is able to infiltrate the ground.

oxbow lake A small lake left when a meander of a stream is cut off and abandoned after a flood.

Pacific decadal oscillation (PDO): Cyclic changes in the eastern part of the northern Pacific Ocean sea-surface temperatures near North America from a positive warm phase to a negative cool phase over a period of 20 to 30 years.

pahoehoe Basalt lava with a ropy or smooth top.

paleoflood analysis Information on previous floods gathered from erosional and depositional features left by such a flood.

paleomagnetism The study of past orientation and strength of Earth's magnetism as preserved in rocks formed at various times in the past.

paleoseismology The study of former earthquakes from examination of offset rock layers below the ground surface.

paleovolcanology The study of former volcanic events from examination of former volcanic deposits.

pali The Hawaiian term for giant coastal cliffs; many of these are now thought to be the headscarps of giant submarine landslides.

Pangaea Supercontinent that began to break up 225 million years ago to form today's continents.

Peléan eruption A major ash-rich volcanic eruption that typically produces pyroclastic flows.

perforated pipes Pipes full of holes that are pushed into a water-saturated landslide to help drain it. Water seeps through the holes into the pipes and then down through the gently sloping pipes to the ground surface.

period The time between successive water waves or seismic waves, or the time between the peaks recorded on a seismograph.

permafrost The condition where water in the ground remains frozen all year.

permeability The ease with which water can move through soils or rocks.

phreatic eruption A volcanic eruption dominated by steam.

phreatomagmatic eruption Similar to a phreatic eruption but with a higher proportion of magma.

pillow basalt Basalt that flows into water or wet mud and chills on its outside surfaces to form elongate rounded fingers of basalt a meter or so across.

piping Water seeping through a dam or levee can produce pipelike openings and begin to cause erosion and excavate sand.

plastic A type of deformation in which a rock deforms gradually without breaking. If the deforming stress is relaxed, then the rock does not return to its original shape. Contrast with elastic.

plate tectonics The theory that lithospheric plates move relative to one another; they can collide, pull apart, or slide past one another. These movements can cause earthquakes, lead to volcanic eruptions, and build mountain ranges. The theory is supported by a wide range of data and is now considered fact.

Plinian eruption Extremely large volcanic eruptions that involve continuous blasts and heavy ash falls. Examples include Vesuvius in 79 AD and Mount St. Helens in 1980.

plume (1) A dense, rising cloud of ash from an erupting volcano. (2) An upward-expanding zone of hot rock deep in Earth's mantle.

plutonic rocks Distinctly grainy igneous rocks such as gabbro and granite that solidified slowly below Earth's surface. They contrast with volcanic rocks, that are fine-grained and solidified rapidly at the surface.

point bar Deposits of sand and gravel on the inside (concave side) of a meander bend of a stream.

polar vortex A mass of air rotating around the North or South Pole. A strong polar vortex correlates with a fast jet stream, a positive NAO and a positive Arctic Oscillation.

poor sorting A sediment with a broad mixture of different grain sizes.

popcorn clay Clay that swells when wet to cause very slippery conditions; on drying it resembles a layer of popcorn.

pore pressure The pressure of water in pore spaces between soil or sediment grains tends to push the grains apart and reduce the contact force between the grains; it can facilitate landsliding.

porosity The percentage of pore space in a sediment or rock.

precipitation Water that falls as rain, hail, or snow.

precursor events Events preceding major natural events such as earthquakes or volcanic eruptions, and which may warn of the coming event.

prediction A statement that a future event will occur at a certain time at a certain place. *Compare with* forecast.

prescribed burn Intentional setting of a fire under controlled conditions to consume fuels and mitigate spread of future large fires.

pressure The force per unit area; a continuously applied force spread over a body.

property rights The right to do what one wishes with one's personal property; it should not be permitted to affect other's rights.

proxy data Long-term records of climate conditions inferred from study of changes in tree rings, sediments, and ice cores.

pumice Frothy volcanic rock dominated by gas bubbles enclosed in glass; typically pale in color and can float on water.

P wave The compressional seismic wave that shakes back and forth along the direction of wave travel.

pyroclastic flow A mixture of hot volcanic ash, coarser particles, and steam that pours at high velocity down the flank of a volcano. Also called nuée ardente or ash flow.

pyroclastic material Fragmental material blown out of a volcano—for example, ash, cinders, and bombs.

quick clay Water-saturated mud deposited in salty water and tends to consist of randomly oriented flakes of clay with large open spaces between the flakes. If the salt is flushed out, the flakes are unstable and may collapse; the mass may flow almost like water.

radiation Electromagnetic energy transferred outward from the sun or other source and converted to heat when it strikes an object.

radiative forcing The difference in sun's radiant energy above Earth's troposphere versus below it. Positive radiative forcing warms Earth. Carbon dioxide is the largest contributor to radiative forcing.

radiometric The measurement of the age of a rock by the analysis of radioactive constituents and their products, along with their known rates of decay.

rain shadow Drier climate on the downwind side of a mountain range.

recurrence interval The average number of years between events of a certain size in a location—for example, floods, storm surges, and earthquakes. Also known as a return period.

refraction *See* wave refraction.

relative humidity The percentage of moisture in air relative to the maximum amount it can hold (at saturation) under its given temperature and pressure.

resisting force The force (such as friction or a load pushing in the opposite direction) that resists downslope movement.

resurgent caldera A huge collapse depression on a volcano that sank into a near-surface magma chamber during eruption of the magma.

resurgent dome A bulge in a caldera where magma rose, and which may or may not develop into a new eruption.

retrofitting Modifying existing buildings to minimize the damage during a strong earthquake.

reverse condemnation Where government restricts a property owner from using his or her property for some potential purpose; that has effectively taken the value of the property for that purpose.

reverse fault Faults (generally, steeply inclined) on which the upper block of rock moved up compared with the lower block.

rhyolite A light-colored, fine-grained volcanic rock that has a silica content of approximately 70% and high viscosity. It generally erupts as volcanic ash and fragments.

Richter Magnitude Scale The scale of earthquake magnitude invented by Charles Richter. *See also* magnitude. It uses the amplitude of waves recorded on a seismograph.

riffles The part of a stream with shallow rapids—generally, the straighter section between meander bends.

rift A spreading zone on the flank of a volcano from which lavas erupt. *See also* rift zone.

rift zone An elongate spreading zone in Earth's lithosphere.

right-hand rule The rule used to remember the direction of wind rotation in a low- or high-pressure cell in the northern hemisphere. With your right thumb pointing in the direction of overall air movement (up away from Earth, or down toward Earth), your bent fingers point in the direction of wind rotation.

right lateral One sense of movement on a strike-slip fault; if you stand on one side of the fault, the opposite side moves to the right during an earthquake.

rip current A short-lived surface current that flows directly off a beach and through the breaker zone. It carries water that had piled up onto the beach back to the sea.

riprap Coarse rock piled at the shoreline in an attempt to prevent wave erosion of the shore or destruction of a near-shore structure or erosion of a stream bank.

risk The chance of an event multiplied by the cost of loss from such an event.

risk map Map showing the probable risk of where and when an earthquake, flood, or other disaster will strike.

river stage Water height in a river, measured relative to some arbitrary fixed height.

rock A mass of interlocking or cemented mineral grains. *Compare with* mineral.

rockbolts Long bolts drilled into and expanded in an unstable rock mass to help keep it from landsliding.

rockfall A rock mass that falls from a steep slope.

rotational landslide A landslide in which the mass rotates (bottom edge outward) as it slides on a concave upward basal slip surface.

rotational slump Same as rotational landslide.

runoff The portion of precipitation that flows off the ground surface.

runout The distance a rockfall will travel, including that beyond the base of the slope.

run-up The height to which water at the leading edge of a wind wave or tsunami rushes up onto shore.

Saffir-Simpson Hurricane Wind Scale The commonly used five-category scale of hurricane intensity based on wind speeds.

sand Particles of rock that range in size from 1/16 to 2 mm in diameter.

sand boil A pile of sand brought to the surface in water expelled from the ground during an earthquake, by compaction or liquefaction at shallow depth, or during a flood when the water pressure under a channel forces groundwater to the surface outside a levee.

sand dune An accumulation of windblown sand, most commonly along the upper beach above high-tide level.

sand sheet Layer of nearly homogeneous, unlayered sand laid down by a tsunami wave that sweeps sand inland from a beach.

Santa Ana winds Southern California hot, dry winds that flow southwest and offshore, bringing dry air from the continental interior. A local name for a Foehn wind.

scientific method A method commonly used by scientists to solve problems. They analyze facts and observations, formulate hypotheses, and test the validity of the hypotheses with experiments and additional observations.

seafloor spreading Where seafloor spreads apart at a mid-oceanic ridge during plate tectonic movement.

seawall Walls of boulders, concrete, or wood built along a beachfront and intended to protect shore structures from wave erosion.

sedimentary rock A rock deposited from particles in water, ice, or air—or precipitated from solution—and then cemented into a solid mass. Examples are sandstone, shale, and limestone.

seismic gap A section of an active fault that has not had a recent earthquake. Earthquakes elsewhere on the fault suggest that the gap may have a future earthquake.

seismic wave A wave sent outward through Earth in response to sudden movement on a fault. *See also* P wave, S wave, surface wave.

seismogram The record of seismic waves from an earthquake or other ground motion as recorded on a seismograph.

seismograph The instrument used to detect and record seismic waves.

ShakeMap Computer-generated maps of ground motion that show the distribution of maximum acceleration and maximum ground velocity during earthquakes.

shatter cone Cone-shaped features, with rough striations, radiating downward and outward from the shock effect of a comet or meteorite impact.

shield volcano An extremely large basalt-lava volcano with gently sloping sides, such as those in Hawaii.

shore profile The seaward slope of the beach.

shotcrete = gunite A fluid cement-type material that is sprayed on a slope to prevent water penetration.

silica tetrahedra An arrangement of four oxygen atoms in four-cornered pyramids around single silicon atoms.

sinkhole A ground depression caused by collapse into an underground cavern.

solar radiation management An approach to geoengineering involving artificial reduction of sunlight by, for example, injecting SO_2 aerosols into the stratosphere.

slip plane, slip surface The sliding surface at the base of a landslide. Also called a slide plane.

slope angle The angle of a slope as measured down from the horizontal.

slump *See* rotational landslide.

smectite A soft "swelling" clay that forms by alteration of volcanic ash, swells when wet, and becomes extremely slippery. It is prone to landsliding and can deform houses built on it.

smokejumpers Firefighters that parachute in to fight wildland fires.

snow avalanche Rapid downslope movement of snow.

"soft" solutions Solutions to a problem that involve avoiding the problem through restrictive zoning and building codes to minimize damage.

soil For engineering purposes, all of the loose material above bedrock.

soil creep Slow downslope movement of near-surface soil or rock, caused by repeated cycles of heating and cooling, freezing and thawing, burrowing animals, and trampling feet.

soluble Able to be dissolved, typically in water.

solution The process of dissolving rocks or minerals in water.

sorted A sediment is well sorted if its grains are all about the same size. *Compare with* poor sorting.

southern oscillation *See* ENSO.

spillway The overflow lip for water from a dam or levee.

spot fire Fires ignited by firebrands which blow through the air ahead of the main fire.

spreading center or rift zone Boundary along which lithospheric plates or volcano flanks spread apart or diverge.

stalactite An elongate deposit of calcium carbonate that grows down from the roof of a cavern by evaporation of carbonate-rich water.

stalagmite A cone-shaped deposit of calcium carbonate that grows up from the floor of a cavern by evaporation of carbonate-rich water dripped from above.

step leader Electrical charges that advance downward from a lightning strike but do not manage to reach the ground.

stony-iron meteorite Essentially a chondrite that contains some nickel-iron.

storm surge The rapid sea-level rise caused by the strong winds of a storm that push water forward and to a lesser extent by the low atmospheric pressure of a major storm.

strain Change in size or shape of a body in response to an imposed stress.

stratovolcano A large, steep-sided volcano consisting of layers of ash, fragmental debris, and lava. *Also called a* composite volcano.

straw wattle A net in the shape of a large hose, filled with straw, and used to hinder surface runoff and erosion on a slope denuded by fire or landslide.

surface wave The third in a sequence of seismic waves of an earthquake; it travels near Earth's surface and causes large ground motions.

streambed mining Excavation of sand and gravel from a stream bed.

stream order A way of numbering streams in a hierarchy that designates a small unbranched stream in a headwaters as first order, the stream it flows into as second order, and so on.

stress The forces on a body. These can be compressional, extensional, or shear.

strewn field Area in which fragments of large meteorites are distributed around the main impact site.

strike The compass direction of a horizontal line on a plane such as a fault or a rock layer.

strike-slip fault A fault (generally vertical) which has relative lateral movement of the two sides.

Strombolian eruption Frequent mild eruptions of basalt or andesite cinders, typically forming a cinder cone.

subduction The process in which one lithospheric plate (usually oceanic) descends beneath another plate.

subduction zone Convergent boundary along which lithospheric plates come together and one descends beneath the other; often ocean floor descending beneath continent.

submarine canyon A deep canyon in the continental shelf and slope offshore and extending outward from the shore.

submarine landslide Subsea-level collapse of sediment or rock, such as the flank of an oceanic volcano, for example, in Hawaii or the Canary Islands.

subsidence Settling of the ground in response to extraction of water or oil in subsurface sediments, drying of peat, or formation of sinkholes. Subsidence of continent or ocean floor occurs by slow cooling of hot lithosphere, such as after passage of a hotspot.

sulfur dioxide A toxic volcanic gas consisting of two oxygen atoms attached to one sulfur atom.

supercell A particularly strong rotating thunderstorm that can spin off dangerous tornadoes.

superoutbreak A large group of tornadoes produced along a major storm front.

surface fire Fires that spread along the ground surface through grass and low shrubs.

surface runoff Water flowing across the ground after a heavy rain.

surface rupture length The surface length of a fault broken during an earthquake.

surface tension The effect by which grains of sand are held together by the thin films of water between them.

surface wave The seismic wave that travels along and near Earth's surface. These waves include Rayleigh waves (which move in a vertical, elliptical motion) and Love waves (which move with horizontal motion) perpendicular to the direction of wave travel.

surge *See* base surge, storm surge.

suspension Sediment grains carried within the water column and buoyed up by turbulent eddies—in contrast to sediment grains transported along a stream bed.

S wave The seismic shear wave that shakes back and forth perpendicular to the direction of wave travel. S waves do not pass through liquids.

swelling soil A soil that expands when wet; generally, a soil that contains smectite, the swelling clay.

talus Coarse, angular rock fragments that fall from a cliff to form a cone-shaped pile banked up against the slope.

tectonic *See* plate tectonics.

tektite Droplets of molten rock—glass—that may have formed by superheated splash from a hypervelocity impact on Earth or possibly the moon.

thalweg The line connecting the deepest parts of the channel along the length of the stream bed.

theory A scientific explanation for a broad range of facts that have been confirmed through extensive tests and observations. In science, it is equivalent to a scientific fact or "law."

thrust fault Fault (usually, gently inclined) on which the upper block of rock moved up and over the lower block. Found in compressional environments.

thunderstorm A storm accompanied by clouds that generate lightning, thunder, rain, and sometimes hail.

tidal current The flow of water through a narrow passage, such as between the ocean and a lagoon, through segments of a barrier island, in response to sea level change between high and low tides, or oceanward flow after storm surge.

tidal wave A colloquial but incorrect name for tsunami.

tide The change in sea level, generally once or twice a day, in response to gravitational pull of the sun and moon.

tiltmeter A device, originally like a carpenter's level, that records any change in slope on the flank of a volcano.

toe The lowest, farthest extent of a landslide.

tornado A near-vertical narrow funnel (generally less than a kilometer or so in diameter) of violently spinning wind associated with a strong thunderstorm.

Tornado Alley The region of the central United States, between Texas and Nebraska that is most noted for frequent tornadoes.

tornado outbreak A series of tornadoes spawned by a group of storms.

tornado warning A warning issued by the weather service when Doppler radar shows strong indication of vorticity or rotation, or if a tornado is sighted.

tornado watch A cautionary statement issued by the weather service when thunderstorms appear capable of producing tornadoes and telltale signs show up on the radar. At this point, storm spotters often watch for severe storms.

trade winds Regional winds that blow from northeast to southwest between latitudes 30° north (or south) and the equator and centered near 15° north (or south).

transform boundary A fault boundary where two tectonic plates slide laterally past one another.

transform fault Boundary along which lithospheric plates slide laterally past one another, for example the San Andreas Fault.

translational slide A landslide with a slip surface approximately parallel to the slope of the ground.

transpiration The passing of water vapor through vegetation pores to the atmosphere.

trench An elongate depression where the descending ocean floor moves under the overriding tectonic plate at a subduction zone and most commonly at the edge of an active continental margin. Most are at the margins of the Pacific Ocean.

trimline The line along a mountainside along which tall trees in the forest upslope are bounded by distinctly shorter trees downslope—sometimes indicating tsunami damage.

triple junction A junction between three lithospheric plate boundaries.

tropical cyclone A large rotating low-pressure cell that originates over warm tropical ocean water. Depending on location, they are called hurricanes, typhoons, or cyclones.

tropics The warm climate area between the Tropic of Cancer (23.5° north of the equator) and the Tropic of Capricorn (23.5° south).

tsunami An abnormally long wavelength wave most commonly produced by sudden displacement of a large mass of water in response to fault movement on the seafloor. It can also form when a landslide, volcanic eruption, or asteroid impact displaces water.

tsunami warning When a significant tsunami is identified, officials order evacuation of endangered low-lying coastal areas.

tsunami watch The alert is issued when a magnitude 7 or larger earthquake is detected somewhere below the Pacific Ocean or some other ocean that may see dangerous tsunami.

tuff A rock formed by consolidation of volcanic ash.

typhoon A large low-pressure weather system that circulates counterclockwise in the northern hemisphere and clockwise in the southern hemisphere. It is equivalent to a hurricane or a cyclone.

Uniform Building Code A national code for construction standards to provide safety for people in high hazard zones for earthquakes, hurricanes, and other natural hazards.

urbanization Change of an area with a concentration of buildings and pavement such as in a city.

VEI *See* volcanic explosivity index.

viscosity The resistance to flow of a fluid because of internal friction.

vog An acidic volcanic smog produced when volcanic gases react with moisture and oxygen in the air, and sun, to produce aerosols.

volatiles Dissolved gases in a volcano.

volcanic ash Any small volcanic particles; formally, particles less than 2 mm across.

Volcanic explosivity index (VEI) A scale of volcanic eruption violence based on volume, height, and duration of an eruption.

volcanic weather Weather generated during a volcanic eruption. The high temperature above an erupting volcano draws in outside air that rises and cools. Moisture in the air condenses to form rain and stormy weather.

volcano A mountain formed by the products of volcanic eruptions.

Vulcanian eruption A style of volcanic eruption that is more violent than Strombolian and less violent than Peléan. They blow out dark eruption clouds of ash and blocks.

wall cloud The rotating area of cloud that sags below the main thundercloud base and from which a tornado may develop.

warm front The boundary between a regional mass of warm air advancing over an adjacent large mass of cold air.

water pressure The pressure of water spaces between grains in the ground.

watershed That part of a landscape that drains water down to a given point on a stream.

water table Roughly the level of the top of the saturated zone of water in the ground; the level to which water would rise in a shallow well.

water vapor The invisible gaseous form of water in the air.

wave base The greatest depth of wave motion that stirs sediment on the bottom—generally, 10 m or so.

wave energy The capacity of a wave to do work—that is, to erode the shore. Because wave energy is proportional to wave height squared, high waves cause almost all coastal erosion.

wave height The vertical distance between a wave crest and an adjacent trough.

wavelength The distance from crest to crest of a wave—for example, in an earthquake wave or a water wave.

wave refraction The bending of water wave crests near shore, where one end of the wave drags bottom and slows down in shallower water.

weather The conditions—temperature, pressure, humidity, and wind—of the atmosphere at a particular place and time.

weather fronts *See* cold front, warm front.

weathered rock or weathering The gradual destruction of rock materials through physical disintegration and chemical decomposition by exposure to moisture, temperature, and chemicals, in the soil and atmosphere.

welded ash Ash from a pyroclastic flow that was hot enough when deposited that the particles fused together to form a solid rock.

westerly winds Regional winds that blow from southwest to northeast and are centered 45° north (or south) of the equator.

wildfire Unplanned and uncontrolled fires in minimally developed areas.

wind The movement of air from an area of higher to lower atmospheric pressure.

wind chill Wind causing evaporation of skin moisture to cause heat loss and make us feel colder.

wind shear The difference in wind direction or velocity across a short distance.

wing dam Segments of walls built to protrude into the current of a river to increase the depth of water in the channel to facilitate shipping.

zoning restrictions Laws that prohibit building in certain areas.

Index

"f" after page numbers indicates figures, photos, or other images. **Bold** page numbers indicate section headings and other important entries. *Italic* page numbers indicate Case in Point features

preplanning failed, *460*
pumps failed, *461*
rebuilding, 451
water rose fast, *461*
Hurricane Sandy, 1
damage, 437, 441
Atlantic City, NJ, 435
atmospheric pressure, *470*
carbon monoxide poisoning, *470*
extreme diameter, *470*
government buys homes, *471*
high tide, surge, *470*
major flooding, *470*
strong onshore winds, *470*
towns lost tax base, *471*
hydrocarbon residue, **485f**
hydrogen
isotope, deuterium (^{2}H), 308
sulfide, volcanic, 157
hydrograph, 357f
debris flow, *374*f
North Dakota flood, *376*f
shape, 357f–358
stream discharge, 357f
stream order, 357
hydrologic cycle, **248f–249f**
hydrophobic soils, **485f, 486f**
hyperconcentrated flow, 364
hypothermia, 271
symptoms, 268
treatment, 268
hypotheses, **28**, 30

I

ice age
extent, 308–309
last, 17
temperature cycles, **308f–310f**
ice, cores, past temperatures, 307–309f, 311f
ice dams, **369f–370f**
ice storm, **270f–272f**
air temperature, 270f
carbon monoxide poisoning, 270, *288*
development, 270f
freezing pipes, *288*
hazards, 270
heating, *288*
humidity, 270f
lake shoreline ice, 270
power lines, *288*
iceberg, 18
Iceland
Mid-Atlantic Ridge, 22, 127
volcanic ash plume, 152–154f
volcanic eruptions, 127, 157
identification, meteorite, **508f–509f**
ignimbrite, 149
ignition, wildfire spread, **482f–483f**
impact crater, **511f–512f**
central uplift, 511–512
meteorite, **510f–512f**
impact(s)
Chesapeake Bay, **513f**
Chicxulub, Mexico, **512f**
dust, 516
effects, immediate, **516**
energy, meteorite, **509f–510f**
melt, shatter cones, **513f–514f**
multiple asteroid, **515f**
plants, animals, humans, **329f–331f**
recognized worldwide, 510
site, flood basalt, 516
size, frequency, 517
trigger other hazards, **516**

impervious carbonate-rich layer, 354
incentives, reduce CO_2 emissions, 337
inclusions, basalt in rhyolite, 126
indefensible locations, **494f**, *500–501*
India
collision with Asia, 25
monsoon rains, 259f–260f, *371*
pollution, 260f
Indian Ocean, tsunami, *114f–116f*
Indonesia tsunami, 2004, *114f–116f*
industrial revolution
climate, 307
CO_2 since, 304
infrared radiation, 305
inhaled volcanic ash, hazard, 152
initiation of tsunami, 97f
injecting fluids, earthquakes, 65f
inlets through barrier islands, **418f**
insect-borne diseases, climate change, 330
insurance, **9–10**
actuarial, 393, 395
disasters, **9**
flood, definitions, **394f**
homeowners, hurricanes, 456–457
lightning damages, 273
natural hazards, 9
policies canceled, 456
premiums, 9
risk formula, 9
sinkhole, 227
wildfire, 494
intensity
earthquake scale, 48f
earthquake shaking, 49f
floods, **356f–359f**
stream runoff, 357
interactions between hazards, 7f
Intergovernmental Panel of Climate Change, 314
intermountain seismic zone, 70
internal pore pressure, 186
internal surfaces, landsliding, 184–185
intraplate earthquakes, Missouri, 43, 44f–45f
intraplate eastern earthquakes, **43–45**
IPCC (Intergovernmental Panel of Climate Change), 314
projected temperature rise, 314f
iridium anomaly, **515**
iron meteorites, **508f–509**
irregular coastline, waves, **413f–414f**
isostasy, **17–18**
Italy
earthquake prediction, 67
volcanoes, map, 162f
Izmit, Turkey, earthquake damage, 54

J

Japan
giant earthquake, 2011, 41, **55f–56**
Sendai earthquake, 2011, 83
tsunami, 2011, *113f*
volcano mudflow, 154f
jet stream, **253f–254f**
dip, 252
drought, 263
meander, 253f
polar vortex speed, 254
shift in summer, 253
shift in winter, 253
jetty, 424
down-coast erosion, 425
negative effects, 425
rip current, 413
sand movement, **426f**

joints, beams fail in earthquake, 76f
Jökulhlaups, **368f–369f**
Joplin tornado, 2011, 278
Juan de Fuca Ridge, 25f

K

kaolinite clay, **185**
karst, **223–224f**, 225
Karst Plain, Missouri, 226
Katmai, Alaska, volcano, 138f
Katrina *See Hurricane Katrina*
Keeling curve, atmospheric CO_2, 312f
Kentucky, ice storm, *288f*
Kilauea, Hawaii, **134f–135f**
eruption, 135f, *175f–176f*
flank collapse, tsunami, 100, 109f–111f
magma gases, 135
potential tsunami, 110f–111f
recent eruptions, *175f–176f*
slowly slumping, 110f
knickpoints, 352
Krakatau, 1883
pyroclastic flow, 151
tsunami, 98
K-T boundary, **513–515f**
Kuiper belt comets, 508

L

L'Aquila, Italy, earthquake, 67
La Conchita, CA, landslide, 188f
La Jolla Canyon, sand loss, 417
La Niña, **256, 257f–259f**
drought, *289*
ocean temperature, 256–257f
southwest weather, 259
ladder fuels, **481f–482f**
wildfires, removal, *492f*
lag, storm, flood crest, 358–359f
lagoon sand, not for beaches, 426
Laguna Beach, CA, landslide, 181, 195f
lahar, **131**, *139f–140f*, **154–156**, 364f–365f
lahar
active volcanoes, 365f
characteristics, 129
Mt. Lassen, 169
lahars, 365f
lake effect
from Great Lakes, 269f
ice-covered lake, 269
ice-free lake, 269
moisture, 269
snow depths, 269f, 270
Lake Nyos, West Africa, 156
Lake Tahoe landslide, tsunami, 99f
Laki flood basalt eruption, Iceland, 157
land cultivation, beach sand, 417
land subsidence, **227–233**
landslide, **181–217**
causes, 186–**189**
causes, overlapping, *209–210*
active fault, 83
after wildfire, 485
becomes debris flow, 375
blasting triggered, ***214–215***
blasting, ***214–215***
causes, 186–189
center of rotation, 194–195f
clays, 185–186
clear-cut logging, 188–189f
coastal areas, *210–211*f
cost, 204
dam reservoir, 203f